TELECOMMUNICATIONS AND DATA COMMUNICATIONS HANDBOOK

BICENTENNIAL
1807
WILEY
2007
BICENTENNIAL

THE WILEY BICENTENNIAL–KNOWLEDGE FOR GENERATIONS

*E*ach generation has its unique needs and aspirations. When Charles Wiley first opened his small printing shop in lower Manhattan in 1807, it was a generation of boundless potential searching for an identity. And we were there, helping to define a new American literary tradition. Over half a century later, in the midst of the Second Industrial Revolution, it was a generation focused on building the future. Once again, we were there, supplying the critical scientific, technical, and engineering knowledge that helped frame the world. Throughout the 20th Century, and into the new millennium, nations began to reach out beyond their own borders and a new international community was born. Wiley was there, expanding its operations around the world to enable a global exchange of ideas, opinions, and know-how.

For 200 years, Wiley has been an integral part of each generation's journey, enabling the flow of information and understanding necessary to meet their needs and fulfill their aspirations. Today, bold new technologies are changing the way we live and learn. Wiley will be there, providing you the must-have knowledge you need to imagine new worlds, new possibilities, and new opportunities.

Generations come and go, but you can always count on Wiley to provide you the knowledge you need, when and where you need it!

WILLIAM J. PESCE
PRESIDENT AND CHIEF EXECUTIVE OFFICER

PETER BOOTH WILEY
CHAIRMAN OF THE BOARD

TELECOMMUNICATIONS AND DATA COMMUNICATIONS HANDBOOK

RAY HORAK
The Context Corporation
Mt. Vernon, WA 98273

WILEY-INTERSCIENCE

A JOHN WILEY & SONS, INC., PUBLICATION

Published by John Wiley & Sons, Inc., Hoboken, New Jersey
Published simultaneously in Canada

For general information on our other products and services or for technical support, please contact
our Customer Care Department within the United States at 877-762-2974, outside the United States at
317-572-3993 or fax 317-572-4002.

Wiley also publishes its books in a variety of electronic formats. Some content that appears in print
may not be available in electronic formats. For more information about Wiley products, visit our web
site at www.wiley.com.

Wiley Bicentennial Logo: Richard J. Pacifico

Library of Congress Cataloging-in-Publication Data:

Horak, Ray.
   Telecommunications and data communications handbook/Ray Horak.
      p. cm.
   Includes index.
   ISBN 978-0-470-04141-3 (pbk.)
   1. Telecommunication—Handbooks, manuals, etc. 2. Digital communications—Handbooks, manuals,
etc. I. Title.
   TK5101 .H6655
   621.382—dc22

                                    2006032496

Printed in the United States of America

10 9 8 7 6 5 4 3 2 1

To Margaret,
For whom my love and devotion are infinite and everlasting.

CONTENTS

PREFACE

I was a civil engineering student in Berlin. Berlin is a nice town and there were many
opportunities for a student to spend his time in an agreeable manner, for instance with
the nice girls. But instead of that we had to perform big and awful calculations.

Konrad Zuse, 1910

Those of us who have been involved in communications technology for any number
of years have witnessed a transformation that truly is revolutionary. The *big and*
awful calculations got even bigger and more awful as the technologies became more
complex. Yet, the old voice network remained much the same from the late 1800s
through the 1960s—Alexander Graham Bell would have recognized it clearly and
understood it completely. Around the time of World War II, however, dramatically
new technologies started to make their presence felt. Microwave radio systems
began to appear and the foundation for electronic computer systems was laid. Soon
thereafter, serious computer systems began to emerge and the need to network
them soon became obvious. Over time, the networks themselves gradually became
computerized. During the 1940s, the groundwork also was laid for the development
of fiber optic transmission systems, which have the ability to transport incredible
volumes of information over very long distances and with crystal clarity. Videoconferencing systems began to emerge in the 1960s, although they became practical
only in the last few years and have yet to gain widespread acceptance. Local Area
Networks (LANs) appeared in the early 1980s with the development and standardization of Ethernet.

The rate of development of the underlying system and network technologies
gathered speed in the last 20 years to the point that it has become difficult for even
the most technically astute to keep pace with the rate of change, much less the depth
and breadth of its impact. Copper wires have yielded to glass fibers. Rather than
flowing through networks in continuous streams over connections, information

often moves in packets, frames, and cells—sometimes on a connectionless basis. Worldwide, the volume of data traffic now exceeds voice traffic. Increasingly, however, the definition of *data* is no longer confined to computer data. Rather, data now include voice data, video data, image data, and even multimedia data. The networks not only connect telephones and computers but also have become networks of computers themselves. Wired networks no longer are just supplemented by wireless network technologies but in many applications are now being challenged by them, especially where portability and mobility are advantageous. In fact, there now are more cellular phones in the world than there are wireline phones, and a great many people now use cellular phones as their primary and even sole telephone.

Government regulation of monopolies has yielded to free market competition, a trend that began in 1984 with the Modified Final Judgment (MFJ) in the United States. Monopolistic companies have been torn apart in the interests of increased competition only to be reconstituted in altered states when the regulators became convinced that the market, indeed, is the best regulator. Competition has become rife in virtually every sector of the communications world, bringing with it the inherent advantages of alternative choice, improved performance, greater creativity, enhanced technology, lower cost, and a bewildering range of options. Community Antenna TeleVision (CATV) providers have entered the fray, offering voice services and Internet access over cable systems originally put in place for entertainment TV, and telephone companies are now beginning to offer cable TV service. The Internet has been commercialized and now is available in every corner of the world, supplanting more traditional means of communication and even threatening more traditional voice and data networks. Underlying Internet technologies and protocols have become the foundation for next-generation networks that are virtually certain to replace the circuit-switched network that served us so well for over 100 years. Audio, images, animated images, and even video clips are attached to electronic mail. Once available only to the technically elite digiterati working in privileged circumstances with dumb terminals connected to mainframe computers, e-mail now is available to all of us, just about anywhere, and through devices as simple and mobile as cell phones.

This book delivers a comprehensive overview of a wide range of communications systems and networks, including voice, data, video, and multimedia. It is written in plain English and provides a commonsense basis for understanding system and network technologies, their origins and evolutions, and the applications they serve. Further, it discusses the origin, evolution, and nature of many relevant standards and explores remaining standards issues. It also provides a view of the evolution and status of regulation and examines a number of key regulatory issues awaiting resolution. From fundamental concepts through the convergence of voice and data networks, this book offers a single source of information for those who need to understand communications networking.

Reserving most discussion of volts, amps, ohms, algorithms, and the like for a later date and another book by another author, this one weaves a fabric of understanding through a complex set of technologies that underlie meaningful contemporary and future applications. Further, this book serves as a language primer, providing a short course in the vocabulary and syntax of the language of telecommunications—having read this book, you will be conversant in telecommunications techno-speak. Finally, you will understand how networks work and why.

HOW THIS BOOK IS ORGANIZED

This book is organized into 15 chapters, each of which addresses closely related areas of telecommunications technology and applications, with an eye toward the development of meaningful and cost-effective solutions to legitimate communications requirements. Taken as a whole, this document is a discourse on the origins, current status, and foreseeable future of the networked world. As this book weaves a bit of a story throughout, it is best read from cover to cover. Yet, those who are either impatient or highly focused will find that each chapter is fully capable of standing on its own.

The first several chapters set the stage for understanding the fundamental nature of systems and networks. Chapter 1 offers a set of basic concepts and provides a set of definitions that apply fairly universally across all communications and networks. Chapter 2 offers a detailed explanation of the essentials of transmission systems, both wired and wireless, including twisted pair, coaxial cable, microwave radio, satellite radio, Free Space Optics (FSO), fiber optics, and Power Line Carrier (PLC). Each transmission system is explained in detail and is compared and contrasted with others along a number of critical dimensions.

Chapter 3 is devoted to discussion of premises-based communications systems that primarily support voice applications, although they increasingly support data and even video communications as well. It begins with an examination of Key Telephone Systems (KTSs) and goes on to examine the several generations of Private Branch eXchanges (PBXs) and their capabilities and applications as well as emerging trends. This chapter then examines Centrex and Automatic Call Distributors (ACDs) and concludes with an examination of Computer Telephony (CT) and Internet Protocol (IP)–based voice systems, recent developments that provide tremendous value and efficiency in the processing and management of voice calls.

Chapter 4 is dedicated to electronic messaging technologies and systems, including facsimile, voice processing, electronic mail, Instant Messaging (IM), Short Message Service (SMS), and Multimedia Message Service (MMS). Increasingly, these technologies are viewed as converging into a unified suite of messaging systems, if not a unified information stream, and ultimately into a suite of unified communications systems. Indeed, we see the beginnings of such unification in the Internet, the World Wide Web (WWW), and next-generation IP-based networks.

Chapter 5 concentrates on the conventional Public Switched Telephone Network (PSTN), addressing the origin, evolution, and contemporary nature of the traditional voice network. The underlying technologies, regulatory and standards domains, carrier/service provider domains, and functional domains all are discussed. This chapter also defines the nature and specifics of the signaling and control systems that manage and control the operation of the various network elements in order to ensure that the network functions as a whole. A wide variety of voice network services are defined and illustrated. The chapter concludes with a discussion of a number of critical PSTN issues, including numbering plan administration, number portability, equal access, and the impact of developing IP-based voice networks.

Chapter 6 begins the discussion of data communications, addressing a number of basic concepts. Data terminal equipment, data communications equipment, communications software, and the network are explained as the various functional domains in a data communications network. Protocols are explained, both as a concept and in terms of certain specific, baseline examples. Additionally, a number of key elements of a data communications protocol are discussed, with specific examples provided. Network architectures are examined, with concentration on layered operations models such as that of the Open Systems Interconnection (OSI) Reference Model, which sets the framework for interconnectivity and interoperability. Discussion follows of the various influential standards organizations, with security issues and techniques serving as the conclusion.

Chapter 7 centers on conventional digital and data networks, which are based on the voice network model. Dedicated and circuit-switched networks are discussed in the context of both private and public data networks. Specific data network options, presented in rough order of deployment, include Dataphone Digital Service (DDS), Switched 56 and classic Virtual Private Networks (VPNs), T-Carrier and E-Carrier, X.25 and packet switching, and ISDN.

Chapter 8 steps back from the traditional Wide Area Network (WAN) to explore the world of the LAN, the concept of which was first formally expressed in 1973. Since the first LAN Standard was published in 1982, LANs have grown to the point that they are virtually ubiquitous. This chapter defines LANs in terms of fundamental dimensions such as acceptable transmission media, physical and logical topologies, baseband versus broadband, and media access control. LAN and internetworking devices are discussed, including hubs, bridges, switches, routers, and gateways. Network Operating Systems (NOSs) are presented in the context of the client/server model, and the fundamentals of LAN internetworking are discussed, with emphasis on the Transmission Control Protocol (TCP)/IP suite. The chapter finishes with a discussion of relatively recent developments including Virtual LANs (VLANs), remote LAN access, high-speed LAN technologies and standards such as 100BaseT and Gigabit Ethernet (GbE), Wireless LANs (WLANs), and Storage Area Networks (SANs).

The next two chapters deal with broadband networking, the high-speed future of communications. Chapter 9 is dedicated to discussion of the physical infrastructure, with the initial focus on recently developed local loop technologies, both wired and wireless. Considerable discussion is devoted to Digital Subscriber Line (DSL), CATV networks, Passive Optical Network (PON), Wireless Local Loop (WLL), and Broadband over Power Line (BPL). Synchronous Optical NETwork/Synchronous Digital Hierarchy (SONET/SDH) and Wavelength Division Multiplexing (WDM) fiber optics, the ultimate in broadband wireline networking, are presented in detail. Chapter 10 focuses on broadband fast packet networks in the forms of Frame Relay (FR) and Asynchronous Transfer Mode (ATM). Discussion of Broadband Integrated Services Digital Network (B-ISDN) and Advanced Intelligent Networks (AINs) conclude the chapter.

Chapter 11 explores the world of wireless communications—not traditional wireless transmission systems such as microwave and satellite, but rather special network alternatives. Popular options examined include Specialized Mobile Radio (SMR), paging, and 2G, 2.5G, and 3G cellular networks. Discussion of Low-Earth Orbiting (LEO) satellites networking rounds out the discussion of wireless communications.

Chapter 12 is devoted to video and multimedia systems and networks. The addition of a visual dimension enhances communications to a very significant extent, although it places great demands on the supporting networks. As video and multimedia networking is highly capacity intensive and as broadband networks are by no means fully deployed, cost-effective applications remain few. But the future beckons—and the future certainly includes video and multimedia.

Chapter 13 profiles the Internet, including its origins, nature, and structure. The TCP/IP protocol suite is discussed in detail, and QoS (Quality of Service) protocols and mechanisms are explained. Internet access options, equipment, and costs are explored, and issues of regulation and security are discussed. A sample of the more interesting and legitimate applications are visited, most especially that of the World Wide Web (WWW). Voice over IP (VoIP) and MultiProtocol Label Switching (MPLS) are examined in detail.

Chapter 14 addresses network convergence, the coming together of voice, data, video, and entertainment networks. As the networked world becomes increasingly deregulated and as users develop an ever more insatiable appetite for ever more exotic and capacity-intensive applications, a host of companies are vying to satisfy that hunger. CATV providers in the United States now deliver not only entertainment TV service but also voice and high-speed Internet access. Telcos now not only offer voice and high-speed Internet access but also are beginning to offer entertainment TV service. The stakes are enormous in magnitude, as the outcome will shape the future of the networked world. The status, the issues, and the likely outcomes are explored in this chapter.

Chapter 15 rounds out the tour of the networked world with a profile of regulation, both domestic United States and international. The origins, evolution, and current status of regulation are tracked through key legislative, judicial, and agency events. Current regulatory issues are discussed with emphasis on deregulation, most especially in the context of the Internet and convergence.

Finally, there are two appendixes. The first is a complete listing of every germane abbreviation, acronym, contraction, initialism, and symbol mentioned in this book, and there are hundreds of them. Consider this appendix to be your secret decoder ring. It also provides you with a tool for finding your way through the highly detailed Index, for I do not list subject matter by acronym or abbreviation there. Rather, I list things in the Index in order of the terms, themselves, spelled out fully. The second appendix is a listing of all the standards bodies and special interest groups that I consider to be of relevance. Included in each listing is full contact information, current as of the time of this writing (June 2007).

RAY HORAK

The Context Corporation
ray@contextcorporation.com

ACKNOWLEDGMENTS

I am a hoarder of two things: documents and trusted friends.
Muriel Spark, "Introduction," Curriculum Vitae, 1992

I owe a great deal to many who gave freely of their time, effort, knowledge, experience, expertise, and technical resources to make this book a reality. This book builds on my best-selling *Communications Systems and Networks*, the first edition of which I wrote for John Wiley & Sons in 1997. Through three editions, that first book sold well over 50,000 copies, which makes it a best seller by any measure.

Communications Systems and Networks began, and continued to evolve, as a course manual for my public seminars which were sponsored by *Network World* as the cornerstone of its Technical Seminar series. Bill Reinstein of *Network World*, with the encouragement of Mark Miller of DigiNet Corporation, had the courage to depart from his highly successful model and to sponsor a series of seminars on network essentials. The result was a seminar designed to introduce datacommunications and its underlying technologies to a new generation of communications professionals. We also updated more than a few old-timers in the process. Bill Reinstein, Deb Mahler, and Bill Bernardi made it all happen for me at *Network World*. They are great, and I'll always be indebted to them.

Mark Miller, further, saw a book in that first seminar manual. As a first-time author, the first edition of *Communications Systems and Networks* nearly killed me, or so it seemed at the time. I thought that the second edition and third editions would be much easier, but that wasn't the case at all. In the intervening years, the basics never changed, of course—a Hertz is still a Hz, a binary digit is still a bit, and a wavelength is still a λ—and the basics remain extremely important. Many of the technologies, applications, service providers, and regulations, however, have changed so much as to be almost unrecognizeable.

So it is with this book, which builds on the original work but goes into much more technical detail and reflects 10 years of change in the business, including a host of new technologies. This book is a condensation of more than 30 years of my knowledge and experience, hundreds of years of the knowledge and experience of my professional associates (who, thankfully, also are my friends), and hundreds of books and thousands of articles written by others over the last 130 years. I am a hoarder of paper, and I forever will owe those authors a debt of gratitude for putting their thoughts and observations on paper and now on the Web.

Mark Miller, president of the Diginet Corporation, was invaluable in the development of this work, and the predecessor works. As consulting editor, he applied his considerable technical expertise to ensure their absolute integrity. Mark Miller put his name on this book, which gives me great pride. His 20 published books are well respected in the industry, and he kindly allowed me to draw from them extensively during the course of this work. Thank you, Mark, for your friendship and guidance over the last dozen years or so.

Bill Flanagan, president of Flanagan Consulting, served as technical editor, providing a great deal of guidance across a wide range of technologies and applications. Bill is perhaps as knowledgeable as anyone across the full range of subject matter covered in this book and has written 11 excellent books, from which I drew extensively. Technically, Bill is absolutely brilliant and totally unyielding. He went through every word of every draft as if his life depended on making me correct every single Hz, bit, and byte. Bill particularly likes to quote George Bernard Shaw: "Beware of false knowledge; it is more dangerous than ignorance." Bill is right, as usual. He also happens to be a patient and skilled collaborator with a great sense of balance and a wonderful sense of humor. I also have worked with him on several consulting projects and know him to be incredibly honest, ethical, fair, and just an all around good guy. I have learned over the years to trust very few people very far, but I trust Bill Flanagan a very long way, indeed.

T.S. Eliot is quoted as having said that "Most editors are failed writers, but so are most writers." He certainly didn't have Mark or Bill in mind, for their success as writers clearly has made them better editors and they, in turn, have made me a better writer.

I also am indebted to the tens of thousands of people who have attended my public and private seminars around the world over the past 20 years. I never taught a seminar that I didn't learn something about a technology, or an application, or a way to phrase a concept to make it more understandable. It is such a pleasure to do something that you love, to work with people who are enthusiastic and giving, to learn at the same time that you teach and write, and to get paid for all of it. Thank you all so very much.

Most of all, I am forever indebted to Margaret Horak, my gorgeous and giving wife. We first met over 25 years ago, but circumstances caused us to lose each other for too many years. Quite by accident, we found each other again through an article I wrote in 1990 for my friend Rick Luhmann, who then was editor-in-chief of *Teleconnect Magazine*. That article, quite clearly, was my greatest literary achievement. Margaret's love and devotion have translated into long hours of graphics development in the late hours of the night and on the weekends. Her graphic interpretations of my words have added immeasurably to my works over the last dozen years or so. Also, her common sense, level-headedness, good nature, and wonderful sense of humor kept me focused and helped me put this all in perspective. Her unyielding love for me has made me work ever harder, every minute of every day, to make her proud. I can only hope that I succeeded, for Margaret truly is my great treasure. Margaret, you are my heart and my soul.

It is not enough to wire the world if you short-circuit the soul. Technology without heart is not enough.

Tom Brokaw in a speech at commencement exercises at the
College of Santa Fe (New Mexico), May 15, 1999

ABOUT THE AUTHOR

Ray Horak is an internationally recognized author, columnist, lecturer, and consultant. *Communications Systems and Networks*, his previous book for Wiley, was a best seller by any measure, with over 50,000 copies sold through the third edition. He has written well over 100 articles for major publications, a number of white papers and case studies, and several regular columns. Ray serves on the editorial advisory boards of several leading technology periodicals and is on the advisory boards of several colleges and universities. Ray lectures before thousands of communications professionals annually around the world. As an author and a lecturer, he is well known for his ability to explain the most complex technologies in a plain-English, commonsense style—and with more than just a dash of humor, just to keep things in perspective.

Ray's 30 plus years' experience in the networked world began with Southwestern Bell Telephone Company (now AT&T), which was part of the AT&T Bell System (which no longer exists) at the time. Toward the end of his nine-year Bell System career, Southwestern Bell loaned him to AT&T (which bears no resemblance to the *new* AT&T) and Bell Telephone Laboratories (now just Bell Labs, which is a part of Lucent, which was just acquired by Alcatel, a French company, if you can believe that) in a failed attempt to make him fit the Bell-shaped mold (which has since been broken, for better or worse). Ray then spent nine years with CONTEL (subsequently acquired by GTE, which subsequently was merged into Bell Atlantic along with NYNEX, and the whole mess has been renamed Verizon), where he founded several successful companies, which were later either sold to or merged with other companies (which no longer exist). He then ran the CONTEL Executone (which no longer exists) operation in Houston, Texas (which is still there) for a short time. When CONTEL failed him, he worked for a software company (which exists but is barely recognizable) for a hard time. Finally, and in a desperate attempt to make a living on his own terms, he founded The Context Corporation, an independent consultancy. That worked so well that he's been independent ever since. Ray hopes for continued success, since he figures that he's been on his own for so long that he now is technically unemployable, which is fine with him. Borrowing and modifying a quote attributed to Groucho Marx, Ray claims that he wouldn't want to work for any company that has the poor judgment to hire him. He also claims no responsibility for the fact that most of the companies for which he worked no longer exist as such.

Ray met his lovely wife, Margaret, during his CONTEL days. They lost touch for years but were reunited when Margaret read an article he wrote for *Teleconnect* magazine in 1990, which fact attests to his skills as an author. Ray and Margaret claim to have a wonderful relationship, although it apparently becomes a bit strained when he is about halfway through writing a book.

RAY HORAK

CHAPTER 1

FUNDAMENTALS OF THE TECHNOLOGY: CONCEPTS AND DEFINITIONS

What is the transmitter? It is an electrical ear which receives the shock of the dancing molecules, just as does the membrane of the human ear. . . . What is the receiver? It is an electric mouth which can utter human sounds.
John Mills, *The Magic of Communication: A Tell-You-How Story*,
Information Department, American Telephone and Telegraph Company

Telecommunications is the transfer of information (*communications*) from a transmitter or sender to a receiver across a distance (*tele*). Some form of electromagnetic energy is employed to represent the data, usually through a physical medium, such as a copper wire or a glass fiber. A wireless medium, such as radio or infrared light, also may be employed. Additionally, a number of intermediate devices are typically involved in setting up a path for the information transfer and for maintaining adequate signal strength.

The information transfer must be established and maintained at acceptable levels in terms of certain key criteria such as speed of connection, speed of information transfer, speed of response, freedom from error, and, finally, cost. The information can be voice, data, video, image, or some combination of these—in other words, multimedia. The information can retain its original, or native, form during transmission. Alternatively, the transmission process can alter the data in some way in order to effect compatibility between the transmit and receive devices and with various intermediate network elements. For example, analog voice often is converted into a digital (data) bit stream for transmission over a digital network and is restored to analog form for the benefit of the analog-oriented human being on the receiving end. Additionally, the information can be compressed in order to improve the efficiency of information transfer and can even be encrypted for purposes of security.

Telecommunications and Data Communications Handbook, By Ray Horak
Copyright © 2007 Ray Horak

The electromagnetic energy employed to carry the data can be in the form of electric impulses or radiated energy in the form of either radio waves or light rays. The media employed can include metallic conductors (e.g., twisted pair or coaxial cable); free space, or airwaves (e.g., radio technologies such as microwave, satellite, or cellular or optical technologies such as free space optics); and glass or plastic fiber (fiber-optic cable). In a network of substantial size that spans a significant distance, a combination of transmission media typically is involved in the information transfer between transmitter and receiver. An intercontinental voice or data call might involve a combination of many media.

Additionally, a wide variety of intermediate devices might be employed to establish and maintain the connection and to support the information transfer. Such devices may include an appropriate combination of modems or codecs, controllers, concentrators, multiplexers, bridges, switches, routers, gateways, and so on.

This chapter examines a number of concepts and defines a fundamental set of elements that apply universally to communications networks. Distinctions are drawn between dedicated, switched, and virtual circuits, with two-wire and four-wire circuits defined and illustrated. The concept of bandwidth is explored in both analog and digital terms, with the advantages and disadvantages of each explained. The concept of multiplexers is discussed in detail, with variations on the theme detailed and illustrated. Finally, this chapter briefly explores the nature and evolution of various types of switches, including circuit, packet, frame, cell, and photonic switches.

1.1 FUNDAMENTAL DEFINITIONS

Developing a solid understanding of communications networking requires that one grasp a number of fundamental definitions, for telecommunications has a language of its own. The following terms, many of which are illustrated in Figure 1.1, are significant and are applied fairly universally across all voice, data, video, and other systems and network technologies. Some of the terms have multiple definitions that can be specific to a technology or application. As this book will use and illustrate these terms many times across a wide variety of technologies and applications, they soon will become part of your everyday vocabulary. (*Note:* This would be an excellent time to pause and warn your family and friends.)

- **Transmitter:** The *transmitter*, also known as the *sender* or *source*, is the device that originates the information transfer. Transmitters include voice telephones, data terminals, host computer systems, and video cameras.
- **Receiver:** The *receiver*, also known as the *sink*, is the *target* device, or *destination* device, that receives the information transfer. Receivers can include telephones, data terminals, host computers, and video monitors. Note that most devices are capable of both transmitter and receiver functions; exceptions include broadcast radio and TV devices.
- **Circuit:** A *circuit* is a communications path, over an established medium, between two or more points, from end to end, between transmitter and receiver. Circuit generally implies a *logical connection* over a *physical line*. Further, the

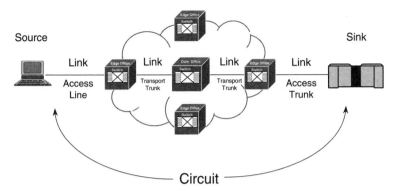

Figure 1.1 Simple circuit between transmitter and receiver across a network involving multiple links and switches.

term circuit often is used interchangeably with *path, link, line*, and *channel*, although such usage can be specific to the underlying technology, the overall context, and other factors. Circuits comprising copper twisted wire are either *two-wire* or *four-wire*, depending on the requirements of the specific application and the fundamental nature of the network. Circuits also may be for purposes of either access or transport. *Access* circuits are from the customer premises to the *edge* of the carrier network, while *transport* circuits are employed in the *core*, or *backbone*, of the network for purposes of long-haul transmission. Circuits may be *simplex* (one-way), *half-duplex* (two-way, but only one way at a time), or *full-duplex* (simultaneous two-way).

· **Link:** A *link* is a two-point segment of an end-to-end circuit (e.g., from terminal to switch or from switch to switch). Typically, a circuit comprises multiple links. Also, a circuit may consist of a single link, as often is the case between a host computer and a peripheral, such as a printer. Link sometimes is used interchangeably with line or circuit.

· **Line:** *Line* has several definitions, which may result in some confusion. In a Private Branch eXchange (PBX) environment, a *station line* refers to the connection between the PBX switch and the station user's terminal equipment, which usually is in the form of telephone, although it could be a computer workstation, a printer, a facsimile machine, or some other device. In rate and tariff terminology, line refers to a *local loop* connection from the telephone company Central Office (CO) switch to the user premises in support of Customer Premises Equipment (CPE) other than a switch. For example, such CPE may be in the form of a single-line residence or business set, a multiline set, or the common control unit of a key telephone system. In any case, line refers to a *voice-grade circuit*, in other words, a circuit serving a single physical location where it terminates in a relatively unsophisticated device. Further, a line has a single associated telephone number and generally is single channel in nature (i.e., supports a single transmission at a time). A line may be thought of as a tributary of a *trunk*. In telephone company (telco) parlance, line describes the user side or local loop side of the central office switch; in other words, the *line side* is the side of the network to which users connect to *access* the network.

The *trunk side* involves the high-capacity trunks that serve to interconnect the various telco switching centers in the core of the carrier network.

- **Trunk:** *Trunk* comes from the Latin *truncus*, meaning *torso*. The trunk is the main body apart from the head or appendages, much as the main channel of a river is apart from its tributaries. In the context of telecommunications, a trunk is a communications circuit, available to share among multiple users, on a pooled basis and with contention for trunk access managed by an intelligent switching device. Therefore, trunks interconnect *switches*. For example, *tie trunks* connect PBX switches in a private, leased-line network, *central office exchange trunks* connect PBXs to telephone company central office exchange switches, and *interoffice trunks* interconnect central office exchange switches. *Trunk groups* are groups of trunks serving the same special purpose, with examples including Direct Inward Dial (DID) and tie trunk groups. Trunks are directional in nature, with the options being one-way outgoing (originating), one-way incoming (terminating), or two-way (combination).

- **Channel:** In formal standards terms, a *channel* is a means of one-way connection between transmitter and receiver—therefore, a one-way *circuit* or signal *path*. In data processing terminology, particularly IBM, a channel is a high-speed two-way connection between mainframe and peripheral. In common usage, a channel is a *logical* connection over a *physical* circuit to support a single conversation. You can configure a physical circuit in such a way as to support one or many logical conversations. Multichannel circuits always are four-wire in nature—either physical or logical four-wire.

- **Switch:** A *switch* is a device that establishes, maintains, and changes logical connections over physical circuits. Common examples of switches include PBXs and Central Office Exchanges (COs or COEs), both of which are *circuit switches*. Circuit switches establish connections between circuits (or links) on demand and as available. While developed to support voice communications, circuit switches can support any form of information transfer (e.g., data and video communications). *Packet switch* is a generic term that actually includes packet, frame, and cell switches. Packet switching evolved in more sophisticated networks, primarily in support of computer-to-computer data and image transfer. In terms of physical placement, there are *edge switches* and *core switches*. Edge switches are positioned at the physical edge of a network; the user organization gains access to an edge switch via an access link. Core switches, also known as *tandem switches* and *backbone switches*, are high-capacity switches positioned in the physical core, or backbone, of a network and serving to interconnect edge switches. Although some switches are very intelligent in many respects, a pure switch makes connection decisions only at the link level. That is to say that a switch has a very limited view and cannot consider the network as a whole. Therefore, switches operate link by link, that is, hop by hop, generally under the control of a centralized set of logic that can coordinate their activities in order to establish end-to-end connectivity across a multilink circuit.

- **Router:** A *router* is a highly intelligent switch capable of making traffic routing decisions based on a view of the network as a whole. This is in contrast to simple switches, which see only an individual link and have no sense of the larger network. Routers are programmable devices that can be quite sophisticated. In determining the route for a given communication, a router can be programmed to consider a number of factors including the addresses of the originating and destination devices, the least-cost route, the least-congested route, and the shortest route. Routers can be capable of connecting dissimilar networks, such as circuit-switched and packet networks, and accomplishing the conversion processes necessary to resolve any issues of incompatibility. Chapter 8 discusses routers in great detail.

- **Network:** A *network* is a fabric of elements that work together much as the fabric of a net to support the transfer of information. In the extreme sense, a network includes everything from the transmitters to the receivers, including all links, switches, and other intermediate devices that can be called upon to support a communication.

  - **Local Area Network (LAN):** A LAN is a local, that is, limited-distance, packet network designed for interconnecting computers, peripherals, storage devices, and other computing resources within a confined area. A LAN may serve an office, a single floor, an entire building, or perhaps a campus of many buildings but generally does not cross a public right-of-way. LANs generally are private networks. LANs can be interconnected, perhaps across a MAN or WAN.

  - **Metropolitan Area Network (MAN):** A MAN is a public network that serves a metropolitan area or perhaps a portion of a metropolitan area such as a city or a suburb. MANs tend to be data oriented and increasingly serve to interconnect LANs. A number of carriers now offer high speed metropolitan Ethernet services, for example.

  - **Wide Area Network (WAN):** A WAN is a network that covers a wide geographic area such as a state, province, region, or country. The *Public Switched Telephone Network* (PSTN) is a voice-oriented WAN that individuals use to connect voice calls. The Internet is a WAN, as are many other data-oriented public networks. WANs can serve to interconnect LANs and MANs. The WAN commonly is depicted as a *cloud* (Figure 1.1), which originated in sales presentations of the 1970s for data communications networks. The thought behind the cloud simply was that the specific internal workings of the networks could be many and various, change from time to time, and vary from place to place. The cloud served to obscure those internal workings from view. The cloud was the consummate conceptual sale—data simply popped in on one end of the network in one format and popped out on the other side of the network in another format. Interestingly, the data network that gave rise to the cloud never worked, but the cloud lives on. Actually, the cloud is entirely appropriate for depicting the Internet, as the specific internal workings are largely unpredictable for any given call.

1.2   DEDICATED, SWITCHED, AND VIRTUAL CIRCUITS

Circuits can be provisioned on a dedicated, switched, or virtual basis, depending on the nature of the application and the requirements of the user organization. Ultimately, issues of availability and cost-effectiveness determine the specific selection.

1.2.1 Dedicated Circuits

Dedicated circuits are distinct physical circuits dedicated to directly connecting devices (e.g., PBXs and host computers) across a network (Figure 1.2). Dedicated circuits make use of access circuits, in the form of local loops, to access the service provider's *Point of Presence* (POP) at the edge of the carrier network. Rather than accessing a switch at the POP, a dedicated circuit terminates in the carrier's *wire center*, where cross connections are made from the short-haul access circuit to a long-haul transport circuit. Dedicated circuits serve a single-user organization only, rather than serving multiple users. They offer users the advantage of a high degree of availability as well as specified levels of capacity and quality. You can *condition* dedicated circuits to deliver specific levels of performance by adding amplification or other signal processing enhancements to the line to optimize its transmission characteristics, whereas you generally cannot do so with switched circuits, at least not from end to end. (*Note:* The local loop links used to access switched networks are dedicated circuits and therefore can be conditioned, but the links within the network cloud vary from call to call and therefore cannot be conditioned.) Additionally, the costs of dedicated circuits generally are calculated on a flat-rate, rather than a usage-sensitive, basis. That is to say you can use them continuously and to their full capacity for the same cost as if you never use them at all.

However, the reservation of a circuit for a specific customer has a negative effect on the network efficiency because that circuit is taken out of shared public use and, therefore, is unavailable for use in support of the traffic of other users. As a result, dedicated circuits tend to be rather expensive, with their costs being sensitive to distance and capacity. Additionally, the process of determining the correct number, capacity, and points of termination of such circuits can be a difficult and lengthy design and configuration process. Further, long lead times often are required for the carrier to configure (i.e., provision) or reconfigure such a circuit. Finally, as dedicated circuits are susceptible to disruption, backup circuits often are required to ensure effective communications in the event of either a catastrophic failure or serious performance degradation.

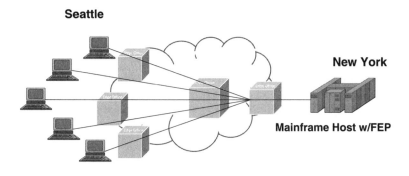

Figure 1.2 Dedicated circuits between Seattle data terminals and a New York mainframe through a Front-End Processor (FEP).

Traditionally, dedicated digital circuits have connected large data centers that communicate intensively. Similarly, many large end-user organizations with multiple locations have used dedicated circuits known as tie trunks to tie together multiple PBXs. In both cases, the advantages of assured availability, capacity, and quality in support of mission-critical, time-sensitive applications often outweigh considerations of configuration difficulty and risk of circuit failure. Dedicated circuits often are known as *nailed-up* circuits because, in days long past, the twisted-pair copper physical circuits were hung from nails driven in the walls of the carrier's wire centers.

1.2.2 Switched Circuits

Switched circuits are connected through the network on a flexible basis through one or more intermediate switching devices. Traditionally, the switches were in the form of the telephone company Central Office exchanges, as illustrated in Figure 1.3. Individual users seeking switched connections through the network connect to the edge switches via dedicated local loops, that is, access circuits, terminating at the premises. Through those local loops multiple users compete for limited core network resources on demand and as available, with each switch serving as a point of contention. This sharing of limited network resources clearly allows the network providers to realize significant operational efficiencies, which are reflected in lower network costs. The end users realize the additional advantages of flexibility and resiliency because the network generally can provide connection between any two physical locations through multiple alternate transmission paths.

In the domain of traditional circuit-switched voice networking, all local, regional, national, and international networks are interconnected. The cost of establishing switched circuits traditionally is sensitive to factors such as the distance between originating and terminating locations, duration of the connection, time of day (prime time vs. nonprime time), and day of the year (business day vs. weekend day or holiday). Yet, circuit-switched connections offer great advantage for calls of short

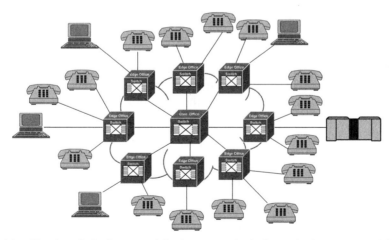

Figure 1.3 Circuit-switched connectivity between single-line telephone sets and between data terminals and a mainframe host computer through edge office and core office switches.

duration, connection between specific locations that communicate relatively infrequently, in cases where network redundancy is important, and at times when a high degree of flexibility is advantageous. Interconnectivity typically is much more selective in the data domain, with the exception of the Internet. Although the Internet is virtually ubiquitous, some governments place restrictions on access and content. The vast majority of voice calls and data calls are carried over switched circuits.

1.2.3   Virtual Circuits

Virtual circuits are logical, rather than physical, circuits. Virtual circuit connectivity is provided over high-capacity, multichannel physical circuits, such as fiber-optic transmission facilities. Virtual circuits are established through the network based on options and instructions defined in software routing tables. *Permanent Virtual Circuits* (PVCs) are permanently defined in routing tables, until such time as the carrier permanently redefines them. *Switched Virtual Circuits* (SVCs) are determined at the moment in time the communication is requested, with relatively sophisticated devices making highly informed decisions about the best path available in support of the specific requirements of the communication. In either case, a virtual circuit provides connectivity much as though it were a physical circuit, with all data traveling the same path. Such a physical circuit often can support a great number of logical circuits, or logical connections. In the high-capacity, fiber-optic backbone carrier networks, dedicated circuits are provided to users on a virtual basis, with the capacity and other performance characteristics of the circuit performing as though the circuit were dedicated.

Now it is worth pausing to further define and contrast the terms *transparent* and *virtual*. Transparent means that a network element (e.g., hardware or software) exists but appears to the user as though it does not. Without special test equipment, the end user may be totally unaware of its existence. Virtual means that the network element behaves as though it were something more than it actually is. So, a user can access a virtual circuit on a transparent basis.

It also is necessary to further define and distinguish between logical and physical circuits and channels. A *logical circuit* refers to the entire range of network elements (e.g., physical circuits, buffers, switches, and control devices) that support or manage communication between a transmitter and receiver. A single logical circuit can support many *logical channels*. In order to establish and support the information transfer, a *physical circuit*, or *physical path*, must be selected for the information transfer. A single physical circuit can support many logical circuits. The transmission facilities in the physical path may be in the form of copper wire (e.g., twisted-pair or coaxial cable), radio (e.g., microwave or satellite), or glass or plastic fiber (fiber optic) [1].

1.3   TWO-WIRE VERSUS FOUR-WIRE CIRCUITS

Telegraph and telephone circuits originally were metallic one-wire, which proved satisfactory even for two-way communications. Soon after the invention of the telephone, however, two-wire circuits were found to offer much better performance characteristics, due largely to their improved immunity from electromagnetic

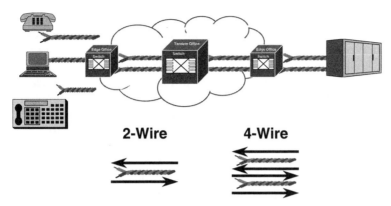

Figure 1.4   Applications for two- and four-wire circuits.

interference. Four-wire circuits offer still better performance, although at higher cost. Both two- and four-wire circuits are widely used.

1.3.1   Two-Wire Circuits

Two-wire circuits carry information signals in both directions over the same physical link or path. Typically, such a circuit is provisioned through the use of a single twisted-pair, copper wire connection. Within such a two-wire circuit, two wires are required to complete the electrical circuit, with the current in one wire opposite to the current in the other, and both wires carry the information signal. A common example is a local loop connection between a telephone company's CO switching center and an individual single-line or multiline telephone set, data terminal, or Key Telephone System (KTS), as depicted in Figure 1.4.

Two-wire circuits generally cover a short distance; the vast majority of two-wire local loops, for example, are less than 18,000 ft in length [2]. Longer loop lengths require some form of amplification in order to maintain signal strength. Additionally, such a circuit offers relatively little bandwidth, or capacity, and is single channel in nature (i.e., supports only a single conversation). Finally, two-wire circuits, generally speaking, are analog in nature; therefore, error performance (quality) is relatively poor. Two-wire circuits often are characterized as *voice grade*, that is, good enough for voice communications between humans, who are reasonably intelligent devices capable of adapting to errors in transmission over a circuit of relatively poor quality. A voice-grade circuit also will support low-speed data transmission through a modem, which has internal mechanisms for dealing with transmission errors. Two-wire circuits of lesser grade serve lesser applications, such as burglar alarms and fire alarms.

1.3.2   Four-Wire Circuits

According to the most basic definition, *four-wire circuits* carry information signals in both directions over separate physical links or paths and in support of simultaneous, two-way transmission. Traditionally, such a circuit was provisioned through the use of two copper pairs, one for transmission (*forward path*, or *upstream path*) and one for reception (*reverse path*, or *downstream path*); such a circuit is known as *physical four-wire*. However, current technology accommodates four-wire transmission over a single physical link or path and over a variety of transmission media, including twisted-pair, coaxial cable, or fiber-optic cable. In other words, the circuit may be physical two-wire (or even physical one-wire) and *logical four-wire*, performing as a four-wire circuit but employing fewer than four wires. In fact, a four-wire circuit can be established without the use of any wires at all, as in the case with a circuit established over microwave, satellite, or infrared transmission systems.

Although the absolute cost of four-wire circuits is higher than that of two-wire circuits, they offer considerably improved performance. As four-wire circuits can accommodate multiple, simultaneous communications in a full-duplex mode, all multichannel circuits are four-wire. By virtue of their multichannel capability, four-wire circuits also are capable of supporting out-of-band signaling and control, which offers the significant advantage of being nonintrusive, that is, nondisruptive. Additionally, such circuits typically offer much greater bandwidth, or capacity, and typically are digital, rather than analog, in nature. As a result, error performance generally improves. Long-haul circuits (traditionally defined as equal to or greater than 50 miles, or 80 km) usually are four-wire [3], as the carriers, or service providers, typically aggregate large volumes of traffic for transport over multichannel facilities. Figure 1.4 illustrates typical examples of cost-effective applications of four-wire circuits, specifically to interconnect PBX, CO, and tandem switches in a voice environment.

1.4 BANDWIDTH

Bandwidth is a measure of the capacity of a circuit or channel. More specifically, it refers to the total frequency on the available *carrier* for the transmission of data. There is a direct relationship between the bandwidth of a circuit or channel and both its frequency and the difference between the minimum and maximum frequencies supported. While the information signal (bandwidth usable for data transmission) does not occupy the total capacity of a circuit, it generally and ideally occupies most of it. The balance of the capacity of the circuit may be used for various *signaling and control (overhead)* purposes. In other words, the total *signaling rate* of the circuit typically is greater than the effective *transmission rate*. The more information you need to send in a given period of time, the more bandwidth you require.

1.4.1 Carrier

Carrier is a continuous signal on a circuit that is at a certain frequency or within a certain frequency range. The primary value of the carrier is in its support of the information-bearing signal (i.e., it carries the information signal), which the transmitter impresses on the carrier by varying the signal in some fashion and which the receiver must detect and interpret. The carrier also can support signaling and control information used to coordinate and manage various aspects of network operations.

1.4.2 Hertz

Hertz (Hz), named after Heinrich Rudolf Hertz, the physicist who discovered radio waves, is the measurement of frequency. Hertz also is the measurement of analog bandwidth, measured as the difference between the highest and lowest frequencies over a circuit or within a channel. Hertz refers to the number of electromagnetic waveforms transmitted per second (i.e., signals per second or cycles per second). Although some applications operate in very low capacity environments, measured in tens of or hundreds of hertz, the frequencies generally are much higher. Hence and by way of example, you can measure analog bandwidth in kilohertz (kHz, or thousands of hertz), megahertz (MHz, or millions of hertz), gigahertz (GHz, or billions of hertz), and terahertz (THz, or trillions of hertz).

1.4.3 Baud

Baud is an olde term that refers to the number of signal events (i.e., signal changes or signal transitions) occurring per second over an analog circuit. The baud rate can never be higher than the raw bandwidth of the channel, as measured in Hz. *Baud rate* and *bit rate* often and incorrectly are used interchangeably. The relationship between baud rate and bit rate depends on the sophistication of the modulation scheme used to manipulate the carrier. The bit rate and baud rate can be the same if each bit is represented by a signal transition. The bit rate typically is higher that the baud rate as a single signal transition can represent multiple bits. Chapter 6 explores the distinction between baud rate and bit rate in more detail.

1.4.4 Bits and Bytes per Second

Quite simply, *bps* (lowercase *b*) is the bit rate, or the number of *bits* transmitted over a circuit per second. It is the measurement of bandwidth over digital circuits and should not be confused with the speed of the electromagnetic signal, that is, the velocity of propagation. In other words, bps refers to the number of bits that pass a given point in a circuit, not the speed at which they travel over a distance. Over an analog circuit, you can manipulate the electromagnetic waveforms to support the transmission of multiple bits per baud. As a result, the bit rate (bps) can be a multiple of the baud rate, even without the application of special compression techniques. A thousand (1000) bps is a kilobit per second, or kbps; a million (1,000,000) bps is a Megabit per second, or *Mbps*; a billion (1,000,000,000) bps is a Gigabit per second, or *Gbps*; and a trillion (1,000,000,000,000) bps is a terabit per second, or *Tbps*.

Bps (uppercase *B*) refers to the number of *bytes* transmitted over a circuit per second. Bps is used exclusively in the context of storage networking, as storage is byte oriented. Storage technologies such as *Fibre Channel* and *ESCON* (Enterprise Systems CONnection) measure the speed of information in bytes per second.

1.4.5 Narrowband, Wideband, and Broadband

Bandwidth levels or ranges fall into three categories: narrowband, wideband, and broadband. These terms are imprecise, as there are no widely accepted, formal standard definitions. Also, the definitions vary, depending on the technological context.

Narrowband: *Narrowband* refers to voice-grade bandwidth. In analog telecommunications terms, a narrowband channel has bandwidth of a nominal 4 kHz, which is the standard for analog voice. In digital terms, a narrowband channel is 64 kbps, which is the fundamental standard for uncompressed, digitized voice. Narrowband sometimes is used to describe a channel or circuit of less than voice-grade bandwidth. Narrowband also is used to describe some number of 64-kbps channels ($N \times 64$ kbps). *Narrowband Integrated Services Digital Network* (N-ISDN), for example, comprises two information-bearing channels of 64 kbps each plus a signaling and control channel of 16 kbps, for a total of 144 kbps. So, narrowband essentially is used to describe a circuit or channel offering relatively little bandwidth. I use the term narrowband to describe bandwidth up to a maximum of 24 channels. In digital terms, this means bandwidth up to the T1 signaling rate of 1.544 Mbps, which supports 24 channels at 64 kbps, at least according to North American standards. The equivalent international standard is E1, which has a signaling rate of 2.048 Mbps and supports 30 channels at 64 kbps. (*Note:* You will find detailed information on narrowband ISDN, T1, and E1 in subsequent chapters.)

Wideband: *Wideband* is used to distinguish a circuit or channel with capacity greater than narrowband. Wideband sometimes is used to describe a circuit or channel that has bandwidth wider than normal for operation. In the radio domain, wideband refers to a radio channel covering a relatively wide range of frequencies. *Ultra-Wideband* (UWB), for example, is defined as a radio system with occupied bandwidth (i.e., the difference between the highest and lowest frequencies in the radio channel) greater than 25 percent of the center frequency. Wideband sometimes is used interchangeably with broadband, as if the terminology were not already confusing enough.

Broadband: *Broadband* is an imprecise, evolving term referring to a circuit or channel providing a relatively large amount of bandwidth. I generally use the term to describe capacity equal to or greater than the nominal T1 rate of 1.544 Mbps, which is the basis for Broadband ISDN (B-ISDN), according to North American standards. European and international standards define B-ISDN at the E1 rate of 2.048 Mbps. Broadband has an entirely different definition in the context of LANs. You will learn about LANs in Chapter 8.

1.5 ANALOG VERSUS DIGITAL

Along one dimension, communications fall into two categories, analog and digital. In the analog form of electronic communications, information is represented as a continuous electromagnetic waveform. Digital communications involves modulating (i.e., changing) the analog waveform in order to represent information in binary form (1 s and 0 s) through a series of blips or pulses of discrete values, as measured at precise points in time or intervals of time.

1.5.1   Analog Sine Waves: Starting Point

Analog is best explained by examining the transmission of a natural form of information, such as sound or human speech, over an electrified copper wire. In its native
form, human speech is an oscillatory disturbance in the air that varies in terms of
its volume or power (amplitude) and its pitch or tone (frequency). In this native
acoustical mode, the variations in amplitude cause the physical matter in the air to
vibrate with greater or lesser intensity and the variations in frequency cause the
physical matter in the air to vibrate with greater or lesser frequency. So, the physical
matter in the space between the speaker's mouth (transmitter) and the listener's
ear (receiver) serves to conduct the signal. That same physical matter, however,
also serves to *attenuate* (weaken) the signal. The longer the distance is between
mouth and ear, the more profound the effect. As a result, it is difficult, if not impossible, to communicate acoustically over distances of any significance, especially
between rooms separated by doors and walls or between floors separated by floors
and ceilings. In order to overcome these obvious limitations, native voice acoustical
signals are converted into electromagnetic signals and sent over networks, with the
compression waves falling onto a microphone in a transmitter embedded in a
handset or speakerphone. The microphone converts the acoustical signals into *analogous* (approximate) variations in the continuous electrical waveforms over an
electrical circuit, hence the term analog. Those waveforms maintain their various
shapes across the wire until they fall on the speaker embedded in the receiver. The
speaker converts them back into their original acoustical form of variations in air
pressure, which can be received by the human ear and understood by the human
brain.

A similar but more complicated conversion process is used to transmit video over
networks. In its native form, video is a series of still images, each comprising reflected
light waves. Transmitted in rapid succession, the series of still images creates the
illusion of fluidity of motion. The transmitter (i.e., video camera) creates analogous
variations in electrical or radio waveforms, which it sends in rapid succession over
a network to a receiver (i.e., monitor), which re-creates an approximation (analog)
of the original information.

Information that is analog in its native form (voice and other forms of audio and
image and video) can vary continuously in terms of intensity (volume or brightness)
and frequency (tone or color). Transmission of the native information stream over
an electrified analog network involves the translation of those variations into amplitude and frequency variations of the carrier signal. In other words, the carrier signal
is *modulated* (varied) in order to create an analog of the original information
stream.

The electromagnetic sinusoidal waveform, or *sine wave*, as illustrated in Figure
1.5, can be varied in amplitude at a fixed frequency using *Amplitude Modulation*
(AM). Alternatively, the frequency of the sine wave can be varied at constant amplitude using *Frequency Modulation* (FM). Additionally, both frequency and amplitude
can be modulated simultaneously to create an analog of the native signal, which
generally varies simultaneously along both parameters. Finally, the position of the
sine wave can be manipulated (actually, can appear to be manipulated), adding the
third technique of *Phase Modulation* [PM, also known as *Phase Shift Keying* (PSK)].
Chapter 6 discusses these modulation techniques in considerable detail.

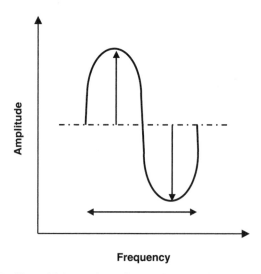

Figure 1.5   Sinusoidal waveform illustrating amplitude and frequency.

Bandwidth, in the analog world, is measured in Hertz (Hz). The available bandwidth for a particular signal is the difference between the highest and lowest frequencies supported by a channel or circuit. For example, a circuit can support a 3.0-kHz voice channel through the use of a *bandpass* (i.e., band-limiting) filter supporting transmission at frequencies between approximately 300 and 3300 Hz. Similarly, a circuit can support a 3.0-kHz channel at frequencies between 7000 and 10,000 Hz. *Passband* refers to the upper and lower cutoff frequencies at which the bandpass filters operate [2].

1.5.1.1 Voice   The signaling rate of a voice-grade channel is nominally (approximately) 4000 Hz, or 4 kHz. The bandwidth in the range 0–300 Hz generally is ignored, suppressed by the equipment's lack of ability to deal with it at those low frequencies. The voice band is approximately 3.0 kHz wide, running at 300–3300 Hz. Signaling and control functions take place in the band 3300–3700 Hz. The lower band of 0–300 Hz and the upper band of 3700–4000 Hz have value, as they are used for maintaining separation between information channels, each of which is supported over a separate carrier frequency range, when analog voice channels are multiplexed using Frequency Division Multiplexing (FDM), which is described later in this chapter. While human speech can transmit and human hearing can receive a much wider range of frequencies, 3.0 kHz is considered sufficient for voice communications and certainly is more cost effective for the service providers than attempting to support full-fidelity voice. [*Note:* At 300 Hz, the cutoff frequency is high enough to reject AC (Alternating Current) electrical hum at 60 Hz in North American networks and 50 Hz in European networks. At 3300 Hz, the frequency is high enough to include all of the important harmonics that make human voice recognizable. Voice bandwidth and frequency range vary, with some filters operating at 200–3500 Hz, some at 300–3400 Hz, and so on.] Band-limiting filters employed in carrier networks constrain the amount of bandwidth provided for a voice application, which certainly conserves bandwidth. Capping the bandwidth at 3300 Hz also prevents

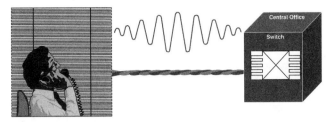

Figure 1.6   Analog voice transmission over a two-wire local loop.

aliasing, a phenomenon that occurs when different continuous signals overlap and become indistinguishable (i.e., becoming *aliases* of one another) when encoded into digital format for transmission over digital facilities [3, 4]. Figure 1.6 illustrates an analog local loop supporting voice communications.

1.5.1.2 Video   An analog cable TV [Community Antenna TeleVision (CATV)] video channel has a width of approximately 6,000,000 Hz, or 6 MHz. Approximately 4.5 MHz is used for transmission of the video signal and the balance is used for *guard bands* to separate the various adjacent channels riding the common, analog coaxial cable system.

1.5.2 Digital Bit Streams: Ones and Zeros

While the natural world is analog in nature, the decidedly unnatural world of contemporary computers is digital in nature. Computers process, store, and communicate information in binary form. That is to say that a unique combination of *1*s and *0*s has a specific meaning in a computer coding scheme, which is much like an alphabet. A *bit* (*bi*nary digi*t*) is an individual *1* or *0*. The output of a computer is in the form of a *digital bit stream*.

   Digital communication originates in telegraphy, in which the varying length (in time) of making and breaking an electrical circuit results in a series of *dots* (short pulses) and *dashes* (long pulses) that, in a particular combination, communicate a character or series of characters. Early mechanical computers used a similar concept for input and output. Contemporary computer systems communicate in binary mode through variations in electrical voltage.

   Digital signaling, in an electrical network, involves a signal that varies in voltage to represent one of two discrete and well-defined states. Two of the simplest approaches are *unipolar* signaling, which makes use of a positive (+) voltage and a *null*, or zero (0), voltage, and *bipolar* signaling, which makes use of a positive (+) or a negative (−) voltage. The transmitter creates the signal at a specific carrier frequency and for a specific duration (*bit time*), and the receiver monitors the signal to determine its state (+ or −). Various data transmission protocols employ different physical signal states, such as voltage level, voltage transition, or the direction of the transition. Because of the discrete nature of each bit transmitted, the bit form is often referred to as a *square wave*. Digital devices (Figure 1.7) benefit greatly from communications over digital transmission facilities, which are not only faster but also relatively free from noise impairments.

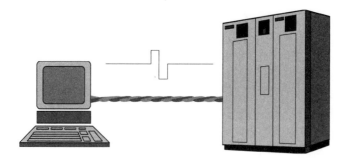

Figure 1.7   Digital communications between a terminal and host.

Digital signaling in an optical network can involve either the pulsing on and off of a light source or a discrete variation in the intensity of the light signal. Digital transmission over radio systems (e.g., microwave, cellular, or satellite) can be accomplished by discretely varying the amplitude, frequency, or phase of the signal.

Bandwidth, in the digital world, is measured in bits per second. The amount of bandwidth required depends on the amount of raw data to be sent, the desired speed of transmission of that set of data, and issues of transmission cost. Compression of data files prior to transmission is fairly routine, as it improves the efficiency of transmission, reduces the transmission time, and thereby reduces transmission costs.

1.5.3 Analog versus Digital Transmission

Transmission systems are either analog or digital in nature. In an analog transmission system, all components operate in analog (continuous-waveform) mode. Similarly, a digital transmission system must be digital from end to end. A network may consist of both analog (A) and digital (D) transmission systems, with A-to-D and D-to-A conversions required to resolve the obvious issues of incompatibility. Analog and digital signal formats each have advantages and appropriate applications.

1.5.3.1 Analog Advantages   Analog transmission offers advantages in the transmission of analog information, in which case it is more bandwidth-conservative than is digital transmission. As analog transmission facilities formed the foundation for telecommunications networking, they were widely deployed and remain virtually ubiquitous.

1.5.3.1.1 Analog Data   Analog has an advantage with respect to the transmission of information that is analog in its native form, such as voice and video. The process of transmission of such information is relatively straightforward in an analog format, as the continuous flow of the native signal is easily and fully represented in the continuous flow of the carrier signal. Although filters may constrain the amplitude and frequency levels of the transmitted signal, the essence of the native signal and its infinite variations is faithfully represented through an analog system and over an analog network. Conversion of analog data to a digital bit stream requires special conversion equipment. Such equipment adds cost, contributes additional points of failure, and can negatively affect the quality of the signal through the conversion process itself. The end result is very much an approximation of the original data. The examination of T-carrier in Chapter 7 includes discussion of the impacts of analog and digital conversions.

1.5.3.1.2 Bandwidth A raw analog information stream, if fully and faithfully transmitted, consumes far less bandwidth in analog form than in digital form. This is particularly evident in CATV transmission, where 50 or more analog channels routinely are provided over a single coaxial cable system. Without the application of relatively sophisticated compression techniques, fewer digital channels could be supported.

1.5.3.1.3 Availability Finally, analog transmission systems are in place worldwide. All aspects of the standards are well understood and easily implemented, and the interconnection of analog systems is routine. As voice traditionally comprised the majority of network traffic and as the vast majority of voice terminals still are analog devices, voice communication largely continues to depend on analog networks at the local loop level. Conversion to fully digital networks would require prohibitively expensive, wholesale conversion of such terminal equipment and local loops.

1.5.3.2 Digital Advantages Digital transmission certainly is advantageous for the transmission of digital information. Additionally, digital data can be compressed effectively and easily. Security of the data can be more readily ensured, and the error performance of digital networks is much improved over their analog counterparts. Finally, the cost effectiveness of such networks is improved by virtue of the greater bandwidth they provide, especially since they can be more easily upgraded and more effectively managed.

1.5.3.2.1 Digital Data Just as it often is better to transmit analog information in an analog format, it is better to transmit digital information in a digital format. Digital transmission certainly has the advantage when transmitting binary computer data. The (modem) equipment required to convert the information to an analog format and send the digital bit streams over an analog network represents additional cost, is susceptible to failure, and can induce errors into the datastream.

1.5.3.2.2 Compression Digital data can be compressed relatively easily, thereby increasing the efficiency of transmission. As a result, substantial volumes of computer data can be transmitted using relatively little raw bandwidth, with the receiving device decompressing the data to reconstitute it in its original form. Transmission of analog voice and video benefits from digital conversion if the analog signals are sampled at appropriate intervals, converted into byte format, and compressed. Several subsequent chapters discuss this process in detail.

1.5.3.2.3 Security Digital systems offer much improved security. While analog systems can offer some measure of security through the scrambling, or intertwining, of several frequencies, you can fairly easily defeat that technique. Digital information, conversely, can be *encrypted* to create the appearance of a single, pseudorandom bit stream. Thereby, the true meaning of individual bits, sets of bits, and the total bit stream cannot be determined without the key to unlock the encryption algorithm employed.

1.5.3.2.4 Error Performance   Digital transmission offers much improved error performance (data integrity) in comparison with analog. This is due to the nature of the devices that serve to boost the signal at periodic intervals in the transmission system to overcome the effects of attenuation. Additionally, digital networks deal more effectively with *noise*, which always is present in transmission networks.

> **Attenuation:** Electromagnetic signals tend to weaken, or *attenuate*, over a distance; this is particularly true of electrical signals carried over twisted-pair copper wire, due to factors including the level of resistance (or impedance) in the wire and the tendency of the signal to radiate, or spread out, from the wire. It also is true of microwave radio and other terrestrial radio systems, due to the physical matter in the air and the tendency of the signal to spread out, or disperse. Attenuation is sensitive to carrier frequency, with higher frequency signals attenuating more than lower frequency signals.
>
> **Noise:** Signals also tend to pick up *noise* as they transverse the network. Again, this is particularly true of twisted-pair copper wire systems. Such wires tend to act as antennas and, therefore, absorb noise from outside sources of *ElectroMagnetic Interference* (EMI) and *Radio Frequency Interference* (RFI). The quality of the signal degenerates as it is distorted by the noise, and the integrity of the data transmission suffers as a result.

1.5.3.2.5 Cost   The cost of the computer components required in the digital conversion and transmission process has dropped to a considerable extent, while the ruggedness and reliability of those components have increased over the years.

1.5.3.2.6 Upgradeability   Since digital networks comprise computer components, they can relatively easily be upgraded, within design limits. Such upgrades might increase bandwidth, improve error performance, and enhance functionality. Certain upgrades often can be effected through software downloads over the network, thereby eliminating the need for a "truck roll," that is, the need to dispatch a technician.

1.5.3.2.7 Management   Generally speaking, digital networks can be managed much more easily and effectively because they comprise computerized *Network Elements* (NEs). Such components can be endowed by their creators with the abilities to determine their status (i.e., on or off), sense their own levels of performance relative to programmed thresholds, isolate and diagnose failures, initiate alarms to upstream management systems, respond to queries, and respond to commands to correct the failure condition. Further, the cost of so enabling these devices is dropping rapidly.

1.6   LOADING COILS, AMPLIFIERS, AND REPEATERS

As noted earlier in this chapter, electromagnetic energy attenuates over a distance, whether the energy passes through a conductor or the air. Therefore, you must place some sort of device at regular spatial intervals in a network to overcome this phenomenon by boosting the signal strength. These boosting units receive a weakened incoming signal and transmit a stronger outgoing signal, which propagates across the network, weakening until it reaches another boosting unit, and so on. Analog networks make use of devices known as *loading coils* and *amplifiers*, which were originally known as relays. Digital networks employ *repeaters*.

> *On long lines the current often becomes so reduced by leakage, and from other causes, that it is insufficient to work an electro-magnet, either to mark paper, or give audible sound. It is therefore usual on such lines to interpose an instrument called a relay. A current weakened by distance, although unable to effectively work the receiving instrument, may have enough force to cause a light armature to be attracted by a small magnet. This movement may be made to bring a local battery into circuit so as to strengthen the current, and such an arrangement constitutes a relay.*
> —*Wonders of the Universe*, The Werner Company. 1899 [5]

1.6.1 Loading Coils

In order to ensure that local loops perform properly, they must be designed in such a way that signal strength is maintained at acceptable levels. The obvious solution to the problem of signal attenuation is to limit the length of the local loop. Therefore, the size of a *Carrier Serving Area* (CSA), that is, the geographical area served by a CO, is generally limited to a radius of about 18,000 ft. Beyond that distance, analog voice-grade signals of 4 kHz sent through Unshielded Twisted-Pair (UTP) copper cables attenuate to such an extent that they become unusable. Now, the attenuation increases substantially across all frequencies in the band because of the high *capacitance* created between the two tightly spaced conductors comprising a cable pair. The performance of long copper local loops can be improved through the use of *induction coils* known as *load coils* or *loading coils*. A load coil is a toroidal (i.e., ring-shaped or donut-shaped) device made up of a powdered iron core or sometimes a soft iron wire core around which copper wire is wound. The coil is then spliced into the local loop at some point where it can be properly sheltered in a weatherproof splice case, underground vault, or some similarly protected environment.

The coil functions as a *lumped* inductor, which is to say that at a specific point in the circuit the process of inductance takes place to compensate for the distributed capacitance. In effect, the load coil tunes the copper circuit, optimizing it for mid-voice-band performance. The load coil also functions as a low-pass filter, increasing loss above the cutoff frequency, which is 4 kHz in this case. The coil can reduce mid-voice-band attenuation by as much as 80 percent. Load coils commonly are placed on local loops that exceed approximately 18,000 ft (5.5 km) in length. The first load coil is placed approximately 3000 ft (0.9 km) from the CO and at intervals of 6000 ft (1.8 km) or so thereafter. Load coils are passive devices, that is, not electrically powered, and generally are limited to use in analog loops. The presence of load coils renders local loops unusable for ISDN, T-carrier, and other loops operating at high data rates, as the load coils filter out the high frequencies that accompany those higher data rates. The existence of so many loaded loops within the existing telephone plant was a significant deployment deterrent when ISDN was introduced some 20–25 years ago, as ISDN requires frequencies above 4 kHz. Neither are load coils acceptable for use on Asymmetric Digital Subscriber Line (ADSL) and other broadband local loops, as the frequency ranges far exceed 4 kHz. Where such services are to be deployed, the local loops must be properly *conditioned*, which entails removing the load coils and other impediments.

The presence of a load coil also has the effects of increasing the impedance of the circuit and reducing the velocity of propagation, that is, speed of signal propagation, to 10,000–12,000 miles per second. This speed penalty is not of particular significance in short voice-grade local loops. If the loops are long, however, load coils can create unacceptable problems with *echo*, or signal reflection. At any point in a circuit where an electromagnetic wave meets a discontinuity, a portion of the wave is reflected back in the direction of the transmitter. Such discontinuities can be caused by impedance mismatches, mismatches between line and balancing networks, irregular spacing of loading coils, and a host of other anomalies that are beyond the scope of this book. Now, echo is not a problem for human-to-human conversations as long as the echo return is not longer that 30–40 milliseconds (ms), which means that a circuit that propagates electromagnetic waves at near the speed of light (186,000 miles per second) can be 3000–4000 miles long without creating an echo problem. However, a circuit of 1000 miles with a signal traveling at 10,000 miles per second returns an echo in 100 ms, which makes conversations nearly impossible. In contemporary networks, high-speed carrier circuits are conditioned, with loading coils removed, so the problem largely disappears except in international communications or where satellite circuits are involved. Contemporary networks also are designed with echo cancelers or echo suppressors to deal with this problem, although it still occurs on occasion when echo suppressors fail.

1.6.2 Amplifiers (Analog)

The active boosting devices in an analog network are known as *amplifiers*. Amplifiers are unsophisticated devices that simply boost, or amplify, the weak incoming signal, much as does an amplifier in a radio receiver or TV set. In addition to attenuating, the signal accumulates noise as it transverses the network; the amplifier boosts the noise along with the signal. This effect is compounded through every step of the transmission system and through each cascading amplifier, thereby creating the potential for significant accumulated noise at the receiving end of the transmission. The resulting *Signal-to-Noise Ratio* (SNR) can produce unacceptable results. Amplifiers are spaced every 18,000 feet or so in a typical analog voice-grade twisted-pair local loop, for example. The exact spacing is sensitive to a number of factors, including the transmission medium and the carrier frequency, which affects raw bandwidth, transmission speed, and attenuation level. According to William Shockley, coinventor of the transistor [6]:

> *If you take a bale of hay and tie it to the tail of a mule and then strike a match and set the bale of hay on fire, and if you then compare the energy expended shortly thereafter by the mule with the energy expended by yourself in the striking of the match, you will understand the concept of amplification.*

The impact of amplification on voice communications generally is tolerable, as humans are relatively intelligent receivers who can filter out the noise or at least adjust to it. In the event of a truly garbled transmission, the human-to-human error detection and correction process simply involves a request for retransmission—the *Huh?* protocol. Should the quality of the connection be totally unacceptable, you can terminate and reestablish the connection. Computer systems, however, are not so forgiving, and garbled data are of decidedly negative value.

There are several exceptions to this broad characterization of amplifiers. *Erbium-Doped Fiber Amplifiers* (EDFAs), used in high-speed fiber-optic systems, amplify light signals falling in a narrow optical frequency range, performing much more cost effectively than optical repeaters. *Raman amplification*, another amplification technique used in fiber-optic systems, makes use of pump lasers that send a high-energy light signal in the reverse direction (i.e., the direction opposite the signal transmission). This technique not only increases the strength of the signal but also serves to improve its clarity. Note that the immunity of fiber-optic systems to ambient noise makes the use of such amplification techniques quite acceptable. A detailed discussion of EDFAs and Raman amplification can be found in Chapter 9.

1.6.3 Repeaters (Digital)

Digital systems generally replace periodic amplifiers with *regenerative repeaters* that regenerate the signal, rather than simply amplifying it. In an electrically based system, for example, the repeater essentially guesses the binary value (1 or 0) of the weak incoming signal based on its relative voltage level and regenerates a strong signal of the same value without the noise. This process considerably enhances the signal quality. Repeaters are spaced at approximately the same intervals as amplifiers, which is approximately 18,000 feet in voice-grade twisted-pair circuits.

The performance advantage of digital networks can be illustrated by comparing the error rates of amplifiers and regenerative repeaters. A twisted-pair, analog network, for example, yields an error rate on the order of 10^{-5} [3]. In other words, digital data sent across an analog network through modems will suffer one errored bit for every 100,000 bits transmitted (Figure 1.8). The very same twisted-pair network, if digitized and equipped with repeaters, will yield an expected error rate of 10^{-7}, or one errored bit in every 10,000,000, which is an improvement of two

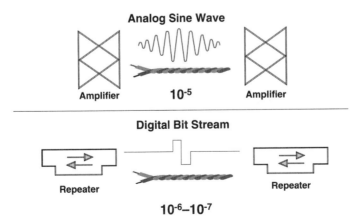

Figure 1.8 Comparative error performance of analog versus digital transmission over twisted pair.

orders of magnitude. Digital fiber-optic systems, currently considered to be the ultimate in transmission systems, commonly yield error rates in the range between 10^{-11} and 10^{-14}, or an error rate as low as 1 bit for every 100,000,000,000,000 transmitted, which is virtually perfect [4].

1.7 CONVERSION PROCESS: MODEMS AND CODECS

Regardless of the relative merits of analog and digital transmission, both technologies are in place. Local loops, which connect the user premises to the COE, generally are analog, at least in residential and small-business applications. Medium- and large-size businesses typically make use of digital local loops in the form of either T-carrier or ISDN over twisted pair. Cellular radio networks in the United States originally were analog, although most now are digital. High-capacity, backbone carrier transmission generally is digital. Analog-to-digital and D-to-A conversions take place routinely in contemporary networks.

1.7.1 Digital to Analog: Modems

As local loops often are analog, computer communications across such circuits are not possible without the assistance of a device to accomplish the D-to-A conversion. Where a digital circuit is available, it generally is considerably more expensive.

The device that accomplishes the D-to-A conversion process is known as a *modem*. Modems *mo*dulate and *dem*odulate the analog carrier wave in order to represent digital bit streams across the analog local loop, reconstructing the digital signal on the receiving end through a process of A-to-D conversion (Figure 1.9). A variety of techniques, explained in Chapter 6, are used to accomplish this process.

1.7.2 Analog to Digital: Codecs

The opposite conversion process is necessary to send analog information across a digital circuit. Certainly, this occurs often in carrier networks, where huge volumes of analog voice are digitized and sent across high-capacity digital circuits. This requirement also exists where high-capacity digital circuits connect premises-based, PBX voice systems to COEs or to other PBXs, assuming that the PBXs or COs have not already performed the conversion. As video also is analog in its native form, a similar process must be employed to send such information across a digital circuit.

The device that accomplishes the A-to-D conversion is called a *codec*. Codecs *co*de an analog input into a digital (data) format on the transmit side of the con-

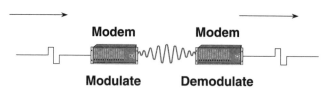

Figure 1.9 Modem: D-to-A and A-to-D conversion.

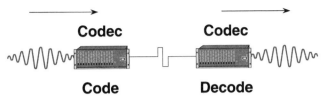

Figure 1.10   Codec: A-to-D and D-to-A conversion.

nection, reversing the process, or *dec*oding the information, on the receive side, to reconstitute an approximation of the original analog signal (Figure 1.10).

Encoding is the process of converting an analog information stream (e.g., voice or video) into a digital data stream. The voice or video signal is sampled at frequent intervals and each sample is expressed in terms of a binary value, usually a four- or eight-bit byte (i.e., data word). The reverse process of *decoding* takes place on the receiving end, resulting in recomposition of the information in its original form, or at least a reasonable approximation thereof.

1.8 MULTIPLEXERS (MUXES)

The term *multiplex* has its roots in the Latin words *multi* (many) and *plex* (fold). Multiplexers (muxes) act as both concentrators and contention devices that enable multiple relatively low speed terminal devices to share a single high-capacity circuit (physical path) between two points in a network. The benefit of multiplexers is simply that they enable carriers and end users to take advantage of the economies of scale. Just as a multilane highway can carry large volumes of traffic in multiple lanes at high speeds and at relatively low incremental cost per lane, a high-capacity circuit can carry multiple conversations in multiple channels at relatively low incremental cost per channel.

> *The modern saying, "Time is Money," is indeed most of all true when applied to tele-graphic signalling; and many endeavours have been made, not only to transmit signals with celerity, but also to transmit more than one communication at the same time along the same wire. This has been successfully done in the duplex system—by which a message is sent from either end of the same wire simultaneously; in the diplex system—in which two messages can be sent simultaneously in one direction; and in the quadruplex system, which combines the two former methods, and by which it is possible to convey four signals along the same wire at the same moment. This last method was invented by Mr. Edison.*
>
> *—Wonders of the Universe* [5]

Contemporary multiplexers rely on four-wire circuits, which enable multiple logical channels to derive from a single physical circuit and permit high-speed transmission simultaneously in both directions. In this manner, multiple communications (either unidirectional or bidirectional) can be supported. Multiplexing is used commonly across all transmission media, including twisted pair, coaxial and fiber-optic cables, and microwave, satellite, and other radio systems.

Traditional multiplexing comes in several varieties, presented in the following sections in chronological order of development and evolution. Included are Frequency Division Multiplexing (FDM), Time Division Multiplexing (TDM) and Statistical Time Division Multiplexing (STDM). Wavelength Division Multiplexing (WDM), a relatively recent development, is used in fiber-optic cable systems.

1.8.1   Frequency Division Multiplexing

Frequency Division Multiplexing (FDM) takes advantage of the fact that a single twisted pair copper circuit, for example, can support much more than the 4 kHz guaranteed for an individual voice conversation. Even in the early days of vacuum tube technology, a set of two copper pairs (a four-wire circuit, with two wires supporting transmission in each direction) could support up to 96 kHz, thereby enabling the support of up to 24 individual voice channels separated by frequency guard bands [3]. In terms of a commonly understood analogy, a single four-wire electrical circuit can support multiple frequency channels through frequency separation much as the airwaves can support multiple radio stations and TV channels.

Through an FDM (Figure 1.11), conversation 1 might be supported over frequencies 0–4000 Hz, conversation 2 over frequencies 4000–8000 Hz, conversation 3 over frequencies 8000–12,000 Hz, and so on. Small slices of frequency within each channel are designated as subchannels, or *guard bands*, which separate the carrier channels used for information transmission. The guard bands serve to minimize the likelihood of interference between conversations riding in adjacent logical information channels over the same physical circuit. This prevents *crosstalk*, in which parties using adjacent channels hear each other unless the filters or frequency converters drift from their proper settings. Of course, the individual channels are not separated spatially; rather, they are overlaid, with all sharing the same physical space on the wires.

Frequency division multiplexers typically are not particularly intelligent. Specific devices or groups of devices often are tuned to using designated frequency bands for communications. As noted in Figure 1.11, the bandwidth associated with those

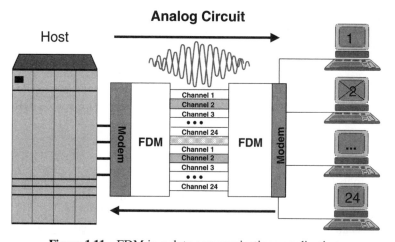

Figure 1.11   FDM in a data communications application.

X'ed-out devices is unused if the communication is inactive for some reason, even though other devices could make effective use of it.

FDM served its purpose well, at the time, for long-haul voice transmission. Data communications over FDM, however, requires sets of low-speed modems, one for each channel, with one set at each end of the facility. FDM currently is used in broadband LANs, which support multiple simultaneous transmissions. FDM also is used in cellular radio networks and in certain digitized voice applications. As noted, however (all else being equal), digital generally is better, especially when data traffic is involved, and data rules in this world.

1.8.2 Time Division Multiplexing

Time Division Multiplexing (TDM) offers all of the advantages of digital transmission, namely improved bandwidth utilization, enhanced error performance, improved security, and upgradability.

At the transmitting end of the connection, the time division multiplexer scans the buffers associated with the ports to which individual devices are attached. Each device port is allocated a channel in the form of a *time slot* for transmission of data. Device 1 transmits through port 1 and over time slot 1, device 2 transmits through port 2 and over time slot 2, and so on, in a serial fashion, as illustrated in Figure 1.12. The transmitting time division multiplexer typically accepts an 8-bit sample of data from each buffered port and byte interleaves those data samples into a *frame* of data. (*Note: Byte* generally refers to an eight-bit value, or *data word*, although some bytes comprise more or fewer bits. The term *octet* is more precise and frequently is used in telecommunications standards to describe an 8-bit unit of data.) As the mux completes a scan of the ports and transmits a set of such data samples, it prepends each frame with some number (1 in the case of T1) of *framing bits*. The

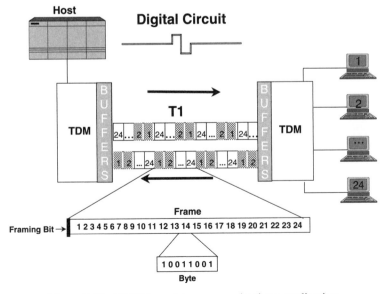

Figure 1.12 TDM in a data communications application.

framing bits are sent in a repeating pattern that delineates one frame from another and are used by the multiplexers and other intermediate devices for purposes of *synchronization*. Sophisticated multiplexers also use the framing bits for various signaling and control purposes.

At the receiving end, the process is reversed. Each channel in each frame is identified, the individual transmissions are demultiplexed, and each is forwarded over the port to which the intended receiving terminal device is attached. Clearly, the muxes must be carefully *synchronized* in time in order for the receiving mux to determine the proper separation of frames and channels of data. The framing bits typically provide the mechanism not only for synchronization between the muxes but also for the attached terminals.

The primary constraint of basic TDM is that of static configuration. In other words, channel 1 is always reserved for port 1, over which terminal 1 always transmits. Terminals that are idle, turned off, unplugged, or otherwise out of action are still allocated valuable bandwidth, which has a negative effect on the cost effectiveness of the facility, as illustrated in Figure 1.12. As their bandwidth allocation capabilities are so limited, basic time division multiplexers are generally considered obsolete.

1.8.3 Statistical Time Division Multiplexing

Statistical Time Division Multiplexing (STDM) is much improved over TDM because the muxes are intelligent. STDMs, or *Stat muxes*, offer the advantage of dynamic allocation of available channels and raw bandwidth. In other words, STDM can allocate bandwidth, in the form of time slots, in consideration of the transmission requirements of individual devices serving specific applications (Figure 1.13). An STDM also can oversubscribe a trunk, supporting aggregate port speeds that may be in the range of 3–10 times the trunk speed by buffering data during periods of high activity. Further, an intelligent STDM can dynamically adapt to the changing

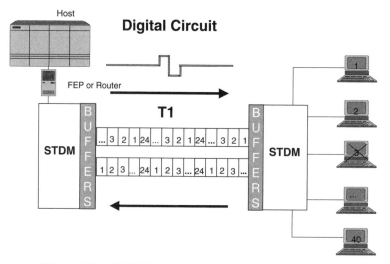

Figure 1.13   STDM in a data communications application.

nature and associated requirements of the load placed on it and in consideration of the available capacity of the circuit.

Stat muxes can recognize active versus inactive devices as well as priority levels. Further, they can invoke *flow control* options that can cause a transmitting terminal to cease transmission temporarily, in the event that the mux's internal *buffer*, or temporary memory, is full, thereby preventing a fast transmitter from overwhelming a slower receiver. Flow control also can restrain low-priority transmissions in favor of higher priority transmissions. Additionally, an STDM may offer the advantages of data compression, error detection and correction, and reporting of traffic statistics.

An STDM typically divides a high-speed, four-wire digital circuit into multiple time slots to carry multiple voice conversations or data transmissions. Channelized T1 (North America), for example, commonly provides 24 time slots to carry 24 conversations, each of a maximum of 64 kbps. Channelized E1 (European and international) commonly provides 30 time slots in support of 30 conversations.

Additionally, the individual channels can be grouped to yield higher transmission rates (*superrate*) for an individual bandwidth-intensive communication such as a videoconference. The individual channels also can be subdivided into lower speed (*subrate*) channels to accommodate many more, less bandwidth-intensive communications, such as low-speed data. Also, many muxes allocate bandwidth on a priority basis, providing delay-sensitive traffic (e.g., real-time voice or video) with top priority to ensure that the presentation of the data at the receiving end is of high quality.

1.8.4 Wavelength Division Multiplexing

Wavelength Division Multiplexers (WDMs) enable a single fiber-optic transmission system to support multiple high-speed channels through the transmission of multiple *wavelengths* of light, or *lambdas*. (*Note*: Physicists denote wavelength with the Greek letter λ (lambda), with wavelength being inversely proportional to *frequency*.) Much as multiple electrical frequencies can support multiple, simultaneous conversations in an FDM transmission system, multiple wavelengths can coexist on a single fiber of the appropriate type. A number of carriers now routinely deploy *Dense Wavelength Division Multiplexing* (DWDM) on fiber-optic systems, with eight or more lambdas introduced into the optical fiber through the use of *tunable lasers* firing through *windows*, or wavelength ranges. Each lambda might run at 2.5 Gbps, for a total yield of 20 Gbps per fiber strand. If each of 48 lambdas on a standard DWDM grid were to run at 10 Gbps, for example, the total yield would be 480 Gbps per fiber strand. At last count, the International Telecommunications Union—Telecommunications Standardization Sector (ITU-T) had defined 160 wavelengths at spacings of 100 GHz and manufacturers currently offer DWDM systems that multiplex as many as 80 lambdas. As some carriers deploy hundreds of fiber strands along a given path, the bandwidth potential is incredible.

1.8.5 Inverse Multiplexers

Inverse multiplexers perform the inverse process of traditional muxes. In other words, they accommodate a single, high-bandwidth data stream by transmitting it over multiple, lower bandwidth channels or circuits. The transmitting mux segments

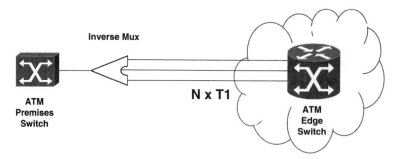

Figure 1.14   Inverse multiplexing over ATM.

the data stream and spreads it across the circuits on a consistent and coordinated basis, and the receiving mux reconstitutes the composite data stream. Clearly, the two devices must synchronize carefully with each other and with the transmission characteristics of the individual paths and channels in order to minimize errors and delays. An individual communication might spread over multiple switched circuits, dedicated circuits, or channels on multichannel circuits. A broadcast quality video-conference, for example, requires a full T1 (1.544 Mbps). Assuming that a full T1 is currently unavailable between two locations, the mux might split the signal across portions of multiple T1s and recombine at the receiving end.

Inverse Multiplexing over ATM (IMA) fans out an *Asynchronous Transfer Mode* (ATM) cell stream across multiple circuits between the user premises and the edge of the carrier network. Where significant levels of ATM traffic are destined for the WAN, a single circuit of appropriate bandwidth may either be unavailable or too costly. In such a circumstance, multiple physical T1 circuits can be used as a single, logical ATM pipe. The IMA-compliant ATM concentrator at the user premises spreads the ATM cells across the T1 circuits in a round-robin fashion, and the ATM switch at the edge of the carrier network scans the T1 circuits in the same fashion in order to reconstitute the cell stream (Figure 1.14). There is a similar Implementation Agreement (IA) for Frame Relay (FR), and Multilink Point-to-Point Protocol (MPPP) serves much the same purpose in the Internet domain.

1.8.6   Data over Voice and Voice over Data

A large number of manufacturers now offer muxes that enable data to be sent over voice lines and voice to be sent over data lines. A digital data circuit, for example, also can accommodate voice (for which it was not intended) through the use of a special mux that digitizes the voice signal and transmits it; the reverse process takes place at the receiving end. The voice and data conversations share the same circuit sequentially, rather than simultaneously. Bandwidth is allocated as appropriate, with priority provided to the delay-sensitive voice traffic.

This approach enables the user to take advantage of excess capacity on a dedicated data circuit. It also can support both voice and data communications over a single circuit-switched analog circuit. There is, of course, an investment required in the multiplexing equipment, although such equipment is quite affordable.

A number of manufacturers have developed muxes and routers that enable voice to share excess capacity on a Frame Relay network. While the quality of *Voice over*

Frame Relay (VoFR) can suffer from delay due to network congestion, the voice conversation essentially is free, as the justification for the Frame Relay network is in support of data communications requirements. *Voice over IP* (VoIP) can be supported over the Internet, although the quality can be poor due to the many uncertainties inherent in the Internet. VoIP quality is much better over high-speed *Internet Protocol* (IP) networks optimized for voice, rather than for the data traffic for which the protocol was originally developed. While the availability of these networks remains somewhat limited at this point, their potential is quite significant. VoIP promises to support toll-quality voice at much lower cost, due to the inherent efficiencies of packet switching as compared to circuit switching. These networks also are designed to support voice, data, video, and image traffic over the same infrastructure, with priority granted to real-time voice and video; such an integrated network offers inherent efficiencies and a single point of service management. You will find considerable detail about VoFR and VoIP in subsequent chapters.

1.9 SWITCHES AND SWITCHING: THE BASICS . . . AND THEN SOME

> *A click of metal against metal as an expert hand thrusts a plug in to a switchboard's face or as mechanisms shift automatically, at the command of electric impulses. Then . . . the simple question "Number, please?" or its mechanical equivalent, both alike in meaning: "Make way for the messages of the people!"*
> *Telephone Almanac,* American Telephone and Telegraph Company, 1934

Switches serve to establish connectivity between terminal devices (transmitters and receivers) on a flexible basis. They effectively serve as contention devices, managing contention between multiple transmit devices for access to shared circuits. In this manner, the usage and cost of expensive circuits can be optimized, based on standard traffic engineering principles. Without switches, each device would require a direct, dedicated circuit to every other device in a *full mesh topology*. Such an approach clearly is extremely resource intensive, highly impractical, and even completely impossible, as early experience proved. This discussion of switches and switching is presented in chronological order of development, beginning with circuit switching and its evolution and progressing through packet, frame, and cell switching. I also introduce the concepts of soft switching and optical switching.

1.9.1 Circuit Switching: Optimized for Voice

Circuit switches establish connections between physical circuits on a temporary, continuous, and exclusive basis. That is to say, on demand and as available, a circuit switch establishes connections through the switching matrix and provides continuous and exclusive access between the physical circuits for the duration of the conversation. From circuit to circuit and through the switching matrix, the switch establishes a transmission path, dedicating the prescribed level of bandwidth to that one transmission. Many switches may be involved in supporting an end-to-end circuit between terminal devices over a long distance. Contemporary circuit switches provide continuous access to a logical TDM channel within the shared internal bus of the switch for each call. A residential or small-business user typically gains access to a port on the switch over a single-channel, voice-grade analog circuit. A larger business user more commonly gains access through a high-capacity, multichannel digital circuit such as T1 copper local loop or perhaps an optical fiber. The high-capacity, multichannel circuits that interconnect the circuit switches in the core of the carrier network are known as *InterMachine Trunks* (IMTs), which generally are in the form of fiber-optic circuits. While circuit switches originally were developed for voice communications, much data traffic also is switched in this fashion. Typical examples of circuit switches include PBXs and COs.

1.9.1.1   Manual Switchboards   These switches involve an operator who manually establishes the desired connection at the verbal request of the transmitting party. The original switch was in the form of a *switchboard*, which literally was a series of small, manual mechanical *jackknife switch*es mounted on a *board*. The operator manually switched the blade of the jackknife switch from one contact to another to establish a unique physical and electrical connection. While the term switchboard is used even today, the technology soon gave way to that of the *cordboard*, another manual switching technology that requires the operator to establish connections on a *plug-and-jack* basis, with the plugs on *cord*s and the jacks mounted on a board (Figure 1.15). Again, the operator establishes a unique physical and electrical connection that remains in place for the duration of the call. When either party disconnects, the operator is alerted and manually disconnects the circuit, which then becomes available for use in support of another call. The size of such switches, the complexity of interconnecting long-distance calls across multiple switches, and the labor intensity of this approach all contributed to their functional obsolescence many years ago. Note, however, that many thousands of such switches remain in service, largely in developing countries.

The term *tip and ring* came from the cordboard plugs that establish the connection between the copper wire pairs. One wire connects electrically through the *tip* of the plug, while the other wire in the pair connects through the *ring*, or seat, of the plug. The New Haven District Telephone Company in New Haven, Connecticut, installed the first such switch on January 28, 1878, at a cost of $28.50. The system connected 21 subscribers referred to by name, rather than telephone number. As

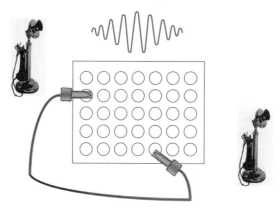

Figure 1.15   Manual switching: cordboard. (Courtesy of AG Communications Systems Corporation.)

was typical in those days, the operators were young boys, most of them experienced as messengers and telegraph clerks. The boys soon proved too boisterous and unreliable and were replaced with women [7].

1.9.1.2 Step-by-Step Switches   Step-by-Step (S × S) switches are electromechanical in nature. Almon B. Strowger, a Kansas City undertaker frustrated with the behavior of the local telephone company operator, invented and patented the first such switch in 1891. According to legend, the operator was directing Mr. Strowger's calls to a competing undertaker, who also happened to be the operator's husband. Building on earlier Bell system work, Strowger invented a system that served 99 subscribers. The telephones that worked with that first automatic switch had two buttons. In order to reach subscriber 99, for example, the caller slowly and deliberately pressed the first button nine times and then the second button nine times. As the caller pressed a button, it would complete an electrical circuit, and as the caller released the button, it would break the circuit. Making and breaking the circuit would cause a mechanical rotary wiper on the switch to rotate from one contact to another. That patent served as the foundation for the company he founded, Automatic Electric Company, which later became the manufacturing subsidiary of General Telephone and Electric (GTE), which is now part of Verizon [7].

More contemporary S × S switches consist of a large number of *line finders* to which groups of individual subscribers are assigned for dial tone. The transmitting party dials a series of numbers, perhaps with a rotary dial telephone terminal, which causes the making and breaking of an electrical circuit. As displayed in Figure 1.16, those electrical pulses cause successive mechanical *line selectors* to step across contacts (one contact per electrical dial pulse) to set up the conversation path as the complete number is dialed [8]. *Note*: A touchtone telephone also can be used if a tone converter is installed on the line at the central office termination.

The S × S switches quickly gained favor, as they largely eliminated human operators, who were expensive and prone to make errors. So many were installed and proved so durable that as late as 1986 approximately 38 percent of all U.S. circuit switches were based on S × S. As they clearly are slow, expensive, large, maintenance intensive, and capacity limited [3], S × S switches are rarely found in the networks of developed countries. Large numbers of S × S switches do remain in service, however, mostly in developing countries.

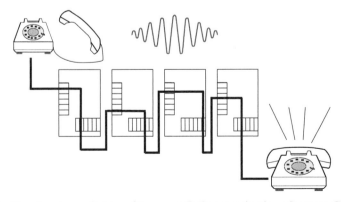

Figure 1.16   Step-by-step switching. (*Source:* AG Communications Systems Corporation.)

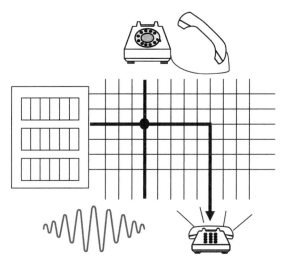

Figure 1.17   XBar switching. (*Source:* AG Communications Systems Corporation.)

1.9.1.3 Crossbar Switches   The first *common control* switches, CrossBar(XBar) switches are electromagnetic in nature. While the original concept was developed at Bell Telephone Laboratories, the Ericsson company in Sweden accomplished much of the early practical development work. The first such switch installed in the United States was a central office exchange in Brooklyn, New York (1938) [6]. XBar switches quickly became predominant.

In an XBar switch (Figure 1.17), a request for dial tone is recognized by a *marker*, which directs a *sender* to store the dialed digits. A *translator* is then directed to route the call, reserving a path through a *switching matrix* [3]. Once the call connects, these various components become available to serve other calls. Compared to the S × S switch, the XBar has relatively few moving parts. XBar switches offer the advantages of increased intelligence, common control, greater speed of connection, smaller physical footprint, lower maintenance, and greater capacity. XBar switches were considered state of the art for nearly 30 years.

1.9.1.4 Electronic Common Control Switches   Electronic Common Contral (ECC) switches reflect the marriage of computer technology and telephony. While the first ECC switches were analog, contemporary switches are fully digital in nature. Voice conversations are digitized and switched over high-speed digital circuits, with all processes accomplished through programmed logic. ECC switches are microprocessor controlled. The first ECC switch was the Electronic Switching System (ESS), developed by AT&T Bell Telephone Laboratories (Bell Labs) with the assistance of Western Electric. It was based on the transistor, invented at Bell Labs in 1948, and involved a development effort that began in earnest in the early 1950s. The first ESS CO began service in Succasunna, New Jersey, on May 30, 1965, connecting 200 subscribers. By 1974, there were 475 such offices in service, serving 5.6 million subscribers. The development effort was estimated to involve 4000 man-years and a total cost of US$500 million [7].

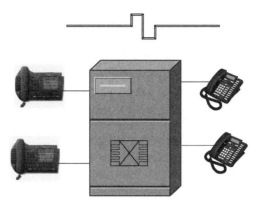

Figure 1.18   ECC switching.

ECC switches (Figure 1.18), as compared to the previous generations of switching technology, offer the advantages of further increased intelligence, greater speed of call setup and overall call processing, and a still smaller footprint. Additionally, they offer lower maintenance costs and can be monitored and managed from a remote location. Many contemporary ECC switches are unmanned, in favor of control from a centralized *Network Operations Center* (NOC). ECC switches offer greater capacity and on a scalable basis; that is, capacity can be increased through the addition of various system modules, or cabinets, with a reasonably graceful relationship maintained between increases in capacity and associated costs. As specialized, software-controlled computer systems, the functionality and feature content of ECC switches often can be upgraded through additional software and/or firmware. Such switches generally possess the ability to switch data and video as well as voice. Finally, they interface with various application processors to further increase the range of services provided. Example application processors include voice processors (e.g., voicemail) and fax servers (fax mail).

1.9.2 Packet Switching: Optimized for Data

First deployed in 1971, *packet switching* grew out of the U.S. Advanced Research Project Agency (ARPA) network. Commonly referred to as *ARPANET*, the network was established to support interactive, asynchronous computer-to-computer communications between the defense and university communities.

Rather than employing circuit switching, which is far too inefficient and expensive for intensive, interactive computer communications, ARPANET and its successors such as the Internet make use of packet switching. Packet switching involves the transmission of data formed into *packets* and sent across a shared network. Each packet, or datagram, is individually addressed in order that the packet switches can route each packet over the most appropriate and available circuit and each packet can survive independently. Each packet may represent an individual set of data or a larger set of data can be fragmented into multiple packets, each of which works its way through the network independently. A packet also can comprise multiple smaller, related sets of data gathered together for transport. Packets offered to the network by large numbers of users make use of the same universe of switches and transmission facilities. Such a highly shared network offers dramatically lower costs of data transmission in comparison to circuit switching.

Traditional packet-switched networks are based on mature and stable technologies, are widely available both domestically and internationally, and are low in cost. *X.25*, an international standard packet-switching interface (Chapter 7), is an excellent example of an early packet-switched network. X.25 offered great advantage in terms of its ability to support the interconnection of virtually any computer system through its ability to accomplish protocol conversion. This highly desirable feature caused X.25-based packet networks to be characterized as the first *Value-Added Networks* (*VANs*). X.25 remains heavily used, especially in developing countries. Disadvantages of X.25 include the fact that it can support only relatively low speed data transmission. Also, as the switches assume a 1960s-vintage analog network environment of twisted pair, each X.25 switch must examine each individual packet for errors created in transmission. In the event that an error is discovered, the switch must resolve the problem through a request for retransmission. These factors, in combination, result in unpredictable, variable levels of packet *latency* (delay). Therefore, packet switching traditionally has been considered to be unsuitable for stream-oriented communications such as real-time voice and video.

Most people, however, immediately think of *Transmission Control Protocol/ Internet Protocol* (TCP/IP) when they think of packet switching. TCP/IP is not only fundamental to the Internet but also serves as the basis for a number of developing high-speed packet networks optimized for voice, currently challenging the traditional circuit-switched PSTN for dominance. These networks, which also are challenging Frame Relay and ATM, are based on the IP suite and various fiber-optic transmission technologies. (*Note*: IP is more prevalent in access networks, while ATM is more prevalent in carrier backbones.) Chapter 12 includes detailed discussion of the IP protocol suite, and Chapter 14 addresses IP-based competitive networks.

1.9.3   Frame Switching: Optimized for LAN Internetworking

A relative newcomer, *Frame Relay* was first offered commercially in 1992 by Wiltel (United States). Essentially a high-speed, streamlined version of X.25, Frame Relay was developed specifically as a protocol for access to a packet network in support of LAN-to-LAN networking. But Frame Relay can be used in support of the transmission of virtually any computer data stream in its native form, with individual *frames* varying in length up to 4096 bytes. Frame Relay became the overwhelming choice for data networking during most of the 1990s and currently is widely available as a domestic service offering in developed nations and on an international basis between them. As Frame Relay, like packet switching, is oriented toward data transmission over a highly shared network, frame latency is variable and unpredictable in duration. While increasingly satisfactory techniques have surfaced for support of voice and video, Frame Relay was not designed with those applications in mind. Detailed discussion of Frame Relay will be provided in Chapter 10.

1.9.4   Cell Switching: Optimized for Everything

Cell switching encompasses both *Switched Multimegabit Data Service* (SMDS) and ATM. Data are organized into *cells* of fixed length (53 octets), shipped across high-speed facilities, and switched through high-speed, specialized switches. SMDS proved very effective for data networking in a metropolitan area, although it was available only in limited serving areas and its popularity was short-lived. At this point, SMDS is but a historical footnote in the evolution of data networking.

ATM was much more successful than SMDS. ATM's value is in its ability to support any type of data stream (e.g., voice, data, video, image, and multimedia), whether real time or non–real time, compressed or uncompressed, and providing each with the appropriate level of Quality of Service (QoS) on a guaranteed basis. ATM is unique in this respect. Chapter 10 includes a detailed discussion of ATM.

1.9.5 Softswitches: Optimized for Flexibility

Softswitches were developed in the late 1990s as a replacement for the *hard-coded* proprietary CO circuit switches used in the traditional PSTN. As packet-based IP, Frame Relay, and ATM networks began to develop in the 1990s to address the explosive growth in data traffic, it became clear that COs were functionally obsolete in such networks. Further, service providers had a strong interest in developing and delivering a wide variety of data and multimedia services in very short periods of time. Softswitches addressed those issues by offloading call processing functions (i.e., signaling and call control) to industry standard server hardware, essentially decomposing the call control logic from the switching platform. This allows the call control logic to reside at some geographically centralized location from which it can control multiple switching platforms, as well as the separate devices that provide for interconnection of circuit and packet networks such as IP and Frame Relay. In addition to controlling this protocol conversion function, softswitches can support multiple QoS and Grade of Service (GoS) mechanisms and levels. Softswitches are software programmable elements that support open *Application Programming Interfaces* (APIs) in consideration of the pressure to shorten the *time to market* for new services. Softswitches tend to be much more flexible and much less expensive and have much smaller footprints than traditional COs.

1.9.6 Photonic Switches: Optimized for Optics

Currently in the early stages of commercial application, *photonic switches* are yet another dimension in the evolution of switch technology. Photonic switches operate at the pure optical level, switching individual lambdas in a DWDM transmission system. Photonic switches do so without the requirement for optoelectric conversion imposed by more traditional electronic common control switching devices with optical interfaces but electronic switching fabrics, thereby offering clear advantages in terms of simplicity, speed, and cost. They are, however, limited to switching at the lambda level. (See Chapter 9.)

1.10 SIGNALING AND CONTROL

Signaling and control comprise a set of functions that must take place within any network to ensure that it operates smoothly. In this context, various elements within the network must identify themselves, communicate their status, and pass instructions. Fundamental examples include on-hook and off-hook indication, dial tone provision, call routing control, busy indication, and billing instructions. Further examples include dialed digits, route availability, routing preference, carrier preference, and originating number or circuit [3].

In more sophisticated, contemporary networks, the responsibility for overall signaling and control functions resides within a separate *Common Channel Signaling* (CCS) and control network. Such a sophisticated CCS network involves highly intelligent devices capable of monitoring and managing large numbers of lower order devices in the communications network that it controls. From a centralized *Network Control Center* (NCC), the network can be monitored, and faults or performance failures can be identified, diagnosed, and isolated. Finally, the lower order devices in the communications network often can be addressed and commanded to correct the condition. Contemporary signaling and control systems include *Signaling System 7* (SS7) in TDM networks and *H.323* and *Session Initiation Protocol* (SIP) in VoIP networks. Chapter 5 explores signaling and control in detail.

REFERENCES

1. Shelly, Gary B. and Cashman, Thomas J. *Introduction to Computers and Data Processing.* Anaheim Publishing Co., 1980.
2. Doll, Dixon R. *Data Communications: Facilities, Networks and Systems Design.* Wiley, 1978.
3. *Engineering and Operations in the Bell System.* Bell Telephone Laboratories, 1977.
4. Bates, Bud. *Introduction to T1/T3 Networking.* Artech House, 1991.
5. *Wonders of the Universe.* The Werner Co., 1899.
6. Augarten, Stan. *Bit by Bit.* Ticknor & Fields, 1984.
7. Brooks, John. *Telephone: The First Hundred Years.* Harper & Row, 1976.
8. *Introduction to Telecommunications.* AG Communications Systems, 1990.

CHAPTER 2

FUNDAMENTALS OF TRANSMISSION SYSTEMS: TECHNOLOGIES AND APPLICATIONS

It was early declared by Professor Morse, and by other distinguished investigators of the nature and powers of the electric current, that neither the ocean itself, nor the distance to be traversed, presented any insuperable obstacle to the laying of submerged oceanic lines from continent to continent, and the confident prophecy that such lines would eventually be undertaken was freely uttered and discussed in learned circles.

R. M. Devens, *Our First Century or the One Hundred Great and Memorable Events in the History of Our Country During the One Hundred Years of Its Existence,* C.A. Nichols & Co., 1876

Information is of considerably increased value if it is shared with others. In this information age of high technology, we understand this principle well and we routinely share great quantities of information across vast distances. The conveyance, or transmission, of information across a distance necessarily involves some form of transmission medium that supports the propagation of the signal. The selection of an appropriate physical transmission medium is critical to the successful conveyance of the information. While the medium is not the message, at least not in the telecommunications domain, it is critical to message communications, particularly if the communication mode is an interactive one.

This chapter addresses all transmission media commonly used in traditional voice, data, video, and image networks, whether analog or digital in nature. Those media fall into two distinct categories, the first of which includes all wired media, also referred to as conducted, guided, bounded, or wireline media. The second category includes all traditional wireless media, also known as radiated, unguided, free space or unbounded.

Wired transmission systems employ tangible physical media. In other words, they are palpable media that human beings can see, touch, and feel. Also known as

Telecommunications and Data Communications Handbook, By Ray Horak
Copyright © 2007 Ray Horak

conducted systems, wired media generally use a metallic or glass *conductor* that serves to *conduct*, or carry on, some form of electromagnetic energy. Twisted-pair and coaxial cable systems, for example, conduct electrical energy, usually employing a copper medium. Fiber-optic systems conduct light, or *optical*, energy, generally using a glass conductor. The term *guided* media refers to the fact that the signal is contained within an enclosed physical path that guides the signal. Finally, *bounded* media refers to the fact that some form of twisting, shielding, cladding, and/or insulating material binds the signal within the core medium, thereby improving signal strength over a distance and enhancing the performance of the transmission system in the process. Twisted-pair (both unshielded and shielded), coaxial, and fiber-optic cable systems fall into this category.

Wireless transmission systems do not make use of a physical conductor to guide or bind the signal. Therefore, they also are known as *unguided* or *unbounded* systems. Rather than relying on electrical energy, such systems generally make use of radio waves; hence the term *radiated* often is applied to wireless transmission. Finally, wireless systems employ electromagnetic energy in the form of radio or light waves that are transmitted and received across space. Therefore, wireless systems often are referred to as *airwave* systems, although *spacewave* is a more accurate term, as the air in the space between transmitter and receiver actually serves to weaken the signal. Microwave, satellite, cellular, and a great number of special-purpose radio systems are wireless in nature. Free Space Optics (FSO) systems are wireless systems using infrared (IR) light signals.

Each specific transmission system has certain unique properties that manifest in advantages and limitations that point to appropriate applications. The application to be supported clearly must be of primary consideration in designing a network and in selecting the most appropriate transmission medium, assuming options are available.

Transmission systems appearing in this chapter include twisted copper wire, coaxial cable, microwave, satellite, FSO, and fiber optics, with this order of discussion being roughly chronological. Chapter 8 discusses cellular radio, packet radio, wireless Local Area Networks (LANs), and other application-specific radio systems.

2.1   ELECTROMAGNETIC SPECTRUM

James Clark Maxwell believed that magnetism, electricity, and light are all transmitted by vibrations in one common ether, and he finally demonstrated his theory by proving that pulsations of light, electricity, and magnetism differed only in their wave lengths. In 1887 Professor Hertz succeeded in establishing proof positive that Maxwell's theories were correct, and, after elaborate experiments, he proved that all these forces used ether as a common medium.

Joseph H. Adams, *Harper's Electricity Book for Boys*, Harper & Brothers, 1907

While human voice frequencies mostly fall in the range of 100–8000 Hz, the energy in the speech spectrum peaks at approximately 500 Hz, with most articulation at higher frequencies. Human hearing can distinguish signals as low as 20 Hz and as high as 20 kHz and is most sensitive in the range of 1000–3000 Hz. Human-to-human voice communication seldom requires technical support over short distances. Voice communication over distances of more than a few meters, however, requires that the acoustical energy be converted into some form of electromagnetic energy and sent over a transmission system of some description. The electromagnetic spectrum comprises all of the frequencies or wavelengths that can be electromagnetically radiated, from the longest electrical or radio waves to the shortest gamma and cosmic rays.

Public Switched Telephone Networks (PSTNs) provide raw, voice-grade bandwidth in channels of 4 kHz, with 3 kHz (300–3300 Hz) used for voice transmission and the balance used for signaling and control purposes and for guardbands for signal separation when multiple analog voice channels are multiplexed. This range of frequencies (i.e., level of bandwidth) is sufficient to support voice communications of reasonable, if not perfect, fidelity. In an electrified telecommunications cable system, the carrier frequency, or range of carrier frequencies, depends on the specific nature of the medium and the requirements of the applications supported. Twisted-pair systems, for example, can support bandwidth of 10–10^6 Hz, and coaxial cable can support signals of up to 10^6–10^8 Hz. Band-limiting filters commonly used in transmission systems allow only the specified *passband* range of frequencies to pass on and stop all others [1–3].

The Institute of Electrical and Electronics Engineers (IEEE) defines *frequency* as the number of complete cycles of sinusoidal variation per unit time, with the unit of time generally being that of 1 s. Plotting $y = \sin x$, where x is expressed in *radians*, yields a sine wave as illustrated in Figure 2.1. [*Note:* From the Latin *radius*, a radian is a unit of plane angular measurement equivalent to the angle between two radii that enclose a section of a circle's circumference (arc) equal in length to the length of a radius. There are 2π radians in a circle.] A complete sine wave entails a cycle as measured from a point of zero amplitude to a point of maximum positive amplitude ($+A$) through zero to a point of maximum negative amplitude ($-A$) and back to a point of zero amplitude. In an electrical network, by way of example, $+A$ could be in the form of positive voltage (e.g., +6 V) and $-A$ in the form of negative voltage (e.g., −6 V). Alternatively, $+A$ could be in the form a relatively high level of positive voltage (e.g., +3 V) and $-A$ in the form of a relatively low level of positive voltage (e.g., +1.5 V). The *wavelength*, or length of the sine wave, can be measured from peak to peak or trough to trough or between the points that cross zero amplitude in the same direction. Wavelength is expressed as the Greek letter λ (*lambda*).

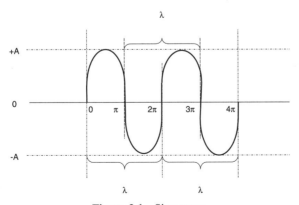

Figure 2.1 Sine wave.

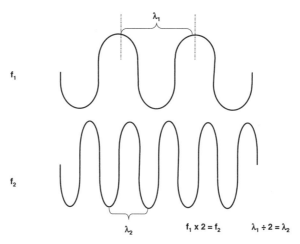

Figure 2.2   Frequency (f) and wavelength (λ).

Frequency (f) and wavelength (λ) are inversely related. As the frequency of the signal (number of cycles per second) increases, the wavelength (length of the electromagnetic waveform) of the signal decreases. In other words, the more waveforms transmitted per second, the shorter the length, or cycle, of each individual wave. Figure 2.2 illustrates the relationship between frequency and wavelength—as the frequency doubles, the wavelength halves.

It is worth noting at this point that signals in electrical and radio networks are described in terms of hertz (Hz). Once the frequency of the electromagnetic signals exceeds the Extremely High Frequency (EHF) of level of 300 GHz and crosses into the infrared light (IR) range of the optical spectrum, hertz no longer has relevance as either a bandwidth measurement or a signal descriptor, as the numbers are just too large and difficult to express. Rather, wavelength is used in the optical domain to describe the nature of the signal. In support of this logic, consider that the upper range of an analog voice channel is 4 kHz. Recalling that the velocity of propagation of all electromagnetic energy in a vacuum is roughly that of the speed of light, or 300,000 km/s, consider that at a frequency of 4 kHz (4000 cycles per second) each cycle is 75 km in length. Once the signal crosses into the optical spectrum, the scale changes in consideration of the preservation of human sanity. Consider that an IR optical signal in a fiber-optic cable at a commonly used wavelength of 1550 nm (0.000001550 m) has a nominal frequency of 193 THz (193,548,387,096,774 Hz). In consideration of the fact that adjacent signals can be spaced at intervals of 200 GHz, or 1.6 nm (at 1550 nm), it is fairly obvious that a lot of time and effort are saved by talking about wavelength rather than frequency (Hz) at this level:

$$\frac{300,000,000\,\text{km/s}}{4000\,\text{cycles/s}} \equiv 75,000\,\text{m}$$

$$\frac{300,000\,\text{km/s}}{193,548,387,096,774\,\text{cycles/s}} \equiv 0.000001550\,\text{m}$$

Table 2.1 defines the frequency and wavelength of various types of radio and light-based communications systems as they relate to the electromagnetic

TABLE 2.1 Frequency Spectrum

Band Designation	Nominal Frequency	Nominal Wavelength $(\lambda)^a$	Example Applications
Audible	20 Hz–20 kHz	>100 km	Acoustics
Direct Current (DC)	0–30 Hz	10,000 km–∞	Not applicable
Extremely Low Frequency (ELF) radio	30–300 Hz	1000–10,000 km	Submarine communications
InfraLow Frequency (ILF)	300 Hz–3 kHz	100–1000 km	Not applicable
Very Low Frequency (VLF) radio	3–30 kHz	10–100 km	Navigation, weather
Low-Frequency (LF) radio	30–300 kHz	1–10 km	Navigation, maritime communications, information and weather systems, time systems
Medium-Frequency (MF) radio	300 kHz–3 MHz	100 m–1 km	Navigation, AM radio, mobile radio
High-Frequency (HF) radio	3–30 MHz	10–100 m	Citizens Band (CB) radio (aka short-wave radio), mobile radio, maritime radio
Very High Frequency (VHF) radio	30–300 MHz	1–10 m	Amateur (Ham) radio, VHF TV, FM radio, mobile satellite, mobile radio, fixed radio
UltraHigh Frequency (UHF) radio	300 MHz–3 GHz	10 cm–1 m	Microwave, satellite, UHF TV, paging, cordless telephony, cellular and PCS telephony, wireless LAN
SuperHigh Frequency (SHF) radio	3–30 GHz	1–10 cm	Microwave, satellite, wireless LAN
Extremely High Frequency (EHF) radio	30–300 GHz	1 mm–1 cm	Microwave, satellite, radiolocation
Infrared (IR) light	300 GHz–400 THz	750 nm–1 mm	Wireless LAN bridges, wireless LANs, fiber optics
Visible light	400 THz–1 PHz	380–750 nm	Not applicable
UltraViolet (UV) light	1–30 PHz	10–380 nm	Not applicable
X rays	30 PHz–30 EHz	0.01–10 nm	Not applicable
Gamma and cosmic rays	>3 EHz	<0.1 nm	Not applicable

Note: k = kilo = 1000, M = mega = 1,000,000, G = giga = 1,000,000,000, T = tera = 1,000,000,000,000, P = peta = 1,000,000,000,000,000, E = exa = 1,000,000,000,000,000,000, km = kilometer (1000 m), m = meter, cm = centimeter (1/100 m), mm = millimeter (1/1000 m), μm = micrometer (1/1,000,000 m), nm = nanometer (1/1,000,000,000 m).

^{a} Figures are for wavelength in a vacuum. Wavelength in a medium will be shorter due to propagation at less than 300 km/s. For example, in glass the speed of light is reduced by the index of refraction, which is about 1.5 in practice, so the speed is 300/1.5 = 200 km/s. Index of Refraction (IOR) is the ratio of speed in a vacuum divided by speed in the medium.

spectrum [2–4]. Note that the higher is the frequency of the carrier signal, the more sine waves are available for manipulation and the greater is the potential level of bandwidth available. If you consider each sine wave to represent raw material for the production of bandwidth and the manipulation of each sine wave to increase the potential of the raw material, you can understand that the more sophisticated the manipulation of the sine waves, the greater the bandwidth created. However, the higher is the frequency of the carrier signal, the greater is the extent to which it suffers from signal attenuation (i.e., weakening). The impact of signal attenuation dictates the maximum allowable spatial separation between various devices such as transmitters, amplifiers or repeaters, and receivers.

2.2   TRANSMISSION MEDIA SELECTION CRITERIA

The selection of the most effective transmission system for a given application must be made in the context of a number of key design considerations. Such considerations include general transmission characteristics such as bandwidth and error performance, both of which affect throughput. Additionally, you must consider the allowable distance between devices as well as issues of propagation delay, security, mechanical strength, physical dimensions, and speed of deployment. Finally, and perhaps most importantly, consider local availability and cost, including cost of acquisition, deployment, Operation and Maintenance (O&M), and upgrade or replacement.

2.2.1   Transmission Characteristics

The basic transmission characteristics of a given medium are of primary importance. Those characteristics include bandwidth, error performance, and distance between network elements. These three dimensions of a transmission system, in combination, determine the effective *throughput*, or the amount of information you can put through, that is, send over, the system.

Bandwidth, in this context, refers to the raw amount of bandwidth, or capacity, the medium supports. *Error performance* refers to the number or percentage of errors introduced in the process of transmission. *Distance* refers to the minimum and maximum spatial separation between devices over a single link, rather than over a complete, end-to-end circuit. Clearly, the attractiveness of any given transmission system increases to the extent that you realize greater available bandwidth, fewer errors, and a greater maximum distance between various network elements such as amplifiers and repeaters.

Note that bandwidth, error performance, and distance are tightly interrelated. In a twisted-pair network, for example, more raw bandwidth requires more raw material in the form of hertz (sine waves), which translates into higher transmission frequencies. The concept of raw material suggests that each sine wave can represent one or more bits of data, depending on the sophistication of the modulation scheme employed. Unfortunately, higher frequencies *attenuate* (lose power) more rapidly than do lower frequencies. This fact results in more errors in transmission, unless the amplifiers/repeaters are spaced more closely together. The following scenarios serve to illustrate the relationship between frequency, distance, and error performance:

- A four-wire ISDN BRI (Integrated Services Digital Network, Basic Rate Interface) circuit, for example, typically runs over a physical two-wire local loop that supports three channels delivering aggregate bandwidth of 144 kbps in both directions at a frequency of approximately 40 kHz. The local loop circuit can span a distance of up to 18,000 ft between the carrier Central Office (CO) and the customer premises. ISDN BRI offers excellent levels of error performance at these distances, with no requirement for repeaters. The twisted-pair local loops must be of excellent quality to support that level of performance.
- A T1 circuit traditionally runs over a four-wire twisted-pair local loop, with each pair providing bandwidth of 1.544 Mbps in one direction at a frequency of approximately 0.75 MHz. T1 error performance is excellent through the placement of regenerative repeaters approximately every 6000 ft to overcome the effects of signal attenuation at the substantially higher carrier frequency. Again, the twisted-pair local loops must be of excellent quality.
- A LAN can support transmission rates of 10 Mbps (16 MHz), 100 Mbps (100 MHz), and even 1 Gbps (600 MHz) between attached devices such as workstations, hubs, and switches and with excellent error performance. The twisted-pair cables that support these levels of performance are of various data grades comprising multiple conductors across which the signals are split. The device separation is limited to a hundred meters in some cases and several hundred meters in others.

Also note that radio systems are naturally limited in terms of bandwidth. As the laws of physics tell us that there is only so much raw bandwidth in the frequency spectrum, various international, regional, and national regulatory bodies apportion frequency ranges on the basis of application and geography. The regulators also limit the power levels at which such systems operate to ensure that the signals do not exceed a certain geographical range or assigned radio cell and therefore do not affect licensed users in adjacent geographic areas. As a result, the licensing of frequency bands is on a site-specific or path-specific basis. Given the limited availability of radio spectrum, it is especially important that it be used with maximum effect. Digital systems are preferable for a number of reasons, including the fact that they support data compression and, thereby, are more efficient in their use of the precious resource of radio spectrum.

Further, consider that radio systems are highly sensitive to the quality of the atmosphere between the transmitter and the receiver. Dust, smoke, haze, and humidity have decidedly negative effects on signal performance. Precipitation (e.g., rain, sleet, snow, and hail) can cause substantial degradation in performance; this phenomenon is known as *rain fade*. These issues of atmospheric quality are more significant for terrestrial radio systems than for nonterrestrial satellite systems, as satellite systems largely transmit through the vacuum of space.

Finally, take note that the nature of the application has significant impact on the selection of the appropriate radio frequency range. Pager networks, for example, generally operate in the 900-MHz band. In this frequency range, the signals can travel relatively long distances at low power levels and can penetrate some amount of relatively dense physical matter (e.g., windows, walls, floors, and ceilings) without serious loss of signal strength. Many cellular radio networks operate in the 800- and 900-MHz bands and enjoy the same benefits. Amplitude Modulation (AM) broadcast radio signals run in the range between 535 and 1605 kHz in the United States and Frequency Modulation (FM) radio between 88 and 108 MHz. At these higher frequencies AM and even FM signals are able to penetrate some number of windows, walls, floors, and ceilings and still maintain acceptable levels of signal strength. (*Note*: There is more to the story, of course. AM and FM radio transmitters operate at power levels of 50 kW in the United States, while cellular transmitters typically operate at radiated power levels of 5–10 W. Further, broadcast radio transmitters also sit on high towers that often are perched on hilltops in order to increase the lookdown view and therefore improve line of sight.) Microwave radio systems operate in much higher frequency bands. As microwave signals cannot penetrate dense physical matter without suffering catastrophic signal loss, micro- wave systems require clear *line of sight* (i.e., unobstructed view) between antennas. As radio frequencies increase in frequency and approach the light spectrum, the signals behave more like light than radio, as we normally think of it. In other words, the signals are absorbed, reflected, and dispersed by physical matter, rather than penetrating it.

2.2.1.1 Propagation Delay

Father Time must be astounded at the extent to which the telephone has challenged him in a domain over which he has long held undisputed sway. The voice travels over tele- phone wires and cables at between 10,000 and 180,000 miles per second. By radio its speed is the same as that of light.
       Telephone Almanac, American Telephone & Telegraph Company, 1937

Propagation delay refers to the length of time required for a signal to travel from transmitter to receiver across a transmission system. Factors impacting propagation delay include the distance between transmitter and receiver and the density of the medium.

The speed of the electromagnetic signal depends on the density of the medium through which it travels. All electromagnetic energy travels at roughly the speed of light, which is approximately 300,000 km/s (actually 299,792,458 m/s), or 186,000 miles per second (actually 186,282.397 miles per second), in a vacuum. As a vacuum is void of all physical matter, there is nothing to impede the signal as it travels from point to point. A signal travels at a slower rate through the atmosphere of earth, as oxygen, carbon dioxide, water molecules, smoke, dust, and other physical matter are present, acting together to increase the density of the medium and impede the progress of the signal. An electrical signal *propagates* (i.e., moves or transmits forward) more slowly as it travels through a copper wire. (*Note*: On loaded loops, i.e., where loading coils are installed, the signal can propagate as slowly as 10,000– 20,000 miles per second.) Similarly, an optical signal slows as it travels through the relatively dense glass comprising an optical fiber. Table 2.2 provides comparative data on signal velocity in various media.

TABLE 2.2 Electromagnetic Signal Propagation Velocity (Approximate) [5, 6]

Medium	Signal Velocity[a] (km/s)	Velocity of Propagation,[a] Vp (Percent of Speed of Light in Vacuum)
Vacuum	300,000	100.00
Air	299,890	99.97
Copper cable[b]	180,000–240,000	60.00–80.00
Water	226,000	75.33
Teflon[c]	210,000	70.00
Optical fiber	205,000	68.33
Polyethylene; polypropylene[c]	200,000	66.67
Polyvinyl Chloride[c]	135,000–180,000	45.00–60.00

[a] Nominal (approximate) values.
[b] Various twisted-pair and coaxial cables.
[c] Cable insulating materials.

Clearly, the total length of the circuit directly impacts the length of time it takes for the signal to reach the receiver. The circuit length between two points can vary considerably in a switched network, as the specific circuit route can vary in length from call to call, depending on the availability of individual links that comprise an end-to-end circuit. Dedicated networks offer the advantage of a reliable and consistent level of propagation delay. In either case, the number of network elements (devices) in the network also affects the level of delay, as each device (e.g., amplifier or repeater, multiplexer, and switch or router) acts on the signal to perform certain processes, each of which takes at least a small amount of time to accomplish. Further, the more complex the processes (e.g., amplification or regeneration, multiplexing, switching or routing, protocol conversion, compression or decompression, and encryption) performed by each device, the greater the level of delay imposed on the signal. The fewer devices involved in a network and the less complex the processes performed, the less delay imposed on the signal. Many applications are intolerant of latency. Many applications also depend on precise and consistent timing of the received signal and, therefore, are intolerant of *jitter*, or variation in latency.

Geosynchronous Earth-Orbiting (GEO) satellite systems illustrate propagation delay very effectively. As the radio signals must travel approximately 22,300 miles (36,000 km) up to the satellite and the same distance on the return leg, the resulting round-trip delay is approximately 0.25 s. (Even at 186,000 miles per second, it takes a while to travel 44,600 miles.) Note that the signal mostly travels through the vacuum of space, contending with only a few miles of atmosphere—and the increased delay and signal distortion it induces—on the uplinks and downlinks. Considering the amount of time required for signal processing on board the satellite, as well as at the Earth stations, the total delay for a round-trip transmission is about 0.32 s. Therefore, the delay between signal origination and receipt of response is approximately 0.64 s, assuming an immediate response. Hence, satellite communications is considered ineffective for highly interactive real-time voice, data, and video applications.

As the data in Table 2.2 make clear, an optical signal propagates more slowly through a glass fiber than an electrical signal travels through a copper cable or a radio signal travels through the air. As a matter of fact, airwave technologies such as microwave radio offer much faster signal propagation than either. So, it follows that the speed of signal propagation is by no means the reason that fiber optics is so advantageous. Rather, it is bandwidth, distance, error performance, and security that make optical fiber so desirable in certain applications.

2.2.1.2 Security   *Security*, in the context of transmission systems, addresses the protection of data from interception as it transverses the network. Clearly, increasing amounts of sensitive data are being transmitted across public networks, well outside the range of physical protection on the user's premises. Therefore, security is of greater concern than ever before, and that level of concern will only heighten as commercial enterprises and nations increasingly seek to gain and protect competitive advantages and as they apply even more sophisticated means to doing so. In hearings before the U.S. Senate in May 1996, a statement revealed that 120 nations either had or were in the process of developing sophisticated computer espionage capabilities. That number undoubtedly has increased during the intervening years. Through its Echelon system, the U.S. National Security Agency (NSA) reportedly eavesdrops on approximately three billion conversations a day in defense of national security. Echelon apparently can tap any electromagnetic transmission system, including fiber optics, anywhere on the globe. (*Note:* While the NSA's activities make most U.S. citizens feel safer from terrorism and other threats to national security, it certainly makes many feel that their privacy is violated.)

Note that airwave systems (e.g., microwave and satellite) are inherently insecure, as access to the signal is easily accomplished and virtually undetectable through an antenna properly tuned and in proximity to the signal path. It is much more difficult to physically tap a wireline circuit. Also note that digital systems offer much greater security potential than analog systems by virtue of the fact that application software can quite effectively *encrypt*, or encode, the data to conceal its true meaning.

2.2.1.3 Mechanical Strength   *Mechanical strength* applies especially to wired systems. Installers must physically manipulate twisted-pair, coaxial, and fiber-optic cables while deploying and reconfiguring them. Clearly, each type of wire, fiber, and cable has certain physical limits to the maximum severity of the bend it can tolerate (*bend radius*) without cracking or breaking and the amount of bending and twisting (*flex strength*) it can tolerate. There also are limits to the amount of weight or longitudinal stress a cable or wire can support (*tensile strength*) without suffering deformation or breaking (*break strength*). Strength members improve the tensile strength of OutSide Plant (OSP) aerial cables and inside riser cables. Strength members can be either metallic or nonmetallic in nature, depending on issues such as the weight of the cable and the need for lightening protection. Aramid fiber such as Dupont's Kevlar is used in fiber-optic riser cables not only for improved strength but also for increased protection of the fragile glass fiber from physical damage.

Wires, fibers, and cables not only stretch and break but also expand and contract due to variations in ambient temperature from season to season and even from day to night, with the latter phenomenon known as *diurnal wander*. These events especially affect cable systems hung from poles, as they are more exposed to the elements than are buried cables and as the weight of the cable magnifies the effect. Copper cables are more susceptible than glass fiber-optic cables, but both expand and contract to some extent. As the cables expand and contract, the length of the path increases and decreases, signal propagation delay increases and decreases, and high-speed digital systems can suffer timing problems that can cause loss of synchronization and, ultimately, temporary system failure.

The issue of mechanical strength also applies to airwave systems, as the reflective dishes, antennas, and other devices used in microwave, satellite, and infrared technologies must be mounted securely to deal with the stresses of wind and other forces of nature. Additionally, the towers, walls, and roofs on which they typically are mounted must be properly constructed and braced in order to withstand such forces and must flex as appropriate.

2.2.1.4 *Physical Dimensions* The physical dimensions of a transmission system must be considered as well. Certainly, you must consider the sheer weight of a cable system. The bulk (diameter) of the cable is important, as conduit and raceway space often is at a premium. The physical dimensions of airwave systems are no less important, as the size and weight of a reflective microwave or satellite dish and mounting system (e.g., bracket and tower) may require support, particularly in locations that experience high winds.

2.2.1.5 *Speed of Deployment* Speed can be of the essence at times. Wired connections take some time, even under the best of circumstances and even using the most pliable wires and quickest connectors. Radio antennas may take some time to install, but once they are in place, the time required to configure and reconfigure the connections between them can take little time, if any. If fact, the antennas can even establish and maintain connections while in motion. Portability and even mobility are key wireless advantages. It is said that time is money, which leads us to consider cost.

2.2.1.6 *Cost* Ultimately, financial considerations rule, and media selection is no exception. Cost considerations include acquisition, deployment, operation and maintenance (O&M), and upgrade or replacement. Without getting involved in a lengthy discussion of each cost issue at this point, it certainly is particularly worth pausing for a moment to compare the deployment costs of wireline and wireless media.

Wired transmission systems require securing legal rights-of-way and digging trenches, boring tunnels, planting poles, placing conduits and manholes, pulling and splicing cables, placing amplifiers or repeaters, and so on. Such costs, clearly, are not trivial. Wireless systems, on the other hand, require securing rights-of-way, erecting towers, mounting antennas, securing spectrum licenses, and so on. While it is difficult to make hard-and-fast generalizations, the deployment of wired systems certainly involves a set of cost issues that can be problematic. Further, wired systems tend to be more susceptible to the forces of man (e.g., cable-seeking backhoes, posthole diggers, and trains) and nature (e.g., earthquakes and floods). Whether caused by man or nature, catastrophic failures add repair costs to the equation.

2.3   TWISTED PAIR: INTRODUCTION TO TELEPHONE WIRE

From iron wire to hard-drawn copper; from overhead to underground circuits. Over 94% of Bell System wire is now in cable.
                    Telephone Almanac, American Telephone & Telegraph Company, 1938

Metallic wires were used almost exclusively in telecommunications networks for the first 80 years, certainly until the development of microwave and satellite radio communications systems. Initially, uninsulated galvanized iron and then steel telegraph wires were used, although it soon became clear that copper was a much better choice for a number of reasons, including its malleability and electromagnetic energy conducting properties. The early metallic electrical circuits were one-wire, supporting two-way communications with each telephone connected to *ground* in order to complete the circuit. In 1881, John J. Carty, a young American Bell technician and one of the original telephone operators, suggested the use of a second wire to complete the circuit and, thereby, to avoid the emanation of electrical noise from the ground [7]. The first long-line copper wire telephone circuits were strung between New York and Chicago. Consisting of uninsulated hard-drawn copper conductors about as thick as a pencil, the two-wire circuit weighed 870,000 lb, filled a 22-car freight train, and cost $130,000 for the copper alone [8]. In certain contemporary applications, copper-covered steel, copper alloy, nickel- and/or gold-plated copper, and even aluminum metallic conductors are employed. The most common form of copper wire used in communications is that of the *Unshielded Twisted Pair* (UTP), which has no shield, or outer conductor, to protect the signal from outside sources of electromagnetic interference.

A twisted pair (Figure 2.3) involves two copper conductors, generally solid core, although stranded wire is used occasionally in applications that require additional flex strength. Each conductor is separately insulated by a *dielectric* (nonconductor of *di*rect *electric* current) material such as polyethylene, PolyVinyl Chloride (PVC), flouropolymer resin, Teflon, or some other low-smoke, fire-retardant substance. The insulation separates the conductors so that the electrical circuit is not shorted and protects the conductors from physical damage. Twisted pair is known as a *balanced* medium as both conductors serve for signal transmission and reception and as each conductor carries a similar electrical signal with identical direct and return current paths. At any given point in the cable, the signals are equal in voltage to ground but opposite in polarity, which has the effect of reducing radiated energy and, therefore, reducing attenuation, which increases signal strength over a distance [1, 3].

Figure 2.3   UTP configuration.

2.3.1 Twisting Process

The manufacturing process involves smoothly twisting the separately insulated conductors in a helix with a constant pitch or distance to make a 360° twist, hence the term *twisted pair*. This twisting process serves to improve the performance of the medium by reducing the radiation of electromagnetic energy and, thereby, improving the strength of the signal over a distance. Reducing the radiated energy also serves to minimize the impact on adjacent pairs in a multipair cable configuration. This is especially important in high-bandwidth applications, as higher frequency signals tend to lose power more rapidly over distances. Additionally, the radiated electromagnetic field tends to be greater at higher frequencies, which impacts adjacent pairs to a greater extent. Generally speaking, the tighter the twist, that is, the more twists per foot, the better the performance of the wire [1].

In applications that involve multiple pairs, the lay length can become an issue. For example, 10/100Base-T Ethernet LANs make use of two pairs of a four-pair Category 5 (Cat 5) UTP cable. The 1000Base-T standard for Gigabit Ethernet (GbE) can run over four pairs in the same Cat 5 cable if the cable is of the proper type and is installed properly. 1000Base-T splits the gigabit signal into four signals, each running over a single pair at 250 Mbps. Each of the four pairs in a Cat 5 cable has a slightly different *twist ratio*, or *twist pitch*, in order minimize *crosstalk*, which can be caused by the unwanted coupling of signals between pairs. (*Note*: You undoubtedly have experienced crosstalk in voice communications as extraneous conversations intruded on yours. Aggravating in voice applications, crosstalk renders data communications difficult, if not impossible.) The difference in twist ratios results in a slightly different *lay length*, that is, physical length if the cable were to be untwisted and laid flat, for each pair. (*Note*: More twists per foot make for a longer physical path per pair foot.) This causes built-in propagation *delay skew* simply because it takes more time for a signal to travel a longer physical path (Figure 2.4). As too much delay skew will cause transmission errors because of timing differences between the signals spread across the various pairs, some cable manufacturers use foamed insulation, rather than solid insulation, on the conductors.

While the specifics of the physics behind this are beyond the scope of this book, note that electrical signals tend to distribute themselves within a conductor so that the current density is greater near the surface of the conductor than at the core. The higher the frequency, the more pronounced is this phenomenon known as *skin effect*.

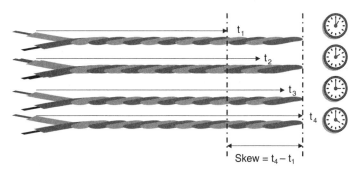

Figure 2.4 Delay skew across four pairs.

For example, and through a 24-gauge conductor with a diameter of 0.0201 in., a signal at 20 kHz travels at a skin depth of 0.0181 in. (which is almost the diameter of the wire, which means there is negligible skin effect), and a signal at 25 MHz travels at a depth of only 0.00052 in. [6]. In fact, much of the signal is in the form of an electromagnetic field surrounding the conductors and traveling through the dielectric insulation. Manufacturers make use of foamed insulation, taking advantage of these phenomena to increase the *velocity of propagation* (Vp), as the foamed insulation contains air and as air supports a higher velocity of signal propagation (see Table 2.2). Further, manufacturers can engineer the foamed insulation on each of the pairs with a slightly different amount of air in order to compensate for the minor differences in the lay lengths and, thereby, to adjust for delay skew. These issues are of particular significance with respect to high-speed LANs such as 100Base-T and 1000Base-T.

2.3.2 Gauge

Gauge is a measure of the diameter of the conductor. The greater is the diameter of the wire, the less the *resistance*, the stronger the signal over a given distance, and the better the performance of the medium. Thicker wires also offer the advantage of greater break strength.

American Wire Gauge (AWG), originally known as Brown and Sharp (B&S) Gauge, is the standard measurement of gauge in the United States for all metals other than iron and steel. The gauge numbers are retrogressive; in other words, the larger the number, the thinner the conductor. The AWG number indicates the approximate number of wires that, laid side by side, span 1 in. (Historically, the AWG number indicated the number of times during the manufacturing process that the copper wire was drawn through the wire machine, with each draw involving a die of slightly smaller diameter in order to reduce the diameter of the wire a bit more. The contemporary process involves many fewer draws.) As an example, a 24-gauge (AWG) wire with a diameter of 0.0201 in. (0.511 mm) has a weight of 1.22 lb/kft (1.82 kg/km), maximum break strength of 12.69 lb (5.756 kg), and DC resistance ohms of 25.7/kft (84.2/km). Twisted pairs commonly employed in telco networks vary from 19 to 28 gauge, with the most common being 24 gauge. England uses Imperial Standard Wire Gauge, also known as British Standard Gauge, which also is retrogressive. Many other countries use Metric Gauge, which expresses the gauge as 10 times the diameter of the wire in millimeters. Still other countries (e.g., Namibia) simply express wire gauge in terms of kilograms per kilometer of bare wire. Table 2.3 provides an abbreviated comparison of the various UTP categories [9].

Category 3 was developed in support of 10Base-T Ethernet LANs, which run at 10 Mbps. Its excellent performance characteristics, coupled with its relatively low cost, have led to its application in OSP applications as well. Telephone companies in the United States commonly use Cat 3 cable for local loops in support of ISDN, T1, and even POTS. Cat 3 UTP has three to four twists per foot.

Category 5e (enhanced) UTP is specified at a signaling rate of 100 MHz over distances up to 350 m in support of data rates of 100 Mbps (100Base-T and 100VG-AnyLAN) and 155 Mbps [Asynchronous Transfer Mode (ATM)]. Some manufacturers have increased the signaling speed as high as 250 MHz in support of

TABLE 2.3   Twisted-Pair (TP) Categories of Performance

Category (Cat) of Performance	Gauge (AWG)	Performance Rating (MHz)	Typical Applications
Cat 1	Various	Unspecified, <1	Analog voice grade, ISDN BRI, low-speed data, speaker wire, alarm cable
Cat 2	24	1	4 Mbps Token Ring LANs
Cat 3	24	16	Plain Old Telephone Service (POTS), ISDN, T1, 10Base-T LAN
Cat 4	24	20	16 Mbps Token Ring LAN
Cat 5	24	100	10/100Base-T LAN
Cat 5e	24	100+	10/100Base-T LAN, 155 Mbps ATM, 1000Base-T (GbE)
Cat 6	23	250	1000Base-T
Cat 7	23	600	10 GbE

Note: As Defined by the Electronic Industries Alliance (EIA), International Organization for Standardization (ISO), and International Electrotechnical Commission (IEC).

1000Base-T (i.e., GbE over UTP). Cat 5e offers improved performance with respect to attenuation and crosstalk. Cat 5e performance is achieved through a tighter twist (three to four twists per inch), electrical balancing between the pairs, and fewer cable anomalies, such as inconsistencies in both conductor diameter and thickness of the dielectric insulation. Cat 5e may be *Shielded Twisted Pair* (STP), which involves a continuous metallic shield protecting each pair as well as a continuous metallic shield surrounding all four pairs in the cable. Cat 5 cable has become the de facto standard for inside wire. *Note*: Most cable sold as Cat 5 actually is Cat 5e.

Category 6 cabling specifications include UTP, STP and *Screened Twisted Pair* (ScTP) rated at 250 MHz over distances up to 220 m, although some manufacturers boast of performance up to 400 MHz. Cat 6 is intended to support GbE, spreading the signals over each of four pairs. Applications include 100Base-T, ATM, and 1000Base-T for GbE applications.

Category 7 specifies STP with a combination foil and braided screen construction. Cat 7 supports signaling rates up to 600 MHz, although the usable spectrum can be up to 750 MHz. Cat 7 is intended to support 10GbE (10-Gigabit Ethernet), spreading the signals over each of four pairs. Due to its STP construction, Cat 7 is relatively expensive to manufacture. It also is bulky and requires larger conduits, ducts, and cable trays.

2.3.3 Configuration

Twisted pair generally is protected by a sheath or jacket made of polyethylene, PVC, Teflon, or some other insulating material that protects the conductors from physical damage and eases the installation process. A cable generally comprises multiple pairs, which is not only cost effective but also provides a measure of redundancy, as some wire pairs invariably suffer damage during the installation process and others may fail over time due to rodent damage or other unpleasant circumstances.

Inside wire and cable systems generally involve relatively small pair counts. For example, Cat 5e cable generally comes in a four-pair configuration for horizontal applications. As 10Base-T and 100Base-T require only two pairs and as PBX telephone sets generally require one or two pairs, a single four-pair Cat 5e cable can support both voice and data requirements for a typical end user. Since 1000Base-T requires four pairs, it requires a dedicated cable system. In consideration of distance limitations and crosstalk issues, inside wire and cable deployments generally are on a *home-run* basis, which is to say that each cable provides an exclusive, uninterrupted physical path between a terminal device such as a telephone, workstation, or printer and a centralized connectivity device such as a hub, switch, or router. In residential or small-business voice applications, inside wire installations commonly are in a *loop* configuration, which involves connecting multiple voice telephone jacks to one or two pairs of wires in a continuous, shared electrical loop. This approach works well enough in support of extension telephones, but at the expense of privacy.

Larger numbers of pairs are bundled into larger cables to serve departments, quadrants of a building, or floors of a high-rise office building. Such cables may contain 25, 50, 100, 250, or 500 or more pairs. As appropriate, the cables and pairs interconnect at *cross-connect* points. (*Note:* As the cables get farther from the switch or router, for example, and closer to the end user, the pair counts tend to get smaller.) In large cables, pairs combine into *binder groups* of 25 pairs for ease of connectivity management. Each binder group is wrapped (bound) with some sort of plastic tape to separate it from other groups. Each pair within a binder group is uniquely color coded for further ease of connectivity management. (*Note:* This is oversimplified, of course. Cable and wire management is a notoriously difficult business and one which is largely lacking in standardization.)

Inside wires and cables are of several types, according to the various national and regional standards such as the National Electrical Code (NEC) in the United States, with the construction of the cables depending on their application. *Plenum cables*, for example, are intended for use in *plenums*, or air-handling spaces, such as those between walls, under floor structures, and above drop (false) ceilings. While plenums are convenient places to run cables, they also are conducive to the spreading of fires within buildings. Therefore, the NEC specifies that the insulation on plenum cables must be fire retardant, low smoke, and low toxicity. *Riser cables*, intended for use between floors of a building, also must be fire retardant. Note that these categories apply equally to all cables, including fiber-optic cables.

While twisted-pair cables of up to 3600 pairs are still used in outside plant (OSP) applications, fiber-optic cables largely have replaced them in contemporary networks. Categories of OSP cables include the following:

- *Overhead* cables hang from poles.
- *Direct burial* cables lie directly in trenches dug in the ground.
- *Indirect burial* cables lie in ducts or conduits placed in trenches dug in the ground.
- *Submarine* cables are underwater, perhaps miles deep.

OSP cables must be extremely rugged and durable, as they variously are exposed to extremes of temperature and pressure, rodents, cable-seeking backhoes and posthole diggers, and other forces of man and nature too numerous to list in this space. Defense mechanisms against man and rodents include lead sheathing and steel armoring. Air pressurization and water-blocking gel help keep cables free of moisture, which not only can cause electrical shorts but also can freeze, thereby causing cable insulation to crack and break.

2.3.4 Bandwidth

The effective capacity of twisted-pair cable depends on several factors, including the gauge of the conductor, the frequency of the signal, the nature of the dielectric insulating material, the length of the circuit, and the spacing of the amplifiers/repeaters. One must also recognize that a high-bandwidth (high-frequency) signal will not only attenuate relatively quickly but also may cause interference with other transmissions on other pairs in proximity. This issue of crosstalk is particularly sensitive in data applications.

While a voice-grade channel is guaranteed at 4 kHz, standard copper circuits can support much greater bandwidth. A single twisted pair in a typical telephone installation may provide up to 250 kHz, or 1–4 Mbps compressed, assuming amplifier or repeater spacing every 2–3 km [1]. Additional examples follow:

- T1 connections provide bandwidth of 1.544 Mbps at approximately 0.75 MHz. T1s are routinely provisioned over specially conditioned, four-wire twisted-pair cable, with repeaters spaced at approximately 6000 ft.
- Category 5e copper, in a LAN environment, provides bandwidth of 100+ Mbps at a signaling rate of 100+ MHz over twisted-pair cable at distances of up to 100+ m.
- Asymmetric Digital Subscriber Loop (ADSL) simultaneously supports as much as 6.144 Mbps in the downstream direction, a bidirectional channel supporting as much as 608 kbps, and an analog voice channel over a single conditioned physical two-wire twisted-pair local loop at distances up to 2 miles. Standards-based ADSL specifications also define considerably higher speeds, although over shorter distances.

2.3.5 Error Performance

Signal quality is always important, especially relative to data transmission. Twisted-pair cable is especially susceptible to the impacts of outside interference, as the lightly insulated wires act as antennas and, thereby, absorb such errant signals. Potential sources of ElectroMagnetic Interference (EMI) and Radio Frequency Interference (RFI) include electric motors, radio transmissions, and fluorescent light boxes. As the carrier frequency of a transmission increases, the error performance of copper degrades significantly, with signal attenuation increasing approximately as the square root of frequency. Further, a high-frequency transmission radiates a strong electromagnetic field, which is absorbed by adjacent pairs in a multipair cable and which affects their error performance.

Error performance in UTP cables also is highly sensitive to proper splicing. Proper grounding and bonding of any shielding is necessary to minimize EMI. The integrity of the outside insulation and shielding of OSP cables is critical. As mentioned above, rodent damage is a constant problem, particularly in older cables employing soy-based insulation, which squirrels, moles, and other critters find to be quite tasty. Water is a serious enemy of electrical cables, in general, and telephone cables often are pressurized to prevent moisture from affecting performance. Water-blocking gels and powders also are used commonly.

2.3.6 Distance

UTP is perhaps the most distance limited of all the media options. As distance between network elements increases, attenuation (signal loss) increases, and error performance degrades at a given frequency. Even low-speed (voice-grade) analog voice transmissions require amplifiers spaced at least every 2–4 miles or so. As a result, local loops generally are 10,000–18,000 ft in length. As bandwidth increases, the carrier frequency increases, attenuation becomes more of an issue, and amplifiers/repeaters must be spaced more closely. T1 transmission, for example, requires repeaters spaced at intervals of approximately 5000–6000 ft. (Remember that there are 5280 ft in a mile and about 3281 ft in a kilometer.)

2.3.7 Security

UTP is an inherently insecure transmission medium. While it is relatively simple to place a physical tap on a UTP circuit, it also is fairly simple to detect the presence of a tap. Through the use of an antenna or inductive coil, you can easily intercept the signal without the placement of a physical tap, as so much of the signal travels in an electromagnetic field around the conductor. With the proper wiretap technology in place, proximity is not as much of an issue as one might think. As is the case with any transmission over any medium, encryption is the best protection.

2.3.8 Cost

The acquisition, deployment, and rearrangement costs of UTP are very low, at least in inside wire applications involving only a few pairs (e.g., between a terminal and a switch or hub). In high-capacity, long-distance applications (e.g., interoffice trunking), however, the relative cost is very high due to the requirements for trenching or boring, placement of conduits or poles, and frequent splicing of large, multipair cables that tend to be relatively short in length. (*Note*: There is only so much cable of a given size and weight that is manageable. The bigger and heavier the cable, the shorter the length of cable a construction crew can put on a truck and pull through a conduit or hang from a pole.) Additionally, there are finite limits to the capacity and other performance characteristics of UTP, regardless of the inventiveness of technologists—hence the popularity of alternatives such as microwave and fiber-optic cable.

2.3.9 Applications

Generally speaking, UTP is no longer deployed in long-haul outside plant transmission systems. Rather, fiber-optic cable, microwave, and satellite are the media of choice in such applications. UTP remains the medium of choice in most inside wire applications due to its overall ease of handling and its excellent overall performance over short distances. Although wireless LANs have replaced many UTP-based LANs, the wireless access points generally are hard wired back to the LAN switches with UTP. Much of the copper embedded in the local loop continues to perform well. In fact, it often greatly exceeds original expectations, as evidenced by the large installed base of ADSL.

Copper has applications other than telecommunications, of course. As the price of scrap copper approximately doubled in 2004 to reach historic highs, UTP cables became attractive targets for theft all over the world. During the first two months of 2006, thieves cut down aerial telephone cables in Kent, Washington (United States), putting hundreds of customers out of service. The Tucson, Arizona (United States), city council is considering an ordinance that would require junk dealers to give police the identities of people who sell scrap metal. While teaching a seminar on telecommunications in Windhoek, Namibia, in early 2006, I read that Telecom Namibia is offering cash rewards for citizens who report thefts of copper telephone cables, with the size of the rewards amounting to thousands of dollars, depending on the amount of cabling stolen. I know that the problem also exists in South Africa, where I teach seminars several times a year.

Unfortunately, UTP also has application in arts and crafts. Several years ago I noticed a *gaily* colored basket in a hotel gift shop in Johannesburg. The colors were familiar and the heft was considerable. The tag on the basket explained that Zulu warriors in times past whiled away their spare time by weaving these *mbenge* from native grasses dyed with the juices of native herbs. Actually bowls rather than baskets, the *mbenge* were covers for clay *ukhamba* (Zulu sorghum beer pots), serving to keep insects and dust out of the beer while still allowing the beer to ferment and breathe. As the Zulus increasingly moved to the cities in search of work, many men became night watchmen, in keeping with the warrior tradition. To while away the long hours while watching over the assets of their employers, the watchmen often wove *mbenge* from materials at hand, including scrap telephone wire. Well, that certainly explained the familiarity of the colors and heft. (*Note*: The insulation surrounding UTP cables is color coded in binder groups of 25 pairs to assist in proper splicing. The insulation colors are blue, orange, green, brown, slate, white, red, black, yellow, and violet and are presented in both solid colors and stripes of various color combinations.) It certainly did not explain the origins of the *scrap* wire, however. While I cringe at the thought of being in possession of stolen property, I just had to buy one, which now adds a dash of color to my living room. I hate to admit it, but I just bought another small *telewire* basket during that same seminar tour in early 2006. [*Note*: Conduct a Web search for *mbenge* and you, too, will find the opportunity to purchase an *mbenge* made of scrap telephone wire. *Mbenge* made of more traditional (and reliably legitimate) materials also are available.]

2.4 SHIELDED COPPER

The simplest form of shielded copper is *Screened Twisted Pair* (ScTP), which involves multiple insulated pairs formed into a core that is enclosed with an overall metallic shield that is encased in a thermoplastic cable jacket. The shield typically consists of helically or longitudinally applied plastic and aluminum laminated solid tape,

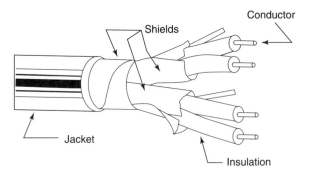

Figure 2.5 STP configuration.

although it may comprise a woven mesh, and steel or copper also may be used. One or more uninsulated steel or tinned copper conductors in contact with the shield serve as *drain wires*, ensuring that the continuity of the shield remains intact in the event that the tape is broken or cracked. *Shielded Twisted Pair* (STP), also known as *Shielded Foil Twisted Pair* (SFTP), is more complex in that a metallic shield surrounds each of the insulated pairs, which may or may not be twisted. The core of shielded pairs is then surrounded by an overall metallic shield of metallic tape or braid, or both, which is encased in a thermoplastic cable jacket, as illustrated in Figure 2.5. Shielding sometimes takes the less expensive form of nickel and/or gold electroplating over the individual conductors, although this approach is less effective.

Shielded copper offers the advantage of enhanced performance through reduction of emission of energy from the subject conductors and reduction of interference from ambient sources of electromagnetic energy such as electric motors, radio systems, and adjacent cables and wires. The shield absorbs ambient energy and conducts it to ground through the drain wire, thereby protecting the signal transmitted through the center conductor. The shield also serves to confine the electromagnetic field associated with the transmitted signal within the core conductors, thereby reducing signal loss and maintaining signal strength over a longer distance. This reduction of emissions also provides additional security and minimizes the potential for causing interference in adjacent pairs or cables.

ScTP and STP also have several disadvantages. First, the raw cost of acquisition is greater because manufacturing costs are higher. Second, the cost of deployment is greater because the additional bulk and weight of the shield and extra insulation increase the difficulty of installation. (*Note:* The insulation used in shielded copper cable systems is significantly thicker than that of UTP of similar gauge and resistance. This is due to the effect of the shield on the electromagnetic fields of the transmitted signal. The closer the shield is to the core conductors, the greater its effect. Hence, the dielectric insulation must be thicker to increase the separation between them.)

Also, the electrical grounding of the shield requires more time and effort during the installation process. As the continuity of the shield must be protected from end to end, either significant flexing or a severe bend radius can compromise the integrity of the shield.

The additional cost of shielded copper historically has limited it to inside wire applications in high-noise environments. It currently also is deployed where high-frequency signals are transmitted and where interference with adjacent pairs presents a concern. Current applications include both Cat 6 SFTP and ScTP and Cat 7 SFTP in support of high-speed LANs (e.g., 100Base-T and 1000Base-T) at signaling rates as high as 750 MHz.

2.5 COAXIAL CABLE

Coaxial cable (Figure 2.6) is a very robust shielded copper cable. The center conductor is much thicker than a twisted-pair conductor (e.g., 20 AWG versus 24 AWG) and is surrounded by an outer shield/conductor that serves to greatly improve signal strength and integrity. A layer of *dielectric* material, either foam or solid, generally separates the two conductors. The entire cable is then protected by another layer of dielectric material, such as PVC or Teflon. The two conductors share a common axis, hence the term *coaxial*. Invented by AT&T Bell Telephone Laboratories in 1934, the first coaxial cables were hollow tubes about one-quarter inch in diameter. Down the center of each pipe ran a copper wire held in place by insulating discs. The pipes were in pairs—one for transmission in each direction. The first coaxial system was placed into service in New York City in 1936. Such a cable was used in New York City to televise the 1940 Philadelphia Republican National Convention at which Wendell Wilke was nominated for president of the United States [27]. While Wilke was unsuccessful in his bid for the presidency, the coaxial cable proved to be popular. By the early 1940s, coaxial cable in commercial service could carry 500–600 telephone channels. By the late 1950s, frequency division multiplexers supported some 1800 conversations over each pair of coax tubes. Specific types of coaxial cables often are referred to by RG number (e.g., RG-6, RG-8, and RG-58). The terminology was established by the United States military in the 1930s, with RG referring to *Radio Guide*, as the Radio Frequency (RF) signal is guided down the center conductor of the cable system. The RG numbering system does not really have any special significance; rather, each RG number is just a page in a book, so to speak. Each RG number does, however, specify the impedance, the core conductor gauge (AWG), and the Outside Diameter (OD) of the cable.

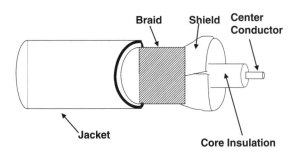

Figure 2.6   Coaxial cable configuration.

2.5.1 Configuration

A coaxial cable, or *coax* cable, typically consists of a single, two-conductor wire with a center conductor and one or sometimes two outer shields, or conductors. The inner conductor generally is solid core, although stranded wire sometimes is used in applications requiring additional flex strength. The outer shield generally consists of a solid metal foil, although a braided or stranded metal screen sometimes is used. The metal used for the inner conductor may be bare copper, silvered copper, tinned copper, copper-clad aluminum, or copper-covered steel. The outer shield generally comprises an aluminum sheath, aluminum braid, bare copper braid, silvered copper braid, or tinned copper braid. *Twinaxial (twinax) cables* contain two thin coax cables contained within a single cable sheath and once were popular for linking IBM terminals to cluster controllers. More recently, the IEEE 802.3ak task force finalized (February 2004) the 10GBase-CX4 standard in support of 10 Gigabit Ethernet (10GbE). That standard yielded a specification for twinax assemblies operating over distances up to 50 ft. The cost of this patch cord technology approximately 10 percent that of comparable fiber-optic solutions.

Regardless of the specifics of the configuration, the center conductor(s) carries the carrier signal. The outer conductor generally is used only for electrical grounding and is maintained at 0 volts. Therefore, coax is described as an electrically *unbalanced* medium. A *balun* (*bal*anced/*un*balanced) connector is used to connect (balanced) twisted-pair and (unbalanced) coax cables.

2.5.2 Gauge

The gauge of the cable is much thicker than a twisted pair. The heavier gauge increases the available bandwidth and increases the distance of transmission as the thicker wire offers less resistance to high-frequency signals. However, the more complex construction of coax also increases the cost of acquisition. Further, coax is more expensive to install as its weight and mass make it harder to pull. It also is difficult to splice and connect, and it must be properly grounded. Traditional coax, such as that used in the original Ethernet LANs, is quite thick, heavy, bulky, rigid, and altogether difficult to manipulate. The subsequent Ethernet generation, sometimes known as *CheaperNet*, made use of *ThinNet* coax of smaller dimensions, but at the expense of performance.

2.5.3 Bandwidth

The effective capacity of coax cable depends on several factors, including the gauge of the center conductor, the nature of the dielectric insulation, the length of the circuit, and the spacing of amplifiers and other intermediate devices. Because the bandwidth available over coax is very significant in comparison to twisted pair, it often was used in high-capacity applications, such as data and image transmission. Note the following examples of coax standards for classic Ethernet LANs; 10Base5 involves a much more substantial cable, with a thicker center conductor than 10Base2:

- **10Base5:** 10 Mbps; *Base*band (single channel); 500 m maximum link length
- **10Base2:** 10 Mbps; *Base*band (single channel); 200 m (180 m, rounded up) maximum link length

The 10GBase-CX4 standard (February 2004) for 10GbE yields a specification for twinax assemblies operating over distances up to 50 ft. Each twinax cable will transmit at a rate of 2.5 Gbps in simplex mode (i.e., in one direction) at a frequency of 3.125 GHz.

Coax is easily capable of supporting data rates of 100 Mbps over much longer distances and with better error performance than twisted pair. In Community Antenna TeleVision (CATV) and other applications, coax routinely supports transmission of multiple channels at an aggregate rate of 500–750 MHz. The CATV trunk cables commonly are made of very thick copper-clad aluminum, and the drops to the premises are much thinner solid-core copper.

2.5.4 Error Performance

Coax offers excellent error performance due to the outer shielding. As a result, coax was used extensively in classic data applications, for example, mainframe to Front-End Processor (FEP) to cluster controller to terminal. But the end-to-end integrity of the shield is crucial; either physical damage or poor splicing will result in awful error performance. The shield also must be grounded properly.

2.5.5 Distance

Coax does not have the same distance limitations as either UTP or STP, because the thicker center conductor offers less resistance to the signal. Further, the outer shield helps confine the signal to the inner conductor, which maintains signal strength. The outer shield also serves to make the signal riding over the center conductor much less sensitive to ambient noise, which is especially important as the signal weakens over long distances. Nonetheless, amplifiers or other intermediate devices must be placed at appropriate intervals to extend high-frequency transmissions over significant distances.

2.5.6 Security

Coax is inherently quite secure as it is relatively difficult to place physical taps on coax without detection. (*Note:* The original Ethernet taps were designed to be easy, but LANs operate in secure office environments.) Little energy is radiated through the outer shield, so radio antennas and inductive coils are of little use in gaining access to the raw signal.

2.5.7 Cost

The acquisition, deployment, and rearrangement costs of coax are very high compared with UTP due to increased bulk and weight as well as the requirement for grounding the outer shield. In certain high-capacity data applications, however, its positive performance characteristics can outweigh that cost.

2.5.8 Applications

Historically, coax often was used in telco interoffice trunking applications as a superior option to twisted-pair cables. The contemporary choices, however, include satellite, microwave, and, especially, fiber-optic cables in such applications. The superior performance characteristics of coax favored its use in many short-haul, bandwidth-intensive data applications. Current and continuing applications include host to host, cabinet to cabinet [e.g., Private Branch eXchange (PBX) and computer], and host to peripheral (e.g., host to FEP). Recent developments in UTP and fiber-optic transmission systems, however, have rendered coax largely obsolete in such applications. Very short lengths of coax are used in a wide variety of devices where high levels of bandwidth are required but fiber optics is impractical due to cost of the optoelectric conversion processes. For example, patching connections at the E-2 (2 Mbps) and T3 (45 Mbps) levels are often coax.

Coax is used extensively in CATV networks due to its overall performance characteristics. Coax-based CATV networks commonly provide bandwidth of either 330 MHz in support of up to 40 channels or 750 MHz in support of up to 116 channels. Such analog, one-way downstream networks support each analog TV signal over a 6-MHz channel, with the channels frequency division multiplexed at the *head end*, or point of signal origin. The signals are demultiplexed at the *set-top box*, or converter. Much coax remains in place in CATV applications, although it is being replaced with fiber optics in the backbone, with the existing coax used for the last leg of the connection to the premise. Such a hybrid network will support one-way entertainment TV, two-way Internet access, two-way voice, and other applications.

2.6 MICROWAVE RADIO

> *The medium through which electrical and magnetic forces act is called "ether." With the theories as to its composition or construction we are not here concerned. It is, however, by virtue of this medium which fills all space that electromagnetic disturbances ... are made manifest at a distance. Of the periodic disturbances thus transmitted light and heat are two classes. The periodic disturbance made use of in wireless telegraphy form a third class.*
>
> John Mills, *Radio Communication*, McGraw-Hill Book Company, 1917

Microwave radio, a form of radio transmission that uses ultrahigh frequencies, developed out of experiments with *radar (ra*dio *d*etecting *a*nd *r*anging) during the period preceding World War II. The first primitive systems, used in military applications in the European and Pacific theaters of war, could handle up to 2400 voice conversations over five channels. Developed by Harold T. Friis and his associates at AT&T Bell Telephone Laboratories, the first public demonstration was conducted between the West Street laboratory and Neshanic, New Jersey, in October 1945. Construction began on the first experimental microwave telephone network in 1947 [7].

Microwave systems are point-to-point radio systems operating in the GigaHertz (GHz) frequency range. The *wavelength* is in the millimeter range, which is to say that each electromagnetic cycle or waveform is in the range of a millimeter, which gives rise to the term *microwave*. As such high-frequency signals are especially

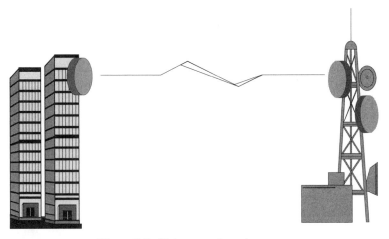

Figure 2.7   Point-to-point microwave.

susceptible to attenuation, they must be amplified (analog) or repeated (digital) frequently. In order to maximize the strength of such high-frequency signals over long distances, the radio beams are tightly focused. Much as a light bulb in a flashlight is centered in a mirror that serves to focus the light beam, the microwave transmit antenna is centered in a concave, reflective metal dish that serves to focus the radio beam with maximum effect on the receiving antenna (Figure 2.7). Similarly, the receiving antenna is centered in a concave metal dish that serves to collect a greater amount of incoming signal and reflect it into the receiver. *Note:* Antennas serve both transmit and receive functions, with transmit and receive frequencies separated to avoid self-interference.

The requirement to so tightly focus the signal clearly limits microwave to application as a *point-to-point*, rather than a *broadcast*, transmission system. Additionally, microwave is a *Line-Of-Sight* (LOS) technology as such high-frequency radio waves will not pass through solid objects of any significance (e.g., buildings, mountains, or airplanes). Actually, line-of-sight is not quite enough, as the signal naturally disperses (i.e., spreads out) in a conical pattern. As a result, portions of the signal reflect off of bodies of water, buildings, and other solid objects and can interfere with the primary signal through a phenomenon known as multipath fading. The impact of multipath fading is that multiple copies of the signal reach the receiving antenna at different levels of strength at slightly different times and slightly out of phase, thereby confusing the receiver and distorting the signal much like the *ghosting* effect that can be so aggravating at times to broadcast television viewers. So, additional clearance is required in the form of a Fresnel ellipse, an elliptical zone that surrounds the direct microwave path. In consideration of LOS and Fresnel zone clearance, antenna positioning and tower height are important considerations in microwave path selection and network design. Clearly, so to speak, antennas atop tall towers positioned on the roofs of tall buildings and the peaks of high mountains tend to provide optimum signal paths. Figure 2.8 illustrates a multihop microwave configuration with consideration given to Fresnel zone clearance.

If a microwave route traverses a smooth-earth path involving no hills, mountains, bulges of earth, tall buildings, or other signal obstructions, the link length is sensitive

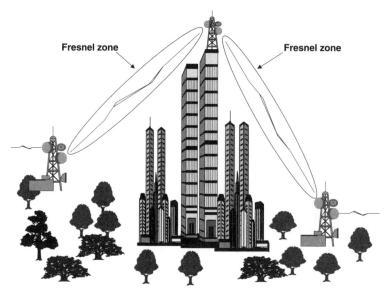

Figure 2.8 Multihop microwave configuration with Fresnel zone clearance.

TABLE 2.4 Example Microwave Frequency Bands (ITU) and Antenna Separation

Frequency Bands (GHz)	Typical Maximum Antenna Separation
2–6	20–30 miles (32–48 km)
10–12	10–15 miles (16–24 km)
18–23	5–7 miles (8–11 km)
28–30	1–2 miles (0.6–1.2 km)

to factors including frequency band, air quality, and curvature of the earth. Higher frequencies suffer more from attenuation than do lower frequencies. In the context of an airwave system such as microwave, air quality and environmental interference issues include dust, smog, agricultural haze, precipitation, fog, and humidity. Table 2.4 lists example international frequency bands allocated by the International Telecommunications Union—Radiocommunications Sector (ITU-R) for commercial microwave and makes clear the relationship between frequency band and antenna separation, assuming typical allowable power levels [10]. These frequency bands are representative of those used throughout the world for microwave applications, although the specifics can vary from region to region and nation to nation.

At the lowest microwave frequencies, attenuation is low enough that the horizon becomes a major consideration, as the curvature of the earth limits LOS. In this scenario, it is necessary to consider the difference between optical LOS and radio LOS. True *optical LOS* is a straight line between the two antennas. *Radio LOS* can be somewhat longer as the density gradient in the atmosphere acts like a lens and tends to bend radio beams back toward the earth, as illustrated in Figure 2.9 [11].

In most countries and regions, regulatory authorities [e.g., the Federal Communications Commission (FCC) in the United States] protect frequency bands

Figure 2.9   Microwave path illustrating differences between optical and radio LOS.

allocated for microwave application. Additionally, the physical path profile, place-ment of the antennas, and the power level of transmission are regulated, with licenses granted to individual carriers and end users. Difficulties, however, have developed over time in certain areas (e.g., Europe and Asia) due to factors that include the small size of the individual nations, lack of regulation, conflicting regula-tions, conflicting commercial and military applications, and unwillingness of national regulators to govern the use of radio frequencies on a coordinated, regional basis.

The range of usable spectrum has increased over the years as technology has evolved. Local Multipoint Distribution Services (LMDS), a Wireless Local Loop (WLL) technology, operates in the 31-GHz licensed band. Some non-standard WLL systems operate in licensed bands as high as 38 GHz. Worldwide Interoperability for Microwave Access (WiMAX) is a relatively recent standard for WLL systems running in the 2- to 11-GHz range. WiMAX operates in both point-to-point and point-to-multipoint topologies and can operate in both LOS and Non-LOS (NLOS) modes. (Chapter 9 discusses WLL in detail.)

The Industrial/Scientific/Medical (ISM) bands are in the ranges of 2.4–2.5 GHz and 5.8–5.9 GHzs. As these bands require no licensing and as signal propagation characteristics are excellent at these relatively low frequencies (particularly in the lower of the two bands), they are employed extensively in a wide variety of applica-tions, including Wireless Local Area Networks (WLANs). (Chapter 8 addresses WLANs in detail.)

2.6.1   Configuration

Microwave radio systems consist of antennas centered within reflective dishes which are attached to structures such as towers or buildings and generally take the shape of either a parabola (bowl) or horn (cornucopia). While the antennas often are quite small, larger horns collect more incoming signals and, therefore, offer greater *gain*, or increase in signal power. Hollow cables or tubes, known as *waveguides*, serve to guide the radio microwaves between the electronic transmit/receive equipment and the antennas.

2.6.2 Bandwidth

Microwave systems offer substantial bandwidth. Digital microwave systems, which account for the majority of contemporary systems, routinely run at signaling rates of T1 (1.544 Mbps) and E1 (2.048 Mbps), with many operating at E3 (34 Mbps) and T3 (45 Mbps) rates and even Optical Carrier (OC) 3 rates of 155 Mbps. [See Chapter 9 for a discussion of the Synchronous Optical NETwork (SONET)/Synchronous Digital Hierarchy (SDH) rate of OC-3.] Digital systems employ sophisticated modulation techniques to increase spectrum efficiency by packing multiple bits into each available hertz. For example, *Quaternary Amplitude Modulation* (QAM) defines two levels of amplitude and four levels of phase shift, for a total of eight possible combinations of amplitude and phase, which allows three bits to be impressed on a single hertz. Some LOS microwave systems employ 256-QAM and 512-QAM, theoretically allowing eight or nine bits, respectively, to be impressed on a hertz. (Chapter 6 discusses modulation techniques in detail.) Note that the issue of overall bandwidth availability is determined not solely by the spectrum allocated for microwave use by the regulatory authority but also by the fact that spectrum is licensed to users on a geographic basis. Once a band of microwave spectrum is licensed to an organization for use along a specific physical route, it becomes absolutely unavailable to other users in proximity. While there is no additional radio spectrum being manufactured these days, technology has continued to develop to the point that some WLL systems run effectively in the range of 38 GHz. Technology certainly will continue to develop and usable spectrum certainly will continue to increase, but it is important to remember that radio spectrum will always be limited.

2.6.3 Error Performance

Microwave, especially digital microwave, offers excellent error performance assuming proper system design and deployment. Physical obstructions must be avoided at all costs, as even grazing the smallest obstructions has a decidedly negative impact on error signal strength due to the phenomenon of Fresnel zones. Microwave radio also is particularly susceptible to environmental interference such as haze, smog, smoke, fog, and precipitation. *Rain attenuation*, or *rain fade*, is a factor at frequencies above 8 GHz and can be especially serious at frequencies above 11 GHz. Rain fade is sensitive to the rate of rainfall, the size of the raindrops, and the length of exposure. Note that microwave operates at very high frequencies, many of which are near the upper limit of the radio spectrum and at the lower edge of the light spectrum. Therefore, microwave behaves more like light than radio, as we humans normally think of it. In other words, it is absorbed, refracted, and reflected by physical matter—the more dense the matter, the worse the effect. Techniques for overcoming signal attenuation include spatial and frequency diversity. *Spatial diversity* involves the use of multiple antennas vertically separated on the tower and with each connected to a separate receiver. As the likelihood is that the signal will not suffer the same level of attenuation as it disperses slightly and propagates along slightly disparate paths, the receiver with the strongest signal assumes control of the transmission. Further improvements in received signal strength can be realized by signal combiners. *Frequency diversity* involves the use of multiple paired transmit and receive antennas operating at different frequencies. As the likelihood is that the signals will not suffer the same level of attenuation at different frequencies, the receiver with the strongest signal assumes control of the transmission. Generally speaking and assuming that the system is designed properly, microwave performs on a par with copper wire networks in terms of error performance.

2.6.4 Distance

Microwave clearly is distance limited, especially at the higher frequencies (see Table 2.4). As a point-to-point LOS radio system, design considerations include topography, antenna height, climate, and curvature of the earth. Distance limitations can be mitigated through larger antennas and antenna arrays incorporating spatial diversity and frequency diversity to increase the quality of the collected signal.

2.6.5 Security

As is the case with all radio systems, microwave is inherently insecure. A radio antenna tuned to the proper frequency range and positioned in proximity to the microwave path can easily capture the raw signal. Security is imposed through signal encryption (scrambling).

2.6.6 Cost

The acquisition, deployment, and rearrangement costs of microwave can be high. But these costs often compare very favorably with those of cabled systems, which require extensive right-of-way procurement processes, trenching and pole placement, conduit systems, splicing, and so on. Additionally, microwave is not affected by *backhoe fade*, as are cabled systems.

2.6.7 Regulation

Microwave systems, generally speaking, operate in licensed frequency bands. Spectrum allocation is the responsibility of the ITU-R at the international level. Regional authorities include Directorate General XIII (DG XIII) of the European Union (EU). National authorities include the FCC in the United States and the Independent Communications Authority of South Africa (ICASA). Within the allocated spectrum, individual microwave transmission systems must be licensed on a case-by-case basis to avoid interference between adjacent systems. Licensing considerations include physical path, tower and antenna placement, tower height, frequency allocation, modulation method, and radiated power level. Local zoning ordinances and health and safety regulations also may affect the placement of antennas. As regulations can be complex, the licensing process can be lengthy and costly. Unlicensed spectrum such as the ISM bands requires no licensing, of course, although radiated power levels are strictly limited to minimize the considerable potential for interference.

2.6.8 Applications

Microwave historically has been used extensively for long-haul voice and data communications. Competing long-distance carriers, first in the United States, found microwave a most attractive alternative to cabled systems, due to the relatively high speed and low cost of deployment. Where technically and economically feasible, however, fiber-optic technology currently is used in most long-haul applications. Contemporary microwave applications include long-haul carrier networks, private networks, carrier bypass, disaster recovery, interconnection of cellular radio switches, and WLL. Microwave certainly is an excellent alternative to cabled systems where terrain is challenging. In nations where regulatory authorities have liberalized telecommunications, emerging competitors find microwave to be an excellent means for deploying competing networks quickly and at low cost, particularly in WLL applications.

2.7 SATELLITE RADIO

Energy ... must obviously have some medium for its action—or, if we like to call it so, for its conveyance.... Wave-motion ... of any kind requires a medium in which to propagate the wave.... The Ether, as this universal medium is called, must be unimaginably rare and subtle.... We can neither see it, nor feel it, nor weigh it; its necessary properties seem in their contradiction to mock all our faculties; yet we must suppose it, or something like it, or all our physical theories fail us, and our known facts fall to the ground as a confused mass.

The Wonders of The Universe, The Werner Company, 1899

Satellite radio is a microwave transmission system utilizing nonterrestrial relay stations positioned in space, where there is no atmosphere and certainly no ether. The concept initially was offered in an article entitled "Extra-Terrestrial Relays: Can Rocket Stations Give World-wide Radio Coverage" published in *Wireless World* in February 1945 by Arthur C. Clarke, then a physicist at the British Interplanetary Society. Clarke is better known as the author of the famous short story, *The Sentinel*, which served as the basis for the movie *2001: A Space Odyssey*, which he cowrote with Stanley Kubrick. The launch of the Earlybird I satellite in 1965 proved the effectiveness of the concept of satellite communications. AT&T Bell Telephone Laboratories subsequently launched Telstar I, the first active commercial communications satellite, in 1962. Since that time, satellites have proved invaluable in extending the reach of voice, data, and video communications around the globe and into the most remote regions of the world. Exotic applications such as the Global Positioning System (GPS) would have been unthinkable without the benefit of satellites [12]. The approximately 200 GEO satellites in operation clearly speak to the popularity of communications satellites. (*Note:* There are approximately 2200 active satellites of various kinds in orbit at any given time.) In a subsequent article entitled "A Short Pre-History of Comsats, Or: How I Lost a Billion Dollars in My Spare Time," Clarke lamented his failure to patent the concept [13].

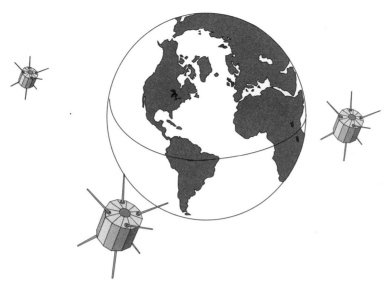

Figure 2.10   Satellites in GEO.

Traditional satellite communications systems involve a satellite relay station that launches into a *geostationary, geosynchronous,* or *geostatic* orbit, also known as a *Clarke* orbit, in honor of Arthur C. Clarke. Such an orbit is approximately 22,235 statute miles (35,784 km) above the equator (Figure 2.10). At that altitude and in an equatorial orbital slot, the satellite is in synchronization with the revolution of the earth. In other words, the satellite rotates around the earth at the same speed as the earth rotates on its axis. As a result, the satellite maintains its relative position over the same spot of the earth's surface. Consequently, transmit and receive earth stations (i.e., terrestrial microwave dishes) point to a fixed spot in the heavens to establish a communications link, secure in the knowledge that the satellite will be there. GEO satellites are also known as *Fixed Satellite Systems* (FSSs) due to their fixed positions relative to the earth's surface.

A GEO can see roughly one-third of the earth's surface from its vantage point; it cannot effectively transmit to such a wide area. Therefore, the downlink transmission is focused on a particular *footprint*, or area of coverage, which might be as wide as an entire continent. *Spot beams,* even more tightly focused downlinks, serve specific applications over smaller regions. The popularity of satellite communications has placed great demands on the limited number of GEO slots, which are spaced at intervals of approximately 1.5°–2.0°. Competition for orbital slots, coupled with the development of new applications (e.g., mobile voice and data) and the technologies to address them, has given rise to new generations of microwave satellite platforms known as *Middle-Earth Orbiting* (MEO) and *Low-Earth Orbiting* (LEO) systems. MEO systems operate at altitudes of 10,062–20,940 km and LEO systems operate at altitudes of 644–2415 km. At these lower altitudes, the satellites circle the earth rather than remaining in a fixed position above and relative to it. Further, the satellites operate along multiple paths, most of which are nonequatorial. Also at these lower altitudes, the satellites have much more restricted views of the earth's surface and, therefore, considerably smaller footprints, as illustrated in Figure 2.11.

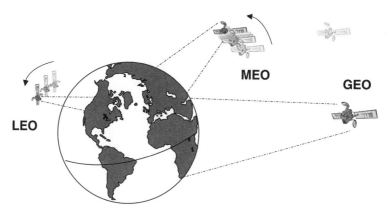

Figure 2.11 Satellites in GEO, MEO, and LEO.

TABLE 2.5 Example Satellite Frequencies, Band Designations, and Applications

Frequency Band	Band Designation	Example Applications
136–137 and 148 MHz	VHF	Weather; military tactical communications
400 MHz	UHF	Military tactical communications
1610–1625.5 MHz 2483.5–2500 MHz	L-band	Global Positioning Satellites (GPS); mobile voice (LEO); Search for ExtraTerrestrial Intelligence (SETI); telemetry
2310–2360 MHz	S-band	Civil defense radio; Direct To Home (DTH) TV; weather; satellite radio (XM and Sirius)
3700–4200 MHz 5925–6425 MHz	C-band	TV broadcast; voice; videoconferencing
5.2–10.9 GHz	X-band	Military (Naval and Air Force); scientific; various comsats
11.7–12.2 GHz 14.0–14.5 GHz	Ku-band[a]	Direct Broadcast Satellite (DBS): TV; Internet access; voice
20 and 30 GHz	Ka-band[b]	Mobile voice and data
36–46 GHz	Q-band	Military
46–56 GHz	V-band	Intersatellite links
75–110 GHz	W-band	Radar and scientific research

[a]The K-band spectrum (10.9–36.0 GHz) is subdivided into the Ku-band and the Ka-band. The Ku-band is so called as it is *u*nder the center of the K-band.
[b]The Ka-band is so called as it is *a*bove the center of the K-band.

The popularity of satellite communications also has placed great demands on the international regulators to allocate and manage radio spectrum. As is the case with terrestrial microwave radio, there are a number of frequency bands assigned to satellite systems, most of which fall in the MegaHertz (MHz) or GigaHertz (GHz) ranges. Due to the wide footprint, or Earth coverage area of a satellite, the frequencies must be managed carefully at the international, regional, and national levels. Generally speaking, geostationary satellites are positioned approximately 2° longitude apart to minimize interference from adjacent satellites using overlapping frequencies [12]. Table 2.5 provides a set of example frequencies and spacecraft serving various applications. Although the C-band is shared with terrestrial microwave, interference is avoided by the use of highly directional antennas. As the K-band is reserved for exclusive satellite use, interference is not a significant issue.

2.7.1 Uplinks and Downlinks

Satellite radio signals theoretically can propagate infinite distances in the vacuum of space with no signal loss if the radio beam can be perfectly focused. (*Note:* Practically speaking, that is not possible.) Such high-frequency signals, however, can suffer considerably from attenuation in the few miles of atmosphere on the *uplink* and *downlink* segments, with the higher frequency bands suffering the most. The uplink and downlink signals run at fairly widely separated frequencies in order to avoid the potential for self-interference between incoming and outgoing signals. As is suggested in Table 2.6, the higher of the two frequencies is used for the uplink as the increased signal attenuation at the higher frequency can be overcome through the application of higher radiated signal power and as power is much more readily available and much less expensive on the surface of the earth than on a satellite that depends on solar power. At a much lower power level, the lower downlink frequency can better penetrate the earth's atmosphere and electromagnetic field, which can act to bend the incoming signal much as light bends when entering a pool of water.

In order to maximize the strength of such a high-frequency signal as well as to direct the uplink transmission to a specific satellite, the uplink radio beams are tightly focused. As is the case with terrestrial microwave, the transmit antenna is centered in a concave, reflective dish that serves to focus the radio beam with maximum effect on the receiving satellite antenna. Table 2.6 provides select uplink and downlink frequency bands and dish sizes. Although the higher frequencies are more susceptible to signal fading through the atmosphere, they offer the advantages of lesser interference and smaller dishes.

Although a GEO system can see roughly one-third of the earth's surface from its vantage point, it cannot effectively transmit to such a wide area. Without shape or focus, the signal would be so weak, particularly at the fringes of the coverage area, that it would be unusable. Therefore, the downlink transmission is focused on a particular footprint, or area of coverage, which might be as wide as an entire continent. Spot beams, even more tightly focused downlinks, serve specific applications over smaller regions. Spot beams are heavily used in Ka-band satellites, as the downlink frequency is so high that a tightly shaped beam is required to overcome the effects of atmospheric attenuation. Spot beams offer the additional advantage of frequency reuse, as a given frequency can be used again in a nonadjacent footprint, much as a frequency can be reused in nonadjacent cells of a cellular telephony network.

TABLE 2.6   Select Uplink/Downlink Satellite Frequency Bands (Approximate)

Frequency Band	Uplink (GHz)	Downlink (GHz)	Dish Diameter (m)
C-band	5.925–6.425	3.7–4.2	2.4
Ku-band	14.0–14.5	11.7–12.2	1.07
Ka-band	27.5–30.0	17.7–21.2	0.61

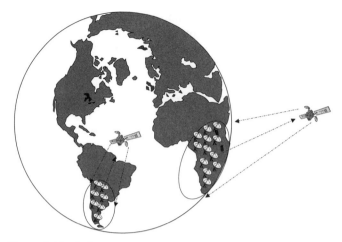

Figure 2.12 Point-to-multipoint GEO networks with footprints.

2.7.2 Footprints

The footprint of a satellite radio system enables a signal to *broadcast* a signal over a wide area. Thereby, any number (theoretically, an infinite number) of terrestrial antennas can receive the signal more or less simultaneously. In this manner, satellites can serve a *point-to-multipoint* network (Figure 2.12) requirement through a single uplink station and multiple downlink stations.

More recently developed satellites can serve a *mesh* network requirement, whereby each terrestrial site can communicate directly with any other site through the satellite relay station. Previously, all such communications were required to travel through a centralized site, known as a *head end*. Such a mesh network, of course, imposes an additional level of difficulty on the network in terms of managing the flow and direction of traffic.

2.7.3 Configuration

Satellite radio systems consist of antennas and reflective dishes, much like terrestrial microwave. The dish serves to focus the signal from a transmitting antenna or to a receiving antenna. The dishes generally are mounted on a tripod or other type of brace which anchors to the earth, a pad, or a roof or attaches to a structure such as a building. Waveguides guide, or channel, the radio signal between the antennas and the transmit/receive electronics. The terrestrial antennas support a single frequency band (e.g., C-band, Ku-band, or Ka-band), while the satellite may support a number of frequency bands for various applications, such as broadcast radio, broadcast TV, paging, voice, mobile voice, telemetry, and data. (Table 2.5) The send/receive dishes that make up the earth segment vary in size, depending on power levels and frequency bands. The higher the frequency band, the smaller the possible size of the dish, as the higher frequency antennas can achieve greater signal gain at a smaller size. (*Note:* The focusing ability of a dish is directly related to the wavelength of the signal.) Further, the higher frequency bands require tighter signal shaping and the antennas must be aligned more tightly, as the signal is so much more susceptible to atmospheric attenuation such as rain fade. Therefore, C-band TV dishes tend to be rather large at approximately 2.4 m in diameter and can tolerate a misalignment of 0.75° before suffering a signal loss of 3.0 decibels (dB), which is significant. (*Note:* Signal attenuation of 3 dB means that signal power is halved, i.e., reduced by 50 percent. Attenuation of 10 dB means that signal power is reduced by 90 percent.) A Ku-band antenna is as small as approximately 0.9 m and can tolerate a misalignment of 0.65°, and a Ka-band dish is smaller still (Table 2.6) at approximately 0.61 m, but can tolerate a misalignment of only 0.4°, which is a very tight tolerance [14]. Also, flat, mechanically passive, phased array dishes are being built in very small sizes and at very low cost for DBS TV and other applications, including live in-flight TV and Internet access for commercial airliners. Such phased-array antennas employ an array of small antennas that work together to logically focus on the point of maximum signal strength, rather than requiring that the reflective dish adjust mechanically. As a point of reference, the Intelsat I (1968) dishes measured 30 m in diameter.

The space segment antennas are mounted on a satellite, of course. The satellite can support multiple transmit/receive antennas, depending on the various frequencies that it employs to support various applications and on whether it covers an entire footprint or divides the footprint into smaller areas of coverage through the use of more tightly focused spot beams such as those required in the Ka-band. A satellite repeater, or *transponder* (*trans*mit and res*pond*), accepts the weak incoming signals, boosts them, shifts them from the uplink to the downlink frequencies, and transmits them to the earth stations in what is known as a *bent-pipe* network configuration. Contemporary satellites commonly support as many as 28–46 transponders. Current generations of broadband satellites are replacing the relatively dumb bent-pipe transponder approach with onboard processing capability in the form of circuit switching and even fast packet switching and statistical time division multiplexing. These satellites variously support IP, Frame Relay, and ATM traffic.

2.7.4 Very Small Aperture Terminals

Very Small Aperture Terminals (VSATs) are terrestrial dishes of very small diameter, or aperture. *Note: Very small* is relative, as typical VSATs are only as *small* as 0.9, 1.2, 1.8, and 2.4 m (approximately 3–8 ft) in diameter, with the specific dish size being sensitive to the placement of the antenna within the satellite footprint. The smallest dishes work well in the center of the footprint, where the signal is strongest. As antenna placements creep farther from the center and closer to the fringes of the footprint contour, larger dishes are required to collect more signal and thereby improve the quality of reception.

Operating in the C-band and Ku-band, VSATs are digital and designed primarily to support data communications on a point-to-multipoint basis for large private networks in applications such as retail inventory management, credit verification and authorization, and general transaction processing. Bandwidth commonly is in channel increments of 56/64 kbps, generally up to an aggregate bandwidth of 512 kbps. Some newer systems support bandwidths of as much as 1.7 Mbps on the downlink and 20 Mbps on the uplink, mesh networking, and compressed voice communications at rates as low as 2.4 kbps. [15]. By far the largest concentration of users is found in North America, claiming about 75 percent of the market. Companies such as Exxon Mobil Corporation and Shell Oil Company have installed networks of as many as 5000 VSAT nodes in support of transaction processing applications.

2.7.5 Bandwidth

Satellites can support multiple transponders and, therefore, substantial bandwidth. Contemporary GEO systems commonly support aggregate bandwidth in the range of 1 GHz per beam, yielding raw bandwidth of 1.344 Gbps, through 28 transponders, each with a channel capacity of 36 MHz [16]. A few GEO systems are based on transponders with channel capacities of 72 MHz, yielding data rates up to 155 Mbps. The level of bandwidth, the number of beams and frequency bands, the number of transponders, and the footprint all influence the size and power requirements of a given satellite.

As in the case of other transmission systems, the higher frequency bands offer greater raw bandwidth. In this sense, the C-band is the most limited, while the Ka-band is the most attractive of the commercial satellite frequency bands. As a point of reference, Intelsat I could accommodate only 240 voice circuits, while Intelsat VI supported 120,000 voice circuits and three TV channels, with total bandwidth of 3.46 GHz [17].

2.7.6 Error Performance

Satellite transmission is susceptible to environmental interference, particularly at frequencies above 20 GHz. Sunspots and other types of electromagnetic interference can have considerable impact on microwave transmission in general and satellite transmission in particular. Error performance also is sensitive to the proximity of the earth stations to the equator, as those nearest the poles and farthest from the equator must deal with more signal absorption and environmental interference as the signal must travel diagonally through more atmosphere. Error performance also is sensitive to the physical location of the receiving antenna within the footprint, as the signal is strongest in the center of the footprint and weakest at the edges. C-band satellite transmission also must deal with competition from terrestrial microwave through highly directional antennas and spot beams. As a result of these several factors, satellite transmission requires rather extensive error detection and correction capabilities [18]. Throughput typically is far less than raw bandwidth, as error control requires either retransmissions or forward error correction, with the latter approach involving the embedding of redundant data in the transmitted data stream.

2.7.7 Distance

Satellite, generally speaking, is not considered to be distance limited, as the signal largely travels through the vacuum of space. Further, each signal travels approximately 22,300 miles or more in each direction, whether you communicate across the street or across the country, and assuming that only a single satellite hop is required. But larger earth stations and additional power are required to serve areas far removed from the equator (e.g., New Zealand and South Africa). In such instances, the signals must travel a longer distance through substantial atmosphere, and they are more likely to be deflected at such severe angles.

2.7.8 Propagation Delay and Response Time

By virtue of their high orbital altitude, GEO satellites impose rather significant propagation delay on the signal and, therefore, doubly affect response time (see Figure 2.13). Given the fact that the radio signals must travel approximately 22,300 miles or more up to the satellite and the same distance on the return leg, the signal propagation delay is about 250 ms (0.25 s). Considering the amount of time required for processing on board the satellite, as well as at the earth stations, the total delay for a one-way transmission is about 320 ms (0.32 s). Therefore, the delay between signal origination (transmission) and receipt of response is about 640 ms (0.64 s), assuming an immediate response requiring only a single satellite hop. Note that the exact level of propagation delay is sensitive to the proximity of the earth stations to the equator.

As a result of such severe propagation delay, highly interactive voice, data, and video applications are not effectively supported via two-way GEO satellite communications. Although satellites commonly support interactive videoconferencing, it generally is accomplished in broadcast lecture mode, with the video signal sent from the central location to the satellite, which broadcasts it to multiple receive-only earth stations. As illustrated in Figure 2.13, the participants at the distant locations interact with the lecturer via landline connections. This hybrid approach mitigates the impact of propagation delay, which would create an intolerable situation if the conference made use of satellite links in both directions. This approach also reduces the overall cost, as receive-only earth stations are much less expensive than transmit/receive stations. Interactive Internet access also is offered via satellite by DBS providers through implementation of this same hybrid approach, known as *telco return*.

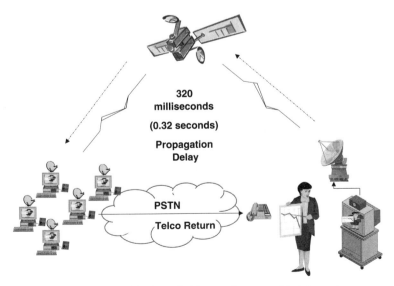

Figure 2.13 Satellite-based videoconference with telco return.

2.7.9   Access Control

Satellite systems are designed to handle multiple simultaneous uplinks and down-links. A great number of earth stations can be contending for access to transmit to a satellite at any given time and a great many more can be in readiness to receive a transmission from the satellite. There are several access protocols that can be implemented in wireless communications in general and in satellite communications systems in specific. While these access control protocols are discussed in detail in Chapter 11, they must be at least mentioned at this early point in the book:

- *Frequency Division Multiple Access* (FDMA): Earth stations have specific assigned uplink and downlink frequencies within the allotted range. Separation of communications is purely by frequency. FDMA is a wireless version of Frequency Division Multiplexing (FDM).
- *Demand-Assigned Multiple Access* (DAMA): In this variation of FDMA, frequency bands are shared by several carriers. An earth station gains access to an available band that is assigned at the time of need, depending on availability.
- *Time Division Multiple Access* (TDMA): Earth stations have specific time slots during which they transmit in short bursts on the assigned uplink frequency and receive in short bursts on the downlink frequency. Thereby, multiple trans-missions can share the same frequency band, separated in time. TDMA is a wireless version of Time Division Multiplexing (TDM).
- *Code Division Multiple Access* (CDMA): Multiple carriers share the same fre-quency band. The individual carriers are separated by specific coded waveforms that distinctly identify each from the others. This highly sophisticated approach allows a great many transmissions to share a limited spectrum.

2.7.10   Security

As is the case with all microwave and other radio systems, satellite transmission is inherently insecure. Satellite transmission is especially vulnerable to interception, as the signal is broadcast over the entire area of the footprint. Therefore, the unau-thorized user must know only the satellite location and the associated frequency range in order to gain access to the raw signal. Only through the use of an encryp-tion or scrambling mechanism can any level of security be implemented.

2.7.11   Cost

The acquisition, deployment, and rearrangement costs of the space segment of satel-lite systems can be quite high (easily $200 million, including the satellite and its launching). But a large number of user networks can share the satellite. User orga-nizations can even share an earth station through a *teleport*. As a result, satellite networks often compare very favorably with cabled systems or terrestrial microwave systems for many point-to-multipoint applications. Cost elements typically include leasing capacity from a satellite provider and the acquisition cost of the terrestrial antennas or the leasing costs associated with shared teleport access. Note that the acquisition cost of receive-only dishes is relatively low, while transmit/receive dishes are considerably more expensive. As is the case with microwave, satellite transmission is not affected by *backhoe (digger) fade*, which plagues cabled systems.

2.7.12 Regulation

National, regional, and international bodies more or less carefully regulate the space segment of satellite communications. Additionally, local zoning ordinances and health and safety regulations may affect the placement of terrestrial antennas. While satellite network technology is widely available in North America and other countries, it is not widely available to end users worldwide. Many countries in Asia, Europe, and Africa have rejected *open-sky* policies, in support of the incumbent carriers. Developing countries, particularly, depend heavily on the imbalance of trade in telecommunications services as a source of hard currency. In fact, many of them intentionally price originating traffic at very high levels in order to suppress outbound traffic and, therefore, to suppress the amount of hard currency outflow. As private satellite transmission bypasses the national network, it has a negative impact on both the incumbent carrier's revenue stream and the trade imbalance (depending on one's perspective) and, therefore, is strongly discouraged.

2.7.13 Applications

Although traditional international voice and data services have been supplanted, to a considerable extent, by submarine fiber-optic cable systems, satellite applications are many and are increasing rapidly. Traditional, and still viable, applications include international voice and data, remote voice and data (e.g., island nations, isolated areas, and sparsely populated areas), television and radio broadcast, maritime navigation, videoconferencing, transaction processing, inventory management and control, and paging. More recent and emerging applications include air navigation, GPSs, mobile voice and data (LEOs and MEOs), Advanced Traffic Management Systems (ATMSs), DBS TV, ISDN, and Internet access. Satellite communications recently have proved to be invaluable in disaster recovery, as evidenced by their extensive use in the aftermath of the tsunami that devastated Southeast Asia (2004), Hurricane Katrina along the Gulf Coast of the United States (2005), and the earthquakes in Pakistan (2005). Terrestrial systems, including microwave and cellular systems, were either totally destroyed or knocked out of service for long periods of time as a result of these natural disasters.

2.8 FREE SPACE OPTICS

Alexander Graham Bell originally developed the concept of using light waves for communications. In early 1880, Bell invented and experimented with the *photophone*, a system utilizing mirrors to focus modulated sunlight onto a selenium cell. He was successful in transmitting voice over a distance of 700 ft on sunny days and was granted four patents for the invention. While the technique was clearly impractical, Bell nonetheless felt the invention to be his greatest achievement. During World War II, the Nazi military experimented with similar but more advanced systems in tank warfare applications, but the technology remained impractical [7].

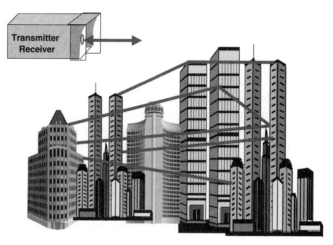

Figure 2.14   IR transmission topology.

Also known as infrared (IR) and sometimes characterized as *wireless fiber, free-space optics* is a relative newcomer to communications transmission systems. At the consumer level, many of us are familiar with IR-based remote controls for TV sets, slide projectors, computer graphics presentations, and the like. Many of us use an IR-based computer mouse, and many laptops have IR networking capability. Since the late 1980s, FSO systems have assumed a position of some, if still limited, importance and appear to have a substantial future in a variety of short-haul applications. FSO systems are point-to-point airwave systems that use focused IR light beams between transmitters and receivers, much as microwave systems use focused radio beams. There are several equipment configurations, with the traditional and most common comprising a focusing lens in the transmitting device to tightly focus a light beam on a collecting lens in the receiving device (Figure 2.14), with both types of lenses contained within each *transceiver* (*trans*mitter/re*ceiver*). Some systems employ dishlike receivers that serve to collect more of a diffused beam for improved reception during foggy conditions. As FSO is a line-of-sight (LOS) system, the transceivers typically are mounted on rooftops or the sides of buildings. Some systems are so sensitive to LOS as to require auto-tracking mechanisms that adjust to the sway of high-rise office buildings due to wind, tremors, and other forces of nature. Some systems allow indoor receiver placement as the light beams will pass through some glass windows, depending on their chemical composition.

FSO transmitters use laser light sources in the form of either Vertical-Cavity Surface-Emitting Lasers (VCSELs) or laser diodes. VCSELs generally operate at a slower speed but are less expensive and less sensitive to environmental damage than laser diodes and require active heating and cooling to maintain signal intensity. FSO VCSELs generally operate in the range of 780–850 nm, while laser diodes operate in the 1550-nm range. As signal power can be higher and attenuation is lower at 1550 nm, the distances can be greater between transmitters and receivers. Also, the 1550-nm range is considered to be eye-safe, even at the higher power levels. The optimization process of balancing cost and performance has led most manufacturers to focus on systems in the range of 800–850 nm, however.

FSO systems suffer from environmental interference, particularly fog, which absorbs, scatters, and reflects the light beam, much as fog affects the beam from the headlights of an automobile. Signal reliability can be affected by scintillation due to differences in air density caused by heated air rising from heat ducts or from the earth on bright, hot days. The impact of this phenomenon can be mitigated through the use of larger aperture receivers and widely spaced redundant receivers. Under optimum conditions, transceiver separation is limited to about 2–5 km, although most tests have shown optimum performance at distances of 500 m to 1 km. In areas where there is a lot of dense fog, links more typically are limited to about 200 m, and some manufacturers offer fog redundancy via microwave systems. On the more positive side, FSO systems are immune to EMI and RFI. As the beams are so tightly focused and beam spread is minimal over the short distances where FSO systems find application, a great many systems can coexist in a small airspace with no concern of saturation and cross-interference. Under optimum conditions and with link lengths constrained to 2 km or so, error performance is in the range of 10^{-8} and link availability in the range of 99.9 percent. When supplemented with microwave backup, FSO systems can deliver a carrier-class availability level of 99.999 percent, which compares favorably with UTP and microwave.

Despite the limiting factors, FSO is enjoying increasing popularity due to its low relative cost when compared to microwave and fiber optics. FSO systems commonly operate at rates of 1.544 Mbps (T1), 2.048 Mbps (E-1), 34 Mbps (E-3), 45 Mbps (T3), 155 Mbps (OC-3), and 622 Mbps (OC-12). (*Note:* Chapter 9 discusses SONET and SDH fiber-optic standards and the associated OC rates.) Some manufacturers offer equipment running at rates as high as 10 Gbps (OC-192), and systems running at up to 160 Gbps have been demonstrated in the labs [19, 20]. FSO systems also can be deployed very quickly as there currently are no FCC licensing requirements and few other regulatory restrictions on its use. Security is a key advantage. Not only is the beam path invisible to the unaided eye, but intercepting the signal involves breaking the beam path, which would drop the connection.

The applications are all short-haul in nature. FSO, for example, is an attractive alternative to leased lines or private cabled systems for building-to-building connectivity in a campus environment or other short-haul point-to-point application, especially in bridging LANs. FSO is also used by carriers as a replacement for various licensed microwave Wireless Local Loop (WLL) technologies. In those WLL applications, FSO typically is deployed in a mesh configuration for purposes of redundancy, in consideration of the potential for link failures due to fog and other environmental or line-of-sight issues. FSO systems also find application in disaster recovery scenarios.

2.9   FIBER OPTICS

I have heard articulate speech produced by sunlight! I have heard a ray of the sun laugh and cough and sing. . . . I have been able to hear a shadow, and I have even perceived by ear the passage of a cloud across the sun's disk. . . . Can Imagination picture what the future of this invention is to be!

   —Alexander Graham Bell (1880) on the success of his experiments with the
          photophone, the precursor to free space optics and fiber optics

While Alexander Graham Bell's *photophone* proved to be impractical, it was the early precursor to the optical technologies that have forever altered the telecommunications landscape. The 1940s saw the first experiments conducted with *waveguides*, involving both microwave radio and optical transmission systems. Such waveguides were rigid, insulated pipes which served to contain the electromagnetic energy and channel it from end to end, while offering protection from outside interference. While significant transmission speeds can be realized through this technique, it is seldom used other than in high-power feeds to broadcast antennas—the physical configuration is obviously impractical.

It was clear that flexible glass fibers offered much more potential as a transmission medium for light. As early as the 1950s, the efforts of the American Optical Corporation resulted in optical fiber cable that could carry light signals a few feet. It was at Standard Telecommunications Laboratories in 1966 that Charles Kao and George Hockham developed the first practical conceptual breakthrough—the purity of the glass was the issue. During the early 1970s, the first practical fiber-optic systems were developed. These systems were made possible by the manufacturing of glass fibers pure enough to permit the transmission of light over long distances with little signal loss. Donald Keck, Bob Maurer, and Peter Schultz of Corning, continued that work with fused silica for both the core and the cladding, adding controlled levels of impurities to the core to make its refractive index slightly higher than the cladding [21]. At roughly the same time, AT&T Bell Telephone Laboratories invented *laser diodes*, which serve as light sources in high-speed optical transmitters. Since then, fiber-optic development has progressed to the point that virtually all high-speed networks are based on fiber-optic technology.

Conventional fiber-optic transmission systems are *optoelectric* in nature. In other words, they involve a combination of optical and electrical electromagnetic energy. The signal originates as an electrical signal which is translated into an optical signal which subsequently is reconverted into an electrical signal at the receiving end. Optical repeaters go through an optoelectric conversion process as they boost the signal strength at various points in long-haul transmission systems. (*Note:* Optical amplifiers are another matter, as discussed later in this chapter.) Chapter 9 presents a detailed discussion of SONET, a set of international standards for fiber-optic transmission systems.

2.9.1 Wavelengths and Windows

At this point, it is necessary to be more specific about the characteristics of the light signals that are used in fiber-optic transmission systems. Those signals are not in the form of white light, as you normally think of natural sunlight or artificial light created by an incandescent bulb or fluorescent tube. *White light* actually is a combination of all of the *wavelengths*, or *lambdas* (λ), in the visible-light spectrum. Rather, the light sources used in transmission systems are much more precise, as they create light signals within very specific and very tightly defined wavelength ranges, measured in nanometers. As discussed previously, wavelength is the inverse of *frequency*. In optical systems, the term wavelength is used and refers to the distance between the peaks or troughs of a sinusoidal electromagnetic signal. In electrical and radio systems, the term frequency is used and refers to the number of waveforms transmitted per second. All optical transmission systems run in the IR

TABLE 2.7   ITU-T Transmission Windows

Band Designation	Wavelength (nm)
850 Band	810–890
O-Band[a]	1260–1360
E-Band[b]	1360–1460
S-Band[c]	1460–1530
C-Band[d]	1530–1565
L-Band[e]	1565–1625
U-Band[f]	1625–1675

[a] Original Band.
[b] Extended Band.
[c] Short Wavelength Band.
[d] Conventional Band.
[e] Long Wavelength Band.
[f] Ultralong Wavelength Band.

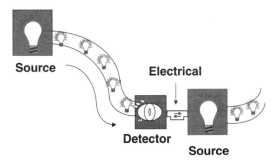

Figure 2.15   Fiber-optic system, consisting of light sources, glass fibers, and light detectors.

range. The most commonly used wavelengths are in the nominal 850-, 1300-, 1310-, and 1550-nm ranges. Contemporary high-speed systems operate in various transmission *windows* specified by the ITU-T and detailed in Table 2.7.

Generally speaking, the higher the transmission window (i.e., the longer the wavelength and the lower the frequency), the less the signal attenuation and the less the signal degradation but the more expensive the associated electronics [22].

2.9.2   Configuration

Fiber-optic systems consist of light sources, fibers, and light detectors, as depicted in Figure 2.15. In a simple configuration comprising a single link, one of each is used. In a more complex configuration over longer distances, many links are involved. Much like any other transmission system, optical signals require boosting at various intervals in order to overcome the effects of attenuation. Traditionally, this signal boosting was accomplished by regenerative repeaters, which essentially are back-to-back detectors and sources. Repeaters accept the incoming optical signal, convert it to electrical format, reamplify it, retime it, regenerate it, and reconvert it back to optical format. Various types of optical amplifiers increasingly are now employed over long-haul links.

2.9.2.1 Light Sources Light sources convert electrical energy into optical energy, that is, convert electrons to photons. Light sources consist of two basic types: *Light-Emitting Diodes* (LEDs) and *laser diodes*. Specific laser types include *Fabry–Perot lasers, Distributed-Feedback* (DFB) *lasers,* and *Vertical Cavity Surface Emitting Lasers* (VCSELs), each of which has unique attributes and associated advantages. Differences include speed (bandwidth), power, physical and optical coupling efficiency, directionality (output pattern), spectral width, coherence, and cost.

- *Speed* is directly related to the *cycle time*, or *rise and fall times*, that is, the length of time it takes for the light source to cycle through a rise to its peak and a fall to its trough of signal intensity (power). Fiber-optic systems use Amplitude Modulation (AM), so the faster the light source can cycle, the higher the bit rate. (*Note:* Most fiber-optic systems are digital, although a few are analog in nature.) Light sources never completely turn off, as that would limit their speed. Lasers are the fastest light sources, followed by VCSELs and LEDs.
- *Power* refers to the optical power, or light intensity. A higher power signal can survive more attenuation and, therefore, can survive over a longer distance without requiring amplification. Lasers are the most powerful light sources and surface-emitting LEDs are the least powerful.
- *Coupling efficiency* refers to the efficiency with which the light source connects to the fiber. The more precisely the light source can inject a tightly focused signal directly into the inner core of a fiber, the stronger the resulting signal and the better the signal performs over a distance. Coupling efficiency is a key advantage of pairing DFB lasers with single-mode fibers, which have an inner core of only 5–10 μm.
- *Directionality* is the extent to which the light beam is lined up with the fiber core. Collimated light beams are lined up in perfect parallel. Divergent light beams spread out as they exit the source. LEDs emit the most divergent light beams, and laser diodes emit the least divergent.
- *Spectral width* refers to the range of wavelengths, or *window*, emitted by the light source. LEDs emit signals of the greatest spectral width, which limits their application. Lasers emit very narrowly defined signals that may be only 1 nm wide or less. Therefore, lasers have application in long-haul systems that multiplex numerous wavelengths.
- *Coherence* describes light signals that are synchronized in phase, with the sine waves rising and falling in unison. Too much coherence is a bad thing, as perfectly coherent light waves can interfere with each other and cause *speckling*, that is, cause the signal to appear grainy, if they travel slightly different paths, or *modes*. Coherence also means the source has a narrow range of wavelengths, ideally only one, so that all photons act identically, which is a good thing.
- *Cost* is always a major consideration. The essence of optimization is balancing cost and performance, and optimization is the goal of any organization geared toward profitability.

Light-emitting diodes are commonly used semiconductor components, found in clocks, calculators, and a plethora of other devices. The LEDs used in fiber-optic transmission are, of course, much more sophisticated. LEDs predominated in early fiber-optic systems largely because of their low costs of acquisition and operation in comparison to diode lasers of that era. LEDs pulse on and off relatively slowly, as specified by the rise and fall times of signal intensity. Therefore, LEDs are relatively bandwidth limited. LEDs also generate broadly defined optical signals; in other words, the signals comprise a relatively broad spectral width, or range of spectrum. Slower LEDs emit light from an area etched into the surface of a semiconductor chip, while the faster LEDs emit light from the edge of the chip. The physical design of LEDs is such that they mechanically couple efficiently only to the relatively broad (62.5-μm) inner core of MultiMode Fiber (MMF). While LEDs no longer are used in long-haul transmission systems, they have found continuing application in LANs, where they support transmission rates of up to 1 Gbps over relatively short distances. LEDs also are used in certain other short-haul transmission systems, including some Passive Optical Networks (PONs). LEDs are matched with the less capable fiber (MMF) and detector [Positive–Intrinsic–Negative (PIN) diode] technologies. LEDs are relatively inexpensive and long-lived.

Laser diodes generally resemble LEDs in structure, although they are much more difficult and expensive to manufacture. They also are associated with more expensive and complex supporting electronics which require careful control of ambient temperature. But they generally are much faster and, therefore, offer much more bandwidth. Diode lasers offer significant mechanical and optical *coupling efficiency*. In other words, they can mechanically couple to a very thin *Single-Mode Fiber* (SMF) and can tightly focus a high-speed optical signal for presentation to its smaller axis, or core. Diode lasers also are capable of generating tightly defined optical signals in very small spectral ranges, or windows. Diode lasers also generate signals at wavelengths longer than 850 nm. In these higher transmission windows (see Table 2.7), the signals attenuate much less and, therefore, can travel much farther without being repeated or amplified. In long-haul carrier-class transmission systems, this tight definition allows the multiplexing of a number of wavelengths through a process known as *Wavelength Division Multiplexing* (WDM). *Dense WDM* (DWDM) systems are particularly intense, multiplexing lambdas closely around a centerpoint of 1552.52 nm (193.1 THz). The spacing between carriers is implementation dependent and includes options of 200 GHz (1.6 nm at 1550 nm), 100 GHz, 50 GHz, and even 25 GHz. Issues of modulation noise and channel isolation currently limit implementations to spacing of 100 GHz. In some relatively short haul carrier-class systems, a similar, but less intense, technique known as *Coarse WDM* (CWDM) is employed. CWDM standards specify 18 wavelengths in the range of 1270–1610 nm, with spacing of 20 nm (2500 GHz at 1550 nm) and targeted at networks with a reach of 50 km or less. All of these WDM techniques essentially are Frequency Division Multiplexing (FDM) at the optical level. As each frequency window is added, the bandwidth of the system increases. A system operating at 2.5 Gbps, for example, enjoys a bandwidth increase of 2.5 Gbps as each DWDM wavelength is added. As diode lasers enjoy a high level of coupling efficiency, they can be used in conjunction with either MMF or SMF. Because diode lasers are the more capable light sources, they generally are matched with the more capable SMF and Avalanche PhotoDiode (APD) technologies.

Fabry–Perot lasers are commonly used general-purpose lasers. Fabry–Perot lasers are more precise than LEDs as they emit an optical signal of relatively narrow spectral width in the range of 3–6 nm. Around the center wavelength, these lasers emit a narrow range of less intense wavelengths. For example, a Fabry–Perot laser operating at a nominal wavelength of 1310 nm might also emit weaker signals at wavelengths ranging from 1307 to 1313 nm. This spectral width causes some amount of chromatic dispersion, which limits bandwidth in SMF systems. Fabry–Perot lasers are moderately fast and moderately priced [23].

Distributed-feedback lasers are high-speed lasers that operate at high output power levels. These lasers have a spectral width under 1 nm, which effectively means that they emit a single wavelength. This narrow spectral width is significant in the context of Wavelength Division Multiplexing (WDM), which is discussed in Chapter 9. Further, DFB lasers can tightly focus a very narrow optical beam from the edge of the semiconductor chip into the center of the thin (5–10-μm) inner core of a SMF. DFB lasers run in the 1300- and 1550-nm regions, which are particularly well-suited for long-haul applications. These characteristics currently make DFB lasers the overwhelming choice of telecommunications carriers and CATV providers. In such high-bandwidth, long-haul applications, their relatively high cost is justifiable.

Vertical Cavity Surface Emitting Lasers (VCSELs) are so named because the lasing cavity runs vertically (from top to bottom) through the chip. The chip is mirrored at the bottom in order to maximize signal output power at the top. VCSELs have capabilities somewhere between LEDs and other lasers, as they have a spectral width somewhere between the two. VCSELs can couple effectively to a MMF with a narrower core (50 μm versus 62.5 μm) than LEDs. They also are faster than LEDs, if somewhat slower than lasers. The first generation of VCSELs operated in the 850-nm regions. The second generation (2005) of VCSELs can run in the 1300- and 1310-nm regions.

2.9.2.2 *Optical Fiber*
While plastic fibers are used in some specialized, low-bandwidth, short-haul applications (e.g., automobiles, airplanes, televisions, and stereo equipment), glass predominates. Generally speaking, fiber cables contain a large number of fiber strands because the incremental cost of redundancy is relatively low. Oftentimes, only a few of the fibers are active, with the remaining ones left *dark* for backup or future use. Major carriers deploy cable systems with as many as 620 fiber strands along a single route, with each fiber supporting as many as eight *lambdas* (i.e., *wavelengths*) through DWDM, with each lambda operating at speeds as high as 10 Gbps. While current technology can support two-way transmission over a single fiber, two fibers generally are used, with one transmitting in each direction.

The mass production of glass fiber employs several similar techniques, including the following *Outside Vapor Deposition* (OVD) technique. All of these techniques, by the way, take place in a vacuum environment, as it is the exposure to oxygen that makes glass so brittle. The process begins with heating silica and germanium to the point that it vaporizes. The glass vapor cools and is deposited as layers of soot on a rotating hollow *bait rod* to create a glass cylinder. The first layer is the core material of germanium-doped silica. On top of the core material are deposited many layers of slightly purer silica soot that form the cladding. (*Note:* Step-index fibers are characterized by an abrupt change in chemical composition between the core and cladding. Graded-index fibers involve many layers of silica of slightly different chemical compositions to yield slightly and successively purer layers of cladding surrounding the axis, much like the arrangement of the annular rings of a tree.) The entire rod is then reheated and collapsed into a *preform* cylinder. The preform

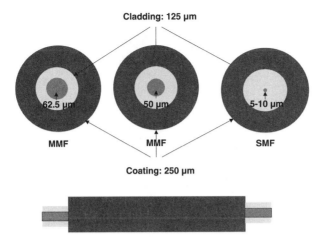

Figure 2.16 Glass fiber-optic cables, side views and cross section.

cylinder is reheated to a temperature of 2500°C in a drawing tower. The molten glass is carefully drawn by gravity, in a process known as *broomsticking*, into fibers with a consistent outside diameter and that can measure 10 km in length. As the fibers cool, an acrylate coating is applied to protect the raw glass from physical damage.

The light pulse is intended to travel through the center *core* of the pure glass fiber. Surrounding the inner core is a layer of even purer glass *cladding*. The difference in purity or clarity of cladding yields a slightly different *Index Of Refraction* (IOR) (*n*) than the glass that comprises the core. The IOR (*n*) is calculated as

$$n = \frac{c}{v}$$

where *c* is the velocity of light in a vacuum, which is a constant value of 186,000 miles per second, or 300,000 km/s, and *v* is the velocity of light in the subject medium, which is a variable sensitive to the chemical composition of the medium.

So, the IOR is the inverse of the velocity of propagation (Vp), as expressed in Table 2.2. The greater the difference in purity between the core and cladding, the greater the IOR. Surrounding the cladding is a thin layer of protective acrylate coating. Glass optical fibers consist of two basic types: *multimode* and *monomode*, or *single mode* (Figure 2.16), each of which has unique properties, operating characteristics, and subtypes. *Note:* The glass used in optical fibers is said to be so pure that you could see through a 3-mile-thick block of it just as clearly as you can see through your living room window, which is made of glass about $\frac{1}{8}$ in. thick.

2.9.2.2.1 Multimode Fiber *Multimode fiber* was the first optical fiber in production and still is extensively used in relatively low speed, short haul applications. The most common form of MMF has a relatively large inner core that has an diameter of 62.5 μm and is suitable for use with LED light sources. The more recently developed version has an inner core with a diameter of 50 μm and is suitable for use with VCSELs. It is important to note at this point that the diameter of the fiber core has quite the opposite effect on optical signal quality as the diameter of a copper wire

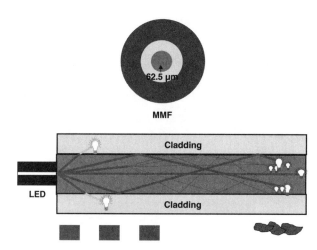

Figure 2.17 Edge-emitting LED light source coupled to 62.5-μm step-index MMF illustrating angle of incidence, scattering, modal dispersion, and pulse dispersion.

has on electrical signal quality. (*Note:* A nanometer (nm) is one billionth of a meter. A micrometer (μm) is one millionth of a meter. A millimeter (mm) is a thousandth of a meter. Your fingernail is about 1 mm thick and it grows at a rate of about 1 nm/s. If 1 nm were scaled up to the width of your little fingernail, your fingernail would be about the size of Delaware, United States.) The following several scenarios elaborate on the relationships between various types of light sources and fibers.

LED AND STEP-INDEX MMF   Consider a system configuration involving a relatively fast edge-emitting LED operating in the 850-nm region and coupled to an MMF with an inner core of 62.5 μm, as illustrated in Figure 2.17. The LED emits a highly divergent pulse of light across a wide range of angles. Some rays of light are injected at angles greater than the *critical angle* (i.e., that are too severe), penetrate the core–cladding interface, and are lost in the cladding. Some rays of light are injected at fairly slight *angles of incidence* and reflect off of the boundary of the core and cladding due to the sharp difference, or *step*, in the IOR. Those signals reflect back into the core, strike the core–cladding boundary on the other side of the fiber, and so on as they propagate across the link. Some rays of light are injected more or less into the center of the step-index fiber core and take a more direct path, but even those can strike impurities or encounter density changes in the glass and spread out to take different paths in a phenomenon known as *modal dispersion*. As some paths are more direct than others and as the time of arrival is directly related to the distance traveled, some portions of the signal arrive before others. (*Note:* This is much like *delay skew* in electrically based copper networks, as discussed earlier in this chapter.) As the distance of the circuit increases, the differences in distances traveled by the various portions of each light pulse become greater as the effects of modal dispersion become more pronounced. As the speed of transmission increases, the *bit time* (i.e., the period of time that each 1 or 0 bit occupies) decreases and the separation between bits is lost. The overall impact is that the pulses of light tend to lose their shape and overrun each other in a phenomenon know as *pulse dispersion* (Figure 2.17), which is a type of *intersymbol interference*. Beyond distances of only a few kilometers, the light detector cannot distinguish between the individual pulses. (*Note:* The diameter of the core clearly is a major factor in pulse dispersion, as it directly affects the length and number of modes. Early multimode optical fibers had a core of 100 μm, and pulse dispersion was much more pronounced.) The signal also has attenuated to a considerable extent due to scattering and absorption as the photons interacted with molecular matter in the fiber and, particularly, as they struck impurities and encountered density changes. The level of attenuation is especially great as the wavelength is so short, that is, the frequency is so high, at 850 nm. Also, some amount of photonic energy was converted to thermal energy during these interactions.

This scenario so far assumes that the fiber lies along a perfectly straight line, which is never the case, of course. In fact, fibers snake around corners and excess fiber is coiled in anticipation of subsequent splicing requirements to resolve future cable breaks. So, light rays injected into the very center of the fiber have no chance of remaining centered from end to end. So, some level of modal dispersion is assured and some level of pulse dispersion is a certainty. As a result, transmission speeds are kept relatively low and link lengths are kept relatively short.

The phenomena of modal dispersion and pulse dispersion in a multimode step-index cable are analogous to a convoy of very large white and black automobiles (white representing the presence of light pulses, or 1 bits, and black representing the absence of light pulses, or 0 bits) traveling at a speed of 50 mph (50 Mbps) and tailgating (small separation, or short bit time) each other down a five-lane (62.5-μm) interstate highway (glass fiber). While each of the automobiles begins in the center lane, some of them drift from shoulder to shoulder, thereby traveling a longer distance from point to point. Not only does the convoy lose its shape over a long (100-m) distance, but also collisions occur frequently.

VCSEL AND GRADED-INDEX MMF   Now consider a system involving a much faster VCSEL operating in the 1300-nm region and coupled to a graded-index MMF with an inner core of 50 μm. The VCSEL emits a pulse that is less divergent than that of the LED and the inner fiber core is considerably smaller in diameter. Some rays of light are injected into the core at angles and enter the cladding, and some find their way into the cladding through the normal course of propagation as modal dispersion occurs at bends in the fiber. Some enter the cladding where the angle of incidence changes due to bends in the cable as it snakes its way around corners. Those errant light rays that stray away from the center of the core do not encounter a core–cladding boundary with a sharp step in IOR. Instead, they encounter very gradual changes in IOR. Rather than either reflecting off the core–cladding boundary or being lost in the cladding, the light rays gradually increase in speed and bend slightly as they enter glass regions of lesser density. The sharper the angle, the greater the increase in speed and the greater the bend. Ultimately, the effect is that the errant light rays gain speed and bend back toward the fiber axis to rejoin the light rays that traveled directly through the center of the core, and the pulse largely regains its shape, as illustrated in Figure 2.18. So, modal dispersion is less of a concern and pulse dispersion is less of an issue even given the faster bit rate and, therefore, shorter bit times associated with VCSELs. The level of attenuation is less as the wavelength is longer, that is, the frequency is lower, at 1300 nm, so the link length can be increased as well.

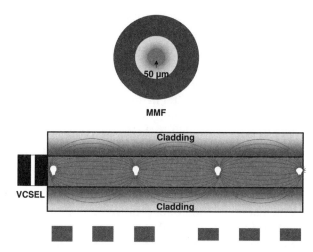

Figure 2.18 VCSEL coupled to 50-μm graded-index MMF.

As graded-index fibers can support transmission rates up to 100 Mbps over distances of 2 km or so, they are commonly used in LAN backbone applications, serving to interconnect high-speed servers, hubs, switches, and routers. Laser-optimized 50-μm MMF also is used in 10 GbE applications for links up to 300 m. As signaling speed or distance increases, it is necessary to use high-powered lasers, which introduce the problem of modal noise. Caused by interactions between the fiber and the connectors, *modal noise* results in power fluctuations at the receivers, increasing the Signal-to-Noise Ratio (SNR) and limiting the length of the fiber link [22]. So, SMF generally is used for distances over 300 m.

Continuing with the convoy analogy, pulse dispersion in a multimode graded-index fiber is like small black and white automobiles traveling at 622 mph (622 Mbps) and *really* tailgating (even smaller separation, or shorter bit time) each other down a four-lane (50-μm) interstate highway (glass fiber). While each of the automobiles begins in the center lane, some of them drift from shoulder to shoulder, thereby traveling a longer distance from point to point. However, the narrower highway is banked (graded) like an oval racetrack, so that the force of gravity (IOR) causes the cars to turn back onto the center portion of the track (core axis) and increase speed. The errant automobiles resume their positions in the convoy, which maintains its shape over a longer distance (2 km), with fewer collisions.

SINGLE-MODE FIBER Also known as *monomode, single-mode fiber* has a thinner inner core of 5–10 μm, effectively providing only a single mode for the light to travel (Figure 2.19). Therefore, neither modal dispersion nor the resulting pulse dispersion is an issue. Single-mode fiber performs better than MMF over longer distances at higher transmission rates. The thinner inner core renders SMF unsuitable for use in conjunction with LED and VCSEL light sources because their lack of *coupling efficiency* produces an unacceptable level; in other words, they are incapable of tightening their focus sufficiently to present the light signal to the axis of the small SMF. Laser diodes, however, are specifically designed to couple efficiently with SMF. Laser diodes operate at much higher rates and, therefore, offer much greater band-

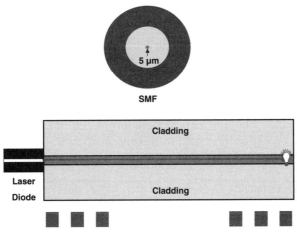

Figure 2.19 Laser diode coupled to 5- μm SMF.

width. Laser diodes also commonly operate in the 1310- and 1550-nm regions. The signal suffers much less from attenuation in the 1550-nm window, in particular, due to its lower frequency, so the signal better retains its shape and better maintains its strength. SMF does suffer from *chromatic dispersion*, which is due to the fact that different wavelengths of light travel through a medium at slightly different speeds. (*Note:* Chromatic dispersion causes the prism effect.) As even the most sophisticated DFB lasers have a spectral width of approximately 1 nm, the emitted light signal is a very narrow wavelength band rather one true wavelength. Hence, there is some amount of chromatic dispersion. Further, some amount of light travels through the cladding as well as the core. Given the difference in IOR between core and cladding, chromatic dispersion is compounded in a phenomenon know as *material dispersion*. Neither chromatic dispersion nor material dispersion is an issue in SMF systems except over very long distances or at very high speeds. In MMF systems, modal dispersion and pulse dispersion render the signal unusable long before chromatic dispersion and material dispersion become issues. Although more costly, MMF is used to great advantage in high-speed, long-haul applications.

Using the same analogy of a convoy of automobiles traveling down a highway, the SMF strand is a 5-μm-wide single lane. Even at 10,000 mph (10,000 Mbps or 10 Gbps) and over a very long distance (200 km), the convoy maintains its shape, with resulting collisions unlikely. There are several types of SMF: Non-Dispersion-Shifted Fiber (NDSF), Dispersion-Shifted Fiber (DSF) and Zero-Water-Peak Fiber (ZWPF). (*Warning*: The following discussion of SMF nuances is a bit detailed. If your interest in SMF is casual, you might consider skipping over the next few points. That said, proceed at your own risk!)

NON-DISPERSION-SHIFTED FIBER *Non-dispersion-shifted fiber*, which variously runs in the range of 1300–1320 nm (O-Band), was the earliest type of SMF, improving considerably on MMF in terms of distance limitations. However, *chromatic dispersion* was discovered to be an issue. Again, chromatic dispersion is the phenomenon by which optical energy at different frequencies travels at slightly different speeds. While some light sources operate in smaller windows than others, all emit signals across a range of wavelengths. So, some portions of the signal can arrive before others. As some errant light signals enter the SMF cladding and propagate through it while the majority of the signals propagate through the core, as the IOR varies between the core and the cladding and as the impact of the IOR varies with the frequency of the waveforms, *material dispersion*, which essentially is a form of modal dispersion, becomes an issue. NDSF runs in the range of 1300–1320 nm, where the effects of material dispersion are lowest in a standard SMF. At 1550 nm, the impacts of dispersion are considerably increased, requiring that the length of the fiber link be shortened in order for the signal to remain intelligible. However, attenuation is much less of a problem at 1550 nm.

DISPERSION-SHIFTED FIBER   *Dispersion-shifted fiber* shifts the optimal dispersion point by adjusting the interface between the core and the cladding. There are two types of DSF. *Zero-Dispersion-Shifted Fiber* (ZDSF) shifts the point of zero dispersion by increasing material dispersion to the point that it cancels out chromatic dispersion at 1550 nm, rather than at 1310 nm. Dense wavelength division multiplexers and Erbium-Doped Fiber Amplifiers (EDFAs) both work in this higher window, which can create yet another noise problem in the form of *Four-Wave Mixing* (FWM), a phenomenon by which wavelengths interact to create additional wavelengths. The EDFAs amplify those signals, and superimpose them on the DWDM channels. *Non-Zero-Dispersion-shifted Fiber* (NZDF) addresses this issue by shifting the optimal dispersion point slightly above the range in which EDFAs operate. A small but finite amount of dispersion remains, which actually helps by providing a means of separating wavelength channels [21] and [22]. Most contemporary SMF is NZDF DSF. (Try introducing that sentence into normal conversation at the next neighborhood barbecue, and see if you are ever invited back.)

ZERO-WATER-PEAK FIBER   *Zero-water-peak fiber* is designed to eliminate issues of attenuation caused by water ions that are residuals of chemical reactions in the manufacturing process or humidity in the environment. This *water peak* causes attenuation of wavelengths and pulse broadening in the general regions of 950, 1380, and 2730 nm. ZWPFs resolve water peak issue in the 1380 nm (1383 nm) region thereby opening the entire spectrum from 1260 to 1625 nm for high-performance optical transmission. ZWPF is the contemporary industry standard for all SMF.

Now, before moving away from what I am sure you must agree (?) is a fascinating discussion of the various forms of dispersion and water peaks and their impact on the error performance of fiber-optic transmission, I must introduce the concept of *Polarization-Mode Dispersion* (PMD). A light signal can travel along two planes through a SMF. If the fiber is perfectly round, light will travel along both planes at exactly the same speed and both planes of light will arrive at exactly the same time, barring other dispersion phenomena. PMD is caused by the fact that fibers are always inherently somewhat asymmetric; in other words, they are slightly elliptical (i.e., not perfectly round) in cross section. Also, some asymmetry is caused as the fibers become somewhat misshapen during installation, as they are bent around corners, twisted, coiled, and so on. Further, transient asymmetry can occur due to vibration and temperature changes at various places along the link or even from aerial fibers swaying in the wind. As the timing difference is so slight as to be measured in picoseconds (10^{-12} s), PMD is not an issue at speeds of 2.5 Gbps or less. At contemporary speeds of 10 and 40 Gbps, however, PMD results in unacceptable bit error rates. Closer spacing of regenerators will overcome the effects of PMD, although that solution tends to be expensive. PMD compensators have been developed to control the effects of PMD at speeds up to 40 Gbps by physically squeezing the fiber to counter stress it [24, 25].

2.9.2.2.2 Plastic Optical Fiber   There are exceptions to every generalization, and *Plastic Optical Fiber* (POF) is the exception to the rule of glass in the domain of fiber-optic transmission systems. There certainly is no debate about the fact that the performance characteristics of *Glass Optical Fiber* (GOF) are far superior to those of plastic, but glass tends to be expensive both to acquire and to install. Also, glass is highly susceptible to catastrophic failure, as it can be so easily broken. Therefore, optical fiber generally is limited to applications where it can be protected from physical damage; even then, redundant fibers often are deployed, which further increases costs. These issues of cost and fragility generally have worked against optical fiber to the desktop and in favor of UTP and even wireless technologies. Recent developments in POF, however, may change that fact. In favor of POF are its ability to withstand extremes of temperature (−40 to +85°C) and its ability to withstand a bend radius of down to 25 mm with no break or damage. Running at a wavelength of 650 nm, POF can support data rates of up to 400 Mbps over distances up to 100 m, which compares very favorably to standardized Cat 5 and Cat 5e UTP, as well as the developing Cat 6 and Cat 7 standards. While its speed rating does not compare favorably with GOF, the ATM Forum has approved POF as a viable medium for use in 155-Mbps horizontal links up to 50 m in length, and the IEEE included POF in the 1394b *FireWire* standard, also for links up to 50 m. Outside of what one would consider to be the normal data networking domain, POF enjoys considerable popularity in applications such as automobile and airplane wiring, where distances are short, temperatures can be extreme, and bandwidth requirements are fairly modest [26]. Also, POF is commonly used in contemporary stereo and television equipment.

2.9.2.2.3 Cords and Cables   Optical fibers are organized into cords and cables which can be indoor or outdoor in nature. Inside fiber systems generally involve one or a very few fibers. As is the case with UTP, inside wires and cables are of several types, according to the various national and regional standards such as the National Electrical Code (NEC) in the United States. *Plenum cables* are intended for use in *plenums*, or air-handling spaces, such as those between walls, under floor structures, and above drop (false) ceilings. As plenums are conducive to the spreading of fires within buildings, the NEC specifies that the insulation on plenum cables must be fire retardant, low smoke, and low toxicity. *Riser cables*, intended for use between floors of a building, also must be fire retardant.

Categories of OSP cables include *overhead cables* that hang from poles, *direct burial cables* that lie directly in trenches dug in the ground, *indirect burial cables* that lie in ducts or conduits placed in trenches dug in the ground, and *submarine cables* that are underwater, perhaps miles deep. OSP cables must be extremely rugged and durable, as they variously are exposed to extremes of temperature and pressure, rodents, cable-seeking backhoes and posthole diggers, boat anchors, trawler nets, sharks, and other forces of man and nature too numerous to list in this space.

Defense mechanisms against humans and rodents include lead sheathing and steel armoring. Air pressurization and water-blocking gel help keep cables free of moisture, which can freeze and cause cable insulation to crack and break. Moisture also can infiltrate tiny cracks in glass fiber cladding and cause the fibers to break if the moisture freezes and expands.

In applications such as Fiber-To-The-Neighborhood (FTTN), six acrylate-coated fibers commonly are contained in a loose-tube buffer filled with water-blocking gel. There may be a number of stranded loose-tube buffers, with the *loose-tube* design allowing the tube, strength member, armoring, cable sheath, and other elements of the cable to expand and contract independently and, thereby, protect the fiber from damage. (*Note:* Tight buffered fibers are used in indoor applications, where temperature variations are more modest.) In more bandwidth-intensive long-haul applications, 6 or 12 fibers commonly are contained in a flat ribbon. There may be 12 or more such ribbons in a single cable.

Strength members are used to improve the tensile strength of outside plant (OSP) aerial cables and inside riser cables. Strength members can be either metallic or nonmetallic in nature, depending on issues such as the weight of the cable and the need for lightening protection. Aramid fiber such as Dupont's Kevlar are used in fiber-optic riser cables not only for improved strength but also for increased protection of the fragile glass fiber from physical damage.

2.9.2.3 *Light Detectors*   Detectors consist of several basic types of photodiodes, the most common being PINs and APDs. The light detectors serve to reverse the process accomplished by the light sources, converting optical energy back into electrical energy, that is, converting photons to electrons. (*Note:* A *diode* is a device that allows a *charge carrier*, and thereby an electric current, to move in only one direction. A diode is analogous to a one-way valve, or check valve, that allows a liquid to flow in only one direction.)

Positive Intrinsic Negative (PIN) diodes comprise three layers of semiconducting material in the forward current-carrying direction. The first layer is chemically doped, that is, infused, to create a *positive (p)* electromagnetic region. The second layer is either undoped or lightly doped to retain its *intrinsic (i)* properties and, therefore, is neither strongly positive nor strongly negative. The third layer is doped to create a *negative (n)* electromagnetic region. So, the diode structure is *P*ositive, *I*ntrinsic, and *N*egative, or *PIN*. A PIN generates a single electron from each photon received and therefore does not provide a significant gain or increase in signal strength. However, PIN diodes are fairly rugged and inexpensive. PIN diodes generally are matched with LED and VCSEL light sources and multimode fibers.

Avalanche PhotoDiodes (APDs) are more sensitive that PIN diodes as they use a strong electric field to accelerate the electrons flowing in the semiconductor. As a result, an APD generates an avalanche of electrons with a multiplication factor that can be in the range of 70; that is, an APD generates 70 electrons from 1 photon [23]. Thereby, a very weak incoming light pulse will create a much stronger electrical effect that can be interpreted more effectively and understood more clearly. So, an APD can be characterized as a very high *gain* photodiode receiver, that is, a one-way photonic receiver with a high ratio of (electrical) output power to (optical) input power in the range of 70:1. Although more sensitive and more effective than PIN diodes, APDs require more electrical power to operate, are more sensitive to extremes of ambient temperature, and are more expensive. APDs generally are used in combination with laser diode light sources and single-mode fibers.

2.9.2.4 Amplifiers and Repeaters

Much like any other transmission system, optical signals require boosting over a distance in order to overcome the effects of attenuation. Traditionally, this signal boosting was accomplished by regenerative repeaters, although various types of amplifiers increasingly are now employed.

Optical repeaters, or regenerators, are optoelectric devices. On the incoming side of the repeater, a light detector receives the optical signal, converts it into an electrical signal, boosts it and adjusts for noise, retimes it, and converts it into an optical signal that it injects into a fiber. This process actually is *Optical–Electrical–Optical* (OEO). Optical repeaters essentially are fiber terminals (light detectors/sources) mounted back to back, with an intermediate electrical element connecting them. There may be many repeaters in a long-haul fiber-optic transmission system, although typically far fewer than would be required in a terrestrial system based on other transmission media. In the last few years, technology has developed to the point that the maximum spacing between repeater sites has increased from 600 km or so to as much as 5000 km. However, an OEO regenerator can repeat only a single wavelength. If multiple wavelengths are multiplexed in a WDM system, they must be demultiplexed, each wavelength must be regenerated independently, and they must then be remultiplexed before being sent on their way.

Optical amplifiers increasingly are used either in place of repeaters or to supplement them. Optical amplifiers are of two types, each of which is a purely optical device—electrically powered, of course. The process of amplification is *Optical–Optical–Optical* (OOO). An *Erbium-Doped Fiber Amplifier* (EDFA) uses a short length of fiber that has been doped (i.e., infused) with erbium, a rare-earth element, and spliced into the operating single-mode fiber in a configuration known as *discrete amplification* or *lumped amplification*. A three-port Wavelength Division Multiplexer (WDM) is used, with one incoming port connected to the operating fiber carrying the primary signal in the range of 1550 nm, one incoming port attached to a *pump laser* operating at 980 or 1480 nm, and the one outgoing port connected to the operating fiber. The pump laser excites the erbium atoms. Weak incoming light from the operating system stimulates emissions from the erbium atoms. As the erbium atoms drop from their excited state, they release the extra energy, which transfers to the primary signal and amplifies it. An EDFA can simultaneously amplify a number of wavelengths in an operating range around 1550 nm (C-Band). A single-pump EDFA involves a pump laser on the upstream (i.e., incoming) side of the erbium-doped fiber section and provides a gain (i.e., increase in signal strength) varying from +10 dB (1000 percent, or 10:1) to as much as approximately +17 dB (approximately 8000 percent, or 80:1). A double-pump EDFA (Figure 2.20) involves one pump laser on the upstream side and another on the downstream side of the erbium-doped fiber section and provides a gain of close to 30 dB (100,000 percent, or 1000:1). *Note:* The pump lasers can operate in either direction. *Optical isolators*, placed on both sides of the EDFA, act like diodes, serving to prevent optical signals from traveling in more than one direction. EDFAs work quite effectively and at lower cost than optical repeaters. But they generally are limited to no more than 10 spans over a total distance of 800 km or so, at which point a repeater must be applied to the signal to filter out the accumulated noise caused by various forms of disper-

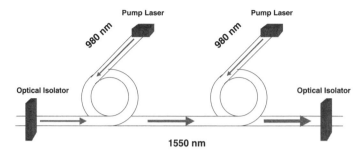

Figure 2.20 Erbium-doped fiber amplifier.

sion. EDFAs are further limited by their inability to amplify wavelengths shorter than 1525 nm [23, 27].

Raman amplification requires no fiber doping and usually is accomplished throughout the length of the transmission fiber itself in a configuration known as *distributed amplification*, rather than in a discrete amplification, or lumped amplification configuration such as that employed by EDFAs. Raman amplification occurs as a high-energy (i.e., high-frequency, short-wavelength) *pump wavelength* is sent in the reverse direction (i.e., in the direction opposite the signal transmission) from the output end of the fiber span, where the incoming signal is weakest. The pump wavelength, which generally is in the 1450-nm range (E-band), interacts with atoms in the crystalline lattice of the fiber core. The atoms absorb the photons and quickly release photons with energy equal to the original photon plus or minus the atomic vibration. In other words, a frequency/wavelength shift occurs as the pump wavelength propagates along the fiber in the reverse direction. The energy lost in the pump wavelength is shifted to longer wavelength (within about 100-nm) signals in the forward direction, thereby serving to amplify them. Raman amplifiers offer the advantage of amplifying signals in the broad range extending from 1300 to 1700 nm. Further, they perform better than EDFAs in terms of Signal-to-Noise Ratio (SNR). Raman amplifiers increasingly are used as preamplifiers to enhance the performance of EDFAs in DWDM systems [27–34].

The concept and effect of Raman amplification is explained very effectively in the following analogy: As you are crossing an old wooden bridge at a walk, you feel the normal slow oscillation of the bridge under the weight of each of your own steps. Then a much larger person is running toward you from the other end of the bridge. The vibrations, or oscillations, you create are waveforms, and they move in both directions. The same is true of the waveforms created by the runner. But those created by the runner are stronger (he is heavier) and longer (he has a longer stride). Your waveforms and those of the runner meet in the middle. Some of the energy from the runner's waveforms is transferred to your waveforms, which are reflected right back to you with greater strength.

2.9.2.5 Optical Switching Chapter 1 introduced various switching technologies, including circuit, packet, frame, and cell switching. That very brief introduction will be expanded considerably in subsequent chapters, as will several other approaches. All of these approaches are implemented at the electrical level. That is to say that, whether the transmission medium is electrical, RF, or optical in nature, the switching system is electrically based. The transmitted signal is interfaced to the switching system, converted into an electrical format as necessary, switched on one basis or another, and sent on its way, converted into an RF or optical signal, as necessary, to interface properly to the outbound transmission system. In the very recent past, however, purely optical switches have emerged. These switches fall into the generic category of MEMSs (*Micro-ElectroMechanical Systems*) and actually are more along the lines of optical cross-connects. There are two types of MEMS switches under development: mechanical and microfluidic.

- *Mechanical switches* are based on arrays of micromachined mirrors, with as many as hundreds of thousands contained on a single silicon chip. Control signals applied to the MEMS chip adjust the position of each mirror to reflect the subject light signal to the correct output port. Some mechanical switches can move mirrors on a two-dimensional basis, and others can operate in three dimensions by swiveling in multiple angles and directions.
- *Microfluidic switches* operate on the basis of the movement of liquid in tiny channels etched into a silicon chip. One such switch comprises a number of silica waveguides with intersecting paths in a grid configuration. A tiny trench etched diagonally at each point of intersection contains a fluid. In default mode, the fluid allows the light to travel through the switch. If it is necessary to switch the signal, bubbles are injected and removed hundreds of times per second in the fluid in order to reflect the signal to the proper output port.

There are several types of actuation mechanisms used in the various mechanical and microfluidic switches—electrostatic, magnetic, and thermal—and there are a variety of fabrication methods used to create them. Unfortunately, these fascinating details are beyond the scope of this book [35, 36].

2.9.2.6 Analog or Digital? Fiber-optic systems can either be analog or digital in nature, although digital is much more common. Analog systems simply vary the intensity (amplitude) of the light wave. Digital systems pulse *on* and *off* (although never totally off) to represent 1s and 0s via Amplitude Modulation (AM). Frequency Modulation (FM) is possible in optical transmission and has been demonstrated but is not considered practical as the complexity of modulating frequencies in the terahertz range is extreme. Further, FM would render WDM difficult, if not impossible, as the wavelengths are so tightly packed, with spacings of only 200, 100, or even 50 GHz. Phase-Shift Keying (PSK) also is possible but not practical due to the complexities of signal manipulation at such high frequencies. Because digital systems offer significant advantages (e.g., error performance and compression), all long-haul fiber systems used in carrier networks are digital. While CATV providers deployed significant amounts of analog fiber for purposes of cable TV delivery, they increasingly are upgrading those systems to digital technology. Clearly, digital fiber is required in a convergence scenario, where the CATV providers support not only TV but also voice and high-speed data.

2.9.3 Bandwidth

Fiber offers by far the greatest bandwidth of any transmission system, often in excess of 2.5 Gbps in long-haul carrier networks. Systems operating at 10 Gbps are now routinely deployed, and some systems in commercial operation now run at 40 Gbps.

Through WDM, carriers routinely introduce 4, 8, 16, or 32 lambdas (wavelengths) into a given fiber. At a bit rate of 2.5 Gbps per lambda, a single fiber strand running 80 lambdas theoretically can support 2.5 million uncompressed voice conversations. Sprint was the first carrier to commercially deploy a DWDM system running 40 channels at 2.5 Gbps, for a total yield of 100 Gbps. The ITU-T has specified 160 lambdas in its DWDM grid, so the potential is staggering. The theoretical limit of fiber is thought to be in the terabit range per wavelength.

2.9.4 Error Performance

As the fiber itself is *dielectric*, it is not susceptible to EMI/RFI; neither does it emit EMI/RFI. Assuming you have properly powered and grounded the associated repeaters and other devices, ambient interference is not an issue. The optical signal does suffer from attenuation and distortion as it transverses the fiber, although not to nearly the same extent as do other transmission systems. Attenuation is a function of a number of factors, primarily scattering and absorption, and is much less of an issue in fiber optics than in other transmission systems. As attenuation is wavelength dependent, long-haul systems focus on the 1550-nm region. Signal distortion is caused by factors including modal dispersion, polarization-mode dispersion, chromatic dispersion, and simple mechanical splices in the cable system. Error performance, depending on factors such as the compression scheme utilized, ranges between 10^{-9} and 10^{-14}, or one errored bit in every 100 trillion [37]. For all practical purposes, fiber is error free, at least in comparison to the alternative transmission systems.

Fiber-optic systems suffer from *diurnal wander*, as do copper and coax systems. Diurnal (daily cycle) wander is a loss of signal synchronization in digital cable systems caused by temperature variations over the course of 24 h. As the ambient temperature can vary considerably from the heat of the day to the cool of the night, the cable stretches and contracts, with the overall length of the cable changing, if only ever so slightly. As the length of the medium changes, the latency of signal propagation is affected, and the number of digital pulses effectively stored in the medium changes. As a result, the network elements (e.g., repeaters and multiplexers) can get out of synch. Diurnal wander most especially affects cables strung on poles, rather than buried under ground, as the exposure to ambient temperatures is greater and as the weight of the cable magnifies the effect.

2.9.5 Distance

Single-mode fiber-optic systems are routinely capable of transmitting unrepeated signals over distances well in excess of 600 miles (1000 km) through the use of Raman amplifiers and EDFAs. As a result, relatively few optical repeaters are required in a long-haul system, thereby both reducing costs and eliminating points of potential failure or performance degradation.

2.9.6 Security

Fiber is intrinsically secure because it is virtually impossible to place a physical tap without detection. Because so little light radiates outside either the strand or the cable, physical taps are the only effective means of signal interception, with the possible exception of exotic technologies employed by a few government intelligence agencies. Additionally, the fiber system supports such a high volume of traffic that it is difficult to intercept and distinguish a single transmission from the tens, or hundreds, or even hundreds of thousands of other transmissions that might ride the same fiber strand. Additionally, the digital nature of most fiber, coupled with encryption techniques frequently used to protect transmissions from interception, makes fiber highly secure.

2.9.7 Cost

While the acquisition, deployment, and rearrangement costs of fiber are relatively high (approximately 130 percent the cost of Cat 5 copper) in the LAN domain, the immense bandwidth can outweigh that cost in bandwidth-intensive applications such as GbE and 10 GbE. *Storage Area Networks* (SANs), which are very bandwidth intensive, typically are designed around fiber-optic transmission. In the *Metropolitan Area Network* (MAN) and *Wide Area Network* (WAN) domains, fiber optics generally is the only realistic option for long-haul transmission employing wireline systems. At Gbps speeds, a single set of fibers can carry much greater volumes of digital transmissions over longer distances than alternative systems, thereby lowering the transport cost per voice conversation to a small fraction of a penny per minute. The cost of transporting a single bit, therefore, is essentially zero; for that matter, the cost of transporting a multimegabyte file is essentially zero.

2.9.8 Durability

While glass fiber certainly does not have either the break strength or flex strength of copper or coax, it does enjoy the same tensile strength as steel of the same diameter. In vertical riser cable applications (i.e., between floors), the integrity of the fiber often is protected through the use of aramid fiber (e.g., Dupont Kevlar) strength members; at some point, neither copper nor glass cables can support their own weight. When covered by a protective jacket or armor, fiber can be treated fairly roughly without damage. Note, however, that you must respect limits of bend radius, as the integrity of the data stream can suffer and the glass fiber can break under a severe bend. Additionally, fiber is more resistant to temperature extremes and corrosion than alternative cable systems. However, notably, and in consideration of the huge number of conversations supported over a typical fiber-optic cable in a WAN application, a train derailment, earthquake, or other traumatic event can have consequences of catastrophic proportions. In LAN backbone applications, the impact of a fiber failure also can be considerable, although train derailments are less likely to be the root cause. Note that while plastic optical fiber (POF) is much more forgiving in terms of flex strength and bend radius, it does not perform at nearly the same speed as does glass.

2.9.9 Applications: Bandwidth Intensive

Fiber-optic transmission systems are most cost effective in bandwidth-intensive applications. Such applications include backbone carrier networks, international submarine cables, backbone LANs, interoffice trunking, computer-to-computer or cabinet-to-cabinet (e.g., mainframes and PBXs) connectivity, distribution networks (e.g., CATV), and certain fiber-to-desktop applications [e.g., Computer Aided Design (CAD)].

Unfortunately, fiber optics has other applications as well. While I have not seen any fiber-optic *mbenge* (see Section 2.3.9), I have heard from reliable sources in South Africa that reels of fiber-optic cable have disappeared. The buffered fiber later turned up at flea markets, where it was being sold as refills for string trimmers. While the flying glass shards could be dangerous, I suppose that you could trim your lawn with the speed of light. A delegate in a seminar I was teaching in Namibia recently told me that fiber cables were disappearing there as well. Apparently, some of the local criminal gangs got the bright idea that they could strip out the Kevlar strength members and stuff them into the door panels of the automobiles to bulletproof them. (Some criminals really are not all that bright!)

2.10   POWERLINE CARRIER

PowerLine Carrier (PLC) is an old technology that uses existing electric power distribution cabling for communications purposes. PLC has gotten new spark, thanks to some fairly recent developments in the area known as *Broadband over Power Line* (BPL). Electric utility companies have used PLC for many years for telemetry applications and controlling equipment at remote substations. Rural telephone companies sometimes use PLC it to provide voice service to extremely remote customers who have electric service but for whom it is too costly to provide telephone service over copper local loops. BPL, the developing version that currently is the object of so much attention, supports not only voice and low-speed data but also high-speed data.

PLC uses existing power distribution cabling and inside wire running at 120 or 240 V, depending on the standards in place for the electric grid. In Europe and most of the rest of the world, the standard calls for communications over the 240-V grid at frequencies from 30 to 150 kHz. In the United States, the standards for the 120-V grid allow the use of frequencies above 150 kHz as well. Power utilities use the frequencies below 490 kHz for their own telemetry and equipment control purposes. There are two variations on the theme of broadband PLC: access BPL and in-house BPL.

2.10.1   Access BPL

Access BPL runs over Medium-Voltage (MV) power lines in the power utilities' distribution networks. Those MV lines generally operate at 7200 V. At the utility substation where the High-Voltage (HV) lines are stepped down to MV for the distribution network, the BPL provider typically terminates a fiber-optic network connection in a device that accomplishes the optoelectric conversion process (Figure 2.21). Inductive couplers wrapped around the power lines without touching them serve as *injectors* for downstream transmissions, injecting the communications signals onto the distribution lines. The same couplers serve as *extractors* to extract upstream signals. The RF carrier supporting the communications signals can share

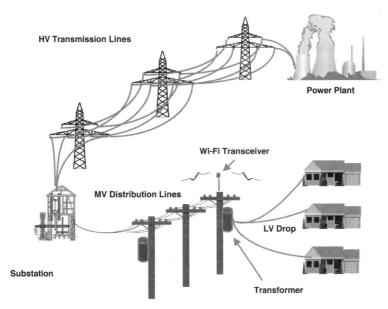

Figure 2.21   Access BPL.

the same line with the electrical signals as they operate at different frequencies, that is, this is Frequency Division Multiplexing (FDM) of telecom and electrical power, with the BPL signal using frequencies between 2 and 80 MHz. Repeaters must be spaced every 300 m or so, which is very tight spacing for a local loop technology. Extractors remove the signals from the power lines just ahead of the transformers that step the voltage down from MV to the 110/220 Low-Voltage (LV) level used within the premises. The connection to the premises can be over the LV lines or via Wi-Fi, the predominant WLAN technology. Access BPL has real potential for broadband service delivery in sparsely populated rural areas where DSL, fiber-optic, and WLL technologies are impractical.

2.10.2   In-House BPL

In-house PLC technologies have been used in key telephone and intercom systems for decades, although not particularly successfully. Standards for in-house BPL, a premises networking technology, however, are a relatively recent development, with HomePlug standards being the most prevalent. Loosely based on Ethernet LAN standards, in-house BPL allows any device to connect to the LAN directly through the LV electric lines (110 V at 50–60 Hz or 220 V at 50 Hz). HomePlug 1.0 supports up to 16 nodes sharing bandwidth up to a theoretical maximum of 14 Mbps. Some proprietary systems support raw signaling rates up to 85 Mbps, which comes very close to 100Base-T performance.

2.10.3   Interference and Other Issues

Interference is a considerable issue with PLC, in general. Since the electric power grid was not designed with telecommunications in mind, it is challenging under even the best of circumstances. As electric distribution cables and inside wires are highly susceptible to EMI and RFI, any reasonably strong source of ambient noise can create problems. Broadcast radio stations, Citizens Band (CB) radios, and various other radio systems can cause considerable difficulty. Vacuum cleaners, electric drills, electric sanders, and other devices generate considerable impulse noise. Television sets create distortion and light dimmers (i.e., rheostats) cause noise. Interference issues work both ways, of course. As HV and MV power transmission lines largely are unshielded and aerial in nature, they emit considerable electromagnetic fields that can interfere with short-wave (e.g., ham) and other radio signals. As interference works both ways, the FCC and other regulatory bodies have established certain excluded frequency bands to avoid interference with amateur and aircraft radio. FCC rules also establish exclusion zones in proximity to sensitive operations such as Coast Guard, Navy, and radio astronomy stations.

Attenuation in PLC is a considerable issue as the signal must deal with inside wire components such as fuse boxes, splices, surge suppressors, and circuit breakers. Security is always an issue, particularly so with technologies using a shared bus topology. As multiple premises typically are served from a common electrical transformer, the physical reach of an in-house BPL network can extend well beyond the walls of an individual home or business. Some BPL systems provide for network encryption. Additional challenges include unresolved regulatory issues and the fact that a loss of electrical power renders PLC in a state of catastrophic failure [38, 39]. Chapter 9 discusses BPL in greater detail.

2.11 HYBRID TRANSMISSION SYSTEMS

While each transmission medium/system has its own unique properties and applications, digital fiber-optic cable clearly offers the most potential in terms of raw performance. Its cost and fragility, however, are limiting factors. The selection of the most appropriate transmission medium is sensitive to the criteria mentioned at the beginning of this chapter. Namely, those considerations include bandwidth, error performance, throughput, distance between elements, propagation delay, security, mechanical strength, physical dimensions, and a number of cost factors. In fact, a given long-haul transmission typically will traverse a number of transmission systems, perhaps both wired and wireless, commonly including twisted pair in the local loop and fiber optics in the backbone.

The true concept of a *hybrid transmission system*, however, generally involves a local loop connection deployed in a well-planned *convergence* scenario. Such a scenario involves one or more providers deploying a communications grid designed to deliver voice, data, and entertainment information to the premises. Hybrid systems usually are described as involving *Fiber-to-The-Neighborhood* (FTTN) or *Fiber-To-The-Curb* (FTTC), with the last link being embedded UTP.

A number of incumbent telephone carriers and CATV providers have made significant investments in fiber-optic local loop technologies. Verizon (United States) has invested over a billion dollars in FTTP and plans to invest billions more. AT&T (United States), previously SBC (nee Southwestern Bell Telephone Company), has made significant investments in FTTN, planning to maximize the use of embedded UTP in a hybrid scenario. Other approaches involve Wireless Local Loop (WLL) technology at various levels. Broadband over Power Line (BPL) is a recently developed technology that is commercially available on a very limited basis. Chapter 9 discusses these technologies in considerable detail.

REFERENCES

1. *Engineering and Operations in the Bell System*, Bell Telephone Laboratories, 1977.
2. Keen, Peter G.W. and Cummins, J. Michael. *Networks in Action*. Wadsworth Publishing Co., 1994.
3. Gelber, Stan. *Introduction to Data Communications: A Practical Approach*. Professional Press Books, 1991.
4. Sherman, Kenneth. *Data Communications: A Users Guide*. Reston Publishing Co., 1981.
5. Steinke, Steve. "Lesson 154: Network Delay and Signal Propagation." *Network Magazine*, May 5, 2001.
6. *Reference Data for Radio Engineers*, 4th ed. International Telephone and Telegraph Corporation, 1956.
7. Brooks, John. *Telephone: The First Hundred Years*. Harper & Row, 1975.
8. Casson, Herbert N. *The History of the Telephone*. A. C. McClurg & Co., 1910.
9. *LAN Cable*. Underwriters Laboratories, 1995.
10. Bates, Bud. *Introduction to T1/T3 Networking*. Artech House, 1992.
11. Freeman, Roger L. *Fundamentals of Telecommunications*, 2nd ed. Wiley-Interscience, 2005.
12. Hudson, Heather E. *Communications Satellites: Their Development and Impact*. Free Press, Division of Macmillan, 1990.
13. Nelson, Robert A. "The Art of Communications via Satellite." *Via Satellite*, July 1998.
14. Careless, James. "Ka-Band VSATs: Blazing the Next Great Frontier." *Via Satellite*, February 2001.
15. Pappalardo, Denise. "GE Spacenet Turbocharges Satellite Network." *Network World*, November 17, 1997.
16. Melnick, Martin and Hadinger, Peter. "Enabling Broadband Satellites." *Satellite Communications*, July 2000.
17. Held, Gilbert. *Understanding Data Communications*. SAMS Publishing, 1994.
18. Paetsch, Michael. *Mobile Communications in the US and Europe: Regulation, Technology, and Markets*. Artech House, 1993.
19. Allen, Doug. "The Second Coming of Free Space Optics." *Network Magazine*, March 2001.
20. Rockwell, David and Mecherle, G. Stephen. "Optical Wireless: Low-Cost Broadband Optical Access." *Lightwave*, June 15, 2001.
21. Keck, Donald B. "Optical Fiber Spans 30 Years." *Lightwave*, July 2000.
22. Greenfield, David. "Lesson 156: Fiber and Optical Networking." *Network Magazine*, July 2001.
23. Sterling, Donald J., Jr. *Technician's Guide to Fiber Optics*, 4th ed. Thompson Delmar Learning, 2004.
24. Fuller, Meghan. "PMD Compensation for 40 Gbits/sec Remains a Question Mark." *Lightwave*, August 2001.
25. Fritschen, John and Lumish, Stan. "40-Gbit/sec Components: Drivers, Issues, and Solutions." *Lightwave*, August 2001.

26. Inoue, Ted. "Whenever and Wherever You Go." *Lightwave*, April 2000.

27. Hecht, Jeff. *Understanding Fiber Optics*, 4th ed. Prentice-Hall, 2002.

28. Lin, Sharlene. "Opening the S-Band, and More, with Raman Amplification." *Lightwave*, March 2001.

29. Nietubyc, Mark. "Raman Amplification in the All-Optical Network." *Telecommunications*, June 2001.

30. Barbier, Denis. "Erbium-Doped Waveguide Amplifiers Promote Optical-Networking Evolution." *Lightwave*, November 2000.

31. Hunt, Peter. "The Search for Ultra-Long-Haul Transmission." *Lightwave*, May 2001.

32. Fuller, Megan. "Raman Amplifiers Combine with EDFAs to Tackle System-Distance Limitations." *Lightwave*, December 2000.

33. Teed, David. "Networking's Horizon for Optical Amplifiers." *Lightwave*, August 2001.

34. Junghohann, James, Schafer, Rick, and Bezoza, Alan. "Raman Amplification—Longer, Wider, Faster, Cheaper." *CIBC World Markets Equity Research*, November 28, 2000.

35. McGarvey, Joe. "Optical Gear: 'MEMS' The Word." *Inter@ctive Week*, March 20, 2000.

36. Bourne, Marlene. "MEMS Switching . . . and Beyond." *Lightwave*, March 2001.

37. Sexton, Mike and Reid, Andy. *Transmission Networking: SONET and the Digital Hierarchy*. Artech House, 1992.

38. Horak, Ray. *Access Broadband over Power Line (BPL)*. http://www.commweb.com/.

39. Horak, Ray. *In-House Broadband over Power Line (BPL)*. http://www.commweb.com/.

CHAPTER 3

VOICE COMMUNICATIONS SYSTEMS: KTS, PBX, CENTREX, AND ACD

If you could make one good invention in the telegraph, you would secure an annual income...and then you could settle that on your wife and teach Visible Speech and experiment in telegraphy with an easy and undisturbed conscience.

Telephone: The First Hundred Years [1], from a letter dated March 13, 1876, to Alexander Graham Bell from Gardiner Hubbard, Bell's future father-in-law

In the beginning, there were only telephone sets and wires between them. The first commercial sets, offered by the Bell Telephone Company in May 1877, consisted of a single piece of wood (black walnut or mahogany) with a single piece of equipment serving as both transmitter and receiver. Power was supplied by the energy in the speaker's audio output and a permanent magnet contained within the device, rather than by a battery or external power source. Telephone sets initially were leased in pairs, and the lesee installed his own telephone wire to connect them. Later, Bell provided the private lines, which were usually leased from Western Union Telegraph Company. The first advertisements offered the use of two telephones and a line connecting them for $20 per year for social purposes, $40 per year for business. Free maintenance was guaranteed. By the fall of 1877, over 600 telephones were in use [1]. As the popularity of the invention grew, visions developed of rooms full of telephones and skies full of wires—and then the visions became reality.

In order to improve the usefulness of the telephone device and reduce the associated costs, some method was required to interconnect telephones on a flexible basis. The first exchange for telephone service was put into service in Boston, Massachusetts, in 1877 by E. T. Holmes, a young man whose father had originated the idea of protecting property by electric wires in 1858. The burglar alarm network proved to be an effective telephone network as well. Holmes obtained telephone numbers six

Telecommunications and Data Communications Handbook, By Ray Horak
Copyright © 2007 Ray Horak

and seven and attached them to a wire in his office. He then placed six box telephones on a new shelf in his office. Any of these telephones could be switched into connection with the burglar alarm wires and any two of the six wires could be joined by a wire cord. It was a simple idea, but a new one [2]. At night, when the telephone operator was off duty, the telephone network reverted to a burglar alarm network.

George W. Coy of New Haven, Connecticut, developed the first practical exchange switch, which was placed into service on January 28, 1878. This manual exchange, or *cordboard*, allowed the flexible interconnection of 21 subscribers. The number of exchanges grew quickly, and although Western Union rather than Bell handled most of the initial installations, the Bell System network grew quickly and soon outpaced the networks of Western Union and other providers [1].

These first exchanges, housed in *Central Offices* (COs), allowed circuits to be connected manually, on demand and as available. Through these central points of interconnection, each subscriber required only one terminal device and one wired connection to the central switch. Central offices or *Central Office Exchanges* (COEs) handled all switching of calls whether they involved parties from different parts of the city, across the hall, or across the office. It soon became clear that a number of factors made this centralized approach less than ideal for serving businesses of any significance and with any appreciable number of terminal devices. Those factors included the following:

- **Labor Intensity:** Manual switchboards were labor intensive, because telephone company operators were responsible for making and breaking all connections. The early electromechanical exchanges were maintenance intensive, as were the later electromagnetic offices and even the early electronic common control exchanges that followed them, although successively less so.

- **Capital Intensity:** COs were costly, and the more subscribers who connected and the more functions performed by the exchange, the larger and more costly it became. Additionally, the cost of individually connecting each telephone terminal to the exchange was considerable, as each local loop connection required an individual pair, usually housed in a multipair cable, which was either placed in conduits and trenches or suspended on a series of poles. Infrastructure is expensive, as a rule, and telephone company infrastructure is no exception.

- **Physical and Functional Limitations:** COs were highly limited in the number of ports and, therefore, the number of local loops they could support. Also, there were finite limits to the number of copper pairs that could be packed together in a cable and either pulled through a conduit or hung from a pole.

- **Personalized Service:** Quite simply, personalized service was not available. In other words, the telephone company operator could not handle the call as courteously and efficiently as would an employee of the user organization. Also, the telephone company was much more likely to misdirect calls due to lack of familiarity with users and their associated station numbers.

Clearly, extending a physical partition of the COE to the customer premises presented a better approach. The subscribing organization could rent the equipment, thereby yielding incremental revenue to the telephone company. Additionally, the cost of labor, cost of switch capital, and responsibility for connection errors would shift to the customer. Further, the subscriber's operators could provide more effective and personalized service to both inbound and outbound callers. Significantly, the equipment also could act as a contention device, allowing multiple calls to share a single, pooled group of local loop connections. This approach would considerably reduce the cost of cabling as well as the number of ports required to connect those local loops to the switch. The shift to the user organization of the functional responsibility for switching both incoming and outgoing calls would also relieve the telco of that burden, allowing the exchange switch to serve more end users. Finally, internal (station-to-station) calls could be connected entirely through the premises equipment, without requiring a connection through the telco's CO. As station-to-station calls are a very large percentage of the total calling traffic in most large organizations, shifting this functional responsibility to the user organization could have considerable impact on the telephone company's capital investment and labor costs.

The first devices to accomplish this feat were *Private Branch eXchanges* (PBXs) in 1879. *Key Telephone Systems* (KTSs) did not arrive on the scene until 1938. Central exchange (Centrex), a CO-based solution, was christened in the 1960s. *Automatic Call Distributors* (ACDs) did not make an appearance until 1973. Rather than exploring these systems in chronological order, this chapter begins with a discussion of KTS, as it is the least complex in many ways. The chapter then progresses through PBX, Centrex, and ACD systems. This chapter concludes with discussions of Computer Telephony (CT) and systems based on the Internet Protocol (IP) that was developed for the Internet.

Originally, the telephone company owned all *Customer Premises Equipment* (CPE) and rented it to the end users. The telco owned all CPE devices connected to the network, including the single-line telephones, KTSs, and PBX systems, and all telephone sets connected behind them. When answering machines and other peripheral devices appeared, they too were rented to the user by the telco. Even cords were telco property rented to the subscriber. In concert, the regulators and telcos maintained that any privately owned equipment of any sort could conceivably cause damage to the network switches and even to telco personnel by introducing uncontrolled levels of electrical current. Further, such devices might well be incompatible with the highly standardized network and, therefore, cause uncontrollable disruption. Interestingly, even acoustically coupled devices were banned, unless provided by the telco. After a series of lengthy legal battles, which I discuss in Chapter 15, the Federal Communications Commission (FCC) in the United States reversed that policy through the 1968 Carterphone decision. That decision permitted user ownership through a special interconnecting device, or *coupler*, that allowed interconnection of registered CPE and protected the network from damage much as a circuit breaker or surge protector protects an electrical network. Over time, the FCC replaced the CPE registration process with a certification process that eliminated the need for couplers. Currently, customers own virtually all voice CPE or lease it from third parties. With few exceptions, the rest of the world has followed this lead and both commercialized and liberalized the telecom environment.

3.1   KEY TELEPHONE SYSTEMS

Key telephone systems are business communications systems intended for small businesses, typically defined in this context as involving no more than 50 stations.

The term *key telephone* dates to the beginnings of telegraphy and telephony when mechanical *keys* were employed to open and close a circuit. The buttons on a key telephone set, also referred to as keys, mechanically opened and closed the line circuit. While contemporary KTSs provide much the same feature content as small PBXs, and while they also act as contention devices for network access, KTSs are not switches. That is, they do not possess the intelligence to accept a call request from a user station, determine the most appropriate circuit from a shared pool of circuits, and set up the connection through common switching equipment. Rather, the end user must make the determination and select the appropriate facility [e.g., local line, tie line, or Foreign eXchange (FX) line] from a group of pooled facilities. KTS control relies on *grayware* (i.e., gray matter, or human brain power), rather than *software* (i.e., computer programs). Therefore, the local loops associated with KTSs are *lines*, rather than *trunks*. This is not just a matter of semantics, for the distinction has a significant impact on monthly cost. A line is a single-channel facility that is associated with a single telephone number and that connects an endpoint to the Public Switched Telephone Network (PSTN). A trunk typically is a multichannel, rather than a single-channel, facility that interconnects switches and is not necessarily associated with a telephone number. Rather, a trunk serves a group of users and a group of telephone numbers through an intelligent switching device that is designed to manage contention between users and channels. Telco rate and tariff logic assumes that a trunk will be used more intensely than a line, so the cost is greater.

Most small KTSs are *squared*, meaning that every key set is configured alike, with every outside line appearing on every set. Thereby, every station user can access every outside line for both incoming and outgoing calls, and all feature presentations are consistent. In larger systems, the physical size of the telephone sets required to maintain the squaring convention would be impractical, but departmental subgroups often are squared.

3.1.1   1A1 and 1A2 KTS

Key systems originally were electromechanical in nature. The common control unit, known as a *Key Service Unit* (KSU), housed multiple circuit packs, known as *Key Telephone Units* (KTUs). Connections between KSU components and between the KSU and the key telephone sets were hardwired. Early KTS feature content was limited to hold, intercom, speakerphones, and autodialers.

- **Hold:** The *hold* feature allows a user to temporarily suspend a call by depressing a designated hold button. This allows the user to engage in another activity such as answering another call. The lamp associated with the first call flashes to indicate the hold status. The user can then retrieve the call, perhaps from another telephone, by depressing the flashing line key in a process known as *recall. I-hold*, or *exclusive hold*, is a privacy feature that prevents recall from any other phone.
- **Intercom:** Key systems include some relatively small number of *intercom* (*intercom*munication system) talk paths that are used exclusively for internal communications between stations belonging to a closed user group, that is, a group or subgroup of privileged users.

- **Speakerphones:** Key systems included separate speakerphone units comprising a combined speaker and microphone. These early speakerphones supported full duplex (i.e., supported simultaneous transmit and receive) communications of exceptional quality.
- **Autodialers:** In the days before software-based speed dialing, telephones sometimes were equipped with *autodialers*. These mechanical card dialers used plastic or thick paper punch cards to dial telephone numbers automatically.

The first standard key system, known as the *1A*, was a hardwired system developed by the Bell System and first marketed in 1938. The *1A1* KTSs, first introduced in 1953 and also truly hardwired, comprised components wired together to form a complete system. 1A1 systems added a few features, including line status lamps that lit steadily to indicate a line in use and flashed to indicate a line on hold. Introduced in 1963, *1A2* systems were modular to some extent, as hardwired circuit packs plugged into a prebuilt chassis that included cable connectors for attaching station equipment [3]. You could add a limited number of enhanced features through common control cards in the form of circuit packs. In either case, wiring between the KSU and the sets required a labor-intensive *home run* (i.e., direct connection) of expensive 25- or 50-pair cable between each voice terminal and the KSU.

3.1.2 Electronic and Hybrid KTS

The electronic KTS entered the small business market in the 1970s, offering many of the same advantages of Electronic Common Control (ECC) as did contemporary PBXs. While these systems add considerable feature content accessible from sophisticated electronic station sets, they remain key systems, rather than switches, at heart. Most key systems currently manufactured are *hybrid KTSs*, which often are marketed as PBXs with the ability to emulate key systems, depending on the software configurations. A hybrid possesses the unique ability to function as either a KTS (direct circuit selection) or a PBX (switched access to pooled facilities). Many hybrids can function simultaneously as both a KTS for one workgroup and as a PBX for another. Just as is the case with contemporary PBXs, hybrids are digital systems comprising microprocessors, printed circuit boards, memory, software, and so on (see Figure 3.1). They also can be upgradeable and expandable, within finite limits determined by the manufacturer. Cabling requirements are reduced from 25 pairs to the same 1–4 pairs required by PBXs. Common KTS configurations are 2×6 (two lines by six stations), 3×8, 6×16, 8×24, 12×32, 16×48, and 24×64.

Hybrids generally are limited to about 256 ports, although some are much larger. Most systems allow individual ports to be configured as either station ports or line/trunk ports, although there may be limits on either. Example hybrid KTS configurations are 48×224 (Avaya Merlin has a maximum of 272 ports, all of which can be configured as either lines or stations) and 80×200 (Nortel Norstar maximum configuration).

Electronic KTSs and hybrids offer a relatively significant set of features that include virtually everything a small or medium-size business might require. Many of those features can be accessed through programmable buttons on feature-rich proprietary station sets. Such *softkeys* commonly are display based and context

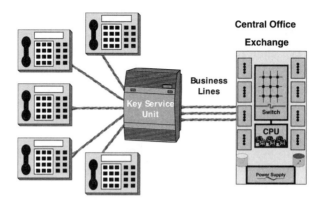

Figure 3.1 KTS configuration.

sensitive. Feature content can be quite simple or very sophisticated. Basic features are numerous and generally include the following:

- **Account Codes:** Users placing outgoing calls, particularly toll calls, can be required to enter an account code to aid in client billing for time and expenses.
- **Automatic Line Selection:** When the station user picks up the telephone receiver, a specific outside line is automatically selected.
- **Barge-in:** An authorized user from an authorized station can barge in on a call in progress (i.e., join an active call without invitation) through the use of an authorization code.
- **Call Forwarding:** A user can forward calls to another station in several ways. By dialing a feature code, the user can forward calls to a specific station on an ad hoc basis. The user can program the station to automatically forward calls to the attendant, a colleague, or voice mail after a selectable number of rings or if the station is engaged.
- **Call Park:** A user can park a call in system orbit. The call can be retrieved by any other user by dialing the associated call park code.
- **Call Pick-up:** Any user can answer any call at any phone where the line appears.
- **Call Transfer:** An incoming or outgoing call can be transferred to another station.
- **Calling Line IDentification (CLID):** The calling number and associated caller name appear on the station display.
- **Camp-on:** The system attendant can forward or extend a call to a station, even if the station is engaged in a call. Camp-on essentially is much like call waiting, although the camped-on call will be routed back to the attendant if not answered within a user-programmable time interval.
- **Conferencing:** The attendant or an authorized user can bridge as many as 2, 4, 8, or perhaps 16 parties on a conference call.
- **Distinctive Ringing:** Distinctive ringing tones distinguish between internal and outside incoming calls.

- **Do Not Disturb (DND):** Depressing a feature button or softkey blocks incoming calls, perhaps automatically forwarding them to the attendant, an administrative assistant, or voice mail.
- **Flexible Station Numbering:** The system administrator can program the station numbering plan for two-, three-, or four-digit dialing to match any existing PBX, Centrex, or intercom plan.
- **Integrated Services Digital Network (ISDN):** CO interfaces typically include digital ISDN lines or trunks.
- **Line Appearance:** Any outside line can appear on any number of station sets and be accessed from any of them. See Privacy below.
- **Music on Hold:** Interfaces to an external audio source to provide background music, promotional messages, or other audio to play while callers are on hold.
- **Off-Hook Voice Announce:** The attendant can announce another call even if the station user is off hook. In some implementations, the announcement is through the telephone speaker, with station microphone muted so that the other party does not hear the announcement. In other implementations, the announcement is through the handset receiver so that the other party was not muted, as the muting was noticeable.
- **Paging:** An authorized station user can access an external paging system, that is, public address system, to make announcements. In larger buildings, paging systems commonly are divided into a number of zones. Unique to key systems is *voice-over paging*, which allows an authorized user to page through the intercom system, which works through the speaker built into the telephone.
- **Privacy:** If a line is engaged, another station user cannot access that line unless the primary station user chooses to override the restriction to allow a conference call. See Barge-in above.
- **Station Message Detail Recording (SMDR):** The system generates a record of incoming and outgoing calling activity by station. These records provide a call accounting system with the data necessary to generate cost allocation reports.
- **Toll Restriction:** Telephones can be restricted from long-distance access through programmable Class-of-Service (CoS) restriction.

More sophisticated features commonly include a nonblocking switching matrix, Automatic Call Distribution (ACD) and T/E-carrier interfaces, and wireless sets. Data communications commonly is supported at various rates. As we traditionally associate many of these features with PBX systems, I discuss them in more detail later in the next section. Note that a hybrid configured as a KTS requires that the users select outside lines manually. Configured as a PBX, the system will make outside channel selections automatically, but over more expensive trunks.

At the opposite end of the spectrum are KSU-less key systems. These systems share no common equipment but rather intercommunicate directly. Feature content is limited to those features that can be contained within the phones, but the systems are inexpensive. Installation is easily accomplished, but the results are highly dependent on the quality of the inside wiring, which is always questionable.

3.2 PRIVATE BRANCH EXCHANGES

While KTSs and hybrids effectively address the communications requirements of small user organizations, they are limited in terms of both feature content and capacity. Clearly, a better approach for the larger user organization involves, in effect, moving a partition of the COE to the customer premises, as portrayed in Figure 3.2. This *Private Branch* of the Central Office *eXchange* significantly benefits both the telephone company and the user organization.

The first PBX was placed into service in the Old Soldiers' Home in Dayton, Ohio, in 1879. While the first systems were little more than nonstandard modifications of telephone company CO switches, AT&T offered a standard PBX (No. 1 PBX) in 1902 [4]. Table 3.1 lists and describes the generations of PBXs. PBXs evolved in the same order as COs with the exception of the last category. IPBXs are a dramatic departure from the conventional approach to voice communications, as they are Local Area Network (LAN) based and, therefore, share a data-oriented infrastructure. IPBXs support packet-based Voice over Internet Protocol (VoIP) rather than TDM-based voice.

Contemporary conventional PBXs are Electronic Common Control (ECC), *Stored Program Control* (SPC), digital computer systems. They commonly support the switching of data calls as well as voice calls. PBXs vary from 10 stations to 10,000 or more stations, or extensions, with the average falling in the range of 200 stations.

3.2.1 PBX Components

A contemporary conventional PBX is a specialized computer system that includes a circuit-switching matrix for the primary purpose of connecting voice calls, although

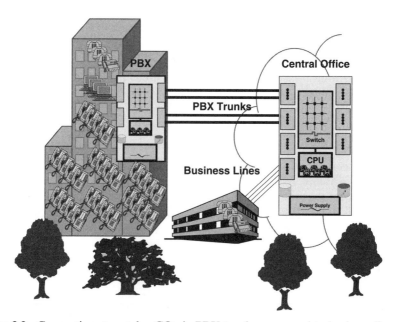

Figure 3.2 Connections to a telco CO via PBX trunks compared to business lines to individual telephone sets.

TABLE 3.1   PBX Generations

Generation	Designation[a]	Nature of Technology
0	Cordboard	Manual switchboard
1	PBX (Private Branch eXchange)	Electromechanical Step-by-Step (SxS)
2	PABX (Private Automatic Branch eXchange)	Electromagnetic Crossbar (XBar) or Crossreed
3	EPABX (Electronic Private Automatic Branch eXchange)	Electronic Common Control (ECC) Analog or digital Stored Program Control (SPC)
4	IP PBX or IPBX (Internet Protocol Private Branch eXchange or Intranet Private Branch eXchange)	Digital SPC LAN-based Voice over Internet Protocol (VoIP)

[a]The terms PBX, PABX, and EPABX often are used interchangeably.

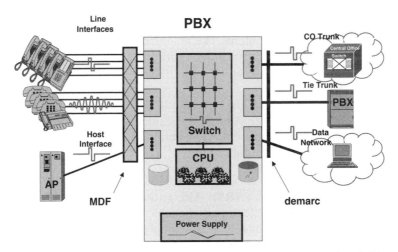

Figure 3.3 PBX components, including power supply, common control, switching matrix, trunk interfaces, line interfaces, and terminal equipment.

many also handle a limited amount of low-speed data. PBXs serve large numbers of station sets, which may include fully featured electronic terminals. PBXs primarily serve voice terminals on a wired basis, although wireless appliqués are commonly available as an option. Otherwise, PBXs are much like other computers, consisting of cabinets, shelves, printed circuit boards, power supplies, and so on. The primary physical and logical components of a PBX include power supply, common control, memory, switching matrix, trunk interfaces, line interfaces, and terminal equipment (Figure 3.3).

3.2.1.1 Common Control Common control is a common set of stored program logic that controls the activities of the system and all of its various elements. In reality, it consists of multiple microprocessors operating under a stored program, which the manufacturers characterize as being so secure as to be *bulletproof*. As in any computer, the control processor consists of memory, input/output equipment, bulk memory equipment, and software.

Central Processing Units (CPUs) are microprocessors that control the operation of the system. They may be centralized but generally are distributed among cabinets, shelves, or even cards (printed circuit boards) for purposes of effectiveness, efficiency, and survivability. The CPUs control functions such as call setup, call maintenance, call release, performance monitoring, system diagnostics, and storage of operational data for analysis and report presentation. *Processor redundancy* is a common characteristic of contemporary PBXs. Not only are the processing functions distributed among multiple microprocessors at the cabinet, shelf, and card levels, but also many PBXs feature *hot standby* processors, available instantaneously should a primary processor fail.

Because the systems largely are software controlled, it is relatively easy to upgrade them. As new software generics are released, associated functions and features become available in the form of programmable functions. These upgrades generally are distributed on either CD-ROM or magnetic tape. Manufacturers typically make fixes for *bugs* (i.e., program errors, also known as *undocumented features*) available for downloading over the public network. Notably, conventional PBXs are highly proprietary, closed computer systems. The manufacturers control all aspects of the systems, leaving the end-user organizations with little ability to affect their operational characteristics. The manufacturers also determine all features and functions, enhancements, and upgrades.

While there are some features that are unique to key systems (e.g., intercom and voice-over paging), PBX feature content is considerably more advanced than that of the typical key system. There are literally hundreds of programmable PBX functions and features, including the following:

- **Automatic Route Selection (ARS):** Also known as *Least Cost Routing* (LCR), this is an optional software feature that enables the user to program the system to route individual calls over the most appropriate selection of carrier and service offerings. Routing factors typically might include the nature of the call; the *Class of Service* (CoS), or level of privilege, of the user; the time of day (e.g., primetime and non-primetime); and the day of the year (e.g., weekdays, weekends, and holidays). In this fashion, executives and other high-priority users with no calling restrictions can connect on a priority basis, even if restricted by availability to a high-cost carrier and service offering. On the other hand, clerks and other low-priority users must queue for a low-cost facility; the system remembers their requests, rings their telephones to alert them when that facility is available, and automatically connects the previously requested call, much like a camp-on call. ARS is most meaningful if the telecom environment is liberalized or deregulated and there are multiple carriers and rate plans from which to choose in a highly competitive environment.

- **Call Pick-up Groups:** Enable users within the same workgroup to answer calls for one another. Authorized users can invoke this option by entering a code on the set keypad or by depressing a designated feature button on an electronic set.

Call Detail Record

Starting Time	Elapsed Time	Calling Party	Called Party	Access Code Dialed	Trunk Used	Date	Account Code
12:30	0:01	431	415-555-1212	9	0003	718	
12:35	0:01	421	360-555-1212	9	0002	718	4538
12:45	0:01	411	415-555-1212	9	0001	718	

Figure 3.4 Example of CDR outgoing call records.

- **Call Forward:** Enables a user to define an extension to which a call will be transferred in the event of a *busy* or *no-answer* condition. The target extension can be predefined or defined on an ad hoc basis.
- **Conference Calling:** Set up either by the user or by the console attendant, this makes use of a special *conference bridge* card in the system. Such a bridge typically supports as many as 3 or 6 parties to a call in a small systems or as many as 8 or 16 in a large system.
- **Call Detail Recording (CDR):** Also known as *Station Message Detail Recording* (SMDR), this information provides details on all incoming and outgoing calls (see Figure 3.4) so the system administrator can develop network usage reports. Information typically includes the originating and terminating station, outgoing and incoming trunk or trunk group, time of connection, time of termination (or elapsed time), access code dialed (e.g., 9 for an outside line or 8 for a private line), telephone number dialed, and account code or authorization code used.

The inclusion of an originating telephone number for inbound calls provides the most complete incoming call detail. This level of detail requires that the incoming PBX trunks support either *Automatic Number Identification* (ANI) or *Calling Line Identification* (CLID), services that carry additional charges from the service provider unless the trunks are ISDN. ANI requires a *trunk-side* connection, also known as *Feature Group D*, which makes the PBX trunk appear to the network as an internal *InterMachine Trunk* (IMT). When so equipped, the PBX trunk receives the same originating call information that is passed internally within the carrier networks for telephone company billing and tracking purposes. Significantly, the calling party cannot block ANI information. CLID, a feature of *Signaling System 7* (SS7), is designed to provide essentially the same information, although the caller can block its transmission. Also, it is possible to spoof the CLID information with the proper (or improper) equipment. CLID *spoofing* is quite easily accomplished in IP telephony, as the header in the call setup packet can be modified much as one might manipulate an e-mail address. The only advantage to spoofing, of course, is in deception for illicit purposes or perhaps preserving anonymity in certain sensitive situations.

Many contemporary PBXs contain adjunct software that allows the generation of a number of reports, including usage costs by cost center (e.g., station, account code, workgroup, department, and division). This critical feature assists in cost control and fraud control as well as permitting costs to bill to clients or projects. As such software generally is limited in its capability, most users prefer to acquire more substantial *telemanagement* software residing on an adjunct computer connected to a special PBX port. Such software typically includes a number of other modules, all of which share a common database for purposes of cost allocation, traffic analysis, equipment inventory management, wire and cable management, service order management, and numerous other critical administrative functions [5].

- **Automatic Call Distribution:** Software that serves to route incoming calls to the most available and appropriate agent. Incoming call centers make extensive use of such specialized software to enhance customer service. In a typical scenario, an *automated attendant* (i.e., front-end voice processor) answers the calls and provides callers with menu selections to guide the call through the system. The system then can queue calls by specialized agent group and route them to available agents. The routing of the call to an agent can be on the basis of next available, longest time since last call, least number of calls answered, or some other fairness routing algorithm. While a specially equipped and partitioned PBX often delivers such capability, the PBX also can be specially configured as a stand-alone ACD. Intensive call center applications typically make use of specialized ACDs that function as highly intelligent switches for the processing of incoming calls.

- **Uniform Call Distribution:** A standard feature of many PBXs; stand-alone units exist, as well. While a Uniform Call Distributor (UCD) serves many of the same functions as an ACD, it lacks the same level of intelligence. Therefore, it is much less capable, although much less costly. A UCD simply routes incoming calls to the next available agent, scanning agent ports in a predetermined, rigid hunt pattern. The unfortunate agent who is first in the hunt pattern definitely earns his or her paycheck, while the lucky agent who is last in the hunt pattern has a very easy time of it.

- **Power Failure Transfer:** Also known as *power failure cut-through*, this is an emergency mode of operation that allows the user organization to maintain some minimal level of access to the outside world in the event of a commercial power failure. This feature allows an analog CO trunk to cut through the PBX switch to a predetermined analog emergency phone in a location such as the receptionist's desk, the security office, or the data center. There may be many such lines if a large PBX serves many users. As analog CO lines and trunks are powered from the CO and as the telco generally has several levels of emergency power backup in place, they are not highly susceptible to commercial power failures. This feature is particularly important if the user organization has no power backup such as an *Uninterruptible Power Supply* (UPS). Large organizations that consider telecommunications to be mission critical generally have UPS systems in place that may include not only battery backup but also backup diesel power generators that serve to recharge the batteries during long outages. Many PBXs offer optional internal battery backup that is sufficient for 1–8h of operation.

- **Centralized Attendant Service (CAS):** Allows the consolidation of PBX console operators for a number of PBXs distributed across a campus, Metropolitan Area Network (MAN), or Wide Area Network (WAN). Thereby, a single group of operators can perform all attendant functions, including placing calls, answering and extending calls, and establishing conference calls.

- **Uniform Dialing Plan (UDP):** Also known as *coordinated dialing plan*, this is PBX software that supports simplified station-to-station dialing in a multisite PBX environment. UDP allows multiple PBXs to share a single numbering plan of four or five digits. The end user need only dial those digits to be connected to another user, regardless of where the two are located. If the two are served by the same PBX, the call connection process is simple. If the two are served by separate PBXs, the call will be sent over a private facility such as a tie trunk if one is available. If the call is sent over a CO trunk for connection through the PSTN, the originating PBX will automatically insert the necessary country code, area code, and CO prefix or otherwise modify the dialed digits as necessary. UDP most commonly is implemented in multisite enterprises over private, leased-line networks connecting PBXs of the same origin (i.e., manufacturer) and software generic. However, it is possible to establish this level of networking between PBXs of disparate origin if they are equipped with a common signaling protocol such as QSIG.

- **Q SIGnaling (QSIG):** A standard that defines services and signaling protocols for interconnecting Time Division Multiplexing (TDM)–based PBXs based on the International Telecommunations Union—Telecommunication Standardization Sector (ITU-T) ISDN standard Q.931. It is a Common Channel Signaling (CCS) protocol that runs over the ISDN D-channel for signaling between nodes in a *Private Integrated Services Network* (PISN). QSIG supports call setup, call teardown, and transparency of features such as message waiting, camp-on, and callback. The International Organization for Standardization (ISO) has adopted QSIG as the *Private Signaling System No. 1* (PSS1).

- **Multitenant Service:** Allows a single PBX to serve multiple tenants or user organizations. Logical software partitions allow each tenant to have its own attendant console, trunk groups, and blocks of telephone numbers. Thereby, each organization has a unique identity through a logical PBX that is part of a single physical PBX.

- **Property Management Interface (PMI):** Intended for the hospitality industry, which includes hotels and motels, dormitories, cruise ships, hospitals, and extended-care facilities such as nursing homes and retirement homes. PMI allows the coordination of front-office and back-office management features with PBX communications functions. For example, a housekeeper can use the telephone in the guest room to indicate room status, such as *clean and ready for occupancy*. Front-desk personnel can disable a guest telephone when a room is vacant or if the guest's credit card has been rejected. Charges for calls through the PBX can be associated with the guest account and a single, all-inclusive bill can be rendered.

- **Simplified Message Desk Interface (SMDI):** Defines a way for a phone system to communicate with a voice messaging system. The PBX sends an SMDI message in advance of each call, advising the voice messaging system of the line that it is using, the type of call that it is forwarding (e.g., call forward, call forward busy, or call forward no answer), and the source (calling number) and destination (called number) of the call. SMDI allows the voice messaging system to process the forwarded call more intelligently by accessing the correct personalized greeting associated with the correct mailbox. If the phone is so equipped, the voice mail system can turn on a *message waiting light*.

- **Automatic Set Relocation:** An administrative feature that simplifies Move, Add, and Change (MAC) activity by allowing the end user to accomplish set relocations without technical assistance. The user simply takes the phone from one location to another, plugs the set into the wall jack, and dials a relocation code. The set identifies itself to the PBX, which changes the station port assignment and reassociates the station number and all assigned features to the new port. Some systems support automatic set relocation across multiple networked PBXs with a Uniform Dialing Plan (UDP), as defined above.

3.2.1.2 Switching Matrix
Contemporary circuit-switched PBXs employ Pulse Code Modulation (PCM) and Time Division Multiplexing (TDM), concepts that are explored briefly in Chapter 1 and are discussed in great detail in Chapter 6. Through these processes, multiple analog conversations are sampled, converted to a digital format, and sent sequentially over a shared electrical *bus*, or common physical path.

3.2.1.3 Trunk and Line Interfaces
Trunk Interfaces are in the physical form of specialized circuit boards that serve to interface the PBX switch to trunks connecting it to other switches. Trunks are directional in nature and can be one-way outgoing, one-way incoming, or two-way (*combination*). Often a PBX will employ all three variations in order to serve various specific applications, to maximize system performance, and to ensure a minimum acceptable level of both incoming and outgoing network access. ISDN offers dynamic call direction on a channel-by-channel basis, which is much more flexible and potentially more cost-effective. I discuss ISDN in detail in Chapter 7.

Trunks generally are multichannel in nature, although single-channel trunks are employed in applications such as power failure transfer and *Foreign eXchange* (FX) service. As high-capacity, multichannel trunks will support multiple conversations, they generally are much more cost effective. Examples of multichannel trunk facilities include T-carrier, E-carrier, and ISDN Primary Rate Interface (PRI), also known as Primary Rate Access (PRA). Trunks of the same directional type and that serve the same purpose are organized into *trunk groups*. The PBX will *hunt* for available channels within a trunk and for trunks within a trunk group, based on a predetermined, user-definable hunt sequence. Trunk interfaces provide access to the following specific types of trunks:

- **Central Office Trunks:** Connect the PBX to the local CO exchange. These trunks serve for access to the local calling area and to all other areas served by the local telco, or Local Exchange Carrier (LEC). They also serve to provide switched access through the LEC to IntereXchange Carriers (IXCs) for long-distance calling.
- **Interexchange Trunks:** Provide direct access to an IXC, bypassing the LEC CO switch. Such trunks generally are intended only for long-distance network access to geographic areas outside the LEC local calling area. In the event of a failure in an interexchange trunk facility, CO trunks can be used for switched access, thereby providing a very effective level of redundancy.

- **Foreign Exchange (FX or FEX) Trunks:** Connect directly to a foreign CO. They are used for more cost-effective access to and from a distant geographic area where a high volume of traffic originates and terminates. A user organization with a high volume of traffic between its Dallas, Texas, offices and the area immediately surrounding Denton, Texas, for example, might choose to avoid long-distance calling charges by installing an FX trunk. Because the FX trunk is priced on a flat-rate, milage-sensitive (it is about 35 miles between the city centers) basis, the user organization avoids long-distance charges to that area. Callers from the Denton area dial a local Denton telephone number associated with that FX trunk, similarly avoiding long-distance charges. As a side benefit, the host organization in Dallas creates the illusion of having established a local presence in Denton. If the level of calling activity is not sufficient to justify the cost of a multichannel trunk, one or more single-channel FX trunks might serve the need at lower cost. FX lines are used with key systems in small- to medium-size system environments.

- **Direct Inward Dial (DID) Trunks:** Intended for incoming traffic only. Each station is assigned a DID number that roughly corresponds to the internal station number. As a call is placed to a seven-digit DID number, the CO recognizes that fact and connects the call over a special DID trunk. The DID number is passed to the PBX in advance of the call by virtue of a special signaling and control arrangement that exists between the PBX and CO. With that information, an intelligent PBX with the proper generic software load can automatically route the call directly to the station, without the intervention of an attendant. The service provider rents DID numbers to user organizations in groups or blocks of 50, 100, or 250, for example. (*Note:* DID calls now can be connected over combination trunks as well if appropriate arrangements are made with the carrier.)

- **Tie Trunks:** Directly interconnect, or tie together, PBXs in a private network configuration. Through the use of ARS software, the systems will route calls between offices over leased-line tie trunks, avoiding toll charges in the process. Coordinated, abbreviated dialing plans programmed within each PBX cause the various PBX systems to interact on a networked basis, appearing to the user as though they were a single system. Tie trunks and other dedicated circuits are sometimes referred to as *nailed-up circuits*, because they literally once were nailed up on the CO wall and specially tagged in order to distinguish them from other circuits. If a key system *sits behind* (i.e., is used in conjunction with) a PBX, tie lines are used to interconnect them. Tie lines also are used to interconnect key systems.

- **Wide Area Telecommunications Service (WATS) Trunks:** Allow for discounted bulk long-distance access. Special-purpose WATS trunks are highly unusual in a contemporary context, having been replaced by discounted long-distance billing plans that are independent of trunk facilities.

- **INward WATS (INWATS) Trunks:** Serve incoming WATS calls. The called party bears the charges at a discounted cost per minute or fraction thereof. Callers originally accessed domestic U.S. INWATS numbers by dialing 1-800-*nnn-xxxx*. When the 800 area code was exhausted some years ago, 888, 877, and 866 numbers gradually were phased in. Outside the United States, the INWATS is known variously as *Freephone* or *Greenphone*, and the dialing patterns differ greatly (e.g., 020, 0500, 00800, 0800, and 900). INWATS often is characterized as *toll free*, which is correct only as far as the calling party is concerned. Dedicated INWATS trunks are rare in a contemporary context. Rather, the INWATS calls are pointed to an incoming or combination trunk or trunk group, which must be properly sized to accommodate the additional traffic. *Dialed Number Identification Service* (DNIS) is a feature that operates much like DID over toll-free lines by passing the dialed digits to the destination PBX in advance of connecting the call. Thereby, the PBX can determine which of perhaps many toll-free numbers was dialed and can route the call accordingly, perhaps to a particular agent group in a call center application. DNIS is heavily used in call centers that service multiple clients or client groups, each of which has a unique toll-free number.

- **ISDN Trunks:** Support *dynamic bandwidth allocation*, also known as *bonding* and *N × 64*. This feature allows the ISDN-compatible PBX to dynamically allocate, or bond, multiple contiguous 64-kbps channels to serve an application that requires more than a narrowband channel. As an example, a videoconference might require 128 kbps (2 channels) or 384 kbps (6 channels). ISDN *Primary Rate Interface* (PRI) trunks are the ISDN equivalent of T1 trunks, providing 24 channels at 64 kbps for an aggregate of 1.536 Mbps, or E-1 trunks, providing 30 channels at 64 kbps for an aggregate of 1.920 Mbps. Dial-up Internet access might benefit from bonding 2 channels for a connection at 128 kbps. (*Note:* Implementing the feature in a PBX is problematic. It is more commonly implemented in a device such as a data router.) Chapter 7 discusses T1, E-1, and ISDN in detail.

- **Direct Inward System Access (DISA) Trunks:** Allow remote access to the PBX system, generally on a toll-free basis. After the PBX answers the call, the caller can gain access to connected resources (e.g., e-mail servers, voice processors, computer systems, and outgoing toll trunks) through the entry of authorization codes. While DISA trunks are useful to authorized users, hackers often targeted them in order to gain unauthorized access to the same resources. Therefore, the integrity of the organization, its toll network, its various networked systems, and the resident databases can be compromised. You should avoid DISA trunks, and, where in place, you should disconnect them and turn off the feature in the PBX unless you secure them properly through techniques such as frequent password changes and the use of intrusion detection systems. Some telemanagement systems include anomaly detection features that will alert a system administrator to unauthorized usage; some will actively disable a DISA port if there are indications that security has been compromised.

- **Analog Trunks:** Necessary to support analog devices. Although more sophisticated fax machines and fax servers are designed with digital line and trunk interfaces, the most common fax machines are simpler and less expensive and have only analog line interfaces. Such machines require an analog line port. If the PBX uses a proprietary digital encoding mechanism rather than the standard PCM, it also is necessary that the analog fax line port connect to an analog trunk port and an analog trunk. Power failure cut-through requires analog trunks and ports as the feature requires line-powered analog phones.

3.2.1.4 Station Interfaces Station interfaces are in the form of printed circuit boards that can support multiple stations of the same general type through multiple *ports* on a single interface card. Analog line cards with analog ports, for example, support analog voice sets, in which case a codec at the line card level digitizes the signal. Digital line cards support digital telephones, in which case a codec embedded in the station set digitizes the analog signal. Digital line cards also directly support computer workstations, printers, and other digital devices. Line cards typically support 4, 8, 16, or 32 ports.

An *Off-Premises eXtension* (OPX) or *Off-Premises Station* (OPS) is a station extension that terminates on a distant telephone, often located on a property that is not contiguous with that on which the PBX is located. An OPX requires a conditioned private line (also known as a burglar alarm circuit), leased from the LEC. OPXs are, and always were, uncommon due to their relatively high cost but sometimes are used to connect a distant guard shack or security building to the PBX or perhaps to provide a technically challenged key executive with a PBX extension at home.

3.2.1.5 Terminal Equipment As PBXs are designed primarily to support voice traffic, the user terminal equipment generally is in the form of a telephone set. However, terminal equipment often includes data terminals as well. The following items are examples of terminal equipment:

- *Telephone sets* can be generic or proprietary in nature. PBXs typically support both dial and touchtone single-line generic sets as well as highly functional electronic sets that offer easy access to a number of system features. Although such sets are proprietary to the system manufacturer and are considerably more expensive, they enable the station user to invoke system features by depressing a single designated button. Programmable sets enable either the station user or the system administrator to program the individual feature buttons, or *softkeys*, which are context sensitive. The proprietary electronic sets generally are digital, with the codec embedded in the set rather than in the station interface card in the PBX cabinet. ISDN sets provide a highly functional ISDN connection to the desktop through an ISDN-compatible PBX connected to an ISDN circuit. ISDN sets generally are proprietary to the PBX manufacturer.
- *Attendant consoles* provide attendants with the ability to answer and extend incoming calls, provide operator assistance to outgoing callers, and establish conference calls. They also often provide alarm indications in the event of a performance failure. When associated with a computer workstation, they often provide electronic directory information and other functionality through an enhanced user interface. In a large, private network scenario, attendants and attendant consoles often are centralized to enhance attendant performance and minimize associated costs. *Centralized Attendant Service* (CAS) is a feature offered by manufacturers of large, network-capable PBX systems.
- *Maintenance and Administration Terminals* (MATs) are PCs connected to the system. The MATs may be connected directly to a maintenance port on the system via an RS-232 connection, although the contemporary method of connection generally is over an Ethernet LAN. In either case, authorized users can

access the system software for purposes that might include *moves, adds,* and *changes, class-of-service* changes, ARS programming, requests for traffic and usage statistics, requests for status reports, and diagnostic testing and analysis. Remote maintenance generally can be accomplished over the PSTN via a modem connection.

3.2.2 System Configuration and Capacity

You must consider carefully the capacity of the PBX in order to ensure that the organization does not outgrow the system over a reasonable period of time. The system initially should be configured and installed to serve near-term as well as current requirements, as subsequent software and hardware additions generally are considerably more costly when installed later. Additionally, the system should have the ability to grow to meet long-term needs with respect to hardware, software, feature content, and so on. The following sections describe the capacity and configuration dimensions of PBXs.

3.2.2.1 Centralized or Distributed PBXs generally are configured as centralized systems, with all cabinets and peripheral devices collocated for ease of administration. An enterprise spread over a large campus environment may opt for a decentralized approach, with multiple intelligent cabinets located in campus quadrants and interconnected with high-capacity circuits. Very large organizations with multiple, widely separated locations may require the capability to network multiple PBXs over a private tie-line network through tie trunks. Alternatively, a *Virtual Private Network* (VPN) can serve the same purpose. A VPN provides much of the functionality of a true private network, but without the requirement for dedicated leased tie trunks. (Classic voice VPNs are discussed in Chapter 5.)

3.2.2.2 System Capacity and Engineering Capacity should be considered along several dimensions: physical and traffic. *Physical capacity* is the measure of the number of lines and trunks that can be supported through additional line and trunk cards and additional cabinets. Every system has a finite or practical limitation to the number of cabinets, shelves, circuit boards, and, ultimately, ports that it will support.

Traffic capacity is the measure of the number of simultaneous conversations that can be supported. This measurement is critical, especially when both voice and data are switched. You must consider the capacities of both the processors and the buses. Processor capacity is measured in terms of the maximum number of *Busy-Hour Call Attempts* (BHCAs) and *Busy-Hour Call Completions* (BHCCs) the system can support. The *busy hour* is the busiest hour of the day, ideally as determined by empirical traffic studies. The capacity of the switch matrix, that is, the bus capacity, usually is evaluated in terms of *Centum Call Seconds* (CCS) capacity, with a *centum call second* being 100 call seconds. One hour contains 3600 call seconds (60s times 60min), or 36 CCS. Traffic engineers use this information to design a proper system, which is to say that they engage in a process of application engineering designed to strike a desirable balance between system cost and performance. Traffic engineering is beyond the scope of this book, but there are a great many good books on the subject. That said, we will do some simple traffic engineering.

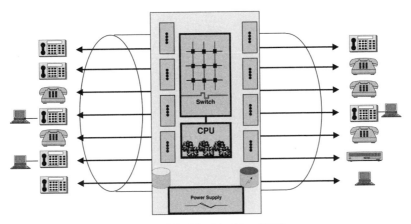

Figure 3.5 Nonblocking PBX.

3.2.2.2.1 Nonblocking Switch   A TDM switch matrix supports a number of connections, or completed circuits. A completed circuit supports a single call over a
single channel. A call that is active for an hour occupies a circuit for 3600 call
seconds, or 36 CCS. Consider a PBX with 200 stations, for example. If all stations
are engaged in station-to-station calls at a given moment, there are 100 calls in
progress. If this continues for an hour, the switch matrix must support 3600 CCS.
A *nonblocking* switch provides a nonblocked (guaranteed) talk path for every
terminal; in other words, there exists a 1:1 (one-to-one) relationship between line
ports and time slots (*Note*: Remember that there are two time slots—one in each
direction—for each conversation.) In this scenario, the PBX is functioning as an
extreme voice intercom system. The scenario is more legitimate if data communication as well as voice communication (Figure 3.5) is supported through the PBX, as
data communication sessions tend to be much longer than voice conversations,
which last 3–4 min on average. Notably, however, traditional PBXs are TDM-based
circuit switches and, as such, are inherently inefficient for many data communication
applications. Just to complicate the scenario a bit, we could provide access to trunks
on a nonblocking basis, but this would make the scenario even less realistic. In days
of yore, PBXs sometimes were configured as nonblocking trunk-to-trunk switches
by long-distance resellers, but that is a highly unusual application scenario.

3.2.2.2.2 Blocking Switch   A much more realistic scenario involves engineering
PBX system capacity to optimize the relationship between cost and availability.
Voice-only systems particularly lend themselves to optimization, as the characteris-
tics of voice traffic are well understood:

- Calls are occasional.
- Not everyone is on the phone at the same time.
- Calls last 3–4 min or so on average.
- Some calls are incoming, some are outgoing, and some are station to station.
- Some people like to listen to themselves talk, but others have better things to do.
- Not all calls are equal.

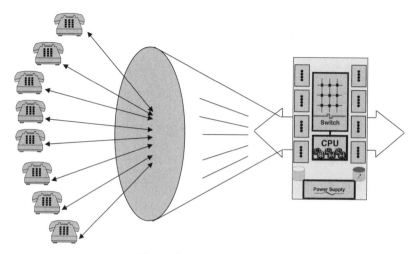

Figure 3.6   Blocking PBX.

Blocking switches (Figure 3.6) are engineered in such a way as to support a reasonable level of traffic, with some call attempts intentionally blocked during the busiest hour of the day, which is a worst-case scenario. This approach involves the concept of *Grade of Service* (GoS), which is the probability of blockage during the busy hour, expressed as a decimal fraction of the calls anticipated to be blocked. The primary technique for designing the system involves the Poisson distribution, which considers not only the traffic load during the busy hour but also the call setup time, including the time required for the user to dial the call and the time the system takes to set the call up. Once the total CCS load is determined, the desired GoS is established. A *P*.05 GoS, for example, means that there is a probability (*P*) that 5 percent (.05) of the offered calls will be denied, and the user will either receive no dial tone or a get busy signal, on the first attempt during the busy hour of the day. A *P*.01 means that there is a probability (*P*) that 1 percent (.01) of the offered calls will be denied on the first attempt during the busy hour of the day. In either case, the likelihood is that those calls will be accepted and processed successfully on the second attempt. However, the difference in circuit cost can be quite significant.

Table 3.2 is an excerpted Poisson distribution table that clearly illustrates the costs savings associated with the number of circuits required to support various traffic levels of traffic in consideration of various GoS levels. If, for example, the desired GoS is *P*.05 and the traffic study yielded a *Busy-Hour Centum Call Seconds* (BHCCS) load of 245, 12 circuits are required. (*Note*: the BHCCS figure of 245 must be increased to the next highest value in the table, which actually yields a slightly better GoS. Reducing the BHCCS to the next lowest value would yield a slightly worse GoS.) If the desired GoS is *P*.01, 15 circuits are required. If a virtually non-blocking GoS level of *P*.001 is desired, 17 circuits are required.

So, the assumption is that the general user population will be willing to accept some relatively minor level of inconvenience during the busiest hour of the day in return for the considerable associated cost savings. (Actually, the assumption is that *management* is willing to *impose* some relatively minor level of inconvenience *on* the general user population during the busiest hour of the day in return for the

TABLE 3.2 Excerpted Poisson Distribution

Number of Circuits Required	Grade of Service				
	P.001	*P*.01	*P*.02	*P*.05	*P*.10
1	0.01	0.04	0.7	1.9	3.8
2	1.6	5.4	7.9	12.9	19.1
3	6.9	16	20	29.4	39.6
4	15	30	37	49	63
5	27	46	56	71	88
6	40	64	76	94	113
7	55	84	97	118	140
8	71	105	119	143	168
9	88	126	142	169	195
10	107	149	166	195	224
11	126	172	191	222	**253**
12	145	195	216	**249**	282
13	166	220	241	277	311
14	187	244	**267**	305	341
15	208	**269**	293	333	370
16	231	294	320	362	401
17	**253**	320	347	390	431
18	276	346	374	419	462
19	299	373	401	448	492
20	323	399	429	477	523
21	346	426	458	507	554
22	370	453	486	536	585
23	395	480	514	566	616
24	419	507	542	596	647

considerable cost savings.) During the balance of the day, the GoS is expected to be much better, of course. The challenge is to strike the most reasonable balance between cost and performance during the hour of the day with the heaviest calling activity. In practice, a *P*.05 GoS generally is considered to be quite acceptable for many PBX environments, while COs more typically are engineered for a GoS of *P*.01.

Importantly, Poisson distribution is a probability theory that expresses the probability of events occurring in a fixed period of time if each event is independent of the previous event. In traffic engineering, that means that each call is completely independent of any previous call. [*Note*: Siméon-Denis Poisson (1781–1840) discovered the distribution, which he published together with his probability theory in 1838 in his work *Recherches sur la probabilité des jugements en matières criminelles et matière civile*, translated as *Research on the Probability of Judgments in Criminal and Civil Matters*.] There are a number of other techniques based on other assumptions underlying other formulas that yield different results. The various Erlang models are based on formulas developed by A. K. Erlang (1878–1929), a Danish mathematician and traffic engineer for the Copenhagen Telephone Company. These formulas calculate GoS based on *Busy-Hour Traffic* (BHT) expressed in hours of traffic, or *erlangs*, presented to circuits:

- *Erlang B* assumes that an offered call is cleared immediately, with no queuing. In other words, it assumes that the call encountering blockage will not appear again. Either the caller will hang up and not attempt the call again or the call will automatically be routed over another circuit if one exists, even if the use of that circuit is more expensive.
- *Extended Erlang B* (EEB) also assumes that an offered call is cleared immediately, with no queuing. However, extended Erlang B assumes that the caller encountering blockage (e.g., busy signal or no dial tone) will hang up and immediately attempt the call again. There is no overflowing of calls to more expensive routes. EEB was developed by Jim Jewitt and Jaqueline Shrago of Telco Research.
- *Erlang C* assumes that calls encountering blockage are queued indefinitely, that is, until a circuit is available. Only circuits in the primary circuit group are considered, with no overflow to more expensive circuits. Erlang C commonly is used to engineer circuit requirements for ACDs in incoming call centers.
- *Equivalent Queue Extended Erlang B* (EQEEB) assumes that calls are queued, but only for a predetermined period of time. If a circuit in the primary group does not become available during that time, either the call is routed over a more expensive circuit or the caller is given the option of trying to place the call again at a later time. A percentage of callers retry their calls until they are successfully completed. Developed by Jim Jewitt of Telco Research, EQEEB was used in incoming call centers in the days when circuits were very expensive and poor GoS levels (e.g., *P*.10) were acceptable [6–8].

Traffic engineering is extremely important, as poor capacity planning can lead to a complete system replacement, which can be very expensive. The term *forklift upgrade* comes to mind. (*Note*: In the days of *heavy metal*, a complete replacement meant driving a forklift into the switchroom, lifting the old system out and bringing in another, bigger one. In the 1980s, a typical 200-line PBX with voice mail was about the size and weight of a pizza oven. A contemporary system—with a lot more functionality—is closer to the size and weight of a couple of family-size pizza boxes filled with double-cheese pizzas with all the toppings. Put anchovies on my pizza, please.)

3.2.2.3 Application Processors Application Processors (APs) often interface with PBXs over high-speed links in order to provide specialized functions such as voice processing, computer host access, e-mail, and facsimile. Generally speaking, the PBX manufacturer supports interfaces either to proprietary APs or to those developed by third parties with which there is a strong, strategic relationship. While this approach is limiting, it ensures that the systems and the interfaces between them function successfully and remain so as the PBX system and the APs evolve. The ability of the PBX to support such processors is critical, as they can greatly enhance the performance of the organization.

3.2.3 PBX Enhancements and Trends

Manufacturers increasingly have positioned the PBX as a communications server for voice, data, video, and image communications. The PBX group at Lucent

Technologies (since spun off to form Avaya) developed the first truly integrated multimedia PBX in October 1995. The MultiMedia Communications eXchange (MMCX) Server supported all customary PBX telephone features on multimedia calls comprising any combination of voice, data, image, and video and supported interfaces to Ethernet, ATM, and ISDN networks. The more typical approach, however, involves high-speed interfaces to a variety of APs. These specialized processors, often developed by strategic partners, perform their various functions to maximum advantage on the most effective platform, while providing application integration through the PBX. The PBX developments and trends discussed in the following sections exclude computer telephony, which I discuss at the end of this chapter.

3.2.3.1 *PBX-to-Host/LAN*   Such links enable users to access computer applications through the PBX, sharing the same cable and wire plant between computer and telephone terminals. Connectivity between the PBX and a host computer or LAN can be accomplished via means such as ISDN or Ethernet links. While such capabilities have existed for a number of years, they never gained any real market share based on the traditional PBX model. Generally speaking, voice communications remained the domain of the PBX and data communications remained the domain of the LAN. In fact, voice has begun to shift to the LAN domain, as I discuss later in this chapter.

3.2.3.2 *Data Communications*   Data communications is increasingly important. PBX manufacturers have long supported modem pooling and low-speed data communications through terminal interfaces. Several manufacturers have supported X.25 packet-switching interfaces for a number of years. Manufacturers increasingly support high-speed connectivity to Ethernet LANs and ATM networks. Again, data communications generally proved too inefficient and expensive through a PBX, so this approach gained little real support over the years. Rather, voice remained switched through the PBX and data remained over the LAN. The shift now is in the other direction, with voice moving to the LAN domain, as I explain at the end of this chapter.

3.2.3.3 *Wireless*   Selected users currently can access wireless communications through a special PBX wireless appliqué. Special PBX ports support wired connections to antennas distributed throughout a large office complex or campus. Selected users who require mobility can take advantage of special cordless telephones to communicate as they roam through the organization, as long as they remain within the limited range of the antennas. Recently, several small manufacturers have released small PBX systems that are entirely wireless in nature; such systems have application in environments that allow the users to enjoy a high level of mobility within the office, campus, or factory. The advantages of wireless PBX technology include increased mobility, increased call completion, decreased call-back costs, and increased productivity due to decreased voice mail messages. Disadvantages include increased cost, lack of feature content at the set level, and concerns about security.

Also known as Wireless Office Telecommunications System (WOTS), wireless PBX technologies comprise various cordless telephony standards. The most recent emergent in this area is VoWi-Fi (Voice over Wireless Fidelity), which makes use of the 802.11 standard for wireless Ethernet in support of VoIP. I discuss these technologies in more detail in several subsequent chapters.

3.2.3.4 *Fax Messaging* Select manufacturers support fax messaging. Through voice processing system prompts, the user can download a fax message stored in a user-specific fax mailbox. The user can access the fax message onscreen at a workstation, much as he or she would access a voice message from a telephone terminal. This enhancement never achieved great popularity in the conventional TDM PBX domain as the demand for fax messaging waned with the advent of e-mail and the cost of the Application Programming Interface (API) was fairly significant.

3.2.3.5 *Asynchronous Transfer Mode* ATM was developed as an outgrowth of ISDN standards in the 1980s and was viewed as the switching solution for broadband ISDN. Over time, ATM became the backbone switching technology of choice in the PSTN and seemed positioned to make significant inroads into the LAN domain. There was a considerable movement in the early and mid-1990s to incorporate ATM into PBXs in various ways, with the ultimate intention of replacing the TDM-based switch fabric altogether with an ATM cell-based fabric. A number of PBX manufacturers incorporated ATM into large PBXs intended for enterprise applications. Avaya's Definity ECS (Enterprise Communications Server) Release 6 software generic, for example, supported an ATM interface card for integration of voice traffic over an ATM campus backbone running at speeds up to 5 Gbps, in support of integrated voice, data, video, and fax traffic. The Siemens Hicom 300E switch supported an integrated ATM InterWorking Unit (IWU), which enabled a connection either to ATM switches in a private network scenario or to public carrier-based ATM services. Nortel's Meridian Passport supported integrated access in support of voice, frame relay, ATM, and transparent data [e.g., native LAN and Synchronous Data Link Control (SDLC)] through a carrier-based ATM network. Numerous manufacturers of next-generation CT (Computer Telephony) PBXs interconnected application servers via ATM over fiber-optic links.

While it remained unusual to find a PBX switch that was fully ATM based, the likelihood was that most manufacturers would incorporate ATM switching matrices alongside the standard TDM-based STM (Synchronous Transfer Mode) matrixes used to support voice traffic and a PBX/ATM switch combination would be positioned as a multimedia communications controller. Despite ATM's technical elegance and considerable advantages, it proved to be too complex and expensive to support in the PBX domain. ATM proved to be an evolutionary dead end, so to speak, and was overwhelmed by the rather inelegant combination of Ethernet and IP to yield the IPBX.

3.2.3.6 *Internet Protocol* The IP is the foundation of the Transmission Control Protocol (TCP)/IP suite developed for use in packet-switched WANs such as the Internet. TCP/IP has found wide application in support of LAN traffic, as well. IP, specifically, has found recent application in next-generation, packet-switched voice WANs. While such nontraditional networks have not yet fully matured, they promise tremendous efficiencies and, therefore, much lower costs than do traditional circuit-switched networks. Further, IP telephony, or VoIP, promises to allow the integration of voice, data, and video over a single set of access and transport facilities—in other words, access through a single high-capacity local loop to a single packet-switched broadband WAN. In anticipation of the popularity of these next-generation WANs, a number of manufacturers of traditional PBXs have developed IP gateway interfaces to resolve the protocol differences between the TDM-based PBX and the packet-based IP network. When presented with an outgoing long-distance call request, these PBXs evaluate the *Quality of Service* (QoS) provided over the IP network at that moment and either present the call to that IP network or route it over a more expensive and more traditional circuit-switched alternative. The PBX automatically routes outgoing local calls over the circuit-switched PSTN, as there is no benefit—cost or otherwise—to be derived from routing them over a VoIP network.

Similarly, station-to-station voice traffic that remains internal to the PBX domain does not benefit directly from VoIP networking. There are, however, considerable merits to this approach. A single inside wire and cable plan for both voice and data clearly is advantageous. A single voice/data LAN-based server network can be accessed through common Ethernet switches. Such an integrated voice/data infrastructure certainly is an attractive concept. The drawbacks include the potential loss of voice quality associated with running IP-based, packetized, compressed voice over a collision-prone shared Ethernet LAN. Another drawback is the requisite prioritization of voice over a LAN and the resulting deprioritization of the data for which the LAN was originally installed. But the speed of contemporary Ethernet networks, routinely operating at a speed of 100 Mbps and currently running as fast as 1 Gbps and even 10 Gbps, combined with various prioritization mechanisms generally translates into little, if any, compromise in terms of QoS. Neither does it generally translate into any appreciable degradation of performance with respect to traditional data traffic over such a high-speed LAN infrastructure. The main items currently at issue in this area concern the reliability of LAN-based voice versus traditional circuit-switched PBXs and the complexity and cost of supporting voice over data networks. Customer support for IP PBXs is a significant issue, as many vendors do not have the requisite skills to support an integrated voice/data environment. I discuss the next generation of PBXs later in this chapter and considerable detail as part of the overall treatment of convergence in Chapter 14.

3.2.3.7 Security Security is of increasing concern, as hackers too easily can penetrate PBXs, voice processors, and other systems to gain unauthorized access to networks and various information resources such as critical databases. PBX, computer, and peripheral equipment manufacturers have added security measures through enhanced software and hardware that can act on either a passive or active basis to deny or terminate such access or, at least, to recognize its occurrence and alert the system administrator [9]. While such products have long been widely available at reasonable cost, the consensus is that far too few PBXs are properly protected. As PBXs move away from the traditional monolithic TDM-based circuit-switched technologies and toward LAN-based systems supporting IP-based packet-switched networking, security risks intensify, particularly as much VoIP traffic traverses the public Internet. Security is very much a *work in progress* and likely always will be.

3.3   CENTREX

Centrex (Central exchange), also known as *CO Centrex*, is something of a conceptual step back in time, providing PBX-like features through a special generic software load in the COE. Typically, each Centrex station is connected to the CO via an individual twisted-pair local loop (see Figure 3.7). While multichannel local loop connections are available, they require an expensive remote CO line shelf to be installed at the customer premises. (This approach literally involves placing a physical partition of the CO on the customer premises.) Therefore, multichannel access is cost effective only in applications that involve large numbers of users at a single site. All switching of calls is accomplished in the CO, which also serves all features to the user population.

Centrex first was made available in the United States and Canada in the early 1960s, through S × S COs. In the United States, AT&T and General Telephone and Electronics (GTE) deemphasized it in the 1970s, in favor of PBXs. This shift in direction largely was due to impending deregulation and divestiture, which took full effect on January 1, 1984. In particular, AT&T reportedly anticipated that it would spin off its local exchange operating companies but would retain its CPE business. In the grand scheme of things, therefore, it was decidedly to AT&T's advantage to deemphasize Centrex in favor of PBXs, and well in advance of divestiture. After all, Centrex requires considerable capital investment in COs, which AT&T would then have to give away to the operating company. PBXs, on the other hand, are either sold or leased to the end-user organization. The comparative merits of Centrex and PBX had nothing to do with the strategy.

Modern Centrex became available in 1984 through the first generation of digital COs [10]. AT&T immediately reversed its position and developed sophisticated

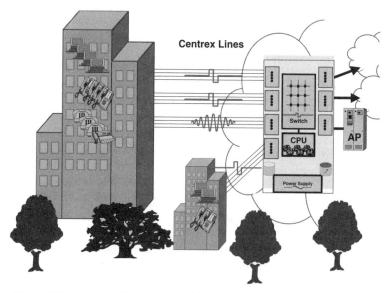

Figure 3.7   Centrex CO, with individual local loops to Centrex stations.

digital Centrex software for marketing to its former subsidiaries, the Regional Bell Operating Companies (RBOCs), and the independent telcos. Through its Western Electric subsidiary, AT&T manufactured and sold both COs to telcos and PBXs to end users and so had all options covered. In addition to Lucent Technologies (nee Western Electric), Nortel and Siemens are notable manufacturers of Centrex COs.

Although Centrex most commonly is provided by Incumbent Local Exchange Carriers (ILECs) such as the RBOCs, it more recently has been offered by Competitive Local Exchange Carriers (CLECs), many of which previously were either IXCs or Competitive Access Providers (CAPs). Traditionally a North American phenomenon, Centrex recently has found market acceptance in the United Kingdom, Europe, Japan, and much of Africa and Asia. Carriers find Centrex attractive because it effectively competes against PBX and KTS systems offered by third parties. Lucent Technologies (5ESS) and Nortel (DMS-100) manufactured the great majority of traditional TDM-based CO Centrex systems installed in the United States.

3.3.1 Features

Centrex is capable of offering the same feature content as PBX technology. Indeed, the platform inherently is much more substantial in terms of traffic capacity, processing power, memory, and virtually every respect. Ultimately, however, the carrier must consider the general market for such features, given an appropriate pricing strategy. Additionally, the impact of features must be considered in the context of limited switch resources such as processing power, memory, and traffic capacity. Finally, the regulators all too often get involved in determining pricing, availability, impact on the rate base, and other factors. Therefore, not all technically feasible Centrex features are available, certainly not on a consistent basis. Centrex features can include the following, all of which were defined previously:

- Direct inward dialing
- Automatic route selection or least cost routing
- Call pick-up groups
- Call forwarding
- Conference calling
- Automatic call distribution
- Call detail recording

ISDN Centrex also is capable of data switching, which more or less effectively delivers low-speed CO-based LAN functionality. Additionally, the Centrex CO generally houses a voice processing system or provides remote access to one, thereby delivering highly effective voice mail and other related functionality [10–12].

The Centrex CO also may offer access to a large number of CLASS (*Custom Local Access Signaling Services*) services, more commonly described as custom calling features. Such services include the following:

- *Calling number delivery*, also known as *calling line identification* (CLID) and *caller ID*, provides the called party with the telephone number of the calling party.

- *Calling number blocking* allows the caller to block the calling number delivery so that the calling telephone is not transmitted to the called party.
- *Calling name delivery*, or *caller name*, transmits the name of the calling party as it appears in the telephone directory. As this feature works in conjunction with CLID, it automatically is blocked if CLID is blocked.
- *Continuous redial*, also known as *repeat dial*, is a programmable feature that continuously redials a telephone number for a period of time, usually up to 30 min, or until the call is successfully completed.
- *Distinctive ringing*, aka, *custom ringing*, allows multiple telephone numbers to route to a single line and associated telephone(s), with a distinctive ring pattern identifying the called number.
- *Call return*, or *automatic callback*, allows the user to determine the originating number of the last missed or unanswered incoming call and to return that call without dialing the number. If the number is busy, Call return will determine when the line is available and ring both parties.
- *Call trace* automatically trades and records a caller's telephone number at the telephone company's security offices. Call trace is used to stop threatening, obscene, or harassing calls.
- *Selective call screening*, aka *security screening*, screens calls that would normally appear as "unavailable," "out-of-area," "unknown," "anonymous," "blocked," or "private" on Caller ID and instructs the callers to identify themselves either by unblocking CLID or entering their telephone number using the tonepad in order to complete their call.
- *Selective call forwarding* allows the user to program some number (e.g., 10 or 15) of originating telephone numbers, calls from which will be forwarded to a cell phone, home phone, or some other local or long-distance number. A distinctive ring identifies a forwarded call.
- *Selective call rejection* blocks calls from some number of programmable originating telephone numbers.
- *Anonymous call rejection* blocks calls from anonymous callers, that is, callers using calling number blocking.

Finally, Centrex offerings increasingly include *customer rearrangement* capability. This feature permits the user organization to manage its Centrex service much as it might manage its own PBX. Through a computer terminal located on the customer premises, access is afforded to the switch database on a logically partitioned basis. Thereby, the system administrator can accomplish MACs, disconnect Centrex stations, change CoS, file trouble reports and work orders and monitor their status, and even perform various system and network diagnostic tests.

3.3.2 Advantages

The carrier provides Centrex, which resides on a very robust computer platform that also is an integral part of the network. As a result, Centrex offers a number of unique advantages, including reduction of both capital investment and operating cost and avoidance of responsibility for system acquisition, maintenance, and management. Additionally, system capacity generally is not an issue, and the carrier often can network the systems to great advantage.

As the capital investment in the switching platform and associated application processors is the responsibility of the carrier, the user organization avoids the financial pain and associated risks of acquiring a potentially substantial capital asset in the form of a premises-based system. Generally speaking, however, the user organization remains responsible for the acquisition and deployment of associated station equipment and inside wire and cable systems. The user organization avoids a host of operating responsibilities and related costs, as those burdens are shifted to the carrier. Such responsibilities and costs can include those of system maintenance, system administration, system upgrade, insurance, physical space, physical and network security, power and backup power systems, fire suppression systems, and redundancy.

Because the risks of ownership are shifted to the carrier, issues of obsolescence and replacement are obviated. Further, COs are relatively unlimited in terms of port, memory, and processor capacities; therefore, it is unlikely that a typical user organization would outgrow the system. This characteristic of Centrex can be extremely important to businesses that tend to expand and contract, perhaps on a seasonal basis. Rather than investing in a PBX switch designed to accommodate the maximum size of the organization during the peak season, additional Centrex lines can be added as required and disconnected as that requirement relaxes, with the associated capacity and telephone numbers perhaps being reserved until such time as the requirement redevelops.

Finally, Centrex permits the user organization to network a number of locations through the serving CO, without the requirement for leasing tie-line circuits to interconnect multiple PBX or key systems. In fact, multiple Centrex COs of the same type can network to provide *virtual Centrex*. This approach offers the advantage of allowing multiple, distant offices to communicate as though they were collocated and served by a single switch.

3.3.3 Disadvantages

While Centrex offers a number of unique and very significant advantages over PBXs, there also exist clear disadvantages, including certain cost issues and considerations of control. The recurring charges for the circuits and features continue forever, although special tariffs for very large customers can serve to mitigate this factor. Further, a *Subscriber Line Charge* (SLC), also known as a *Customer Access Line Charge* (CALC), applies to each local loop circuit or defined group of loops. Because Centrex generally involves a separate voice-grade local loop for each Centrex station, the typical Centrex user pays a higher price for local access than does a PBX user organization. Finally, rate stability has always been a Centrex issue, as all local rates are subject to regular review and adjustment. In many jurisdictions, rate stability plans have served to mitigate this concern, at least in the short term. Rate changes affect PBX users to a lesser degree because relatively fewer circuits of relatively higher capacity are shared on a pooled basis.

As noted above, standard Centrex feature content is determined by the service provider based on market demand and in consideration of the resources available in the serving CO switches. The state regulatory agencies oversee the ILECs in terms of capital investment, operations expenses, and rate levels. Because the offered features are geared toward the average user, the overall feature set of the typical Centrex offering tends to be much less complete than that of the typical PBX. Therefore, user organizations with requirements for more exotic features may find Centrex an unacceptable alternative. Additionally, feature content and feature access may well vary by carrier, regulatory jurisdiction, exchange CO, switch manufacturer, and generic software load. Features such as automatic call distribution are unusual, while certain features such as message waiting may not be presented well. Still other features, such as voice-over paging, are not available at all. As a result, some users still place small PBXs, hybrid KTSs, and other CPE systems on premises to interface the station equipment to the Centrex lines, thereby extending more exotic features to select workgroups.

Perhaps the most significant disadvantages of Centrex revolve around issues of control, as the carriers determine the nature of the offerings, with participation by the regulators where Centrex is a regulated service offering. Special features may require lengthy tariff approval processes, assuming that the user organization can successfully negotiate the support of the carrier and assuming that the regulators are sympathetic. Certainly, the user is dependent on the carrier for switch administration, maintenance, and enhancement. Considering that many ILECs have a less than stunning reputation for responsiveness, this absolute level of dependency is of great concern to many companies.

3.3.4 Customer Premises Equipment

Proprietary voice terminals (*P-phones*) are required for ease of feature access, although most systems support generic sets as well. As P-phones are switch specific, a multilocation company can face additional inventory costs associated with the requirement to stock multiple set types. Confusion also results when users must deal with multiple user interfaces.

The inherent cost and flexibility limitations of Centrex have given rise to a new breed of Centrex CPE that significantly enhances Centrex capabilities. Although the equipment must be matched carefully to the Centrex switch, such CPE includes Centrex-compatible electronic voice terminals (i.e., telephones), ACDs, voice processors, paging systems, attendant consoles, and message lamp controllers. A number of manufacturers have developed *set handlers*, also known as *mediation devices*, that act as protocol converters to effect proper interfaces between high-performance third-party electronic station sets and the Centrex system.

3.3.5 Applications

Centrex service providers have responded to heightened competitive pressure by deploying more attractive features on a more consistent basis, extending limited network management capabilities to the end user, designing rate plans to attract smaller businesses, and generally becoming more responsive. Typical applications include multilocation organizations within a single metropolitan area, seasonal businesses, and businesses that tend to locate only on a temporary basis (e.g., construction companies and contractors). Additionally, government agencies, educational institutions, and not-for-profit organizations are heavy users of Centrex due to capital and expense budget limitations and restrictions.

3.3.6 Trends and Futures

Centrex will increase in popularity in most developed nations for reasons that include increased competitive pressure, enhanced feature content, improved pricing, and enhanced networking capability. Additionally, Centrex manufacturers and providers increasingly are applications focused, developing and offering capabilities that include enhanced data communications, videoconferencing, and integrated messaging. *IP Centrex* systems, also known as *virtual Centrex* and *hosted PBX*, were developed in the 2000–2001 time frame and deployed by a select few CLECs. I discuss IP Centrex toward the end of this chapter.

3.4 AUTOMATIC CALL DISTRIBUTORS

ACDs essentially are highly sophisticated PBXs designed specifically to switch incoming calls in call center applications. As call centers are highly active, with relatively large numbers of callers queued for a much smaller number of agents, ACDs generally are nonblocking systems. Call centers may be specific to the user organization but often are set up on a service bureau basis. A service bureau might answer calls for a large number of clients, on either a primary or overflow basis, with software and scripting specific to the individual client's requirements. The system identifies the target client by the telephone number called, with a separate telephone number or set of telephone numbers associated with a specific client, a special promotion, a specific service offering, or a special subset of each client's customers through *Dialed Number Identification Service* (DNIS), a service offered by the carriers and generally associated with toll-free numbers (800, 888, 877, and 866 in the United States and 0800 and 0500 in much of the rest of the world). The DNIS information is passed to the ACD in advance of the call.

The process, as depicted in Figure 3.8, typically involves a front-end voice processing system that prompts the caller through a menu for call routing purposes. Based on factors such as the originating telephone number, the specific number dialed, the menu option selected, and other information input by the caller (e.g., account number and password), the system then can route a call to an appropriate agent group, queue it if an agent is not available, and deliver it to the first available qualified agent.

Multiple call centers can be networked, with calls routed to the most appropriate call center closest to the caller, in consideration of network costs. In the event that the closest call center's queue length exceeds user-definable parameters, the call is then served to the next nearest call center with a queue of acceptable length. Recent developments allow the first call center to examine the queue lengths of all call centers in the network, determine each call center's ability to handle a call in consideration of service-level parameters, and forward the call to the call center most likely to satisfy those objectives based on *look-ahead routing* logic. This approach reduces unnecessary levels of congestion and unacceptable numbers of unhappy customers.

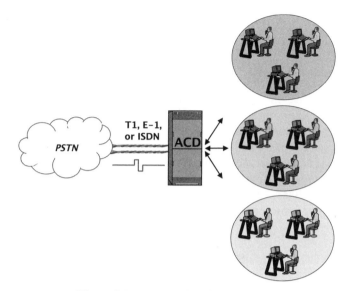

Figure 3.8   Automatic call distributor.

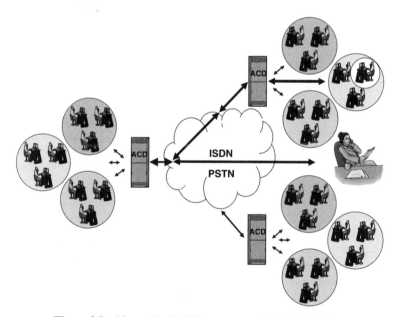

Figure 3.9   Networked ACDs connected via ISDN links.

A more recent, much improved alternative allows multiple call centers to logically connect as portrayed in Figure 3.9. The first appropriate available agent in any of a number of call centers can be identified and reserved via SS7 technology, which is fundamental to ISDN. IP-based ACDs now accomplish all of this over the Internet or other IP-based networks.

ACDs employ various call routing techniques, including keypad input, voice input via *Interactive Voice Response* (IVR), calling number, or called number. Regardless of the technique, the process of customer-programmable call handling and routing

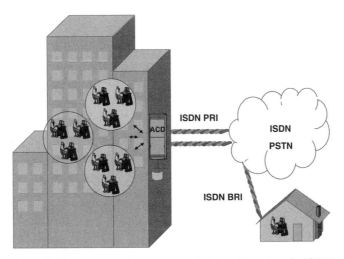

Figure 3.10   Agent-at-home, connected to call center via ISDN.

is known variously by terms such as *call vectoring* and *custom control routing*. Once a call center accepts the call, the identification of the caller can be determined through several means, including a *Personal Identification Number* (PIN), *Calling Line Identification* (CLID), or *Automatic Number Identification* (ANI). The identification number then can be matched against a computer database in order to establish the caller's profile. At that point, the profiles of the available agents are analyzed in order to identify those most capable and available to handle the call. Once an agent is selected, the caller's profile is presented to the agent in advance of the connection of the call in what is known as a *screen pop*. In this manner, the most available and capable agent has access to full account information and, therefore, can provide the highest possible level of service. That agent might have special language skills or a positive historical relationship with the caller. This capability is known as *skills-based routing*. Should the agent need to transfer the call to another agent or to a supervisor, the screen pop typically travels with the call, thereby providing each successive party with the same information to assist in processing the call in the most effective manner.

In addition to agents working at formal call centers, remote agents and even at-home agents can network to an ACD, often over ISDN connections. ISDN technology enables the remote agent to engage in a voice conversation with the caller, while simultaneously maintaining a data session with the centralized host computer and its associated databases (Figure 3.10). Service bureau call centers in the United States even make use of inmates in correctional facilities (e.g., federal prisons)—security concerns aside, inmates apparently make extremely good employees, as they have relatively few distractions and seldom call in sick. The ACD commonly interfaces to a PBX, so calls can be transferred to other departments [13].

3.4.1 Benefits

Benefits of ACD technology include increased productivity, because the system serves incoming calls to groups of agents from a queue of waiting callers. As soon as an agent concludes one call, another is available to be connected from the queue. Additionally, the incoming callers realize enhanced customer service because they are directed to the call center with the shortest queue and held in queue until a properly skilled agent is available. While the holding time can be long during periods of heavy calling load, this approach is preferable to encountering busy conditions that require repeated attempts to reach an agent. In fact, holding time can be used for advertising, promotional, or service-related messages. Should calls be queued from both local trunks and toll-free long-distance trunks, the ACD can place the toll-free callers at the head of the queue in order to minimize long-distance costs to the user organization. ACDs provide rather substantial workforce management reporting software that enables the system administrator to view and trend calling patterns, average queue lengths, average holding times, and so on. Through the use of such a tool, management can better anticipate future incoming call loads and, therefore, can more accurately schedule the number of agents required to handle that load satisfactorily, in consideration of customer service objectives.

Customer service is enhanced further when the ACD supports skills-based routing. A credit card company, for example, might capture the originating telephone number of an incoming caller or request that the caller enter his or her account number. Through matching that information to a database of customers, the company might determine that a particular caller holds a platinum card, prefers to deal with a Mandarin-speaking agent, and has failed to make a payment in the past two billing cycles. Based on that information, the ACD might direct the call to the closest call center with a readily available, multilingual agent capable of initiating collection action involving such a privileged customer [14–16].

Applications for ACDs in call center environments include reservation centers such as hotels, auto rental agencies, and airlines. Financial institutions such as stock brokerages, commodities traders, and banks make heavy use of ACDs, as do ticket agencies. Call centers are used extensively for customer service and technical support.

3.4.2 Enhancements and Trends

ACDs no longer are intended only for large, isolated incoming call center applications. In response to general organizational trends toward downsizing, ACDs now are highly cost effective in relatively small configurations. In fact, a number of manufacturers offer highly scalable ACD systems based on a client/server computing concept. Organizations increasingly network call centers. Further, they increasingly take advantage of agents-at-home, remoting access to the ACD and associated databases via various devices generically known as *call extenders*. Such devices support integrated voice and data communications between the ACD and the agent-at-home and can operate over ISDN links, analog links, or even ADSL or cable modem networks through the Internet. Skills-based routing often is employed to direct the incoming caller to the most capable and available agent, whether situated in the call center or working from home [17].

Call-back messaging enables an incoming caller to register his or her desire to be called back by an agent should he or she grow weary of waiting in the queue. Once the queue has been satisfied and an agent becomes available, the system will call the customer back and connect him or her to an agent automatically.

Call blending permits call centers to improve their cost effectiveness by serving both incoming and outgoing calling functions. Through the use of a *predictive dialer*, the system will monitor the status of incoming calling activity and the level of availability of the agent pool. The system will introduce outgoing calls when it determines that the level of incoming calling activity has dropped to the point that the quality of the service to incoming callers will not be affected adversely. The predictive dialer will search a database of customers to be called, statistically predict the availability of an agent, dial the associated telephone number, detect when the call is answered, and connect the call to an available agent. (You certainly have experience with a predictive dialer. If you answer the telephone at home and experience a noticeable pause before the caller comes on the line, a predictive dialer likely called you. When the system detected your answer, it searched for an available agent, presented the agent with a screen pop, and then connected the call.) If the predictive dialer senses an answering machine or voice processor at the target telephone number, it can simply hang up and dial the next number in the database. Predictive dialer software logic may be contained within the ACD or it may reside on a LAN-attached server with a network connection to the ACD [18]. Call blending and predictive dialing have combined to revolutionize the call center, reportedly doubling the average agent's productive talk time to as much as 54 min per hour [19, 20]. There also is a decidedly negative side to call blending.

Telemarketing is the use of the telephone to solicit prospective customers to sell products or services and dates at least to the 1970s, when AT&T promoted long-distance telemarketing under the name *Phone Power* to promote WATS. As predictive dialers and other technologies developed and calling costs dropped, telemarketing reached new heights for marketers—and new lows for consumers—in the 1980s and 1990s. It seemed as though it was impossible to sit down for a meal without experiencing a constant barrage of telemarketing calls. Under pressure from consumers and consumer groups, the U.S. Congress passed the *Telephone Consumer Protection Act* (TCPA) in 1991. The TCPA specifically mentions automatic telephone dialing systems, or autodialers, and prerecorded messages and includes rather substantial penalties for telemarketers found guilty of violating the privacy of residential, fax, and certain other categories of users. In 2003, the FCC and Federal Trade Commission (FTC) finally addressed the problem in a proactive way with the *National Do-Not-Call Registry*. A number of state governments already had such registries in place. The Do-Not-Call Registry was extended to cellular phones in late 2004. The Controlling the Assault of Non-Solicited Pornography And Marketing (CAN-SPAM) Act was passed in 2003, extending the ban on certain unsolicited marketing to e-mail. In 2005, that ban was further extended to text messaging directed toward wireless phones and other mobile devices.

Speech recognition software has dramatically improved the ease of customer access to the modern call center. Callers accessing the contemporary call center can work their way through menu options via voice, rather than a touchtone keypad, commands. This approach, which supports multiple languages, is especially useful when the caller is using a cellular telephone, as the distraction of dialing while driving is generally recognized as a contributing factor in many automobile collisions. It also is highly valued as a user interface by many callers who are visually and physically impaired. Many sighted people with nimble fingers who do not happen to be driving during the call find speech recognition to be an incredible annoyance, especially when no other option is available, myself included. It is akin to being in voice mail jail with a dull, dim-witted, and overly talkative cellmate.

Networked ACDs have become increasingly common since 2000 for several reasons, including outsourcing, offshoring, and network costs. There has been a general trend towards *outsourcing*, that is, replacing full-time employees with contract employees, for some years. Most recently, much of the outsourcing emphasis has been on *offshoring*, that is, replacing domestic workers with foreign workers. These trends are evident across a wide range of industries and job functions. At the professional and technical levels, a great number of computer programming, engineering, and accounting positions have moved offshore from the United States. The trend is highly visible in the call center industry, including sales and technical support positions. The trends are evident in Europe as well, although to a lesser degree due to government restrictions. The reasons for these trends include cost reduction, liability avoidance, and improved quality, at least theoretically. Much of the shift has been to countries such as India, Pakistan, and South Africa, where there are large numbers of highly educated, highly motivated, low-wage workers with good (especially English) language skills. Traditionally, limited bandwidth and high network costs kept these jobs onshore. The tremendous growth in the Internet, coupled with the trend toward liberalization and deregulation of the telecommunications industry, led to a glut of bandwidth worldwide. So, the network cost of routing a customer service call from Milwaukee to Mumbai became insignificant, while the disparity in wages for a call center worker remained highly significant. While opinions differ dramatically, some suggest that the quality of customer service improved as a result of the shift. Closer to home, McDonalds, the world's largest restaurant chain, recently began a limited outsourcing trial involving the use of remote call centers for taking drive-through orders. "If you're in L.A. and you hear a person ... with a North Dakota accent taking your order, you'll know what we're up to," McDonald's Chief Executive Officer Jim Skinner said during a presentation to analysts in New York in March 2005 [21]. In my humble opinion, the next phase of the trial may involve a call center agent with excellent American English skills but with a slight Pakistani (or Indian or perhaps Filipino) accent.

The World Wide Web–enabled call center represents an exciting ACD development. Customers accessing a company's website through the Internet increasingly have the option of clicking a button to request that an agent call back, with the website perhaps indicating the approximate length of time until that return call can be launched. The customer can receive the call while remaining connected to the website if a second line is available. Therefore, both the customer and the call center agent can view the same information onscreen during the ensuing telephone call. In the not-too-distant future, the caller will be able to connect directly to an agent during the data session established with the website. Assuming that the customer has a *softphone* (i.e., voice-enabled PC) and a multifunction voice/data modem [e.g., Digital Subscriber Line (DSL) modem] and the call center is properly enabled, a VoIP conversation can be established and maintained while both parties simultaneously view the same data presented on the website. Multimedia queuing logic will allow incoming calls to be queued, regardless of how they call the center, but prioritized based on the method. Toll-free callers, for example, might be placed in a priority queue, as those calls cost the call center sponsor; VoIP calls over the Internet might be placed in a queue of lower priority based on the assumption that those calls are free to both parties. I discuss this recent development in more detail in the following section.

3.5 COMPUTER TELEPHONY

The discussion up to this point has been of traditional voice communications systems. Such systems can be characterized as circuit-switched, TDM based, and highly proprietary. The underlying technologies are very mature and the systems built on them are highly reliable. Since the mid-1990s, a great deal of effort has been expended on the development of the next generation of voice systems. Such systems take several forms but generally can be characterized as being computer based, programmable, and client/server in nature. They also, generally speaking, support IP-based packet voice and data over an Ethernet LAN.

Computer Telephony (CT), a term coined by Howard Bubb of Dialogic Corporation (subsequently acquired by Intel) and a concept later championed by my old friend Harry Newton, can be characterized as a 3.5G technology, as it was an intermediate step in the evolution from the third generation of digital ECC systems to the fourth generation of IP-based systems. CT can be defined as the blending of telecommunications switching with computer processing power and programmable logic. Computers have long been used to program and manage PBXs, hybrids, and COEs, which themselves are specialized computers. Yet the system interfaces were proprietary and tightly linked, with one computer associated with one set of databases and programs residing on one switching system, one voice processor, and so on. The system manufacturers ruled this world of monolithic *heavy metal*. During this period, users received only the features the manufacturers deemed worthy and in a presentation mode they felt appropriate.

More recently, it became clear that a single computer system could control and integrate multiple telephony devices, taking advantage of the programmed (and programmable) PC's intelligence and the real-time call processing power of the telephony switch. By way of example, a PBX, ACD, and voice processor could integrate to provide a more effective set of solutions for the processing of incoming calls; additionally, *call blending* could be achieved through the application of a *predictive dialer* to improve the productivity of the call center.

While users and third-party software developers have long coveted access to PBX, ACD, and COE switches, the manufacturers had no real incentive to provide it on an *open* basis. Claiming the risk of potential corruption of the switch databases, their real reason was more one of erosion of their very lucrative market for application software, proprietary station sets, and so on. Over time, they did enable software developers to access expensive and difficult Application Programming Interfaces (APIs) to accomplish these feats. More recently, and under pressure from end users, these manufacturers opened the systems to APIs developed and standardized by various industry groups. Those APIs became available at very low cost and were much easier to use. In fact, every major PBX vendor began shipping products that included one or more APIs.

At a minimum, the concept of CT involves the use of PCs to facilitate user access to PBX and Centrex switch features (e.g., call answer, call transfer, conference calling, call hold, and call hang-up), placing call control functions in the hands of

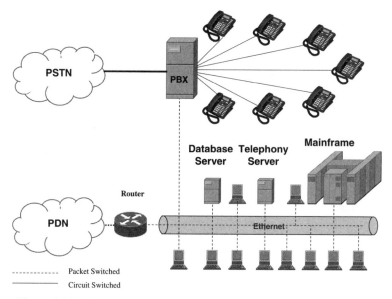

Figure 3.11 CT voice/data system based on client/server architecture.

the individual station users through a *Graphical User Interface* (GUI) on the workstation. Taken a step further, the concept involves stripping the switch of much of its intelligence and placing it in an adjunct computer system. The adjunct computer then effectively becomes the PBX or Centrex feature server and call control platform, while the *switch* is reduced to a high-performance switching matrix. Additionally, users gain access to complex information and can invoke complex features through a workstation. In other words, the PBX, ACD, or Centrex system is rendered *dumb*, or *dumber* (i.e., *programmable*). The adjunct computer contains all the generic feature content, which either a third party or the user organization can customize. Service providers write the applications in accordance with interface standards or specifications. Those applications then are uploaded to the switch through a CT link, which usually is in the form of a serial cable link. The PBX makes procedure calls to the adjunct computer controller, as required [22–24].

Taken further, a CT system can be based on a true client/server architecture (Figure 3.11). A single CT server supports internal voice and data switching, provides access to and from the voice and data WANs, and provides call control functions. Voice communication is provided over telephone sets, although *softphones* (i.e., multimedia voice-enabled PCs equipped with headsets or handsets) also can be used. The PC data terminals are connected to the server via an Ethernet LAN. In addition to supporting normal data communications between data terminals, mainframe and midrange hosts, servers, and peripherals, the LAN also supports communications between the PCs and the server for voice call control purposes.

The call flow depicted in Figure 3.12 illustrates the enhanced functionality CT delivers in call center applications. Not only can the call be transferred and conferenced through computer keyboard and mouse commands, but the system can match incoming call identification (calling number or PIN) to a caller profile residing in a computer database. Through a *screen pop*, the system presents the caller's profile to

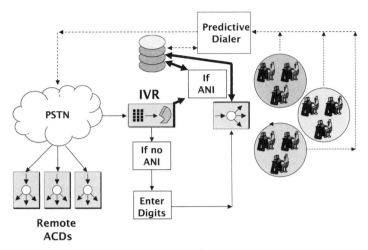

Figure 3.12   CT employed in conjunction with an ACD in a call center environment.

the agent in advance of the call, thereby providing the agent with sufficient information to personalize the contact and improving customer service as a result. Should the original agent need to transfer the call to another agent, the user profile and all other relevant data transfer along with the voice call. The earlier discussion of ACDs presented this application in greater detail.

3.5.1   Technology, Standards, and Specifications

The technology concept involves ECC switches, computer systems, and application software. The switches can be KTSs, hybrids, PBXs, ACDs, or Centrex COs. Third-party developers write the application software in accordance with *application programming interfaces*. Defined and supported by the equipment manufacturers, such interfaces permit a great deal of flexibility, incorporating user-definable call handling parameters. The computers can be in the form of mainframes, midranges, or PCs; LAN-based client/server configurations are commonplace. IBM's CallPath Systems Architecture (CSA), DEC's Computer Integrated Telephony (CIT), and Hewlett-Packard's Applied Computer Telephony (ACT) comprised the first generation of APIs. (*Note:* DEC and Hewlett-Packard both subsequently merged with Compaq.) Early implementations involved little more than switch-to-host request and status links. Subsequently, CT progressed to the workstation level. The specifics of both the internal buses and the links between the adjunct computer and the switch were many and various. Standards initiatives and specifications included the following:

- **Computer-Supported Telephony Applications (CSTA):** Developed by the European Computer Manufacturers Association (ECMA), CSTA was the first truly open CT development standard for link-level protocols. CSTA was improved and formally standardized on an international basis by the ITU-T, incorporating the U.S.-developed Switch-to-Computer Applications Interface (SCAI). CSTA is a full protocol stack that requires an open-system interface to a PBX, ACD, or Centrex COE.

- **Telephony Services Application Programming Interface (TSAPI):** Developed jointly by AT&T and Novell and released in March 1994, the TSAPI specification is strongly oriented toward PC platforms. TSAPI is a lesser protocol stack that does not require that the switch manufacturer fully open the switch interface. Although a number of PBX manufacturers supported TSAPI, the cost of TSAPI PBX drivers is quite high; as a result, its widespread use has been discouraged. Additionally, many suggest that it is limited in capability and cumbersome to use; TSAPI received criticism for being a CSTA translator for AT&T Definity switches and as a means of promoting AT&T's and Novell's interests. Release 2 (1995) extended the API to other PBXs and computer systems, including OS/2, Macintosh, UnixWare, and Windows NT [25, 26]. Newer versions of TASPI extended compliance with CSTA and supported a Windows NT or 2000 client and a TCP/IP network [27].
- **Telephony Application Programming Interface (TAPI):** Microsoft and Intel jointly developed this specification in response to the problems associated with TSAPI. As an integral part of Microsoft's *Windows Open Services Architecture* (WOSA), TAPI runs in a Microsoft Windows NT/2000 environment. Because it is Windows client/server based, TAPI involves limited additional cost to the user organization [25, 26, 28]. More recent versions also supported third-party call control.
- **Java Telephony Application Programming Interface (JTAPI):** JTAPI is a Java-based solution developed jointly by Sun Microsystems, Lucent Technologies, IBM, and Nortel. Similarly characteristic of Java, JTAPI is a cross-platform, multivendor solution that uses highly efficient applets, which are small sets of application program code, for network-based CT operation. JTAPI adds Internet/intranet functionality to CT, thereby enabling the creation of Web-based applications that integrate browser applications with call center functionality [27].

Each of these initiatives enjoyed some level of success at some point in time, but their overall complexity created implementation difficulties. Manufacturers largely have ceased to support these specifications, with the exception of TAPI.

3.5.2 Forums and Consortia

Several forums and consortia exist that actively promote open-system architectures and standard APIs for the development of CT technologies and applications. In addition to the ECMA, they include the Enterprise Computer Telephony Forum (ECTF) and the International Multimedia Teleconferencing Consortium (IMTC).

- **Enterprise Computer Telephony Forum:** Now under the umbrella of the Computing Technology Industry Association (CompTIA), ECTF promotes interoperability and standard approaches to CT. The ECTF has developed a highly sophisticated framework, or model, for both carrier-class and end-user-level systems.
- **International Multimedia Teleconferencing Consortium:** IMTC is a not-for-profit organization with the stated mission of promoting, encouraging, and

facilitating the development and implementation of interoperable multimedia teleconferencing and telecommunications solutions through open standards. IMTC focus is on the T.120 and H.320 standards suites for data conferencing and video telephony, respectively.

3.6 IP SYSTEMS

Sometime around the turn of the twenty-first century (i.e., 2000) and just as CT was gaining market acceptance, the term computer telephony slipped out of the popular technology vocabulary. About that time, VoIP reached adolescence and fast Ethernet had largely displaced its competition in the LAN domain. These factors converged and CT morphed into IP CPE. The IPBX was born. IP Centrex emerged soon afterward. New system installations increasingly are IP based and it now seems clear that VoIP will overwhelm the voice market, both on the premises and in the WAN domain.

An IP-based system is fundamentally quite different from one based on TDM. A conventional TDM system is monolithic in nature, much like a special-purpose mainframe computer associated with a circuit switch. Such a system is designed as a stand-alone computer with a proprietary operating system and closed APIs to interface adjunct computers for voice mail and other messaging applications. Trunk-side signaling and control protocols are SS7 and QSIG. As the intelligence and databases associated with a conventional TDM system resides in the centralized CPU, the telephones essentially are dumb voice terminals. The telephones connect to the centralized host via dedicated home runs of Unshielded Twisted-Pair (UTP) cable. An IP-based system distributes intelligence across one or more microprocessor-based *telephony server*s and databases across multiple *database servers*, and the telephones themselves are intelligent terminals that can be in the form of *hardphone*s (i.e., hardware-based telephones) or *softphones* (i.e., software-based telephones residing on desktop, laptop, tablet, hand-held, or other computer platforms). The operating system is commercial rather than proprietary and generally in the form of Microsoft Windows or LINUX with open APIs that encourage third-party software developers (note that some manufacturers offer open-source IPBX operating systems—the basic systems are free, but more feature-rich versions must be licensed) [26]. Trunk-side signaling and control protocols are *H.323* and *Session Initiation Protocol* (SIP), which I will explain in detail in subsequent chapters. The architecture is client/server in nature and the infrastructure is that of a high-speed switched Ethernet LAN that supports voice and data, with voice prioritized in consideration of QoS requirements. The shared infrastructure involves a common transmission system that can incorporate various combinations of UTP (Cat 5 or better), fiber optics, and RF-based wireless media. Table 3.3 provides a side-by-side comparison of TDM and IP CPE voice systems along these dimensions.

There is, of course, a huge installed base of TDM CPE in the form of KTS, PBX, and ACD equipment. There also is an enormous base of TDM COs, many of which support conventional TDM Centrex. Those systems will survive for many years as carriers and end-user organizations seek to maximize the return on the substantial embedded investment in TDM technology, which remains quite functional for voice communications. As TDM systems reach the end of their functional lives and begin

TABLE 3.3   Comparison of TDM and IP CPE Voice Systems

Description	TDM	IP
Architecture	Monolithic, centralized, host based	Manifold, distributed, client/server LAN
Voice terminals Form Intelligence	Hardphones, dumb	Hard, and Softphones intelligent
Operating system	Proprietary	Commercial
Line-side protocols	Proprietary, closed	Ethernet, TCP/IP, SIP; open
Trunk-side protocols	SS7, QSIG	H.323, SIP
APIs	Proprietary, closed	Proprietary, open
Switching technology	Circuit switched	Switched Ethernet (packet)
Orientation	Voice (data incidental)	Data and voice
QoS	Voice QoS guaranteed	Voice prioritized
Transmission system: configuration media	Star; dedicated home run UTP (Cat 3)	Bus, shared UTP ($\geq$Cat 5e), fiber optics, RF wireless

to experience failures, as manufacturers cease to support legacy systems, and as new IP-based features increase in number and luster, IP systems will replace them. In fact, that replacement cycle has already begun. The shift does not necessarily take the form of a *forklift upgrade*, however, as there are several variations on the theme of the IPBX. Many of the new generation of the CPE voice systems are primarily TDM switches with IP gateways and are generically known as IP enabled. This approach serves to extend the life of the embedded investment in station equipment, especially given the fact that not all users require the functionality of an expensive new and highly intelligent IP phone. Some implementations are pure IP, and still others are hybrids, supporting both TDM and IP.

3.6.1   IP-Enabled PBX

The simplest approach, in many respects, to incorporating IP into the PBX domain is a phased approach that involves coupling IP onto a TDM PBX platform. As illustrated in Figure 3.13, the intelligent IP phones can take the form of either feature-rich hardphones or softphones that connect either over a shared switched Ethernet LAN or over an embedded cable plant deployed for telephone sets. The LAN-attached data terminals interconnect as usual. The IP phones connect to the PBX through an Ethernet port on a line card that includes an IP gateway, which resolves the interface issues between the TDM bus and the Ethernet frames that contain VoIP packets. Calls between the LAN-attached IP phones are conducted on a peer-to-peer basis using their LAN addresses and are confined to the LAN. Calls between TDM phones also are on a peer-to-peer basis through the TDM switching matrix. Calls between a LAN-attached IP phone and a PBX-attached TDM phone go through the gateway, where protocol issues are resolved, including address translation between PBX extension numbers and Ethernet LAN addresses. The PBX is responsible for all voice call control in this scenario. (*Note:* The path used for signaling and control purposes may differ from the talk path.)

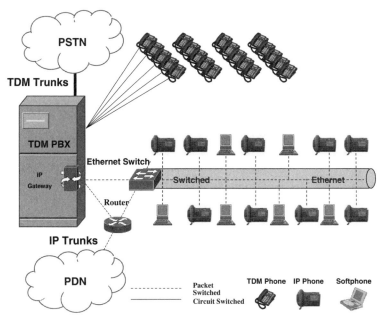

Figure 3.13   IP-enabled PBX.

Calls between the PBX and WAN can follow several scenarios. Calls originating from LAN-attached data terminals go through an Ethernet LAN switch and through a router to a Public Data Network (PDN) such as the Internet, as usual. Calls between LAN-attached IP phones and IP phones located elsewhere can follow the same general route. Calls originating from a PBX-attached voice terminal can go through an IP gateway, over an IP trunk, and to an IP PDN or can connect over a TDM trunk (e.g., T1 or E1) to the PSTN, as usual.

3.6.2   IP PBX

A pure IP PBX, or IPBX, steps completely away from TDM technology developed for voice communications. A pure IPBX is based on a distributed client/server architecture that generally is implemented on a switched Ethernet LAN running at 100 Mbps or better. Client software residing on intelligent IP hardphones and soft-phones runs against one or more servers that can be distributed across an enterprise, perhaps in geographically diverse locations. One or more *telephony servers* are responsible for all call control functions such as call setup and teardown.

The infrastructure can be shared between voice (and even video) and data applications. Therefore, there also typically are database and applications servers against which the client workstations run. The transmission media are shared between all LAN-attached devices. In a large-enterprise application, there commonly are multiple LAN switches. The switches serve to segment traffic geographically, confining traffic associated with a given call to only those segments (and associated geographic areas) necessarily involved in the communication and, thereby, mitigating issues of congestion that traditionally plagued early Ethernets. (*Note:* Many, if not most,

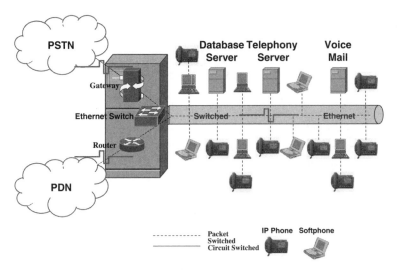

Figure 3.14   IP PBX.

contemporary IPBX manufacturers insist on fully switched Ethernets, with one segment per desktop.)

In a simple scenario involving only one segment and one switch, as illustrated in Figure 3.14, data calls are switched to a router and over a PDN, as usual. Pure VoIP calls can be switched and routed in the same manner. Calls originating on LAN-attached IP phones and destined for the PSTN must traverse a gateway router in order to resolve basic protocol issues such as formatting differences between IP packets, Ethernet frames, and T1 frames. In a larger enterprise, a single router and gateway can serve a great many users interconnected by many LAN switches. In an even larger and more complex multisite enterprise, geographically dispersed routers can be interconnected across an IP WAN through a Virtual Private Network (VPN). (Chapters 12 and 14 discuss IP VPNs in detail.) In such a networked IPBX scenario, a single telephony server can support all sites, with call control typically centralized at the site in proximity to the largest number of users. This approach eases the administrative burden and lowers infrastructure costs but increases network costs and creates issues of interdependence. In the event of congestion in the WAN or a catastrophic failure, the remote sites would not have the ability to function independently. A more resilient approach is to place a telephony server, or *media gateway*, at each site, perhaps synchronizing their databases so that any given server can assume call control responsibilities and each site can function autonomously in the event of a failure either of another server or in the network interconnecting them.

Note that an IP phone, whether in the form of a softphone or hardphone, essentially is a computer terminal with an Ethernet LAN address associated with a Network Interface Card (NIC) and an IP address assigned by the network administrator. Like any laptop or tablet PC, the IP phone is highly portable. In other words, it can be moved from room to room, building to building, and even site to site across national boundaries. Once plugged into an Ethernet port associated with an IPBX company and initialized with a proper password and user ID, the IP phone is active.

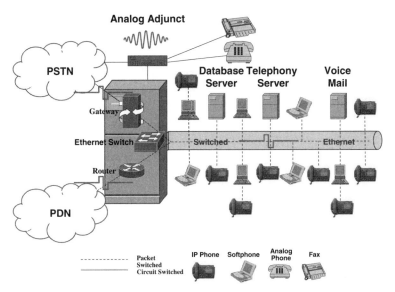

Figure 3.15   IP PBX with analog adjunct.

Further, the user can connect through the Internet from a public wireless hotspot in a coffee shop, hotel, airport, or other public place. If equipped with wireless capability, an IP softphone becomes not only portable but also mobile as the user can establish and simultaneously maintain both a VoIP and a data session while moving around, at least at pedestrian speeds, within range of the same wireless LAN Access Point (AP). This level of flexibility is not only convenient but also adds a significant measure of resiliency to VoIP networking. In a catastrophic failure scenario, if at least one site is operational and the IP phone can access it over the highly redundant Internet, the corporate voice network remains viable.

Speaking of system failure scenarios, conventional PBXs make use of power failure cut-through that connects analog trunks to analog phones via a copper talk path between analog trunk ports on special analog trunk cards and analog line ports on special analog line cards. IPBXs do not support such an arrangement, as a single Ethernet LAN switch port serves a number of IP phones and data terminals on an unchannelized basis. Neither do IPBXs necessarily support analog phones or analog fax machines. To resolve this issue, PBX manufacturers and third parties offer adjunct gateways, or subsystems, that interface analog phones to an IPBX. They also provide a connection between analog fax machines and analog trunks and support power failure cut-through, as illustrated in Figure 3.15. *Note*: There are standards for IP fax (Chapter 4), but implementation can be complex for an end-user organization and involves additional cost.

As pure IPBXs are based on client/server architecture and commercial operating systems residing on commercial hardware platforms, they naturally are highly flexible in terms of their physical and logical configurations. Call control, telephony applications, and associated databases can reside on a single server or multiple servers, with redundancy if so desired. The servers can be centralized (clustered in a single location) or distributed (one cluster per location) for enhanced performance and resiliency. The systems also can be highly scalable. Small systems, sometimes known as *media gateways*, are sized for *Small-Business Enterprise* (SBE) applications up to a dozen or so extensions or perhaps SME (*Small-to-Medium Enterprise*) applications up to 50 or so extensions. Media gateways also can be networked with a larger *media server, media manager*, or *communication manager* that serves a regional office of 400 or so extensions. The Avaya Communication Manager is an excellent example of scalability, as it is designed to support as few as 100 extensions or as many as 36,000 on a single system or as many as 1 million on a single network [27].

3.6.3 Hybrid TDM/IP PBX

A number of manufacturers offer hybrid PBXs, with TDM and IP components coexisting side by side. This approach blends the best of both. The TDM component comprises TDM line and trunk cards and ports and a TDM bus. The IP component comprises Ethernet ports, an Ethernet switch, a router, and IP trunk ports. A gateway interconnects the TDM and IP components, both of which are under the control of a telephony server running a commercial operating system, as illustrated in Figure 3.16.

The TDM component accommodates fax machines, modems, and power failure cut-through. The TDM component also supports digital and analog phones for areas such as lobbies and other public places, hospital rooms, dormitory rooms, and hotel rooms where the cost of IP phones cannot be justified and where the feature content of an IP phone would be lost on the untrained user [28]. As local T1 and E1 CO trunks are TDM trunks, they are assigned to the TDM ports on the hybrid IPBX. The IP component serves the more fully featured—and more expensive—IP phones and terminates the IP trunks used for access to the public Internet and other IP networks.

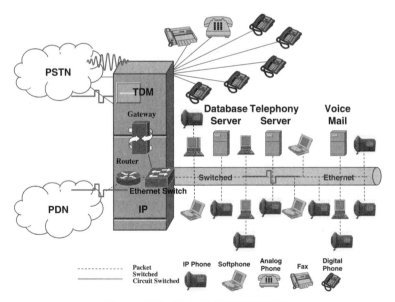

Figure 3.16 Hybrid TDM/IP PBX.

3.6.3.1 IPBX Features and Benefits Although still in its adolescence, this fourth generation of PBXs offers considerable feature content. While many of the features have been available on digital TDM PBXs for years, some are enhanced. Some significant features are absolutely unique to IPBXs:

- **Shared Infrastructure:** The IPBX inherently involves a shared transmission infrastructure, including transmission media, Ethernet switches, routers, and perhaps servers. This shared infrastructure reduces capital investment and lowers administrative and maintenance costs.

- **Scalability:** Many IPBXs are designed in a modular fashion. As memory, port, or traffic capacity is exhausted and additional capacity is required, another system module can be added at a fairly modest cost. So, a reasonably graceful relationship is achieved and maintained between performance and cost. Some systems are designed to grow in such a fashion from only a few hundred devices or users to many thousands. Generally speaking, the systems can grow in this fashion within a single site or spread across perhaps a great many sites in a distributed enterprise, with all sites networked across the Internet or other IP network.

- **Networking:** Internetworking multiple sites is simplified with IPBXs. Rather than the complex and expensive private leased-line tie trunks required to inter-connect monolithic TDM PBXs, low-cost IP access trunks into the Internet, or a carrier's IP network, simply and effectively network IPBXs. The client/server model distributes resources across a network, which can be a WAN just as well as a LAN.

- **Redundancy and Resiliency:** The Internet is inherently redundant and, there-fore, highly resilient. IP networks, in general, tend to share these attributes. VoIP call control can be centralized or distributed among multiple sites, with multiple backups and mirrored databases.

- **User Interface:** CT systems in general and IPBXs in specific offer improved user interfaces. Alphanumeric and softkey displays are characteristic of IP hardphones. Softphones are common and bring with them an inherently user-friendly, highly familiar, and even intuitive GUI that can be navigated easily by mouse and keyboard and touch screen.

- **Messaging:** Voicemail, electronic mail (e-mail), Instant Messaging (IM), and even fax mail are highly and easily accessible through a single user interface on a single softphone. Client software on a workstation running against match-ing software on a server allows a user to filter, organize, and manage e-mail by virtue of the header information such as originating address(es), destination address, subject line, date and timestamp, message length, and attachments, if any. Voice mail and fax mail can be managed just as easily, using similar header information. The user interface on an IP hardphone is not quite as robust but is still functional. The end result is *unified messaging*, which Chapter 4 addresses in some detail.

- **Presence:** Users can manage their own *presence*, or availability status, much as they do with instant messaging. In other words, a user can post or advertise his or her availability as online and available for e-mail or IM but unavailable for telephone calls, unavailable for IM or telephone calls, out to lunch, out of the office on business or vacation but returning on a certain date, and so on. Presence notification can be linked to a calendar program such as that contained in Microsoft Outlook.

- **Find-Me/Follow-Me Call Forwarding:** Users can establish their own rules for call forwarding, with distinctive options for friend and family, co-workers, and clients, for example. In the event of a *no-answer* condition, calls from one class of callers might go directly to voice mail and others perhaps to a cell phone or home phone. If an IPBX is based on the SIP signaling and control protocol, *forked ringing* allows a call processor to ring multiple phones at once, rather than in turn, and serve the call to the first phone that answers.

- **Multimedia Conferencing:** A user can organize a conference using a calendaring program, but with the added ability to determine which conferencing capabilities each participant has access to at a given moment. Thereby, the host can best match a multimedia presentation to the group, both as individuals and as a whole. Media options could include voice, video, text, graphics, and whiteboarding.

- **Voice/Data Integration:** As the infrastructure is shared between voice and data and as IP softphones are intelligent, multitasking, voice-enabled computer workstations, access to data during the course of a voice call is easily available. The data can be in the form of a caller profile or any combination of related or unrelated information. For example, using *Lightweight Directory Access Protocol* (LDAP), a user can search for a person and click on a telephone number to dial, then click an address to e-mail or instant message.

- **Moves, Adds, and Changes:** MAC activity is easily accomplished in an IPBX environment, just as in a data-only Ethernet. Moves, in particular, are often *plug 'n play*. Many other MACs can be performed by the system administrator, or even by the end user, through a password-protected intranet portal.

- **Portability and Mobility:** Softphones are highly portable, as users can move them anywhere within the company that an Ethernet port is available, plug them in, and be online virtually instantly. When equipped with wireless interfaces, a hard wired port is unnecessary and the softphones even become portable.

- **Remote Access:** Softphones are portable to the extent that they can easily connect to the IPBX over the Internet from home, a hotel room, a coffee shop, or anywhere they can gain Internet access. *Teleworkers*, or *telecommuters*, are particularly well supported by IPBX technology, assuming that they have broadband connections [e.g., Asymmetric Digital Subscriber Line (ADSL) or cable modem] and the hardware and software necessary to establish a VPN tunnel to the main office. Telework offers real advantages, including improved productivity as commuting time and associated stress are reduced or eliminated, as are traffic congestion and associated pollution. Telework results in greater employee satisfaction and, therefore, improved retention and provides the company with a larger pool of prospective employees at lower hourly wages and greater flexibility of scheduling to respond to peaks and valleys in calling traffic. Research from IDC indicates that there were about 112,000 outsourced home-based call center agents in the United States in 2005 and that those ranks will swell to over 328,000 by 2010, for a 24 percent annual growth rate [29]. Not every job is a candidate for telework, however, and some people just do not have the discipline to work at home. A recent study indicated that 1 in 8 male teleworkers and 1 in 14 female teleworkers say they do their jobs in the nude. Further, more than half the women do not shower on work-at-home days, as compared to about two-thirds of the men [30]. Personally, I find that disgusting as I sit at the computer writing this book in my bathrobe.

3.6.3.2 IPBX Issues While IPBXs offer a number of attractive features, including some that are unique to this fourth generation of PBXs, there are issues of some significance. Some of these issues are of the kind associated with any adolescent technology; others are integral to the platform. Issues include reliability, technical support, QoS, and security:

- **Reliability:** Conventional PBXs are highly stable platforms with bulletproof Operating Systems (OSs) and no direct Internet access for hackers. Although general-purpose microcomputer platforms are increasingly stable, they—and the applications they support—do tend to experience a much higher failure rate than conventional PBXs. Also, the IPBX OS and applications software have yet to prove themselves capable of measuring up to that of conventional PBXs. Carrier-class platforms used in IP Centrex systems certainly are more reliable but have yet to prove themselves at the level of conventional CO Centrex. Hardened, or fault-tolerant, computer platforms are essential if voice communication is considered time sensitive and mission critical. A fully redundant approach with systems (hardware and software) distributed across multiple sites is optimal. There are a variety of high-availability storage solutions for consideration with respect to IPBX databases, including database mirroring, replication, and failover and failback, all of which are beyond the scope of this book.
- **Technical Support:** The integration of voice and data naturally poses some support issues, particularly with respect to IP-enabled PBXs, which are a kluge of legacy voice systems and VoIP over high-speed switched Ethernet LANs. IPBXs certainly pose support challenges to in-house technicians as well as technical support personnel at the manufacturer and distributor levels. As an example, technicians can experience great difficulty attempting to isolate the cause of poor audio quality, which could lie in a gateway, transmission line, router, switch, or some other network element.
- **Quality of Service:** QoS is an important consideration for any type of communication, but especially so for voice. Noise and latency are always concerns, but VoIP increases latency due to compression and decompression processes, packet queuing, and packet serialization. Packetizing voice and routing it over a public Internet add to latency and jitter or variability of latency. The Internet and Ethernet LANs combine to raise issues of VoIP prioritization and data loss. Although these and other QoS parameters, issues, and resolutions are beyond the scope of this immediate discussion, they are covered in this book in subsequent discussions of LANs, the TCP/IP protocol suite, the Internet, and VoIP. At this point, however, it is worth mentioning that broadband local loops are a requirement for high-quality VoIP over the Internet and broadband LANs

are a similar requirement on the premises. Also, VoIP QoS generally benefits considerably from the separation of voice and data if voice is prioritized through the use of *Virtual LANs* (VLANs).

- **Security:** Security is a long-standing issue in the domain of conventional PBXs and voice mail systems, even though the PBX operating systems are considered virtually bulletproof. Security issues are quite significant in the domain of the Internet and systems that network through it. Therefore, IPBXs connecting through the Internet put the integrity of both voice and data systems at risk. Hackers, crackers, Trojan horses, Denial-of-Service (DoS) attacks, spyware, malware, and various other Internet plagues make encryption, authentication, firewalls, VPNs, and numerous other security measures as important for voice communications as they long have been for data.

3.6.4   IP Call Centers

In the IP domain, the term automatic call distributor has fallen out of favor. *Call centers* use IPBXs with the addition of ACD programmed logic in a highly functional blend of the two systems. IP call center systems offer all the features and benefits of an IPBX, including shared infrastructure, scalability, networking, redundancy and resiliency, and remote access. IP call center systems also perform all of the functions of a traditional ACD, with some notable improvements. Dozens or even hundreds of skill profiles can be defined not only for incoming and outgoing voice calls but also for e-mail and Web-based text chat exchanges in real time. The Web-enabled IP call center also supports form sharing and page sharing so that the customer and agent can look at the same information during the voice conversation.

The multimedia contact can be initiated by the end user by clicking a "Talk to an Agent" button on the website, thereby launching a VoIP call to a call center agent over the Internet. The call center agent's screen pop includes both the screen from the website and the caller's profile data. While the agent and the caller are engaged in the voice conversation, they simultaneously view the same website screens. The caller must have a suitably equipped multimedia PC and a voice/data modem, DSL, or cable modem connection and must be running the appropriate protocols in the TCP/IP suite. The yield is that the call center agent can handle the call in a much more productive fashion. A decidedly low-tech alternative involves the caller's clicking a "Have an Agent Call Me Back" button on the website. This approach makes use of the PSTN to launch the return voice call and requires that the caller have two lines available, one for the Internet data call and one for the return voice call, unless he or she has broadband access such as ADSL or cable modem access. While such exotic solutions are not yet widely available, they clearly are technically feasible.

3.6.5   IP Centrex

IP Centrex essentially is a network-based carrier-class IPBX to which the user organization connects as though it were a remote office. *Incumbent Local Exchange Carriers* (ILECs) such as AT&T, BellSouth, Qwest, and Verizon tend to favor the term *IP Centrex*, in keeping with their legacy telco image of stability and reliability. *Competitive Local Exchange Carriers* (CLECs) and competitive service providers, which essentially function as *Application Service Providers* (ASPs), generally prefer *hosted PBX* or *virtual PBX*, although the terms have fallen out of favor in the last few years. Much like traditional Centrex, the user organization avoids all the risks and costs of ownership of the main body of the system, although at the expense of control. Notably, however, some IP Centrex service providers provide a Web portal that allows end-user system administrators to perform MACs and allows individual users to access their own call logs and configure many preferences and options. Much like an IPBX, the user organization realizes advantages such as shared infrastructure, voice/data integration, unified messaging, presence, portability and mobility, and remote access, although there are concerns of reliability, technical support, QoS, and security surface. As IP Centrex is a server-based telephony platform, feature content is on parity with that of IPBXs, which is quite a departure from the considerable disparity in feature content between conventional Centrex and PBX systems [31, 32].

The physical positioning of the main body of the system, including call control, application software, and associated databases, depends on the service provider. An ILEC is likely to place the IP Centrex at the edge of the network, where it may be in the form of an IP-enabled CO or perhaps a true carrier-class hybrid TDM/IP platform, with a broadband local loop such as a DSL circuit or a T1 or E1 providing the connection to the network edge. An ASP is more likely to place a hybrid TDM/IP platform in some (perhaps distant) location central to its customer base and in proximity to its technical support staff. In this scenario, a broadband local loop connects the user sites to an Internet Service Provider (ISP), which provides access to the ASP through the Internet, as illustrated in Figure 3.17. Note that the broadband local loop will support a number of simultaneous VoIP and data sessions. The integration of voice and data at the local loop level consolidates circuit requirements and thereby reduces associated costs. This approach also compares most favorably with traditional Centrex, which required either a local loop per station or an expensive remote line shelf on the premises, with expensive T1 or E1 TDM local loops back to the Centrex host.

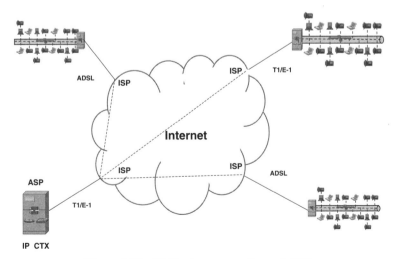

Figure 3.17   IP Centrex access through ISPs.

3.7 FUTURES

Voice communications systems traditionally fell into neat categories that gradually have become less meaningful. A KTS was generally considered to be an ideal solution for small businesses of 50 lines or less. PBXs were available in line sizes as small as 10–20 and as large as 1000 or so. At that line size, the PBX essentially was a premises-based CO that could grow to a line size of 10,000. CO Centrex was an option for those end-user organizations that wished to avoid the costs and responsibilities of ownership. Hybrid PBXs blurred the lines a bit when they arrived on the scene, but they really were just PBXs that could emulate key systems.

CT and IP not only have blurred the lines but also have essentially erased them. Manufacturers design systems for certain line sizes, of course, but many of them are modular and highly scalable. Indeed, some can grow from several hundred lines located at one site to tens of thousands of lines distributed across a great many sites networked through the public Internet. Further, IP-based systems are multimedia in nature, so they integrate voice, data, whiteboarding, and video. While it will be many years before the last TDM PBX connects its last call, there seems to be little doubt that TDM-based systems have a bleak future in the face of that sort of competition.

REFERENCES

1. Brooks, John. *Telephone: The First Hundred Years*. Harper & Row, 1976.
2. Casson, Herbert N. *The History of the Telephone*. A.C. McClurg & Co., 1910.
3. Costello, Rich and Lassman, Jay. "Key/Hybrid Systems and PBX Systems: Technology Overview." *IT Continuous Services*. Datapro Information Services, August 2, 2001.
4. Gasman, Lawrence. *Manager's Guide to the New Telecommunications Network*. Artech House, 1988.
5. Horak, Ray. "Telemanagement Systems and Software: Overview." *Datapro Communications Analyst*. Datapro Information Services Group, January 1995.
6. Parkinson, Richard. "Traffic Engineering Techniques in Telecommunications." http://www.tarrani.net/mike/docs/TrafficEngineering.pdf.
7. Freeman, Roger L. *Fundamentals of Telecommunications*, 2nd ed. Wiley, 2005.
8. Bates, Regis J. and Gregory, Donald W. *Voice & Data Communications Handbook*. McGraw-Hill, 2000.
9. Rohde, David. "Northern Telecom to Bring Security, Savings Back to PBXs." *Network World*, May 1, 1995.
10. Abrahams, John R. *Manager's Guide to CENTREX*. Artech House, 1988.
11. Horak, Ray. "PBX versus Centrex Comparison." *Datapro Communications Analyst*. Datapro Information Services Group, December 2001.
12. Costello, Richard A. "Centrex: Overview." *Datapro Communications Analyst*. Datapro Information Services Group, May 1996.
13. Doren, Donald van. "CTI: Moving into High Gear." *TeleProfessional*, February 1996.
14. Bodin, Madeline. "ACD Features at Work." *Call Center Magazine*, May 1995.
15. Klenke, Maggie. "ACDs Get Skills-Based Routing." *Business Communications Review*, July 1996.

16. Klenke, Maggie. "ACDs Get Skills-Based Routing." *TeleProfessional*, March 1996.

17. Lassman, Jay. "Automatic Call Distribution Systems: Technology Overview." *IT Continuous Services*. Datapro Information Services Group, April 4, 2001.

18. http://www.usatoday.com/tech/products/services/2005-03-11-mc-call-centers_x.htm.

19. Dawson, Keith. "Out of the Box, onto the Phones." *Call Center Magazine*, May 1995.

20. Dawson, Keith. *Call Center Savvy*. Telecom Books, 1999.

21. Dawson, Keith. *The Call Center Handbook*. CMP Books, 2001.

22. Krupinski, David, McConnell, Brian, and Schick, Charlie. *Next Generation Phone Systems*. CMP Books, 2001.

23. Grigonis, Richard. "PBX Survival Techniques." *Computer Telephony*, July 1995.

24. Lau, Delina, Xue, Cindy, Paetamai, Poonpon, and Ghani, Muhammad Usman. *Computer Telephony Integration*. University of San Francisco, December 4, 1995.

25. King, Rachael. "CTI Up Close: Analyzing the API Angle." *Data Communications*, October 1995.

26. Taylor, Kieran. "Distributed PBXs: Big Benefits, Little Boxes." *Data Communications*, October 1995.

27. Rosenberg, Arthur M. "Call Center Computer Telephony: Technology Overview." *Datapro IT Continuous Services*. Datapro Information Services Group, December 10, 1997.

28. Burton, James. "CTI: The View from Windows 95 and NT." *Business Communications Review*, January 1996.

29. Bednarz, Ann. "Call Centers Are Heading for Home." *Network World*, January 30, 2006.

30. Network World Staff. *Study: The Dirty, Naked Truth about Teleworkers*. http://www.networkworld.com/news/2006/031306-teleworkers.html.

31. Horak, Ray. "IP Centrex." January 7, 2004. http://www.commweb.com/17200585.

32. Galitzine, Greg. "Hybrid IP: the Best of Both Worlds." *Internet Telephony*, October 2005.

CHAPTER 4

MESSAGING SYSTEMS

I see no reason why intelligence may not be transmitted instantaneously by electricity.
Samuel Morse, 1832

Samuel Morse was absolutely correct, although there is a slight technical issue in the form of propagation delay. Also, we must assume that he meant information, rather than intelligence. That said, there is great value in communicating information instantaneously by sending messages. Also, and it would not have occurred to him at the time, there is great value in storing and later forwarding certain messages. Electronic messaging systems comprise facsimile, voice processing, electronic mail, instant messaging, and short message service. Each system deals with information in a different native form, and each serves its own unique purpose to great advantage. While these technologies differ greatly in their basic characteristics and certainly in their applications, they do share some commonalities. They all reside on computer platforms of various descriptions, and they all yield the greatest benefits when widely networked. Perhaps most importantly, all of these electronic messaging systems are characterized by their abilities to support *store-and-forward* communications. In other words, a system can accept a message and store it in temporary memory in a mailbox from which it can be delivered or retrieved at a later time. The user often can access the mailbox remotely and may be able to forward the message to another user, perhaps with annotations. Several other of these technologies support instant, or more correctly near-real-time, message communications.

Telecommunications and Data Communications Handbook, By Ray Horak
Copyright © 2007 Ray Horak

Store-and-forward technology adds significant value because it overcomes the requirement for a real-time communication between people or machines. The specific benefits of store-and-forward technologies include the fact that differences in time zones are mitigated, since you can create the message during business hours in one location and forward it to the recipient who can access it during normal business hours at the distant location, with the transmission perhaps taking place when network traffic loads are lightest and costs are lowest. Whether the parties are across the street, at opposite ends of the country, or separated by oceans, communication can be accomplished on a non-real-time basis. Additionally, you commonly can access systems and messages over a Wide Area Network (WAN) from remote locations (e.g., field offices, hotels, and client sites) at any time of the day or night, thereby offering tremendous benefits to the contemporary *road warrior*. Finally, the messages tend to be abbreviated, containing only necessary information; this is especially true of voice mail, which replaces the normal human-to-human conversational mode of communications. As social animals, humans tend to socialize before getting down to the business at hand; however, you are much less likely to attend to interpersonal niceties via a machine-enabled, one-way communication. Therefore, communications by messaging systems tends to be highly efficient, if not particularly personal, in nature. Note that this store-and-forward capability is effective only for non-real-time messaging applications. Real-time voice and video communications, for example, are highly sensitive to tight and immediate timing considerations.

While all of these messaging devices initially were of a proprietary nature, contemporary systems are computer-based messaging systems that generally are networked and are based on standards that ensure both their interconnectivity and their interoperability at some minimal level. You variously can access them from a telephone, computer workstation, or even a cellular phone or other mobile device. Additionally, you often can access them remotely across a WAN through a Key Telephone System (KTS), Private Branch eXchange (PBX), Automatic Call Distributor (ACD), Central exchange (Centrex), or Central Office (CO) system. They also increasingly are enhanced through the application of Computer Telephony (CT) technology; in fact, voice processing inherently is CT. Taken to the technically feasible extreme, CT enables the user to access a variety and combination of messages through the switch from a Wireless Local Area Network (WLAN)–attached, multimedia tablet PC workstation using a single, intuitive Graphical User Interface (GUI). Finally, the separate technologies of facsimile, voice processing, e-mail, and instant messaging currently integrate to yield *unified messaging* systems which allow voice, audio, text, facsimile, image, and even video messages to blend, thereby enhancing the aggregate effectiveness and impact of the individual messages. The technology currently exists to support unified messaging, and a number of such systems are commercially available, although they have been slow to penetrate the market. The next step in the evolution of messaging systems is that of *unified communications*, which will support the translation of message formats in such a way that an end user can receive any message or combination of messages in virtually any native message format using virtually any device (e.g., cellular telephone or computer), with the formats translated for optimum display or presentation. Further, the end user can respond to that message with the translations working in reverse.

4.1 FACSIMILE (FAX) SYSTEMS

The equipment required is simple to use. An office boy has been trained to operate the transmitter in 1 week. An executive's wife learned not only to adjust a home receiver but also to replace paper and the printer blade that serves as ink in 1 hour.
Lee Hills and Timothy J. Sullivan, *Facsimile*, McGraw-Hill Book Company, 1949

Facsimile comes from the Latin *facere simile*, which translates to *make similar*. Traditional facsimile systems are unique as they communicate information in graphic form, rather than audio or data form. Edward Davy invented the first practical facsimile machine in 1837 but abandoned the invention shortly thereafter. Alexander Bain (1811–1877), a Scottish clockmaker and inventor, revived the concept and patented the *recording telegraph* in 1843. Bain's primitive transmitting device used a stylus attached to a clock pendulum that passed over metal type, sensing dark and light spots on a metal-plated document. The dark spots were transmitted via electrical signals over a telegraph circuit to a synchronized clock pendulum on the receiving device, which passed over chemically treated paper, making a dark stain at a corresponding point where the transmitter sensed a dark spot. The Bain device was used commercially in the United States and England, where it competed with the Cooke–Wheatstone telegraph, which could transmit images through etching metal with a stylus. The first commercial facsimile service was established in 1865 by Giovanni Casselli with a circuit between Paris and Lyon. Circuits were added to other cities, and Casselli sent 5000 faxes in the first year using his patented Pantelegraph machine, which was based on the Bain recording telegraph. The service was discontinued in 1870. A number of other inventors developed various wireline facsimile devices over the next 50 years or so, but none achieved any great level of success, for they were overshadowed by the much more functional and practical electric telegraph (1844). The electrical telegraph system was invented by Samuel F. B. Morse (1791–1872) and Alfred Vail (1807–1859) and began operation in 1844. That system involved a transmitter in the form of a traditional manually operated telegraph key for sending alphanumeric data using Morse code. The receiver recorded the code symbols with an armature that scratched a paper tape. The stock ticker (1870), which was the predecessor of the teletype, and the telephone (1876) further discouraged usage of facsimile systems.

The next wave of facsimile development was in the early 1920s, with work on both wireline systems and Amplitude Modulation (AM) and Frequency Modulation (FM) broadcast radio systems. The Associated Press started a wire photo facsimile service in 1934, stimulating newspapers and law enforcement agencies all over the world to use fax for photo transmission. Although the Radio Corporation of America (RCA) developed a radiophoto system in 1926, it was not until 1936 that 1000 homes in the United States were experimentally equipped with facsimile radio receivers that could print newspapers transmitted overnight, when normal audio broadcasts were off the air. Although commercial developments were stalled by World War II, planning continued and further trials of radio-based facsimile newspaper transmission continued into the early 1950s. Needless to say, this application was impractical and failed miserably.

Facsimile was not widely deployed until the 1970s, when the technology matured sufficiently and the International Telecommunications Union—Telecommunication

Standardization Sector (ITU-T) set interoperability standards. At that point, the devices became sufficiently affordable for the technology to find relatively significant market acceptance in commercial, educational, and government applications. Facsimile became truly widespread after Group III standards were established in 1980. The low cost of current fax technology renders it cost effective even for widespread consumer use [1,2]. Estimates of the installed base vary but generally suggest that there are somewhere in the range of 200 million fax devices of one sort or another installed worldwide.

4.1.1 Technology Basics

Facsimile transmission typically involves a pair of stand-alone fax devices that serve to both transmit and receive image documents through built-in modems that interface these inherently digital devices to the analog Public Switched Telephone Network (PSTN). The transmitting fax scans the image document from top to bottom and from left to right, looking for dots of black and white; some systems will also support 256 levels of grayscale and some will support color. Through the modem, these various dots are translated into data bits, the data bits are compressed in order to reduce transmission time, and the resulting compressed data file is translated into modulations of analog sine waves, which travel over a voice-grade analog local loop to the edge of the PSTN. Alternatively, the fax transmission is converted from analog into digital format for transmission over a digital local loop, such as a T1. Within the core of the PSTN, transmission commonly is in digital format. At the egress edge of the PSTN, the process is reversed. The receiving device reads the analog signal through a matching modem and either prints a facsimile of the original document or stores the data in memory for printing or distribution at a later time. The most basic and least expensive fax machines use an old *electrothermochemical* printing technology that varies the temperature of a print head to cause the image to be reproduced on chemically treated paper. Plain paper fax machines use ink jet or laser printer technology. Early fax machines were expensive and proprietary in nature, which served to discourage their widespread use. Beginning in 1966, various standards initiatives from the Electronic Industries Alliance (EIA) and the ITU-T allowed manufacturers to build to standard sets of specifications, which enabled interoperability of machines, encouraged widespread adoption of the technology, and thereby lowered costs due to greater manufacturing volumes. The generations of fax machines defined by these standards are as follows:

- *Group I* (G1) standards were published by the EIA in 1966 as EIA RS-328 (Recommended Standard). The EIA standard was accepted by the ITU-T in 1966 as the T.2 mode of operation and became known as Group I (G1). That standard, now considered obsolete, specified analog transmission with modems using double-sideband modulation. (*Note:* The process of *Amplitude Modulation (AM)* results in the creation of two *sidebands*. An *upper sideband* is above the carrier frequency and a *lower sideband* is below the carrier frequency.) Group I machines conforming to international standards use Frequency Modulation (FM), transmitting at two frequencies, with 1500 Hz pegged as the white frequency and 2300 Hz as the black frequency. The North American standards peg 1500 Hz as white and either 2300 or 2400 Hz as black. As Group I machines

TABLE 4.1   Facsimile Generations and Characteristics

Generation (ITU-T Group)	Transmission Speed (approx. per page)	Compatibility	Resolution (lpi)
Group I	4–6 min	Group I	100
Group II	2–3 min	Group II	100
Group III	3–30 s	Groups I–III	Variable horizontal and vertical:
Group IV	3–4 s	Groups I–IV	100, 200, 300, and 400 lpi

used no compression mechanism, transmission was slow at about 4–6 min per page, even at the relatively poor resolution of about 100 scan *lines per inch* (lpi). Group I machines used an obsolete electrochemical printing process. Table 4.1 compares fax Groups I–IV.

- *Group II* (G2) fax machines conform to the ITU-T T.3 (1978) black-and-white mode of operation, which accomplishes modest bandwidth compression through the use of encoding and *Vestigial SideBand* (VSB) transmission. (*Note:* Vestigial sideband is a technique involving the transmission of the carrier, one complete sideband and only a portion of the other sideband. The VSB assists in demodulation of the signal.) Group II machines use AM and Phase Modulation (PM) at a carrier frequency of 2100 Hz. Group II machines use compression to improve transmission speed to approximately 2–3 min per page, but resolution remains relatively poor at 100 lpi. Group II machines use the same obsolete electrochemical printing process as Group I machines and are, themselves, considered obsolete.
- *Group III* (G3) fax machines conform to ITU-T Recommendation T.4 (1980) and are backward compatible with Group I and Group II devices. Group III devices convert a document to digital form and employ a run-length encoding algorithm, as described below, that compresses the document prior to transmission.
- *Group IV* (G4) fax machines conform to ITU-T Recommendations T.563 and T.6. Group IV machines are highly specialized and relatively expensive fax computer systems designed to make use of digital circuits to improve quality and improve transmission speed at rates up to 64 kbps. Circuit options include switched 56 kbps service and Integrated Services Digital Network (ISDN) at 64 kbps. Group IV fax machines also are Group III compatible and connect to analog circuits as required. Group IV machines are unusual, particularly in North America, where ISDN penetration is relatively light. Table 4.1 provides a view of the basic characteristics of each ITU-T fax generation [2–5].

4.1.2 Compression

Compression mechanisms employ mathematical algorithms in a process that serves to reduce the amount of data to be transmitted or stored. Compression is possible since there always is some amount of data redundancy or there may be a predictable flow to the data. These characteristics of a set of data or a stream of data (i.e., datastream) allow the use of a sort of mathematical *shorthand* to represent or describe the original data in fewer bits. A matching decompression process reverses the compression process and restores the data to its original form, or a facsimile thereof, so to speak. Such reduction serves to improve the efficiency of data transmission and storage and is especially valuable if bandwidth and memory resources are limited. (*Note:* Resources are always limited.) There are trade-offs, of course. Compression and decompression processes involve mathematical calculations that require computational resources in the form of Central Processing Unit (CPU) cycles to act on program logic. Some amount of cost is involved in the CPU cycles, the programs, and the associated storage, and some amount of time is required to accomplish the processes. So, compression takes time and costs money, like so many other things in life, and the more complex the compression algorithm, the more time it takes and the more money it costs. As applications for fax began to grow and technology evolved in the 1970s, compression made its appearance with Group III standards. Before discussing compression, it is necessary to understand that Group III standards also specified a number of options for document *resolution*, which refers to the level of detail of reproduction of an image and which is directly related to the density of the dots of color (black, white, and perhaps other colors). Group III specifications provide a number of options, expressed as *lines per inch* (lpi) in terms of scanning and *dots per inch* (dpi) in terms of sensing and printing, along both the horizontal and vertical axes in the format H × V (spoken as "H by V"). Of those options, the actual (and nominal) industry standards are as follows:

- *Standard*: 98 × 203 (100 × 200)
- *Fine*: 196 × 203 (200 × 200)
- *Superfine*: 392 × 203 (400 × 200)

Resolution clearly has an impact on the size of an image file, as greater resolution involves more lines per inch in scanning and dots per inch in sensing, which increases the number of bits required to express the image in digital terms. As the resolution increases, the number of bits required to express the image increases, and the amount of time required to transmit the set of data increases proportionately, assuming that the available bandwidth is constant. The yield, of course, is more dots per inch in printing, which improves image quality. Compression offers a solution to this bandwidth issue without compromising print quality. There are three forms of compression used in Group III/IV fax systems:

- *Modified Huffman* (MH) is a relatively simple compression algorithm that eliminates signal redundancy using a one-dimensional *run-length* encoding (i.e., digitizing) process that serves to compress a document prior to signal modulation and transmission. The transmitting machine scans a document from top to bottom and left to right, sensing dots of color, which in this case are black or white, at some interval that depends on the resolution setting: standard, fine, or superfine. Rather than transmitting a set of bits identifying value (black or white) of each dot of each line, the scanning machine looks for redundancy, or runs, of dots of the same value. The machine then can transmit a set of bits identifying that value and the length of the run before the value changes from black to white, for example, then transmit the length of the run of white, and so on. The receiving machine reverses the process, decompressing the data in order to reconstruct a facsimile of the original image. Compression at this point serves to reduce the amount of data that must be transmitted, thereby improving the efficiency with which the limited bandwidth offered by an analog circuit is used. Further efficiencies can be realized if the circuit is of good quality and, therefore, a fax modem can employ a relatively sophisticated modulation technique. Transmission rates over analog lines range from 2400 bps to 33.6 kbps over analog lines, depending on the quality of the circuit at a given moment, through the use of an ITU-T standard modem protocol. The modems test the quality of the circuit during the *handshaking* process that precedes the operational phase of the transmission. The modems negotiate a sustainable transmission rate in consideration of circuit quality and can dynamically adapt to changing conditions. Group III devices operating at 9.6 kbps transmit documents at "business letter" quality at a rate of approximately 30 s per page using the MH compression algorithm. MH is supported by all Group III devices as the lowest common denominator and yields a compression ratio of about 20:1 in black and white.

· *Modified Read* (MR), a Group III option used in some machines, scans and compresses the first line using MH. Subsequent lines are scanned and compared to the first line, and only the differences (*deltas*) are encoded and transmitted. This process continues for some predetermined number of lines in a group, at which point the process is reset and the first line in another group is encoded using MH, and so on. MR is an optional compression algorithm that is particularly effective if there are few differences between lines.

· *Modified Modified Read* (MMR), a compression algorithm used in Group III machines operating at 14.4 kbps or better, supports transmission at a rate of as low as 3 s per page. Specified in ITU-T T.6, MMR uses a two-dimensional compression technique that permits the transmitting modem to view and consider multiple lines of data during the encoding process. At 28.8 kbps, Group III machines can transmit a page in about 4 s using a V.34 modem employing *Quadrature Amplitude Modulation* (QAM). At 33.6 kbps, Group III machines can transmit a page in about 3 s using a V.34bis (aka V.34+) modem employing *Trellis-Coded Modulation* (TCM). (*Note:* Modulation techniques were discussed briefly in Chapters 1 and 2 and will be covered in greater detail in Chapter 6.)

4.1.3 Computerized Fax

Fax boards and *fax software*, both Group III/IV compatible, exist for computer systems ranging from PCs to mainframes. When sending a computer-based fax document, the fax software instructs the fax board to print the document to a remote facsimile machine, rather than a printer. The computer fax board contains a fax modem, thereby enabling any computer file to be transmitted to another similarly equipped computer or to a fax machine through *fax emulation* (imitation). Fax software and fax boards once commonly accompanied an applications suite packaged with a PC. Client/server versions involve fax software residing on the client workstations, with a relatively small number of fax boards residing on the fax server; the server may be a stand-alone (dedicated) fax server or a partition of a multifunctional server. The fax software in the client workstations sends the fax documents

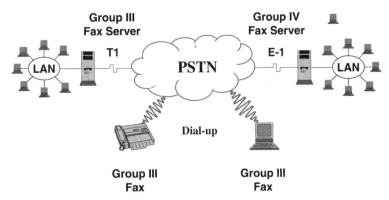

Figure 4.1 Facsimile transmission in a fax server environment.

across a Local Area Network (LAN) to the fax server (see Figure 4.1), which queues them as required before transmission, thereby considerably reducing hardware and circuit costs [6–8]. The fax server, which also may support other applications, accepts and queues faxes from multiple workstations on the basis of a *print-to-fax* option. The fax server takes faxes from the queue, *rasterizes* them (i.e., converts them from text to image format), adds either a default or customized cover page, compresses the data, perhaps selects the Least Cost Route (LCR) for transmission, and sends the fax. International faxes can be held in queue for transmission after normal business hours, when traffic loads are lightest and calling costs are lowest. If the fax devices are Internet Protocol (IP) enabled, the LCR options may include the Internet or a special IP-based carrier. The fax server also provides accountability, as records can be kept of fax traffic through employee numbers and department codes in order that the cost of fax transmission can be billed back to the responsible cost center [9].

These computerized approaches have several drawbacks, however. First, the transmitted documents either must exist as computer files or must be scanned by peripheral equipment and converted to computer files through a rastering process accomplished by application software. Second, the receiving fax computers must be turned on and networked continuously in order to be accessible. Third, in a stand-alone PC environment, the process of fax receipt may interrupt other applications in progress at the receiving computer workstation. Finally, the received documents are memory-intensive image files, which must be converted to text files in order to achieve full effectiveness. Also, so many people now have broadband access [e.g., Asymmetric Digital Subscriber Line (ADSL) and cable modems] and so seldom use their modems for dial-up access that PC-based fax is just too troublesome for a large percentage of users.

On the positive side again, the converted documents can be fully editable text files that can be forwarded to other workstations or servers without suffering the loss of quality characteristic of documents refaxed via the traditional approach. You also can archive the converted incoming faxes and burn 100,000 or so onto a single CD-ROM disk without having to scan them first. An additional advantage to this approach is that you can retrieve the faxes over the Internet, thereby relieving the recipient of the requirement to have access to a conventional fax machine.

To resolve this computerized fax dilemma, applications suites such as Microsoft Office 2003 support fax over the Internet through a fax service provider. The service provider offers downloadable fax software for signing, editing, and managing faxes. Incoming faxes are delivered as e-mail attachments. No fax modem, fax machine, or separate telephone line is required. The fax service may be bundled with Voice over Internet Protocol (VoIP) and other Internet or Web-based services.

4.1.4 Fax-On-Demand

Fax-On-Demand (FOD) most commonly is an integration of voice processing and facsimile. The traditional FOD approach involves a voice processing front end that answers the telephone call, then prompts the caller to select a document from a menu of options, enter a return fax number, and perhaps enter a credit card number for billing purposes. The system also may have the capability to automatically verify the credit card number on a machine-to-machine basis. FOD has integrated with websites for access over the Internet. Visitors can click a *fax back* button, select the requested documents, and enter a return fax number and billing information.

FOD systems can be delivered on a turnkey basis, or the user organization can build one with a component toolkit. FOD systems also may include broadcast capabilities which support high-speed outgoing fax transmission to large numbers of receivers through the entry of a distribution list. In larger applications, FOD systems are in the form of fax servers, typically residing on dedicated computer platforms and accessible by multiple client workstations across a LAN.

While e-mail and the Internet currently may be the preferred methods of electronic communications, fax continues to offer advantages as the lowest common denominator, which is particularly significant in developing countries and other circumstances where Internet access to e-mail is not available. Like e-mail and the Internet, FOD is fairly instantaneous. Although typically in black and white, cleverly formatted documents rich in graphic content can be faxed easily and quickly. Further and very much unlike most e-mail systems, fax provides instantaneous reporting of the results of transmissions, whether successfully delivered or not.

4.1.5 Conventional Fax Standards

Conventional fax is based on a number of international standards, which naturally increased in number and complexity over time and by generation. Those standards include T.30 and T.434.

4.1.5.1 T.30   T.30 is an ITU-T standard (1996) that describes the handshaking protocol used between two Group III/IV devices for establishing and maintaining communications. T.30 also provides for routing faxes to users via subaddresses or fax mailboxes. Message security is included, so that only those responsible for certain manual routing processes can view even the cover page. The routing can be accomplished in several ways, including DTMF, DID, OCR, and manually:

- *Dual-Tone MultiFrequency* (DTMF) routing requires the sender to enter the appropriate fax extension via a telephone tonepad.

- *Direct Inward Dial* (DID) routing requires a PBX or fax server which is so equipped, and each fax extension must have a separate DID number.
- *Optical Character Recognition* (OCR) software permits the server to recognize the name or special identification of the intended recipient; OCR currently is more expensive and less reliable than the other options.
- *Manual* routing is the most common approach, with an individual viewing only the cover page and then routing the fax as appropriate.

In a client/server computing environment, each LAN-connected client workstation is assigned a fax extension number. As the fax server receives inbound fax messages, it automatically routes them to the specific workstation associated with the intended recipient. Through entry of the appropriate security password, the recipient then accesses the facsimile message. Assuming that the message is T.434 compliant, the recipient can edit and annotate it prior to either responding or forwarding the fax to a user-definable distribution list. Software exists that allows the server to recognize the dialed telephone number with the trailing subaddress; the fax then can be routed to a fax machine, another client workstation, or an e-mail address. Further, the server may be intelligent enough to determine whether to send the message via the PSTN or the Internet or another IP-based network as an e-mail attachment [10].

4.1.5.2   T.434   T.434 is an ITU-T industry standard (1999) for *Binary File Transfer* (BFT) that permits compliant facsimile devices to send any file type, reproducing the original quality at the receiving end. Additionally, the received document is in the form of an editable file, if allowed by the sender. T.434 provides for interoperability among BFT products from disparate manufacturers, allowing data files to be sent much as e-mail messages so that *fax-on-demand* essentially becomes *file-on-demand*. Benefits of T.434 include increased throughput and reduced document storage requirements through data compression. Additionally, the specific file attributes (e.g., image format as in .eps, .pcx, or .bmp files) are maintained. The standard works with computer-based facsimile systems and Group IV fax machines. The standard allows the linking of fax systems to photocopiers, scanners, e-mail gateways, and PCs and invites integration with PBXs and voice mail systems in a unified messaging scenario.

4.1.6   Fax over Internet Protocol

Fax over IP (FoIP) is a relatively recent (1998) development that offers significant cost savings compared with the traditional method of transmission over the circuit-switched PSTN. The traditional method certainly has advantages in terms of ease of use through a standard telephone interface, ubiquitous access through modems over analog local loops connected to the PSTN, and low cost of terminal equipment since the development of Group III machines. Group IV devices extended that capability to digital circuits, adding a significant level of intelligence in the process. The contemporary fax document almost always originates as a data file that increasingly is sent from computer to computer, rather than printed out and manually fed into a conventional fax machine. It is difficult to argue the logic behind sending a fax data file over a highly efficient packet-switched network optimized for data, rather than over the circuit-switched network optimized for voice. The trick is to somehow provide a mechanism for interfacing the huge installed base of contemporary fax machines with a network for which they clearly were not designed.

The Internet and emerging special-purpose IP networks are built on the concept of packet switching and the underlying TCP/IP suite for packet data. Without getting too deep into the specifics of packet switching and the TCP/IP suite, which are presented in several subsequent chapters, suffice it to say that the two concepts work together very nicely to support reliable fax transmission over a highly shared packet data network. Packet networks require that a file of data be fragmented into multiple, discrete units, or packets, of data that flow independently over the network from the originating edge to the terminating edge, where they are linked together and where the original file of data is reconstituted. Packets of all sorts flow into, across, and out of a network that may support thousands or even millions of simultaneous transmissions, each in packet form, and all of which contend for limited resources in the form of switches and transmission facilities. Under load, this packet-by-packet contention for limited network resources results in congestion, which imposes variable and unpredictable levels of delay on the individual packets. Contemporary fax machines rely on an internal timing mechanism between the transmitting and receiving terminals. The carefully timed PSTN supports this approach beautifully. Because packet networks violate this timing mechanism, however, the devices simply cannot transmit effectively over such a network.

In order to send a fax document over a packet network (Figure 4.2), the terminal equipment must adapt to the inherent nature of that network. You can typically accomplish this adaptation process through a fax *gateway*, which serves as a physical gate between the circuit-switched and the packet-switched networks. Just as importantly, the gateway runs gateway protocols that convert from the carefully timed PSTN to the TCP/IP-based packet network. IP-enabled fax devices (e.g., fax machines, PCs, and servers) do not require the services of such a gateway. Relevant IP fax standards include T.37 and T.38.

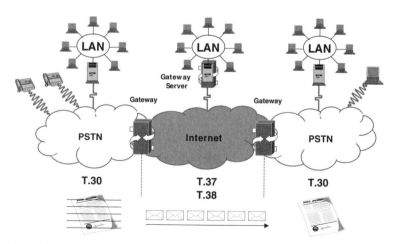

Figure 4.2 IP fax illustrating relationship between conventional PSTN and Internet fax domains through gateways.

4.1.6.1   T.37   T.37 (June 1998) is a joint ITU-T Recommendation and Internet Engineering Task Force (IETF) standard [Requests for Comment (RFCs) 2301–2306] for store-and-forward fax via e-mail through the incorporation of SMTP (*Simple Mail Transfer Protocol*) and MIME (*Multipurpose Internet Mail Extension*). SMTP is an application layer extension of TCP/IP that governs electronic mail transmissions and receptions. MIME is an SMTP extension that supports compound mail; in this context, MIME provides for the attachment of a compressed fax image to an e-mail. Fax image documents are attached to e-mail headers and encoded in the TIFF-F (*Tagged Image File Format-Fax*) compressed data format using the Modified Huffman (MH) technique. In *simple mode*, T.37 restricts fax transmission to the most popular fax machine formats (e.g., standard or fine resolution and standard page size); this restriction is effected through limitation of TIFF-F encoding to the S-profile. Simple mode provides no confirmation of delivery. *Full-mode* extensions include mechanisms for ensuring call completion through negotiation of capabilities between transmit and receive devices. Full mode also provides for delivery confirmation. Extensions also have been developed for color fax.

4.1.6.2   T.38   T.38 is an ITU-T Recommendation (June 1998, Amendment 4 September 2001) for store-and-forward fax via e-mail. Derived from X.25 packet standards, T.38 addresses IP fax transmissions for IP-enabled fax devices and fax gateways, defining the translation of T.30 fax signals and *Internet Fax Protocol* (IFP) packets. The specific methods for various T.38 implementations include fax relay and fax spoofing. *Fax relay*, also known as *demod/remod*, addresses the demodulation of standard analog fax transmissions from originating machines equipped with modems and their remodulation for presentation to a matching destination device. Fax relay depends on a low-latency IP network (i.e., 1 s or less) in order that the session between the fax machines does not *time out*. *Fax spoofing* is used for fax transmissions over IP networks characterized by longer and less predictable levels of packet latency that could cause the session with the conventional fax machines to time out. Packet transmission over such a network can result in variable latency, or *jitter*. T.38 compensates for both increased latency and jitter by padding the line with occasional *keep-alive packets* to keep the session active, rather than allowing it to time out. Thereby, T.38 *spoofs*, or fools, the receiving device into thinking that the incoming transmission is over a real-time, carefully timed voice network. Delays up to 5 s can be tolerated in this manner. T.38 improves on T.37 in a number of ways, including immediate confirmation of receipt.

T.38 provides for two transport protocols, *User Datagram Protocol* (UDP) and *Transmission Control Protocol* (TCP). UDP is the faster of the two, but the less reliable, due to the lack of error detection and correction within the network. T.38 overcomes this shortcoming either through redundant transmission of the image data packets, which is inherently inefficient at the network level, or through a *Forward Error Correction* (FEC) technique, which is inherently inefficient at the device level. TCP includes an error correction mechanism employed at the router level, with the routers typically positioned only at the edges of the network. (*Note:* Switches, rather than routers, typically are positioned in the core of the network.) Although T.38 strips this process from consideration for the IP fax packets, the level of delay nonetheless is increased; therefore, spoofing techniques are required to maintain fax sessions.

The inherent efficiencies of fax transmission over the Internet or other packet-based IP networks can lead to lower network costs as the incremental cost of one more packet transmission generally is negligible, if not zero, and generally is not distance sensitive. This cost structure compares favorably with a relatively expensive fax call over the PSTN. IP-enabled fax devices essentially incur no usage-sensitive transmission costs other than those possibly imposed by an IP fax service provider; the costs of so enabling a device vary widely but generally can be justified for fax-intensive environments. Devices not so enabled must make use of an IP gateway from a service provider, with the costs to the end user of transmission in this environment varying widely and not necessarily comparing favorably with the traditional approach.

In either case, access to the packet network is on the basis of a local call, which does not carry a per-minute charge in many countries. A number of telcos in the United States and abroad support IP fax, as do some Internet Service Providers (ISPs) and most fax service bureaus. IP fax–capable routers have the ability to transmit a fax over an IP network assuming that the level of delay is acceptable and to default to the more conventional means of transmission over the PSTN when delays are deemed unacceptable [11–14]. Such routers also have the ability to secure the fax document during transmission via the IPsec (*IP security*) encryption mechanism, thereby providing a substantial level of security over the inherently insecure public Internet. In the absence of a defined relationship between fax routers running matching encryption software, the IP fax user is at risk in transmitting over the Internet. As testimony to issues of IP fax security, I just received an e-mail response from a hotel chain asking me to fax some information. The following cautionary statement appeared at the end of the message: "Please be advised that our fax machines transmit through the internet. [sic] For your protection, please block out any non-essential information such as the last twelve numbers of your credit card account."

4.1.7 Fax Features

Fax capability comes in a variety of forms from simple and inexpensive stand-alone Group III fax machines to integrated devices that combine the capabilities of a fax machine, printer, copier and scanner and culminating in Group IV fax servers with IP networking capabilities. Available features include the following:

- **Blocking:** If caller ID is available on the fax line, the user can build a list of junk fax numbers to be blocked.
- **Broadcasting:** Also known as *group fax*, broadcasting allows the user to program a distribution list of fax telephone numbers intended to receive a given fax. The machine will transmit the document to the target numbers, in sequence.
- **Color:** Color fax was specified in T.30E. While color lengthens transmissions times considerably and reduces the number of pages that can be stored in memory, it certainly is an aesthetic improvement over black and white. Some fax machines using bubble jet, inkjet, or laser print technology offer color, as well as black-and-white, fax capabilities.

- **Delayed Fax Send:** The user can program the machine to transmit a document at a future time and date, in consideration of factors such as time zone differences and discounts for non-prime-time calling.
- **Distribution List:** The user can program distribution lists for fax broadcasting.
- **Duplex:** Fax servers may be full duplex, supporting simultaneous fax transmission and reception.
- **E-Mail Gateway:** Fax servers may include an e-mail gateway for direct IP fax transmission.
- **Forward:** A machine can forward incoming faxes to another machine if the user programs it to do so while the user is away from the office, for example.
- **Line Sharing:** If the telco offers line sharing, two telephone numbers can coexist on the same line, with one number for the fax machine and another for voice telephones. Distinctive ringing patterns distinguish fax calls from voice calls. The fax machine can be programmed to recognize the fax ringing pattern and to autoanswer those calls.
- **Phone Book:** The user can build an alphabetical phone book for speed dialing by name rather than speed dial code. The phone book can be imported from or exported to other applications.
- **Polling:** Polling allows a machine to download fax documents stored in a remote machine, with security provided on the basis of a programmable poll code.
- **Queuing:** Fax servers queue documents for faxing just as print servers queue documents for printing. Multiple queues can be established with priority-level distinctions.
- **Redial:** The user can program the fax machine to automatically redial a number if the first attempt encounters a busy or no-answer condition. The machine will make a predetermined number of attempts at predetermined intervals.
- **Reporting:** Fax machines typically offer reports of incoming and outgoing attempts, including fax number, start time, duration, mode [*Error Control Mode* (ECM) yes/no], number of pages, and result (e.g., *OK, send error, stop pressed,* and *no answer*).
- **Resolution:** *Standard* (98 vertical × 203 horizontal dpi), *fine* (196 × 203 dpi), and *superfine* (392 × 203 dpi) settings typically are available. *Note:* Higher resolution improves document quality, but at the expense of transmission time.
- **Security:** High-end fax machines and servers may require that the target recipient enter a password in order to retrieve a protected fax. Without some form of security mechanism, fax is inherently insecure, as anyone walking by a fax machine has access to the document. Also, correct fax transmission is entirely dependent on the sender's entering the correct telephone number.
- **Stamping:** The user has the ability to stamp an outgoing fax with a time and date stamp as well as sender name and fax telephone number.
- **Speed Dialing:** The user can build a fax speed-dial list.

4.1.8 Fax Applications

Application of facsimile technology traditionally has focused on document transfer. A key advantage, of course, is that any document can be transmitted by fax. Whether it is a letter, an invoice, a blueprint, or even a photograph, a facsimile of any original paper document can be transmitted successfully. Numerous sales-oriented enterprises still rely heavily on fax broadcast systems, in place of more traditional direct mail or other forms of advertising, at least in developing countries where Internet access is not widely available or is expensive and bandwidth is capped. Additionally, a faxed ad is more likely to garner personal attention than a similar ad in the media, a direct mailer, or an e-mail message. While users cannot block unwanted faxes quite as easily as they can block unsolicited e-mail, or *spam*, there are laws in some countries that prohibit *junk fax* and stipulate penalties for violations. In the United States, the Telephone Consumer Protection Act of 1991 prohibits using a telephone facsimile machine, computer, or other device to send an unsolicited advertisement to a telephone facsimile machine. The Junk Fax Protection Act of 2005 clarified and strengthened that 1991 legislation.

4.1.9 Future of Fax

Rather than being totally replaced by e-mail and other forms of messaging technology, facsimile appears to have a long, if not highly prosperous, remaining life. Dependable, inexpensive, standardized, and virtually ubiquitous, fax promises to continue its role as a valuable communications tool. While e-mail offers a number of advantages over fax, it is not universally available in developing countries. It also is worth noting that fax, unlike e-mail, does not present any barriers to communication in languages that involve complex alphabets. Further, facsimile is an inexpensive and highly effective complement to other messaging systems. Fax messages also have long enjoyed the advantage of being recognized as legally admissible in court as, unlike e-mail messages, they cannot easily be modified. (*Note:* Telex transmissions and telephone records also are considered legally admissible in court, as the trusted carrier guarantees the identity of the sender.) Traditional, stand-alone fax machines continue to grow in number even as e-mail replaces fax as the preferred method of communication for the technically privileged with access to the Internet. As a highly effective, lowest common denominator, traditional fax transmission will survive well into the future.

As costs continue to drop, fax penetration will increase. But outside of developing countries and other situations where Internet access is not available or where fax offers unique legal and other advantages, fax usage will continue to decrease. IP fax will grow, but not at the significant rate predicted several years ago. In the face of competition of e-mail over the very same IP networks, IP fax just does not do well. FOD usage also will continue to decrease in the face of increasingly intense competition from e-mail and the Internet and the Web.

4.2 VOICE PROCESSING SYSTEMS

Press one for sales. Press two in a hopeless effort to get technical support. Press three for answers to questions you don't have. Press four if you're gullible and pessimistic. Press five if you're willing to buy something just so you can talk to a human being.

The comic strip *Dilbert* by Scott Adams

Voice processing originated as voice mail (aka voicemail, v-mail, or vmail), the invention of which a number of sources attribute to Gordon Matthews (1937–2002). Matthews certainly filed the basic patents and first commercialized the systems. However, my recollection is that voice mail systems were developed at Bell Telephone Laboratories in the mid-1970s and trialed by AT&T *Bell Operating Companies* (BOCs) on an *alpha test* basis in 1977–1978 as a replacement for answering machines. (I was at Southwestern Bell in those days, so I know I am right.) Those systems were never commercially deployed, however, as the BOCs were prevented from doing so under the terms of the *Modified Final Judgement* (MFJ) that broke apart the Bell system. Under those terms, voice mail was considered an *enhanced service*, and the *Regional Bell Operating Companies* (RBOCs) formed by the MFJ initially were prohibited from offering any enhanced services. AT&T seemed to lose interest in voice mail, along with cellular telephony and a few other stellar technologies, for reasons known only to certain nearsighted AT&T executives who made stunningly bad decisions and whose names now, thankfully, are forever lost in the mists of time.

In any event, the Gordon Matthews story is that he was on a business trip in the 1970s and was having trouble reaching his office at Action Communications, the second company he founded. He was having time zone problems and playing telephone tag while trying to pick up his messages. He mentioned the problem to his wife, who suggested that he invent a computer so that he and his employees could leave messages for each other. Matthews went to work on the project, left Action Communications, and founded VMX (Voice Message eXpress) in Dallas, Texas (United States) in 1979. In 1992, Matthews retired and sold VMX to Octel. (*Note:* Octel subsequently merged with the Enterprise Networks Group of Lucent Technologies, an AT&T spin-off. VMX and Octel systems are now a product line of Avaya, a Lucent spin-off.) Shortly thereafter, 3M bought the first commercial system, a stand-alone voice mail system with an interface to the PBX, from which the call was forwarded in the event of a busy or no-answer condition at the user station. Gordon Matthews' wife, Monika, recorded the first greeting on this first commercial system. Since that time, a great number of manufacturers have entered the voice processing business, changing the nature of the systems considerably to include increased feature content, integrated messaging, PC platforms, application development toolkits, and networking [15, 16].

4.2.1 Technology

Voice processing systems are specialized computer systems consisting of ports, processors, an operating system, codecs, and storage. Although typically a special-purpose computer, the platform may be a general-purpose computer with special application software. The operating system typically is UNIX, MS-DOS, or Windows NT or better, although it may be proprietary. Similarly, the processors may be Intel or another industry standard or they may be incorporated into proprietary ASICs (*Application-Specific Integrated Circuits*) that are under the skin of (i.e., contained within) a multifunction box that tightly integrates the functions of a small PBX and voice processor. The codecs serve to convert incoming voice analog signals to digital format, compressing the data in the process in order to conserve system storage capacity. Disk drives or ASICs may contain application programs and provide for storage of data, including digitized voice greetings and incoming messages that store in individual mailboxes in the form of memory partitions. The amount of storage is sensitive to factors such as the number of mailboxes, the number of messages to be stored, the average message length, the nature of the analog-to-digital conversion process and compression algorithm, and the number of messages to be archived.

A default or customized voice greeting provides the calling party with menu options that are exercised through either DTMF (i.e., touchtone) selection or speech recognition with the latter option having become commonplace. The audio input can be either analog or digital, depending on the nature of the connecting circuit. The voice processing system digitizes the signal, as required, and compresses the voice data, generally employing a proprietary code format and compression scheme.

The voice processor interfaces to a KTS, PBX, ACD, or CO and may perform a number of functions. The voice processing system can act as an *automated attendant*, in which case the switch answers the call and forwards it to the voice processor, which enables the caller to directly access a department or station through interaction with a menu of options on the basis of touchtone or speech input. In the event that the caller does not know the desired station number, an automated directory can provide that information on the basis of a name search. Once the station number is identified, the voice processor signals the switch, instructing it to connect the call. In a voice mail application, the switch directs the caller to the voice processing system in the event of a busy or no-answer condition at the target station. Interfaces to general-purpose computer systems also are becoming commonplace. In such an application, the system provides access to a database such as bank accounts or perhaps an e-mail mailbox, with the data being converted from text to synthesized speech.

Platforms for voice processing no longer are limited to proprietary hardware, except in cases of larger systems for intensive applications. Many smaller organizations and those with certain, specific applications currently make use of special software and voice cards (printed circuit boards) that reside on PC platforms in a client/server environment. A number of manufacturers offer client/server application software suites that include an automated attendant, voice mail, customized mailboxes, and audiotex. Such an approach is highly functional and cost effective, although concerns remain relative to reliability and database integrity and fault-tolerant platforms and security mechanisms go a long way toward addressing those concerns. Additionally, a voice message requires substantial amounts of memory, even when highly compressed [17].

4.2.2 Applications

The applications for voice processing have increased dramatically, with penetration having reached the point that it is virtually ubiquitous. Typical applications fall into the categories of audiotex, voice mail, call processing, and Interactive Voice Response (IVR).

4.2.2.1 Audiotex *Audiotex*, also known as *audiotext*, is a simple technology that enables callers to select prerecorded messages from a menu. Essentially a voice bulletin board, audiotex generally is incorporated into a more substantial suite of voice processing applications such as call processing. The audiotex feature commonly enables callers to work through multiple tiers of a hierarchical menu to gain information.

4.2.2.2 Voice Mail *Voice mail*, the most common application, involves the direction of the incoming call to a voice mailbox associated with a particular user or application. Voice mail digitizes, compresses, and stores the voice incoming message in the mailbox and then advises the user of the fact that a message is waiting, most commonly through message-waiting lamp indication or stuttered dial tone. When the user accesses the system and enters the proper command and password, the message is resynthesized and played back, ultimately in analog audio form in order to be compatible with the human ear. Features may include the following:

- **Annotation:** A recipient can annotate a message to add a comment before archiving it or forwarding it to another station.
- **Archiving:** Messages can be archived, or saved, perhaps for a limited amount of time unless they are resaved. Messages may be archived on an external storage medium for legal reasons. As examples, telephone companies sometimes must archive messages for long periods of time when effecting a change of carrier, as must stockbrokers when confirming a solicited trade, and telemarketers when confirming a solicited sale of certain types.
- **Attendant Access:** The system provides the caller with the ability to reach an attendant or alternative answering point if the caller does not wish to leave a message. This important feature, if turned on, allows the caller to speak to a real, live person, rather than being trapped in *voice mail jail.*
- **Broadcasting:** An authorized user can broadcast a message to a user-definable distribution list. The targeted user must listen to the message in its entirety before playing any other message. This feature can be used to make important announcements.
- **Certification:** An authorized user can request a certification or confirmation, which will generate a notification that a message has been read by the target system user.
- **Class of Service (CoS):** The system administrator can define multiple classes of service, each of which provides a different level of privilege for features such as incoming message length, archival storage capacity, broadcasting, and privacy.
- **Find Me:** The system provides the caller with a *find-me* locator option. If invoked, the system will try several (perhaps three) preprogrammed find-me numbers (e.g., home office phone, cell phone, and home phone) in sequence. If the system is unable to find the target user, it will so advise the caller and offer the opportunity to leave a message.
- **Forwarding:** A recipient can forward a message to another system user. E-mail forwarding allows the system to forward a voice message as an audio attachment to an e-mail, with the e-mail address being user definable.
- **Off-Site Notification:** The system will dial a user-definable outside telephone number (e.g., cell phone or home phone) to advise of a message waiting.
- **Personalized Greeting:** A user can record personalized greetings, perhaps for specific situations such as a *busy* condition or a *no-answer* condition.

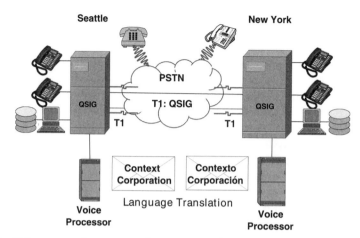

Figure 4.3   Voice processing systems interconnected through Time Division Multiplexing (TDM)–based PBXs running QSIG in a private, leased-line network.

- **Prioritization:** Messages can be marked *urgent*, thereby sending them to the head of a message-waiting queue.
- **Privacy:** A sender can mark a message *private* or *confidential*, thereby denying the recipient of the ability to forward it to another user.
- **Purge:** The system will automatically purge archived messages after a predetermined period of time established by the system administrator. Prior to doing so, the system will leave the affected user a voice message, offering the user the opportunity to resave those messages to avoid their being purged.

Networking of voice processing systems is fairly routine in large user organizations, although generally limited to systems of the same origin (manufacturer) and generic software load. An employee based in New York who also has an office in Seattle, for example, would have voice mailboxes in both locations. Internal and external callers in each city dial a local number over a DID CO trunk and, in the event of a busy or no-answer condition, are directed to a local voice mailbox. The voice processing systems are interconnected through the private corporate PBX network. The PBXs run the Q SIGnaling (QSIG) protocol (Chapter 3) and are interconnected through the PSTN via dedicated, leased tie lines, as illustrated in Figure 4.3. On a scheduled basis, the voice processing systems engage in a computer-to-computer dialog, transferring messages to the system in the city where the employee is working that day, perhaps performing language translation just to test the language skills of the staff member. (*Note:* Language translation is an unusual feature that is available on some systems.)

This non-real-time transfer of voice data also can be accomplished cost effectively over packet data networks such as Frame Relay (FR) and the Internet. Also, unlike systems can network successfully if they both are compliant with the AMIS (Audio Messaging Interchange Specification) standard, published by the Industry Information Association in 1992. AMIS specifies message file formats, addressing conventions, and message transmission. As illustrated in Figure 4.4, this networking scenario requires the deployment of gateways that accomplish protocol conversion

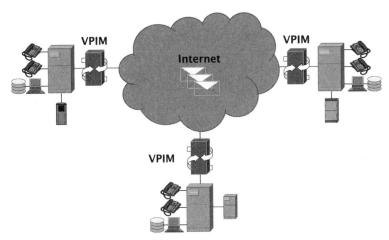

Figure 4.4 Voice processing systems interconnected through the Internet via TDM-based PBXs and VPIM gateways.

to resolve issues between the TDM-based PBXs and the IP-based Internet. Those gateways typically are compliant with the Voice Profile for Internet Mail (VPIM) published by the IETF as RFCs 2421 and 2421v2 (1998). Frame relay networks also can be used to internetwork voice processing systems.

In addition to what might be described as more legitimate applications, increased use of voice mail advertising has recently developed. Solicitors have built voice mail databases that enable them to broadcast advertisements and various other messages to thousands of current and potential customers and contributors. While such use is aggravating to many voice mail recipients, recipients do have the option of calling the advertisers and requesting that they remove their names from such a database. Taking voice mail advertising a step further, some service bureau voice mail providers offer discounted or even free service to customers willing to listen to advertising messages each time they access the system. A more legitimate and common application is for school districts to notify parents of their children's absence from school, notify parents that report cards have been mailed via the postal service, schedule sporting events and practice sessions, or announce school closings or delays due to inclement weather.

4.2.2.3 Call Processing *Call processing* applications position the voice processing system as a front-end call processor to a PBX or ACD. The *automated attendant* feature of the system will answer the call and provide menu options that can be invoked through either DTMF input or speech recognition software. Features may include automated call routing, Automatic Number Identification (ANI), and Fax-On-Demand (FOD) access. Once the caller selects a menu option, the call can be directed to the appropriate agent, extension, or computer resource. Should the system be unable to connect the call immediately, either the call can be placed in queue or the caller can leave a call-back message in a general system mailbox, rather than simply abandoning the call.

Speech recognition, also known as *Automatic Speech Recognition* (ASR), recently has been the subject of a great deal of interest. This surge in interest is largely due to the development of the *VoiceXML* specifications, which are based on the industry standard XML (*eXtensible Markup Language*) developed by the W3C (*World Wide Web Consortium*) for use on the Web. VoiceXML is a high-level programming interface to speech and telephony resources for application developers, equipment manufacturers, and service providers. Speech recognition technology, in general, allows a caller to interact with a computer system in natural conversational voice mode, that is, by speaking to the machine just as you would speak to another person. This technology, in effect, takes call processing to the level of speech-driven call processing [18].

United Airlines, for example, provides toll-free access to its reservation centers. Mileage Plus members are provided with different numbers than the general public, and those numbers are sensitive to the caller's Mileage Plus level (e.g., Premier, Premier Executive, and Premier Executive 1K). Through the use of DNIS (*Dialed Number Identification Service*), callers are segregated by level and directed by the ACD to the logical partition customized for their use. The voice processor then prompts the caller to select either domestic or international reservations. Finally, the processor places the caller into the proper queue to speak to an appropriately skilled agent.

4.2.2.4 *Database Access: Interactive Voice Response*   In a *database access* application, the voice processing system is positioned as a front end to a general-purpose computer on which reside appropriate databases (Figure 4.3). As a means of imposing a level of security to the transaction, various means of caller authentication, such as DTMF input of an account number and *Personal Identification Number* (PIN), can be employed. More unusual and expensive techniques include *speech recognition* (recognition of voice commands) and *voice print matching* (verification of the caller's identification by matching the characteristics of the voice input to a stored sample). That information (e.g., bank or credit card account information) then either is played back from prerecorded *sound bites* or is voice synthesized for the benefit of the caller, yielding what is commonly known as *interactive voice response*. Access to multiple databases often is provided across a LAN. Features may include speech recognition (i.e., the recognition of speech) and even voice recognition (i.e., the recognition of a specific voice), voice print matching, and *Text-To-Speech* (TTS) capability. Speech recognition combined with TTS can allow a complete transaction to be accomplished on a voice basis without human intervention. Even should it ultimately be necessary for a human being to get involved, this initial interaction can considerably reduce operator talk time.

Reservation centers and financial institutions, for example, make heavy use of such capabilities in support of routine transactions, thereby reducing staffing levels and providing enhanced customer service on a 24 × 7 basis. Telephone companies increasingly use voice recognition technology to provide automated access to local directory databases. Major banks commonly provide automated access to account information without human intervention. A caller is prompted to enter his or her account numbers, followed by a user-definable PIN, Social Security Number (SSN), or federal tax ID number. At that point, menu options can include the account balance, last five checks posted, and last three deposits posted. Airline reservation centers make use of IVR for automated flight status information. The caller dials a special number set aside for that purpose, enters a series of commands (e.g., date and flight number) via tonepad signaling, and is provided the required information without the costly involvement of an agent. Among the main benefits of an IVR system is that it eliminates the need for human employees to answer repetitive questions. While an IVR system does not totally eliminate the need for human call center agents, it does relieve them of some of the less challenging aspects of their jobs. IVR systems also provide the significant benefit of being available to work 24×7, with no breaks, no lunch hours, no sick time, no vacation time, and no whining [19].

My financial institution, for example, provides account information through IVR, with the following hierarchy of menu options:

For English, press 1 (one)

Para Español, marque 2 (dos)

> I pressed 1. Although my English might be poor, it is better than my Spanish.

For information or services on an existing account, press 1

Open a new account or apply for loans or credit cards, 2

ATM or branch locations, 3

Year-end tax information, 4

Merchant verification of funds, 5

At any time during this call, you may speak with a specialist by pressing 0

To return to this menu any time, press *

> I pressed 1.

For account inquiries, press 1

To transfer funds, 2

For Internet banking or bill payment by phone, 3

To order checks, 4

For account maintenance, 5

To speak to a specialist, press 0

> I pressed 1.

For deposits, press 1

Checks, 2

All account activity, 3

> I pressed 1.

Using the touchtone keypad, please enter your 15-digit account number followed by the star (*).

> I entered my account number and pressed the star.

Using the touchtone keypad, please enter your four-digit password.

> I entered my password. Sadly, there have been no deposits made to my account.

While I am disappointed over the lack of deposits, particularly as I was expecting a royalty check for this book, my bank has saved a good deal of money by automating the account inquiry process and making me do all the work. However, I take solace in the knowledge that the bank will pass those savings on to me in the form of lower service charges. I then find humor in my own extreme gullibility in believing that my bank actually will pass the savings on to me. Once again, I have amused myself with IVR.

4.2.3 Voice Processing Developments and Futures

The future of voice processing is very bright, indeed. While such systems initially were a North American phenomenon, they now are commonplace around the world. Specific technology futures include archival applications, multimedia/unified messaging, voice-to-text, voice-to-fax, and language translation.

Computer telephony has impacted voice processing to a considerable extent, with the application software residing on a LAN-attached server in the form of an industrial-strength PC running a commercial operating system. (See Chapter 3.) The server runs at high clock speed, has a high-speed internal bus, and includes substantial Random Access Memory (RAM) and hard-drive memory. CT systems may allow the front-end voice processor to actually process the call, connecting the caller to the target station without the involvement of the switch. Typical of the evolution of program logic, some manufacturers of smaller telephone systems have reduced much of the voice processing software to the chip level. Embedding the core voice processing functions in ASICs, this firmware approach enables much faster system operation than a software approach, although it sacrifices some level of programmability at the user level. In order to overcome this limitation and to permit the user organization to customize the application, a combination of firmware and software is necessary to drive system operation.

Archival technology has been developed to allow voice messages to be saved on an external archival system in the form of digitized audio files, although this technology is not widely available. While voice processing systems currently enable the user to archive a message within the system, internal memory limitations generally tightly restrict the number and length of stored messages.

Multimedia messaging, also known as *unified* or *integrated messaging* and *unified communications*, enables multiple messages in multiple formats to be developed, coordinated, and networked. A voice message can, thereby, be associated with a text or image document. A voice processor acting as a front end, for example, might recognize a fax tone and store the fax along with the voice message; subsequently, the target user can listen to the voice message while viewing the fax document. The ability to store voice, textual (e-mail), image (fax), and even video messages in a single mailbox and to access them from a single point of interface has obvious advantages. Unfortunately, the underlying technologies required are complex and generally remain unaffordable at this time. Although CT systems provide the user with the ability to accomplish this feat relatively easily through coordinated access to multiple media from either a telephone set or a computer monitor, such systems currently are not widely implemented. While text-to-speech and speech-to-text systems enable the remote worker to access a range of message formats from a single device, the messages typically must be presented sequentially, rather than simultaneously if a telephone is used. Personal Digital Assistants (PDAs) and more substantial tablet and laptop computers allow the user to view message lists. Advanced pager technology, while impressive, is particularly limited in this regard.

While the emerging IP-based WANs promise to overcome some of these limitations at the network level, the underlying technologies and architectures at the system level currently are proprietary in nature.

Voice-to-text and *voice-to-fax* technologies have been developed that will convert voice messages to text or fax messages, and vice versa. While the technology is still maturing at this time, a number of commercially available products will convert text to speech for applications such as remote e-mail and fax access from a telephone.

Speech recognition software enables the caller to speak commands to the voice processing system, rather than entering commands via the telephone tonepad. Such systems are speaker independent, although the speaker must speak clearly and the network connection must be relatively free of noise. While all but the most expensive systems currently are limited in vocabulary, such technology will become increasingly affordable and, therefore, prevalent in the foreseeable future. For example, some years ago IBM developed speech recognition technology to the point that computer system vocabularies exceed 50,000 words [20, 21]. Regardless of the level of sophistication, many users find the current technology too slow and cumbersome and the application annoying.

Language translation systems ultimately will find their way into voice processing systems. Some years ago, AT&T and other IntereXchange Carriers (IXCs) had rather complete plans to implement this developing technology as a replacement for the human translators who currently provide such services. It is just a matter of time until the technology progresses to the point that it becomes sufficiently advanced and affordable to be implemented on a wide scale.

Voice processing systems are destined to become even more commonplace, to the chagrin of the many who bemoan the fact that they no longer can talk to a live human being. While users often hide behind voice mail in order to be more productive in the workplace, callers understandably get frustrated at their inability to reach them. Additionally, many systems are set up intentionally to defeat the caller's attempts to reach any live person—this phenomenon is known as *voice mail jail,* and callers do not appreciate it.

4.3 ELECTRONIC MAIL (E-MAIL)

Amazingly, 70 percent of all first-class mail in the United States is generated by computers. Much of this is destined for other computers. It is printed, split into separate sheets, fed into envelopes, sent to a mail room, stamped, carried manually to a post office, sorted, delivered to planes and sent to another post office, sorted again, delivered to a mail room, sorted, distributed, and then laboriously keyed for entry into another computer. All of this ought to take place electronically.

James Martin, *Telematic Society*, Prentice-Hall, Inc., 1981

Once upon a time, people communicated by talking over the fence or the dinner table, but we do not do that much anymore. Over longer distances people text messaged by writing letters, which were sent through the highly reliable but very slow Post Office Department in the United States and similar government agencies around the world, but we do not do that much anymore either. The first telegraph message was sent in 1844 and the first commercial telegraph service was established in 1850, as best I can determine. The Pony Express began service in April 1860 with the promise that it could deliver a letter from St. Louis, Missouri, to Sacramento, California, in 10 days or less. Eighteen months later, the completion of the Western Union transcontinental telegraph network doomed the Pony Express. Although fax machines date to 1837, the first commercial facsimile network was not built until 1865 and fax did not pose much of a threat until much later. The telephone arrived in 1876 and soon replaced telegraphy as the primary means of electronic communications. Western Union introduced *teletypewriter* service in 1923 so that companies could link branches and even join other companies in private text messaging over leased private-line networks. Beginning in about 1935, telegraph companies began to use *telex* (*tele*typewriter *ex*change), a rotary dialing system much like that used in telephone networks. Telex was an effective means of routing telegraph calls, and it ran at the amazing signaling speed of 45.5 bps and, later, at 50 bps, or 66 words per minute (wpm). Altogether, telex services eventually reached some 190 countries and 3.5 million machines, by some estimates. Telex was the first true electronic mail service and, at one time, was undoubtedly the most heavily used textual telecommunication system in the business world. Telex service is still in use, largely in developing countries.

Federal Express (1971) and other overnight courier services have since become a routine means of sending written documents. Fax machines became affordable and popular in the 1980s, enabling instantaneous transmission of image documents of any sort, not just those text based. Then came *electronic mail* (*e-mail* or *email*), an application software system originally intended for textual messaging. E-mail now permits the attachment of other forms of information, including binary files, images, graphics, and even digitized voice and video. Nonetheless, its primary use remains for textual messaging, at least for the time being.

E-mail has it origins in the mid-1960s as a means for multiple users of time-share computer systems to communicate. E-mail quickly became popular for government and military communications in the late 1960s and early 1970s, especially as an application on the ARPANET (Advanced Research Projects Agency NETwork), which was the predecessor to the Internet. E-mail was popularized in the late 1970s and early 1980s as part of the office automation concept designed to lead us toward the paperless office, which was about as successful as the personal helicopter. Typically residing on a midrange computer, early systems included IBM PROFS (PRofessional OFfice System), Digital Equipment Corporation DECmail (included in the All-in-One applications suite), and the Wang Office. Network-based e-mail services were available from providers including AT&T Information Systems (AT&T Mail), CompuServe (EasyPlex and InfoPlex), MCI International (MCIMail), and Western Union Business Services (EasyLink). The proliferation of PCs and PC LANs in the 1980s and 1990s stimulated the growth of e-mail, which then became affordable and, therefore, accessible by a much larger community of potential users.

According to a 2005 study by IDC, 84 billion e-mail messages will be sent each day worldwide in 2006, about 33 billion of which messages will be *spam*, that is, unsolicited e-mail. That equates to about 30,660,000,000,000 (thirty trillion, six hundred and sixty billion) messages per year, of which 12,045,000,000,000 (twelve trillion, forty-five billion) will be spam. IDC further projects that business e-mail volumes sent annually worldwide will exceed 3.5 exabytes in 2006, more than doubling the amount over the past two years. (*Note:* 1 exabyte = 1 thousand petabytes = 1 million terabytes = 1 billion gigabytes) [22]. That compares with about 212 billion pieces of *snail mail* handled by the U.S. Postal Service (USPS) in 2005. For many, e-mail is the first thing checked at the start of the business day and the last tool touched at day's end. It has become a fundamental, if not the primary, mode of communications for the electronically privileged *digiterati*, that is, the digitally literate.

E-mail provides the ability to distribute information to large numbers of people virtually instantaneously and very inexpensively. A 50-page document, for example, can be transmitted coast to coast in approximately 30 s by using a high-speed modem and in a fraction of that time by using ADSL, a cable modem, or another broadband access technology. Attach a portfolio of photographic images, an audio file, or even a video clip. The cost of such a transmission is in the range of pennies, rather than dimes or dollars for facsimile, or tens of dollars for overnight delivery via the USPS, Federal Express, or a similar courier service. The dark side of e-mail, however, includes the fact that once you click the mouse the message is sent immediately and may be irretrievable. Gone are the days when you dictated a sensitive letter and let it rest overnight in a desk drawer while emotions cooled so you could reread and edit it in the light of a brighter day. Getting darker, still, e-mail is an easy way for reckless or malicious employees to instantaneously expose company secrets, damage reputations, offend co-workers, spread computer viruses, and otherwise create a host of business, personal, legal, and technical problems [23].

4.3.1 Technology

E-mail involves application software that resides on various computer platforms. The contemporary e-mail network generally is in the form of a *client/server* configuration involving a dedicated server to which client workstations gain access for messaging purposes. Clients are the workstations that create, transmit, and receive the messages. A client includes software in the form of a *Mail User Agent* (MUA) that effects compatibility with the mail server. The server provides message storage and transport through a *Mail Transfer Agent* (MTA). The server also provides directory services to the client, perhaps in conjunction with another server that performs Internet address lookups and *Network Address Translation* (NAT). Various clients can be linked to a server through Application Programming Interfaces (APIs), messaging *middleware* (software that logically sits between client and server), and a driver package. Examples include Microsoft Messaging Application Programming Interface (MAPI) and Apple Computer's Open Collaboration Environment (AOCE). Such APIs permit client applications such as word processing and spreadsheets to communicate with an e-mail server. In other cases, special drivers must access servers running proprietary access protocols.

A server may be in the form of a mainframe, midrange, or PC platform which may be logically partitioned in support of multiple applications. Many client/server e-mail systems run equally effectively in several Operating System (OS) environments, such as Windows, Macintosh, and UNIX. On premises, the server and the clients generally are LAN-attached in business environments, and access generally is provided to the Internet. The server also may be provided on a service bureau basis, such as with specialized carriers and national, regional, and international online services including Africa Online, America Online, AT&T Worldnet (Webmail), Google (Gmail), MSN (Hotmail), and Yahoo! (Yahoo! Mail). Even the smallest ISPs also offer e-mail services.

4.3.1.1 Networking Although some e-mail networks are closed, these messaging islands are increasingly rare. E-mail servers typically support multiple protocols, with internetworking between disparate e-mail systems adding great value to messaging technologies. Most provide for wide area networking over a *backbone*, or centralized high-capacity network, in order to provide for both outbound and inbound message communications with the outside world through the Internet. Remote access to the internal e-mail resources generally is provided to employees, vendors, clients, and others within the inner circle.

4.3.2 Protocols

Each e-mail network is governed by a set of *protocols*, that is, the rules and conventions by which the network and its various component elements operate. Such protocols are embedded in the application software and the specific OS or Network Operating System (NOS) that governs how the computer or computer network functions. An element of the software that achieves compatibility between the client workstation and the server over the network generally resides on the client as well as on the server (Figure 4.4). E-mail protocols govern such characteristics as addressing conventions, routing instructions, message structure, and message transfer. Protocol categories include document exchange, directory management, and e-mail access. In addition to standard protocols, there exist a number of proprietary protocols.

4.3.2.1 Message Handling and Document Exchange *X.400* is the ITU-T international standard protocol for e-mail message handling and document exchange. Created in 1984 and updated in 1988 and again in 1992, X.400 is compatible with the International Organization for Standardization (ISO) Open Systems Interconnection (OSI) model, functioning at layer 7, the application layer. X.400 is a complex protocol that gained considerable popularity in Europe, largely for in-house implementations. In the United States, the preference developed for the *Simple Mail Transfer Protocol* (SMTP) of the TCP/IP protocol stack, upon which the Internet is based. While most carriers do not use X.400 in native implementations, they commonly use it as a gateway protocol for X.400-to-SMTP gateways, particularly for international networks. As depicted in Figure 4.5, X.400 acts as a gateway protocol, permitting disparate e-mail systems to interoperate at a minimal level over either packet networks or asynchronous dial-up circuits [24].

4.3.2.2 Directory Protocols *X.500* is an international standard developed jointly by the ISO and the ITU-T, also in 1988, to support the requirements of X.400 name and address lookup. X.500 provides for global directory services that theoretically enable network managers to store information about all users, machines, and applications in a distributed fashion. The current version of X.500 provides for directory replication functions, which permit multiple copies of the directory information to be stored throughout the network, rather than reside on a centralized

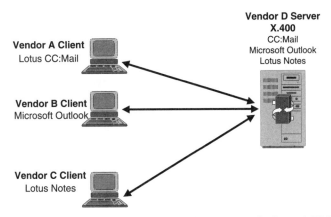

Figure 4.5   Client/server e-mail, supporting multiple protocols through X.400 gateway.

server. The current version also provides improved security through authentication access controls. While X.500 is a very robust global directory standard, it requires significant computational resources to implement and often receives criticism for being overengineered Although X.500 has yet to be fully embraced by software developers—and it is highly questionable as to whether it ever will be at this late date—a number of proprietary e-mail products have incorporated X.500.

Lightweight Directory Access Protocol (LDAP) was developed to simplify the demands of the X.500 DAP. Developed at the University of Michigan in the 1993 time frame and described in the IETF's RFC 1777, LDAP is a lean subset of X.500 that can run over TCP/IP networks. LDAP can run as a stand-alone directory system or can be used as a means of accessing an X.500 directory or other directory. LDAP has been extremely popular and currently is in version LDAPv3, as described in RFC 3377. LDAPv3 supports non-ASCII (American Standard Code for Information Exchange) and non-English characters for international directories and can sort through multiple directories on the basis of a single request. Directories follow the X.500 format, which is a hierarchical tree of entries, each of which consists of a set of attributes, each of which has a name. An LDAP directory request might be to provide all available contact information on all persons in the Mount Vernon, Washington, U.S.A. office of The Context Corporation with the first name of Ray who have an e-mail address. That search of the corporate directory would return complete contact information on Ray Horak (that would be me), including full name, title, division and department, physical mailing address, telephone and fax numbers, and any other information available to be shared with the requestor, based on his or her level of privilege.

4.3.2.3 *Access Protocols*   Access to e-mail within the LAN domain is a relatively straightforward process of logging onto the e-mail server over the LAN. As the user is physically located on premises, physically connected to the LAN (with the exception of wireless techniques), and must enter one password to boot the PC, another to log on to the LAN, and another to access the e-mail server, security is not a significant issue. Remote access to the corporate e-mail server also is a relatively straightforward process accomplished either through a dial-up connection over a VPN (Virtual Private Network) through the public Internet or perhaps over a DSL circuit or a cable modem network. VPNs are inherently secure, at least as secure as communications over an inherently insecure packet data network can be at this point. Dial-up access generally is through a *Remote Access Server* (RAS) that is located on the premises and that is associated with a *Remote Authentication Dial-In User Service* (RADIUS) server that performs authentication and accounting functions to ensure network security. Dial-up access generally is accomplished through the PSTN with the assistance of a modem, although *Integrated Services Digital Network* (ISDN) connectivity may be supported.

Access to Internet-based e-mail is a bit more complex. Such e-mail is stored in a network-based mailbox under the control of a national service provider such as AOL or CompuServe; a Web-based service such as MSN, Netscape, or Yahoo!; or a smaller ISP such as Isomedia, in my case. The larger service providers have hundreds, or even thousands, of POPs (Points Of Presence) that allow my daughter, for example, to access her Gmail from virtually anywhere in the country on the basis of a local call. The POPs are in the form of access routers and servers located variously in telco COs or third-party facilities at strategic locations across the country and, for that matter, across the world. The protocols run in such networks often are proprietary in nature.

Remote access through smaller ISPs is more problematic, although based on standard protocols. Such access commonly is on the basis of a dial-up modem connection over the PSTN, which usually is in the form of long-distance calls involving toll charges if the call is placed from outside the local calling areas served directly by the local or regional ISP. Toll-free calling may be offered at an additional charge. Large numbers of local and regional ISPs have formed nationwide consortia linking the POPs in their respective local calling areas through the Internet. These networks generally use one of two access protocols:

- *POP3 (Post Office Protocol version 3)* is an IETF standard (RFC 2449) used to retrieve e-mail from a remote server over an IP network. POP3 essentially is a store-and-forward service that runs on both the client workstations and the mail server. Prior to accessing a mailbox at the remote server and downloading all mail to the client, the user can elect either to delete that mail from the server or leave a copy on the server. The second option provides advantages such as checking mail remotely while on the road or from another computer, as the user always can access the same mail later from a home location and personal computer. After downloading that mail in either case, the user can disconnect from the remote server and work with the mail offline. Note that POP3 is involved only when downloading mail from the mailbox. When uploading mail, your access is to an SMTP server, which merely forwards mail after looking up the proper IP addresses on a *Domain Name Server* (DNS). POP3 is widely deployed.

- *IMAP (Internet Message Access Protocol)* is a more recent and much improved IETF standard (RFC 2193) protocol for accessing e-mail. IMAP acts as something of a remote mail server, allowing the client to manage mail much more effectively than POP3. IMAP, for example, allows the user to view the heading and sender of each mail message and then decide whether to download it, delete it, or take other action. IMAP also allows the user to create, manipulate, and delete individual mail folders and mailboxes on the server. IMAP requires that the connection be maintained between client and server continuously while working with mail. Security is an issue with IMAP, as the remote client takes on the appearance of a remote virtual server.

4.3.2.4 Application Protocols E-mail running over the Internet makes use of several extensions of the TCP/IP suite. Although I discuss these protocols and many more aspects of TCP/IP in great detail in Chapter 12, I cannot leave this treatment of e-mail without briefly mentioning SMTP and MIME.

4.3.2.4.1 Simple Mail Transport Protocol Connectivity likely will remain an issue for some time. A minimum set of agreements must be reached within the e-mail development community in order to resolve issues including commonality of header fields, end-to-end messaging services (e.g., priority transfer, confirmation notification, and security), addressing schemes, and management standards. Most systems, for example, use Simple Mail Transport Protocol (SMTP), defined by the IETF in RFC 821 to pass e-mail to and through the Internet. Yet each SMTP gateway may deal with addressing schemes in a different way, causing conflicts during the process of message routing. Additionally, each e-mail system places different limitations on the size of file attachments, and neither delivery confirmation nor user authentication are supported between such systems through SMTP. Further, binary files typically get converted to text and then converted back to binary files on the receiving end, resulting in a complete loss of formatting.

Notably, SMTP was developed to support the seven-bit ASCII code for simple text. This coding works just fine for relatively simple alphabets (e.g., 26-character English), supporting uppercase and lowercase characters. ASCII also supports numbers 0–9, punctuation marks, and a reasonable set of control characters (e.g., paragraph break and page break). ASCII, however, does not support the rich text formatting (e.g., italics, bold, and color, other than the default black in which this document is printed) supported by most contemporary word processing programs. This seven-bit format also prevents the transmission of eight-bit binary data found in executable files.

Some e-mail systems, such as Microsoft Outlook, also support *rich text* and *HyperText Markup Language* (HTML) formatting. Developed for the Web, HTML supports very rich text formatting and embedded graphics. Unfortunately, communication with e-mail clients not supporting rich text or HTML creates considerable formatting incompatibilities.

4.3.2.4.2 Multipurpose Internet Mail Extensions Multipurpose Internet Mail Extensions (MIME) is an SMTP extension that was developed to overcome SMTP's ASCII-based limitation. MIME standards, as defined in the IETF's RFC 2045, include a number of *types* and *subtypes* that support a range of data formats. Those types include the following:

- *Text type* for textual messages. Subtypes include *plain text* for seven-bit ASCII and *rich text* for enhanced text formatting.
- *Image type* for image files. Subtypes include *Graphics Interchange Format* (GIF) and *Joint Photographic Experts Group* (JPEG).

- *Video type* for time-varying picture images. The *Moving Pictures Experts Group* (MPEG) subtype is defined.

- *Audio type* for basic audio data at 8 kHz.

- *Application type* for executable code and any data that do not fit neatly into any of the other types. Subtypes include *octet stream* for binary data and *postscript* for PostScript files.

- *Message type* for encapsulated messages within e-mail. Partial subtype permits a long e-mail message to be fragmented at the transmitter and reassembled at the receiver.

- *Multipart type* supports the combination of multiple types into a single e-mail message [25].

4.3.2.5 Proprietary Systems and Protocols   Proprietary e-mail systems and protocols include cc:Mail (Lotus Development), Lotus Notes (IBM/Lotus), Microsoft Outlook (Microsoft), Eudora (QUALCOMM), and Pegasus (David Harris). Online service providers such as America OnLine, Google, Netscape, and Yahoo also use various proprietary e-mail systems. Through *gateway* technology, such as is in place on the Internet, unlike mail systems can interoperate, at least at a basic messaging level. Gateway software acts as a protocol translator, or interpreter, converting from one native environment to another. In this manner, Outlook and Lotus Notes software, for example, can pass messages. For instance, several editors, the publisher, and I were able to accomplish much of the editing process for this book via e-mail. From my home office, using a desktop computer running Outlook e-mail software and with Internet access via ADSL through Verizon to Isomedia, a local ISP, I sent the editors electronic Word and PowerPoint drafts as e-mail attachments, with absolutely no idea what e-mail program they were running internally. The editing process continued on the road via a laptop running Outlook software, with dial-up Internet access through hotel PBXs in the United States and South Africa. The recipients of this mail were able to respond with edits until the document was deemed satisfactory. Incompatibility was not even an issue, thank goodness. (I had to deal with plenty of other issues.)

4.3.3 Features

E-mail systems offer a growing number of features that commonly include time and date stamping and the ability to read, write, reply, and archive messages. Archived messages can be sorted by criteria such as status, priority, sender, date, size, and subject. A variety of file types (e.g., text, binary, audio, video, and image) and file formats can be attached or appended. Features also can include encryption for additional security, receipt confirmation, automated response, and automated forwarding. The ability to build customized distribution lists enables a user to send a single message to large numbers of user mailboxes with the click of a mouse. The following is a list of typical features:

- **Address Books:** Personal and corporate address books can be built and searched for addressing mail. Links may be provided to a personal contact list that contains greater detail about each individual, including perhaps physical address, detailed contact information, and free-form comments. LDAP Internet directories can be searched. Address books can be imported from and exported to other applications.

- **Confirmation:** The sender can request that the recipient send a receipt to confirm that the message has been received and, perhaps, send another receipt to indicate that the message has been read.
- **Folders and Files:** Mail received can be organized into user-definable folders and files.
- **Formatting:** Mail format options include plain text, rich text, and HTML. A number of stationary options may be available. A signature can be affixed to outgoing mail, including perhaps sender name, mailing address, contact numbers, and even a quote. For example, I currently affix the following quote to my e-mail: "Whenever I'm asked what kind of writing is the most lucrative, I have to say ransom notes" (originally stated by literary agent H. N. Swanson, who once represented F. Scott Fitzgerald and Ernest Hemmingway).
- **Mailboxes:** The *Inbox* contains all received mail; the *Outbox* contains all outbound mail not sent; the *Junk* mailbox contains all mail designated as junk (i.e., spam), either manually or by message rules; the *Sent* mailbox contains all sent mail; and the *Deleted* mailbox contains all mail that has been deleted. Mail read is distinguished from mail unread. Mail can be sorted by priority level, status (read or unread), attachment size, sender, subject, date received, flag status (follow up: yes or no), and even color.
- **Recall:** While there is no *Unsend* button, it is possible to recall mail under certain circumstances. Microsoft Office Outlook 2003, for example, supports recall if both the sender and recipient use a Microsoft Exchange Server 2000 (or later) e-mail account. It is not possible to recall a message sent to someone's personal ISP POP3 e-mail account. (*Note:* Outlook is client software; Exchange Server is the Microsoft messaging and collaboration server. IBM offers Lotus Notes on the client side and Domino on the server side.)
- **Search:** All mail is searchable by key word.
- **Security:** Mail can be encrypted through the use of a digital certificate.

4.3.4   Internet E-mail in Practice

E-mail is perhaps best understood through an example scenario, which is illustrated in Figure 4.6. *Note:* some of the specifics of this scenario are discussed in later chapters. For example, the details of Ethernet frames and TCP/IP packets and the interrelationships between them are discussed in Chapters 8 and 12, respectively.

1. Margaret Horak (my lovely bride) at the Evergreen Group composes an e-mail message on her PC client workstation using a MUA. She selects the name *Ray Horak* (me) from an address book of correspondents, and the client e-mail software addresses the mail to ray@contextcorporation.com. She attaches a text document and a graphic file, which she has compressed with a commercial software program. Margaret clicks on the *Send* button.

2. The MUA formats the message in Internet e-mail format. The MUA organizes the data into IP packets, with UDP also used to provide some level of reliability of data stream transport across the inherently unreliable networks

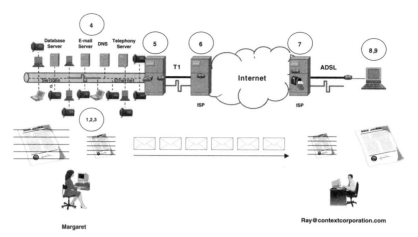

Figure 4.6 E-mail in practice.

of the local area Ethernet and the wide area Internet. SMTP is used to support the e-mail message and MIME is used to support the attached text and graphic files. The packetized mail exits the MUA.

3. The mail passes through the Network Interface Card (NIC), which organizes it into Ethernet frames, adds the originating Ethernet address of the client workstation and the destination address of the e-mail server, and places it on the corporate Ethernet LAN.

4. The MTA software on the e-mail server receives the Ethernet frames through the NIC. The MTA consults a DNS to find the name of the Internet-attached mail exchange server that accepts mail for ray@contextcorporation.com. If the address is not on the LAN-attached DNS, the DNS of the ISP may be consulted. The mail exits the MTA, headed for the Internet, addressed to a mail server at Isomedia.com, the ISP where Ray maintains the virtual domain of www.contextcorporation.com.

5. The mail traverses the Ethernet, passing through a router that strips off the Ethernet frame and presents the e-mail packets to the ISP across a digital T1 trunk. [*Note:* The T1 trunk also runs one of several protocols, including Frame Relay (FR), Asynchronous Transfer Mode (ATM) and Point-to-Point Protocol (PPP). This level of detail is outside this discussion and, therefore, these protocols are explained in subsequent chapters.]

6. The ISP receives the packets and presents them to the Internet for transport.

7. Isomedia, Ray's ISP, receives the packets from the Internet, routes them to the MTA on the e-mail server, and places them into his mailbox, which is in the form of a logical memory partition.

8. Ray, who is an independent consultant and teleworker, checks his mailbox on Isomedia's POP3 MTA e-mail server and downloads them to his MUA client software over a broadband ADSL local loop and through an ADSL router. (*Note:* The ADSL link also runs either FR or the ATM protocol.)

9. Ray opens the e-mail and attachments using a matching decompression program.

4.3.5 Applications

While e-mail traditionally has been text oriented, it is now often appended with binary, audio, video, and image files. As these files can be extremely large, they often are compressed prior to transmission in order to improve the efficiency of transmission. This is especially important if the sender or one or more of the recipients do not have broadband Internet access but rather rely on voice-grade dial-up access. The compression algorithms associated with attached files require that both the sender and the receiver have compatible software for compression and decompression. While much of that software is available at no charge, it is nonstandard and can be cumbersome to use. Nonetheless, e-mail has become essential for the conduct of business. Indeed, it is virtually universal, having found its way into widespread consumer use through online information services, connected via the Internet and mail *gateways*, which serve as protocol converters. If your experience is at all similar to mine, you receive a small fraction of the postal mail that you received 10 or even 5 years ago and you receive very few, if any, faxes. You do, however, receive dozens, if not hundreds, of e-mails daily, of which at least 75 percent are spam, offering you prescription drugs from Canada, pornography from Russia, and (counterfeit) Patek Philippe watches from The Philippines. A half-dozen of the spam messages have compressed, virus-laden e-mail attachments and another dozen offer you great wealth if you will only help to launder some ill-gotten gains for a Nigerian diplomat who has somehow gotten crossways with the national tax authority. (*Note:* My advance apologies to all Canadians, Russians, Filipinos, and Nigerians, but I don't make this stuff up.)

4.3.6 Spam and Freedom of Speech

Junk mail plagues the e-mail community, just as it does the worlds of facsimile, voice mail, and snail mail. Advertisements, once considered a breach of e-mail etiquette, have become a fact of life on the Internet. Those of us who subscribe to technology magazines almost always are requested or even required to include our e-mail addresses so vendors can deluge us with junk mail. Now that many technology magazines are no longer even available in print form, but only online, e-mail addresses are the only delivery option. Unsolicited e-mail, or junk mail, has justifiably earned the nickname *spam*, reportedly after Hormel's ever-popular canned meat product made of leftovers from the processing of pork plus lots of additives. The analogy supposedly is that junk e-mail is broadcast all over, just as the Hormel meat product spatters when hurled against a solid object with sufficient force. The U.S. federal government and a number of states have attempted to either limit or eliminate spam but always run up against arguments of the Constitution's guaranteed freedom of speech. Many consumers find unsolicited e-mail annoying and time consuming. In addition, unwanted messages sent to wireless phones and other devices can be intrusive and costly. A number of studies have shown that the costs of dealing with such mail have reached incredible proportions. Ferris Research, for example, conducted a study in 2003 that showed an associated cost of $8.9 billion for U.S. corporations, $2.5 billion for European businesses, and $500 million for U.S. and European ISPs. The cost estimates include lost time, help desk support, and infrastructure [26]. A more recent study conducted by the Radicati Group in

September 2005 indicated that 33 percent of corporate e-mail is spam and nearly 25 percent is personal in nature. In combination, spam and personal mail account for well more that half of the e-mail working its way through corporate networks [27].

Finally, in 2003, the U.S. Congress enacted the Controlling the Assault of Non-Solicited Pornography And Marketing (CAN-SPAM) Act to curb spam. As required by the act, the Federal Communications Commission (FCC) adopted rules that prohibit sending unwanted commercial e-mail messages to wireless devices without prior permission. This ban took effect in March 2005 and includes substantial penalties for violations. While there have been notable successes in terms of arrests and prosecutions under the law, many spammers have managed to elude detection and prosecution, in many cases by moving offshore and therefore outside the jurisdiction of U.S. law enforcement. While other countries face the same spam issues, many have not passed laws or otherwise taken strong measures against spammers. Perhaps the best solution, at least in the short term, remains the *block sender* or *junk mail filter* feature included in all e-mail programs.

4.3.7 Scams and Frauds and Viruses . . . and the List Goes On

E-mail is a great way to communicate, but it also is a favorite tool of scam artists and fraudsters. A popular scam is the *Nigerian Scam* or *419 Advance Fee Fraud*, named after a relevant section of the Criminal Code of Nigeria, where the scam originated and where most activity still originates. The target receives an unsolicited e-mail indicating something along the lines that some member of a previous government or royal family of Nigeria or some other West African nation (or now Iraq or Saudi Arabia or some other *exotic* place) has had substantial funds frozen by the current government and cannot access or expatriate those funds. The recipient of the e-mail can help by wiring $5000 or so as an advance fee or transfer fee for which the target will receive a much larger sum of money. Sometimes the offer is for the target to accept a cashier's check for $100,000 and return $75,000 of that via wire transfer but keep the rest as a fee. The $100,000 cashier's check is bogus, of course, while the $75,000 wire transfer is real. This sort of fraudulent activity has gone on at least since the 1970s through the postal service and via fax machines but is much more prevalent these days due to the widespread reach and ease of use of Internet e-mail.

Phishing, a recent favorite, involves unsolicited e-mail contact in which the scam artist attempts to get valuable information from the target by gaining that person's confidence through various social engineering techniques. The term *phishing* was coined in the 1996 time frame by *crackers* (malicious computer *hackers*) to describe the process of fishing for suckers by using some sort of lure or bait. (Hackers commonly replace *f* with *ph*, "phor" reasons that I still cannot entirely "phathom".) Phishing commonly involves phony e-mails from banks or other financial institutions warning that your account has been subjected to fraud or perhaps that your credit card is due to expire and that you must confirm certain information such as account number and password to prevent that occurrence. The mail includes a link to a phony website that quite closely matches the legitimate website. If the scam is successful in enticing the target, the recipient clicks on the link and divulges information necessary for the scam artist to perhaps wipe out a bank account, max out a credit card, or even steal a person's identity, incur extraordinary debts in his name, and generally ruin his credit. According to the Gartner Group, between May 2004 and May 2005, roughly 1.2 million U.S. computer users suffered phishing losses valued at $929 million. *Pharming* is a similar scam in which pop-up boxes appear at reputable websites and hijack the user, who is directed to a phony website and enters sensitive financial data there. The Gartner projection is that companies in the United States lose more than $2 billion annually as their clients fall victim [28].

Trojan horses, viruses, spyware, and other *malware* (*mal*icious soft*ware*) are more direct attacks on the host computer system that often are spread through e-mail attachments. They also commonly are spread through contact with websites, often as a result of e-mails that entice the recipient to visit those sites. Countermeasures include firewalls, e-mail filters, antivirus software, and a liberal dose of common sense. Gullibility has no place in cyberspace. Trust no one there.

4.4 INSTANT MESSAGING

Then you suddenly tore the market wide open by offering fantastic bargains, trading the six-year rights for seventh-year and eighth-year. Your computer made such lavish use of Instant Messages to Earth that the Commonwealth defense office had people buzzing around in the middle of the night.

Cordwainer Smith, *John Fisher*[100], *Norstrilia*, ibooks, 1964

The term *instant message* was first used in the science fiction classic *Norstrilia* by Cordwainer Smith, the pen name of Paul Linebarger, at least as best I can determine. In the book, Rod McBan's computer stays up late one night trading stocks on the New Melbourne Exchange on his behalf using instant messages in his recorded voice. The computer makes him enough money to buy Earth. I personally have never owned a computer that made me any money on Instant Messaging (IM), although the bank, my wife, and I do own a little less than an acre of the Earth and my computer does help me in my work. Also, I confess that I do enjoy sending the occasional instant message.

Instant messaging, at least in earthly application, is much like e-mail, but in real time. Instant messaging originated in the 1970s on PLATO, a private online instructional system for schools and universities in the United States [29]. Instant messaging was popularized in 1996 by ICQ, an Israel-based company later acquired by AOL. (*Note: Instant Message* is a service mark of AOL.) Public Web-based IM services include AOL's Instant Messenger (AIM) and ICQ, Google's Google Talk, Microsoft's MSN Messenger, and Yahoo! Messenger. A number of stand-alone *chat* clients that support all of these, as well as IRC, are commercially available. Enterprise systems include IBM's Lotus Sametime, Microsoft's Live Communications Server (LCS), and Novell GroupWise.

Instant messenging is a client/server application that fundamentally is much like e-mail, with several notable exceptions. The following is a comparison of IM and e-mail characteristics:

- **Text Based:** Like e-mail, IM is essentially for textual communications over the Internet or other IP network.
- **Multimedia:** E-mail protocol enhancements (i.e., the MIME extension to SMTP) support multimedia attachments, including text, graphics, image, video, and audio. VoIP coexists with e-mail and other applications on IP networks. Some IM systems now support multimedia attachments as well. Some systems also support VoIP and even videoconferencing.
- **Temporal:** E-mail is a store-and-forward technology designed to support communication in non–real time. IM communications, however, take place in near real time, in conversational mode, much like an *Internet Relay Chat* (IRC). IM therefore requires that both correspondents be online at the same time. Some IM systems now support one-way messaging if the recipient is not online. In this mode, the recipient can access the message at a later time, much like an e-mail communication.
- **Presence:** As IM requires that both parties be online, it is essential that some form of presence mechanism be in place to advertise the status (e.g., available or unavailable) of users. As a non-real-time mode of communication, e-mail is insensitive to presence.
- **Standards:** Although enterprise-level e-mail application software is proprietary, gateways exist to resolve protocol issues and allow intercommunication between proprietary networks. Also, e-mail is highly standardized at the Internet level through the TCP/IP suite. IM is totally proprietary, although standardization is proposed.

According to research conducted by the Radacati Group in 2005, 85 percent of businesses use IM for either business or personal reasons. The study indicates that there were over 300 million IM users globally in 2004 and that number will grow to nearly 600 million by 2008 [30]. Additional research conducted by the Radicati Group projects that the IM volume will increase from the 2005 level of 13.9 billion per day to 46.5 billion in 2009 [31] (Figure 4.7). Research conducted by analyst Michael Osterman in 2005 indicates that about 90 percent of organizations have some consumer IM operating in their networks and that 25 percent of e-mail users also are IM users [32]. Further Osterman research in March 2005 indicates that more than 50 percent of business organizations are using IM for business purposes, that virtually all organizations eventually will use some form of IM, and that 87 percent of e-mail users will also use IM [30]. There are numerous other studies that offer different statistics based on different methodologies, but all seem to suggest that the IM market is huge and that market penetration is significant and growing rapidly.

4.4.1   Features

IM systems, much like e-mail systems, offer a growing number of features that commonly include time and date stamping and the ability to read, write, reply, and archive messages. A variety of file types (e.g., text, audio, video, and image) and file formats can be attached or appended. The following is a list of typical features:

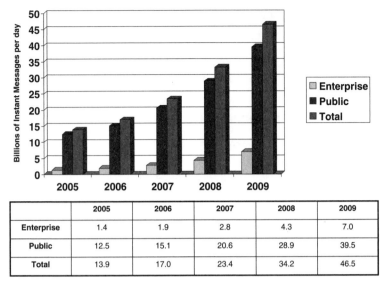

	2005	2006	2007	2008	2009
Enterprise	1.4	1.9	2.8	4.3	7.0
Public	12.5	15.1	20.6	28.9	39.5
Total	13.9	17.0	23.4	34.2	46.5

Figure 4.7   Projected growth in IM sent per day in billions: 2005–2009. (*Source:* The Radicati Group.)

- **Presence:** Status options advertised to correspondents on the contact list may include *Online, Busy, Be Right Back, Away, On The Phone, Out To Lunch,* and *Appear Offline.*
- **Privacy:** The user can block presence status to all but those on the contact list.
- **Contact List:** A contact list can be established and maintained and imported from files, e-mail contact lists, or other LDAP-compliant lists
- **Attachments:** Files and photos can be attached to instant messages.
- **Message History:** A history of messages can be maintained and organized by correspondent, in chronological order, by session.
- **Multimedia:** Voice and video communications may be supported as well as IM textual messaging.
- **Conferencing:** Some enterprise IM systems will support real-time collaboration, including audio- and videoconferencing and whiteboarding, with digital ink (i.e., handwriting).
- **Integration:** Enterprise systems feature application integration, offering access from IM to word processing, spreadsheet, graphics, and other applications.
- **Polling:** IM can include a polling feature that allows authorized users to conduct multiple-choice polls to gauge response to questions of general interest to the universe of IM correspondents or a subset of correspondents in a *buddy list,* i.e., closed correspondent group.
- **Security:** Filters are available to reduce or eliminate SPIM (*SPam over Instant Messaging*), which has become a considerable problem, especially for users connected to public services. Such filters may reject messages from senders not on an approved list, messages containing a *Uniform Resource Locator* (URL)

(i.e., website address), or messages attempting to initiate a file transfer. As is the case with e-mail spam, SPIM can be a vehicle for viruses and other malware. Enterprise-level systems typically support strong security measures, including *authentication* and *encryption.* *(Note:* SPIM is covered under the same federal CAN-SPAM Act that addresses e-mail spam.)

4.4.2   Standards and Interconnectivity

Instant messaging systems all were developed as proprietary systems without thought to interconnectivity. As the popularity of IM grew, however, it became clear that there were advantages to linking disparate systems, both public and enterprise. Gateways were developed to support protocol conversions and thereby support interconnectivity between systems, if not full-featured networking.

Microsoft's Live Communications Server (LCS), for example, is an enterprise-level system primarily designed for stand-alone use in internal messaging applications. However, users who are licensed for public IM connectivity can now add contacts, send instant messages, and share presence information with users of MSN Messenger, AOL Instant Messenger (AIM), and Yahoo! Messenger, all of which are public, Web-based services. LCS also supports *federation,* which is the ability to establish secure IM relationships with trusted business partners. In January 2006, IBM announced plans to allow users of its Lotus Sametime IM system to work with AIM, Yahoo! Messenger, and Google Talk, as illustrated in Figure 4.8. MSN Messenger and Yahoo! Messenger interconnected late in 2005, bringing together some 275 million IM users, representing approximately 44 percent of the online IM community. As both MSN Messenger and Yahoo! Messenger are based on the *Session Initiation Protocol* (SIP) *for Instant Messaging and Presence Leveraging*

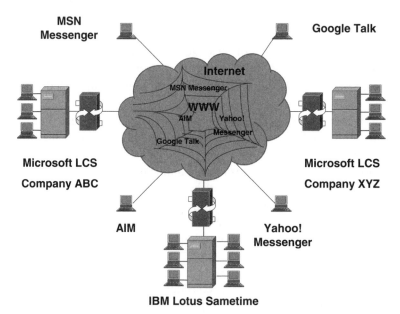

Figure 4.8   IM between enterprise and Web-based systems through gateways.

Extensions (SIMPLE) protocol, interconnection was relatively simple, so to speak. AOL's AIM remains based on proprietary protocols [33].

The gateways that will link these systems and others are based on various standards initiatives from the IETF. Those initiatives are as follows:

- **IMPP:** *Instant Messaging and Presence Protocol* (IMPP) is an effort to define the protocols necessary to build an IM system, including presence awareness and notification, and that will scale to Internet size. The IMPP group has published the following RFCs: "A Model for Presence and Instant Messaging" (RFC 2778), "Instant Messaging/Presence Protocol Requirements" (RFC 2779), and "Common Profile for Instant Messaging," CPIM (RFC 3860). The CPIM defines common semantics and data formats for IM to facilitate the development of gateways between services.
- **SIMPLE:** The predominant IM standard, SIMPLE is an extension of SIP (RFC 3261) to the suite of services the IETF refers to as *Instant Messaging and Presence* (IMP). SIMPLE is compliant with RFC 2779 and the CMIP specification. Microsoft's Live Communications Server and IBM's Lotus Sametime, both enterprise-level systems, are based on SIMPLE. Web-based MSN Messenger and Yahoo! Messenger also are based on SIMPLE.
- **XMPP:** *eXtensible Messaging and Presence Protocol* (XMPP), as specified in RFCs 3920–3923, is based on the *eXtensible Markup Language* (XML) to provide the functionality specified in RFC 2770. XMPP is deployed in the open-source Jabber system for which the initial protocols were originally developed.

While some analysts project that IM will soon eclipse e-mail as the primary mode of business communications, most find those projections unrealistic. IM requires all parties to be online, and it is often intrusive. IM just does not have the depth of support to replace e-mail, but it clearly will have a strong position for many years in a suite of messaging technologies.

4.5 MOBILE MESSAGING: SMS AND MMS

Mobile messaging is messaging in motion, which has several technological implications. First, mobile messaging must be untethered, as wires and fibers limit freedom of motion. Only wireless networks are wireless, of course. Only cellular networks currently support high-speed mobility, although *Wireless Fidelity* (Wi-Fi, the wireless LAN standard) hotspots permit limited freedom of motion. *Worldwide Interoperability for Microwave Access* (WiMAX) is a developing technology, the mobile extension of which is awaiting standardization. Second, the wireless terminal devices must be small and light and, therefore, highly portable. In contemporary terms, only cellular telephones, *personal digital assistants*, tablet PCs, and assorted other devices match those requirements. The small sizes of all of those devices limit the user interfaces. Specifically, the displays and keyboards or keypads are small on tablet PCs, smaller on PDAs, and tiny on cell phones. Improvements in display technologies such as font-size options, backlighting, brightness controls, and color serve to mitigate display size issues. Icons, shortcuts, softkeys, predictive text entry, abbreviations, multiple text entry modes, templates, voice recognition, and even digital ink (i.e., on-screen handwriting with a stylus) ease the system navigation and data entry processes. Regardless of the cleverness of the engineers and the state of the technology, however, the market for these devices is limited to those with good eyesight and small and nimble fingers.

Limitations aside, mobile messaging offers great benefits. The ability to communicate in text mode from virtually anyplace at any time is not only a convenience but also a time saver and even a life saver. Multimode communications only enriches the experience.

Mobile messaging over cellular networks takes the form of Short Message Service (SMS) for text messaging, which has evolved into Multimedia Message Service (MMS) as cellular networks have evolved to offer greater bandwidth and device technologies have evolved to support image and video capture and transmission. SMS and MMS have considerable legitimate business applications and have become so incredibly popular for personal use in the younger (i.e., preteen, teen, and young adult) market to the point that they have assumed the status of a cultural phenomenon. While is more pronounced in Asia and Europe than in the United States, the phenomenon is undeniable. Further, it likely is a permanent communications mode, rather than a passing cultural flirtation with a technological curiosity.

SMS volumes are incredible. In the Philippines, where cell phones outnumber landlines 10:1, studies show that more than 200 million SMS text messages are sent daily—that works out to roughly 73 billion messages per year [34]. According to the Mobile Data Association (MDA), SMS volume in the United Kingdom grew from 1.1 billion in 1999 to approximately 32.5 billion in 2005 [35]. In the United States, the Cellular Telecommunications and Industry Association (CTIA) estimates that 9.8 billion SMS messages were sent in December 2005 and 81.2 billion in the calendar year 2005 [36]. SMS spam is beginning to be a huge problem, which is hardly a surprise. Although the National Do-Not-Call Registry offers consumer protection against SMS spam in the United States, there are exceptions if the recipient has given prior consent to the caller or has an established business relationship with the caller. In any case, the protection is apparently only theoretical, as SMS spam is increasingly an issue. The carriers actually have a disincentive to control the problem, as SMS messages are an optional service that carries additional charges of as much as US$0.10 per message.

4.5.1 Short Message Service

Short Message Service (SMS) is a text-messaging service available on most digital cellular telephone networks. SMS was originally designed to support one-way information transfer for applications such as weather reports, sports scores, traffic reports, and stock quotes as well as short e-mail-like messages, which may be entered through the service provider's website. Most service providers also allow the cellular user to respond to e-mails via two-way SMS using the cell phone keypad. Personal digital assistants and other wireless-enabled devices often have much more functional keyboards, of course. Contemporary SMS supports two-way communication between cell phones, other wireless devices, and computers connected to the Web.

Initially SMS was defined in the *Global System for Mobile Communications* (GSM) standards, but it also has been available on a number of other digital cell networks since the late 1980s or so.

SMS is a store-and-forward messaging technology that generally includes a chat option that operates in near-real-time synchronous mode, much like IM. In fact, many IM systems support mobile communications via interfaces to SMS systems. As SMS messages use the same SMTP as is specified in the TCP/IP suite for e-mail, messaging interconnection issues are relatively modest. Interconnection of cellular networks for SMS applications presents something of a challenge in the United States as there are so many transport protocol variations. Second-generation (2G) network protocols include *Time Division Multiple Access* (TDMA), GSM and *Code Division Multiple Access* (CDMA). In every case, and whether the SMS message originates or terminates in a mobile phone, the message travels through a centralized message center, also known as a *Short Message Service Center* (SMSC). Also in every case, the SMS data are packetized, but in TDMA and GSM networks the data packet transport is over the signaling channel via *Signaling System 7* (SS7), while data packet transport in CDMA networks is over the normal data channels. [*Note:* The SS7 protocol limits the packet size to 140 octets, which translates to 160 ASCII seven-bit characters: $140 \times 8 = 1120/7 = 160$. Where other coding schemes are employed, the number of characters per packet varies. For example, a double-byte scheme such as that used in support of complex alphabets (e.g., Chinese and Japanese) limits the packet size to 70 characters. Content exceeding this limitation must be fragmented prior to transmission and reconstituted upon reception.] In most countries, GSM is the sole 2G network standard, so gateway issues are much less significant. All cellular networks are limited in terms of bandwidth, of course, although the newer 2.5G and 3G networks are much less so. The user interface is always an issue with mobile devices as display size and keyboard size are naturally limited.

Much of the SMS traffic is as a result of partnerships with public IM providers such as AOL, Google, MSN, and Yahoo!. Through these partnerships, computer users are able to correspond with mobile users on cell networks from their desktop, laptop, and tablet PCs. Mobile users also are able to access Web content on these networks from their mobile devices. Gaming is another popular application.

4.5.2 Multimedia Messaging Service

Multimedia Messaging Service (MMS), the next step in SMS evolution is currently available from many service providers and has been widely available in Japan and select other countries since 2001. MMS allows the cell phone user, for example, to create a text message, add a photo or perhaps a graphic downloaded from the Internet, add a video clip, and add a recorded audio file.

MMS makes use of the *Wireless Access Protocol* (WAP) standardized by the *3rd Generation Partnership Project* (3GPP) and *3GPP2*. 3GPP sets WAP standards for 3G cellular telephone systems based on the GSM standards, and 3GPP2 sets similar standards for systems based on CDMA2000. Although cellular technology is discussed in considerable detail in Chapter 11 and, therefore, is beyond the scope of this chapter, note that SMS and MMS will work with any digital cellular system technology, although performance is much improved if the underlying technology is either 2.5G or 3G, as available bandwidth is so much greater. The standard technologies, by generation, are as follows:

2G
- Digital Advanced Mobile Phone System (D-AMPS)
- Global System for Mobile Communications (GSM)

2.5G
- High-Speed Circuit-Switched Data (HSCSD)
- General Packet Radio Service (GPRS)
- Enhanced Data Rates for Global Evolution (EDGE)

3G
- Code Division Multiple Access 2000 (CDMA2000)
- Universal Mobile Telecommunications System (UMTS), aka Wideband CDMA (W-CDMA)
- Time Division-Synchronous Code Division Multiple Access (TD-SCDMA)

4.6 UNIFIED MESSAGING AND UNIFIED COMMUNICATIONS

Unified messaging extends across e-mail, voice mail, and fax technologies. E-mail messaging increasingly includes binary computer files, graphics files, image files, video files, and even voice and audio files. While such attachments can be very bandwidth intensive, they clearly enhance the effectiveness of the communication. But unified messaging goes well beyond the world of simple attachments, ultimately providing a single mailbox for fax mail, voice mail, e-mail, and even video mail—all accessible from a single terminal device. Text-to-speech (TTS) and fax-to-speech conversion will allow all current forms of mail to be accessed from a wired or wireless voice terminal, and speech-to-text will allow all current forms of mail to be accessed from a single wired or wireless data terminal. Video mail will be more of a challenge, even though the basic technologies have been developed and archived streaming video presentations are commonly transmitted over the Internet and even over cellular networks. Particularly in the wireless domain, unified messaging can offer significant advantages for the mobile worker when the technologies reach the point that your messaging system can find you, wherever you are, and deliver your messages to you on whatever sort of terminal device you might have, converting the message formats as required.

While speech-to-text, text-to-speech, and fax-to-speech technologies are fairly immature at this point, they are developing quickly and certainly within the next 5–10 years will be considered reliable and inexpensive enough to become commonplace. Parallel developments in network, system, and terminal technologies, particularly in the wireless domain, will greatly enhance the functionality and, therefore, the acceptance of unified messaging. The general appeal of being able to access e-mail from your cellular telephone and voice mail from your laptop or PDA is just too great for unified communications to be anything but successful. Early implementations of these capabilities have proved functional but have drawn mixed reviews because of lingering issues involving display and other interface shortcomings.

Unified communications, at least according to my definition, takes unified messaging to a higher level, allowing you to receive messages in any format you choose and to respond in the same way, regardless of the format of the original message. In other words and for example, you will be able to respond to the voice message that was converted to text before being sent to your PDA. You will enter a text response that will be sent via e-mail over a wireless network and converted to voice by your unified communications system. That voice response then will be delivered by phone to the caller or deposited in his voice mailbox unless he now must be reached through his PDA, in which case his unified communications system will recognize that fact, find him, convert the message as appropriate, and deliver the response in text format. This scenario, by the way, is one of many convergence scenarios in which voice and data intertwine over circuit-switched and packet-switched networks that are both wired and wireless, with mobility being a critical driver of the application. Taking the scenario one step further, there technically is no reason that the archives and databases of the multiple devices cannot be synchronized so that the fax that was converted into a voice message sent to a cell phone cannot be saved as a text message in all three formats on one or more servers and one or more devices, along with any annotations. In this fashion, messages can be sent from any device to any device, anywhere, anytime, and one time.

While such an ideal unified communications system does not exist at the moment, the fundamental technologies do exist and will develop to the point that such a scenario will be practical in the foreseeable future. For that matter, this scenario may well be understated, assuming that compression, display, power, and a variety of other technologies develop to even greater levels than I have assumed here. Video mail, for example, by no means is out of the question. In fact, all of these variations on the theme of unified messaging are available in early versions, which currently are being promoted heavily.

Several formal standards bodies and industry consortia have begun work on unified messaging. The *Electronic Messaging Association* (EMA) spent considerable effort in the late 1990s toward the development of a mechanism for passing voice mail between voice processors over the Internet. That effort led to the development of the *Voice Profile for Internet Mail* (VPIM), which is based on MIME and which was published by the IETF as RFC 2421 in September 1998. The IETF continues that work towards the development of RFC 2421v2 and v3, although at a slow pace. There also was some interest at the IETF in loosening the restrictions of VPIM and making it the voice component of a unified messaging protocol suite, although it appears that little progress has been made in the regard in the last few years.

REFERENCES

1. Hills, Lee and Sullivan, Timothy. *Facsimile*. McGraw-Hill, 1949.
2. Kauffman, Maury. *Internet and Computer Based Faxing*, 2nd ed. Telecom Books and Flatiron Publishing, 1998.
3. Green, James Harry. *The Irwin Handbook of Telecommunications*, 2nd ed. Irwin Professional Publishing, 1992.

4. Glossbrenner, Alfred and Glossbrenner, Emily. *The Complete Modem Handbook*. MIS Press, 1995.

5. http://www.atis.org/tg2k/_facsimile.html.

6. Grigonis, Richard. "Beyond Fax and Beyond." *Computer Telephony*, August 1995.

7. "How Network Fax Systems Work." *Datapro Communications Analyst*. Datapro Information Services Group, November 1995.

8. Coleman, Ken and Meley, Mimi. "Implementing Network Facsimile Successfully." *Datapro Communications Analyst*. Datapro Information Services Group, February 1995.

9. Sullivan, Kristina B. "Faxing Still Alive and Well." *PC Week*, March 16, 1998.

10. "Evolution of the Fax." *Information Week*, April 3, 1995.

11. Adelson, Josh. "Beyond Cost Savings: IP Fax Benefits Today." *CTI*, October 1998.

12. Arnum, Eric. "Fax Via the 'Net.'" *Business Communications Review*, October 1998.

13. "Fax Over IP—Opportunities and Options." Natural Microsystems, http://www.naturalmicrosystems.com.

14. Liebmann, Lenny. "Rethinking IP Fax." *Network Magazine*, July 1, 1999.

15. Commweb. "Gordon Matthews—Voicemail Inventor—Dies." www.commweb.com/article/COM20020225S0012, February 25, 2002.

16. "The History of Voicemail— 'Press One to Find Out . . . '" http://www.thehistoryof.net/the-history-of-voicemail.html.

17. Kalman, Steve. "Beyond Voice Mail." *Network World*, August 7, 1995.

18. Harler, Curt. "Speech Recognition Tools Get Standards, Respect." *BCR's Voice 2000*, May 2000.

19. Kinsey, Graeme. "How Interactive Voice Response (IVR) Systems Can Boost Business Efficiency." *Telecommunications*, April 1995.

20. Wolf, Robert. "A Primer to Using Voice Technologies." *TeleProfessional*, March 1996.

21. Foster, Peter and Schalk, Thomas B. *Speech Recognition*. Telecom Library, 1993.

22. "IDC Examines the Future of Email As It Navigates Security Threats, Compliance Requirements, and Market Alternatives." IDC Press Release. http://www.idc.com/getdoc.jsp?containerId=prUS20033705, December 22, 2005.

23. McCune, Jenny C. "This Message Is for You." *Beyond Computing*, June 1996.

24. Myer, Ted. "Straight Talk about E-Mail Connectivity." *Business Communications Review*, July 1995.

25. Chae, Lee. "Tutorial. Lesson 110: E-Mail and MIME." *Network Magazine*, October 1997.

26. "Study Says Junk E-mail Costs Billions a Year." *Seattle Post-Intelligencer*, January 3, 2003.

27. Gross, Grant. "Study: One Quarter of Corporate E-mail Is Personal." http://www.computerworld.com/printthis/2005/0,4814,106460.00.html.

28. Kerstein, Paul L. *How Can We Stop Phishing and Pfarming Scams?* http://www.csoonline.com/talkback/071905.html.

29. http://www.platopeople.com/about.html.

30. "What You Don't Know Can Hurt You: How to Protect and Manage Instant Messaging." White paper. Postini, http://www.postini.com/, September 12, 2005.

31. Kim, Gary. "Google Talk Makes Noise." *Fatpipe*, September 2005.

32. Osterman, Michael. "Allowing Your Users to Use IM While You Protect the Corporate Net." *Network World*, June 7, 2005.

33. Vaas, Lisa and Charny, Ben, "Yahoo, MSN to Form World's Largest IM Community." *eWeek*, October 12, 2005, http://www.eweek.com/article2/0,1759,1870365,00.asp.

34. http://www.tegic.com/press_view.html?release_num=19.

35. http://www.esendex.com/uk/news-2006-q1-another-record-year-for-sms.aspx.

36. http://files.ctia.org/pdf/Wireless_Quick_Facts_April_06.pdf.

CHAPTER 5

PUBLIC SWITCHED TELEPHONE NETWORK

Of the many wonderful scientific discoveries and inventions which have made the nine-teenth century remarkable, certainly none is of more popular interest than the simple little piece of apparatus known as Bell's articulating telephone. By this instrument it becomes possible to transmit ideas between far-distant places, not in the form of signs afterwards to be deciphered, but as actual articulations, an echo of those produced by the human voice at the point of transmission.

The Wonders of the Universe, The Werner Company, 1899

The old networks were designed around basic voice communications, or *Plain Old Telephone Service* (POTS). While the first telephone networks used transmission lines leased from Western Union, telegraphy was conducted over a separate, pre-existing network. Since data, image, and video systems did not exist until many years later, there certainly was no requirement to network them. The characteristics of voice traffic were, and still remain, well known and easily understood. Specifically, voice calls are occasional, short, bidirectional, continuous, stream oriented, and analog in their native form.

Voice calls typically are occasional, because talking on the telephone is not the focus of most people's personal and professional lives. Clearly there are exceptions, such as agents who work in call center environments. Human-to-human voice calls also tend to be short in duration–3–4 min or so, on average. While business calls tend to be short and to the point, purely personal calls are more social in nature and tend to last longer. Additionally, some people tend to express themselves in very concise terms, while others tend to be very wordy. (My late mother, rest her soul, was one of the wordy ones. I am not.)

Clearly, voice calls are conversational, or bidirectional, in nature. (Conversations with my mother tended to be one-way.) Further, the usage of the circuit is fairly

Telecommunications and Data Communications Handbook, By Ray Horak
Copyright © 2007 Ray Horak

constant, or potentially so. (See the previous parenthetical comments.) Regardless of the direction of the conversation (talking versus listening) and the number and length of pauses in the conversation, the network must support the communication continuously. Voice communication is characterized as *isochronous* and stream oriented. In other words, voice (and video) information flows at a constant and regular pace, with each utterance, or lack thereof, being of equal importance. Therefore, all elements in the network must be available continuously to accept, switch, transport, and deliver the data—this effectively means 100 percent of the time. Further, each element of the voice communication must be delivered to the receiver in exactly the sequence in which it was presented to the transmitter and with no significant level of either *latency* (i.e., delay) or *jitter* (i.e., variation in delay). Finally, the native information signal is analog, rather than digital, in nature.

Analog transmission over copper twisted pairs was quite suitable and remains so. But the inherent advantages of other media (Chapter 2) often make them preferable in contemporary applications. Circuit switching also was quite suitable and still is. Manual switchboards eventually gave way to Step-by-Step (S × S), CrossBar (XBar), and then Electronic Common Control (ECC) switches. Local switches were interconnected with other local switches through intermediate network switches for longer distance connections. Contemporary networks, of course, are much more sophisticated, supporting incredible volumes of not only voice but also data and even some level of video and image transmission. Analog has given way to digital, copper has yielded to fiber, wireline networks have given way to wireless in many applications, and so on.

A number of developing next-generation public networks are based on packet switching, rather than circuit switching. While packet switching was developed specifically for data communications, it offers both lower costs and increased functionality in support of voice. But quality issues can be considerable in packet-switched networks. While this chapter focuses on conventional voice networks, it briefly explores *Voice over IP* (VoIP) as a developing replacement for the Public Switched Telephone Network (PSTN). I discuss VoIP and *Voice over Frame Relay* (VoFR) in considerable detail in later chapters. While both VoIP and VoFR networks make use of highly sophisticated compression techniques to deal with quality issues, reliable *toll quality* voice (i.e., the traditional level of quality you have come to expect) is difficult to achieve in such networks.

As more people became comfortable with the new telephone technology, more people subscribed to the service, which made the technology more useful. Businesses came to depend on telephony, and usage increased. When costs began to drop, the technology became generally affordable and usage increased even more. Long-distance usage, which was highly profitable to the carriers, proved to be extremely price sensitive.

Around the turn of the twentieth century, the Bell System had approximately 800,000 telephones in service—that compared to about 600,000 for the nearly 6000 independent, privately owned telephone companies that sprang up in the United States [1]. In many cities and towns, multiple telephone companies operated in direct competition under franchises granted by the local government. In sparsely populated rural areas and isolated towns and cities (e.g., in Alaska) where telephone service was not commercially viable, telephone cooperatives and municipally owned telcos formed, and a number of them still exist. Approximately 1300 U.S.

independents (non-Bell companies) still exist according to the USTA (United States Telecom Association). Notably, the telephone was not accepted elsewhere quite so quickly. Reportedly, a group of British experts stated that the telephone "may be appropriate for our American cousins, but not here, because we have an adequate supply of messenger boys." As late as 1937, half of the telephones in the world were in the United States, which accounted for approximately 6 percent of the world's population. At that time, there were about 14 telephones for every 100 population in the United States, 11 per 100 in Canada, and 2 per 100 in Europe [2].

Initially, these networks were isolated islands of local service. The telephone companies soon interconnected with each other and with the long-distance network, usually under government and regulatory pressure. In the United States, the Bell System avoided providing interconnection, preferring to aggressively acquire the independent telcos and even acquiring a large block of stock in Western Union Telegraph Company, which had entered the telephone business in 1877 and was its chief competitor at the time. Increasingly pressured by the federal government under antitrust laws, Nathan C. Kingsbury, an AT&T vice president, wrote a letter to the U.S. Attorney General in December 1913 to resolve various issues. Known as the *Kingsbury Commitment*, that letter committed AT&T to dispose of its holdings in Western Union, to purchase no more independent telephone companies without the approval of the Interstate Commerce Commission (ICC), and to make interconnection with the independent telephone companies. Eventually, national standards were established to govern the nature and rules of interconnection. The ITU-T governs standards recommendations at the international level.

Many nations acquired, or even confiscated, the telephone networks from the private owners, forming *Post, Telegraph, and Telephone* (PTT) agencies. French telephony, for example, was originally conducted by private interests but was virtually confiscated in 1889 by the government, "which proceeded to inflict upon it the worst evils of bureaucracy. Rates were inequitable; there was no national plan or research effort; subscribers were required to buy their telephones; and operators were subject to baroque bureaucratic rules, such as being forbidden to marry policemen, cashiers, foreigners, or mayors of towns, lest they betray the secrets of the switchboard" [1] (*Note:* The French company Alcatel is now in the process of acquiring Lucent Technologies, which includes the remnants of Bell Labs, the research and development arm of the Bell System prior to divestiture in 1984. At the time, Bell Labs was on the order of a national treasure. Let us all hope that the French do not "betray the secrets of the switchboard" once they gain access to them all.) On January 12, 1912, the British government assumed full control and ownership of the national telephone system, leaving the United States as the only major nation in which the network was privately owned. Eventually, people viewed access to telephone service as a basic human right in many industrialized nations. In 1935, an act of Congress defined the concept of *universal service* and created the Federal Communications Commission (FCC), which was chartered with accomplishing this goal. The goal (admittedly paraphrased and based on my best recollection from my days at Southwestern Bell) of the Bell System, for example, became "universal service of the highest possible quality at the lowest possible cost." In order to ensure that high-quality basic service was available universally, a complex set of cross-subsidies formed to fund the deployment of the network in high-cost areas, with particular emphasis on single-line residential service. The Telecommunications Act of 1996 amended the Communications Act of 1934 to require that all telecommunications companies that provide interstate telecommunications services contribute money to preserve and advance universal service.

Many governments have discovered that they cannot easily fund the capital-intensive upgrades their networks require, and they cannot effectively serve the communications needs of their constituents at reasonable cost. They also have come to recognize that the networks are immensely valuable assets that can be sold to fund various social programs. Therefore, the networks have been *liberalized*, that is, opened to competition, in many nations. The networks also rapidly are being *privatized*, either partially, with the government continuing to hold controlling interest for some period of time, or completely. As a result, service has become more universally available around the world, the quality of service has improved, the range of services has increased, and the overall cost to the user has dropped considerably. *Note:* This is a trend, rather than a universal fact, and progress sometimes is slow.

In the Republic of South Africa (RSA), for example, Telkom South Africa was government owned until May 1997, when 30 percent of the company was sold to the Thintana consortium of SBC Communications (United States) and Telekom Malaysia. Telkom South Africa remained a state-enforced monopoly until 2003, when the search began for a qualified Second Network Operator (SNO). Also in 2003, the company went public and was listed on the Johannesburg and New York stock exchanges in March. The government held competitions for the SNO and actually awarded the licenses a time or two, only to retract them for various political reasons. The SNO licenses were awarded again in early 2006 and SNO Telecommunications prepared to launch services in selected areas of the country. The hope is that SNO will take a more creative approach to its service offerings than it took in the selection of its name.

5.1 NETWORK CHARACTERISTICS

Each type of network can be described in terms of a number of key characteristics that define its basic nature and application. The PSTN, also known as the Global Switched Telephone Network (GSTN), can be characterized as designed for voice communications, primarily on a circuit-switched basis, with full interconnection among individual networks. The network is largely analog at the local loop level, digital at the backbone level, and generally provisioned on a wireline, rather than a wireless, basis.

5.1.1 Voice (Primarily)

The original network was designed to carry voice communication only. At the time, the only other form of telecommunication was telegraphy, which was the province of Western Union Telegraph Company in the United States. The contemporary PSTN still exists primarily to support voice communication. While much data traffic continues to tranverse the network, intensive data traffic largely travels special-purpose data networks, or physical and logical partitions of the PSTN. The PSTN also can support the transfer of image information, with facsimile being an excellent example. Additionally, special videoconferencing equipment and interfaces support video traffic.

Note that much Internet traffic travels over the PSTN, at least through the *Local Exchange Carrier* (LEC) access networks. For example, a great many users access an Internet Service Provider (ISP) over the PSTN through a modem embedded in a desktop or laptop computer. Over an analog local loop, the modem dials the telephone number of the ISP, and the connection is provided through a circuit-switched Central Office (CO) designed for voice communications. The ISP's local loop is in the form of a channelized T1 circuit designed for voice. Should you need a connection to the Internet, it is supported from the ISP to the Internet backbone over an unchannelized T3 leased from the LEC and designed for voice. This traditional method of Internet access is giving way to new technologies such as *Digital Subscriber Line* (DSL), *cable modem*, and *Wireless Local Loop* (WLL) technologies. I discuss these Internet access scenarios and the underlying technologies in detail in later chapters.

5.1.2 Switched (and Dedicated)

Traditional voice networks are largely circuit switched, thereby providing great flexibility to the end users and significant economies to the carriers. (*Note:* While packet-switched networks are rapidly replacing circuit-switched networks in the United States and some other highly developed regions of the world, we will defer that discussion.) Large user organizations that communicate intensively between well-defined physical locations (e.g., headquarters, region, and field locations) often employ dedicated circuits in order to mitigate network costs. Such dedicated circuits generally are leased from the carrier(s), making use of existing PSTN transmission facilities. In relatively rare instances, large end-user organizations deploy their own facilities, often in the form of private microwave, to extend connectivity to areas that the telcos find to be economically unattractive or where telco construction charges would be excessive. Also in rare instances, large end-user organizations purchase or lease from the carrier a set of fiber-optic facilities that can be used to connect the user's locations on a dedicated basis. Such optical fibers are contained within a larger set of facilities originally deployed for the carrier's own use and sold or leased to the user on an as-available basis. As *Wavelength Division Multiplexing* (WDM) made its appearance, carriers began leasing individual wavelengths, or *lambdas*, to end-user organizations and other carriers.

5.1.3 Analog (and Digital)

The traditional voice network originally was analog in nature. The local loop connection terminating at the residence or small-business premises generally is still analog, with the exception of Integrated Services Digital Network (ISDN), which is digital from end to end, and, very recently, Fiber-To-The-Premises (FTTP). ISDN never achieved any appreciable level of penetration in the United States, although it was widely deployed in many other countries. Large user organizations often have digital access circuits in the form of T-carrier or E-carrier, which may be formatted as leased lines or ISDN.

Backbone circuits in the internal telco networks and internetworks largely have converted to digital for reasons that include increased bandwidth, improved bandwidth utilization, superior error performance, and enhanced network management capabilities. This conversion process largely is complete in most major industrialized nations and regions, including Australia, Hong Kong, Japan, New Zealand, North America, Singapore, and Western Europe. Developing nations and regions (e.g., Eastern/Central Europe, Sub-Saharan Africa, The Philippines, and Thailand) either are converting the PSTN to digital or are abandoning obsolete analog facilities in favor of constructing new digital networks.

5.1.4 Interconnected

In 1903, only thirty-nine per cent of the world's telephones could be connected with the Bell System. Today, every telephone served by the Bell System can reach about ninety-two per cent of all the telephones in the world.
           Telephone Almanac, American Telephone and Telegraph Company, 1934

Complete interconnection between all local and long-distance and national and regional networks is fundamental to the PSTN. Islands of telephony were interconnected many years ago in order that any subscriber might have the ability to connect to any other, subject to availability of transmission and switching capacity and issues of national or regional security. In other words, any-to-any connectivity is fundamental.

5.1.5 Wired (and Wireless)

While the traditional networks largely are wired (i.e., twisted pair and optical fiber), much wireless technology also is employed. Wired networks generally perform better in terms of bandwidth, error rate, security, and other dimensions, as is discussed in Chapter 2. Fiber-optic transmission systems are particularly notable in these respects. Note that wired networks, and especially those employing optical fiber, are very much under the control of the owner, because they are not so susceptible to environmental interference (e.g., precipitation, sunspots, and static electricity). Further, each wire or fiber is a self-contained world of bandwidth, whereas wireless systems are limited by spectrum availability and frequency allocation. While wired networks often are more difficult, time consuming, costly, and sometimes impossible to deploy and reconfigure, they can be advantageous in the long term.

A limited radio spectrum, electromagnetic interference, lack of security, and an assortment of other limitations all plague wireless networks. Therefore, they generally are avoided in the architecture of the PSTNs of developed nations. Microwave, however, is used extensively in many networks, either as a primary means of connection where wired networks are impractical or as backup in the event of network overload or failure. Satellite communications also is used extensively for certain international communications and in order to provide access to island nations and remote areas. *Wireless Local Loop* (WLL) technology has gained in popularity in recent years as an alternative means of network access. WLL also finds application as a *facilities bypass* technology that enables the user organization to bypass the facilities of the *Incumbent Local Exchange Carrier* (ILEC) in order to gain access to an *IntereXchange Carrier* (IXC) or other service provider. WLL often compares quite favorably with ILEC access circuits in terms of both speed of deployment and cost. However, issues of security and interference remain. (*Note:* I consider cellular networks to be a special case and not part of the PSTN as such. I spend a good deal of ink discussing cellular networks, in specific, and wireless networks, in general, in Chapter 11.)

Developing nations, in particular, make extensive use of satellite and microwave systems and currently deploy WLL aggressively. The advantages of wireless communications, as is noted in Chapter 2, include both the rapidity and low cost of deployment, particularly in difficult terrain (e.g., mountainous areas where the soil is rocky), where islands must be interconnected, or where remote or sparsely populated areas require service. India, for example, makes extensive use of satellite communications to link thousands of remote villages. Similarly, remote towns in the outback of Australia are linked to the PSTN via satellite. Carriers use microwave extensively to link cities in island nations such as the Philippines and Malaysia, where WLL also has found considerable application.

5.2 NUMBERING PLAN ADMINISTRATION

In order for any telephone to connect to any other telephone in the world, a carefully developed and administered numbering plan, or *logical addressing* scheme, must be in place. The ITU-T is responsible for Numbering Plan Administration (NPA) at the international level, with each nation or region having similar responsibility within its domain. In 1947, AT&T and Bell Telephone Laboratories established the *North American Numbering Plan* (NANP) as a means of integrating the *area codes* and Central Office Exchange (COE) codes in the area loosely known as North America. This area, officially known as World Zone 1, excludes Mexico and includes the Continental United States, Hawaii, Canada, Puerto Rico, the Virgin Islands, and parts of the Caribbean. Within each state in the United States, the dominant LEC (read the *Bell Operating Company*) administered the NANP under the direction of the FCC [3].

Coincident with the breakup of the Bell System, in 1983 *Bellcore* (*Bell Communications Research*) assumed responsibility for the NANP. The Canadian Radio-television and Telecommunications Commission is responsible for the numbering plan in that country. In the Caribbean, some of the various governments delegate administration, while others retain that responsibility. In 1995, the *North American Numbering Council* (NANC) was chartered as an impartial body to assume oversight responsibility for the NANP. In January 1998, NANP direct administration was removed from Bellcore (now Telcordia Technologies) and placed in the hands of Lockheed-Martin, thereby enhancing the administrative function with a presumed level of impartiality. Lockheed Martin formed NeuStar as an independent business unit specifically to handle that function and subsequently spun off NeuStar as a separate entity. The U.S. number format comprises fixed-length, 10-digit national numbers which fit into the international NPA dialing scheme (+CC.NPA.*NXX*. *xxxx*).

The ITU-T *E.164* recommendation (*The International Public Telecommunication Numbering Plan*) specifies the current international NPA convention at a maximum of 15 digits, although the number of digits required for calling within a nation varies. In many cases, numbering schemes vary within the same country; 6- and 7-digit telephone numbers, for example, coexist in Namibia and many other countries.

When dialing a telephone number, the user effectively instructs the network to establish a connection between two physical addresses based on a logical address, which is a series of numbers following a specific pattern. Using an example with which I am familiar (it's my company), the following logical addressing convention functions in establishing a connection originating from a physical address in Johannesburg, South Africa, and terminating at another associated with the Context Corporation at 1500A East College Way, Suite 443, Mount Vernon, Washington 98273, United States (and in reverse):

- *0* is the access code for an outside line from behind a PABX. (9 is the access code for an outside line from behind a PBX in the United States.)
- *09* indicates that the call is international. (The international access code for calls originating in the United States is 011.)
- *1* is the *country code* for the United States. (27 is the country code for South Africa.)
- *360* indicates the *area code*, or NPA. Specifically, 360 is the geographic area of western Washington surrounding the Greater Seattle metropolitan area. *Note:* 360 is in Local Access and Transport Area (LATA) 674. (11 indicates a specific area in Johannesburg, Guateng Province, and is akin to an area code.)
- *428* is the *Central Office* (CO) *prefix*, which indicates a specific Mount Vernon COE of Verizon, the ILEC. These three digits are associated with an exchange that resides in a wire center, which may house several such exchanges. The exchange also is a *rate center*, meaning that it is identified by V&H (*Vertical and Horizontal*) coordinates that are used to calculate the distance between that rate center and another so that rates involving distance-sensitive call rating algorithms can be calculated correctly. (The dialing pattern is similar in South Africa from this point, as local telephone numbers are seven digits.)
- *5747* is the *line number*, which indicates the port and circuit ID, which is associated with a local loop, which in turn is associated with terminal equipment at the physical address of The Context Corporation.

Through such a series of dialing steps, the logical address of the Context Corporation is translated, in steps and across carrier domains, by several networks in order to route the call to the target telephone system. Changes in the dialing scheme require changes in switch logic (e.g., PBXs and COs) in order that the switches can recognize the legitimacy of the dialing pattern. The Context Corporation also has an 800 number (i.e., IN-WATS, or toll-free, number) that the network translates into 360.428.5747 in order to route domestic calls on a toll-free basis. As the 800 number points to a regular telephone number associated with a multifunctional local loop rather than a separate INWATS line, the number technically is in the form of *virtual WATS*.

Demand for telephone numbers increased considerably beginning in the early 1990s due to the popularity of cellular telephony, fax machines, pagers, and modems for computer access to the Internet—all of which require telephone numbers. The Telecommunications Act of 1996 added considerably to the pressure on the NANP as CLECs (*Competitive Local Exchange Carriers*) began to request the assignment of telephone numbers for their potential customers. The NANP convention called for numbers to be assigned by CO prefix, each of which contains 10,000 numbers (0000–9999). So, a LEC requiring 10 or 100 numbers in a given geographic area was assigned a full block of 10,000 numbers, the vast majority of which went unused. Further, requests for number blocks were approved fairly automatically, regardless of their legitimacy. Under that sort of pressure, the FCC approved 640 new area codes in 1995 and expected another 88 existing area codes to be exhausted by 2010 or so [4–10].

Subsequently, however, common sense prevailed as consumers put pressure on the regulators to reconsider the introduction of new area codes. As a result, LEC requests for number blocks were scrutinized and were rejected if their existing number blocks were considered to be less than fully utilized. Further, the NANC introduced number pooling, whereby multiple carriers are each assigned blocks as small as 1000 numbers within a pool of 10,000 per CO prefix. The FCC also instituted a number reclamation policy to ensure that unused numbers are returned to the pool and a 60 percent utilization threshold, increasing to 75 percent over time, that carriers must meet before getting additional numbers in their service area. As of January 1, 2006, 328 area codes were in service out of a possible 800 available combinations that fit the acceptable numbering convention. Current projections are that the remaining supply of area codes will last until at least 2039 [10]. Similar factors in Europe have pressured the numbering plans in that part of the world as well— hence the expansion of the international dialing convention to 16 digits and the eventual expansion of the NANP to 11 or 12 digits once the supply of existing CO prefixes and area codes is exhausted.

The changes in the NPA have impacted the network to a considerable extent, as all network switches must be reprogrammed to recognize and honor dialing instructions under the new conventions. At least as importantly, NPA changes have significant impacts on end users, who must keep current and prospective customers and other contacts informed of telephone number changes. Among other things, advertising, stationery, business cards, and signage must be changed and at considerable cost.

In order to understand the implications of NPA changes, you must understand the difference between the old and new conventions. Traditionally, the U.S. area code was a three-digit, *NNX number*. In other words, the first and second digits were limited to specific numbers (*N*), while the third digit consisted of any number (*X*). Specifically, the first digit could be any number other than 1 or 0, either of which would confuse the network, which would interpret it as an instruction to provide access to an operator, connect a call across an area code boundary, or connect an international call. The second digit was required to be a 1 or 0, and the third digit could be any number. As the existing area codes were exhausted, the FCC opened the second position in 1995. Consequently, The Context Corporation (my company) found itself removed from the Seattle 206 area code and established in the new 360 area code on the basis of a geographic split. In order for callers to reach the company, all involved switching devices had to be reprogrammed to recognize the validity of the new area code and, thereby, process the call, rather than reject it. End users were required to make software changes in their PBX systems if they wanted to be able to call the new area code, and the carriers had to make software changes to all network switches. Not only were such software enhancements expensive, but also some older switches could not be upgraded. The end users were denied access to the new number unless the system was replaced or peripheral equipment was added to effect compatibility. I had to advise all my existing and prospective clients of the number change and had to go to the expense of changing my advertising, letterhead stationery, business cards, website, and so on. For my small independent consultancy, the costs and inconveniences were relatively minor. For a large multinational company, the costs easily can run into the millions of dollars, the inconveniences can be significant, and the loss of business can be considerable.

Notably, both area codes and CO prefixes are geographically specific, with the exceptions of 500, 700, 900, and toll-free (e.g., 800, 888, 877, and 866) services. When an existing area code is exhausted and a new one is required, there are two choices. The traditional choice includes a geographical split of an existing area code (e.g., 206 and 360), but this causes considerable disruption. The contemporary choice increasingly is that of an overlay area code, which requires that the caller always dial a 10-digit telephone number, even to place a local call. Neither solution is particularly attractive.

Note, however, that this geographical limitation disappears or at least can be extended with Internet telephony, that is, telephony over the IP-based Internet. As an example, an Internet telephony provider can subscribe to a telephone number in the financial district of Manhattan, New York. The end user can connect to that telephone number from a computer-based IP softphone through the Internet from anyplace in the world. So, a call placed from Dallas, Texas, to the Manhattan telephone number can be connected through the Internet to the subscriber, who may be in a hotel room in Mumbai. The subscriber in Mumbai can call through the Internet, connect to that same number in Manhattan, and place an outgoing call to Dallas, with the Manhattan telephone number appearing on caller ID at the Dallas telephone. Whether the call is incoming or outgoing, the Dallas party has no knowledge of the fact that the other party is anywhere other than Manhattan. This technology may well contribute to shortages of telephone numbers associated with major financial centers such as New York and Hong Kong, for example, and may further enable financial scams and other fraudulent activities.

Note also that in September 2000 the Internet Engineering Task Force (IETF) approved RFC 2916, which documents what is commonly known as ENUM (*Electronic NUMber*). ENUM bridges the gap between the logical addresses specified in the PSTN by E.164 and the logical addresses specified for IPv4 in RFC 791 and IPv6 in RFC 2460. Clearly, it is important to translate between the addressing schemes in a scenario in which voice calls travel both the PSTN and the Internet or other IP network. Therefore, both E.164 and IP addresses are registered with the ENUM *Domain Name Service* (DNS), which can be consulted by gateways that interconnect the two disparate networks.

5.3 DOMAINS

Perhaps the most effective means of examining the fundamental nature of the network is to consider it in terms of *domains*, or spheres of influence. The traditional voice network can be organized into functional, regulatory, and carrier domains.

5.3.1 Functional Domains

Functional domains address the various functions performed in the PSTN. Customer premises equipment, switches, and transmission facilities all are physical elements of the PSTN, each performing specific functions and all supported by a signaling and control system. At a higher level, you can view the PSTN in terms of providing the functions of network access, transport, switching, and service delivery. According to the 1934 AT&T *Telephone Almanac:*

> *A telephone instrument, taken by itself, is merely a thing of ingeniously fabricated insulating materials and metal. A telephone instrument, capable of interconnection with other telephone instruments, not only locally but with a nation-wide and international communication system, becomes an agency of human service whose horizons extend beyond those on commonwealth or continent. ... So measured, its value is limitless.*

5.3.1.1 *Customer Premises Equipment*

Customer Premises Equipment (CPE) is the term for transmit and receive equipment in the voice realm, including voice terminals (telephone sets), key equipment, PBXs, Automatic Call Distributors (ACDs), and peripheral equipment such as answering machines. Historically, the local telephone company owned the entire network, from transmitter to receiver, including all voice CPE, which the end user was required to lease from the telco. Beginning in the United States with the Carterfone decision in 1968, end users were allowed to purchase terminal equipment and connect it to the network as long as it was connected through a properly registered interface device rented from the telephone company. In 1975, the FCC's *Part 68* registration program allowed for the manufacturers of *foreign* (i.e., nontelco) equipment to certify that it would cause no harm to the network, thereby eliminating any need for the interface devices.

Data Terminal Equipment (DTE) is the parallel term in the data realm. As computer equipment historically comprised islands of mainframes and attached terminal and printers confined within a data center, issues of interconnection across public networks were modest until the 1970s. So, DTE historically was owned by the end-user organization. Further, the telcos considered DTE to be too complex and, therefore, had no interest in it. A notable exception, of course, was *teletypewriters*, which the telco owned and interconnected through *telex* and TWX networks. There were small numbers of terminals connected to mainframes across separate networks through modems, which were leased to end users by the telcos, but the numbers were so small that the issues of ownership were of relatively little concern. The Carterfone decision applied to this equipment as well, so end users were allowed to purchase and interconnect DTE beginning in the late 1960s.

5.3.1.2 *Inside Wire*

Inside wire includes all wires and cables located inside the customer premises. In a midsize or large business enterprise, such wires and cables might connect the terminal equipment to the voice common equipment, such as a PBX or a Key Service Unit (KSU), and from there to the *demarcation point* (*demarc*)—the point of delineation between the customer premises and the carrier network. A Centrex customer typically has P-phones that connect directly to the demarc, with no intermediate common equipment. A small business or residence typically uses single- or multiline telephones, also connected directly to the demarc.

The demarc generally is in the form of a *Network Interface Unit* (NIU), also known as a *Network Interface Device* (NID), that includes some form (carbon fuse, gas tube, or electronic chipset) of lightning protector that insulates the premises from potentially disastrous high-voltage current caused by lightning strikes on the outside copper cable plant. Contemporary NIDs also contain a chipset that supports remote testing of the local loop from a centralized *Network Operations Center* (NOC). In the case of some older business installations, the demarc is merely a logical point of demarcation, tagged as such by the ILEC.

Historically in the United States, the inside wire was owned by the telco, which had end-to-end responsibility for the network. Under the terms of the *Modified Final Judgment* (MFJ) in 1984, the responsibility shifted to the customer for all new inside wire. Further, the customer was given the opportunity to take ownership of and assume full responsibility for that already in place. Therefore, the demarc also became the point at which the carrier's responsibility ends and the user's begins. Such responsibilities include installation and management, such as upgrade, maintenance, and security. Many other countries have since followed that same path of deregulation.

5.3.1.3 Switches Switches are the devices that establish connectivity between circuits through an internal switching matrix. In a traditional PSTN environment, such switches are circuit switches, which include COs, tandem exchanges, access tandem exchanges, and International Gateway Facilities (IGFs). On demand and as available, circuit switches set up such connections between circuits through the establishment of a *talk path*, or *transmission path*. The switches set up and maintain those connections and provide the associated bandwidth temporarily, continuously, and exclusively for the duration of the *session*, or call.

5.3.1.3.1 Central Offices The *Local Exchange Carriers* (LECs) own *Central Offices* (COs), also known as *Central Office Exchanges* (COEs) and just *exchanges*. COs provide local access services to end users via local loop connections within a relatively small area of geography known as an exchange area, or *Carrier Serving Area* (CSA). In other words, the CO provides the ability for a subscriber within that neighborhood to connect to another subscriber within that neighborhood by dialing a 7-digit number (*NXX-XXXX*) in the United States. (*Note:* Dialing patters can vary in other countries.) In cases where overlay area codes are in place, a 10-digit (*NXX-NXX-XXXX*) dialing pattern is required. The FCC has since mandated 10-digit dialing for many geographical areas, to put new carriers and users in new overlay area codes on equal footing with incumbents. COs also provide a number of services, such as custom calling features (e.g., call waiting, call transfer, and three-way calling) and Centrex. Also through the CO, a subscriber typically gains access to the LEC metropolitan calling area, which does not involve a long-distance (toll) charge, by dialing a 7-digit number (*NXX-XXXX*). In some cases zone charges apply for calling, generally between noncontiguous zones, with each *zone* comprising multiple CSAs. Finally, most subscribers also gain access to the various long-distance networks through the CO, with a domestic long-distance call requiring dialing 1 + a 10-digit number (1 + *NXX-NXX-XXXX*) in the United States.

COs, also known as *end offices*, reside at the terminal ends of the network. In other words, they are the first point of entry into the PSTN and the last point of

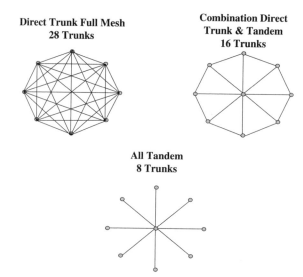

Figure 5.1 Network configurations: direct trunk full mesh, all tandem, and combination direct trunk and tandem.

exit. They also are known as *Class 5 offices*, the lowest of the five classes in the switching hierarchy, and *edge offices*, as they are at the very edges of the service provider's network. Manufacturers of COs include Lucent Technologies (5ESS), which previously was AT&T; Nortel (DMS), which previously was Northern Telecom; Siemens (EWSD); and Ericsson (AXE).

5.3.1.3.2 Tandem Switches *Tandem switches* are network switches that serve in partnership with lesser switches, linking them together. In other words, tandem switches serve no end users directly; rather, they serve to interconnect lesser switches. At the lowest level, tandem switches serve to link together CO switches over dedicated interoffice trunks. This approach can be used to form a fully interconnected and toll-free metropolitan calling area, for example. There are a number of basic network topologies, including full mesh, full tandem, and a combination tandem and direct trunk plan.

- **Full Mesh:** If all COs are interconnected through direct trunking in a full-mesh network topology, a large number of trunks and trunk groups are required, as calculated by the formula

$$X = \frac{n(n-1)}{2}$$

where n is the number of nodes to be interconnected. If there are eight nodes, as illustrated in Figure 5.1, the number of trunks required is calculated as

$$28 = \frac{8(8-1)}{2}$$

So, a full-mesh configuration of eight COs requires 28 trunks, if all trunks are full duplex (i.e., two-way). If all trunks are simplex (i.e., one-way), then twice that number, or 56, trunks are required to implement a full-mesh, full-duplex configuration. This approach is the most trunk intensive but provides a direct path between any two switches. This approach also presents the lowest risk of catastrophic failure, as each switch can exercise multiple paths to every other switch in the network, under the direction of a centralized signaling and control system.

- **Full Tandem:** A full-tandem configuration is the most efficient in terms of trunking, as it reduces the total number of trunks to eight. However, this approach places a great traffic load on the tandem, which much provide all interconnections between the COs. The tandem also becomes a single point of failure, which may present an unacceptable level of risk to the carrier. Further, the tandem adds some amount of call setup time and propagation delay. This configuration also is a physical and logical *star configuration*.

- **Combination:** A combination configuration often is the most reasonable approach, as it strikes a balance between trunking and switching. Adjacent COs interconnect directly; nonadjacent COs connect through the tandem. If there is a failure at the tandem level, nonadjacent COs can interconnect via alternate paths under the direction of a centralized signaling and control system.

Depending on the design philosophy of the service provider and the size of the network, there may be as many as four levels of tandem switches in the backbone of the circuit-switched PSTN. At each level, the tandem switches serve to interconnect lesser network switches. Contemporary network switches often are multifunctional, with one physical Class 4/5 hybrid switch partitioned to serve both as a Class 4 tandem and a Class 5 CO.

5.3.1.3.3 Access Tandem Switches *Access tandem switches* serve to connect the LECs (telcos) to the IXCs (long-distance carriers) over dedicated interoffice trunks, known as *access trunks*. In this manner, the local service providers originally were interconnected to the long-distance providers. As access tandem switches represent an additional point of potential network failure, involve additional costs to the IXC, and impose additional delay on the process of call setup, the larger IXCs often terminate high-capacity trunks directly in the LEC COs or tandem switching centers. Thereby, the access tandem arrangement is bypassed. Where the physical space is available and can be securely partitioned, the LECs must lease space to the IXCs in order that they can collocate their termination facilities in the LEC CO. The cost to the IXC of such leased space is based on actual costs to the LEC plus a reasonable profit margin, at least theoretically. (*Note:* The LECs maintain that, in practice, the cost to the IXC is heavily discounted.)

5.3.1.3.4 International Gateway Facilities *International Gateway Facilities* (IGFs) are the switches owned by the international carriers. Located at *landing points* on each end of the international connection, they provide connectivity between the international carriers and the national and local carriers on the originating and terminating ends. The IGFs provide physical gates between the international and national networks. They also serve as protocol converters, converting between T-carrier and E-carrier, for example.

5.3.1.4 *Transmission Facilities*

Transmission facilities, which are explored in Chapter 2, are the physical transmission media and associated electronics that provide the circuits in all domains of the PSTN. Oftentimes, some combination of twisted-pair, coaxial cable, microwave, satellite, infrared (IR), and optical fiber is employed.

5.3.1.4.1 *Access*

The local loop functions to provide access to the carrier-provided *Wide Area Network* (WAN). Access facilities typically extend from the customer premises to the LEC CO exchange, with the *demarcation point*, or *demarc*, serving as the point of separation between the CPE and LEC domains. Generally, the *incumbent* LEC (ILEC) provides the access facilities, which terminate in the ILEC CO. A *Competitive* LEC (CLEC) may provide its own access facilities, typically in the form of optical fiber to areas where there is a significant concentration of high-volume commercial traffic. Access increasingly is provisioned through Wireless Local Loop (WLL) technologies, with various microwave systems typically used for this purpose, although infrared (IR) is an option. Alternatively, the CLEC may lease existing facilities from the ILEC in those states that support competition in the local loop. Access to voice networks increasingly is provided by *Community Antenna TeleVision* (CATV) providers using coaxial cables or *Hybrid Fiber/Coax* (HFC) transmission facilities to connect from the premises to the service provider's head end, where connection is established either to the LEC for access to the PSTN or to the Internet or other IP-based network for VoIP. In the United States as of June 30, 2005, end users obtained local telephone service by utilizing approximately 144.1 million ILEC switched access lines and 34.1 million CLEC switched access lines. Of the 34.1 million CLEC lines, approximately 9.1 million were provided over their own local loop facilities, of which 4.6 million were provided over CATV coaxial cable connections [11].

Access facilities also may provide direct connection from the customer premises to the IXC networks, bypassing the LEC switching systems in the process. The IXCs may provide those access loops themselves or they may lease them from the ILEC. Alternatively, end-user access to IXCs may be through loops leased from *Competitive Access Providers* (CAPs), also known as *Alternative Access Vendors* (AAVs). An access environment employs transmission facilities, switches, and signaling and control systems.

5.3.1.4.2 *Transport*

Transport is information *transport*ation in the backbone of the PSTN, that is, within the *cloud*. Transport can be in the LEC domain or the IXC domain for long-haul applications. Either individually or in concert, and depending on the geographic scope of the communication, both LECs and IXCs may participate in information transport. Also, a third-party carrier, that is, a *carrier's carrier*, may provide the transport facilities. The transport domain also employs transmission facilities, switches, and signaling and control systems. In the United States as of June 30, 2005, ILECs were the presubscribed interstate long-distance carrier for about 49 percent of the switched access lines they provided. For CLECs, the percentage was 74 percent [11].

5.3.1.5 Signaling and Control Signaling and control systems and networks are used to signal (i.e., alert and incite to action) various network elements and to control (i.e., guide or manage) its operation. Examples include status indication (i.e., on hook and off hook), dial tone provision, call routing control, busy indication, ringing, and performance monitoring. The traditional PSTN makes use of *Signaling System 7* (SS7) for this purpose, and networked digital PBXs make use of the *Q Signaling* (QSIG) protocol. IP-based networks variously make use of *H.323* and *Session Initiation Protocol* (SIP). Analog signaling remains in place for embedded analog PBXs.

5.3.1.6 Services Services include a wide variety of options provided by LECs and IXCs. Such services include various discounted calling plans, custom-calling services, Centrex services, and toll-free (e.g., 800, 888, 877, and 866) calling. I discuss a number of services later in this chapter.

5.3.2 Regulatory Domains

A community is known by its public utilities. The life of our country is built around them. . . . That such services may be extended and developed to be of the greatest use to the greatest number, the Federal Government and practically all of the states have appointed Public Service Commissioners as permanent tribunals to regulate public utilities with fairness to all concerned.

Telephone Almanac, American Telephone and Telegraph Company, 1922

There exists a complex set of domains that address regulatory and standards areas of responsibility. While it would be much simpler if these domains were discrete, they overlap nevertheless.

5.3.2.1 International Regulation is largely nonexistent at the international level, although it is heavily influenced by the ITU-T, which is chartered by the United Nations (UN) and which previously was known as the *Consultative Committee for International Telephone and Telegraph* (CCITT). The original predecessor organization was the *International Telegraph Union* (ITU), which was formed in 1865 to ensure the interconnectivity of national telegraph networks. The ITU-T primarily is responsible for setting standards recommendations intended to ensure the interconnectivity of national networks. Specific standards recommendations also address voice, data, fax, and video applications. The *International Telecommunication Union—Radiocommunication Sector* (ITU-R), previously known as the *Consultative Committee for International Radio* (CCIR), governs over-the-air communications. Intelsat and other consortia are responsible for allocating and managing satellite orbital slots. The *International Telecommunication Union—Development Sector* (ITU-D) works to further telecommunications development around the world, especially in developing countries.

The *International Organization for Standardization* (ISO) comprises the national standards institutes of 156 countries, with the *American National Standards Institute* (ANSI) representing the United States. (*Note:* ISO is not an acronym. Rather, it is from the Greek *isos*, meaning *equal*, implying that each nation has an equal voice in setting ISO standards.) The ISO has great influence over a wide range of international standards. For purposes of this book, the ISO is best known for the development of the *Open Systems Interconnection* (OSI) *Reference Model*, which I explore in Chapter 6. The *World Trade Organization* (WTO) also gets involved in telecommunication rate and tariff issues as part of its role in attempting to resolve trade disputes between member nations. For the most part, the WTO's involvement in telecommunications has been in the realm of international long-distance rates, where it has been only marginally successful.

5.3.2.2   *Regional* Regional authorities, although few in number, have the same responsibility as the ITU, but within a more compact region. *Directorate General XIII* (DG XIII), for example, is responsible for dealing with such issues of regulation and standardization within the *European Union* (EU), which largely comprises Western Europe. Previously known as the *European Community* (EC), the EU has real authority in the areas of regulation and standards, establishing and enforcing policy matters such as network competition within a member nation's network.

5.3.2.3   *National* National regulation is critical in most areas of the world, although less so in the case of the EU member nations, as they have ceded a good deal of this responsibility to the regional authority. Many nations have well-defined and tightly enforced rules, regulations, and standards. Areas of influence often include such issues as competition, rates and tariffs, radio frequency spectrum allocation, and characteristics of electrical local loops. National authorities also determine the basis on which international carriers can establish a presence at landing points for purposes of interconnection with the national networks. In the United States, the FCC is the national authority. Some countries have taken a much more light-handed approach. New Zealand is most unusual in that the national network was entirely deregulated and the regulator abolished some years ago, although the Telecommunications Act of 2001 placed the Commerce Commission back in that role as a competition authority with very limited power. The Australian Telecommunications Authority (AUSTEL) was disbanded in 1997 in favor of market self-regulation, although the Minister for Communications, Information Technology & the Arts has some responsibilities in this area, and two ombudsmen are appointed to deal with consumer complaints. Limited legislative and judicial controls remain in place in these countries.

5.3.2.4   *State or Province* In the United States and Canada, state or provincial regulation is considered important. Issues of intrastate competition and rates and tariffs are managed at this level. In 1937 the state governments began to form *Public Utility Commissions* (PUCs), also known as *Public Service Commissions* (PSCs), in the United States. In some cases, the state PUCs have taken the lead in terms of deregulation and competition. Illinois, for example, for years and well before the federal Telecommunications Act of 1996, permitted some level of competition in the local exchange environment. In Chicago, specifically and for a number of years, CAPs have provided local and Centrex service in competition with the LECs. As the Telecommunications Act of 1996 became tied up in the courts, the state regulators assumed a preeminent role in the introduction of competition into the local exchange. Currently, most states permit some level of competition. In late 2005, the states of Indiana and Texas passed bills to allow for a single statewide franchise provision for companies entering the cable television (CATV) business. Traditionally, CATV franchises in the United States were granted by municipal and other local authorities. The state of Virginia passed a similar, but weaker, bill in early 2006. As of this writing (June 2007), other states are considering similar measures.

5.3.2.5 Local Local regulation enters the picture relative to local zoning ordinances that variously address wireless tower placements, height limitations, and even aesthetics that might require a cellular telephone tower to masquerade as a pine tree or bell tower, for example. Local regulators in the United States also generally control CATV franchises and cable rights-of-way. In Milpitas, California, for example, the local regulators demanded the right to levy franchise taxes against Pacific Bell when the LEC sought right-of-way permission to lay fiber-optic cables to deliver CATV programming in competition with the incumbent CATV provider. PacBell refused to submit, arguing that it was subject only to state and federal regulation and not local regulation. As a result, the Information Superhighway ground to a halt in Milpitas in 1994.

In some instances, local regulation predominates. Alaska and certain Scandinavian countries are unusual in that they have many municipally owned LECs. Commercial carriers had no interest in providing service in such sparsely populated areas, given the high costs of provisioning and the low potential for revenues and profits. Note that many cooperatives were established for the same reasons.

5.3.3 Rates and Tariffs

Carriers in the United States traditionally have been required to file a set of *tariffs* with each appropriate regulatory authority. Those tariffs describe the services the carrier intends to offer in that domain, the rates it proposes to charge for them, and the proposed obligations, rights, and responsibilities of both the carrier and the customer. The regulatory authority examines the tariff filing, holds public hearings on the proposal, and renders a binding decision, which the carrier can appeal to the regulator or through the judicial system [12].

In the United States, national regulation is the domain of the FCC, which has the responsibility to determine all interstate issues, including radio frequency allocation. At the state level, the regulatory authority is in the form of a Public Utilities Commission (PUC) or Public Services Commission (PSC), which has the authority to regulate all exclusively intrastate matters not delegated to the municipalities and other local authorities.

Regulators evaluate carrier tariff filings and determine rate schedules based on complex and varying sets of considerations including allowable costs of service provisioning and support, involving both capital investment and operating expenses. Particularly at the state level, regulators attempt to restrict rates to reasonable and affordable levels. At the same time, the regulator must permit the carrier a reasonable rate of Return on Investment (ROI) that positions the company well to secure necessary funds for expansion and enhancement of its facilities through both the equity (stock) and debt (bond) markets. Additionally, the regulators generally attempt to ensure that residential service is universally available at reasonable cost. Regulators traditionally have set rates for PBX trunks, tie lines, and other services consumed by large-business users well above the costs of providing service, thereby establishing a complex set of cross-subsidies that supports lower rates for residences and small businesses. Basic rates for basic local exchange lines generally are kept at low levels, especially for residential consumers. Enhanced services (e.g., voice mail and custom-calling features) and long distance are considered to be optional and, therefore, are priced to yield more profit in support of basic services.

As the network monopolies were dismantled and the trend toward competition developed, the various regulators focused on the incumbent carriers, who were considered dominant in their respective domains. Beginning with the Carterfone decision in 1968, carrying through the MFJ in 1982 and continuing into the Telecommunications Act of 1996, the emphasis therefore was on AT&T, GTE, and the ILECs. The intentions of the FCC and the PUCs are to encourage competition by providing the new entrants with an advantage. Once the incumbent carriers (i.e., AT&T and the ILECs) have demonstrated that they no longer hold dominant positions, the regulators relax restrictions on them in favor of permitting market forces to prevail. This intent was demonstrated with respect to the IXC market, as the FCC gradually relaxed its requirements of AT&T. This trend continued and extended to the ILECs as a result of the Telecommunications Act of 1996, which is discussed in detail in Chapter 15. Ultimately, this waxing and waning of regulation had unforeseen consequences, as competition became so severe as to force mergers between incumbent IXCs and LECs, that have led to what could be considered a reconstitution of the Bell system, albeit much weaker and considerably disfigured.

Regulation has focused on basic voice network services provided in the traditional fashion by the incumbent carriers. Such services certainly are more basic and necessary than data and other services. This posture now can be argued, particularly with the advent of the Internet. In fact, voice communication over the Internet is possible in a number of ways, and some service providers have developed commercially available international service offerings based on various VoIP approaches using the Internet for transport. Considered by many as a threat to the traditional PSTN and the concept of universal service, a number of interested parties have requested several times that the FCC examine the issue, with the intent to regulate voice over the Internet or even to ban it altogether. Interestingly, the major incumbent IXCs were not parties to this request. Rather, they took strong positions as ISPs and encouraged its use. In the late 1990s and early into the twenty-first century, a number of competitive carriers constructed fiber-optic IP-based networks separate from the Internet and heavily promoted their use for VoIP, further adding pressure to the traditional PSTN. Commercial Frame Relay (FR) services, which are highly popular in corporate data networking, often support *Voice over Frame Relay* (VoFR), which also competes with the PSTN. I discuss the Internet, VoIP, Frame Relay, and VoFR at length in subsequent chapters.

Local exchange competition, VoIP, and VoFR all threaten the concept of universal service, which has been a cornerstone of the PSTN since the formation of the FCC in 1934. In order to ensure the universal availability of voice service at affordable cost to the subscriber, a complex structure of *settlements* (cross-subsidies) developed between incumbent IXCs and LECs. Also, rates in urban areas were a bit higher in order to subsidize basic service in rural areas. Thereby, a subscriber in a *high-cost* area (i.e., an area in which the cost of providing basic service is defined as *high*) such as Hackberry, Arizona (population 1), could gain affordable network access, single-party service, access to emergency services, access to operator services, and so on, just as could a subscriber in New York, New York, despite the obvious cost differences in the carriers providing service. Generally speaking, high-cost areas are remote and sparsely populated. Unless the integrity of the universal service fund is maintained, with all carriers contributing, the concept of universal service may be relegated to a historical footnote.

5.3.4   Carrier Domains and Network Topology

Some years ago, and certainly prior to AT&T's divestiture of the Bell Operating Companies (BOCs) in 1984, the network in the United States was relatively simple in terms of its ownership and topology. Each operating telephone company provided service in its franchised serving areas and gained access to the AT&T long-distance network on a fairly straightforward basis. Beginning in the late 1920s, the network was organized on a layered basis, with five levels of hierarchy, known as *classes* [12].

- *Class 5 offices* are the *local exchange offices*, or Central Offices (COs), that serve end users through local loop connections. As recently as 2001, there were approximately 19,000 Class 5 offices in the United States, although many of them have been consolidated. Each of the remaining 16,000 or so is geographically positioned to address a *Carrier Serving Area* (CSA), as illustrated in Figure 5.2. The CSA has a radius of approximately 18,000 ft, which is the typical maximum length of a voice-grade local loop without special conditioning provided by loading coils, amplifiers, or repeaters. (*Note:* Approximately 90 percent of the local loops in the United States are 18,000 ft or less in length.) The carrier can

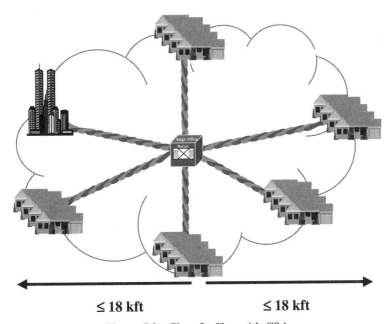

$\leq$ **18 kft** $\leq$ **18 kft**

Figure 5.2   Class 5 office with CSA.

extend the radius of the CSA through the deployment of intelligent remote COs, unintelligent remote line shelves, or *Digital Loop Carrier* (DLC) systems, which essentially are remote time division multiplexers. The various remotes connect to the centralized CO through high-capacity circuits. Should significant volumes of traffic be exchanged directly between COs, they may be directly interconnected. More commonly, they are interconnected through tandem switches.

- *Class 4 offices* are *tandem toll centers*, which serve to interconnect Class 5 offices not interconnected directly. As the lowest class of toll center, these also serve as the first point of entry to the long-distance, or toll, network. Class 4 offices are interconnected within a relatively local toll network and provide access to higher order toll centers. In many instances, a Class 4 office also serves as a Class 5 office; in other words, a hybrid Class 4/5 switch serves as both a Class 4 tandem toll office and a Class 5 CO, with the separate functions provided through logical and physical partitioning within the switch. Approximately 1500 tandem toll centers existed in North America prior to AT&T's divestiture of the BOCs. It is uncertain how many of them remain.

- *Class 3 offices*, or *primary toll centers*, are higher order toll centers, generally serving to connect Class 4 offices for intrastate toll calling. Class 3 offices typically serve to interconnect independent telcos and BOCs. Approximately 200 such offices existed prior to divestiture. It is likely that most, if not all, have now been decommissioned.

- *Class 2 offices*, or *sectional toll centers*, serve to interconnect primary toll centers, largely for interstate calling within a geographic region such as the northeast or the southwest. Approximately 67 sectional toll centers existed in the AT&T network prior to divestiture. It is likely that most, if not all, have now been decommissioned.

- *Class 1 offices*, or *regional toll centers*, serve to interconnect sectional toll centers in support of interregional calling. There were 10 regional toll centers in place in the United States prior to divestiture. At the end of 2001, 7 remained in the United States and 2 in Canada. It is likely that most, if not all, have now been decommissioned.

As illustrated in Figure 5.3, the offices traditionally were interconnected on a hierarchical basis, with end offices residing at the bottom of the network food chain. As a user places a long-distance call, the Class 5 switch examines and analyzes the destination telephone number in the context of the geographic area it serves. Based on that information and relying on programmed logic, the CO processes and routes the call. Local long-distance calls (e.g., within the San Francisco Bay Area) are handled either by directly connected Class 5 offices or through a Class 4 tandem toll office that interconnects multiple Class 5 offices. A coast-to-coast call of only a few years ago, on the other hand, might have involved all five classes of the hierarchy.

For example, a call from Turlock, California, to New York City originated in the Class 5 switch of Evans Telephone Company, an independent telco, and was handed to a nearby AT&T tandem toll center. The call then worked its way up the hierarchy until it reached the Class 1 regional toll center in San Francisco. High-capacity,

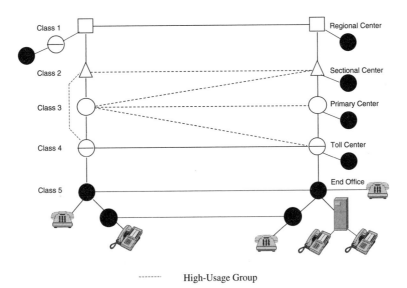

Figure 5.3 Traditional network hierarchy.

coast-to-coast tandem toll trunks carried the call to New York City, where it worked its way down an abbreviated hierarchy and was delivered to the target number in Manhattan. Note that each switch acted on the call more or less independently. In the above classic scenario, each switch looked at the call, determined whether or not it could serve the connection request, and handed the call off to another switch, either higher or lower in the hierarchy. The sole exception was the terminating Class 5 switch, which finally served to establish the connection to the target telephone number. While the switches worked in series to set up the connection, each acted independently and forwarded the connection request blindly along, never knowing what would happen end to end. The same set of processes took place in each switch, the same logic was exercised, and a talk path was set up through each switch and across each interconnecting transmission link.

This approach was quite sensible in the days when AT&T dominated the local and long-distance networks. As calls worked their way through the network, ever larger volumes of traffic were aggregated by ever more capable switches and shipped over trunks of ever greater capacity, taking advantage of the economies of scale.

This network topology has flattened over the years as the cost of transport over fiber-optic facilities has dropped, as the cost of switches has decreased, as hybrid Class 4/5 switches have been replaced by separate Class 4 and Class 5 switches, as competition has increased, and as traditional circuit-switched infrastructure has been replaced with Asynchronous Transfer Mode (ATM) and especially Internet Protocol (IP) infrastructure. As a result, there exist fewer switches, many of which are more intelligent and multifunctional and are interconnected by higher capacity transmission facilities in a *sparse network configuration*. As the various TDM–based offices have been decommissioned, many of them have been replaced with *softswitches*, which are software-based and highly flexible, whereas TDM circuit switches are hard coded and highly inflexible. Softswitches can support multiple protocols, including perhaps TDM, Frame Relay, and IP.

Contemporary carrier and service provider domains fall into three categories: local exchange, interexchange or national, and international. Competitive Access Providers (CAPs) sprang to life in the United States as a result of the Telecommunications Act of 1996 (Chapter 15), focusing on the provisioning of direct access from the customer premises to the IXC through ILEC facilities bypass. Overlay carriers and wireless carriers also are relatively new entrants spawned by liberalization and deregulation movements that began in the mid-1980s and intensified considerably beginning in the mid-1990s.

5.3.4.1 Customer Premises Equipment

CPE, as is noted earlier in the chapter, is the terminal and switching equipment located on the customer premises. While such equipment traditionally was rented to the subscriber by the LEC, deregulation generally places both the privilege and the responsibility of ownership on the end user in most countries. Similarly, inside wire and cable generally are deregulated.

5.3.4.2 Demarcation Point (demarc)

The demarc constitutes the boundary between the end user and carrier domains. In a residential environment, the demarc is in the form of a *Network Interface Unit* (NIU). A NIU includes a *protector*, which serves to protect the premises wiring and equipment from aberrant voltages possibly induced by carrier power supplies, power utility transformers, or lightning strikes. Contemporary NIUs are intelligent, enabling telco technicians or automatic test systems to regularly test the integrity of the local loop from the CO to the customer premises. In the United States, the NIU is located outside the residence, perhaps on an outside wall or in the garage, because the regulated telco cannot place equipment inside the premises under the terms of the MFJ. (*Note:* On special request, the telco can install the demarc in a basement.)

In a business environment or a multitenant building, the demarc is positioned at the *Minimum Point of Entry* (MPOE), defined as the closest logical and practical point within the customer domain. In a high-rise office building, for example, it typically is defined as a point of the entrance cable 12 in. from the inside wall. Newer entrance cable facilities involve a physical demarc, while older facilities typically are tagged by telco technicians in order to indicate a logical point of demarcation. From that point inward, the cable and wire system is the responsibility of the user or building owner, as appropriate.

5.3.4.3 Local Exchange Carriers

LECs provide local telephone service, usually within the boundaries of a metropolitan area, state, or province. Their primary charter is to provide local voice services through a network of local loops and COs, which can be connected either directly or through a tandem switch, as depicted in Figure 5.4. The LECs also provide short-haul, long-distance service, Centrex, certain enhanced services such as voice mail, and various data services. In the United States and many other developed nations, both *Incumbent LECs* (ILECs) and *Competitive LECs* (CLECs) exist. In many developing countries, only a single ILEC exists.

ILECs are the original LECs, each of which was awarded by the regulatory authority a franchise to provide service within a given geographic area. In the United States, a municipal or county government originally awarded those franchises. Over time, the *Bell Operating Companies* (BOCs), owned by the AT&T Bell

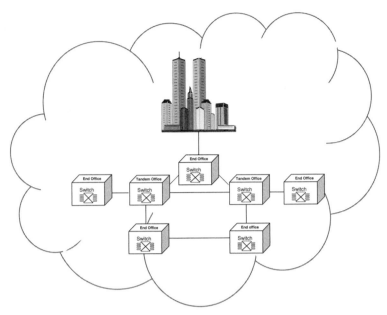

Figure 5.4   ILEC metropolitan serving area, with COs interconnected directly and through a tandem switch.

System and reporting directly to AT&T general headquarters, came to dominate the ILEC landscape. Effective January 1, 1984, those 22 operating telephone companies were spun off from AT&T as a result of the *Modified Final Judgement* (MFJ). Also known as the *Divestiture Decree*, the MFJ was rendered by Judge Harold H. Greene of the Federal District Court in Washington, DC, on January 8, 1982. The MFJ, in fact, was a negotiated settlement representing the culmination of the U.S. Justice Department's long efforts to break up what it characterized as an oppressive monopoly. As a result, the BOCs were reorganized into seven *Regional Bell Operating Companies* (RBOCs), also known as *Regional Holding Companies* (RHCs), as noted in Table 5.1. Over time, each RBOC fully absorbed its component BOCs, creating a single legal entity with a centralized management structure. Cincinnati Bell and Southern New England Telephone (SNET) were not affected in this manner, because they were not wholly owned subsidiaries of AT&T.

In addition to causing AT&T to divest the BOCs, the MFJ effectively limited the BOCs to providing basic voice and data services within defined geographical areas, known as *Local Access and Transport Areas* (LATAs). These services included *intraLATA* toll service, also known as *local long distance*, within the confines of the 196 defined LATAs. Additionally, the BOCs and RBOCs could not engage in certain other activities such as manufacturing communications equipment and providing enhanced services such as voice mail; subsequently, the latter restriction was lifted. LATAs now serve primarily as reference points for call routing.

AT&T Long Lines, which became AT&T Communications and then AT&T Corporation (and was acquired by SBC in November 2005 to become AT&T, Inc.), was restricted from providing intraLATA toll, as were MCI (later MCI Worldcom, acquired by Verizon in January 2006 to become Verizon Business), US Sprint (later

TABLE 5.1   Bell System Operating Company Organizational Structure Before and After MFJ and to Present

Bell Operating Companies (Primary States of Operation), Predivestiture	Regional Bell Operating Companies (Headquarters), Postdivestiture
Illinois Bell (Illinois), Indiana Bell (Indiana), Michigan Bell (Michigan), Ohio Bell (Ohio), Wisconsin Telephone (Wisconsin)	Ameritech (Illinois); acquired by SBC Communications (October 1999), now AT&T
Bell of Pennsylvania (Pennsylvania), Diamond State Telephone (Delaware), The Chesapeake and Potomac Companies (District of Columbia, Maryland, Virginia, and West Virginia), New Jersey Bell (New Jersey)	Bell Atlantic (Pennsylvania); now Verizon Communications
South Central Bell (Alabama, Kentucky, Louisiana, Mississippi, and Tennessee), Southern Bell (Florida, Georgia, North Carolina, and South Carolina)	BellSouth (Georgia); acquired by AT&T (December 2006)
New England Telephone (Massachusetts, Maine, New Hampshire, Rhode Island, and Vermont), New York Telephone (New York)	NYNEX (New York); acquired by Bell Atlantic (August 1997); now Verizon Communications
Pacific Bell (California), Nevada Bell (Nevada)	Pacific Telesis (California); Acquired by SBC (April 1997); now AT&T
Southwestern Bell (Arkansas, Kansas, Missouri, Oklahoma, and Texas)	Southwestern Bell Corporation (Texas); then SBC Communications; now AT&T (November 2005)
Mountain Bell (Arizona, Colorado, Idaho, Montana, New Mexico, Utah, and Wyoming), Northwestern Bell (Iowa, Minnesota, North Dakota, Nebraska, and South Dakota), Pacific Northwest Bell (Oregon and Washington)	US West (Colorado); aquired by Qwest (June 2000)

Sprint Corporation, merged with Nextel in August 2005 to become Sprint Nextel), and the balance of what were known as the *Other Common Carriers* (OCCs). Those companies, which became known as interexchange carriers (IXCs, or IECs), were limited to providing long-distance service on an interLATA basis (across LATA boundaries). Further, AT&T could not enter the local service business, although the OCCs had no limitations in that respect.

AT&T Technologies (now Altacel-Lucent Technologies) was formed of Western Electric, the manufacturing arm of AT&T, and AT&T Bell Telephone Laboratories (Bell Labs), the research and development organization. Of all the AT&T entities, only Bell Labs could retain the Bell name. The RBOCs, in consortium, formed Bell Communications Research (Bellcore) as an R&D entity. Bellcore focused on the needs of its RBOC clients/owners in terms of software R&D, standards development, and other requirements; under the terms of the MFJ, it initially was precluded from involvement in the physical sciences [13, 14]. The RBOCs divested Bellcore in 1998, when it became clear that they would be competitors, rather than collaborators, in a deregulated environment. Science Applications International Corporation (SAIC) acquired Bellcore and changed its name to Telcordia Technologies, Inc., in March 1999. SAIC sold Telcordia to investment bankers in November 2004. Telcordia's stated focus is on next-generation networks.

In the United States, LECs traditionally had the exclusive rights to local and intraLATA markets, although many states either eroded or modified this policy; Connecticut, for example, has permitted local exchange competition since July 1, 1994. Although the Telecommunications Act of 1996 set the stage for complete and open competition for virtually all services, as it bogged down in the courts, the various state regulators took the initiative in terms of local service competition.

In addition to the RBOCs and their LEC subsidiaries, approximately 1300 independent telephone companies exist. GTE was the largest independent until Bell Atlantic acquired it in July 1999 to form Verizon Communications. Southern New England Telephone Company (SNET), which was partially owned by AT&T prior to the MFJ, was acquired by SBC in October 1998.

At this point, it is worth noting that the Telecommunications Act of 1996 has changed this landscape in a very dramatic way. Casting aside the MFJ, the act permitted the IXCs to begin competing immediately with the ILECs for local and intraLATA service. The act also allowed the RBOCs to offer interLATA long-distance service outside the states in which they operate as ILECs. Once the RBOCs satisfied certain local competition requirements, they were allowed to offer interLATA toll services within their home states as well. As competition intensified and both the ILECs and IXCs lost their dominance, interesting mergers began to take place, as noted above and below:

- SBC acquired AT&T in November 2005 to become AT&T, Inc.
- Verizon acquired MCI Worldcom in January 2006 to form Verizon Business.
- AT&T, Inc. acquired BellSouth in December 2006.

Competitive Local Exchange Carriers (CLECs) can compete for local service in most states. As of June 30, 2005, at least one CLEC was serving customers in 85 percent of U.S. postal ZIP codes, which include approximately 98 percent of the population [11]. While many CLECs originated as CAPs, others were IXCs, and still others were created specifically as CLECs. Metropolitan Fiber Systems (MFS), for example, originally was a facilities-based CAP. Subsequently, MFS acquired UUNET, a very large Internet access and backbone provider. MFS then was acquired by MCI, which then was acquired by Worldcom, with the combined company assuming the name MCI Worldcom. While MFS has lost its identity, the network that it constructed served MCI Worldcom well for IXC access and positioned it well as a CLEC. MCI Worldcom subsequently changed its name to Worldcom and then back to MCI after it was bankrupted by senior management. MCI later was acquired by Verizon, as previously noted.

AT&T constructed fiber-optic facilities for direct access to the premises in areas where there existed high concentrations of large user organizations. In Spokane, Washington, AT&T Wireless, spun off from AT&T in July 2001, tested a Wireless Local Loop (WLL) solution based on its Personal Communications Services (PCS) network technology and licenses. While the trials were successful, negative market conditions forced the closure of that business unit in December 2001. AT&T also amassed wireless licenses in over 300 markets in order to bypass the ILEC to reach its business customers without the investment and delay of constructing wireline facilities. (*Note:* In October 2004, AT&T Wireless was acquired by Cingular. In December 2006, the new AT&T, formed of the acquisition of the remnants of the old AT&T by SBC, acquired BellSouth. That acquisition included Cingular, which SBC and BellSouth formed in 2001 from a conglomeration of 11 regional companies. So, Cingular became AT&T Wireless.) In July 1998, AT&T also acquired TCGI (Teleport Communications Group, Inc.), a large CAP with both wireline and wireless access facilities. In March 1999, AT&T acquired TCI (Tele-Communications, Inc.), a large CATV provider, to gain access to its Hybrid Fiber/Coax (HFC) cable network. AT&T made several other CATV acquisitions and formed AT&T Broadband as a full-service provider of entertainment TV, high-speed Internet access, and voice communications services. [*Note:* In November 2002, AT&T Broadband merged with (read *was acquired by*) Comcast Corp.] In combination with, and through some significant investments in, network upgrades, these fiber, coax, and wireless networks provide access to a very large potential customer base. Further, they variously support voice, data, and video communications and high-speed Internet access.

5.3.4.4 Interexchange Carriers IXCs are responsible for long-haul, long-distance connections across LATA boundaries. The IXC networks are connected to the LECs through *Points of Presence* (POPs), which typically are in the form of tandem switches. An interLATA call originating in a LEC serving area is recognized as such and is passed to the access tandem switch, which in turn passes the call to the IXC POP via dedicated trunks leased from the ILEC (see Figure 5.5). Alternatively, the IXC may collocate network termination equipment in the LEC exchange office, assuming that space is available and that secure physical separation can be established and maintained.

Alternatively, large users can gain direct access to the IXC POP, bypassing the LEC switching network in the process. Such a bypass arrangement typically is

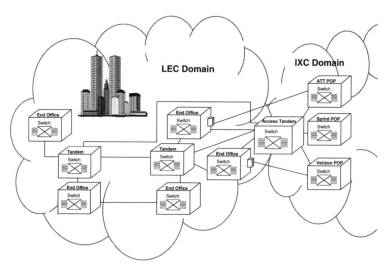

Figure 5.5 Interconnection between LEC and IXC via access tandem switch and a collocated termination equipment.

effected through a leased line provided by the ILEC. Alternatively, the ILEC can be completely bypassed through a direct link, which can be wireless in nature and provided directly by the IXC.

IXCs, by definition, are facilities-based carriers, or *heavy carriers*. While they prefer to contract directly with large end users, they also often work through non-facilities-based carriers (*light carriers*), including resellers, aggregators, and agents. Of the 400 or so heavy carriers in 2000, AT&T, MCI Worldcom, and Sprint dominated the market. Ultimately, however, they succumbed to various competitive forces. The tattered remnants of AT&T and MCI Worldcom eventually were absorbed by ILECs, and Sprint decided to focus on the wireless business.

5.3.4.5 Competitive Access Providers
Competitive Access Providers (CAPs), also known *as Alternative Access Vendors* (AAVs), provide an alternative means of connection between large user organizations and IXCs, completely bypassing the LEC network in the process. CAPs generally deploy high-capacity, fiber-optic facilities from IXC POPs to areas where there exist high densities of large user organizations (e.g., commercial office parks and urban business areas). In many cases, the CAP will extend the fiber-optic connection directly to the user's CPE, such as a PBX, as depicted in Figure 5.6. Wireless CAPs provide access on the basis of either various licensed microwave or infrared technologies.

An optical fiber approach presents numerous benefits, including the fact that the performance characteristics of the fiber-optic facilities are far superior to those of the traditional LEC copper local loop. Additionally, the CAP network automatically provides a measure of local loop redundancy; the CAP facilities are used for IXC access, while the LEC local loops can still be used for IXC access in the event of a failure in the CAP network. Further, the fiber-optic network generally is redundant, thereby minimizing the likelihood of a service-affecting network failure. As the fiber-optic network also offers substantial, and even elastic, bandwidth, the user organization typically can increase its level of access to the IXC much more quickly and easily than through a LEC connection. Call setup time (speed of connection)

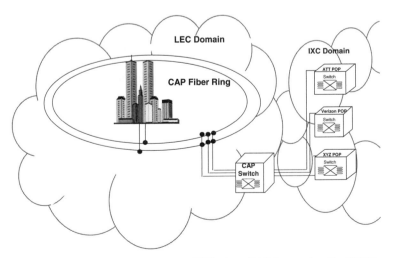

Figure 5.6   End-user access to an IXC via a CAP, bypassing the ILEC.

also is improved, as the LEC CO, tandem switch, and access tandem switch all are bypassed. Finally, the costs of IXC access often are much reduced because the CAP's unregulated rates generally are highly competitive.

CAPs operate most especially in the United States, where many now function as voice/data CLECs, providing Centrex service, local calling service, high-speed data services, and certain enhanced services.

5.3.4.6 International Carriers   International carriers provide communications transport services across national borders, based on access from the LEC or IXC domain by connection to an *International Gateway Facility* (IGF). Contemporary international carrier transmission facilities are largely satellite and submarine fiber-optic cables, the latter often jointly owned through consortia. Examples of international carriers include AT&T, Cable & Wireless (C&W), Global Crossing, and Verizon. *International Record Carriers* (IRCs) offer *record communications services*, which are services designed or used primarily to transfer information that originates or terminates in written or graphic form. Examples of record communications services include telex and TWX.

5.3.4.7 Overlay Carriers   Overlay carriers build networks that overlay the traditional PSTN. While such networks are unusual in highly developed countries, they are common in developing nations. In some countries of eastern and central Europe, for example, private carriers have been franchised to build overlay digital microwave networks to provide service to large governmental, educational, and commercial organizations. In such cases, the PSTN simply could not be upgraded quickly enough to provide satisfactory communications, which are considered vital to economic growth. Overlay carriers generally deploy microwave systems that effectively overlay the outdated wireline facilities of the incumbent carriers.

5.4 SIGNALING AND CONTROL: EXPANDED VIEW

Only in rare instances can a telephone operator know how momentous are the spoken messages for which she established the connections. Life and death may depend upon the swiftness and sureness with which she handles any call. All Bell System telephone calls are treated as if marked "urgent." Methods of operation and mechanisms are so planned that every connection may be established with all possible speed. Bell System service was the fastest in the world ten years ago [1925], *when the average time required to complete a long distance call was seven minutes. It is almost five times as fast today* [1935].

Telephone Almanac, American Telephone & Telegraph Co., 1935

The two basic types of information transfer are user data (content or payload) and signaling and control. In order for the network to function properly, the various devices, components, or elements of the network must have the capability to signal (i.e., alert and inform) each other, indicating their status and condition. Typical status indications include available (dial tone), unavailable (busy), and alerting (ringing signal). The terminal devices also must pass identification information, as well as certain instructions through the network, perhaps as far as to the receiving terminal. Such information and instructions might include the originating number or circuit and the target number, based on the dialed digits. Within the carrier network, such information includes route preference and route availability [12]. Additionally, the LEC network must determine and honor the end user's IXC designation in order that it can hand off a long distance to the carrier of choice.

Signaling and control systems and networks also handle billing matters, perhaps querying centralized databases in the process. Billing options might include bill to originating number (e.g., DDD and WATS), bill to terminating number (e.g., INWATS, or toll-free), bill to third number (third party), and bill to calling card with verification of a PIN against a database.

Finally, certain network management information often is passed over signaling and control links. Such information is used for remote monitoring, diagnostics, fault isolation, and network control. In this fashion, a centralized *Network Operations Center* (NOC) can monitor the network, and faults or degradations in performance can be determined and isolated. Diagnostic routines can then be invoked in order to determine the specific nature of the difficulty. Finally, a network management system or network manager can instruct the network element in difficulty to resolve the problem, perhaps by resetting or reinitializing itself, or by disabling a failed port and activating a standby port.

5.4.1 In-Band Signaling and Control

In-band signaling and control functions take place over the same physical path as the conversation and occupy the same frequency band (analog) or time slots (digital). The impact of simultaneous data transfer and signaling over the same frequency range or time slots disrupts the information stream and, therefore, can be characterized as disruptive or intrusive. As a result, in-band signaling and control seldom is employed in contemporary networks, with the exception of analog local loops.

By way of example and in years past, I frequently called long distance to talk with my young son in East Texas. During those conversations, Barrett (my son) had the annoying habit of "accidentally" depressing the buttons on the tonepad with his chin. The resulting Dual-Tone MultiFrequency (DTMF) signals were interpreted by the network as a priority instruction set. Since the signaling tones occupy the same range of frequencies at the same moment in time as the conversation, they override and interfere with it. In contemporary networks, this in-band signaling and control technique actually is disruptive only over the local loops, which extend from the CPE to the edge of the network cloud; internally, the networks make use of out-of-band signaling and control. Nonetheless, Barrett got a set of tones in his ear and I got a set of tones in my ear; Barrett found this to be highly amusing, while I found it to be highly aggravating, which Barrett found to be even more amusing, and so on, and so on. (*Note:* James Barrett Horak is now a Sergeant in the United States Marine Corps. No longer is he easily amused, as it is against Marine Corps policy.)

Notably, in-band signaling and control goes wherever the call goes, at least at the edges of the network. You can use this fact to your advantage. You can, for example, signal and control a voice processor through a system feature known as *cut-thru*. When checking your voice mail, you access the voice processor, which begins to *talk* to you. Because you know exactly what the machine is going to say, you simply send DTMF tones to the machine through the dial pad, overriding the data transmission.

You enter your password while the machine *talks* to you. You enter option *1* to play the message, enter *2* to repeat the message, enter *3* to save the message, and so on—all while the machine *talks* to you. In other words, you override the downstream data transmission with higher priority signaling and control data, saving time and money in the process.

5.4.2 Out-of-Band Signaling and Control

Out-of-band signaling and control, in the simplest analog application, takes place over frequencies separate from those that carry the information. In a contemporary digital network, signaling and control data generally occupy separate, specifically designated time slots. In either case, there is no interference between the two functions. In other words, out-of-band signaling and control is termed to be nondisruptive or nonintrusive. Out-of-band signaling and control is the standard approach in digital networks and, most certainly, in the core of the carrier networks.

5.4.3 Common Channel Signaling and Control

In 1916, the average time required to establish a Bell System long distance connection was approximately eleven minutes. By 1926, it had been reduced to 5.6 minutes. The average for 1936 was 1.4 minutes.
Telephone Almanac, American Telephone & Telegraph Company, 1937

The carriers use *Common Channel Signaling* (CCS) and *control* systems to carry large volumes of signaling and control information in support of high-traffic networks. The CCS links are digital in nature, based on packet switching, and often are in the form of a dedicated T1 channel over a high-speed optical fiber. Essentially, the CCS network is a highly robust subnetwork that supports the operations of the primary communications network. The CCS subnetwork connects the various network switches to centralized computer systems of significant intelligence and on which reside very substantial databases. Through the use of centralized intelligence supported by carefully synchronized databases, you can control the operations of an entire network and monitor its performance, from end to end. While complicated and expensive to design and deploy, CCS networks are more effective and less costly than the alternative of placing lesser levels of intelligence in each of the various network switches, each of which is required to perform redundant processes in complete harmony with the others.

Signaling System 7 (*SS7*), the current version, was developed and deployed based on ITU-T standards recommendations. Thereby, all carriers can achieve and manage interconnection on a standard basis (Figure 5.7). SS7 significantly speeds call setup and call completion processes. (Call completion takes only a few seconds in 2006, which compares quite favorably with 1.4 min in 1936.) Additionally, SS7 is responsible for the delivery of many enhanced custom-calling features often associated with ISDN. These CLASS (*Custom Local Access Signaling Services*) include caller ID, which has been enhanced as name ID, providing the name of the calling party as listed in the telephone directory. Other services include selective ringing (or priority ringing), selective call forwarding, call block (or call screen), repeat dialing, call trace, and automatic call-back (call return). SS7 is fully deployed in all major

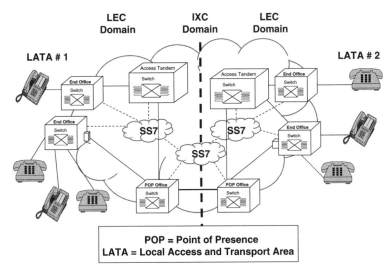

Figure 5.7   SS7 in support of LEC and IXC networks, interconnected.

TDM-based IXC networks. While SS7 is largely deployed in the major ILEC networks in developed countries, older Class 5 switches do not support it.

5.5 NETWORK SERVICES

> *By means of what is known as Conference Service, a number of telephones, at widely scattered points, may be connected together so that their users may converse with each other as if they were all seated in the same room. Widely used for business purposes, it has its social uses also—as an antidote for loneliness.*
>
> *Telephone Almanac*, American Telephone & Telegraph Company, 1938

You can group network services into several categories that define their basic nature. Those categories include access services, dedicated services, switched services, and virtual services.

5.5.1 Access Services

Access services are those services that provide circuit access to the network, which may be a LEC CO or an IXC POP. The circuits may be analog or digital and single-channel or multichannel. Access services include residential and business lines and PBX trunks.

5.5.1.1 Residential Lines   Residential lines are local loop connections between the residential premises and the CO. Although generally analog in nature, the market penetration of digital ISDN Basic Rate Interface (BRI) local loops is relatively high outside the United States. Contemporary residential lines also can be provided over analog or digital channels of a DSL local loop or even over a CATV network. According to the 1951 Bell Telephone System *Telephone Almanac*:

Enjoy "Monopoly in the rumpus room, but not on a party-line telephone, please....
Everybody benefits ... when each person uses the service courteously and with consider-
ation toward the others. This includes answering your telephone promptly—allowing an
interval between your calls so that others may use the line—giving up the line quickly.

Residential lines usually are *private lines*. A very few *party lines* remain in the
United States, although they are not unusual in developing countries where infrastructure is limited in remote areas. A party line involves a local loop shared
between perhaps 2, 4, 8, or as many as 16 residences. Some form of distinctive ringing
comprising various combinations of short and long rings distinguishes a call intended
for each individual residence on the shared line. Clearly, there is no privacy on a
party line, as any party can pick up the phone and answer the call or listen in on it.
Placing outgoing calls is a free-for-all, as the caller must pick up the phone to determine if the line is available before placing the call. If someone else is using the line,
the process must be repeated at another time, and perhaps again and again in hopes
that the line eventually will be available. Good telephone etiquette is a positive
quality in a party line subscriber. (*Note:* As we will discuss in Chapter 8, Ethernet
is much like a party line, as the media access control protocol is contentious and
nondeterministic.)

5.5.1.2 *Business Lines* Business lines are local loop connections between the
business premises and the CO. Business lines provide access to single-line and
multiline terminal sets as well as to Key Telephone Systems (KTSs). Business lines
generally are analog in nature, although digital ISDN BRI has increased in popularity. Contemporary business lines also can be provided over analog or digital channels of a DSL local loop or even over a CATV network. Business lines are almost
always private lines, although there are very rare instances of party line service in
developing countries.

5.5.1.3 *PBX Trunks* PBX trunks are local loop connections between PBX
switches and network switches. As noted in Chapter 3, PBX trunks may be incoming
only, outgoing only, or bidirectional (combination) in nature. Trunk connections may
be provisioned individually and may be analog in nature. More typically, trunks are
provisioned on a high-capacity, digital, multichannel basis (e.g., T1, E1, and ISDN
Primary Rate Interface, or PRI). Trunks that serve a specific, common purpose are
grouped into a *trunk group*.

5.5.1.4 *Dedicated Transport Services*   *Dedicated circuits*, in the traditional
sense, are leased-line circuits dedicated to a specific use by a specific user organization. Again, they may be either analog or digital and either single channel or multichannel. Typically, the LEC provides dedicated circuits on an intraLATA basis,
although CAPs also provide such circuits to the extent that their facilities match
user requirements. An IXC typically provides interLATA leased lines, with the
ILEC or CAP providing the local loop connection to the IXC POP and the IXC
providing the interLATA portion to the destination POP, where the circuit interfaces with the ILEC local loop serving the target location. (Note that LATAs are
regulatory boundaries that exist only in the United States. Also note that the distinction between LECs and IXCs is largely a regulatory distinction in the United States

and in most cases both LECs and IXCs are allowed to carry both intraLATA and interLATA domestic traffic. In the contemporary PSTN, the significance of LATA boundaries is largely limited to call routing.)

5.5.1.5 Foreign Exchange *Foreign exchange* (FX or FEX) lines or trunks connect the user premises more or less directly with a foreign exchange (i.e., with an exchange other than the local exchange that normally provides local dial tone). As is the case with all leased lines, FX circuits generally are billed on a distance-sensitive basis and without any consideration of usage. As is described in Chapter 3, a user organization can lease an FX circuit from Dallas, Texas, to Denton, Texas, to support outgoing and incoming traffic between those general areas; the circuit terminates in Dallas CPE but has a Denton local telephone number. Neither long-distance charges nor any other usage-sensitive charges apply to the traffic.

5.5.1.6 Tie Lines and Tie Trunks Tie lines, as is described in Chapter 3, are circuits that connect KTS together directly and across the PSTN. Tie trunks connect PBXs together. Because they are leased circuits, in neither case are PSTN switches involved. Between two PBX systems, for example, the carrier(s) provide dedicated circuits that pass through wire centers housing network switches. The circuits bypass the switches, which are unnecessary in this application, yet take advantage of amplifiers or repeaters, multiplexers, and other systems and elements embedded in the carrier network. Tie trunks may be either analog or digital but usually are digital T-carrier or E-carrier. Note that the core of the network is typically digital, so an analog tie trunk (or tie line, for that matter) usually is analog only at the local loop level.

5.5.1.7 Off-Premises Extension An Off-Premises eXtension (OPX) circuit is a dedicated PSTN circuit that connects a PBX or KTS to an extension voice terminal located off premises, that is, at another location some distance away. The terminal appears to the PBX exactly as though it were an on-premises extension, providing the user with the same level of functionality. OPXs are unusual, as both the circuit and the special-purpose system interface card are expensive. Additionally, and as contemporary users often are not office bound, voice mail, e-mail, cellular telephony, pagers, and other technologies are adequate alternatives for most situations.

5.5.2 Switched Transport Services

Switched services include all typical local and long-distance voice traffic, whether inbound or outbound. Specific services in the United States include DDD, WATS, virtual WATS, INWATS, 500, and 900/976. Many other telcos around the world offer similar services and a great many other services under a wide variety of names. Deregulation, divestiture, and resulting competition have significantly reduced the costs of these services over the years. The emergence of specialized VoIP carriers has further intensified the price wars, and many long-distance services truly have become commodities, with any differentiation limited to pricing.

5.5.2.1 Message Telecommunications Service The Bell System introduced *Message Telecommunications Service* (MTS), also known as *Direct Distance Dialing*

(DDD) or *1 + dialing*, in November 1951. MTS enables the user to place long-distance calls on a dialed basis, without the intervention of an operator. Such calls traditionally were billed on the basis of a combination of distance, duration, and time of day; contemporary calls generally are billed on the basis of a flat, blended rate based solely on duration. Discounts may apply to long-duration and off-peak calls. DDD calls may be domestic or international in nature.

While the last manual exchange in the United States converted to dial many years ago, there are still a great number of manual exchanges in developing countries. Operators must still act on long-distance calls in some developing countries, particularly in the case of international calling where *International DDD* (IDDD) is not in place. In some developing countries it is still necessary to make an *appointment call*, which involves making an appointment for the international operator to seize an international trunk and place an international call at an appointed time. *Message service*, also known as a *messenger call*, is available in developing countries. This service involves the telephone company's sending a messenger to a remote village or other location with a message advising that person to expect an appointment call to be placed to a particular pay station, telco office, or agency in another village with telephone service; the cost of the messenger is added to the cost of the call.

In remote, rural areas of the United States, there remain a large number of *nondialable toll points*. These toll points are six-digit numbers in the 88*X-NXX* format. Operator intervention is required to access these telephone numbers, which terminate in locations that are beyond the reach of cable systems and are too low and sheltered to be reached via satellite. (The bottom of the Grand Canyon is one good example.)

5.5.2.2 Wide Area Telecommunications Service WATS resembles DDD service but is billed according to a variety of discount plans for large user organizations. WATS originally required special-purpose outgoing trunks that had access to specific areas of the country. Intrastate WATS provided intrastate coverage only. Interstate WATS was organized according to mileage bands 1–5, which were presented as crudely concentric areas of increasing geographic coverage. Band 1 reached adjacent states, band 2 included the next concentric ring of states, band 3 the next ring, band 4 the next ring, and band 5 provided full coverage of the 48 contiguous states. The greater the area of coverage was, of course, the higher the cost of the WATS service was. Traditional WATS was provided on a full-time or a measured basis. You could categorize full-time WATS as "all-you-can-eat" WATS because it was billed at a flat rate, with no usage monitoring or billing. Part-time, or measured, WATS was billed on a flat rate for the first 10 or 20h of usage, with overtime charges applying to traffic over the threshold. Clearly, the process of analyzing traffic patterns and configuring an optimum WATS network, with overflow to DDD, could be a fairly complex process.

Banded WATS in the United States has been abandoned in favor of *1+ WATS*, also known as *Virtual WATS*, which was enabled by increased network intelligence and encouraged by increased competitive pressure. Virtual WATS simply involves a discounted billing arrangement, including a small monthly fee and discounted usage, with the usage discount being sensitive to calling volume and volume commitment. The traditional requirement for special-purpose circuits no longer exists because the originating circuit and responsible user entity are identified by the billing systems, with the appropriate rating algorithm applied at the time the bill is rendered. A great many telcos around the world offer discounted long-distance plans under a wide variety of names.

5.5.2.3 *Inward WATS* Also known as *800 Service, Inward WATS* (INWATS) resembles virtual WATS, with the charges billed to the called party. Therefore, the call is toll free, at least to the caller. INWATS (1.800) service became so popular in the United States that the supply of numbers was exhausted several years ago. Therefore, 888 numbers were added, and 877 and 866 numbers were added soon afterward. Future additions will follow this existing convention of 8*NN*, with the last two numbers being identical. Outside the United States, INWATS service is known by various terms, including *FreeCall, Freephone, Greenphone*, and *Green Number*, and is accessed through various dialing conventions, including 080, 0800, and 0500. In 1997 the ITU-T implemented a *Universal International Freephone Number* (UIFN) convention, which involves dialing 800 + 8 digits.

INWATS is an effective means of encouraging current and prospective customers to call, because they incur no toll charges. Incoming call centers, therefore, are heavy users of INWATS, often terminating large numbers of INWATS numbers in a single call center. An ACD in a call center often is equipped to recognize the INWATS number called, with the called number being delivered to the system through the *Dialed Number Identification Service* (DNIS), which generally is provided by the carrier at additional cost, although it may be a no-cost feature of ISDN PRI trunks. Based on the DNIS and other information, the ACD can route the incoming call to the optimal agent group and agent with greater effectiveness and efficiency, delivering a screen pop in the process.

INWATS also is a cost-effective means for telecommuters and mobile employees to access various company offices for both voice and data applications. With the billing reversed to the target number, the company is relieved of the cost of processing expense vouchers for telephone charges. Further, the cost of the call often is far less than if it were charged to a calling card.

INWATS was the first instance of an *Intelligent Network* (IN) service. The INWATS numbers dialed do not represent real telephone numbers, at least not in the classic sense. In other words, they do not conform to the standard numbering scheme, wherein a telephone number (logical address) relates directly to a physical location (physical address) based on a standard numbering convention. Rather, an INWATS number can be directed to any location within the carrier's domains. In order to direct the call to the proper physical location, the network must query a database of INWATS numbers to translate that number into a conventional number associated with a specific physical location. Additionally, the database advises the originating LEC as to how the call should be routed. In the case of an interLATA call, the call routes to the IXC with which the target user has an INWATS relationship [15].

5.5.2.4 *500 Services* In the United States, *500 numbers* support premium *follow-me* personal communications services, which are defined as "as set of capabilities that allows some combination of personal mobility, terminal mobility, and service profile management" [16]. From a remote location, the subscriber may access the network logic in order to program (or reprogram) a priority sequence of numbers to which calls should be forwarded, with such a sequence perhaps including a cellular number. A single number can be used for both voice and fax, with facsimile routing invoked through a special dialing instruction such as * or # (an asterisk or number symbol), after a voice prompt. The ultimate plan is for 500 numbers to provide the capability to selectively forward calls received only from those that have knowledge of a PIN. 500 numbers promise to offer the first set of capabilities envisioned as *Personal Communications Services* (PCS), whereby one number theoretically can be retained for life—transportable across carriers and carrier domains.

AT&T offered a similar service some years ago in the form of *700 service*. The service provided follow-me call forwarding, although only AT&T long-distance subscribers could access a 700 number. No companies currently market 700 services [17–19]. *Note:* In some countries, 700 and 0700 numbers are designated other purposes such as for Internet calls, which may be billed at a different rate than voice calls.

5.5.2.5 900/976 Services   The *900/976 services* are premium information services that carry either a flat cost per call or the cost per minute that the called party (sponsoring party) determines. The revenues are divided among the sponsoring party receiving the call and the various carriers involved. Originally intended for applications such as telethons and informational services, 900/976 services have gained a bad reputation because many providers of telephone sex and other questionable services make extensive use of them. As a result, 900/976 services have fallen out of favor and are not widely used any longer. Fraud schemes actively use 900/976 numbers. In such a scheme, a caller will leave a message asking for a return call to a 900/976 number, with the call carrying an exorbitant charge.

5.5.3 Virtual Private Network Services

Virtual Private Network (VPN), or *Software-Defined Network* (SDN), services are intended for use by very large user organizations. Classic voice VPNs are circuit switched in nature, creating the effect of a private, leased-line network but without the associated issues of design complexity, long deployment time, high recurring cost, and vulnerability to failure. VPN services generally are interexchange, and often international, in nature because they find their greatest application in large multisite enterprises that transcend local boundaries. To configure a VPN, it is necessary to identify each terminating location in the multisite enterprise and the level of bandwidth required by each. Dedicated access circuits are established between each point of termination and the closest VPN-capable POP (Figure 5.8). Rather than interconnecting the various sites with dedicated circuits, the carrier routes the traffic over high-capacity transmission facilities on a priority basis, with the paths identified in switch routing tables. This ensures that the level of service provided is roughly equivalent to that of a true private network. The carrier realizes the benefit of sharing the involved network with other users, and at least a portion of those cost savings are reflected in lower network costs to the VPN customer.

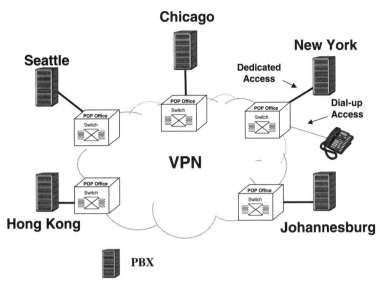

Figure 5.8   Virtual private network.

VPNs offer the advantage of scalability because new sites can be added and bandwidth to individual locations can be increased relatively easily and quickly, while maintaining a graceful relationship between the associated incremental cost and the incremental functionality. Additionally, configuration and reconfiguration effort and expense are reduced, as the only significant requirement is that the points and level of access be considered. Clearly, the time frame associated with carrier provisioning or reconfiguring such a network also is reduced, as dedicated circuits need not be provided between the various points of termination. Further, such a network is more resilient than a truly dedicated network, since the carrier network is highly redundant, and the carrier, therefore, can quickly route traffic around a point of blockage or catastrophic failure [14]. VPNs may be either domestic or international in scope.

5.5.4   Value-Added Services

You can define *value-added services*, also known as *enhanced services*, as those that alter the form, content, or nature of the information, thereby adding value to it. Examples include store-and-forward services such as voice mail, e-mail, and fax mail. In the data realm, networks that accomplish the process of protocol conversion are considered as providing value-added services.

5.6   PORTABILITY: A SPECIAL ISSUE

Portability of numbers is an issue of real significance, and increasingly so. Tradition-ally, all numbers have been associated with a geographic area (e.g., NPA and exchange code), a carrier (e.g., 800, 500, and 900 numbers), or a service offering (e.g., DID). Users prefer to retain the same number (logical address), regardless of physical location, carrier, or service offering. The intensity of the issue increased considerably with the advent of competition in the local exchange domain.

In the United States, 800 numbers (currently 888, 877, and 866 as well) have been portable across carriers since 1992 (although subject to restriction by LATA boundary). Portability of local numbers became an issue of great significance in the mid-1990s, when a small number of states began to permit competition in the local exchange. As the CLECs began to provide local service, they attacked the installed base of large ILEC customers. They were forced to require that those customers change their telephone numbers to fit into a block of DID numbers leased by the CLEC from the ILEC. Regardless of the attractiveness of the CLEC service offerings, potential customers were understandably reluctant to undergo a number change, which could involve a potentially significant loss of business, potentially considerable costs for reprinting stationary and otherwise advertising the new number, and an obvious disruption in the business of the enterprise. Further, customers would have to change numbers again if they were unhappy with the new provider and, therefore, choose either to return to the ILEC or to switch to another CLEC. The issue grew exponentially with the drafting of the Telecommunications Act of 1996 and the expectation that the CLEC business would experience dramatic growth.

Therefore, the act mandated the establishment of the *Local Number Portability Administration* (LNPA) to oversee the development and deployment of a mechanism for *Local Number Portability* (LNP). The first implementation of LNP was in the state of Illinois, which was among the first to permit competition in the local exchange and which has served as the model for the current method. That method involves the use of a *Local Routing Number* (LRN) of 10 digits and makes use of both the SS7 signaling and control network and the *Advanced Intelligent Network* (AIN) that SS7 supports. When a caller dials a telephone number, the originating CO consults a *Service Control Point* (SCP), which dips into a regional database. The database, as appropriate, provides the LRN, as well as the *Carrier Identification Code* (CIC) of the CLEC, in order that the call can route to the competitive carrier. *Number Portability Administration Centers* (NPACs) serve as clearinghouses for all local operators. Neustar, which currently is responsible for administration of the North American Numbering Plan (NANP), operates the NPACs. Note that LNP, as the name suggests, is local in nature; in other words, the number is portable between LECs only within the local calling area supported by the serving LEC (i.e., not across ILEC domains and not between calling areas involving toll calls).

As a footnote, 500 numbers and 900 numbers, both of which are geographically independent, are not portable across carriers, although the FCC has determined that such portability lies in the public interest [19, 20].

The ultimate in portability, as we see it today, is the concept of *Personal Communications Services* (PCS), which is explored in Chapter 11. In its full form, PCS involves inexpensive wireless phones that will offer two-way access anywhere and anytime for voice, data, video, and image communications. PCS, at least theoretically, will enable an individual to retain a single number (i.e., logical address) for life. That number will serve many devices (e.g., voice telephones, fax machines, and computer modems), with the caller prompted to make the appropriate selections and with the called party controlling options for the restriction of such privileges. *Note:* PCS is only conceptual at this point.

5.7 EQUAL ACCESS: ANOTHER SPECIAL ISSUE

Equal access is intended to ensure that the end user can access any IXC with equal ease. In other words, a user can dial a long-distance call from the residence or business premises simply by dialing the telephone number. Equal access is intended to facilitate a competitive environment through removing unnecessary technical barriers. Prior to its implementation in the United States, access to an IXC other than AT&T required dialing a lengthy carrier access number, a lengthy authorization code, and the target telephone number. This requirement clearly placed other carriers at a competitive disadvantage.

The implementation of equal access required that users in a specific geographic area be surveyed and afforded the right to choose a carrier on the basis of preselection. Users who did not respond were assigned a default carrier; such defaults were selected randomly and spread across the available carriers based on their respective local market penetration. All user choices or default selections are compiled in a centralized database residing on a database server, which is queried as each call is placed. Based on the originating circuit number, the database is consulted and the call connected through the designated IXC. This same process can apply equally to all outgoing calls, regardless of distance (i.e., local, intraLATA, interLATA, interstate, and international), subject to regulatory approval and deployment of the technology. New customers similarly have the right to choose a long-distance carrier; if they choose not to do so, they are assigned a carrier based on the same random selection process.

Alternatively, the user can access the carrier through dialing an access number (1010*XXX*). Such a technique would be used in order to place a call through another carrier in the event of a failure or blockage of the network of the primary carrier. The technique also is used to access a carrier that advertises special rates, which are generally to their great advantage, not the caller's. Equal access is 100 percent implemented in the United States.

Access charges are intended to compensate the LEC for the costs of connecting the call across expensive local loop facilities, conducting the preselection survey, investing in the database server, and administering the equal-access database. While the structure of access charges varies from country to country, all include some combination of Subscriber Line Charges (SLC) and Carrier Access Charges (CACs).

The *Subscriber Line Charge* (SLC) is billed to the user by the LEC on a monthly basis. The SLC is a flat-rate, recurring charge that generally varies by type of facility (e.g., residence line, business line, PBX trunk, and FX line). The SLC applies to all users of LEC loops, whether or not they use the LEC network for IXC access. The FCC subsequently allowed the LECs to collect an additional SLC, in the form of a *Digital Port Line Charge* (DPLC), for all digital circuits, including ISDN.

The *Carrier Access Charge (CAC)* is billed by the LEC to the IXC in two forms. First, flat-rate, recurring charges apply for tandem exchange termination. Second, the *Carrier Common Line Charge* (CCLC) is a minutes-of-use charge that applies to each call connected to the IXC [21].

5.8 VoIP: NEXT-GENERATION PSTN

The future of the telephone holds forth the promise of a service, growing always greater and better, and of continued progress—the end of which no one can foresee.
Telephone Almanac, American Telephone and Telegraph Company, 1927

The incumbent voice carriers all built their networks around circuit switching. As they converted from analog to digital technology, they introduced TDM, ISDN, and SS7, but circuit switching remained the technological network foundation. The carriers introduced Asynchronous Transfer Mode (ATM) into the backbone in the mid-1980s, and by the early 1990s ATM was touted as the ultimate network-switching technology in the WAN and even in the Local Area Network (LAN) domain. Ethernet overwhelmed ATM in the LAN and the Internet Protocol (IP) is doing the same thing in the WAN. While I discuss the details of Synchronous Optical NETwork (SONET) fiber optics, ATM switching, IP, data compression, and other elements of the VoIP mix in other chapters, I find it appropriate to discuss the overall concept now as VoIP seems certain to be the foundation for the next-generation PSTN. More correctly, IP will be the dominant protocol in the next generation of networks. These networks will support a multimedia blend of voice, audio, data, fax, image, and video at broadband speeds.

As noted previously, voice originates as an acoustic signal. In order to transmit voice over a network, it must first be converted to an analog electrical signal format. To send that voice signal over a digital network, it must be encoded into a digital (data) format, and it must be decoded back into an analog signal on the receiving end. The standard encoding technique is Pulse Code Modulation (PCM), which requires that the analog signal be sampled 8000 times per second at precise and regular intervals of 125 µs (microseconds), which represents 1/8000th of a second. In other words, each sample represents exactly 125 µs of a voice information stream. Each of the PCM samples comprises eight data bits, or one data byte (think of it as a *sound byte*). The string of sound bytes in a voice conversation, according to the normal conventions, requires a 64-kbps channel, also known as a DS-0 (Digital Signal level Zero), which is the fundamental building block of all digital telephony. A sound byte might represent an utterance or it might represent a moment of silence (i.e., a *silence byte*). A voice transmission contains lots of moments of silence. Further, the human conversational convention generally involves only one active direction of the conversation at a given time. In other words, we take turns talking, rather than overtalking each other. Therefore, a circuit supporting a voice conversation generally is silent in one direction or the other.

In a conventional PSTN scenario, each of the sound bytes (and silence bytes) associated with one voice transmission is interleaved with those of other transmissions through a process of Time Division Multiplexing (TDM) and flows across the network from end to end. At every step of the way, every network element must maintain a very tight synchronization of the individual sound bytes that comprise the information stream. Essentially, all the circuits and switches must ensure that each originating sound byte in a stream of sound bytes is received in exactly the same order and at exactly the same pace as it was created. Only in this fashion will the voice stream retain its fluidity and reasonably natural sound. If a number of voice samples were to suffer loss or error in network transit, the quality would suffer noticeably. If some voice samples were to arrive in rapid succession and others were to be delayed, the voice stream would lose its fluidity and sound "herky-jerky," *if you k n o w wha t I me an*. This variability in the data arrival rate, also known as *jitter*, is most unpleasant. All of the elements in the network (e.g., access circuits, end offices, multiplexers, transport circuits, repeaters, and tandem switches) must work together in a highly synchronized fashion in order to support *toll quality* voice communications. To ensure that the quality of the voice transmission is preserved, conventional TDM networks commit a time slot for each sampled byte, whether a sound byte or a silence byte. This approach works beautifully. This entire process is coordinated under the control of signaling and control, which is in the form of SS7, at least in contemporary digital voice networks.

Voice over IP, like fax over IP (Chapter 4), involves a totally different type of network. Voice over IP supports the transmission of voice data over a highly shared packet data network running the Internet Protocol and at times involving advanced compression techniques. The first step in VoIP is the collection of a number of PCM voice samples in a buffer at the IP *gateway*, a protocol converter responsible for converting the PCM datastream to a compressed IP packet stream. The gateway also resolves signaling and control issues between the circuit-switched PSTN and the IP-based network. Specifically, the conventional PSTN relies on SS7, and IP-based networks variously employ *H.323 or SIP*, both of which I discuss in subsequent chapters. Physically, the gateway can be under the skin of a PBX or ACD or can be a stand-alone CPE device situated between the PBX or ACD switch and the local loop. The gateway can be in the form of software in a softphone that connects to the Internet over a broadband local loop such as ADSL. Alternatively, the gateway can be in the form of a router situated at the edge of the carrier network, where a CO is positioned in a conventional PSTN. Multifunctional network-based gateways commonly are known as *softswitches*, which are software-based switches that act not only as interfaces between circuit-switched and packet-switched networks but also as interfaces between SS7 and H.323 or SIP signaling and control systems. In any case, the gateway contains a number of *Digital Signal Processors* (DSPs), which are silicon chipsets on *Printed Circuit Boards* (PCBs) that fit into slots in a cabinet.

Using G.723.1, one of the standard compression algorithms, the buffer in the gateway gathers 160 voice samples, each representing $125 \mu s$ of the voice stream. Therefore, the set of 160 samples represents $20 \mu s$ (1/50th of a second). The set of 160 samples is evaluated as a discrete set of binary data, and any redundant data (generally quite a lot) is identified. Human speech, for example, often contains long pauses. Even during a continuous stream of utterances, there are relatively long pauses that are not noticeable to human beings but which are quite discernable to computer systems. These pauses are noted and compressed out of the data set, with the beginning and the length of the pause being noted. An utterance also can contain a lot of redunnnnnnnnnnnndancy. This is noted in a similar fashion. The DSPs are responsible for the process of analog encoding to the digital format through a built-in codec and for voice compression.

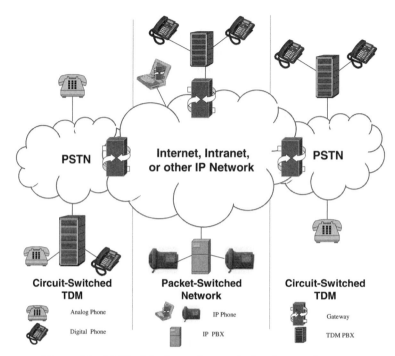

Figure 5.9   VoIP, with alternative methods of network access.

Also, at the gateway, the compressed set of binary data is formed into an IP data packet (Figure 5.9). Each packet is a discrete, independent unit of data, or *datagram*, that is presented to the network. Each packet wends its way through a packet network, comprising various combinations of routers and switches and interconnecting links, until it reaches the terminating edge gateway, as identified by the IP address contained in the packet header. (Note that, once again, a logical address is translated into a physical address. In this case, the IP address is a set of bits in the first field of data, and the physical address is a gateway, rather than a local loop.) The key advantage to a packet data network is that all of the resources (i.e., circuits and switches) are highly shared. That is, the packets are presented to the gateway and queued up in a buffer until such time as the packet switch has computational resources available to process it. In other words, the packets are queued until the packet switch can read the address information, consult a routing table, and make a decision as to the specific route across which the packet should be forwarded. Once processed, the packet might queue in a buffer until the selected circuit is available to send it on its way. If there are many switches and links en route, the packet must endure this process many times. At the terminating gateway, the packet is received, the data packet decompressed (reinserting the periods of silence and the redundancies) and decoded, and an approximation of the original acoustic signal is reconstructed.

This compression process essentially reduces the bandwidth in the network required to support the voice transmission. To the extent that bandwidth demands are reduced, efficiency is increased, more data can be sent over the same circuits and through the same switches in the same period of time, and costs are lowered.

If this sounds too good to be true, it is because it is too good to be true—and the reasons are many:

1. The process of compression and decompression takes some time—not much time, but some.
2. The packets can encounter considerable delay, or *latency*, as they transverse the network. If the network is under a relatively light load at a particular instant, the routers and switches can fairly instantly process the packet, and the links can become available almost immediately. If the network is under relatively heavy load at a particular instant, the packet might queue up for a much longer period of time. The more routers, switches, and links that are involved along a chosen path, the greater the latency.
3. The level of delay is variable and unpredictable in nature. This variability in latency is known as *jitter*.
4. The stream of packets might take a relatively short physical path on one call and a relatively long path on the next call, all depending on the load on the network at the time. The longer the path taken, the greater the propagation delay involved. The more devices (e.g., routers and switches) involved, the more processes involved, and the greater the cumulative delay as the processes are performed.
5. Individual packets in a stream of packets might take different physical paths. Again, the longer the path taken, the greater the propagation delay involved. The more devices (e.g., routers and switches) involved, the more processes involved, and the greater the cumulative delay as the processes are performed. Note that each packet is treated individually. The final result can be an unacceptable level of jitter, even if the packets arrive in the proper sequence. If the packets arrive out of sequence, some process must be invoked to resequence them in order that they can play in the proper order. If some packets are delayed too long, they may not arrive in time to play in the proper order, in which case they are rejected.
6. Individual packets might suffer errors in transit. Errored data are of no value. In fact, errored data are of decidedly negative value. Once the packet arrives, if ever, it must either be accepted or rejected. There is no time for a retransmission of an errored voice data packet.

So, the inherent nature of the packet network is problematic. While this approach offers great efficiencies, it also imposes variable and unpredictable levels of delay on the packets. In other words, VoIP is a classic design trade-off between cost and performance. Next-generation VoIP networks propose to optimize this trade-off in several ways:

1. Voice over IP carriers deploy gateways only at the edges of the network, where the most complex decisions must be made and where the most involved (read time-consuming) processes must be invoked. They deploy high-speed ATM switches in the core of the networks, where speed is of the essence. The routers and switches currently can run at internal bus speeds in the range of 1 Tbps, at least according to the design specifications.

2. The carriers define paths from each gateway to every other gateway, thereby ensuring that all packets travel the same route between any two given points. Therefore, all the packet voice transmissions from San Francisco to Dallas, for example, likely will suffer reasonably similar levels of delay, at least.

3. The carriers deploy incredibly high-speed optical fiber transmission systems in the core. Along any given path, there may be hundreds of fibers, each of which can support numerous lambdas, each of which can run at rates in the Gbps range.

4. The compression algorithms are extreme. Traditional PCM voice requires 64 kbps. Voice over IP requires as little as 5.3 kbps, depending on the specific algorithm selected, plus IP overhead. (*Note:* With the IP overhead, the brings the bandwidth requirement back up to approximately 64 kbps, so clearly VoIP is not all about reducing bandwidth requirements.)

To illustrate the concept of compressed, packetized voice, consider the following analogy. You have become concerned about your health. You finally decided to take your grandmother's advice and include a bowl of stewed prunes in your daily diet. (Your grandmother has lived a long and healthy life but is not much of a gourmet!) You call your grandmother and ask her to pick some prunes from her orchard and send them to you on a regular basis. Daily, your grandmother picks the ripest prune plums (bytes) at an exact pace (samples at precise intervals), puts 160 of them (data set) in a bowl (buffer), dehydrates (compresses) them, removing the water (silence) in order to reduce their weight and volume (bandwidth) and lower the cost of postage (network cost). The result is prunes. Every day she sends you a small, numbered box (packet) of prunes through the mail, along with specific instructions that you eat them every day, at exactly the same time of the morning, and in exactly the same order in which she mailed them to you. In that fashion, you always will have fresh prunes for breakfast and you will remain regular (perfectly timed). You follow her instructions precisely for some number of days. Each day you open a box of prunes, put them in a bowl (buffer) and soak the contents in order to reinsert the water (silence), thereby reconstituting (decompressing) them to an approximation of their original form. Then disaster strikes. One box (packet) of prunes (sound bytes) arrives in sequence but crushed (errored). As the contents (compressed voice payload) are inedible (unintelligible), you throw away (discard) the box and its contents. The next day, the postal system fails you, and no box of prunes arrives (latency, or delay). The next day, two boxes arrive (jitter); you eat two bowls of prunes for breakfast and regret it. You call your grandmother and ask her to send the daily prune ration via Federal Express (high-speed network). They almost always arrive every day and at exactly the same time. Each box (packet) still wends its way through the network independently, but the overbuilt (broadband) network of high-speed access circuits (couriers), high-speed jumbo jets (transport circuits), and high-speed airports and processing centers (routers and switches) provides a good balance of cost and performance. Federal Express does such a good job that it lowers its costs. You are such a good customer that Federal Express offers you a discount; the total cost compares favorably to that of growing your own prune plums.

Now, back to the bottom line, so to speak. Circuit switching and TDM are inherently wasteful for bursty data applications but perform well for stream-oriented transmissions such as voice. Packet networks are inherently efficient but are not intended to support voice. The VoIP carriers overcome this quandary through a combination of effectively overengineering the packet network and using complex compression algorithms and buffers to improve performance. Note that, at least for the foreseeable future, the circuit-switched PSTN remains available as a backup in the event that the performance of an IP-based voice network is less than satisfactory at any given moment. In fact, VoIP service providers use the PSTN for backup, and the most capable VoIP PBXs have the ability to choose between the two networks based on user-definable performance parameters. Note that not all VoIP is alike. Much consumer-level VoIP, particularly peer-to-peer services, is over the Internet, where quality is always questionable.

This discussion of VoIP is obviously very brief, and VoIP is much more complex than is portrayed here. Among other things, VoIP requires a signaling and control mechanism, and the earlier discussion of signaling and control in this chapter was specifically oriented toward the conventional circuit-switched PSTN. The IP-based networks supporting VoIP entail entirely different signaling and control mechanisms. Where those two networks meet, signaling and control issues must be resolved through a gateway that accomplishes protocol conversion. I discuss this and other VoIP issues in subsequent chapters.

REFERENCES

1. Brooks, John. *Telephone: The First Hundred Years*. Harper & Row, 1976.
2. *Telephone Almanac*. American Telephone & Telegraph Company, 1937.
3. Greene, Tim. "Bells Bond to Spread the Word on New Area Codes." *Network World*, May 20, 1996.
4. Potter, David C. "Cracking the NANP Code." *Telecommunications*, October 1995.
5. Potter, David C. "Surviving the North American Numbering Plan." *TeleProfessional*, March 1996.
6. Toth, Victor J. "Winners and Losers in FCC's New NANP System." *Business Communications Review*, October 1995.
7. Toth, Victor J. "Preparing for a New Universe of Toll-Free Numbers." *Business Communications Review*, November 1995.
8. Greene, Tim. "Phone Numbers Are Running Out." *Network World*, May 20, 1996.
9. Rohde, David. "Those #$ (&%*!% Area Codes." *Network World*, September 18, 1996.
10. http://www.nanc-chair.org/docs/nowg/Jan06_NANPA_Report.doc.
11. Federal Communications Commission. "Federal Communications Commission Releases Data on Local Telephone Competition." *FCC News*, April 3, 2006.
12. *Engineering and Operations in the Bell System*. Bell Laboratories, 1977.
13. Tunstall, W. Brooke. *Disconnecting Parties*. McGraw-Hill, 1985.
14. Elbert, Bruce R. *Private Telecommunication Networks*. Artech House, 1989.
15. Rebber, Roger. "Parade of Carrier Options." *Call Center Magazine*, February 1996.
16. "500-NXX Assignments." http://www.nanpa.com/number_resource_info/500_codes.html.
17. Taff, Anita. "Follow-Me Phoning." *Mobile Office*, November 1995.
18. Greene, Tim and Wexler, Joanie. "Follow-Me Phone Numbers Available." *Network World*, February 15, 1995.

19. Seybold, Andrew M. "Follow-Me Phone Numbers." *Mobile Office*, August 1995.

20. Toth, Victor J. "The FCC's Complex Plans for Local Number Portability." *Business Communications Review*, September 1995.

21. Foster, Mark, McGarry, Tom, and Yu, James. "Number Portability in the GSTN: An Overview". Internet Draft, February 9, 2001.

CHAPTER 6

FUNDAMENTALS OF DATA COMMUNICATIONS

A favorite subject for cartoonists used to be the mechanical man. Next came the electronic computer or mechanical brain that could solve extremely complicated problems with lightning quickness. Last year the Bell System demonstrated a service that may give the artists a new inspiration—a device that lets machines "talk" over telephone lines to other machines. Known as the Dataphone service, this arrangement lets customers send up to 800 words a minute over telephone lines from a business machine to another machine, or machines, throughout the country. Any type of information that can be recorded on tapes or cards may be transmitted.... The new system is about ten times faster than previous methods [telex and TWX] *and will be a great help to businesses that send large amounts of business information.*

Telephone Almanac, Bell Telephone System, 1959

Electronic data communications was born in 1844 with the transmission of the first telegraph message, "What hath God wrought?" sent by Samuel F.B. Morse in 1844. Development of telegraph networks was fastest in the United States. At the beginning of 1846, the only working line was Morse's experimental line, running 40 miles between Washington, D.C., and Baltimore, Maryland. In 1848, there were approximately 2000 miles of telegraph lines. The first commercial telegraph service began in 1849 between New York, New York, and Philadelphia, Pennsylvania. In 1850, there were over 12,000 miles of line operated by 20 different companies offering commercial telegraph service [1]. Western Union dates to 1851 through its predecessor, the Mississippi Valley Printing Telegraph Company, which became the Western Union Telegraph Company in 1856 through a series of acquisitions. As telegraphs were manually operated, telegraph services were labor intensive and expensive. Telegraph services also were regional in nature and interconnection, if any, of the regional networks involved a manual process. A telegraph operator would have to

Telecommunications and Data Communications Handbook, By Ray Horak
Copyright © 2007 Ray Horak

receive the telegram transmission over one network, write it down, and rekey it into the next network. The Pony Express began service in April 1860 with the promise that it could deliver a letter from St. Louis, Missouri, to Sacramento, California, in 10 days or less. Eighteen months later, the Western Union transcontinental telegraph network was completed and the Pony Express was decommissioned shortly thereafter. The first telephone networks appeared in 1877 and soon replaced telegraphy as the primary means of electronic communications. Long-distance voice telephone service became so inexpensive in the 1970s and 1980s that the use of telegrams declined precipitously, never to recover. Telegraphy died a slow death, however. Western Union sent its final telegram on January 27, 2006, after 150 years in the business. At its peak in 1929, Western Union sent some 200 million telegrams. In 2005, it sent about 20,000. According to MSNBC, "the last 10 telegrams included birthday wishes, condolences on the death of a loved one, notification of an emergency, and several people trying to be the last to send a telegram" [2]. That's a far cry from "What hath God wrought?" Western Union introduced *teletypewriter* service in 1923 so that companies could link branches and even join other companies in private text messaging over leased private-line networks. (*Note:* The terms *teletype, teletypewriter*, and *telewriter* are used interchangeably.) Beginning in about 1935, telegraph companies began to use *telex* (*tele*typewriter *ex*change), a rotary dialing system much like that used in telephone networks. Telex was an effective means of routing telegraph calls, and it ran at the amazing signaling speed of 45.5 bps and, later, at 50 bps, or approximately 66 words per minute (wpm). At that low signaling rate, one analog voice-grade channel could support 24 or 25 telex transmissions through Frequency Division Multiplexing (FDM). In 1958, Western Union introduced its Telex service, a direct-dial consumer-to-consumer teleprinter service. Initially, consumers could direct dial numbers only in the United States, Canada, and Mexico; operator assistance was required for dialing to other countries. In 1930, AT&T released its TWX (TeletypeWriter eXchange, and pronounced "twix") service, a high-speed telex service that ran at the blazing speed of 75 bps and later 150 bps [3]. Western Union operated the TWX service in the United States for many years until AT&T acquired Western Union's Telex network in 1990, which then reached over 190 countries. Altogether, telex services eventually reached some 3.5 million machines, by some estimates. Telex was, at one time, undoubtedly the most heavily used textual telecommunication system in the business world. Lest you dismiss this bit of history as relating only to simple times and primitive technologies, you should take note of the fact that telegraphy was considered quite complex at the time and that telegraphers were in great demand. Dodge's Institute of Telegraphy (Valparaiso, Indiana) claimed to be "the most completely equipped school in the world and the only institution in which a student can become entirely qualified for a position" (as a telegraph operator). The institute was "endorsed by Officials of the Western Union Telegraph Company. Situations secured for graduates. Demand for operators greatest in the history of the telegraph. We have for some time been unable to fill demand made upon us for qualified students," or so read the advertisement in 1903. Training was on site. "The cost of a six month's course, including tuition (telegraphy and typewriting), table board and furnished room need not exceed $87" [4]. This is reminiscent of the advertisements by schools teaching courses for certified network administrators and systems engineers only a few years ago, although the classes did not last quite six months and room and board was not included in the tuition. Those advertisements will seem equally quaint and amusing 100 years from now. (They are certainly not that amusing today, at least not to many certified network administrators and systems engineers who have switched careers for lack of work.) Telex service is still in use, largely in developing countries. Tele-TYpewriter (TTY) service, also known as *Telecommunications Device for the Deaf* (TDD) in the United States, *textphone* in Europe, and *minicom* in the United Kingdom, is heavily used by those with hearing or speech impairments.

Many of the early chapters of this book have explored the world of voice communications. During that exploration, it became clear that many of the devices and circuits in the Public Switched Telephone Network (PSTN) are digital in nature. In large part, the contemporary PSTN is a data network, transporting and switching voice data in digital format. After all, voice can be considered as the first in a long string of what, in a contemporary context, are considered to be data applications. This chapter introduces the basic concepts of the communications of computer data. Reflecting the reality of contemporary systems and networks, the balance of this book largely focuses on computer-to-computer data communications.

In order to comprehend the intricacy of contemporary data communications systems and networks, it is necessary to develop a solid understanding of certain basic concepts. These definitions and concepts extend across all technologies and service offerings, from the historical to the most contemporary. The historical is more than a footnote; it is important for purposes of understanding. Indeed, the contemporary and future networks are built on those fundamental concepts.

This chapter first addresses the concept of functional domains. I explain terminal equipment, communications equipment, and communication software and provide a detailed discussion of modems and DSUs/CSUs. I explore the concept of protocols at length, with discussion of basic protocol dimensions and issues. I also discuss computer network architectures, with emphasis on IBM's *Systems Network Architecture* (SNA) and the *Open Systems Interconnection* (OSI) reference model. Finally, I pause to consider the importance of security in data systems and networks.

6.1 FUNCTIONAL DOMAINS

Functional domains comprise the spheres of influence exerted by the various network elements that perform specific tasks in a data network. Data terminal equipment, data communications equipment, communications software, switches, and transmission facilities all are physical elements of such a network, with each performing specific functions and all supported by a signaling and control system. Since the general nature and specifics of both circuit switches and signaling systems are explored in Chapters 1 and 5, I will not repeat those discussions here. Please seek out the discussions in those earlier chapters for more information on those topics.

6.1.1 Data Terminal Equipment

The data equivalent of Customer Premises Equipment (CPE) in the voice world, *Data Terminal Equipment* (DTE) comprises the computer transmit and receive equipment. DTE includes a wide variety of *dumb* terminals, or terminals without embedded intelligence in the form of programmed logic. Dumb terminals (e.g., Hewlett-Packard HP2521P and Televideo 950) are devices that merely provide a user interface to a more capable host computer. Semi-intelligent terminals (e.g., IBM 317x and 327x) possess a limited amount of intelligence, enabling them to perform certain, limited processes, independent of the intelligence contained in the host computer. Intelligent terminals generally are in the form of Personal Computers (PCs) that are networked to a host computer. Such devices are highly capable hosts in their own right, although and in this context they often are linked across a network to an even more capable host. At the top of the terminal food chain are client workstations, highly intelligent and capable host computing devices that access a more capable server in a *client/server* environment. In such an operating environment, clients' requirements for access to files, applications, and network communications software are satisfied by one or more servers, which typically are accessed across a Local Area Network (LAN). As a result, the client workstation can perform certain appropriate functions (e.g., screen formatting) related to the specific user task at hand, while the servers' memory and processing power are dedicated to the performance of tasks such as file storage, database management, and security management that are accomplished more effectively on a centralized basis.

DTE also is in the form of *host computers* such as mainframes and midrange (mini) computers. Host computers, also known as *host nodes*, are highly capable devices with substantial processing power and storage memory. Hosts also, at least theoretically, are carefully administered to ensure that they operate successfully and reliably. Hosts also serve as highly effective information repositories, with the data backed up and archived on external storage media such as magnetic tapes and CD-ROMs or on networked storage devices such as RAID (*Redundant Array of Inexpensive Disks*) systems. *Note:* In the DTE context, host computers and nodes are end-user endpoint (i.e., transmit and receive) equipment, rather that the various types of computer nodes that might be employed in networking capacity, such as the computers that run the IP protocol suite on the Internet.

Also note that the lines have blurred between categories of host computers. The PC on which I am writing this book is much more capable (and much less expensive) in many respects than was a typical mainframe only 15–20 years ago. Most of you would not even recognize a mainframe if you saw one. I would, but then I come from the days of *heavy metal*, when a *forklift upgrade* meant using a forklift to move out the old computer and move in the new one. The old one became a boat anchor or artificial reef. (This is true, except for the part about the boat anchor and artificial reef.)

6.1.2 Data Communications Equipment

Also known as *Data Circuit Terminating Equipment* (DCTE), *Data Communications Equipment* (DCE) is the equipment that interfaces the DTE to the network and resolves any issues of incompatibility between those domains in the process. Incompatibility issues can include digital versus analog, voltage level, signaling speed, and bit density. DCE includes modems, DSUs and CSUs, and Front-End Processors (FEPs), all of which I discuss in greater detail later in this chapter.

6.1.3 Communications Software

Communications software often is required and generally is embedded in the computer operating system; alternatively, it can take the form of a systems task under the control of the computer's operating system. The role of communications software is to assist the operating system in managing local and remote terminal access to host resources, to manage security, and to perform certain checkpoint activities. The remote terminals interface to the operating system access methods, which contain the specific code required to transfer data across the network channels between the devices. An example of an access method includes IBM's *Virtual Telecommunications Access Method* (VTAM).

Alternatively, commercial communications management software can control and manage access to the host. IBM's *Customer Information Control System* (CICS) is such a product. This software resolves contention issues between diverse applications without impacting programs or terminals. It handles polling, selection, and program interrupts, thereby ensuring minimum response time. It also resolves error conditions at both the data and line levels. *Random-Access Memory* (RAM) maintains CICS and other *Terminate-and-Stay-Resident* (TSR) software.

6.1.4 Networks

Networks provide the connections between computer resources in order to accommodate the flow of information. Networks support the logical transfer of data during a communications session through the establishment of paths, circuits, or channels over a physical medium. The network can be in the form of a Local Area Network (LAN), a Metropolitan Area Network (MAN), or a Wide Area Network (WAN). These networks support communications over areas of increasing geographic scope. Data LANs, MANs, and WANs, which I discuss in later chapters, also can be interconnected.

6.1.5 Switches

Developed in support of voice communications, circuit switches serve for the flexible interconnection of circuits. Once the circuit is established, a circuit switch dedicates a communications path for the duration of the session in support of a data transmission that is presumed to be in the form of a continuous stream. Because of their ubiquity, Central Office (CO) circuit switches remain widely used in support of dial-up data communications for Internet access. With that exception, most data communications takes place over networks that variously employ packet, frame, and cell switches.

These switches are highly advanced computerized switching devices that have substantial capacity as well as the ability to share high-capacity transmission systems among large numbers of individual user transmissions. In capsule, such switches read the destination address of each packet, frame, or cell of data and forward it through the switch and across the network, perhaps on the basis of priority, in recognition of the underlying application being supported. As all network resources are highly shared, the network operates with much greater efficiency than does a network based on circuit switching, at least for bursty, low-volume data communications applications.

Switches make forwarding decisions independently, link by link, although their activities commonly are coordinated by higher level system. *Routers* are highly intelligent switches that are capable of making path selection decisions across a network, from end to end and in consideration of perhaps a wide range of factors such as priority, protocol, and underlying application. Routers also can operate at much higher levels than switches, as they can be equipped to accomplish protocol conversions, manage security, and so on. I discuss packet, frame, and cell switching and routing in detail in later chapters. (*Note:* A *switch* is beneath a router, and network purists will find my characterization of routers as *highly intelligent switches* to be offensive, even in a descriptive sense. I reckon they'll get over it, eventually.)

6.2 DCE: EXPANDED VIEW

While I discuss data networks and switches in considerable detail in other chapters, this chapter is a convenient and meaningful place to pause and explore the concept and detail of several types of data communications equipment. Specifically, I want to examine modems, codecs, terminal adapters, CSUs and DSUs, and FEPs.

6.2.1 Modems

*Modems mo*dulate and *dem*odulate signals. In other words, they change the characteristics of the signal in some way. Modems, as discussed in this chapter, are of several basic types: line drivers, short-haul modems, and conventional PSTN modems. The term also is used to describe a wide variety of other devices such as ISDN Terminal Adapters (TAs), ADSL modems, and cable modems used in CATV networks, each of which I explore in subsequent chapters.

6.2.1.1 Line Drivers Line drivers actually are interface converters, rather than modems in the classic sense. Line drivers extend the distance of a digital connection, within limits, by converting the digital signal to a low-voltage, low-impedance signal that can transmit more effectively and over longer distances on dedicated, specially conditioned twisted-pair circuits. The *Recommended Standard* (RS) *232* specification [more correctly known as *Electronic Industries Alliance* (EIA) *232*], for example, generally limits the distance between devices to 50 ft at transmission rates of 56 kbps. At lower speeds, line drivers can reshape the digital pulses to extend that distance considerably. At speeds of up to 9.6 kbps, for example, line drivers can extend that limitation to 500–5000 ft over Category 3 Unshielded Twisted Pair (UTP). Line drivers can support speeds up to 19.2 kbps over distances up to 10,000 ft over good-quality Shielded Twisted Pair (STP) or Screened Twisted Pair (ScTP). You can extend the distance further through cascading line drivers. *Note:* Line drivers are unidirectional and work over simplex circuits. Bidirectional communications requires separate sets of line drivers operating over separate twisted-pair circuits.

6.2.1.2 Short-Haul (Limited-Distance) Modems   *Short-haul modems* are used where line drivers fail in terms of either capacity or distance. Short-haul modems can work at distances between 5000 and 10,000 ft, with distance sensitive to signaling speed; that is, the higher the speed, the shorter the allowable distance. Also known as *limited-distance modems*, they usually are used for private-line and hardwired links but can operate over nonloaded local loop facilities.

6.2.1.3 Conventional Modems   Conventional modems allow digital devices to communicate across an analog circuit, accomplishing the digital-to-analog conversion in order to resolve that dimension of incompatibility between the DTE and the network. AT&T set the original de facto standards for modems with the introduction of the DataPhone service in 1959 [5, 6]. Currently, the ITU-T sets modem standards at an international level. Those standards fall into the *V* series, which includes all standards recommendations for *data communications over the telephone network*. The digital input to the modem is in the form of a baseband signal of varying electrical voltage levels that represent binary 1s and 0s. The output from the modem is a modulated analog carrier wave, which can be modulated in terms of its amplitude, frequency, phase, or some combination thereof. Through this process, the 1s and 0s of the data stream output by a digital computer can be sent over an analog voice network. While a network that is digital from end to end is preferred for reasons that include error performance and bandwidth, there often are analog components or links involved. In developed countries, of course, the core of the carrier network generally is fully digital in nature. In consideration of lower costs and greater availability, however, the local loop often is analog in residential and small-business applications. If the local loop is analog, the network is presented to the computer as analog. As the lowest common denominator rules, so to speak, the high-speed digital computer must adjust to the voice-grade analog local loop, and that is done through a modem.

It certainly is worth mentioning, and even emphasizing, that modems are unique in the wired world in terms of their portability, as modems work virtually anywhere. Contemporary *road warriors* largely would be lost without the advantage of modem access to the Internet from hotel rooms, airports, and client sites. While the modem speed may be constrained by the internal processes of an intermediate PBX, the device is absolutely indispensable for remote access to e-mail and other applications through the Internet and World Wide Web. In the United States and many other developed countries since 2000 or so, broadband access has become widely available in hotels, airports, coffee shops, and other public and private places through either hardwired or wireless Ethernet LANs. However, conventional modem access is an essential backup technique in the event that those options are not available. So, I travel with a conventional international modem (the modem standards are the same, but the physical and electrical connections vary from country to country), an Ethernet adapter, and a wireless Ethernet modem.

Now, I want to consider the way conventional modems work by using a simplified example. While researching this book through the Web, sending drafts to various editors and to my publisher, and otherwise fooling around, I sometimes used a modem connection while on the road. To set up the remote modem connection to my Internet Service Provider (ISP), I click an icon and the modem card I have inserted in my laptop computer "goes off-hook," gets an internal dial tone from the hotel PBX, and dials the telephone number of the local ISP in partnership with my ISP back home. [*Note:* Conventional modems can be external or internal. They also can take the form of a PCMCIA (*Personal Computer Memory Card International Association*) card, which fits into a slot on a laptop computer. Also note that the *PCMCIA card* generally is abbreviated as *PC card*.] The dialed number is preceded by a *dial access code* of *9* or *0*, depending on the PBX convention in that particular country. As instructed by the dial access code, the PBX sets up an outside line, draws the dial tone from the local CO and dials the number of the ISP. The ISP's modem answers, and the two modems pass a set of control signals back and forth in a process known as *handshaking*, to negotiate the basis on which the communication will be conducted. Specifically, each modem identifies itself and its capabilities to the other through a set of data organized into a frame. The modems pass the frames simultaneously, in full-duplex mode, with one using a relatively high range of frequencies and the other a relatively low range. Each modem knows exactly the format of the data bits and fields within the frame, as each is based on standards in the form of ITU-T Recommendations. The data are passed at the rate of 2400 baud, which is the maximum expected available baud rate, internationally, over an analog network. Referring back to Chapter 1, *baud rate* is defined as the number of signal events, or signal transitions, per second and *bit rate* is defined as the rate of information transfer. Also referring back to Chapter 1, the baud rate cannot exceed the bandwidth of the channel, which in this case is nominally 3000 Hz. So, the rate theoretically could be as high as 3000 baud, but that would be extremely optimistic and would leave no margin for error. So, 2400 baud has been established as a reasonable level of maximum expectation for a voice-grade local loop anywhere in the world. Assuming that everything is just right in the network (i.e., the analog local loops are in good condition and there are no issues of interference at the moment), the modems then start to pass data at 2400 baud over the analog local loops by modulating the sine waves.

Note that modem networks are balanced and symmetrical. In other words and as illustrated in Figure 6.1, there must be two modems, either standalone or in

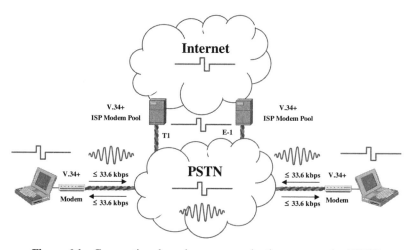

Figure 6.1 Conventional modem communications across the PSTN.

modem pools, and they each must be able to communicate on the same basis in order to establish a connection. Note also that the highest common denominator always rules. In other words and in this case, the more capable modem must adapt to the capabilities of the lesser, and both must adapt to the condition of the local loop circuit, the processes of any intermediate PBX, and so on. So, the modems fall back, as necessary, to adjust to the line conditions and other factors, until they find a baud rate and a bit rate that satisfy their performance requirements. In this scenario, the modems agree that the maximum of 2400 baud works well, so they proceed. Modems variously employ three basic modulation techniques: Amplitude Modulation (AM), Frequency Modulation (FM), and Phase Modulation (PM). More sophisticated techniques include Quadrature Amplitude Modulation (QAM) and Trellis-Coded Modulation (TCM). Remember that each baud can support multiple bits, which qualifies a modulation technique as a compression technique, and some of them are quite extreme.

6.2.1.3.1 Amplitude Modulation   Also known as *Amplitude Shift Keying* (ASK), *Amplitude Modulation* (AM) involves the modulation of the amplitude (i.e., strength or voltage) of the analog carrier sine wave. The transmitting computer outputs a baseband signal, which is the digital transmission of electrical pulses, with 1 bits and 0 bits defined as discrete voltage levels. (*Note:* A *baseband* signal is a signal in its original form, without being altered in any way, whether by modulation or conversion.) Using a (single-bit) AM technique, each 1 bit entering the transmitting modem is expressed as a relatively high-amplitude sine wave, or series of sine waves. Each 0 bit is expressed as one or more low-amplitude sine waves, as illustrated in Figure 6.2. The high and low levels are defined in terms of a reference level or by the relative difference between the levels. At 2400 baud, this *unibit* technique yields a transmission rate of 2400 bps, with one bit transmitted per baud. Note: A bit has one of two possible values, a 1 or a 0. To express one bit with each baud, therefore, there must be $2^1 = 2$ possible signal states.

It is possible to express two bits with each baud by defining $2^2 = 4$ possible signal states. In a *dibit* (two-bit) coding scheme using AM, for example, the lowest level of amplitude represents a 00 bit pattern, the next highest a 01 bit pattern, the next a

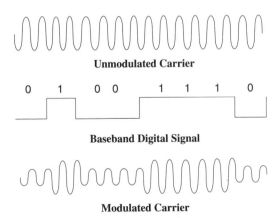

Figure 6.2   Amplitude modulation: unibit.

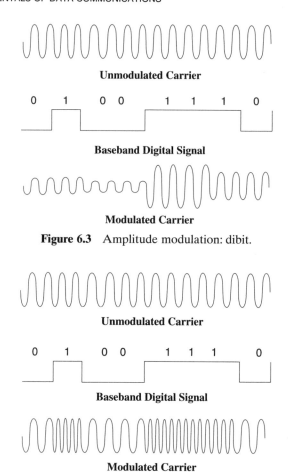

Figure 6.3 Amplitude modulation: dibit.

Figure 6.4 Frequency modulation: unibit.

10, and the highest a 11, as illustrated in Figure 6.3. In this fashion, two bits are impressed on each baud. Thereby, the speed of data transmission is doubled at the same analog line speed; that is, at 2400 baud, the transmission rate is 4800 bps. Thereby, the connection time is halved, and the cost of transmission is halved, or at least is reduced considerably. Amplitude modulation rarely operates independently because it is highly sensitive to the impacts of attenuation and line noise.

6.2.1.3.2 Frequency Modulation Also known as *Frequency Shift Keying* (FSK), *Frequency Modulation* (FM) is the sole technique used in low-speed, Hayes-compatible modems. FSK involves the modulation of the frequency of the analog carrier sine waves (Figure 6.4). When no bits are transmitted, the carrier is at a reference frequency of 1700 Hz. A unibit FM technique impresses one bit on each baud by shifting the carrier to 2200 Hz when transmitting a 1 bit and to 1200 Hz when transmitting a 0 bit. At 2400 baud, therefore, the transmission rate is 2400 bps. The benefits of dibit transmission can be realized by defining four frequencies, with each sine wave or set of sine waves representing a two-bit pattern (00, 01, 10, and 11). Thereby, at 2400 baud, the transmission rate is 4800 bps.

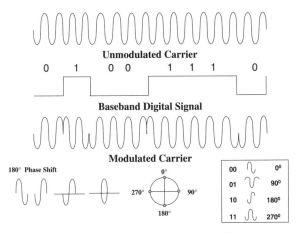

Figure 6.5   Phase shift keying: unibit.

6.2.1.3.3 Phase Modulation   *Phase Modulation* (PM), or *Phase Shift Keying* (PSK), involves the carefully synchronized shifting of the position of the sine wave (Figure 6.5). *Binary Phase Shift Keying* (BPSK) is a unibit technique in which the continuous sine wave pattern is interrupted and restarted at the baseline with a 180° phase shift to indicate a change in value (e.g., from a 1 bit to a 0 bit). *Quadrature Phase Shift Keying* (QPSK), also known as *Quaternary Phase Shift Keying* and *Quadriphase Keying,* yields dibit transmission and can be achieved by defining four phase shifts separated by 90° (0°, 90°, 180°, and 270°). Figure 6.5 provides several ways of looking at phase shift. The first illustration is of a 180° shift of an entire sine wave. The second illustration flips half of the sine wave horizontally, to represent a 180° shift. The third illustration represents the four degrees of phase shift as positions on a circle. The fourth illustrates the four degrees of phase shift as their signatures might appear on an oscilloscope.

Through the definition of eight phase shifts separated by 45° (0°, 45°, 90°, 135°, 180°, 225°, 270°, and 315°), contemporary modems can affect *tribit* transmission, achieving three bits of data per signal. (*Note:* $2^3 = 8$.) Thereby, at a signaling rate of 2400 bps, the transmission rate is 7200 bps.

Differential Phase Shift Keying (DPSK) is a variation on the unibit PSK theme. With DPSK, as illustrated in Figure 6.6, each 1 bit triggers a 180° phase shift, but 0 bits have no effect. A great number of applications specify various PSK techniques. For example, Wi-Fi5 (802.11a) wireless LAN standards, which have nothing at all to do with the PSTN, call for BPSK at 6 Mbps and QPSK at 12 Mbps.

6.2.1.3.4 Quadrature Amplitude Modulation   High-speed modems combine multiple modulation techniques. *Quadrature Amplitude Modulation* (QAM), for example, splits the carrier into two waveforms that are 90° out of phase and specifies two possible amplitude values for each of four phase shifts separated by 90° (0°, 90°, 180°, and 270°). This yields eight distinct signal states, as illustrated in the signal constellation graph in Figure 6.7. Thereby, each signal impulse, or *symbol*, carries one of eight possible signal combinations and represents three bits. (*Note:* $2^3 = 8$.) At a signaling rate of 2400 baud, this tribit modulation scheme yields a transmission rate of 7200 bps.

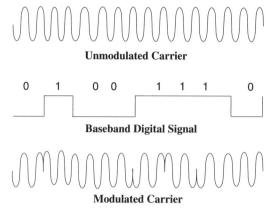

Unmodulated Carrier

0   1   0   0   1   1   1   0

Baseband Digital Signal

Modulated Carrier

Figure 6.6   Differential PSK: unibit.

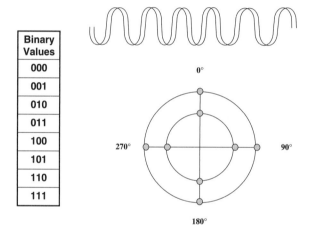

Binary Values
000
001
010
011
100
101
110
111

Figure 6.7   Quadrature amplitude modulation.

The ITU-T V.29 recommendation is for 16-QAM. This approach splits the carrier into two waveforms that are 90° out of phase and specifies two possible amplitude values for each of eight phase shifts separated by 45° (0°, 45°, 90°, 135°, 180°, 225°, 270°, and 315°), as illustrated in the signal constellation graph in Figure 6.8. Thereby, each *symbol* carries one of 16 possible signal combinations and represents four bits. (*Note:* 2^4 = 16.) At a signaling rate of 2400 baud, this quadbit modulation scheme yields a transmission rate of 9600 bps.

It is possible to achieve still higher modulation rates. A 64-QAM technique yields 64 possible signal combinations, with each symbol representing six bits (2^6 = 64). A 128-QAM technique yields 128 possible signal combinations, with each symbol representing seven bits (2^7 = 128). A 256-QAM technique yields 256 possible signal combinations, with each symbol representing eight bits (2^8 = 256). A 512-QAM technique yields 512 possible signal combinations, with each symbol representing nine bits (2^9 = 512). As the modulation technique increases in sophistication, the number of bits per symbol increases, and the transmission efficiency increases. However, the signal points are brought closer together, which increases the

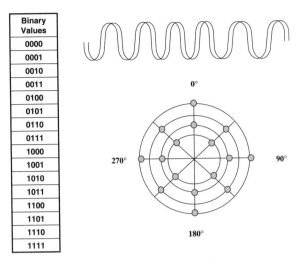

Binary Values
0000
0001
0010
0011
0100
0101
0110
0111
1000
1001
1010
1011
1100
1101
1110
1111

Figure 6.8   The 16-QAM scheme.

susceptibility to signal impairments such as line noise. A great number of standards specify the use of QAM. In the United States, 64-QAM and 256-QAM are used in digital CATV applications. Wi-Fi5 (802.11a) calls for the use of 16-QAM and 64-QAM. ADSL and Local Multipoint Distribution System (LMDS) standards also call for QAM.

6.2.1.3.5 Trellis-Coded Modulation   *Trellis-Coded Modulation* (TCM) uses the same modulation scheme as QAM but adds a sophisticated error correction technique known as *Forward Error Correction* (FEC) to overcome the increased susceptibility to signal impairments. TCM is so named because the plotting of the signal points resembles the latticework of a trellis such as that used in a rose garden, only four dimensional, which explains the lack of a figure accompanying this text. TCM employs a convolutional (i.e., error-correcting) coding scheme. The scheme involves adding an extra bit to every symbol for error control purposes. For example, recall that a 128-QAM technique yields 128 possible signal combinations, with each symbol representing seven bits ($2^7 = 128$). As TCM uses one bit for error control, only six payload bits remain ($2^6 = 64$). So, the modem accepts six bits at a time. The two *Least Significant Bits* (LSBs) are separated from the six-bit payload and are analyzed and a parity bit is added that describes the mathematical value (odd or even) of the sum of the LSBs. The resulting three bits and the other original four bits are recombined into a seven-bit symbol prior to transmission. The receiving modem reverses the process, analyzes the parity bit describing the LSBs, and accepts the data as correct, adjusts the data to correct for an error if possible, or requests a retransmission. The LSBs are the rightmost bits in value. As they change rapidly if the total value changes even slightly, they are highly sensitive to errors and very telling in the event that errors occur. As the symbols are plotted onto the logical trellis by the receiver, there are only 64 ($2^6 = 64$) legitimate states, or positions, plus the two for the error control bit, for a total of 66 states. If the indicated plot point is one of the other 62 ($2^7 = 128 - 66 = 62$), the received symbol is assumed to have been errored in transit. TCM was invented by Gottfried Ungerboeck, who published the theory in an article with the exciting title "Channel Coding with Multilevel/ Phase Signals" in 1982. In 1984, the ITU-T published modem standards incorporating TCM for speeds of 19.2 kbps and higher. ITU-T recommendations for dial-up modems (and maximum speeds) specifying TCM currently include V.32 (9600 bps), V.32bis (14.4 kbps), V.32ter (19.2 kbps), V.34 (28.8 kbps), and V.34bis (33.6 kbps), aka V.34+. (*Note:* The term *bis* comes from Latin, meaning *second*; in other words, the second and enhanced release of the standard. Third releases are designated *ter*, translated from Latin as *third*.) All of these modem standards provide for fully symmetric data transfer, which is to say that the data transfer is at the same speed in both directions.

6.2.1.3.6 General Modem Characteristics Conventional modems can be characterized along a number of dimensions, including asynchronous versus synchronous, compression, diagnostics, error control, equalization, gain control, and band limitation:

- *Asynchronous and synchronous* modems both are available. Asynchronous modems transmit one character at a time, with the receiving device relying on start and stop bits to separate transmitted characters. Synchronous modems are much faster, as the signal is *synchronized* (timed) at the bit rate of the connection by a *Transmit Clock* (TC) in either the transmit modem or the transmit terminal. The paired modems synchronize on that clocking pulse in order to distinguish between blocks of data being transmitted, rather than identifying each individual character in a transmission and surrounding it with a start and stop bit. Particularly when transmitting large amounts of data, synchronous modems increase the efficiency of data transfer, resulting in increased speed of transfer and lower associated transmission cost. Synchronous modems faithfully transmit any bit sequence, rather than just ASCII characters.
- *Diagnostic* modems can test their internal clock and transmit and receive circuits. Additionally, such modems may have the capability to monitor their performance and even diagnose certain conditions contributing to performance degradation. Further, they can respond to loopback tests and, therefore, are manageable through higher level *Element Management Systems* (EMSs). Such management systems typically are located remotely and are capable of managing large numbers of modems and modem pools (i.e., groups of modems to which access is shared among multiple users).
- *Error correction* capabilities are included in most contemporary modems. The proprietary *Microcom Networking Protocol* (MNP) was among the first to include error correction and subsequently was incorporated into the ITU-T V.42 standard. While proprietary error correction software remains embedded in certain modems, the ITU-T V.42 and subsequent generations of modems have standardized this function.
- *Compression* is a characteristic of high-speed modems. There are a wide variety of compression techniques. ITU-T Recommendation V.42bis, for example, eliminates unused bits in ASCII bytes used to express numerical values. Fax modems make heavy use of various run-length encoding algorithms to compress data

prior to transmission, as discussed in Chapter 4. Sophisticated modulation techniques such as QAM are widely used to further improve on the efficiency of transmission. *Lemple–Ziv* (LZ) compression in some modems attains compression ratios of better than 5 : 1 for some forms of text and numerical data. LZ compression allows the DTE to operate at speeds up to 128 kbps while the analog link between the modems remains at 38.4 kbps or less over a 4-kHz channel.

- *Equalizers* reduce frequency and phase distortion on a circuit by selectively introducing a small amount of delay to compensate for variations in attenuation and latency at different frequencies in the transmission band. *Adaptive equalizers* continuously monitor the signal and adjust the equalization process to optimize performance at all times.

- *Automatic Gain Control* (AGC) amplifiers serve to adjust for amplitude variations of the input signal and to ensure that the outgoing signal is of a constant strength. (*Note:* Gain is the opposite of attenuation. *Gain* defines an increase in strength between the incoming and the outgoing signals. *Attenuation* is a loss in signal strength.)

- *Band-limiting filters* improve error performance by managing the frequencies of the incoming signal, filtering out any extraneous frequencies.

- *Dynamic rate adaptation* enables modems to dynamically adjust the speed of data transfer to varying line conditions in order to ensure the integrity of the data stream. The actual transmission speeds that modems can realize in either direction depends on the attributes of the analog local loop and various transient interference issues. As the modem speed increases, the signal modulation techniques become more complex and the signal points are brought closer together, which increases the susceptibility to signal impairments such as line noise. The germane local loop characteristics include loop length, wire gauge and presence of mixed wire gauges, number and quality of splices, *bridged taps* (i.e., multiple appearances of the same cable pair, usually as a result of old and unused connections to other customer premises), bonding and grounding, and the integrity of cable sheathing and splice casings. Modem speeds are affected negatively if any of these attributes is outside acceptable parameters. Whether persistent or transient in nature, ElectroMagnetic Interference (EMI) caused by electrical storms, radio transmissions, electric motors, and other sources of electromagnetic energy also clearly impact modem performance. Additionally, high-speed services such as T1 and ADSL running on cable pairs in proximity within the same cable can cause difficulty. Such services involve relatively high frequencies and, therefore, radiate a relatively strong electromagnetic field, potentially causing interference and even crosstalk. Modem speeds are stated in terms of their maximums, assuming that conditions are optimal. Dynamic rate adaption enables fallback modems to negotiate a lower rate of transmission using a less sophisticated modulation technique and perhaps adjusting the baud rate when line conditions are less than optimum and to ratchet up the transmission rate when conditions improve. For example, V.34bis modems use QAM and can transmit at 33.6 kbps at 2400 baud when line conditions are optimal. If that is not the case, they can fall back to 31.2 kbps and then 28.8 kbps, if necessary, and ratchet back up as the opportunity presents itself.

6.2.1.4   56-kbps Modems: V.90 and V.92   Modem technology took a giant leap forward with the introduction of 56-kbps modems in the mid-1990s. US Robotics (now 3Com) was the first to develop 56-kbps technology with its proprietary x2, referring to the fact that the modems were about twice the speed of the then-current V.34 (28.8 kbps) modems. Lucent, Rockwell, Motorola, and others soon followed with equally proprietary *K56flex* (Kbps 56 *flexible*) modems. The two incompatible approaches were standardized in November 1998 with the ITU-T recommendation V.90. Generally, the modems are combined data/fax modems because they have the ability to emulate a fax machine. They also generally consist of a combination of firmware and hardware, although some are entirely software based. (Note that firmware is faster, but software is more flexible.) Both x2 and K56flex modems were upgradable to V.90 through a software downloads.

V.90 modems are asymmetric in nature, providing a maximum of 56 kbps downstream (i.e., from the network to the modem) and 33.6 kbps upstream (i.e., from the modem to the network). In order to achieve this level of performance, a 56-kbps modem configuration (Figure 6.9) requires that only one transmission link be analog. The end-user dial-up connection is through a V.90 modem over an analog local loop to the PSTN. The connection through the PSTN must be entirely digital, including the originating and terminating COs, all tandem offices, and all transmission facilities. At the terminating end (e.g., corporate intranet site or ISP), the local loop connection must be digital (e.g., T-carrier, E-carrier, or ISDN). Matching 56-kbps modem technology must be in place at the terminating device, typically in the form of an access server or router. This configuration limits the transmissions to only a single D-to-A-to-D (Digital-to-Analog-to-Digital) conversion process, which limits the amount of quantizing noise associated with the D-to-A process. This results in higher speed transmission without sacrificing error performance.

The A-to-D conversion process in the PSTN uses the Pulse Code Modulation (PCM) algorithm, as specified in ITU-T recommendation G.711. As this standard for voice-grade, A-to-D conversion specifies a 64-kbps channel, the theoretical transmission rate for V.90 modems is 64 kbps, symmetric. In the United States, however, intrusive signaling and control are assumed to consume 8 kbps, thereby limiting the theoretical effective transmission rate to 56 kbps. (See Chapter 7 for detailed discussions of T-carrier, PCM, and *bit robbing*.)

The asymmetric nature of these modems is due to several factors. First, the upstream relationship between the modem and the network is not highly precise. In other words, the digital network cannot interpret the modulated sine waves without the introduction of some quantizing noise, which limits the effective throughput. Actually, it can be done; it is just that you need a more sophisticated (read more

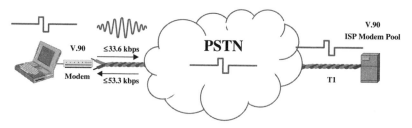

Figure 6.9   The 56-kbps modem configuration.

expensive) modem to do it. Therefore, asymmetry is less expensive. As the manufacturer makes its profits on end-user modems in volume, the price point is important. If enough end users are attracted to high-speed, inexpensive modems, the lower volume of the much more expensive software for 56-kbps servers yields profits as well. Second, the end user generally does not require full 56-kbps bandwidth on a symmetric basis, as most of the bandwidth is required downstream (i.e., downstream from the network). After all, the bandwidth-intensive graphics generally are downloaded from the Web or a corporate intranet site; upstream, the user often sends only a few mouse clicks or keyboard commands or, perhaps, a textual e-mail. If users need to send bandwidth-intensive files upstream, they are no worse off than with a V.34+ modem, which is exactly how the 56-kbps modem works upstream.

Notably, 56-kbps modems also are V.34+ modems. Assuming that the terminating modem is V.34+ or lesser, the 56 kbps modem "falls back" to that standard, which supports symmetric transmission at speeds up to 33.6 kbps. (Note that the lowest common denominator always determines the maximum level of performance.) Also notably, the fine print on the boxes of the 56-kbps modems indicates quite clearly that they currently are limited to 53.3 kbps because they cannot exceed FCC amplitude (signal strength) limitations. (Note, again, the highest common denominator rule, referring to the network, in this case.) The FCC established this limitation many years ago to minimize the likelihood that transmissions on a cable pair might cause interference on adjacent pairs in the same cable sheath. While the FCC is considering relaxing this restriction, it is clear that the modem manufacturers, rather than the FCC, are responsible for the fine print on the modem boxes [7–11].

It was fully anticipated that V.90 would be the last in a long line of modem standards, but modem technology got a boost with the ITU-T standardization of *V.92* (November 2000). Based on work done at Hughes Network Systems for its DirecPC (now HughesNet) satellite Internet service, V.92 offers several key advantages over V.90:

- **Faster Connection:** *QuickConnect* is a feature that cuts the time required for handshaking approximately in half, to about 10–15 s. QuickConnect trains the modem on the first call and remembers the characteristics of the circuit. Assuming that the circuit is the same on the next call, the circuit characteristics do not have to be relearned, which results in faster connect times, which has obvious advantages to the end user. It also offers advantages to the ISPs, which may handle thousands, or even millions, of dial-up calls a day—the faster you connect, the faster you take care of your business and disconnect, the shorter the period of time you tie up an expensive port, and the quicker it is available for another user.

- **Faster Upstream Speed:** Upstream transmission speed is increased from 33.6 to 48 kbps under optimum conditions using a variation of PCM that allows the upstream data stream to use the same clocking source as the downstream data stream.

- **Improved Compression:** V.44 replaces the V.42bis compression algorithm used in V.90. The V.44 is a string-coding algorithm that offers compression in the range of 6:1, improving throughput by 20–60 percent and as much as 200 percent for certain kinds of highly compressible data. That translates into theoretical downstream throughput rates as high as 300 kbps, compared with the maximum rates of 150–200 kbps possible with V.90 modems. (*Note*: These figures are highly optimistic. Actual performance can vary widely, depending on the type of files involved and the line conditions. Performance also depends on whether the ISP decides to turn on the compression option—many do not.)

· **Modem on Hold (MoH):** A V.92 modem can put a data session on hold when it detects a voice call, either incoming or outgoing, through a call waiting indication and gracefully resume that session when the voice call is terminated. Thereby, V.92 allows a single analog line to be used for both voice and data. MoH requires that the line be equipped with *call waiting*, a telco PSTN service.

The announcement of V.92 certainly is not as dramatic as that of V.90, which nearly doubled downstream transmission rates. It does, however, offer considerable advantages for dial-up users, including not only those without broadband access via ADSL, cable modem networks, or satellite but also those of us who use modem connections on the road or as a backup in the event of a broadband access failure. Essentially, that means that all users stand to benefit at one time or another once V.92 modems are widely available and widely supported by ISPs. Unfortunately, there seems to be no market driver significant enough to compel the ISPs to expend the capital necessary to upgrade V.90 servers to V.92. Some V.90 modems can be upgraded to V.92 via a network download [12–14].

6.2.2 Codecs

The reverse conversion of A-to-D is necessary in situations where analog information is to be sent across a digital circuit. Certainly, this often is the case in carrier networks where huge volumes of analog voice are digitized and sent across high-capacity digital circuits. This requirement also exists where high-capacity digital circuits connect premises-based, analog voice PBXs or Key Telephone Systems (KTSs) to Central Office Exchanges (COEs) or to other PBXs or KTSs.

The device that accomplishes the A-to-D conversion is known as a *codec*. Codecs *co*de an analog input into a digital (data) format on the transmit side of the connection, reversing the process, or decoding the information, on the receive side to reconstitute the analog signal. Codecs are used widely to convert analog voice and video to digital format and to reverse the process on the receiving end. There exists a wide variety of codecs for use in wired and wireless networks of all descriptions. Many of those codecs are discussed in detail in subsequent chapters. [*Note*: In contemporary systems, a codec commonly is incorporated into a *Digital Signal Processor* (DSP).]

6.2.3 Terminal Adapters and NT-Xs

Terminal Adapters (TAs) are interface adapters for connecting one or more non-ISDN devices to an ISDN network. Also known as *ISDN modems*, TAs are ISDN DCE that performs protocol or interface conversion for equipment that is not ISDN compatible. *Network Termination* (NT), in ISDN networks, is a function accomplished through the use of programmed logic embedded in the carrier network and the user equipment. NT2 is an interface to an intelligent ISDN-compatible device (e.g., PBX or router) responsible for the user side of the connection to the network, performing such functions as multiplexing and switching. NT1 is responsible for interfacing to the carrier side of the connection, performing such functions as signal conversion and maintenance of the local loop's electrical characteristics. These functions resemble those provided by DSUs and CSUs.

6.2.4   Channel Service Units and Digital Service Units

Channel Service Units (CSUs) and *Digital Service Units* (DSUs) are devices that, in combination, serve to interface the user environment to an electrically based, digital local loop. In contemporary systems, CSUs and DSUs generally combine into a single device known variously as a CSU/DSU, CDSU, or ISU (*Integrated Service Unit*), which typically appears in the form of a chipset on a printed circuit board found under the skin of another device such as a channel bank, multiplexer (mux), switch, or router. They are used in a wide variety of digital voice and data networks, including DDS, T-carrier, and E-carrier, which I discuss in detail in Chapter 7.

6.2.4.1   Channel Service Unit   Channel service units are circuit-terminating equipment that provide the customer interface to the circuit, as illustrated in Figure 6.10. They also permit the isolation of the DTE/CPE from the network for purposes of network testing. CSU functions include electrical isolation from the circuit for purposes of protection from aberrant voltages, serving the same function as a protector in the voice world. Additionally, the CSU can respond to a command from the carrier to close a contact, temporarily isolating the DTE domain from the carrier domain. This enables the carrier to conduct *a loopback test* in order to test the performance characteristics of the local loop from the serving CO to the CSU and back to the CO. Many contemporary CSUs also have the ability to perform various line analyses, including monitoring the signal level. Such intelligent CSUs also often have the ability to initiate loopback tests, although arrangements must be made with the carrier in advance.

The CSU also serves to interface the DTE domain to the carrier domain in an electrical environment. Within the DTE, for example, 1 bits commonly are represented as positive (+) voltages and 0 bits as null (zero) voltages. The network requires that 1 bits be alternating positive and negative voltages and that the 0 bits

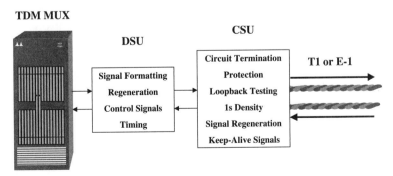

Figure 6.10   CSU and DSU.

be zero voltages. Further, the network requires assurance that *1s density* is achieved. Depending on the carrier network, 15–80 zeros can be transmitted in a row as long as the density of 1s is at least 12.5 percent (1 in 8) over a specified interval of time. CSUs insert, or *stuff*, 1 bits on a periodic basis in order to ensure that the various network elements maintain synchronization.

The CSU also serves to provide signal regeneration and generates *keep-alive* signals to maintain the circuit in the event of a DTE transmission failure. Finally, the CSU stores various performance data in temporary memory for consideration by an upstream network management system.

Smart CSUs increasingly are positioned as *Integrated Access Devices* (IADs). These multiport devices support interfaces to voice, data, and video devices such as PBXs, routers, and videoconferencing units. The programmable IAD supports bandwidth allocation for the various devices, enabling them to share a single T1 or other digital facility.

6.2.4.2 Data Service Unit
Data service units convert the DTE unipolar signal into a bipolar signal demanded by the network. DSU functions variously include regeneration of digital signals, insertion of control signals, signal timing, and reformatting. Some of these functions can be ceded to either the CSU or the terminal equipment [15]. In any case, the functions must be performed, even though the CSUs and DSUs lose their identity.

6.2.5 Front-End Processors

Front-End Processors (FEPs) combine the functions of concentrators and message switches. In other words, they have the ability to concentrate and switch traffic between multiple terminals and among groups of terminals in order to share a single circuit for access to mainframe resources. They also serve as an interface to *Wide Area Network* (WAN) circuits to provide remote terminals access to mainframe resources. Most FEPs are midrange computers that have their own databases to support assigned functions such as error detection and correction, queuing, editing validation, and limited application processing. While the mainframe clearly could perform such tasks, it is more cost effective to apply a lower order computer to the performance of such mundane and highly repetitive tasks, thereby reserving the power of the mainframe for more difficult and demanding tasks in support of user-oriented applications.

In a more contemporary client/server environment there is no exact equivalent of an FEP. Although the same functions must be performed, they are spread across multiple and various devices such as an access router with a firewall for security purposes and a *Remote Access Server* (RAS) or *Virtual Private Network* (VPN) server for performance of security functions and granting of access privileges.

6.3 PROTOCOL BASICS

Protocols are rules of behavior. In the context of data communications, protocols are the procedures employed to ensure the orderly exchange of information between devices on a data link, data network, or system. Protocols comprise conventions that, at a basic level, commonly include the dimensions of line setup, transmission mode, code set, and non–data exchanges of information such as error control. Protocols have two major functions, handshaking and line discipline [16]:

- *Handshaking* is the sequence that occurs between the devices over the circuit, establishing the fact that the circuit is available and operational. The handshaking process also establishes the level of device compatibility and determines the speed of transmission by mutual agreement. Devices accomplish the process of handshaking by passing frames of data back and forth in order to negotiate the basis on which they will communicate, in consideration of the performance characteristics of the circuit. As always, the rule of the lowest common denominator applies and, therefore, communication is limited by the least capable device. Actually, the rule is more like the highest common capability of the least capable device determines the maximum level of performance.
- *Line discipline* is the sequence of network operations that actually transmits and receives the data, controls errors in transmission, deals with the sequencing of message sets (e.g., packets, blocks, frames, and cells), and provides for confirmation or validation of data received.

Several hardware and software solutions are designed to specifically deal with protocol analysis and conversion:

- *Protocol converters* are software-based devices that translate from one native protocol into another [e.g., from ASCII to EBCDIC, from TCP/IP to IBM SNA/SDLC, or from PCM voice to VoIP].
- *Gateways* in ARPANET terminology, were packet-forwarding devices that, in contemporary data communications terminology, are referred to as *routers*. Contemporary gateways are hardware/software combinations that connect devices running different native protocols. In other words, gateways are packet-forwarding devices that run *gateway protocols* for purposes of protocol conversion, perhaps at all layers of the OSI Reference Model (i.e., from the Physical Layer to the Applications Layer). In addition to protocol conversion, gateways provide a point of physical interconnection between incompatible networks, with examples including X.25-to-Frame Relay (FR) gateways and T-carrier-to-E-carrier *International Gateway Facilities* (IGFs).
- *Protocol analyzers* are diagnostic tools for displaying and analyzing communications protocols. Analyzers enable technicians, engineers, and managers to test the performance of the network to ensure that the systems and the network function according to specifications. LAN managers, for example, use protocol analyzers to perform network maintenance and troubleshooting and to plan network upgrades and expansions.

6.3.1 Line Set-Up: Connectivity

A very basic protocol issue involves the manner in which the circuit is set up between devices. There are three alternatives: simplex, Half DupleX (HDX), and Full DupleX (FDX), as illustrated in Figure 6.11 and defined below.

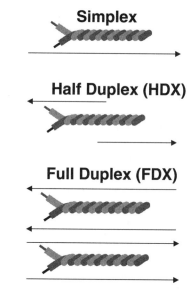

Figure 6.11 Simplex, HDX, and FDX transmission.

6.3.1.1 Simplex Transmission *Simplex* transmission is unidirectional. The information flows in one direction across the circuit, with no capability to support a response in the other direction. Simplex circuits are analogous to escalators, door-bells, fire alarms, and security systems. Contemporary applications for simplex circuits include remote station printers, card readers, and alarm systems (e.g., fire, smoke, and intrusion alarms). Generally speaking, simplex transmission is con-ducted across dedicated circuits of low capacity. An intrusion alarm, for example, requires very little bandwidth because the only information sent across the circuit indicates that an electrical contact has been broken. The intrusion alarm circuit must be dedicated from end to end (e.g., contact to central alarm station) to avoid the possibility that either failure or congestion in a switch or other intermediate device could prevent the alarm from being registered. The circuit must be simplex to prevent the alarm from being reset remotely, rather than investigated locally. Tele-phone company tariffs sometimes refer to simplex circuits as burglar alarm circuits. Broadcast TV and radio operate in simplex mode.

6.3.1.2 Half-Duplex Transmission *Half-duplex* (HDX) transmission operates in both directions, although not simultaneously. A HDX circuit is perhaps best illustrated by a walkie-talkie, *Citizens Band* (CB), or other *Push-To-Talk* (PTT) radio link, over which the speakers must take turns talking. Many speakerphones operate in HDX. In data communications applications, HDX generally is used for relatively low-speed transmission, usually involving two-wire, analog circuits pro-vided on a circuit-switched basis through the PSTN. As the circuit must be turned around in order to support the change in direction of the transmission, the line turnaround time tends to limit the speed of conversational data communications. HDX example applications include line printers, polling of remote buffers, and modem communications (many modems can support FDX as well). HDX is used extensively in transaction-based communications, such as credit card verification and Automatic Teller Machine (ATM) networks. Such applications are not impacted seriously by delays associated with line turnaround. In order to control the direction of the circuit, some sort of control mechanism must be employed to identify which device will transmit and which will receive. The most common approach involves a device sending a *Request-To-Send* (RTS) control signal. If the request is granted, the other device, or perhaps an intermediate switch or router, sends a *Clear-To-Send* (CTS) signal. Wi-Fi (802.11) Wireless Local Area Networks (WLANs) use this same approach to transmission over a radio channel. In Europe, a HDX circuit is commonly referred to as a simplex circuit.

6.3.1.3 Full-Duplex Transmission *Full duplex* (FDX) is a fully bidirectional transmission mode in which communication is supported in both directions simultaneously. The first working FDX communications circuit was invented by Joseph B. Stearns of Boston and installed in 1872 on a one-wire telegraph system using a ground return. This effectively doubled the traffic capacity of the network, and the duplex equipment could be purchased and installed at much lower cost than stringing another wire [1]. Human beings, including telegraph operators, are born fully equipped for FDX communications, but computers require software to enable this capability. While we would rarely employ a one-wire approach these days, there are several different ways to configure an FDX circuit.

- **Physical Four Wire:** The most commonsense FDX configuration involves a physical four-wire circuit with one pair of wires supporting transmission in one direction and another pair supporting transmission in the other direction. This approach is commonly used for high-capacity, dedicated circuits, of which most are multichannel in nature. T-carrier and E-carrier circuits traditionally were configured in this way. All wideband and broadband circuits are FDX in nature.
- **Physical Two Wire:** FDX modems use another approach in order to achieve FDX data transmission on a physical two-wire circuit. Modems cause a two-wire circuit to emulate a four-wire circuit through a split-channel approach, using Frequency Division Multiplexing (FDM) to create two frequency channels. This process allows the creation of two carrier waves that are then modulated using Frequency Modulation (FM). Figure 6.12 illustrates this technique over a voice-grade circuit. In this example, one channel uses an unmodulated carrier wave at 1170 Hz, modulating it at 1070 Hz to represent 0 bits and at 1270 Hz to represent 1 bits. The reverse channel uses an unmodulated carrier wave at 2125 Hz, modulating it at 2025 Hz to represent 0 bits and at 2225 Hz to represent 1 bits. Other modulation techniques can be used within these channels as well, including AM, PSK, and QAM.
- **Optical:** FDX transmission in optical fibers is achieved either by using separate fibers for transmission in different directions or by using *Wavelength Division Multiplexing* (WDM), which essentially is FDM at the optical level.
- **POTS:** This discussion of FDX begs the question as to how FDX is achieved over a physical two-wire circuit for *Plain Old Telephone Service* (POTS), that is, analog voice communications, since there is only one voice-grade channel of

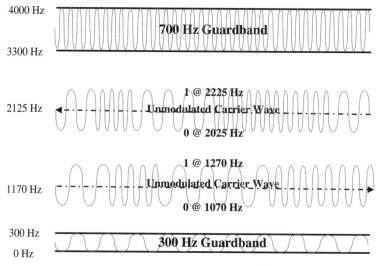

Figure 6.12 FDX communications over a two-wire circuit via FDM and FM.

4 kHz. Remember that there are two conductors and current flows in both directions across the circuit. So, signals can travel in both directions as well. Voice communications is not highly precise and, therefore, not highly demanding of the circuit, so a modest amount of signal distortion is acceptable—actually it is not even noticed.

FDX circuits sometimes are used to connect HDX terminals in order to avoid issues of line turnaround time. More typical examples of FDX applications include channel links between host processors, channel links between controllers/concentrators and hosts, and other applications involving the interconnection of substantial computing systems. Carrier services that deliver FDX capabilities include DDS, E/T-Carrier, and broadband services such as Frame Relay, SMDS, and ATM. I discuss all of these in later chapters.

6.3.2 Transmission Mode: Transmission Method

There are two basic methods of data transmission: asynchronous and synchronous. As long as we are on the theme of timing, we will briefly explore the concepts of isochronous and plesiochronous communications as well.

6.3.2.1 Asynchronous *Asynchronous*, or *character-framed*, transmission is a method that grew out of telegraphy and teletypewriting. From Latin and Greek, it translates as *not together with time*—in other words, not synchronous. Asynchronous transmission is a start–stop method of transmission in character mode that traditionally is used when keyboarding (nee typing). In this mode, characters are transmitted one at a time with variable intervals between characters, as determined by the timing of the fingers as they strike the keys. When no keys are being struck and the transmitter is sending no data, it sends a constant stream of signals in the *mark*, or 1, state. When the transmitter begins to transmit data, a character (letter, number, punctuation mark, or control character) at a time, each is preceded by a *start bit* at the 0, or *space*, level. The start bit alerts the receiving terminal to the transmission across the circuit of something worthy of its attention, which generally is presented in the form of an eight-bit byte. A *stop bit* at the mark, or 1, level, succeeds the transmitted computer value, advising the receiving terminal that the transmission of that set of information has ended. Some asynchronous protocols make use of two stop bits.

PCs, teletypes, and other devices that make use of asynchronous transmission *frame*, or surround, each byte of information with start and stop bits, which are interpreted by the receiving terminal and subsequently stripped away in order to get to the actual data *payload*. The inclusion of start and stop bits adds two or three bits of *overhead* to the transmission of each eight-bit byte. Additionally, asynchronous transmission commonly involves the addition of a *parity* bit as an error control mechanism, which happens to be relatively poor, as we shall see later in this chapter. The framing of the data with these three or four bits of control information yields an overhead, or inefficiency, factor of 20–30 percent.

Asynchronous transmission can be characterized as start-stop (not synchronized) transmission of one character at a time at a variable speed. Additionally, overhead is high and error control is poor. (*Note*: Asynchronous Transfer Mode (ATM) also is asynchronous in nature, although in a much more complex manner than discussed in this baseline example. I discuss ATM in Chapter 10.)

6.3.2.2 *Synchronous*   From Latin and Greek origins, *synchronous* translates as *together with time*. Such transmission is *message framed* and overcomes the inefficiencies of asynchronous, start–stop transmission for high-speed data communications applications. Rather than surrounding each character with start and stop bits, a relatively large set of data is *framed*, or *blocked*, with one or more synchronization bits or bit patterns used to identify the beginning and end of a logical block of data. Both analog synchronous modems and digital DTE synchronize the receiving terminal on the rate of transmission of the data from the sending terminal. Through the receipt of the synchronizing bits, or *clocking pulses*, the receiving device can match its speed of data receipt to the rate of data transmission across the circuit. Thereby, each bit of data and control information can be distinguished at the physical layer. Higher layers sort out when to expect what information, in which data fields, and in what sequence, based on an agreed-upon protocol such as Frame Relay IP. Because only a few framing bits and synchronizing bits surround a large block of data, the overhead is much reduced, the efficiency of transmission is much increased, and the effective throughput is much improved. As stated in Wonders of the Universe (The Werner Company, 1899):

> The Morse printing instrument . . . is a beautiful, but rather complicated piece of mechanism, for besides the printing and electric arrangements it is furnished with clockwork to keep the paper tape in motion whilst the message is being delivered. But lately these accessories have in many cases been dispensed with, and the operator depends upon his ear for the translation of the message sent.

Error control in synchronous communications protocols is quite sophisticated and reliable, involving statistical sampling techniques and mathematical calculations performed on the set of data. Synchronous transmission can be characterized as transmission of multiple characters at a time organized into character sets and presented in blocks or frames. The transmission bit rate is synchronized between transmitter and receiver and takes place at a predetermined and relatively high rate of speed. Further, error control is excellent and overhead relatively low.

6.3.2.3 Isochronous   *Isochronous* (Isoc) data are synchronous data transmitted without a clocking source. From the Greek isochronos, translating as *equal in time*, all bits are of equal importance and are anticipated to occur at regular intervals of time. Bits are sent continuously, with no start–stop bits for timing. Rather, timing is recovered from transitions in the received data stream, with a whole number of bit-length intervals between characters. Bit integrity is preserved, with no modifications (i.e., bipolar conventions). The transparent isochronous transmission does not recognize control characters. Some T-carrier nodes operate isochronously on the links, syncing up with several lines operating at slightly different speeds. Isoc often is used in secure military applications that require encryption.

Real-time, uncompressed voice communication is a type of isochronous data because human conversation is not synchronized and can be presented in a continuous stream. If voice were synchronous, we would all talk at a precise and common rate of speed and not overtalk each other. Similarly, real-time, uncompressed video communication is isochronous, or stream oriented. The traditional circuit-switched, PCM-based PSTN supports isochronous datastreams beautifully. This network, from end to end and in both directions, commits time slots to the real-time, uncompressed voice transmissions, whether sound or silence is being carried. Further, those time slots appear at regular, precisely timed intervals (i.e., every 125 μs, or every 8000th of a second). So, the native voice conversation is isochronous, but the network is synchronous. Any confusion created here will be sorted out in the discussions of PCM and digital carrier (i.e., T-carrier and E-carrier) in Chapter 7.

6.3.2.4 Plesiochronous   From the Greek *plesio*, meaning *near*, and *chronos*, meaning time, *plesiochronous* communications involves devices running at nominally the same rate. Plesiochronous networks comprise subnetworks and devices that are free running, although at approximately the same rate and within defined parameters of tolerance for variation. Much like clocks and watches run at approximately the same rate, devices in a digital network are free running, with some running at a slightly faster or slower pace. So, the network is almost, but not perfectly, synchronized across all devices.

The *Plesiochronous Digital Hierarchy* (PDH) is the digital network hierarchy of T-carrier and E-carrier systems. Originally, the PDH involved a master clocking source in the form of a highly precise *master clock* off of which all digital network elements *slaved* in a hierarchical fashion to *sync up*, that is, take their initial timing, after which time they ran independently. In the United States, Local Exchange Carrier (LEC) and IntereXchange Carrier (IXC) networks took their initial timing from a master Stratum 1 clock, which, as best I recall, was positioned in St. Louis, which is roughly the center of the country. Those clocking pulses were passed down the hierarchy to the regional and local networks of switches and T3 subnetworks that included multiple T3 and T1 links. Roughly in the 1980s, multiplexers also had the option of mutual synchronization, in which a pair of free-running muxes would simply sync up independently of the network. In either case, it was virtually ensured that there would be some level of slippage, or lack of synchronization. In order to adjust for that slippage and thereby resolve minor timing issues in order to allow multiple T1s to be multiplexed into a T2, multiple T2s to be multiplexed into a T3, and so on, *stuff bits* were added to the overhead at each higher level. The same logic was used in E-carrier and J-carrier networks. In contemporary T/E/J-carrier networks, the switches and muxes sync up with a Global Positioning System (GPS) master clock. As a result, timing issues are relatively modest, although they still exist and some small number of stuff bits remain used to resolve those issues.

6.3.3 Code Sets

Analogous to alphabets, *code sets*, or *coding schemes*, are employed by all computer systems to create, store, and exchange information. While code sets vary, they all rely on a specific combination of 1s and 0s of a specific total length in order to represent something of value, such as a letter, number, punctuation mark, or control character (e.g., carriage return, line feed, space, blank, and delete). Contemporary standard coding schemes include Baudot, ASCII, EBCDIC, and Unicode.

The first widely accepted standard coding scheme was Morse code, invented by Samuel Morse sometime prior to 1844 for use in telegraphy, which, of course, is a human-to-human digital data communications method. Friedrich Clemens Gerke invented the International Morse Code in 1848 out of necessity, as some of the spaces in letters created difficulty in radiotelegraphy. The international version was standardized by the International Telegraph Union (ITU) in 1865 and is still used today by amateur radio operators, or *hams*. Morse code uses series of short and long marks in the form of *dots* (short marks) and *dashes* (long marks), with spaces between them, to represent letters, numbers, punctuation marks, and *procedural signals (prosigns)*. The length of the spaces varies, with a short space between letters, a longer space between characters, a longer space between words, and a still longer space between sentences. In order to speed transmission, the fewest number of dots and dashes represent commonly used letters (e.g., *E* is •, *T* is –, *A* is • –). Commonly used words are abbreviated (e.g., *calling* is abbreviated *CG*, or – • – • – – •, as are commonly used phrases (e.g., *love and kisses* is abbreviated *88*, or – – – • • – – – • •. Table 6.1 provides the International Morse Code for English letters, numbers, and select punctuation marks and prosigns. (*Note:* The terms *ham* and *plug* originated in telegraphy to describe a telegraph operator lacking in proficiency [2]. *Ham* alludes to a *ham-fisted* operator.)

6.3.3.1 Baudot Code (ITA 2)

Morse code was the primary communication code for many years, until Emile Baudot invented the Baudot Distributor in the 1870s. That device provided for the transmission of values in a five-bit coding scheme over a line between two synchronized electromechanical devices. The Baudot Distributor soon gave way to the teletype, which also was based on the Baudot coding scheme, subsequently known as International Telegraph Alphabet 2 (ITA 2). *Telephone Devices for the Deaf* (TDDs) still use ITA 2.

Baudot code, updated in 1930, is limited to 32 (2^5) characters. Considering that each bit has two possible states (1 or 0), five bits in sequence yield 2^5 (32) possible combinations. Because 32 values is not sufficient to represent all 26 characters in

TABLE 6.1 International Morse Code Character Set

Letter	Code	Letter	Code	Number	Code	Punctuation, Prosigns	Code
A	•–	N	–•	1	•––––	Period (.)	•–•–•–
B	–•••	O	–––	2	••–––	Comma (,)	––••––
C	–•–•	P	•––•	3	•••––	Question (?)	••––••
D	–••	Q	––•–	4	••••–	Colon (:)	–––•••
E	•	R	•–•	5	•••••	Semicolon (;)	–•–•–•
F	••–•	S	•••	6	–••••	Hyphen (-)	–••••–
G	––•	T	–	7	––•••	Dollar ($)	•••–••–
H	••••	U	••–	8	–––••	At sign (@)	•––•–•
I	••	V	•••–	9	––––•	Stop	•–•–•
J	•–––	W	•––	0	–––––	Wait	•–•••
K	–•–	X	–••–			Invitation to	–•–
L	•–•–	Y	–•––			transmit	•–•
M	––	Z	––••			Received	

the English alphabet, plus the 10 decimal digits, necessary punctuation marks, and the space character, the shift key operates to shift between letters and other characters. An arrow pointing down represents letters (LTRS). Lowercase (LTRS shift) means that all following characters are alpha characters (LTRS). An arrow pointing up represents figures (FIGS). When a FIGS character is recognized, all succeeding characters are recognized as FIGS numbers and special characters until the shift key activates another arrow [16, 17].

Baudot employs asynchronous transmission, as start and stop bits separate characters. Error detection and correction require human editing. Therefore, Baudot is a human-to-human, rather than a machine-to-machine, communication technique. Detected errors must be corrected through retransmission.

Clearly, Baudot is a highly limited coding scheme. The limited range of letters expression, at 32 characters, is barely enough to accommodate the relatively simple English, French, and Spanish alphabets. To stay within that range, all letters must be in uppercase, which is just as well, as the shift key is used for other purposes. The shift key provides another 32 characters for numbers and control characters, but Baudot is still very limited in its range of expression. Additionally, the asynchronous requirement for start and stop bits makes Baudot overhead intensive. Finally, the error detection and correction technique is far less than desirable. As a result, Baudot currently is limited to use in teletypewriters and very old telex machines.

As a footnote, limited coding schemes are not necessarily overly limiting. Proprietary five- and six-bit codes, for example, have been used in the airline reservation systems (e.g., American Airlines' SABRE System and United Airlines' APOLLO) for many years. The airlines still use the IBM *Airline Control* (ALC) protocol, or the Unisys version, *P1024B*, both of which used a six-bit coding scheme. Such applications involve a limited character set that is easily accommodated by a six-bit code; in fact, a seven- or eight-bit coding scheme would be excessive and inefficient.

6.3.3.2 *Extended Binary Coded Decimal Interchange Code* Developed by IBM in 1962, *Extended Binary Coded Decimal Interchange Code* (EBCDIC) was the next standardized code used extensively. An improvement over earlier (1950) Binary-Coded Decimal (BCD) and (1951) extended BCD, EBCDIC was developed to enable different IBM computer systems to communicate based on a standard coding scheme. Although EBCDIC is standardized today, users have the ability to modify the coding scheme [16].

EBCDIC involves an eight-bit coding scheme, yielding 2^8 (256) possible combinations and, thereby, significantly increasing the range of expression. As a result, more complex alphabets can be supported, as can upper- and lowercase letters, a full range of numbers (0–9), and all necessary punctuation marks. Equally importantly, if not more so, the eight-bit coding scheme supports a large number of control characters, which is critical in the coordination of communications between complex mainframe computers.

The EBCDIC-based machines communicate on a synchronous basis, thereby improving on the speed of transmission. Since start and stop bits do not surround each character, overhead reduces, efficiency of transmission improves, and more payload bits can transmit per unit of time. Further, a more complex, machine-to-machine error detection and correction technique yields improved performance in that regard. Detected errors may require retransmission, although *forward error correction* is often employed, with the receiving system identifying, isolating, and correcting the errored bits.

6.3.3.3 American (National) Standard Code for Information Interchange

Developed in 1963, the *American Standard Code for Information Interchange* (ANSCII or ASCII) was specifically oriented toward data processing applications. It was modified in 1967 by the American National Standards Institute (ANSI) to address modifications found in contemporary equipment; that version, originally known as ASCII II, is now known simply as ASCII [16].

ASCII employs a seven-bit coding scheme, supporting 128 (2^7) characters, which is quite satisfactory for the English alphabet, Arabic numerals, punctuation marks, a reasonable complement of special characters, and a number of control characters, as displayed in Table 6.2. (*Note:* Actually, the Indians invented numbers 1–9. The Arabs invented only 0 but introduced 0–9 to Europe.) As ASCII was designed for use in asynchronous computer systems (non-IBM, in those days), fewer control characters were required, making a seven-bit scheme acceptable.

Figure 6.13 illustrates ASCII in the context of asynchronous communications, with start and stop bits framing each character, without employing synchronization bits. Asynchronous transmission makes use of a simple error detection and correction scheme known as parity checking, also illustrated in Figure 6.13. Parity checking is error prone with detected errors often going unnoticed or requiring retransmission, although forward error correction may be employed.

6.3.3.4 Universal Code

Universal Code (Unicode) is an attempt by the Unicode Consortium to standardize coding schemes, of which there are a great number, worldwide. Further, there are duplicate numerical codes for different characters, depending on the scheme involved. Unicode provides a unique numerical code for every character, regardless of the computing platform, the application program, or the language, whether human or machine. The Unicode Consortium developed the original standard, UTF-16 (*Unicode Transformation Format* 16), in 1991 as a standard encoding scheme to support complex alphabets such as Chinese, Japanese, and

TABLE 6.2   ASCII Character Set with Definitions of Example Control Characters

Bit positions 1–4	Bit Positions 5–7							
	000	100	010	110	001	101	011	111
0000	NUL	DLE	SP	0	@	P	'	p
1000	SOH	DC1	!	1	A	Q	a	q
0100	STX	DC2	"	2	B	R	b	r
1100	ETX	DC3	#	3	C	S	c	s
0010	EOT	DC4	$	4	D	T	d	t
1010	ENQ	NAK	%	5	E	U	e	u
0110	ACK	SYN	&	6	F	V	f	v
1110	BEL	ETB		7	G	W	g	w
0001	BS	CAN	(	8	H	X	h	x
1001	HT	EM	)	9	I	Y	i	y
0101	LF	SUB	*	:	J	Z	j	z
1101	VT	ESC	+	;	K	[	k	{
0011	FF	FS	,	<	L	\	l		
1011	CR	GS	-	=	M	]	m	}
0111	SO	RS	.	>	N	~	n	~
1111	SI	US	/	?	O	—	o	DEL

Abbreviations:

NUL: *Nu*ll character. A transmission control character used to serve a media-fill or time-fill requirement, i.e., a *stuff* character or *padding* character.

SOH: *S*tart *O*f *H*eader. A transmission control character indicating the start of a message heading.

STX: *S*tart of *T*e*X*t. A transmission control character to start the reading, transmission, reception, or recording of text.

ETX: *E*nd of *T*e*X*t. A transmission control character to terminate the reading, transmission, reception, or recording of text.

EOT: *E*nd *O*f *T*ransmission. A transmission control character to terminate a transmission that may have included one or more texts or messages.

ENQ: *EN*Quiry. A transmission control character used to request a response from a station to which a connection has been established. The request may be for the station identification, type of equipment, and station status.

NAK: *N*egative *A*c*K*nowledegment. A transmission control character sent by the receiving device to the transmitting device to indicate that a received block of data contained one or more errors. A NAK will trigger the transmitting device to retransmit that errored block.

ACK: *ACK*nowledgment. A transmission control character sent by the receiving device to the transmitting device to indicate that a received block of data contained no errors.

BEL: *BEL*l. A transmission control character that causes a bell to ring or activates some other audio or visual device to gain the attention of the operator at the receiving station.

ETB: *E*nd of *T*ransmission *B*lock. A code extension character used to indicate the end of the transmission of a block of data.

CAN: *CAN*cel. A transmission control character indicating that the associated data are in error or are to be ignored.

EM: *E*nd of *M*edium. The physical end of a data storage medium, or the usable portion of the medium.

SUB: *SUB*stitute. Used in place of a character that is known to be invalid, i.e., in error. Also used to indicate a character used in place of one that cannot be represented on a given device, e.g., *e* may be used in place of ε (epsilon) or *d* may be used in place of δ (delta).

ESC: *ESC*ape. A code extension character used to indicate a change in code interpretation to another character set, according to some convention or agreement. This is much like the use of the shift key in Baudot code to indicate a shift between figures and characters.

CR: *C*arriage *R*eturn. A format control character that causes the print or display position to move to the first position, or left-hand margin, of the screen or print medium. Now often associated with an LF (Line Feed), which moves the print position down to the next line.

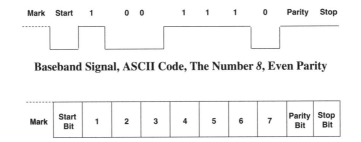

Baseband Signal, ASCII Code, The Number *8*, Even Parity

Mark	Start Bit	1	2	3	4	5	6	7	Parity Bit	Stop Bit

Asynchronous Mode, Character Framing, Even Parity

Figure 6.13 ASCII example of the number *8*, in asynchronous mode with character framing.

Korean. In the Japanese language, for example, even the abbreviated Kanji writing system contains well over 2000 written ideographic characters; Hatakana and Katakana alphabets are also used, further adding to the complexity. As seven- and eight-bit coding schemes cannot accommodate such a complex alphabet, computer manufacturers traditionally have taken proprietary approaches to this problem through the use of two linked eight-bit values. UTF-16 supports 65,536 (2^{16}) characters, in what is known as the *Basic Multilingual Plane* (BMP), which accommodates the most complex alphabets; in fact, multiple alphabets can be satisfied simultaneously. Further, Unicode standardizes the coding scheme so computers of disparate origin can communicate information on a standard basis. Since the transfer of Unicode data does not require translation of proprietary coding schemes, speed of transfer is improved, errors are reduced, and costs are lowered.

Unicode accommodates preexisting standard coding schemes using the same byte values for consistency. In Unicode terms, ASCII, for example, is known as UTF-7. There even is a UTF-EBCDIC, specifically for IBM mainframes. UTF-8 supports any universal character in the Unicode range using one to four octets (eight-bit bytes) to do so, depending on the symbol. UTF-32 uses four octets for each symbol but is rarely used due to its inherent inefficiency. The Unicode standard has been adopted by companies such as Apple, HP, IBM, Microsoft, Oracle, SAP, Sun, Sybase, and Unisys. Unicode is required by modern standards such as CORBA 3.0, European Computer Manufactures Association ECMAScript (JavaScript), Java, Lightweight Directory Access Protocol (LDAP), Wireless Markup Language (WML), and eXtensible Markup Language (XML).

Unicode is developed in conjunction with the International Organization for Standardization (ISO). ISO 10646 defines the *Universal Character Set* (UCS), into which the UTF code sets map. UCS-4 is a four-octet code set into which UTF-32 maps and UCS-2 is a two-byte code set into which UTF-16 maps. UCS-1 encodes all characters in byte sequences varying from one to five bytes.

6.3.4 Data Format

Data formatting is a critical part of a communications protocol. Data formats enable the receiving device to logically determine what is to be done with the data and how to go about doing it. Data formats include code type, message length, and

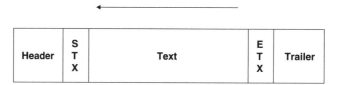

Header	S T X	Text	E T X	Trailer

Figure 6.14   Generic data format, with header, text, and trailer.

transmission validation techniques. A data format generally involves a header, text, and a trailer (refer to Figure 6.14 and the bulleted list that follows shortly), with the actual data content contained within the text field. While the header and trailer are overhead, they serve critical functions in support of the successful transfer of the data content. Generally, both a header and trailer frame the data content, or text. In total, the header, text, and trailer comprise what is known variously as a packet, block, frame, or cell, with the specific terminology being sensitive to the specific protocol involved. (*Note:* Some protocols, such as ATM, do not involve a trailer.)

- **Header:** A *communications header* precedes the data to be transmitted, often serving to establish the fact that the transmission link exists both physically and logically. The header may include synchronization bits that provide for synchronization between the devices and the link. Address fields in the header often include both source address and destination address. The source address identifies the originating device so that a response can be directed correctly and so that a retransmission can be requested if the data suffer errors during transmission. The destination address enables the receiving device to identify data intended for it and enables intermediate devices (e.g., switches and routers) to route the data correctly. Certain data protocols also use fields in the header to identify the length of the text field and the type of data, to indicate the level of tolerance for delay or loss during network transit, and any optional headers that might follow. The *user header* includes user-definable information such as system access (*password*), organization or department ID, operator ID, terminal ID, database or application ID, destination address, message sequence number, date/time ID, and message priority.

- **Text:** The *text* portion of the data set is the information to be communicated. It may contain either a fixed or a variable amount of information, depending on the specific protocol involved. The text may be preceded by *Start-of-TeXt* (STX) and succeeded by *End-of-TeXt* (ETX) control characters so the receiving device can determine the location of the message data. The *text field* also is known generically as the *data field*, or the *payload*.

- **Trailer:** The *trailer, tail,* or *trace* portion of the data set contains information relative to the analysis of the message, including message tracking and diagnostics. Trailing the text, the trailer information may contain the originating ID, the data block number and total number of blocks being transmitted, and identification of system processing points involved in the transmission. The trailer often includes an error detection and correction mechanism to manage the integrity of the transmitted data.

6.3.5 Error Control: Data Transmission Integrity

The integrity of the transmitted data is of prime importance. Several techniques exist for error detection and, ideally, correction. The three basic modes of error control are recognition and flagging, recognition and retransmission, and recognition and forward error correction:

- *Recognition and flagging* provide for no automatic means of correction of errors. Used primarily in networks involving dumb terminals with no means of *buffering* or retaining information transmitted; retransmission of errored data is not possible. Detected errors simply are flagged (identified) as such by the receiving device; error correction requires a human-to-machine request for retransmission. Parity checking is an example of recognition and flagging.
- *Recognition and retransmission* are used in more sophisticated networks where the transmitting device has *buffer memory* and, therefore, can retransmit a set of data that a receiving device, or perhaps an intermediate router or other intelligent network node, has determined to have been errored in transit. Serious failures in the devices and/or the circuit can result in repeated errored retransmissions, which lower the throughput of the communication link. In other words, recognition and retransmission are network intensive. Recognition and retransmission commonly are used in wireline networks where circuit quality is good and bandwidth is highly available. Some protocols, such as X.25 (Chapter 7), make use of recognition and retransmission on a link-by-link basis where circuit quality is poor and error performance is critical. Block parity is an example of recognition and retransmission.
- *Recognition and forward error correction* involve the addition of enough information that the receiving device can make the required corrections without requiring retransmission. While the addition of this redundant information automatically increases overhead relative to the data and, therefore, has a negative effect on the efficiency with which the network resources are used, it enables the receiving device to correct for most errors without requesting a retransmission—which might also be errored. FEC, however, places a load on the computational resources of the receiving device. FEC can be characterized as system intensive, rather than network intensive. FEC often is used in networks where link quality is poor and bandwidth is limited or where latency is high. FEC is used, for example, in cellular and other wireless networks in support of e-mail and Internet access to the Web. Satellite communications often make use of FEC, as the quality of the link is uncertain, bandwidth is limited, and latency is definitely an issue.

6.3.5.1 Echo Checking *Echo checking* is one of the earliest means of error detection and correction. The receiving device *echoes* the received data back to the transmitting device. The transmitting operator can view the data as received and echoed, making corrections as appropriate. But errors also can occur in the transmission of the echoed data, making this approach highly unreliable.

You can characterize echo as very slow and overhead intensive because characters are transmitted one at a time, in asynchronous mode; therefore, the process is bandwidth intensive as well. Further, the error detection and correction process is manual (human to machine) and decidedly unreliable. As a result, contemporary data communications seldom use echo checking.

6.3.5.2 Parity Checking *Parity checking* is by far the most commonly used method for error detection and correction because it is used in asynchronous devices such as PCs. Parity refers to the number of marks, or 1 bits, in a character. The network can be set for either *odd parity* or *even parity*. Once set, the networked devices always create odd or even data values, character by character or set by set. This less than ideal approach is implemented easily and offers reasonable assurance of data integrity. Parity checking has two dimensions: Vertical Redundancy Checking (VRC) and Longitudinal Redundancy Checking (LRC).

Vertical redundancy checking entails the appending of a parity bit at the end of each transmitted character or value to create an odd or even total mathematical bit value. The letter *C*, for example, in ASCII, is coded as a bit sequence of 1100001, which is an odd number of marks, or 1 bits. So the parity bit would be a 0 if the network or link is set for odd parity, as that would create an eight-bit byte with the sequence 11000010, thereby retaining the odd parity value. Alternatively, the parity bit would be a 1 if the network is set for even parity, as that would create an eight-bit byte with the sequence 11000011, thereby creating an even parity value. The receiving device executes the same mathematical process to verify that the correct total bit value was received—hence the use of the terms *redundancy* and *checking*. Speaking in terms of the logical manner in which humans add numbers physically positioned in columns, the two devices sum the bit values vertically, as represented in Figure 6.15—hence the use of the term *vertical*. While inexpensive and easily implemented in computers employing asynchronous transmission, this approach is highly unreliable, as two errored bits in a character can yield an undetectable error in a character. Further, VRC provides no inherent means of error correction; VRC often is characterized as *send and pray*. According to Maria Price La Touche (1824–1906):

> *There is no greater mistake than to call arithmetic an exact science. There are . . . hidden laws of number which it requires a mind like mine to perceive. For instance, if you add a sum from the bottom up, and then again from the top down, the result is always different.*

LRC

VRC

Bit/Value	C	O	N	T	E	X	T	P
1	1	1	0	0	1	0	0	0
2	1	1	1	0	0	0	0	0
3	0	1	1	1	1	0	1	0
4	0	1	1	0	0	1	0	0
5	0	0	0	1	0	1	1	0
6	0	0	0	0	0	0	0	1
7	1	1	1	1	1	1	1	0
P	0	0	1	0	0	0	0	

Figure 6.15 Example ASCII with VRC and LRC odd parity checking.

Longitudinal redundancy checking, or *Block Checking Character* (BCC), adds another level of reliability because data are viewed in a block or data set. Again, this approach is characterized in terms of the manner in which human beings add numbers in rows across columns, as though the receiving device were viewing the data set in a matrix format. This additional technique of checking the total bit values of the characters on a longitudinal (i.e., horizontal) basis employs the same parity (i.e., odd or even) as does the vertical checking technique (Figure 6.15). While remaining relatively inexpensive and easily implemented in devices employing asynchronous transmission, LRC/BCC adds a significant measure of reliability. Still, it is less than completely reliable, as compensating errors still can occur in nonadjacent characters. Also known as *checksum*, the LRC is sent as an extra character at the end of each data block [17].

6.3.5.3 *Block Parity*

The technique of *block parity* improves considerably on simple parity checking. While *Spiral Redundancy Checking* (SRC) and interleaving improved on the detection of errors due to increased transmission speeds and more complex modulation techniques, they gave way to *Cyclic Redundancy Checking* (CRC), which is commonly employed today.

CRC validates transmission of a set of data, formatted in a block or frame, through the use of a statistical sampling process and a unique mathematical polynomial, both of which are known to the transmitter and receiver. The transmitting device statistically samples the data in the block or frame and applies a 17-bit generator polynomial based on an Euclidean algorithm. The result of that calculation is a description of the text field, which is appended to the block or frame or text as either a 16- or 32-bit value. The receiving device executes the identical process, comparing the results of its process to the CRC value appended to the data block. The result is an integrity factor of 10^{-14}; in other words, the possibility of an undetected error is 1 in 100 trillion. By way of example and at a transmission speed of 1 Mbps, one undetected error is expected approximately every 30 years!

An unerrored block or frame is *ACK*nowledged by the receiving device through the transmission of an *ACK*, whereas an errored block or frame is *N*egatively *AcK*nowledged with a *NAK*. A NAK prompts the transmitting device to retransmit that specific block or frame, which has been stored in buffer memory. The transmission of an ACK by the receiving device cues the sending device that the block or frame of data can be erased from buffer memory and the next block or frame of data then can be sent.

CRC is relatively memory and processor intensive, but it is easily accommodated in high-order computers that benefit from synchronous transmission techniques. As CRC ensures that data transmission is virtually error free, it is considered mandatory in most sophisticated computer communications environments.

6.3.5.4 *Forward Error Correction*

Forward Error Correction (FEC) involves the addition of redundant information embedded in the data set so the receiving device can detect errors and correct them without requiring a retransmission [18]. The two most commonly employed techniques are *Hamming* and BCH (Bose, Chaudhuri, and Hocquengham).

While even more memory and processor intensive than CRC, the costs of CPU cycles and gigabytes of memory are so low in contemporary computers that FEC really is not much of a cost issue. FEC enables the receiving device to correct for errors in transmission, thereby avoiding most requirements for retransmission of errored blocks or frames of data. As a result, FEC improves the efficiency, or throughput, of the network, reducing transmission costs in the process, and without sacrificing data integrity. So, FEC is used in applications where bandwidth is at a premium and errors are common, with text messaging over cellular networks being a prime example.

6.3.6 Data Compression

As the length of the data sets increases, the distances over which they travel increase, and the likelihood of errors in transmission increases accordingly, *data compression* becomes sensible. Additionally, data compression can significantly reduce the bandwidth required to transmit a set of data. Regardless of the level of bandwidth available in even the most capable networks, bandwidth always has an associated cost. Data compression techniques can include formatting, redundant characters, commonly used characters, and commonly used strings of characters:

- *Formatting* of the data need not be transmitted across the network. In a basic example, data compression might involve the removal of formatting from a commonly used form, such as an expense report. Such formatting can involve a large amount of redundant data because the receiving device can reformat the data easily, placing the various fields of data in the appropriate places on the form, which resides in memory. An excellent example is that of access to the Internet and the Web through client/server software such as America Online, Netscape Navigator, or Microsoft Internet Explorer. In each case, many of the graphic-intensive screens are stored on the client workstation. When accessing the various Internet *portals*, therefore, you do not need to download the full set of graphics. This process is extremely bandwidth intensive and, therefore, ultimately translates into long delays and higher costs. Rather, only the updated information must be downloaded.
- *Redundant data* can be identified easily by the transmitter and communicated to the receiver. This approach is also known as *string coding*, yielding compression factors of as much as 4:1. An excellent example is that of fax modem compression algorithms, which use various methods of *run-length encoding*. As transmitting fax modems scan a document from left to right and from top to bottom, they can quickly sense a run of whitespace. Then, the transmitting modem notes that "nothing" is being transmitted and notes the length of the run of nothing, all in a few bits stored in an internal buffer. Once some "real" data appear, the modem notes this fact and begins to send corresponding bits to the internal buffer. After a specific number of bits are stored in the buffer, the modem packs them into a frame that it transmits. As runs of nothing and runs of real data reoccur, the modem recognizes that fact and adjusts accordingly. Very quickly, therefore, the transmitting modem can transmit a document with lots of whitespace, with that transmission requiring very little bandwidth through the supporting network. (Try faxing a white sheet of paper. Then try faxing a document of very dense text. You will immediately see the difference in transmission time.)

- *Commonly used characters* are easily identified and abbreviated through the use of an identifier and a smaller set of bits, similar to the technique used by Samuel Morse in the development of Morse code. *Huffman coding* is commonly used in this instance, yielding *compaction factors* of 2:1 or 4:1.
- *Commonly used strings of characters* similarly can be identified and transmitted in abbreviated form. Such an approach relies on the probability of character occurrence following a specific character (e.g., *Q* is generally followed by *U*). *Markov source* and other techniques address this potential.

6.3.7   Asynchronous Data Link Control Protocols

Asynchronous Data Link Control (DLC) protocols are used primarily for low-speed data communications between PCs and other very small host computers. *Framing* occurs at the byte level, with each byte surrounded by a start bit (a 0 bit) and a stop bit (a 1 bit). A parity bit often accompanies each character as well. *Telex* transmission incorporates an additional stop bit.

Kermit and XMODEM are asynchronous protocols, organizing information into 128-byte packets. Kermit also uses CRC error control. The data also can be *blocked* at the application level, and adding the technique of LRC can complement VRC for improved error control.

6.3.8   Bit- versus Byte-Oriented Synchronous Protocols

Two general types of data communications protocols exist—byte oriented and bit oriented. While the performance characteristics of byte-oriented protocols are acceptable for many applications, bit-oriented protocols are much more appropriate for communication-intensive applications in which the integrity of the transmitted data is critical.

Byte-oriented protocols require an entire byte to communicate a command signal to the target station. Byte-oriented protocols communicate value strings in byte formats, generally of eight bits per byte. Control characters are embedded in the header and trailer of each byte or block of data. As byte-oriented protocols are overhead intensive, they are used exclusively in older computer protocols at the second layer, or *link layer*. Byte-oriented protocols generally are asynchronous and HDX, operating over dial-up, two-wire circuits. One example includes *Binary Synchronous Communications* (*Bisync*, or BSC).

Bit-oriented protocols can change a single bit with a frame control byte to send a different command to a target station. Bit-oriented protocols transmit information in a much larger bit stream, with opening and closing flags identifying the separation of the text from the control information, which addresses control issues associated with the entire data set. The much less overhead-intensive, bit-oriented protocols are usually synchronous and FDX and operate over dedicated, four-wire circuits. Examples include IBM's *Synchronous Data Link Control* (SDLC) and the ISO's *High-level Data Link Control* (HDLC).

6.3.8.1   *Binary Synchronous Communications*   IBM developed Bisync in 1966 as a byte-oriented protocol that frames the data with control codes that apply to the entire set of data. Bisync organizes data into blocks of up to 512 characters,

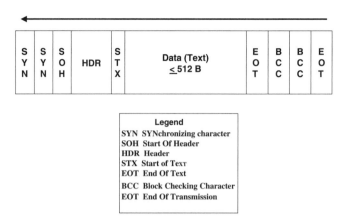

Figure 6.16 BSC block.

which are sent over the link sequentially (one at a time). An ACK or a NAK is transmitted from the receiving terminal to the transmitting device following the receipt of each block. Error control is based on a Block Checking Character (BCC) that is transmitted along with the data; the receiving device independently calculates the BCC and compares the two calculations.

The Bisync block consists of synchronizing bits, data, and control characters sent in a continuous data stream block by block. While there are six basic Bisync block formats, the elements of a generic Bisync block, as illustrated in Figure 6.16, are as follows and in sequence [19–21]:

- **PAD:** A *PAD*ding character (hexadecimal 55) may be sent as the first character to alert the receiving device of the transmission of a block of data and to ensure that the receiving device is in step with the data bits.
- **SYN:** *SYN*chronizing characters (usually two) establish character synchronization between the transmission and the receiving devices. There must be enough bit transitions to allow the receiver to confirm the bit rate.
- **SOH:** A *S*tart-*Of-H*eader control character precedes the routing information.
- **Header:** The *Header* field contains one or more octets of information indicating the address of the transmitting device.
- **STX:** A *S*tart-of-*TeX*t control character indicates the beginning of the data.
- **Text:** The *Text* field, aka payload, is the data being transmitted. This field can be up to 512 octets in length.
- **ETX:** An *E*nd-of-*TeX*t control character indicates the end of the data.
- **BCC:** *B*lock *C*heck *C*haracters detect errors. There are one or two BCCs.
- **EOT** or **PAD:** An *E*nd *O*f *T*ransmission character or *PAD*ding character (hexadecimal FF) character trails the transmission to ensure the receipt of all previous characters and to indicate the end of the block.

In terms of raw efficiency, Bisync is approximately 98 percent efficient if the text field is fully packed with 512 octets of data, as there are only 10 or so octets of control information. As Bisync is a HDX protocol, however, it is considered to be very inefficient by today's FDX standards due to the turnaround times involved.

6.3.8.2 *Synchronous Data Link Control* *Synchronous Data Link Control* (SDLC), developed in the mid-1970s, is at the heart of IBM's System Network Architecture (SNA). SDLC is a bit-oriented, point-to-point protocol that uses bit strings to represent characters. SDLC uses CRC error correction techniques—specifically known as *Frame Check Sequence* (FCS) here. SDLC supports high-speed transmission and generally employs FDX, dedicated circuits. SDLC works either in HDX or FDX, supports satellite transmission protocols, and works in point-to-point or multipoint network configurations.

Up to 128 *frames* can be sent in a string, with each frame containing up to seven blocks, each up to 512 characters. Each block within each frame is checked individually for errors. Errored blocks must be identified as such to the transmitting device within a given time limit or they are assumed to have been received error free. As a carefully timed, point-to-point protocol, SDLC depends on high-performance circuits, usually in the form of dedicated leased lines.

The SDLC frame consists of synchronizing bits, data, and control characters sent in a continuous data stream frame by frame. The specific elements of the SDLC frame (Figure 6.17) are as follows and in sequence [19–21]. The same format applies to High-level Data Link Control (HDLC) frames and X.25 packets:

- **Flag (F):** Flag bits, in a specific eight-bit pattern, alert the receiving device to the transmission of the frame, thereby initiating the error-checking procedures. The most commonly used flag character is 01111110 (7E in hexadecimal).
- **Station Address (A):** This address field of eight bits identifies the specific target device for which the frame is intended, a group address for multiple target terminals, or a broadcast address to all terminals. This field also can be used to distinguish commands from responses.
- **Control (C):** The eight-bit control field identifies the type of frame being transmitted. An *information frame* is used for the transfer of messages, frame numbering of contiguous frames in a message, and so on. A *supervisory frame* is used for purposes such as to indicate a detected error in transmission, acknowledge that frames have been received without error, request the transmission of specified frames, and order the transmitting device to stop sending.

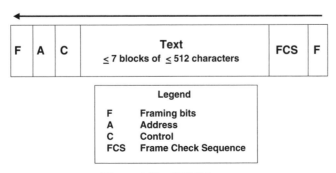

Figure 6.17   SDLC frame.

- **Information (I):** This text field (aka, message field, data field, or payload) of variable length contains the information (data) or request being transmitted. This field also can include a format identifier, logical channel group number, packet-type identifier, and packet sequence numbers. In total, the information field can contain as many as seven blocks, each of which can contain as many as 512 octets of data, for a total of 4096 octets.
- **Frame Check Sequence:** This 16- or 32-bit field contains the CRC character sequence used to check the integrity of the transmitted address and control information as well as the data.
- **Flag (F):** Flag bits, in a specific eight-bit pattern, alert the receiving device to the end of transmission of the frame, thereby terminating the error-checking procedures. The most commonly used flag character is 01111110 (7E in hexadecimal).

Considering that the information field can contain as many as 4096 octets of payload and that the control fields involve only 6 or 8 octets (8 at full payload), the SDLC frame can be as much as 99.8 percent efficient, at least theoretically. There is some overhead in the information field, of course, but even then the SDLC frame is 95–97 percent or so efficient, which is remarkable.

6.3.8.3 *High-Level Data Link Control* The ISO developed HDLC as a superset of IBM's SDLC and the U.S. *National Bureau of Standards'* (NBS) (now *National Institute of Standards and Technology*, or NIST) *Advanced Data Communications Control Procedures* (ADCCP). A version of HDLC is the *Link Access Procedure— Balanced* (LAP-B), which is used in packet-switched networks conforming to the ITU-T X.25 Recommendation. While HDLC was built on SDLC and is very similar, the two generally are not compatible, depending on the framing conventions in the specific HDLC implementation.

6.4 NETWORK ARCHITECTURES

A *network architecture* defines the communications products and services that ensure the various components can work together. Early on, even the various systems of a given manufacturer did not interoperate, let alone afford connectivity with the products of other manufacturers. While IBM's Systems Network Architecture (SNA) and the Digital Equipment Corporation's Digital Network Architecture (DNA), aka DECnet, architectures solved these internal problems, they still did not interoperate. Truly *open*-system architectures still remain in the distant future, although great strides have been made in this regard through the Open Systems Interconnection (OSI) model fostered by the International Organization for Standardization (ISO).

A number of standard computer network architectures have been defined, many of which segregate various functions into discrete layers of responsibility for ease of development and management. In addition to SNA and DECnet, network architecture examples include Xerox Networking System (XNS); Advanced Research Projects Agency Network (ARPANET), the U.S. government–sponsored predecessor to the commercial network we now call the Internet; U.S. Department of

Defense's Defense Data Network (DDN); and the Open Systems Interconnection (OSI) model.

Layered models serve to enhance the development and management of a network architecture. While they primarily address issues of data communications, they also include some data processing activities at the upper layers. These upper layers address application software processes, presentation format, and establishment of user sessions. Each independent layer, or level, of a network architecture addresses different functions and responsibilities. The layers work together, as a whole, to maximize the performance of the process. The various functions address the functions of data transfer, flow control, sequencing, error detection, and notification.

Data transfer enables the transfer of data from one node to another. Included are such issues as normal or expedited data flow; packet, block, or frame sizing; and data assembly and segmentation. *Flow control* controls the pace of packets through the network to prevent a device or link from being overwhelmed with data and to reduce congestion and the resulting degradation of network performance. *Sequencing* of the data packets is required where packets may take different routes between nodes. *Error detection* is required to ensure data integrity. *Notification* provides for the advice from receiver to transmitter of the receipt of packets and their condition.

6.4.1   Systems Network Architecture

Developed in 1974 by IBM, SNA was a five-level design architecture that has grown into a seven-layer model. Although similar, SNA is not compatible with the OSI model. SNA comprises software and hardware interfaces that permit various IBM systems and software to communicate. SNA includes network nodes, physical units, and logical units:

- *Nodes* are physical devices in the SNA network. Nodes can include computers, communications processors (e.g., FEPs), terminal controllers, and terminals.
- *Physical Units* (PUs) manage the communications hardware and software, participating in the controlling and routing of network communications. All physical devices are assigned a PU type (1, 2, 3, or 5) that identifies the level of the device (i.e., terminal, controller, communications processor, or host node) and its origin (i.e., IBM/SNA or non-IBM/SNA).
- *Logical Units* (LUs) manage communications software for communications with end users. A logical unit *session* is an end-to-end communication between an end-user terminal and the originating application residing in the host. LU 6.2, for example, supports peer-to-peer communications between intelligent devices, without requiring the host to assume responsibility for communications support activities. LU 6.2 is also known as *Advanced Program-to-Program Communications* (APPC).

6.4.2   Open Systems Interconnection Model

The *OSI Reference Model* is a layered architecture (see Figure 6.18) consisting of a set of international networking standards known collectively as X.200. Developed

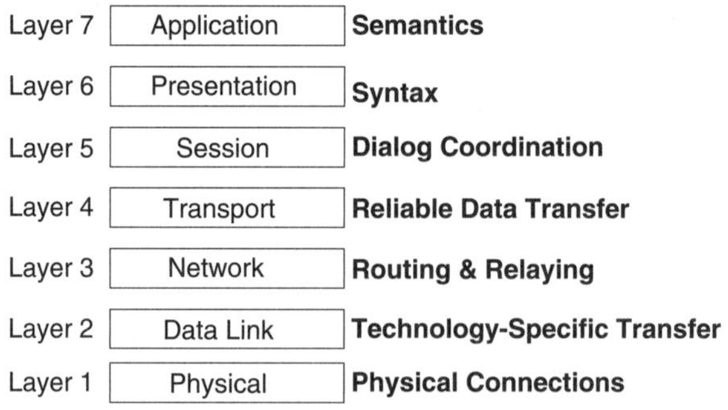

Layer 7	Application	**Semantics**
Layer 6	Presentation	**Syntax**
Layer 5	Session	**Dialog Coordination**
Layer 4	Transport	**Reliable Data Transfer**
Layer 3	Network	**Routing & Relaying**
Layer 2	Data Link	**Technology-Specific Transfer**
Layer 1	Physical	**Physical Connections**

Figure 6.18 OSI Reference Model.

by the ISO, the basic process began in 1977 and was completed in 1983. *Note*: At that point the ISO heavily promoted the model as a full standard. The U.S. government spent billions of dollars on *Government Open Systems Interconnection Profile* (GOSIP), a specification that essentially required all government networking products to be OSI compliant. Digital Equipment Corporation, IBM, Unisys, and other large and reputable organizations invested billions of dollars in total to gain OSI compliance. Despite all the time, money, and energy devoted to OSI, the standard never gained any real traction. At that point, DECnet, SNA, TCP/IP, and a few other standards seemed to have satisfied most people's appetite for standards initiatives. The Europeans were certainly an exception, however. The *European Computer Manufacturers Association* (ECMA) was very instrumental in the development of the OSI model, in large part because of the fact that the multinational nature of Western Europe had led to substantial difficulty in the interconnection of computers and computer networks. Not only was the OSI model seen as a solution to this problem, but the European Union (EU) actually legally imposed the model for some applications. Eventually, TCP/IP pushed OSI aside as a standard, but the reference model remains valuable.

The OSI model defines a set of common rules that computers of disparate origin can use to exchange information (communicate). As is the case with SNA and other such proprietary architectures, the model is layered to segment software responsibilities, with supporting software embedded in each node to provide an interface between layers. Specific levels of service can be negotiated between nodes.

The transmitting device uses the top layer, at which point the data are placed into a packet, prepended by a header. The data and header, known collectively as a *Protocol Data Unit* (PDU), are handled by each successively lower layer as the data move across the network to the receiving node. At the receiving node, the data work way up the layered model; successively higher layers strip off the header information. The seven layers of the OSI Reference Model are as follows:

- **Layer Seven (Application Control):** Provides support services for user and application tasks. File transfer, interpretation of graphic formats and documents, and document processing are supported at this level. X.400 e-mail messaging, for example, takes place at Layer Seven. TCP/IP application protocols such as

Simple Mail Transfer Protocol (SMTP), *Telnet,* and *File Transfer Protocol* (FTP) also take place at layer 7.

- **Layer Six (Presentation Control):** Performs functions related to the format and display of received data by terminals and printers. Functions herein include data formatting, code set (e.g., Baudot, ASCII, EBCDIC, and Unicode), code set conversion [e.g., HTML to ASCII], text compression and decompression (e.g., WinZip), and encryption (e.g., DES).

- **Layer Five (Session Control):** Formats the data for transfer between end nodes, provides session restart and recovery and general maintenance of the session from end to end. This layer is considered by some to be of little or no consequence and is often disregarded and considered to be a function of layer 6.

- **Layer Four (Transport Control):** Responsible for maintaining the end-to-end integrity and control of the session. Data are accepted from the Session Control layer and passed through to the Network Control layer. Long message blocks are divided into shorter message blocks for transmission, sequence numbers are added, checksums are calculated and appended, retransmissions are generated in the event of errored message blocks or timeouts, and security is added. Example protocols that can be used at this layer include *Transmission Control Protocol* (TCP), *User Datagram Protocol* (UDP), and the five classes of the OSI *Transport Protocol* (TP). These protocols ensure end-to-end integrity of the data in a session. The X.25 packet-switching protocol operates at layers 1–4.

- **Layer Three (Network Control):** Comprises software that addresses and sequences the PDUs and transports them to the ultimate destination, setting up the appropriate paths between the various nodes. At this layer, message routing, error detection, and control of internodal traffic are managed. The Internetwork Protocol (IP) operates at this layer.

- **Layer Two (Data Link Control):** Establishes the communications link between individual devices over a physical link or channel. At this level, framing, error control, flow control, data sequencing, timeout levels, and data formatting occur. HDLC, LAP-B, LAPD are at this level. Frame Relay is a layer 2 protocol. LANs operate at layers 2 and 1, and Network Interface Cards (NICs) cut across portions of these two layers.

- **Layer One (Physical Control):** Defines the electrical and mechanical aspects of the interface of the device to a physical transmission medium, such as twisted pair, coax, or fiber. Communications hardware and software drivers are found at this layer, as are electrical specifications such as EIA/TIA-232 (nee RS-232). T/E-carrier runs at layer 1. Synchronous Optical NETwork (SONET) and Synchronous Digital Hierarchy (SDH) run at layers 1 and 2.

6.5 SECURITY

The only secure computer is one that is turned off, locked in a safe, and buried 20 feet down in a secret location . . . and I'm not completely confident of that one either.
Bruce Schneier, security technologist and author

Security is an issue of prime importance across all dimensions of communications and networks, but perhaps most importantly in the world of data communications. In the traditional data world of *mainframes in glass houses*, security was controlled very tightly. In the contemporary world of distributed processing and networked computer resources, security is much more difficult to develop and control. Perhaps the greatest strength of networks is that they enable information to be shared; perhaps the greatest weakness of networks is that they enable information to be shared. The trick, of course, is to permit only legitimate users to share. Security encompasses a number of dimensions, including physical security, authentication, authorization, port security, transmission security, and encryption.

6.5.1 Physical Security

Physical security involves access control, that is, control over the individuals who have access to the facilities in which the systems reside. Methods of access restriction include security guards, locks and keys, electronic combination locks, and/or electronic card key systems that require additional input, such as a Personal Identification Number (PIN). The latter is preferable because the system can maintain a record of specific access. Physical security also entails some decidedly low-tech tools such as document shredders and burn bags, which jointly serve to make paper documents and electronic media unusable after they serve their purposes.

6.5.2 Authentication

Authentication provides a means by which network managers can validate the identity of those attempting access to computing resources and the data they house. Authentication consists of password protection and intelligent tokens. *Password protection* can restrict individuals on a site, host, application, screen, and field level. Passwords should consist of an alphanumeric value of reasonably long length and should be changed periodically. A current trend points toward the use of dedicated password servers for password management. *Intelligent tokens* are one-time passwords generated by hardware devices and verified by a secure server on the receive side of the communication. They often work on a cumbersome *challenge–response* basis.

Remote Authentication Dial-In User Service (RADIUS) is a highly popular public network authentication service. Developed by Livingston Enterprises and based on a model defined by the Internet Engineering Task Force (IETF), RADIUS comprises an authentication server installed on the user's host computer and client protocols. Remote users are authenticated through a series of encrypted communications between the remote client and the centralized server. RADIUS is an open approach that can be modified easily to work with any security system and virtually any communications device. Access via a RADIUS *Remote Access Server (RAS)* commonly not only requires a password but also the remote computer must be equipped with a *smart card reader* into which a *smart card* must be inserted. The smart card must match the password, user ID, and a PIN before access is granted.

6.5.3 Authorization

Authorization provides a means of controlling which legitimate users have access to which resources. Authorization involves complex software that resides on every secured computer on the network; ideally, it provides *single sign-on* capability. Authorization systems include Kerberos, Sesame, and Access Manager:

- *Kerberos*, the best-known authorization software, makes use of private-key authentication. Developed by the Massachusetts Institute of Technology (MIT), Kerberos is available free, although commercial versions exist. IBM's Kryptoknight, for example, is a Kerberos variant. Kerberos was named for the three-headed dog, Cerberus, that guarded the gates of Hades in Greek mythology.
- *Sesame (Secure European System for Applications in a Multivendor Environment)* was developed by the ECMA (European Computer Manufacturers Association). It is flexible, open, and intended for large, heterogeneous network computing environments. It also is highly complex and not effective for smaller applications.
- *Access Manager* uses an API for applications, employing scripting. *Scripting* involves a process of mimicking the log-on procedures of a program, providing basic levels of security for small networks.

6.5.4 Port Security

Port security is essential to deny unauthorized remote access. Passive devices report on unauthorized access, usage anomalies, and so on. Active devices, which are preferable, act to deny access to unauthorized users and disable ports if user-definable parameters (e.g., number of access attempts) are exceeded.

6.5.5 Transmission Security

Transmission security is critical to ensure that unauthorized entities are not permitted to intercept the information as it traverses the network. Transmission of data is especially insecure over analog links because analog transmission does not lend itself to encryption, although encryption can be performed in the Data Terminal Equipment (DTE). Wireless transmission is inherently insecure, although digital wireless systems generally support signal encryption. Transmission security is virtually ensured over coaxial cable and other shielded and screened copper media and, especially, over fiber-optic cable because these media cannot easily be tapped. In order to maximize security, however, it is necessary that the data be *encrypted*.

6.5.6 Encryption

Encryption involves scrambling and perhaps compressing the data prior to transmission; the receiving device is provided with the necessary logic to decrypt and decompress the transmitted information. Encryption logic generally resides in firmware included in stand-alone devices, although it can be built into virtually any device.

Encryption logic, for example, often is incorporated into routers, which can encrypt data on a packet-by-packet basis. Encryption comes in two basic flavors: private key and public key. *Private key* is a symmetric encryption method that uses the same key to encrypt and decrypt data and requires that the key be kept secret. *Public key* is an asymmetric encryption method with two keys—an encryption (encoding) key that can be used by all authorized network users and a decryption (decoding) key that is kept secret. Data encryption standards include *Data Encryption Standard* (DES), *Triple DES*, and *Advanced Encryption Standard* (AES).

6.5.7 Firewalls

Firewalls comprise application software that can reside in a communication router, server, or some other device. That device physically and/or logically is a first point of access into a networked system. On an active basis, the device can block access to unauthorized entities, effectively acting as a *security firewall*. Firewalls can use one or more basic approaches to access control. A *packet-filtering firewall* examines all data packets, forwarding or dropping individual packets based on predefined rules that specify where a packet is permitted to go and in consideration of both the authenticated identification of the user and the originating address of the request. *Proxy firewalls* act as intermediaries for user access requests by setting up a second connection to the resource. That second connection can be established at the application layer (layer 7) by an *application proxy* or at the session (layer 5) or transport (layer 4) layer by a *circuit relay* firewall. A *stateful inspection* firewall examines packets, notes the port numbers that they use for each connection, and shuts down those ports once the connection is terminated. Firewalls are the subject of much continuing interest, especially as organizations seek to protect their data from the ravages of hackers and other less than honorable creatures that prowl the Internet [22, 23].

REFERENCES

1. Standage, Tom. *The Victorian Internet*. Berkley Books, 1998.
2. http://www.msnbc.msn.com/id/11147506/.
3. Martin, James. *Telecommunications and The Computer*, 2nd ed. Prentice-Hall, 1976.
4. Dodge, G. M. *The Telegraph Instructor*. G. M. Dodge, 1903.
5. Brooks, John. *Telephone: The First Hundred Years*. Harper & Row, 1976.
6. *Telephone Almanac*. Bell Telephone System, 1959.
7. English, David. "V.90 Modems: The End of the Line." *PC World*, January 1999.
8. Cray, Andrew. "New Juice for Analog Modems?" *Data Communications*, December 1998.
9. Fratto, Mike. "V.90 Modems Burn Up the Wires with Standards-Based 56-Kbps Access." *Network Computing*, November 15, 1998.
10. Wexler, Joanie. "56k Modems: A Bandwidth Bird in the Hand." *Business Communications Review*, October 1998.
11. Shah, Deval and Holzbaur, Helen. "56K? No Way." *Data Communications*, August 1997.

12. Pappalardo, Denise. "ISPs Reluctant to Offer Relief for Dial-Up Users." *Network World*, July 30, 2001.

13. Spangler, Todd. "Modems Get an Extra Bit of Oomph." *Inter@ctive Week*, July 17, 2000.

14. Ambion, Jason. "V.92 Modems Will Help Dial-Up Users." *Network World*, September 11, 2000.

15. Flanagan, William A. *T-1 Networking*. Telecom Books, 1997.

16. Sherman, Kenneth. *Data Communications: A User's Guide*. Reston Publishing Co., 1981.

17. Held, Gilbert. *Understanding Data Communications*. SAMS Publishing, 1994.

18. Doll, Dixon R. *Data Communications: Facilities, Networks and Systems Design*. Wiley, 1978.

19. *Data Communications Concepts*. IBM (GC21-5169-5), September 1985.

20. Gelber, Stan. *Introduction to Data Communications: A Practical Approach*. Professional Press Books, 1991.

21. Fitzgerald, Jerry. *Business Data Communications*, 4th ed. Wiley 1993.

22. Schultz, Keith. "Taming the Flames." *Communications Week*, March 10, 1997.

23. Newman, David, Holzbaur, Helen, and Bishop, Kathleen. "Firewalls: Don't Get Burned." *Data Communications*, March 21, 1997.

CHAPTER 7

CONVENTIONAL DIGITAL AND DATA NETWORKS

In 1816, Mr. Ronalds (afterwards Sir Francis Ronalds) showed that an electric telegraph was possible, and endeavoured to persuade the Government of the importance of his system. The official reply to his appeal was as follows "Mr. Barrow presents his compliments to Mr. Ronalds, and acquaints him, with reference to his note of the third instant, that telegraphs of any kind are now wholly unnecessary, and that no other but the one now in use will be adopted. Admiralty Office, Aug. 5, 1816." The "one in use," here indicated, was the semaphore, . . . which, it may be mentioned, was quite useless during the night, or when fog prevented the signals being seen.

Wonders of the Universe, The Werner Company, 1899

Data communications began in 1835 with the invention of the first practical telegraph by Samuel F. B. Morse and with his first long-distance message, *What hath God wrought!* sent from Baltimore, Maryland, to Washington, D.C., in 1844. This simplex (in this context, *simplex* means one-way, single-channel) device used start and stop signals of varying lengths over uninsulated iron, and later copper, wire. Subsequently, the technology improved to *diplex* (one-way, two-channel) and *quatraplex* (one-way, four-channel). In 1850, there were over 12,000 miles of line operated by 20 different companies offering commercial telegraph service [1]. Western Union dates to 1851 through its predecessor, the Mississippi Valley Printing Telegraph Company, which became the Western Union Telegraph Company in 1856 through a series of acquisitions. Telegraphy enjoyed a monopoly on electronic communications until 1877, when the first telephone networks appeared. Telegraph networks were not only the first data networks but also the only telecommunications networks for 30 years or so. Western Union had the opportunity to acquire American Bell, Alexander Graham Bell's original telephone company, and all its patents for $100,000 in 1876 but did not see the value of them. When presented with the

Telecommunications and Data Communications Handbook, By Ray Horak
Copyright © 2007 Ray Horak

opportunity, William Orton, president of Western Union at the time, asked what must most certainly be one of the least profound business questions of all time, *What could this company make of an electrical toy?* [2]. (*Note:* I figure that $100,000 at an annual interest rate of 3 percent, compounded monthly is $4,916,260.18 in 2006 dollars, which is a bargain. At 6 percent, that $100,000 investment would be worth $239,362,685.57 today, which, although a considerable sum of money, is still a bargain.)

According to the minutes of a Western Union meeting (ca. 1876):

> *Bell expects that the public will use his instrument without the aid of trained operators. Any telegraph engineer will at once see the fallacy of this plan. The public simply cannot be trusted to handle technical communications equipment. Bell's instrument uses nothing but the voice, which cannot be captured in concrete form . . . we leave it to you to judge whether any sensible man would transact his affairs by such means of communications. In conclusion, the committee feels that is must advise against any investment whatever in Bell's scheme.*

After all, the telephone was far too technical for the average person. Somehow, the average person was able to overcome that obstacle, and the telephone networks quickly overtook the telegraph networks in terms of size, traffic, revenue, and virtually every other measure. But the telephone networks were limited to voice, and the telegraph networks carried all the data traffic—at least for 100 years or so.

Bell's invention proved itself and his company thrived over the next few years. Western Union continued to lease telegraph lines and supplied its customers with various kinds of telegraph keys, printing telegraphs, and dial telegraphs, some of which could transmit 60 words a minute. Western Union believed that these instruments could never be replaced by a scientific oddity such as the telephone until the Gold and Stock Company, one of its subsidiaries, reported that several of its instruments had been superseded by telephones. Western Union quickly realized its mistake, challenged Bell's patents in court, and formed the American Speaking-Telephone Company. Western Union announced that it had "the only original telephone" and was ready to supply "superior telephones with all the latest improvements made by the original inventors" (Dolbear, Gray, and Edison). Those ridiculous statements lifted Bell's invention from the status of scientific *toy* to article of commerce as businesses began to take the telephone seriously [2]. Western Union's patent attorney finally convinced the company that it could not win the lawsuit and encouraged them to settle out of court. In the final treaty (1879) between the two companies, Western Union agreed to admit that the Bell patent was the original, to admit that his patents were valid, and to retire from the telephone business. The Bell Company agreed to buy the Western Union telephone system (56,000 telephones in 55 cities), to pay the Western Union a royalty of 20 percent on all telephone rentals associated with that system, and to keep out of the telegraph business. The agreement remained in force for 17 years [2].

As noted in Chapter 6, Western Union first offered teletypewriter service in 1923 and telegraph companies introduced rotary dial telex services in 1935. AT&T introduced TetetypeWriter eXchange (TWX), a high-speed dialup telex service in 1930. Telegraph, telex, and TWX networks were all separate and distinct from, or at least were distinct physical partitions of, the Public Switched Telephone Network (PSTN).

That approach worked well for a number of years. After all, there were not a lot of computers and certainly not much need to network them. In 1943 Thomas Watson, Sr., Chairman of IBM, said, "I think there is a world market for maybe five computers." Although his vision of the future might have occasionally been blurred and the quotation is questionable, Thomas Watson clearly was an extraordinary businessman. He and later his son, Thomas Watson, Jr., led IBM to a position of leadership in the data processing industry, worldwide—and they sold a lot more than five computers. Increasingly, it became necessary to network those computers, and the PSTN seemed a logical way to do that. At this point, it is appropriate to revisit the PSTN and reexamine its characteristics, particularly as they relate to its application to data communications in the 1950s:

- **Voice Oriented:** The PSTN was designed for voice, only. Although data communications precedes voice, telegraphy (and later telex and TWX) always took place over separate networks, or at least separate physical network partitions. Early data communications relied largely on having a solid copper path for the entire length of the circuit, so the end devices could use the *loop current*. As voice over the PSTN might be amplified or go through transformers or other devices, there might not be a hard copper path end to end. So, any data communications taking place over the PSTN would have to make some adjustments.

- **Analog:** The PSTN was entirely analog in nature. Although the first digital computer, the complex-number calculator, was invented in 1939, digital transmission systems were not invented and trialed until the 1950s and not placed into commercial use until 1962. Analog transmission and switching systems are error prone and relatively slow in data communications applications as compared to digital systems.

- **Voice Grade:** Bandwidth was limited in the PSTN, with the entire network oriented to voice-grade communications in channels 4 kHz wide.

- **Transmission Media:** Copper predominated in the PSTN of the 1950s. Unshielded Twisted Pair (UTP) was widely used in the local loop, coaxial cable was heavily used for interoffice and long-haul trunking, and some amount of analog microwave was used in long-haul applications. Neither fiber optics nor Free-Space Optics (FSO) were available at the time.

- **Duplex:** The PSTN was Full-Duplex (FDX) in nature, which is consistent with the requirements of conversational voice. Certainly, not all data communications applications took advantage of FDX or even Half-Duplex (HDX) transmission mode in the 1950s, but they certainly did not suffer from the network's ability to support it. The core of the network was four-wire and the local loop largely two-wire, much as it is today.

- **Circuit Switched:** The PSTN was entirely circuit switched at the time, with the exception of leased lines, of course. Circuit switching is entirely appropriate for uncompressed, real-time voice communications and for data communications applications such as large file transfers that involve continuous circuit usage for long periods of time.

- **Ubiquitous and Affordable:** Since the Federal Communications Act of 1934, the PSTN rate and tariff structure in the United States has included a complex set of cross-subsidies to ensure that basic telephone service is available almost universally and at reasonable cost, even in remote, rural, and *high-cost* areas. So, the PSTN in the United States is virtually ubiquitous and was so in the 1950s. The PSTN was widely available in many other developed countries as well.

- **Interconnected:** Interconnection is fundamental to the PSTN, which provides for flexible connection between voice terminals, virtually regardless of location. Local Exchange Carriers (LECs), IntereXchange Carriers (IXCs), and international carriers all have been interconnected for many years. Many data networking applications do not demand full interconnection at this level but do not suffer because of it, issues of security aside.

In the late 1950s, large organizations (initially in North America) desired the ability to move data over telephone lines. This was first accomplished with a variation of the IBM 729 tape drive, which interfaced with the analog PSTN through a matched pair of *Bell datasets* (1957) [3], or *DataPhones* (1959) [4], via acoustic couplers and telephone sets. These early modems operated over the analog PSTN at 300 bps and, later, 1200 bps, which was incredibly fast at the time. At that point, voice and data networks began to merge. Datasets quickly spread around the world, rented by the telcos and PTTs to end users until deregulation afforded users the option of acquiring and interconnecting such equipment. The original datasets connected to the PSTN through a *Data Access Arrangement* (DAA) device that served as a coupler, or protector, to protect the network from high signal levels, out-of-band frequencies, and aberrant voltages. This protection is incorporated into contemporary modems and other devices, which are standardized and regulated by the ITU-T on an international basis and by the U.S. Federal Communications Commission (FCC) and other national and regional regulatory bodies.

The telcos began to digitize their networks in the 1960s, as digital technology became reliable and inexpensive enough to support telecommunications applications and as the requirement surfaced for increased bandwidth in the carrier networks. Digital transmission facilities, in the form of T-carrier (North America) and E-carrier (Europe), increased the traffic capacity of existing facilities. In the 1970s, analog Electronic Common Control (ECC) switches began to be replaced with fully digital switches. Data transmission at relatively high speeds and over fully digital networks became a reality, although a number of years passed before such capability became widely available.

Clearly, digital transmission offers significant advantages, especially for data transmission. Those advantages include increased bandwidth and bandwidth utilization, improved error performance and increased throughput, and enhanced management and control. This chapter focuses on conventional digital data networking options, which include dedicated leased lines, circuit switching, and packet switching. Specifically, those technologies and service offerings include Dataphone Digital Service (DDS), Switched 56, classic Virtual Private Networks (VPNs) on the PSTN, digital carrier systems (T-carrier and E-carrier), X.25 and packet switching, and Integrated Services Digital Network (ISDN). In later chapters I provide a detailed introduction into more recent data networking options, including Frame Relay (FR), Asynchronous Transfer Mode (ATM), and Transmission Control Protocol/Internet Protocol (TCP/IP).

7.1   DATAPHONE DIGITAL SERVICE

AT&T introduced *Dataphone Digital Service* (DDS), also known as *Digital Data Service* and *SubRate Digital Loop* (SRDL), in 1974 [3], in response to the increasingly obvious need to interconnect mainframe computers over a wide area. The term DDS now is used generically to describe an end-to-end, fully digital, dedicated service provided by most incumbent carriers. DDS is widely deployed in the United States and Canada and many other developed countries and is intended for relatively high speed data transport applications between purely digital devices (i.e., computers). Employing specially conditioned, dedicated, leased-line circuits provided to user organizations by the carriers, a DDS configuration may be either point to point or multipoint. In either event, all network control is the responsibility of a designated *head-end* system. The head end, traditionally in the form of a *Front-End Processor* (FEP) or, in more contemporary terms, a *communications server*, controls all access to the network through a process of polling the remote devices. Additionally, all communications must pass through the head end; in other words, devices cannot communicate directly as they can in a *mesh* network, where all locations are interconnected directly.

DDS is intended for FDX synchronous communications provided over four-wire circuits between computing systems that communicate intensively (i.e., frequently and passing significant volumes of data). The DDS network provides network timing and synchronization through a master clock, which ensures that all clocks in all slaved network nodes operate at the same bit rate, or clock speed, and at the same clock phase. The Data Communications Equipment (DCE) extracts timing from the received signal. DCE is in the form of a Data Service Unit/Channel Service Unit (DSU/CSU) that operates at the full line rate or on a subrate basis (lower speed), as required. *Note:* While the DDS circuit operates in FDX, HDX and simplex applications are supported. In fact, a large percentage of the DDS applications were for HDX polling.

Transmission rates vary, within limits, according to the user organization's requirements. Bandwidth generally is available at line rates of 2400 bps, 4800 bps, 9600 bps, 19.2 kbps, 56 kbps, or 64 kbps and digital carrier rates of 1.544 Mbps (T1) and 2.048 Mbps (E-1). Note here that the DDS signals actually are carried inside T-carrier or E-carrier channels in the backbone carrier networks. (I discuss T-carrier and E-carrier later in this chapter.)

While the cost of DDS circuits varies according to specific carrier tariffs and pricing strategies, cost is sensitive to both the distance between the points of termination and the level of bandwidth. Such is the case with all dedicated leased-line services. A traditional rule of thumb is that DDS generally is cost effective in applications that require communications between two locations, for a total of one hour per day or more, at a rate of 56 kbps. That equation clearly is sensitive to local rates and the availability of alternative services, such as Frame Relay.

The cost equation changes, of course, where there is a requirement to interconnect multiple locations in a multipoint network configuration. A *multipoint circuit* also is known as a *multidrop circuit* in telco parlance because local loop connections historically are dropped from poles. Multipoint circuits also are referred to as *fantail circuits* because they fan out at the tail end, that is, the end distant from the head end. As noted in Figure 7.1, a headquarters data center in New York might be con-

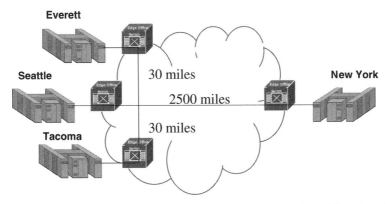

Figure 7.1 DDS leased lines connecting data centers in New York and Seattle, with drops to Everett and Tacoma.

nected to a regional data center in Seattle, some 2500 miles west of New York. Field offices in Everett (approximately 30 miles north of Seattle) and Tacoma (approximately 30 miles south of Seattle) can take advantage of the same coast-to-coast circuit very cost effectively, since the incremental circuit mileage is short and the associated cost of those additional drops therefore, is low. [*Note:* The circuits are interconnected in Central Offices (COs) but bypass the CO switches, as the service is dedicated, rather than switched. DDS and other dedicated services use these COs simply as *wire centers* rather than switching centers.]

In such a multipoint network, the head end addresses each remote system connected to the circuit on the basis of a unique logical address and in a user-definable and variable polling sequence. The target device recognizes its address and responds across the network, while all other devices remain silent. At that point, the two computer systems conduct a dialog until such time as either the data transfer is complete or the head end truncates the communication in order to address other devices according to its programmed polling schedule. Automatic teller machines traditionally were connected to the central bank in such a manner, and many such networks remain in place. In a typical scenario, the central computer polls the individual automatic teller machines, downloads the user's request for a cash withdrawal, and matches the account number and PIN for authentication purposes. It then queries the centralized database to determine the assigned level of withdrawal privileges and the current account balance and authorizes or denies the cash withdrawal.

DDS provides excellent reliability, which generally is in the range of *five nines*, that is, 99.999 percent. You should note, however, that all dedicated services are susceptible to catastrophic failure from such causes as *cable-seeking backhoes*. Therefore, network redundancy must be considered in the form of either backup DDS circuits or some alternative network service such as modem dial-up, Switched 56, or ISDN.

From an applications perspective, DDS is used for relatively intensive data-only communications applications between devices at fixed physical addresses. In such an environment, it can be highly cost effective since usage charges do not apply to network traffic over dedicated circuits. Typical applications include connecting data centers for purposes of file transfer or data backup. Image transfer and other band-width-intensive applications such as *Computer-Aided Design* (CAD) can make cost-effective use of DDS circuits, also benefiting from the bandwidth and excellent error performance offered by the dedicated digital circuits. DDS also serves to connect e-mail and Group IV facsimile servers in a messaging network. DDS traditionally is used in intensive transaction processing environments, as in the automatic teller machine example stated earlier. Department store chains traditionally deployed extensive DDS networks to support transaction processing applications between the centralized data processing center and the retail stores. Many of those networks remain in place. Similarly, oil companies traditionally had large and complex DDS networks in place at gas (petrol) stations with pay-at-the-pump options, with the tail circuits connecting large numbers of retail outlets to a long-haul circuit that terminated in the data center. The oil companies still commonly make use of short-haul, multidrop circuits to connect multiple outlets to a central retail outlet that is equipped with a satellite dish. The long-haul portion of the connection to the data center is provided over satellite facilities.

7.2   SWITCHED 56

Switched 56 (kbps service) is the popular term for *Digital Switched Access* (DSA), even though 64-kbps service is available in some areas. Switched 56 is a circuit-switched digital service intended generally for the same applications as is DDS, although it is more cost effective for less intensive communications. Although the service is switched, rather than dedicated, most of the general characteristics and all of the components closely resemble those of DDS, with the exception that digital COs are involved in setting up the DSA connections. Data Terminal Equipment (DTE) is in the form of computer systems, which connect to digital local loops through DCE in the form of a DSU/CSU. Digital exchanges serve to switch the connections (see Figure 7.2), which are provided through digital carrier transmission facilities on the basis of special routing logic.

The key difference between DDS and Switched 56 is that the calls are switched between physical locations on the basis of a logical address, which is the computer

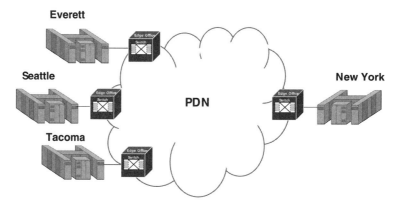

Figure 7.2   Switched 56-kbps service.

equivalent of a voice telephone number. In fact, Switched 56 is the digital data equivalent of a circuit-switched voice call through the PSTN. Based on specific routing instructions contained in programmed logic, the *Public Data Network* (PDN) switches establish the end-to-end connection over entirely digital circuits. (*Note:* In this case, the PDN really is just a physical and logical partition of the PSTN.) The call is set up, maintained, and torn down much like a voice call. Further, the call is priced similarly. In other words and depending on the pricing strategy of the carrier, the cost of the call either is priced on a blended, flat-rate basis or is priced sensitive to distance, duration, time of day, and day of year. As the carriers' Switched 56 service networks typically are not interconnected, calling generally is limited to each specific carrier domain unless the user has made arrangements otherwise.

While DDS is more cost effective for applications in which communications are intensive between specific physical locations, Switched 56 service is more cost effective for communications between locations that communicate less frequently or communicate lesser volumes of data. As Switched 56 calls are switched through the highly redundant carrier networks, rather than relying on vulnerable dedicated circuits as with DDS, Switched 56 services traditionally were employed as a backup to DDS facilities. They are still used for this purpose in some situations, although alternatives such as ISDN are more common in a contemporary context.

7.3 VIRTUAL PRIVATE NETWORKS: IN THE CLASSIC SENSE

Virtual Private Networks (VPNs), also known as *Software-Defined Networks* (SDNs) and *Software Defined Data Networks* (SDDNs), grew out of the voice world, as did the majority of network technologies. In a purely data context, VPNs are incumbent IXC offerings (e.g., AT&T, Cable & Wireless, and Verizon) that operate much like a voice VPN or Switched 56, although the level of bandwidth provided can be much greater in support of intensive data communications, videoconferencing, or multimedia conferencing. Depending on the carrier, VPNs support the following bandwidth levels on a circuit-switched basis:

- 56/64 kbps
- $N \times 64$ kbps
- 384 kbps
- 768 kbps
- 1.544 Mbps (T1) or 2.048 Mbps (E1)
- 44.736 Mbps (Ts) or 34.368 Mbps (E3)

Access to a VPN can take a number of forms, including dedicated digital loops, Switched 56, and ISDN. The IXC VPN service provides priority access and data transport between privileged sites on the basis of a private dialing plan. A wide variety of features are supported, including managed security at the network level. A VPN provides performance similar to that of a dedicated leased-line network, with the additional advantages of flexibility and redundancy. In other words, a Virtual Private Network is not a private network but is virtually so, at least in some respects. These classic VPNs are rapidly being replaced by IP VPNs, which I discuss in later chapters.

7.4 DIGITAL CARRIER SYSTEMS AND NETWORKS

As far back as 1882 Mr. Frank Jacob, Technical Adviser of Messrs. Siemens Bros., designed a plan by which the principle of the Wheatstone Bridge is utilized. . . . The wires forming the metallic loop should also balance in conductivity, insulation, and capacity. When these conditions are attained, T may speak to T', and TT to T'T', without the faintest sound of overhearing being apparent.
        William Henry Preece and Juius Maier, *The Telephone,* Whitaker & Co., 1891

Carrier systems are defined as systems that derive multiple logical channels from a single physical communications path, thereby supporting multiple communications. Initially developed for use within public carrier (i.e., LEC and IXC) networks, the systems provided increased traffic capacity between exchanges without requiring additional transmission facilities. As voice traffic grew dramatically in the post–World War II period, new generations of Central Office Exchanges (COEs) were developed to relieve the strain, and digital carrier systems were developed to relieve the strain on the transmission facilities connecting them [4]. Before exploring digital carrier, we now dedicate a little ink to the history of analog carrier, just to put things in context.

Analog carrier systems first appeared in telephone networks many years before digital carrier systems and actually appeared in telegraph networks years before that. Alexander Graham Bell had been experimenting with the *multiple telegraph,* or *harmonic telegraph,* as early as 1872. In 1874, that work led to his interest in *electric speech* and to the relationship with Thomas A. Watson, his able assistant. The concept of using different tones (i.e., frequencies) for sending different signals simultaneously across the same wire was well appreciated as early as the 1870s, and a number of schemes for accomplishing this were suggested by Bell, Edison, and others. At about the same time in the early 1890s, three different inventors developed practical analog carrier systems for telegraphy based on what we now term *Frequency Division Multiplexing* (FDM). Early work on analog carrier multiplexing for telephony networks was done in the laboratories at American Bell Telephone Company, the predecessor to AT&T, as early as 1894. The first commercial carrier system was installed between Baltimore, Maryland, and Pittsburgh, Pennsylvania, in the United States, in 1918. This Type A system provided four carrier channels above the 4-kHz voice band on short-haul open-wire circuits in the frequency range from 5 to 25 kHz and used the same frequency for transmission in both directions, which caused some amount of crosstalk. Type B (1920) systems reduced the number of channels to three above the voice band and used different frequencies for transmission in different directions over short-haul open-wire loops, thereby reducing crosstalk. This Type B system used *equivalent four-wire* transmission; that is, it was physical two-wire and logical four-wire. Type C (1924) carrier systems increased the bandwidth to support more channels and repeaters could be spaced as far as 120–125 miles, making them the first long-haul carrier systems. Type D (1926) systems were short-haul systems developed for inexpensive rural open-wire applications in support of one two-way channel above the voice band. Type E (1928) systems were developed for voice communications over commercial power lines. This early version of *PowerLine Carrier* (PLC) proved unreliable, although a few of them were employed by the power utilities for their internal use. Type F systems never made it into commercial application, but Type G (1936) and Type H (1937) systems were developed for various other open-wire applications [5].

The original *N-carrier* system (1950) was a short-haul carrier system that employed four-wire twisted-pair cable to deliver *groups* of 12 frequency division multiplexed voice-grade channels for connecting CO exchanges. ON1- and ON2-carrier systems later provided 16 and then 24 channels, respectively. Technology developed further to provide *supergroups* of 60 channels and *master groups* of 600 channels [3, 6, 7].

Type J (1938) systems were intended for long-haul open-wire circuits. Type K (1937) systems were the first long-haul systems developed for cable, rather than open wire; nominal repeater spacing was 17 miles.

L-carrier (1941) was quite an improvement, employing coaxial cable and an analog transmission scheme. L5E (1978), the last L-carrier system, used 22 coaxial *tubes*, in pairs, to carry a total of 132,000 simultaneous voice-grade conversations. Although this was an impressive improvement over N-carrier, the inherent problems of analog transmission were still present. Additionally, the coax cables were expensive and bulky, and the analog Radio Frequency (RF) amplifiers were expensive and prone to failure [3, 6]. At that point, it seemed as though Bell Labs engineers were intent on using up the entire English alphabet on analog carrier. When they developed the first digital carrier, they skipped a few letters. (That's a joke.)

The United States Bell System turned on the first commercial digital carrier system in 1962 under the streets of Chicago, Illinois, where electrical noise from high-tension lines and automotive ignitions interfered with analog systems. The system was designated *T1*, with the *T* standing for *Terrestrial* to distinguish the land transmission from satellite transmission. Bell Laboratories also launched Telstar I, the first communications satellite, in 1962 [7].

The impetus for development of digital carrier and *Time Division Multiplexing* (TDM) was due to the following factors:

1. Terminal multiplexing equipment could be made compact at low cost and could take maximum advantage of advances in solid-state digital circuit technology through system software and firmware upgrades.
2. Highly reliable techniques could be employed to provision a rugged transmission circuit that did not require complex design procedures and elaborate adjustments. In other words, the system could truly be standardized and would not require tuning in either the design or implementation phases.
3. A variety of services could share the same circuit and without the requirement to base the design on the most sensitive service.
4. Error performance would be improved considerably and would not be sensitive to circuit length, as regenerative repeaters, rather than amplifiers, would be employed [3].

T1 refers to a specific set of cable pairs and digital repeaters spaced every 6000 feet or so. T-carrier was rapidly and extensively deployed throughout the carrier networks, initially for short-haul interexchange trunking. Also known as *digroup* (*digital group*), T1 was first offered commercially to end users by AT&T in 1977 on the basis of a *special assembly tariff* and was added to the interstate DDS Tariff #267 in December 1981 [8]. In 1983, AT&T tariffed T1 under the name *Accunet 1.5*. T1 provides 24 channels, based on a convenient multiple of 2×12 channels, which formed the basis for the original analog N-carrier [3]. (*Note:* The 24-channel limitation was due to the nature of the ON-carrier vacuum tube technology at the time. Those vacuum tubes were able to support total bandwidth of 96 kHz. Because each voice-grade channel runs at 4 kHz, a vacuum tube capable of 96 kHz could support no more than 24 analog channels. As backward compatibility was an issue and connectivity between new digital T-carrier and legacy analog ON2-carrier was critical, T1 was limited to 24 channels.)

Before proceeding with the discussion of T-carrier, I have to pause for just a few words to reinforce the fact that digital communications goes back to telegraphy and to mention that time division multiplexing dates at least to 1874, although in a crude form. In that year, Jean Maurice Emile Baudot of the French telegraph administration devised an automatic telegraph that involved synchronized rotating distributor arms that switched the use of the telegraph line between four or six sets of equipment at each end of the line. As the distributor arms revolved two or three times a second, they switched the line between the operators, each of whom had access to the line for a fraction of a second, which was just long enough to transmit a five-bit word in Baudot code. On the receiving end of the line was an electromechanical device that printed the messages in Roman type on a paper tape [1]. With that bit of history to put things in context, we can now fast-forward to 1962, and the next major development in digital transmission technology—T-carrier.

7.4.1 T-Carrier Concept

T-carrier is a dedicated, digital, leased-line service offering that employs TDM in order to derive multiple channels from a single four-wire circuit operating in FDX transmission mode. In capsule, T-carrier offers the advantages of digital error performance, increased bandwidth, and improved bandwidth utilization. As is the case with digital services, in general, T-carrier also delivers increased management and control capabilities to the carriers and end users alike. Additionally, T-carrier is medium independent. In other words, it can be provisioned over any of the transmission media (i.e., twisted pair, coax, microwave, satellite, free space optics, or fiber-optic cable), at least at transmission rates of T1 (1.544 Mbps) and below. At the higher rate of T3 (44.736 Mbps), twisted pair is not a suitable transmission medium, except over very short distances, due to issues of signal attenuation.

As is the case with any dedicated service offering, T-carrier cost is sensitive to distance and bandwidth. While T-carrier initially was deployed in support of voice transmission, it supports data, image, and video as well. Further, T-carrier supports any and all such information streams on an unbiased basis. In other words, all bits and bytes are afforded the same level of treatment, which is the uncompromising level of performance demanded by uncompressed voice. As a result, T-carrier offers the advantage of supporting integrated communications across all information types, whether or not they expect that same high level of treatment. As noted in

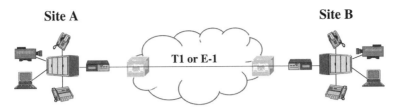

Site A   **Site B**

T1 or E-1

Figure 7.3 Digital carrier as a replacement for multiple dedicated, leased-line networks.

Figure 7.3, T-carrier can obviate the need for multiple voice, facsimile, data, video, and image networks [6, 9].

The significance of T-carrier extends well beyond its practical advantages. Specifically, and as the first digital carrier system, it set the standards for digital transmission and switching, including the use of *Pulse Code Modulation* (PCM) for digitizing analog voice signals. T-carrier not only set the basis for the North American digital hierarchy but also led to the development of similar standards, such as E-carrier in Europe and J-carrier in Japan. Ultimately, the CCITT (now ITU-T) developed international standards recommendations to ensure interconnectivity of national networks. Although T-carrier, E-carrier, and J-carrier are very different in terms of certain specifics of the protocols employed (e.g., transmission rates, encoding techniques, and signaling and control methods), their basic characteristics are much the same.

7.4.2 Channelized T1

The fundamental building block of T-carrier is a 64-kbps channel, referred to as *DS-0* (*Digital Signal level Zero*). Digital carrier is a channelized service, at least in a standard voice implementation. In other words, a single high-capacity digital circuit supports multiple logical channels, with each channel supporting a separate conversation. A T1 circuit, for example, operates at 1.544 Mbps, supporting the standard 24 *time division multiplexed* information-bearing channels, each with a bit rate of 64 kbps (Figure 7.4). E-1 supports 30 TDM channels of 64 kbps plus 2 separate signaling and control channels; J-1 supports 24 channels, as does T1.

The American National Standards Institute (ANSI) set the T-carrier hierarchy standards (see Table 7.1) in its T1.107 specifications. Beginning at the T1 level, the hierarchy progresses up to T4, which provides bandwidth of approximately 274 Mbps in support of 4032 channels. Most end users subscribe to T1 services, because one or more T1s generally satisfy their bandwidth requirements.

The process of transmitting data (voice, data, video, or image) in designated and consistently repeated channels, or *time slots*, is known as *byte interleaving*. For example, each voice conversation to be transmitted is accepted by the multiplexer, assuming capacity is available, and is assigned a time slot. The eight-bit bytes associated with that voice conversation are sent in the designated time slots 8000 times per second, for a total of 64 kbps, which is a voice-grade channel. Those time slots are reserved for that conversation, with the multiplexer providing the transmitting device with regular and repeated access to them for the duration of the communication. Time slots are reserved in both directions, because both real-time voice and T-carrier are FDX in nature.

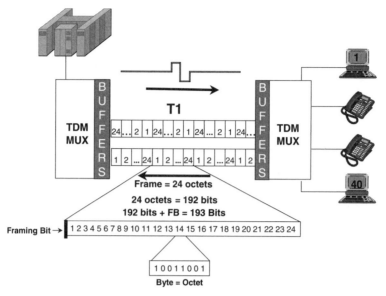

Figure 7.4 Channelized T1, framing convention.

TABLE 7.1 North American Digital Carrier Hierarchy (T-Carrier)

Digital Signal (DS) Level	Data Rate	Number of 64-kbps Channels (DS-0s)	Equivalent Number of Tx's
DS-0	64 kbps	1	Not applicable
DS-1 (T1)	1.544 Mbps	24	1 T1
DS-1C (T1C)	3.152 Mbps	48	2 T1
DS-2 (T2)	6.312 Mbps	96	4 T1, 2 T1C
DS-3 (T3)	44.736 Mbps	672	28 T1, 14 T1C, 7 T2
DS-4 (T4)	274.176 Mbps	4032	168 T1, 84 T1C, 42 T2, 6 T3

7.4.3 Unchannelized T1

Unchannelized T-carrier can support bandwidth-intensive services that do not lend themselves to 64-kbps channelization and standard framing conventions [7]. In other words, the traditional convention of 64-kbps channels can be abandoned in favor of carving the T1 pipe into any combination of segments of bandwidth of any usable size or increment. Additionally, any combination of bits can be transmitted, including an infinite number of zeros, without concern for the violation of the *ones density* rules (discussed later in this chapter)—in other words, a *clear channel* of 64 kbps or more, rather than a 56-kbps channel [8, 11]. A very high speed data communication or a full-motion videoconference, for example, might require a full T1 pipe. A less intensive communication might require 512 kbps, that is, eight channels, or one-third of a T1 facility. Such services are supported through customer equipment in the form of highly intelligent time division multiplexers, routers, or data switches (Figure 7.5). In a private, dedicated, leased-line network, this is easily

Host

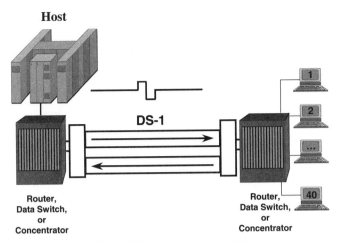

DS-1

Router,
Data Switch,
or
Concentrator

Router,
Data Switch,
or
Concentrator

Figure 7.5 Unchannelized T1.

accomplished. In a switched network application, however, the carrier must be aware that such use will be made of the facility, so the entire facility can be allocated and managed properly.

Unchannelized T-carrier commonly is used for access to a packet-based network such as Frame Relay or ATM. Chapter 10 discusses how these services switch data in packet format, specifically in the form of frames or cells. The switches employed in such networks multiplex frames or cells over an unchannelized T-carrier circuit. Carving the packets or cells into eight-bit samples for transmission over 64-kbps channels only to re-form them on the other end of the link serves no purpose. In fact, such a process only introduces additional levels of complexity, opportunities for timing errors, additional latency, and so on.

7.4.4 Encoding

While T1 is a digital service, it also supports the transmission of data such as voice and video that is analog in its native form. Codecs must be used to convert such analog signals to digital format prior to their being transmitted over a T1 circuit. The standard digitizing technique for voice, known as *Pulse Code Modulation* (PCM), was developed as an integral part of T-carrier. It also became the standard technique for digitizing voice in Private Branch eXchanges (PBXs) and other devices, for the obvious reason of providing seamless transmission between such devices and the network. The quantizing techniques typically employed in contemporary T1 networks include PCM and Adaptive Differential PCM (ADPCM); there also are a number of nonstandard approaches, including CVSD, VQL, VQC, and HCV.

7.4.4.1 Pulse Code Modulation Pulse Code Modulation (PCM), formally known as ITU-T G.711, is based on the *Nyquist theorem* developed by Harry Nyquist in 1928. Nyquist established the fact that the maximum signaling rate achievable over a circuit is twice the number of signal elements, or hertz [12]. In consideration of the Nyquist theorem, PCM specifies that the analog voice signal be sampled at twice the highest frequency on the line. As a voice-grade analog line

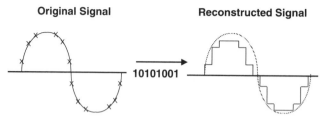

Figure 7.6 PCM encoding of analog voice signal, with reconstruction of approximate analog voice.

is defined as providing bandwidth of 4000 Hz, the Nyquist theorem requires the signal to be sampled 8000 times per second. Each sample is a measurement of the amplitude of the sine wave. (*Note:* Technically, the sampling process must detect every change in direction, i.e., up or down, of the analog waveform.) The individual samples comprising the compressed voice stream are *encoded* (*quantized,* or quantified) into eight-bit binary (digital) approximate values, based on a table of 256 (2^8) standard values of amplitude (Figure 7.6) according to the nonlinear PCM scale. The individual samples then are transmitted in designated time slots over the T-carrier circuit at the very precise pace of 125 μs (1/8000th of a second). The process is reversed on the receiving end of the connection as the encoded amplitude samples are expanded (i.e., decoded, or decompressed) to reconstitute an approximation of the original analog voice signal. The twin processes of *com*pressing and ex*panding* are jointly known as *companding.* The sampling rate and the eight-bit coding scheme yield very high quality voice.

It should be noted that sampling that is too infrequent results in a phenomenon known as *aliasing,* in which the digitized points can be used to represent more than one analog signal. As a result, the reconstructed analog voice signal is less than smooth and accurate, or even totally unintelligible. Similarly, sample encoding that yields bit values that are too approximate yields low-fidelity voice. Even PCM yields some amount of *quantizing noise,* which results from the inexact representation of a smoothly varying analog signal by a digital value restricted to 256 discrete steps. PCM also intentionally introduces some amount of quantizing noise, as it places the greatest emphasis on the amplitude levels in the low and normal volume range, where most speech activity takes place. PCM is nonlinear, much like a sliding-scale or graduated-scale voice ruler with the gradations spaced much closer together near zero volume and farther and farther apart as volume levels increase. This nonlinear approach sacrifices some voice quality at the higher amplitude levels, but the effect is masked by other distortions created by the telephone microphone (transmitter), speaker (receiver), and circuit when the volume is at such high levels. So, PCM can be considered to be a form of compression [3, 11]. (*Note:* The frequency of signal change, or tonal quality of the voice signal, is automatically taken into account.)

$$4000 \text{ cycles/s} \times 2 = 8000 \text{ samples/s} \times 8 \text{ bits/sample}$$
$$= 64,000 \text{ bps}$$

The above calculation shows that 8000 eight-bit samples per second yield a bandwidth requirement of 64 kbps for a PCM-encoded digital voice signal. As PCM was the first standard technique widely used in digital carrier systems, the channel width of 64 kbps became the worldwide standard for all forms of digital networking.

There are two different PCM companding techniques specified in G.711. In North America and areas under North American influence, *μ-law*, often printed as *mu-law*, is used. (ASCII does not conform to an expanded character set that includes Classic Greek letters, so you will please understand if the *μ* does not typeset correctly.) A slightly different nonlinear encoding technique known as *A-law* is used elsewhere in the world. The two techniques are similar, but different enough to require a gateway to convert from one to the other.

This standard approach of channelized T1, as noted previously, was developed and optimized for voice communications using PCM and TDM. Notably, T-carrier was developed for use in the carrier networks, and subsequently the carriers made it available to end-users. As a carrier-provided service, T1 must conform to all of the expectations of the carrier network, unless special arrangements are made. This is true of leased-line T-carrier networks, as depicted in Figure 7.3. It is particularly true of T-carrier when applied as an access service for purposes of access to, rather than through, the carrier network. For example, an end-user organization that employs a T1 local loop from a PBX to a circuit-switched PSTN must conform to the requirements of the carrier, which typically specifies 24 channels of 64 kbps and PCM-encoded voice. But large end-user organizations with large carrier-provided VPNs may take advantage of more efficient encoding techniques such as ADPCM, assuming that the carrier supports them. The following discussion explains how ADPCM supports toll-quality voice at 32 kbps, thereby yielding more efficient use of available bandwidth.

7.4.4.2 *Differential Pulse Code Modulation*   *Differential Pulse Code Modulation* (DPCM) makes more efficient use of bandwidth than PCM, as only the changes in signal level are encoded and transmitted. Based on the logical assumption that the change, or differential, in the voice signal occurs relatively gradually, fewer bits can be used to represent each sample. While DPCM will work with various numbers of bits, a four-bit approach generally is used in this technique, which yields a 2:1 *compression ratio*. This level of compression enables a T1 circuit to support 48 channels of 32 kbps, rather than the PCM standard of 24 channels of 64 kbps. DPCM generally provides voice quality comparable to that of PCM. However, noise (distortion) may result on occasions when the signal varies significantly from one sample to another. A common example is that of a modem transmission, as the amplitude (i.e., volume) and frequency (i.e., pitch, or tone) levels can vary abruptly. This issue can be resolved through several workarounds. One approach is to run PCM on some channels of an intelligent TDM PBX or multiplexer and DPCM on others. If the PBX or multiplexer is intelligent enough, it will sense the modem transmissions and route those calls through the PCM channels. Another approach is to incorporate modems into the voice interface modules on the multiplexer. The modems will recognize the modem signals and demodulate the analog signal into a digital signal in PCM format transparently [11]. DPCM is unusual, as this set of issues and workarounds is generally troublesome.

The following calculation shows that 8000 four-bit samples per second yields a bandwidth requirement of 32 kbps for a DPCM-encoded digital voice signal:

$$4000 \text{ cycles/s} \times 2 = 8000 \text{ samples/s} \times 4 \text{ bits/sample}$$
$$= 32,000 \text{ bps}$$

7.4.4.3 Adaptive Differential Pulse Code Modulation *Adaptive Differential Pulse Code Modulation* (ADPCM), formally defined in ITU-T G.722 and G.726, can improve the quality of DCPM further, without increasing the number of bits required. Through increasing the range of signal changes that can be represented by a four-bit value, DPCM adapts to provide higher quality for voice transmission. Because ADPCM does not interface with a COE based on PCM, it is necessary that special equipment in the form of a *Bit Compression Multiplexer* (BCM) be used to insert two compressed voice conversations into a single PCM channel. A BCM generally is in the form of a printed circuit board that fits into the T1 multiplexer [11].

Notably, ADPCM overcomes the deficiency of DPCM with respect to support of modem transmissions over T1. As you undoubtedly notice when dialing into the Internet, modem tones are very different from voice tones in that both the amplitude and frequency shifts are extreme. As noted previously, DPCM cannot accommodate these shifts. ADPCM, however, can distinguish the presence of a modem tone and can adapt by reverting to a channel width of 64 kbps or by forcing the modem to adapt to a lower speed. Although 32 kbps is the most commonly used ADPCM bit rate, ITU-T standards specify bit rates of 64, 56, 48, 40, 32, 24, and 16 kbps.

7.4.4.4 Digital Speech Interpolation *Digital Speech Interpolation* (DSI) is rooted in a voice compression algorithm known as *Time-Assigned Speech Interpolation* (TASI) developed by Bell Telephone Laboratories in the 1950s for transatlantic telephone cable systems [13]. DSI makes the legitimate assumption that there are predictable pauses in normal human speech. During those pauses, additional voice signals are inserted through a technique known as *silence suppression*. As DSI works on the basis of statistical probabilities, it is employed effectively only when there are a significant number of voice conversations supported. For example, 72 channels yield additional compression of 1.5:1, and 96 channels yield an additional 2:1 (32 kbps). Newer implementations can provide as much as 4:1 (16 kbps), although voice quality is compromised at this level.

DSI suffers the disadvantage of degradation of the signal quality during periods of heavy use. If the parties in a conversation speak rapidly, with few pauses, the voice signal can be *clipped*, or truncated, as the system attempts to detect valid speech and struggles to allocate bandwidth. The more conversations supported, however, the more predictable the average pauses and the lower the statistical probability of such degradation. DSI is commonly used in satellite communications, where bandwidth is precious.

7.4.4.5 Quantizing Variations Variations in the quantizing method are sometimes employed, although they are neither generally accepted nor widely deployed. Those variations include the following [11]:

- *Variable Quantizing Level* (VQL): compression ratio 2:1 (32 kbps)
- *Continuously Variable Slope Delta* (CVSD): compression ratio 4:1 (16 kbps), or 8:1 (9.6 kbps)

- *Vector Quantizing Code* (VQC): compression ratio 4:1 (16 kbps)
- *High Capacity Voice* (HCV): compression ratio 8:1 (8 kbps)

7.4.5 Framing

T-carrier employs a very specific set of conventions to transmit information. Framing is one example. Using T1 as an illustration, each channel of input is time division multiplexed into a T1 *frame*, or set of data. In other words, conversation 1 might be allocated time slot 1 (channel 1), conversation 2 might be allocated time slot 2 (channel 2), and so on, through conversation 24 and channel 24. That set of sampled data is inserted into a frame, which is prepended by a framing bit, as illustrated in Figure 7.4, to distinguish it from subsequent frames of data. The process is repeated for frame 2, frame 3, and so on.

The combined processes of voice encoding and framing yield total T1 bandwidth of 1.544 Mbps. Of that total, 1.536 Mbps is available for information transfer, as noted in the following calculation; the remaining 8 kbps is required for framing and other transmission overhead:

$$4000 \text{ cycles/s} \times 2 = 8000 \text{ samples/s} \times 8 \text{ bits/sample}$$
$$= 64,000 \text{ bps} \times 24 \text{ channels}$$
$$= 1,536,000 \text{ bps} + 8000 \text{ bps}$$
$$= 1,544,000 \text{ bps}$$

There exist several generations of framing conventions, which are designated as D1, D2, D3, D4, and Extended SuperFrame (ESF). Additionally, the ITU-T has developed an international set of recommendations for framing digital carrier signals:

- *D1 framing*, developed in 1962, *robbed* the *Least Significant Bit* (LSB)—the eighth bit—in each channel of each frame in order to insert a signaling bit in the form of alternating 1s and 0s. Although T1 and PCM are designed around the interleaving of eight-bit bytes, *bit robbing* does not affect the quality of digitized voice because seven bits are satisfactory for reconstructing a high-quality approximation of the analog voice input. By truncating an eight-bit value, however, data are seriously impacted; the integrity of the data stream is violated as well. (Imagine the impact of a bit change that converts a decimal point to a comma during the transmission of a financial transaction.) In order to avoid this impact, the data always avoid the eighth bit in every channel, thereby limiting data transmission to 56 kbps. While D1 framing no longer is used, the LSB still is robbed, even in the contemporary D4 framing technique. Therefore, data transmission remains constrained to 56 kbps in many carrier networks.
- *D2 framing*, was used to create a superframe, a 12-bit pattern in the *F* (Framing) bit position, that is, for framing locators. Information was transmitted in a 12-frame sequence or *superframe*. D2 framing is considered obsolete.
- *D3 framing*, which is still in use, assumes that all inputs—whether voice or data—are analog. It uses a *superframe* format and sequence bits.

- *D4 framing*, also known as *M24 Superframe*, uses a 12-bit sequence (1000 1101 1100) of the F bit, repeated every 12 frames, to define the frame locations and enable the receiver to find the channels. D4 enables robbing of the LSBs of the sixth and twelfth frames only. Voice and data are accommodated; data are treated as digital input. This approach improves available signal capacity and yields better voice transmission. Data transmission, however, remains limited to 56 kbps, as even the slightest level of bit robbing negatively affects the integrity of the data stream. Additionally, *ones density* (i.e., the density of 1 bits required to the receiver to recover timing) must be maintained through the insertion of *stuff bits*. Considered together, the 12 frames are designated a superframe.

- *Extended superframe* was originally tariffed by AT&T in 1985 and now heavily used. Extended superframes are 24 frames in length; signaling is performed in frames 6, 12, 18, and 24. It offers the advantages of nondisruptive error detection [six-bit Cyclic Redundancy Check (CRC)] and an Embedded Operations Channel (EOC) for network management using only 8 kbps of overhead [3, 6, 9]. This is accomplished because the highly intelligent ESF channel banks require only 2 kbps for purposes of synchronization. Thereby, the remaining 6 kbps of the framing bits are liberated for other purposed. Specifically, 2 kbps is used for error detection (CRC) and 4 kbps is used for end-to-end diagnostics, network management, and maintenance functions.

- *ITU-T international framing conventions* differ greatly from those described above, which are used in North America and Japan (modified). ITU conventions call for level 1 (E-1) to employ 32 DS-0 channels, 30 for information and 2 specifically designated for signaling and control. The first such DS-0 channel carries the functional equivalent of framing bits, while the sixteenth carries signaling bits [11], as illustrated in Figure 7.7.

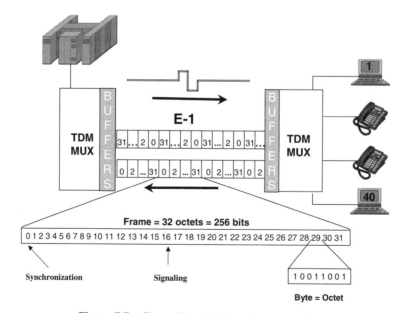

Figure 7.7   Channelized E1, framing convention.

7.4.6   Transmission

Digital carrier transmission facilities can include unshielded twisted pair (22 or 24 gauge), shielded copper, coaxial cable, microwave, satellite, infrared, or fiber-optic cable. Therefore, digital carrier is said to be medium independent. Regenerative repeaters reshape and boost the signal at regular intervals. In a twisted-pair T1 circuit, also known as a *T-span* or *T1 pipe*, the repeaters are spaced at intervals of approximately 6000 ft, which corresponds with the spacing of loading coils in analog circuits. As digital technology replaced analog, the loading coils were removed from their protected housings and replaced with repeaters, which were designed specifically to operate at those intervals. While T1 and E-1 were originally developed as short-haul carrier systems, they can operate over twisted pair for distances up to 200 miles (325 km) or so, with proper spacing of up to 50 repeaters.

The repeaters are *line powered*, that is, powered from the CO exchange over the same transmission line that they serve at levels up to 100 V. The repeaters maintain their synchronization through the transmission bit stream; therefore, bipolar transmission is critical, as is the *ones density* rule.

7.4.6.1   *Alternate Mark Inversion*   Bipolar transmission refers to the fact that the electrical signal has two active states, positive (+) and negative (−) voltage, plus an inactive state of zero voltage. More specifically, *Alternate Mark Inversion* (AMI) reverses the polarity of alternate *marks*, or 1 bits, expressing the first as a positive voltage of +3 V, the second as a negative voltage of −3 V, the third as +3 V, and so on. Between any two successive marks, the signal returns to and pauses at zero voltage in order to maintain a zero reference point for the receiving multiplexer, as illustrated in Figure 7.8 [11].

Since 0 bits are represented by a zero voltage, they provide neither clocking pulses nor power for the repeaters. Neither do 0 bits provide any clocking for the receiving multiplexer. A string of too many successive 0 bits will cause synchronization issues, which can translate into timing slips, or jitter, as the receiver loses the ability to measure the bit time and determine where one bit ends and the next begins. This, in turn, can jeopardize the integrity of the bit stream. As a result, there is a requirement for a certain ones density. Over time, there have been several

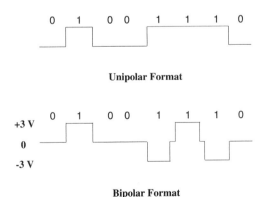

Figure 7.8   Bipolar signal format.

specifications for ones density. AT&T originally specified a limit of 15 consecutive zeros and at least one 1 bit in every eight bits (i.e., byte or, more correctly, octet), for an average density of at least 12.5 percent. As newer equipment was installed in the network, the FCC later relaxed the rule to 80 consecutive zeros but retained the requirement for a ones density of 12.5 percent.

7.4.6.2 *Bipolar with Eight-Zeros Substitution*   Although T1 and E1 were developed specifically for voice communications, they are used intensively in support of data, video, and all variety of applications. Some of those applications, and video in particular, can involve long strings of zeros as legitimate data. In order to accommodate this requirement, extended superframe (ESF) for T-carrier and the ITU-T framing conventions for E-carrier both provide clear channel communications without bit robbing. They also support long strings of zeros through a technique known as *Bipolar with Eight-Zeros Substitution* (B8ZS). When B8ZS encounters a string of eight zeros (00000000), it substitutes a specific bit pattern that intentionally violates AMI, that is, includes an intentional *BiPolar Violation* (BPV). If the preceding mark (1 bit) was represented as a +3 V, the substituted bit pattern is 000+−0−+, as illustrated in Figure 7.9. If the preceding mark was represented as a −3 V, the substituted bit pattern is 000−+0+−. Since the bit pattern is known to both the transmitting and receiving multiplexer, the receiving multiplexer can restore the original 00000000 bit pattern.

7.4.6.3 *Transmission Media*   Note that twisted pair typically is not used at transmission rates above T1 or E-1. At those higher speeds, the native carrier frequency is so high as to make twisted pair unusable due to issues of signal attenuation, at least in a Wide Area Network (WAN) environment. Note that twisted pair performs well at very high frequencies in the Local Area Network (LAN) domain. As the UTP (Category 3, 4, or 5) cable is specifically designed for such frequencies, the cable runs are short, and the environment can be controlled to minimize issues of ambient interference. For reasons that are discussed in Chapter 2, optical fiber is the preferred medium, although infrared and wireless systems offer significant benefits where cabled systems are not practical or where portability is desirable.

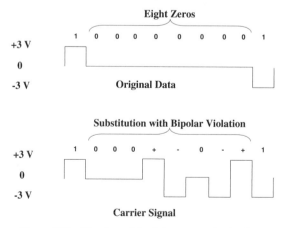

Figure 7.9   Bipolar with eight-zero substitution.

7.4.7 Hardware

DS-1 equipment is required both for end-user organizations and for carriers; that equipment must be of the same generation in order to effect compatibility. Ideally, the user organization should use equipment of the same origin and software generic as the carrier in order to ensure access to all of the functions and features. Hardware includes channel banks, channel service units and digital service units, multiplexers, and digital cross-connect systems.

7.4.7.1 Channel Banks Channel banks were among the first DS-1 devices. Designed for voice-only service in analog applications, channel banks interface analog switches (PBXs and COs) to DS-1 circuits. Channel banks perform two functions in sequence. First, they multiplex up to 24 analog signals on a common *Pulse Amplitude Modulation* (PAM) electrical bus. Second, they encode the individual PAM channels into a digital format, using PCM, for transmission over a DS-1 circuit [6, 9].

Channel banks also accommodate digital data. As relatively unintelligent devices, channel banks place each conversation on a separate channel; for example, a 9.6-kbps data conversation occupies a 64-kbps channel, just as does a 56-kbps data transmission or a digitized voice conversation. Therefore, channel banks do not make efficient use of available bandwidth. Combined channel banks and CSUs often are in the form of printed circuit boards that fit into PBX card slots for seamless interface to a network T1 circuit.

7.4.7.2 Channel Service Units and Digital Service Units Developed circa 1974, Channel Service Units (CSUs) and Digital Service Units (DSUs) are discussed at length in Chapter 6. As a brief recap, they are devices that, in combination, interface the user environment to the digital network at the physical level, corresponding to layer 1 of the OSI model. In contemporary systems, they generally combine into a single device, known as a *Channel DSU* (CDSU) or an *Integrated Service Unit* (ISU), which may reside under the skin of another device, such as a multiplexer. They are used in a wide variety of digital data networks, including DDS and T-carrier.

7.4.7.3 Multiplexers (Muxes) Multiplexers (muxes) are a significant step up from channel banks in terms of intelligence, capability, and cost. Originally based on channel banks and containing CSUs and DSUs, contemporary time division multiplexers offer a tremendous range of flexibility and capability. Muxes typically offer capabilities that include support for both channelized and nonchannelized service, support for multiple medium interfaces (e.g., twisted pair, coax, and fiber), support for multiple trunk types [e.g., Direct Inward Dial (DID) and combination trunks], support for superrate transmission (i.e., channels of a width greater than 64 kbps), and support for subrate transmission (i.e., channels of a width less than 64 kbps). Additionally, they offer the advantages of user-definable configuration, internal diagnostics capability, voice compression, and T-carrier-to-E-carrier protocol conversion. Intelligent muxes also have the ability to allocate bandwidth on a priority basis for specified users and applications, and even to reserve bandwidth, perhaps for a scheduled videoconference or large file transfer associated with a regular data backup. Intelligent muxes can allocate bandwidth on a dynamic basis, assigning channel capacity as required to meet the demands of traffic. A videoconference, for example, may require superrate capacity for a short period of time; multiple, low-speed data communications may require subrate channels for a brief moment; and, at other times, the entire capacity of the circuit may be in support of 32-kbps voice conversations. Finally, contemporary multiplexers commonly are capable of being remotely configured and managed.

Recently, a new breed of multiplexer has emerged in the form of *Integrated Access Devices* (IADs). These devices, which can be quite small and very low in cost, support multiple interfaces, perhaps to a small PBX for voice and a router for Frame Relay. Substantial economies can be realized through an IAD, which enables a single T1 or E1 circuit to be shared by multiple data types in support of multiple applications and services.

7.4.7.4 Nodal Multiplexers Nodal muxes, a step further up the mux food chain, are truly intelligent network nodes acting as T/E-carrier network switches. In addition to serving as traditional muxes for the resident site, they also serve as true networking devices, much like a combined CO/tandem switch in the voice carrier world. Nodal muxes provide the additional function of dynamic alternate routing, which enables them to switch traffic over an alternate path in the event of a condition of blockage or failure in the primary circuit. Figure 7.10 illustrates a fully meshed private leased-line T1 or E-1 network. Recall from Chapter 5 that a full-mesh configuration requires $N(N - 1)/2$ circuits, where N is the number of nodes. So, a four-node configuration requires $4(4 - 1)/2 = 6$ circuits. As the number of nodes increases to 5, the number of circuits increases to 10. As the number of nodes increases to 6, the number of circuits increases to 15. The mathematical relationship between nodes and circuits in a fully mesh configuration is such that a fully meshed private leased-line network rarely becomes too complex and expensive beyond three or perhaps four nodes, despite the advantages of redundancy and resiliency

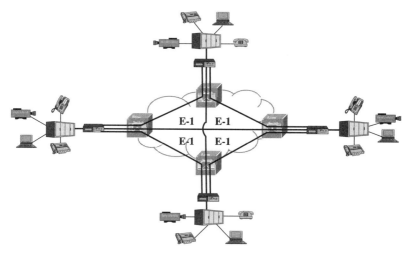

Figure 7.10 Private E-1 network with nodal multiplexers to provide dynamic alternate routing in a fully meshed network configuration.

that a full mesh offers. A partial mesh can be almost as effective, with far fewer circuits.

7.4.7.5 Digital Access Cross-Connect System *Digital Access Cross-Connect Systems* (DACSs, or DCCSs) are nonblocking, ECC switches that serve to cross-connect digital carrier bit streams on a buffered basis by redirecting individual channels or frames from one circuit to another. They provide an electronic common control means of cross-connection that replaces the traditional manual method of physical cross-connection of wires. A DACS can redirect traffic to better manage the capacity and performance of the T-carrier network [13]. A DACS does not switch traffic call by call in the sense that you normally think of a switch. Rather, it switches data circuit to circuit or frame by frame on a preprogrammed basis. Although originally developed for carrier use, DACSs also are deployed in large user organizations to support private digital carrier networks. Smaller versions, residing on a PC, are available for less communications-intensive environments. Typically of significant port capacity, DACSs provide support for DS-0, DS-1, and DS-3 [6] and can accomplish conversions between T-carrier and E-carrier.

7.4.8 Variations on the Theme

While the United States set the theme for digital carrier, the concept was quickly adopted by the CEPT (Committee on European Post and Telegraph) and the CCITT and is now specified in ITU-T G.703 and G. 704. The resulting E-carrier standard differs greatly in its implementation, and there exist some further differences in various national implementations. The Japanese version, J-carrier, resembles T-carrier, but with differences sufficient to cause incompatibility. The DS-0 channel is universal, and some number of DS-0 channels are organized into frames that are repeated 8000 times a second at a precise rate of $125\,\mu$s.

7.4.8.1 E-Carrier *E-carrier* involves a different voice-encoding technique using A-law rather than μ-law. E-carrier is characterized by an entirely different digital hierarchy (refer to Table 7.2), beginning with E-1 at 2.048 Mbps. E-1 supports 30 clear information channels, with 2 channels set aside for framing and nonintrusive signaling and control. Rather than using a framing bit for frame synchronization, E1 uses time slot 0. Specifically, time slot 0 begins with an *International* (I) bit in bit position 1. E1 frames involve a fixed seven-bit pattern (0011011) in bit positions 2–8 of time slot 0 for even-numbered frames and, in odd-numbered frames, a single 1 bit in position 2. In those alternate frames, bit position 3 is used for frame

TABLE 7.2 International (ITU-T) Digital Carrier Hierarchy (E-Carrier)

Level	Data Rate (Mbps)	Number of 64-kbps Channels (DS-0s)	Number of E-1s
1	2.048	30	1
2	8.448	120	4
3	34.368	480	16
4	139.264	1920	64
5	565.148	7680	256

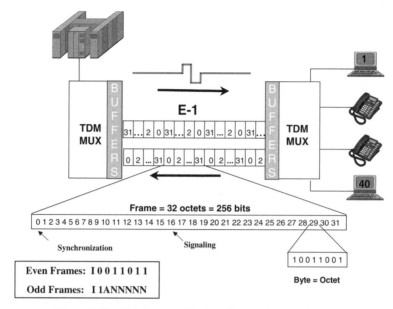

Figure 7.11 E-1 frame with signaling and control detail.

synchronization *Alarms* (A), with a 1 bit indicating a yellow alarm and a 0 bit indicating normal alarm status. Bit positions 4–8 are set aside for *National* (N) use, to be determined by the national carrier, as illustrated in Figure 7.11. All signaling takes place in time slot 16. This signaling and control convention results in E-carrier providing clear channel communications of a full 64 kbps per channel, with no concerns for bit robbing, which restrains T1 channels to 56 kbps where older equipment is in place in the network. E1 line coding employs *High-Density Bipolar 3-zeros* (HDB3). Similar to AMI in T1 networking, HDB3, however, imposes a limit of three successive 0 bits. A fourth 0 bit triggers zeros suppression, substituting a known bit pattern with an intentional *bipolar violation*. The E-carrier *multiframe* of 16 frames roughly corresponds to the T-carrier *superframe* of 12–24 frames [11, 14]. In multiframe format, the I bit in time slot 0 of even-numbered frames can be used for error correction employing the CRC-4 mechanism and the N bits can be relabeled as *Spare* (S) bits.

7.4.8.2 J-Carrier *J-carrier* closely resembles T-carrier, although the hierarchy is slightly different. Line coding and framing also vary considerably from the ANSI approach developed in the United States. Although diversity can be interesting, the advantages of these differences are questionable as, not surprisingly, incompatibility is ensured. The J-carrier digital hierarchy begins at 1.544 Mbps and proceeds to 6.313 Mbps, 32.064 Mbps (J1), 97.728 Mbps (J3), and 397.20 Mbps (J4) [6, 8]. Table 7.3 provides a clear comparison of the various DS levels.

7.4.9 T2 and Above: The Rest of the Hierarchy

T1 certainly is not the end of the story. Generally, the next step is T3, at a nominal rate of 45 Mbps. The T1C (T1 Concatenated), T2, and T4 levels are very unusual

TABLE 7.3 Digital Hierarchy: T-Carrier, E-Carrier, and J-Carrier Based on ITU-T G.702

DS Level	Number of Data Channels	Total Signaling Rate (Mbps)		
		T-Carrier (North America)	E-Carrier (International)	J-Carrier (Japan)
DS-0	1	0.064	0.064	
DS-1	24	1.544	—	1.544
	30	—	2.048	
DS-1C	48	3.152	—	3.152
DS-2	96	6.312	—	6.312
	120	—	8.448	
DS-3	480	—	34.368	32.064
	672	44.736		
DS-3C	1,344	91.053		
DS-4	1,440	—	—	97.728
	1,920	—	139.264	
	4,032	274.176		
DS-5	5,760	—	—	397.200
	7,680	—	565.148	
DS-6	30,720	—	2200.00	

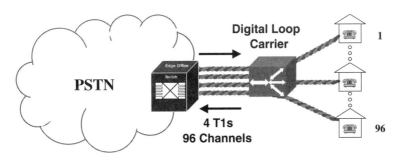

Figure 7.12   T2 circuit connecting a DLC to a CO.

in end-user implementations, and the carrier networks rarely employ them any longer. There does remain some T2 in place in the local loop, where Bell companies used it in *Digital Loop Carrier* (DLC) applications. The *Subscriber Line Carrier 96* (SLC-96) system introduced by Western Electric (now Lucent) in 1979, for example, comprises four T1s multiplexed to support 96 channels, commonly provisioned over a single T2 facility from the CO. The SLC-96 essentially is a remote line shelf and TDM mux that allows a single four-wire twisted-pair circuit (or a fiber circuit) to serve as many 96 channels and, therefore, 96 single-line residences or small businesses. As the channels from the premises to the CO are dedicated, the DLC is not particularly intelligent and is not positioned as a contention device of any sort. As illustrated in Figure 7.12, the SLC contains a channel bank that accomplishes the Analog-to-Digital (A-to-D) or D-to-A conversion processes, multiplexes the signals, and so on. A later Lucent system (1985) supports up to 192 channels over four T2 copper circuits or a duplex optical fiber configuration. As the Incumbent Local Exchange Carriers (ILECs) replace UTP local loops with fiber optics, the embedded DLCs will be decommissioned and replaced either with newer versions that have both copper and fiber interfaces or with fiber optics to the premises.

While the data rate increases at the T1C and T2 levels, the carrier reference frequency increases, and issues of signal attenuation and crosstalk over copper twisted pair increase. This fact creates special engineering problems, which can be resolved by various means, including spacing repeaters ever more closely together. At the higher levels of T3 and T4 and the higher associated signaling speeds, these issues cannot be resolved satisfactorily. Therefore, these higher levels run over fiber optics, Free Space Optics (FSO), or microwave.

Above T1, additional multiplexing is required. In the case of T2, the mux is termed an *M12* (*M*ultiplex T*1* to T*2*). As mentioned in the above discussion of DLC and SLC-96, a T2 runs at 6.312 Mbps and comprises four T1s at 1.544 Mbps apiece, plus 132 kbps of overhead and *justification*, or *bit stuffing*, to adjust for variations in the clocking rates of the incoming T1s. The following set of simple equations builds the logic for the signaling speed from a single DS-0 channel to a T1 and then a T2:

$$64 \text{ kbps} \times 24 \text{ channels} = 1.536 \text{ Mbps payload}$$

$$1.536 \text{ Mbps} + 8 \text{ kbps framing} = 1.544 \text{ Mbps signaling speed}$$

$$1.544 \text{ Mbps} \times 4 = 6.176 \text{ Mbps}$$

$$6.176 \text{ Mbps} + 136 \text{ kbps overhead} = 6.312 \text{ Mbps}$$

At the T2 level, the overhead comprises a total of 168 kbps, including T1 framing. That comes to approximately 2.7 percent, leaving the effective payload at 97.3 percent assuming that the DS-0s are clear 64-kbps channels packed with raw user data. If, however, bit robbing comes into play, each of those channels is reduced to only 56 kbps and the effective payload drops to 5.376 Mbps (56 kbps × 24 × 4), or 85.17 percent. T2 originally was developed for digital carrier applications of 500 miles or less, primarily between metropolitan areas. Traffic levels long since have grown well beyond the capabilities of T2, and it has been replaced with T3 or higher speed Synchronous Optical NETwork (SONET) facilities, which I discuss in Chapter 9.

At the T3 level, an M13 mux gets involved. Actually T3 begins by multiplexing four T1s into a T2, as discussed above and illustrated in Figure 7.13. Then seven

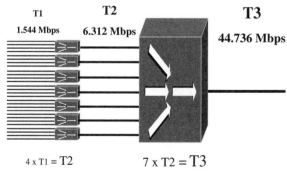

Figure 7.13 M13 Multiplexer.

T2s are multiplexed to yield a rate of 42.924 Mbps. Stuff bits are added to adjust for variations in the clocking rates of the incoming T2s, bringing the signaling rate up to 44.736 Mbps for 672 channels ($24 \times 4 \times 7 = 672$), according to the following formula:

$$1.544\,\text{Mbps} \times 4 = 6.176\,\text{Mbps}$$

$$6.176\,\text{Mbps} + 136\,\text{kbps} = 6.312\,\text{Mbps}$$

$$6.312\,\text{Mbps} \times 7 = 44.184\,\text{Mbps}$$

$$44.184\,\text{Mbps} + 552\,\text{kbps} = 44.736\,\text{Mbps}$$

At this level, overhead includes 555 kbps for T3, 882 kbps for T2 (136 kbps × 7), plus 32 kbps for T1 (8 kbps × 4), for a grand total of 1.469 Mbps, or approximately 3.28 percent. That yields a payload factor of 96.72 percent, assuming clear channel communications. If bit robbing is involved in old North American networks, another 5.376 Mbps drops away (8 kbps × 672 channels), for a grand total of 6.845 Mbps, leaving a theoretical payload of only 37.891 Mbps for an efficiency factor of 84.7 percent. If the payload contains data blocks, frames, packets, or cells with their own overhead issues, it gets even more overhead intensive. Now all of this overhead may seem wasteful, and it is at some level. At another level, however, we must realize that each bit of overhead has a purpose somewhere in some piece of equipment developed at some point in time to be backward compatible with something else that had some legitimate reason for doing what it did some years before, perhaps in a way that we would do entirely differently if we knew then what we know now. So, we deal with it, things interconnect to form networks that interconnect, and life is good. With that bit of wisdom having been painfully extracted from the dark recesses of my mind at this late hour and firmly implanted in this work, let us proceed.

7.4.10 Fractional T1

Fractional T1 (FT1), originally offered in Canada, first was tariffed in the United States in 1987 by Cable & Wireless. Now offered by many LECs and IXCs, FT1 provides T1 functions and features but involves fewer DS-0s. It is offered in fractions of T1 channel capacity, generally at 1, 2, 4, 6, 8, or 12 DS-0 channels. Subrate transmission also is available at speeds of 9.6 kbps. FT1 is particularly applicable where relatively small branch locations connect to a more significant location such as a regional office. There they connect to a full T1 mux or nodal processor, which aggregates the traffic with that of the larger site over a full T1 or T3 backbone network. Figure 7.14 provides a graphic view of an FT1 in an example private T-carrier network configuration. As is the case with T-carrier in general, FT1 serves not only voice applications but also videoconferencing, data communications, and other applications that require more than 56/64 kbps but less than a full T1. It should be noted that FT1 is no less resource intensive than a full T1. The same four-wire circuit is required, as is the same CSU/DSU and other equipment. Essentially FT1 is a T1 with some number of channels deactivated. Therefore, some carriers charge the same for an FT1 local loop as for a full T1, although there may be some savings

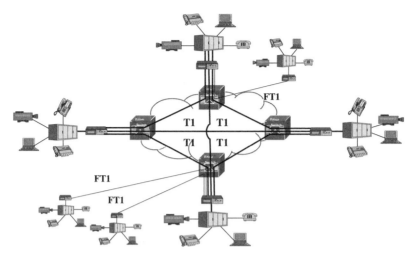

Figure 7.14 Private T-carrier network with nodal multiplexers to provide dynamic alternate routing in a partially meshed network configuration, including FT1 tail circuits.

in terms of port charges or other charges. Fractional E1 service is available in some countries.

7.4.11 Digital Carrier Applications

The applications for digital carrier are many. Large user organizations find digital carrier services to be highly cost effective for local loop access, typically replacing multiple, single-channel PBX trunks. Large corporations find T/E-carrier effective for private, leased-line networks or access to VPNs. The ability of T/E-carrier to accommodate voice, facsimile, data, video, and image information on an unbiased basis and, therefore, to eliminate or reduce the number and variety of specialized circuits offers great advantage.

Internet Service Providers (ISPs) commonly use channelized T/E1 to provide modem-based access to small users requiring channel width of no more than 64 kbps. This channel width is perfect for even the highest speed conventional modems (i.e., V.90 and V.92 at 53.3 kbps upstream), which gain access to the ISP on a dial-up basis through the circuit-switched PSTN. ISPs commonly make use of unchannelized T-carrier (T/E1 or T/E3) for access to an Internet backbone provider. The unchannelized approach is preferred for this application because data typically move between the ISP and the Internet backbone provider using the Frame Relay or ATM protocol, both of which I discuss in Chapter 10.

Although private leased-line T/E-carrier networks largely have given way to newer LAN internetworking technologies such as Frame Relay and IP networking, embedded T/E-carrier networks continue in place and are expanded at times. Incidentally, Frame Relay and T-carrier naturally coexist. A large number of user organizations now run Frame Relay data over the dedicated, leased-line networks they put in place years ago for PCM-based voice traffic. Also, access to ATM, Frame Relay, and IP networks commonly is accomplished over unchannelized T/E-carrier circuits.

7.5 X.25 AND PACKET SWITCHING

Paul Baran and his research associates for the RAND Corporation invented packet switching in the early 1960s. Interestingly enough, the concept first was published in 1964 as a means of transmitting secure voice for military application. In the late 1960s, the U.S. General Accounting Office (GAO) issued a report suggesting that there existed a large number of data centers supported, at least in part, by the federal government. Further, the report indicated that many of those data centers were underutilized and others were severely overloaded. The imbalance was due largely to the lack of a WAN technology that would permit the sharing of those resources on a cost-effective basis.

As a result of that study, the Advanced Research Project Agency NETwork (ARPANET) was developed. ARPANET, the first sophisticated packet-switched network architecture, was born in 1971. ARPANET was intended to link computers on a time-share basis in order to share computer resources more cost effectively [15]. Specifically, ARPANET was designed to support various defense, higher education, and Research and Development (R & D) organizations [16].

Packet switching soon was commercialized and made widely available in North America by companies including Telenet, Tymnet, and Graphnet (a facsimile-like service) in the United States and Datapac in Canada. In Europe, packet switching was offered early on by the pan-European Euronet. In fact, packet switching quickly became available in most countries and currently is virtually ubiquitous. The CCITT (now ITU-T) internationally standardized X.25 as the interface into a packet-switched network in 1976 and subsequently revised the standard in 1980, 1984, 1988, and 1993.

In 1983, the majority of ARPANET users spun off to form the Defense Data Network (DDN), also called MILNET (MILitary NETwork), which included European and Pacific Rim continents. Locations in the United States and Europe that remained with ARPANET then merged with the Defense Advanced Research Project Agency Network to become Defense Advanced Research Projects Agency (DARPA) Internet [17].

The wide availability of packet switching has made it consistently popular over the last 20 years or so. Additionally, packet networks are highly cost effective for applications that require many-to-many connectivity and involve relatively low data volumes. That popularity is growing and is ensured well into the future, largely through its historical deployment as the network technology of the Internet. You should note that X.25 is an interface specification, and does not define the internal operational characteristics of the data network.

7.5.1 Concept of Packet Switching

The basic concept of packet switching is one of a highly flexible, shared network in support of interactive computer communications across a public WAN. Previously, large numbers of users spread across a wide area and, with only occasional communications requirements, had no cost-effective means of sharing access to host computer resources, that is, *time share*, from their remote asynchronous terminals. The issues were several:

- **Bursty Communications:** Asynchronous communications are *bursty* in nature. In other words, data transmission occurs in bursts of keystrokes or data file transfers. Further, lots of idle time occurs on the circuit between transmissions of relatively small amounts of data.
- **Analog Technology:** Additionally, those early networks consisted of analog facilities in the form of twisted-pair local loops, with various combinations of twisted pair, coax, and microwave in the core. Analog technology offered very poor error performance and relatively little bandwidth.
- **Cost:** Calling costs were quite high across the WAN. This was especially true of data calls in support of asynchronous communications, as connect times are long even though data volumes tend to be low.

Existing circuit-switched networks certainly offered the required flexibility, as users could dial up the various host computers on which the desired database resided. Through a low-speed modem, which was quite expensive at the time, data could be passed over the analog network, although error performance was less than desirable. But the cost of the connection was significant because the calls were billed based on the entire duration of the connection, even though the circuit remained idle much of the time. Dedicated circuits could address the imbalance between cost and usage, because costs are not usage sensitive and dedicated circuits can be shared among multiple users through a concentrator. But dedicated circuits were expensive, especially in long-haul applications, and involved long implementation delays. Further, users tended not to be concentrated in locations where they could make effective use of dedicated circuits on a shared basis. Finally, large numbers of dedicated circuits were required to establish connectivity between clusters of users and the various hosts.

Packet switching solved many of those problems in the context of the limitations of the networks existing at the time. Packet-switched networks do a very cost-effective job of supporting low-speed, asynchronous, conversational, and bursty communications between computer systems in full-duplex mode. The low-speed, bursty nature of interactive asynchronous terminal-to-host applications enables large numbers of users to engage in simultaneous data sessions across a highly shared network. Rather than establishing connections across physical circuits to transmit data, the various network devices organize user data into packets and interleave them with packets generated by other users in a manner much like TDM muxes interleave bytes, although on a much less rigid basis. As packet-switched network usage can be billed to the user on the basis of the number of packets transmitted during a session, rather than billing for the duration of a call, packet networks are very cost effective for low-volume, interactive data communications. Further, packet-switched networks can perform the process of error detection and correction at each packet switch, or node, thereby considerably improving the integrity of the data from end to end.

Understanding the concept and nature of packet switching requires the examination of a number of dimensions and characteristics of such networks. The following is an exploration of the X.25 protocol suite, including the packet layer, access procedure, and frame format. There is considerable detail on switching and transmission, error control, connectionless service, latency, permanent virtual circuits versus switched virtual circuits, and protocol conversion.

7.5.2   X.25 Protocol Suite

The X.25 protocol suite maps into the Open Systems Interconnection (OSI) Reference Model at the three lowest layers:

- *Packet Layer Protocol* (PLP) maps into layer 3, the network layer.
- *Link Access Procedure—Balanced* (LAP-B) is the X.25 bit-oriented protocol for encapsulating the PLP packet. LAP-B maps into layer 2, the link layer.
- *X.21bis* defines the mechanical and electrical parameters for the physical, which is at layer 1, the physical layer. Layer 1 options include EIA/TIA-232, EIA/TIA-449, EIA-530, and G.703.

7.5.2.1   *Packet Layer Protocol*

The X.25 network layer protocol is the *Packet Layer Protocol* (PLP), which manages packet exchanges between physical DTE across a network of virtual circuits. PLP also can run on LANs and ISDN interfaces running *Link Access Procedure—Data channel* (LAP-D). PLP is responsible for call setup, synchronization, data transfer, and call clearing (i.e., call teardown). In data transfer mode, PLP transfers data between DTE across both *Permanent Virtual Circuits* (PVCs) and *Switched Virtual Circuits* (SVCs). Data transfer mode is responsible for data segmentation on the transmit side and reassembly on the receive side of the communication. This mode also handles bit padding, flow control, and error control.

An X.25 packet network transports and switches data through the network on the basis of *packets*, each of which is of a finite maximum size and of a specific structure, including a header and a payload. As illustrated in Figure 7.15, PLP packet fields include the following:

- *General Format Identifier* (GFI) is a 4-bit field that identifies packet parameters, which can include payload type (e.g., user data or control data), windowing information, and whether or not delivery confirmation is required.
- *Logical Channel Identifier* (LCI) is a 12-bit field that identifies the logical channel group and channel number of the virtual circuit that connects to the destination DTE.
- *Packet-Type Identifier* (PTI) is an 8- or 16-bit field that identifies the PLP packet type, of which there are 17. Packet types include various call setup and call clearing, data and interrupt, flow control and reset, restart, and diagnostic

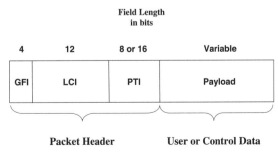

Figure 7.15   X.25 packet structure.

packets. The exact composition of this field varies slightly, depending on whether the network is set for *modulo 8* or *modulo 128*, as identified in the GFI field, with *modulo* referring to *windowing* information that identifies the number of packets that can be sent in a string before an acknowledgment must be returned by the receiving device across a link. If the network is set for modulo 8 and the packet type is either data or flow control, the PTI field includes a three-bit ($2^3 = 8$) *packet-receive sequence number* and a three-bit *packet-send sequence number*. If the network is set for modulo 128 and the packet type is either data or flow control, the PTI field includes a seven-bit ($2^7 = 128$) packet-receive sequence number and a seven-bit packet-send sequence number. If the packet is a data type, there also is included a one-bit field indicating whether the packet is part of a sequence of packets to be treated as a logical whole. If the packet is a call setup, call clearing, or registration type, the PTI field replaces the sequence numbers with originating and destination DTE addressing information.

- *User or Control Data* comprise the payload. If the packet is a data packet, the payload is encapsulated higher layer application information. If the packet is a control packet, the payload comprises various information relating to call setup and clearing, flow control and reset, and so on. The default maximum payload size is 128 octets, which every network must support. Public X.25 networks variously accommodate packets with maximum payloads of 16, 32, 64, 128, 256, 512, and 1024 octets. Airline reservation networks commonly use packet sizes of 1024 octets, although some custom networks can accommodate packet sizes of up to 4096 octets.

7.5.2.2 Link Access Procedure—Balanced

Link Access Procedure—Balanced is a derivative of the High-level Data Link Control (HDLC) protocol, which in turn is based on the IBM Synchronous Data Link Control (SDLC) frame. LAP-B is a bit-oriented protocol running at layer 2, the data link layer, of the OSI Reference Model. (*Note:* Refer to Chapter 6 for a discussion of all of these terms.) The LAP-B frame comprises a header and trailer that encapsulate the PLP packet and provides a mechanism for transporting that packet across a link, ensuring that frames of data are ordered correctly and are free from error. LAP-B is a balanced protocol that operates in *Asynchronous Balanced Mode* (ABM), which refers to the fact that the devices have a *balanced*, rather than a *master/slave*, relationship. Therefore, a device at either end of the link can initiate a dialogue at any time. As illustrated in Figure 7.16, the LAP-B frame format includes the following fields:

- **Flag:** The flag field is a one-octet field that delimits (i.e., establishes the limits or boundaries of) the beginning and end of the frame. The flag is always the specific bit pattern 01111110 (7E in hexadecimal), which is known to all transmitters and receivers.
- **Address:** The address field is a one-octet field that contains no address information whatsoever. Since the frame moves across a point-to-point link between one DTE and one DCE, it is hardly likely that addressing would be an issue. Rather, addressing is the responsibility of the LCI field contained in the PLP packet. So, the address field is used simply to distinguish between commands

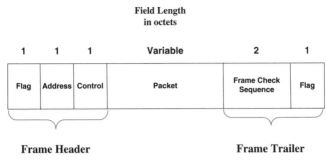

Figure 7.16   X.25 LAP-B frame format.

from DTE to DCE and the associated responses and commands from DCE to DTE and the associated responses.

- **Control:** The control field is a one-octet field that identifies the frame type. An *Information* (I) frame carries upper layer (e.g., application) information and some control data. A *Supervisory* (S) frame carries control information such as I-frame acknowledgment, request for retransmission, and flow control. An *Unnumbered* (U) frame carries control data such as disconnection request, acknowledgment frame, and frame reject.
- **Data:** The data field is a variable-size field that contains upper layer information in the form of an encapsulated PLP packet.
- **Frame Check Sequence (FCS):** The FCS field is a two-octet CRC field that provides excellent error detection. The CRC provides meaningful information for error correction, which is performed at a higher level.

7.5.2.3   X.21bis   *X.21bis* is a Layer 1, or Physical Layer, specification used in X.25. X.21bis defines the mechanical and electrical parameters for cables and connectors in support of FDX transmission at speeds from 9600 bps and 64 kbps for point-to-point connections over four-wire circuits. As a pure Physical Layer specification, X.21bis addresses the movement of electrical bits across a wire that connects DTE and DCE, and that's about all. In addition to moving bits, of course, there are provisions for call control, which entails the devices using signaling leads to signal when they are *ready* or *not ready* to receive calls. There also is a provision for signal timing, or synchronization, at 8 kbps. Figure 7.17 illustrates the relationship between the Layer 3 PLP packet, the Layer 2 LAP-B frame, and the Layer 1 X.21bis bit stream. Note that the physical layer circuits connecting nodes in the network core can take any number of forms, from analog to digital, T-carrier to E-carrier, copper to fiber, and anything in between.

7.5.3   Error Control

X.25 provides for error control through the use of a CRC contained in the FCS field of the LAP-B frame. Should an error occur in transmission from the DTE through the DCE to the originating network node, that node will recognize the error and correct for it by requesting a retransmission of the corrupted packet. Through the core of the network, each node acting on the packet repeats that process. At the

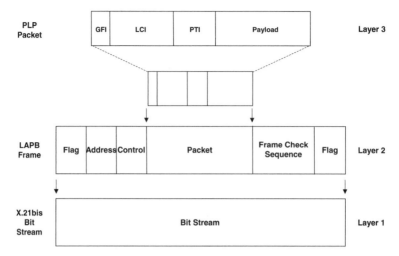

Figure 7.17 X.25 packet encapsulation in LAP-B frame.

destination network node, the sequence numbers of the packets are checked, any missing packets are identified, and requests for retransmissions are made, as required. Once all packets are present and accounted for, they are resequenced as required, and the data stream is presented to the destination host. *Note:* There is no error control at the PLP packet layer, which is Layer 3, the Network Layer. There may be an error control function at a higher layer, but that is an issue under the control of the user host computers and is outside the scope of X.25.

Error control was extremely important in early packet networks because the facilities consisted of analog modems on twisted pair, which clearly is error prone. A cascading error control process, therefore, was developed to ensure the integrity of the individual packets and of the entire packet stream. Notably, the process of retransmission has a negative effect on overall throughput because errored packets that require retransmission consume bandwidth. As error control is accomplished at the link level, X.25 discovers and corrects for errors link by link, thereby limiting this throughput issue. If error control were end to end, retransmissions would be end to end, as would be throughput issues. However, this process is demanding of the computational resources of each of the nodes, thereby adding to their cost. Additionally, the process is time consuming because each packet must be checked for errors prior to being forwarded to the next node. As the time consumed during the error-checking process imposes some level of latency on each packet, the level of latency from end to end increases as the number of nodes involved increases. As is discussed in Chapter 10, the next major evolutionary step in packet services was Frame Relay, which shifted the error control process to the end-user domain in order to improve network efficiency and reduce cost.

7.5.4 Datagram Mode: Connectionless

The most basic mode of X.25 operation is that of *datagram* mode. In this mode, each packet, together with its destination address and usually originating address, can be exchanged between host computers over the packet-switched network, inde-

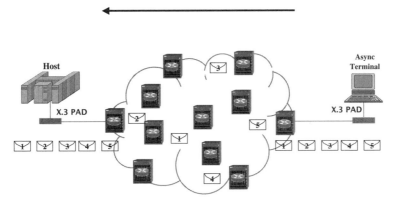

Figure 7.18   X.25 packet-switched network, supporting transmission in datagram mode.

pendently of all other datagrams. The datagram mode deals with each packet individually, without any consideration that each packet is one of a stream of packets associated with a session supporting a file transfer, for example. In other words, each packet works its way through the network, entirely on its own, and either makes it to the receiver or not, which has no affect whatsoever on any other packet trying to do the same thing, whether it is part of the same session or not. The datagram mode is *connectionless*, meaning that there is no predetermined path set up for the packets to go through the network. Rather, each packet can take an entirely different route from originating host to destination host, as illustrated in Figure 7.18. In a *connection-oriented* mode, a path is set up through the network for all packets associated with an originating and destination address pair, either permanently or perhaps just for a given session.

In a typical scenario, the transmitting terminal, equipped with a modem, dials a telephone number to gain access to a local packet node on a circuit-switched basis through the LEC COE. Alternatively, a short-haul, dedicated circuit might connect the user location directly to the packet node. Once the connection to the packet node is established, the transmitting device sends a control packet across the network to establish a data session with the target host computer. That node sends the packet to the target device, which acknowledges its receipt and establishes a session by responding with a control packet to the originating device. Then the originating device begins sending a stream of data, segmented into packets, with each packet numbered sequentially. The originating node receives each packet, checks for transmission errors, reads the address, and forwards the packet toward the destination, across the most direct and available link. The process repeats at each node until the data reach the packet node serving the target host. Each packet routes through the network independently, from node to node, in the direction of the target device, taking the most direct and available path at that instant. Should a reasonable route or the computational resources of the node not be available immediately, the packet will queue in buffer storage at a node for a defined length of time, until a link becomes available. Some packets may encounter little or no congestion and therefore work their way through the network quickly across a relatively direct path involving relatively few nodes. Other packets may encounter considerable congestion and therefore queue for relatively long periods of time and take indirect routes involving a relatively large number of nodes. All of these factors contribute to latency. When the communication session is complete, a control packet is sent across the network to terminate the data call. ARPANET pioneered the concepts of *locally adaptive routing, network message segmentation*, and *datagram transmission mode*. The datagram mode is the lowest common denominator, and all national carriers are expected to support it at a packet size of 128 octets.

X.25 internodal links originally were dedicated analog trunks leased from the various local and long-haul carriers. Over time, those analog facilities were replaced by digital circuits, usually in the form of 56-kbps DDS circuits. Many of those circuits subsequently were replaced with T-carrier and E-carrier facilities. Currently, the facilities generally are high speed and fiber optic in nature, although a variety of analog and digital media are employed in consideration of specific network economics, all of which are sensitive to the region.

7.5.5 Virtual Circuits: Connection Oriented

Datagram mode aside, X.25 packet switching is a *connection-oriented* service. That is to say that a call is set up over a shared physical path, or *Virtual Circuit* (VC), before the first packet is sent and over which all packets may travel in support of a logical connection. In datagram mode, X.25 is connectionless, with each packet perhaps traveling a different path, depending on the availability and performance of the various network links at any given moment in time. In either case, each packet of data is addressed separately and, therefore, is capable of working its way through the network independently of the other packets in a stream of packetized data. This characteristic of packet networks is a critical advantage because the network and all of its elements are shared among a large number of users. Hence, the cost of transmission across such a network is very low in the context of an appropriate application. There are two types of virtual circuits: PVCs and SVCs.

7.5.5.1 Permanent Virtual Circuits Packet switching supports a large number of transmissions riding over the same previously designated circuit or path. While the individual packets of the typical user may travel different paths, those of large user organizations that use the network intensively commonly travel over *permanent virtual circuits* (PVCs). In this scenario, all packets always travel the same path between two host computers, as illustrated in Figure 7.19. The path is established on the basis of routing instructions programmed in the routing tables of the involved nodes and is invoked based on the logical channel group and logical channel number contained in the LCI field of the PLP packet. As the path is predetermined and programmed in a routing table, it can be identified and exercised quite quickly.

The links that comprise the route are defined by the service provider on a *permanent* basis until such time as they are permanently redefined, perhaps when the service provider rebalances the network to improve overall performance in consideration of changing usage patterns. Because the physical circuits are shared by large numbers of users and large numbers of packets and packet streams, rather than being committed to a single data stream, the customer's connections are *virtual* in nature. While latency always is an issue in packet networks, a virtual circuit at least provides some assurance that the level of latency will remain fairly consistent from packet to packet. As all packets travel the same path, they will arrive in sequence.

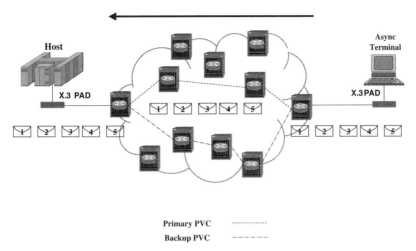

Primary PVC   ------------
Backup PVC   ---·---·---

Figure 7.19 X.25 packet-switched network, supporting transmission over PVCs.

Further, a virtual circuit that creates significant levels of packet errors can be identified more easily, the specific link or switch causing the problem can be isolated, and
the problem can be corrected more readily than if packets whizzed around the
network over multiple changing paths and circuits in datagram mode. Because
PVCs are permanently defined, however, they are subject to catastrophic failure.
Should an individual switch or circuit fail along a given PVC, the network provider
either must correct the problem or redefine the PVC. In the meantime, the user
organization cannot use the network, unless a backup PVC is provided at additional
cost.

7.5.5.2 Switched Virtual Circuits   Alternatively, the network nodes may select
the most available and appropriate path on a call-by-call basis using *switched virtual
circuits* (SVCs) that are set up on the command of (i.e., signaling from) the user
equipment before the first packet is sent. Once selected, all packets in a given session
travel the same path, just as they do in a PVC scenario. SVCs differ from PVCs as
the path is set up in consideration of both the condition and the load at the instant
the connection is required, rather than being defined well in advance. Therefore,
failed and congested switches and circuits are bypassed, and overall performance is
improved. This process of automatic load balancing offers benefits to both the user
organization and the carrier. SVCs, however, demand a greater level of network
intelligence, which adds to total network cost to the carrier and ultimately translates
into higher cost to the end-user organization. The establishment of an SVC also
involves some level of delay in the call setup process, since the network nodes must
examine multiple paths prior to making an optimal path selection.

7.5.6 Protocol Conversion

As an option, packet-switched networks accomplish protocol conversion, including
any protocol that is well established, well understood, widely deployed, and therefore supported by the carrier. As this process of protocol conversion adds value,

packet networks (X.25) are widely recognized as the first *Value-Added Networks* (VANs) [18]. This capability certainly added great value some years ago when protocol conversion was considered to be quite demanding of limited and expensive computational resources. In many cases, the carriers could accomplish the process more cost effectively than the end-user organizations. The limited number of supported protocols included asynchronous, IBM Binary Synchronous Communications (BSC), and IBM SDLC.

In this contemporary world, however, the cost of protocol conversion is relatively minor and the number of protocols requiring support is both considerable and dynamic. Further, protocol conversion adds to overall packet latency, thereby affecting other users of a shared packet network. Therefore, and in a contemporary setting, protocol conversion generally is best accomplished by intelligent devices in the end user, rather than the carrier, domain. As I note in Chapter 10, Frame Relay is a packet service that shifts the protocol conversion process to the end-user domain to improve network speed and reduce cost.

7.5.7   Latency

Latency, or delay, is a troublesome and limiting characteristic of packet networks. As each packet may take a different route though the network in datagram mode, each may travel a route of a different length; therefore, propagation delay may vary from packet to packet. Additionally, each packet may travel through a different number of packet nodes, each of which must act on the packet to read its address, check for errors, request retransmissions of errored packets, and so on. Each of these processes compounds the issue of packet delay. Further, each packet may encounter a different level of congestion in the network and, therefore, can spend a different amount of time in queues. Finally, delay is imposed on each packet if protocol conversion is required; while this process adds value, it also adds to the latency factor. The end result is that some level of latency not only is assured but also is variable (jitter) and uncertain in magnitude [8].

The level of magnitude is measured in milliseconds (ms, or thousandths of a second), which does not seem like much on the surface. But while this characteristic of packet switching does not affect many applications, it renders others ineffective. Many data communications applications, such as e-mail, are not seriously impacted by latency or jitter. Real-time uncompressed audio, voice, and video, however, are affected quite seriously and, therefore, traditionally have not been considered good candidate applications for packet switching.

7.5.8   Access

X.25 actually is the ITU-T recommendation for a standard describing the physical, link, and packet-level protocols between the user DTE/DCE and the network [17, 19]. Host computers connect to the network over an X.25 link. The user data are packetized by DCE in the form of a *Packet Assembler/Disassembler* (PAD), for which *X.3* is the standard. The PAD also may be responsible for password protection and performance reporting [10].

Occasional or casual users typically access a packet network on a dial-up basis from asynchronous PCs through modems, as described above. In such a scenario, the actual packetizing of the data can be performed at the originating network node or by software on the host. Individuals accessing the Internet through an online information service, for example, may use this approach.

Large user organizations often access the network via a dedicated, leased-line link to the closest network node. Such access often is in the form of an unchannelized T1/E1 facility. In either case, the PAD organizes the user data into PLP packets and encapsulates each in an LAP-B frame before presentation to the network [20, 21].

7.5.9   Network Interconnection: X.75

X.25 networks are widely available as a public data network (PDN) service offering, generally using packets of 128 or 256 bytes. Certain applications, however, are supported more effectively by transmission of larger packets. The airline reservation systems (e.g., American Airlines' SABRE and United Airlines' APOLLO), for example, deploy custom packet networks that use packet payloads of 1024 bytes. As this application involves the frequent transmission of relatively large sets of data (e.g., flight schedules, fares, and seating availability), a larger packet size is more appropriate. The larger packet size improves efficiency because the payload is very large, while the overhead information is roughly the same as it is for a smaller packet. As a larger packet is more likely to contain an errored bit, require retransmission, and therefore reduce throughput, the custom reservation networks typically employ digital facilities to minimize this exposure. In the United States and other highly developed countries, the reservation networks largely have shifted to Frame Relay, although X.25 remains heavily used elsewhere, in consideration of poor link quality.

The interconnection of such disparate networks is accomplished through ITU-T Recommendation X.75. Through an X.75 gateway that serves as a network-to-network interface (Figure 7.20), issues of packet size are resolved. This relatively

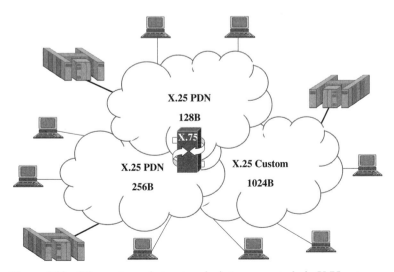

Figure 7.20   Disparate packet networks interconnected via X.75 gateway.

simple level of protocol conversion occurs at a network node that examines the packet for errors and either segments the payload of a large packet into multiple smaller packets or combines the payloads of multiple smaller packets into a single larger one. After the resulting packet payload(s) form, the node encapsulates each with the necessary control data in the form of a header and trailer, modifies the addressing scheme as required, and presents each to the target network.

7.5.10 Packet-Switching Hardware

The user of an X.25 packet network may require no hardware other than a PC and modem, with the packetizing performed at the local X.25 node. Alternatively, a hardware- or software-based PAD conforming to the X.3 standard can perform that function at the host. (*Note:* The advantage of the hardware PAD is that if it misbehaves, you can hit it with a hammer.) The PAD performs the packet *assembly* (i.e., aggregation of many individual characters) for the transmitting device and *disassembly* (i.e., disaggregation) for the receiving device in order to reconstitute the data in its native format.

Packet carriers, of course, must invest in packet nodes rather than circuit switches. Such packet nodes are intelligent devices capable of supporting complex routing tables, buffering packets in temporary memory, resolving packet errors, and accomplishing protocol conversions. Where X.25 networks are interconnected or where protocol conversions of any sort are required, gateways must be deployed.

7.5.11 Packet-Switching Standards

The ITU-T sets standards recommendations for packet switching. Those standards include the following [8, 10, 21]:

- **X.3:** Packet Assembly/Disassembly (PAD) functions
- **X.25:** Interface between DCE and DTE for public packet networks
  Packet Layer Protocol (PLP): Network Layer
  Link Access Procedure—Balanced (LAPB): Link Layer
  X.21bis: Physical Layer
- **X.28:** Terminal-to-PAD communications formats
- **X.29:** Host-to-PAD communications formats
- **X.31:** Packet-mode services over ISDN
- **X.32:** Defines X.25 synchronous dial-up mode
- **X.75:** Internetwork call control procedures

7.5.12 Packet-Switching Applications and Futures

X.25 packet switching originally was intended for interactive time sharing, which involves long connect times and low data volumes. While X.25 still supports such applications effectively, contemporary applications include online interactive processing (e.g., reservations systems), messaging (e.g., e-mail), batch file transfer (e.g., data backup), and Internet access.

X.25 offers the advantage of being a highly mature, if limited, network technology. Therefore, it is relatively inexpensive to deploy and is highly cost effective in support of applications that require many-to-many connectivity and involve relatively low volumes of data transport over error-prone circuits. Additionally, it is virtually ubiquitous, having been deployed in every corner of the globe. X.25, however, is limited in terms of speed and is characterized by significant levels of latency. As a result, most applications and service providers have moved or are moving toward newer network technologies. The airline reservations systems, for example, have largely transitioned to Frame Relay, at least for the domestic networks in the United States; in less developed countries and regions where network reliability and performance may be at issue, X.25 is still heavily used. Internet Service Providers prefer various combinations of dial-up modem access, Frame Relay, ATM, ISDN, and TCP/IP. The Internet backbone network largely has shifted to Frame Relay or ATM, operating at minimum speeds of T/E1 or T/E3. Much of the Internet backbone has been upgraded to fiber-optic facilities operating at speeds of 155 Mbps or much more. While the future of X.25 is past its prime, the future of packet switching, in general, is quite bright. *Note:* Tymnet, one of the early commercial data networks, began operations in 1966 and at one point was reputed to be the largest commercial X.25 PDN network in the world. Tymnet finally ceased operations in 2003.

There are several bright spots in the future of X.25. As it is the packet protocol used in SS7, it is heavily used in the contemporary PSTN for signaling and control purposes. Therefore, it also is used in the ISDN D channel, not only for user-to-network signaling but also for user-to-user signaling in private ISDN networks and for certain user-to-user data communications applications. However, the contemporary PSTN is in a state of decline due to the popularity of Voice over IP (VoIP) and the Internet. ISDN is not exactly on the upswing either. Actually, I reckon these are more like dim spots. I'm sorry. I didn't mean to get your hopes up.

7.6   INTEGRATED SERVICES DIGITAL NETWORK

A CCITT study group first explored *integrated services digital network* (ISDN) as a concept from 1968 to 1971. A more focused conceptual study took place during the 1981–1984 CCITT study period. The first set of published standards recommendations appeared in 1984 in the form of a CCITT *Red Book*, which provided the basic framework for the concept, network architecture, UNI (*User Network Interface*) protocols, and common channel signaling protocols. As a result of the 1985–1988 study period, a *Blue Book* was published that provided descriptions of supplementary services, rate adaptation, ISDN frame relay, and the initial set of B-ISDN (*Broadband ISDN*) recommendations. (*Note:* The color of the books has no significance, other than the fact that a different color represents each study period.)

Rather than being a technology, ISDN is described as a suite of services based on a set of technologies, including transmission, switching, and signaling and control. It is a set of international standards recommendations that permits the provisioning of a wide range of services intended to be available on a ubiquitous basis. Additionally, the ISDN network is accessible through a standard set of interfaces—one for low-bandwidth applications and another for high-bandwidth applications.

The specific characteristics of ISDN include its entirely digital nature—Customer Premises Equipment (CPE), transmission facilities, and switching systems all are fully digital in nature. The three identified channel types include Bearer (B) channels that bear the end-user information; Data (D) channels (aka Delta channels) for signaling and control, low-speed end-user packet data, and telemetry; and High-speed (H) channels for channel aggregation to accommodate bandwidth-intensive applications. The UNI is defined differently at two levels: Basic Rate Interface (BRI) for low-speed access and Primary Rate Interface (PRI) for high-speed access. Common Channel Signaling System 7 (SS7) is a fundamental requirement of ISDN.

Announced to the world amidst great fanfare, ISDN quickly captured the interest of carriers, manufacturers, and user organizations worldwide. ISDN offered the compelling advantages of increased bandwidth, enhanced flexibility, improved error performance, greater reliability, broad availability, and interconnection to a wide range of services. Unfortunately, it then stalled and progressed at a glacial pace for the next few years. Among the many reasons for its slow development are long delays in standards development, lack of adherence to standards, lack of availability, regulatory hurdles, circuit and equipment costs, and poor marketing, particularly in the United States.

Standards development at the ITU-T is infamously slow. Standards traditionally were released every four years, in monsoon fashion and with total droughts in the interim. Over time, the various committees attained the privilege of developing and releasing certain standards recommendations on an intermediate basis. Standards then came in showers. ISDN standards include three layers. The ISDN Physical Layer (Layer 1) addresses mechanical and electrical issues including connectors, signaling rate, and line coding. At the Data Link Layer (Layer 2), ISDN specifies the *Link Access Procedure—D Channel* (LAP-D). Network Layer (Layer 3) specifications include user-to-user and network-to-network signaling protocols for both circuit-switched and packet-switched networking.

Standards from the ITU-T actually are in the form of recommendations. Individual member nations can implement ISDN options as they see fit or they can deviate from the standards as long as international interconnectivity is accomplished at some reasonable level. The most notable international difference is that of the basic ISDN hierarchy. The North American version follows the T1 hierarchy, with PRI including 24 channels; the European (ITU-T) version is based on E1, providing 30 information-bearing channels. While this difference is understandable in the context of maintaining backward compatibility with existing networks, it also perpetuates issues of basic protocol incompatibility.

Systems manufacturers of COs and PBXs have a strong interest in maintaining the proprietary nature of their systems architectures. Therefore, they implement ISDN in distinctly different ways. ISDN *compatibility* became ISDN *compliance*, a decidedly lower level of conformance.

Additionally, carriers have implemented nonstandard versions of ISDN. For example, Pacific Bell (now SBC) initially offered ISDN at a rate of 56 kbps per channel, rather than the standard 64 kbps. This limitation was due to the presence of older channel banks in the Pac Bell ISDN carrier network and the fact that SS7 was not fully deployed. Therefore, in-band signaling and control consumed 8 kbps of channel bandwidth due to bit robbing.

Islands of ISDN resulted from these various implementations. A given carrier using the hardware and software of a given manufacturer could not easily achieve full connectivity with another carrier deploying another version of ISDN. Weary of delays in the standards process, some carriers (e.g., Southwestern Bell and Ameritech, both now merged into SBC) developed and implemented proprietary versions of ISDN, further contributing to the problem. In recent years, this problem has been mitigated through cooperation of the manufacturers and carriers, with the active involvement of Bellcore (now Telcordia) [18, 22].

Availability of ISDN was slow to develop in the United States because the carriers were reluctant to invest in the technology unless they were convinced that a market existed for the services or that the technology offered internal cost savings. ISDN is not inexpensive to deploy, a fact that unfortunately has been reflected in relatively high installation charges, recurring circuit charges, and equipment costs.

Regulators in the United States generally have required that the LECs pass on the cost of ISDN infrastructure to ISDN users, rather than averaging those costs across the entire rate base. In other words, they have viewed ISDN as an optional service that must pay its own way, or the carriers must absorb any associated losses. Because the carriers were unwilling to do so, ISDN rates remained high. Coincident with the development of competition in the local exchange and encouraged by the regulators, the ILECs have exercised more freedom in pricing ISDN attractively. Each ILEC, of course, has its own pricing strategy involving installation charges, monthly rates, and usage charges. The CLECs (Competitive LECs) are free to price ISDN (and any other service offering) as they see fit and generally price it at very attractive levels.

Rates for ISDN access historically were not tariffed at attractive levels in the United States, compared with the cost of basic services. Again, the regulators were largely responsible. The ILECs bear responsibility, as well, because they were not willing to absorb initial losses in order to stimulate the growth of the service offering. Currently, the charges for ISDN vary widely, from carrier to carrier and from state to state. Generally speaking, ISDN charges are somewhat higher than those for two analog lines and for monthly recurring charges as well as installation charges. Some carriers also charge for usage per channel and sometimes charge for packet traffic based on packet volume. To encourage ISDN usage, some carriers waive or lower those usage charges during evening and weekends, when calling activity is light.

Equipment costs were high because the manufacturers constantly were investing in R&D to maintain ISDN compliance with developing standards. Additionally, the limited demand for ISDN caused the manufacturing runs to be small, which tends to increase equipment prices. While equipment costs subsequently came down considerably, they remain an additional expense.

Marketing by the ILECs in the United States proved ineffective. Not only were costs maintained at unattractive levels, but advertising and promotion were minimal and availability was highly limited. Further, meaningful and cost-effective applications were not identified and stressed. With typical lack of foresight, the LECs placed heavy emphasis on the low-speed BRI version, which is suitable only for residence, small-business, and SOHO (Small Office/Home Office) application. High-speed PRI was not emphasized heavily as a replacement for T1 trunking.

ISDN Centrex was touted heavily, but with limited success. Centrex ISDN marketing was heavily slanted toward CO-based local area networking, which proved to be about as successful as the paperless office.

ISDN frustrated the industry in the United States, in general, and its less than stunning level of success eroded gradually in the face of V.90 modems. Since 2000 or so, *Digital Subscriber Line* (DSL) and *cable modems* have virtually wiped ISDN off the map, at least in the residential and small-business markets. At least one ILEC in the United States has removed all mention of ISDN from its website, closed its ISDN support office, and left ISDN customers with only a telephone number connected to an answering machine on which they can leave messages that (some say) are never returned. (I'll not name that company, in order to save them the embarrassment. At the same time, it will save them the trouble of filing a civil lawsuit against me for damages that they can't prove but that I can't disprove. *Note:* The United States has approximately 1,000,000 lawyers, which is about 70 percent of the world's lawyers, and a total population of about 300,000,000, which is less than 5 percent of the world's population.)

ISDN matured much more quickly and completely in Europe and certain parts of the Pacific Rim and Africa than in the United States. In those regions, the regulators encouraged deployment. Additionally, marketing was much more effective, focusing on PRI, rather than BRI [18, 23]. Outside the United States, ISDN pricing also has positioned it favorably in comparison to analog lines and modem-based Internet access. In South Africa, for example, ISDN BRI is priced very attractively for low-to-moderate Internet access. In many regions of the world, ISDN also benefits from the lack of competition from DSL and cable modems.

7.6.1 ISDN Devices and Reference Points

ISDN specifications include a number of functional groupings at layer 1, the physical layer. ISDN hardware, at the end-user side of the connection, includes Terminal Equipment (TE), Terminal Adapters (TAs), and Network Termination (NT) devices, line termination equipment, and exchange termination equipment, as depicted in Figure 7.21. The following discussion begins at the subscriber premises and works toward the carrier networks.

7.6.1.1 Terminal Equipment   *Terminal Equipment* (TE) is the term for a device that connects a customer site to ISDN services. In more traditional non-ISDN terms, TE includes all *Customer Premises Equipment* (CPE), which category includes voice terminal equipment and *Data Terminal Equipment* (DTE). TE also includes premises-based switching equipment, including PBXs and routers. There are two types of TE:

- *TE1* is ISDN-compatible equipment, that is, equipment that can interface directly to an ISDN circuit via a four-wire twisted-pair interface. Examples 2, 3, and 4 in Figure 7.21 illustrate TE1.
- *TE2* devices do not enjoy native ISDN compatibility. TE2 equipment must connect through a *Terminal Adapter* (TA) that resolves issues of incompatibility. Example 1 in Figure 7.21 illustrates TE2.

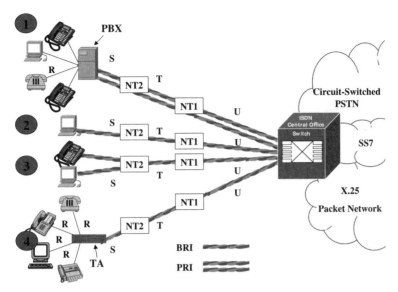

Figure 7.21 ISDN devices, reference points, and networks.

7.6.1.2 Terminal Adapters *Terminal Adapters* (TAs), also known as *ISDN modems*, are interface adapters for connecting one or more TE2 (non-ISDN) devices to an ISDN network. A TA acts as ISDN DCE, providing a function equivalent to that of a protocol or interface converter for equipment that does not have ISDN capability built in. A TA generally is in the form of a stand-alone unit, as illustrated in example 4 of Figure 7.21. A TA also can be in the form of a printed circuit board that fits into an expansion slot of the TE2. If the TE2 and TA are separate units, they connect on the basis of a standard interface such as EIA/TIA-232, V.24, or V.35.

A key function of the TA is that of *rate adaption*, which operates in several ways. In one case, rate adaption effectively throttles down the transmission rate from 64 kbps to the rate at which the non-ISDN device is capable. Rate adaption also serves to bond multiple B channels into H channels for more bandwidth-intensive applications. *V.110* is the ITU-T Recommendation that specifies support for data terminal equipment (DTE) with asynchronous or sunchronous serial interfaces over an ISDN network through rate adaption. *V.120* specifies support for data terminal equipment (DTE) with asynchronous or sunchronous serial interfaces over an ISDN network through data encapsulation. V.120 includes specifications for allowing multiple terminals to share a 64-kbps B channel through *Statistical Time Division Multiplexing* (STDM).

7.6.1.3 Network Terminations *Network Termination* (NT) devices are physical devices that operate to interface the four-wire customer wiring to the two-wire UTP local loop. As illustrated in Figure 7.21, there are two types of NTs:

- *NT1* operates at Layer 1, the Physical Layer, serving to provide physical and electrical connection between the customer wiring and the carrier local loop.

If there are multiple TEs, the NT1 provides for multidrop termination, managing physical layer contention issues between the TEs and the carrier circuit. The NT1 also performs such functions as signal conversion, synchronization, Layer 1 multiplexing, frame alignment, and line maintenance and performance monitoring of the local loop. Echo cancellation is performed at this level. In North America, the NT1 is in the form of CPE. In international implementations, the NT1 is the carrier responsibility.

- *NT2* operates at Layer 2, the Data Link Layer, and Layer 3, the Network Layer. NT2 is an intelligent device responsible for the user's side of the connection to the network, performing such functions as Layer 2 multiplexing, switching, or ISDN concentration. An NT2 commonly is actually an NT1/2 device, performing the combined functions of an NT1 and an NT2 and operating at Layers 1, 2, and 3 of the OSI Reference Model. Such a device likely would be in the form of a PBX, router, or data switch.

7.6.1.4 Reference Points These functional groupings of TE, TAs, and NTs are logically distinguished by *reference points*, which are points of reference, interface, or demarcation. As illustrated in Figure 7.21, the reference points are as follows:

- *Reference Point R* corresponds to the interface between TE2 and the TA.
- *Reference Point S* refers to the point of interface of an ISDN terminal and the NT2, serving to distinguish between terminal equipment and network-related functions. The S interface is defined as a passive bus for up to eight NT2 devices.
- *Reference Point T* references a minimal point of termination at the customer premises. It is the reference point between the NT1 and NT2 devices.
- *Reference Point U* is the reference point between NT1 devices and line termination equipment in the carrier CO. The reference point describes the full duplex data signal on the physical two-wire subscriber line, including line coding and framing conventions. The U reference is relevant only in North America, as the NT1 function is provided by the carrier elsewhere.
- *Reference Point V* is the point of interface at the network side of the connection between the line termination or loop termination and the exchange termination. In other words, it is the point between the circuit-terminating equipment and the ISDN CO. As the V interface exists only if the CO does not have embedded circuit-terminating equipment, it is unusual in contemporary ISDN-compatible COs.

7.6.2 Standard Interfaces and Channel Types

The current version of ISDN is Narrowband ISDN (N-ISDN); I discuss Broadband ISDN (B-ISDN), which is still on the drawing boards where it will probably remain forever, in Chapter 10. ISDN currently is available in essentially two interface varieties (see Figures 7.21 and 7.22): *Basic rate interface* and *primary rate interface*. In each case, the ITU specifies the electrical characteristics, signaling, coding, and frame formatting. Regardless of the specifics of the interface and channel type,

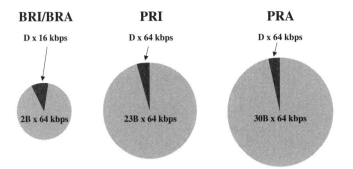

Figure 7.22 ISDN BRI and PRI.

ISDN advantages include digital technology, clear-channel communications, and symmetric bandwidth.

7.6.2.1 *Basic Rate Interface*

Basic Rate Interface (BRI), also known as *Basic Rate Access* (BRA) and *2B+D*, provides two *Bearer* (B), or information-bearing, channels, each operating at the clear-channel rate of 64 kbps by virtue of SS7 non-intrusive signaling. Each B channel can carry digital data, digitized voice (PCM encoded at 64 kbps or a lower rate), or a mixture of low-speed (subrate) data as long as it all is intended for the same destination. BRI also provides a *Data* (D) channel at 16 kbps, which is intended primarily for purposes of signaling and control, messaging, and network management. The D channel also generally is made available for X.25 packet data transmission and low-speed telemetry when not in use for signaling purposes; cost-effective applications include credit card authorization, which involves very small bursts of data [24]. BRI is used primarily for residential, small-business, Centrex, and telecommuting applications that are not particularly bandwidth intensive. The B channels can be *aggregated*, or *bonded*, to provide up to 128 kbps to a given conversation, such as a videoconference or Internet experience; additionally, multiple BRIs can be bonded for even greater capacity. Whether bonded or not, ISDN BRI provides multiple channels over a single physical loop, which is a great advantage.

A single BRI line can support up to 16 devices that contend for access to the BRI channels through a *Terminal Adapter* (TA). The devices can be in a variety of forms, including telephones, facsimile machines, computers, and video cameras. Additionally, ISDN can support up to 64 individual *Service Profile IDentifiers* (SPIDs), which are equivalent to directory numbers, one per terminal device [25]. The SPID is used in the initialization procedure when the device goes *off hook* to establish a connection with the network through the CO. While BRI supports as many as three simultaneous calls, the specifics of the carrier offering determine how they can be used. Some carriers offer a confusing array of ISDN BRI packages that either restrict particular B channels to voice or data application or enable them both to be used for voice or data. Some carriers offer application-specific BRI variations such as *1B+D* and *0B+D*, with the latter targeted at applications such as credit card verification, as illustrated in example 3 of Figure 7.21.

BRI uses an eight-pin connector as defined by the International Organization for Standardization (ISO) in *ISO 8877* and commonly known by the industry term RJ-45. Full-duplex (FDX) connectivity is accomplished over a digital twisted-pair local loop through the application of special carrier electronics, with four-wire connectivity accomplished over one or sometimes two physical pairs, depending on the specifics of the carrier's implementation. An NT1 device provides for compatibility with network protocols.

7.6.2.1.1 Line Coding The BRI interface between the CO and the customer premises is known as the *U Reference Point* (Figure 7.21), and it runs at an actual transmission line rate of 160 kbps, carrying two 64-kbps B channels, one 16-kbps D channel, and 16 kbps of overhead for framing, echo cancellation, and an *Embedded Operations Channel* (*EOC*) that is used for line testing and monitoring. In order to support this transmission rate in full duplex mode over a physical two-wire, logical four-wire UTP local loop, the 2B1Q (*2 Binary 1 Quaternary*) encoding technique, with echo cancellation, is used in North America. 2B1Q is a form of *Pulse Amplitude Modulation* (PAM) that uses four (i.e., quaternary) levels of amplitude (i.e., voltage), each of which represents two adjacent bits in a bit stream, and is accomplished by varying the voltage at nominal levels of ±1 (actually 0.833) and ±3 (actually 2.5) volts, as illustrated in Figure 7.23. Specifically, −3V represents a 00, −1V a 01, +1V a 11, and +3V a 10. [*Note*: This approach is similar to the dibit Amplitude Modulation (AM) technique discussed in Chapter 6.] Because two bits are impressed on each baud (i.e., signal or signal change), the baud rate is halved, and a baud rate of 80 baud will support a transmission rate of 160 kbps. Because statistics force the line voltage to be positive half the time and negative half the time, on average, the signal power lies at a frequency of 40 kHz, which is half the baud rate. This frequency is well above the standard voice-grade rate of 4 kHz over a local loop of up to 18,000 ft. Therefore, the loop must be of excellent quality, which often means that it must be specially conditioned to perform at this level. As 2B1Q scales well, it is the electrical line-coding technique used in *High-bit-rate Digital Subscriber Line* (HDSL), which is a DSL version of T1/E1.

In European and many other countries, the line-coding technique employed is *4 Binary 3 Ternary* (4B3T), a *block code* that combines four bits to represent one

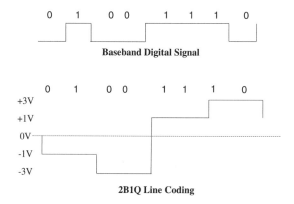

Figure 7.23   2B1Q line coding.

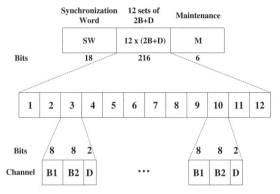

Figure 7.24 ISDN BRI framing at U interface.

ternary signal state. As a result, the baud rate is three-fourths of the transmission rate, and ISDN BRA at 160 kbps requires a baud rate of 120 kbaud. 4B3T yields shorter ISDN transmission distances than 2B1Q, but distances in Europe and elsewhere often are much shorter between the CO and the customer premises [19, 26].

7.6.2.1.2 Framing All TDM-based services require framing, and ISDN BRI is no exception. The 2B1Q-coded transmission must be organized and synchronized in some fashion in order for the B and D channels to be interleaved without losing their identity. Signaling and control functions also must be performed. The ISDN BRI involves several framing conventions, with one at the U interface and the other between the S and T interfaces:

- **U Interface:** The framing at the U interface, as illustrated in Figure 7.24, begins with a *synchronization word* of 18 bits that provides the receiver with the means to synchronize on the beginning of each frame. That is followed by 12 sets of 2B+D samples, with each B sample comprising 8 bits and each D sample comprising 2 bits, for a total of 18 bits per set. The frame ends with 6 bits of *overhead* in the *M* channel for *M*aintenance and other purposes. Much as is the case with T1 and E1, sets of eight frames are organized into superframes. It is at that level that the M channels find their purpose in addressing, error control, power management, and various network management functions.

- **S/T Interface:** The framing at the S/T interface is a bit more complex. The BRI/BRA framing structure comprises the two B channels (2 × 64 kbps = 128 kbps) plus the D channel (16 kbps) multiplexed into repetitive frames at 192 kbps. Each frame is 48 bits long, including two samples for each B channel, for a total of 32 bits of user data (2 × 8 = 16 × 2 = 32), plus 4 bits from the D channel, for a grand total of 36 bits. Added to that are 12 bits of overhead to bring the total up to 48 bits. Given the fact that the signaling speed of the link is 192 kbps, frames repeat at the rate of one frame every 250 μs (48 bits/192 kbps = 1/4 ms, or 250 μs). The B-channel samples are byte interleaved, as they are in T1 or E1. The four D-channel bits are spread out among the four B-channel byte samples. The overhead bits also are spread out and include bits for framing [first (F) bit and last (L) bit], channel activation (A bit), DC balancing (L bit), contention resolution when multiple TEs contend for a channel on a passive bus (E bit), and various other purposes. The line-coding format is *pseudoquaternary coding*, which differs from AMI only with respect to the polarity of the framing bit (F) and the requirement for BiPolar Violations (BPVs). Remember that the S and T interfaces are reference points involving NT1s and NT2s, which are functional groupings that often are not presented in the form of physically distinct devices. So, all of this typically takes place under the hood of a TA or some sort of TE.

7.6.2.2 Primary Rate Interface *Primary Rate Interface* (PRI) also is known as 23B+D in the United States and Japan. The European or international version is known as *Primary Rate Access* (PRA) or 30B+D. PRI offers 23 B (Bearer) channels plus 1 D (Data) channel and is backward compatible with T1 and J1 transmission systems, respectively. PRA offers 30 B channels plus 1 D channel and is backward compatible with E1 transmission. PRI and PRA both provide a full-duplex (FDX) point-to-point connection through an NT2-type intelligent CPE switching device, such as a PBX or router, for interfacing with the carrier CO switch. The DS-0 is the basic building block of both PRI and PRA, as both the B and D channels operate on clear channels at 64 kbps. As is the case with BRI, the B channels can be used individually or can be bonded for voice, data, video, facsimile, any other data and any multimedia combination, but the D channel is reserved exclusively for signaling. As signaling and control functions are fairly light, the standards provide for *Non-Facility-Associated Signaling* (NFAS), which allows a D channel to support up to five PRI connections. The first PRI in a PBX trunking application, for example, would be provided at 23B+D, and the next four PRIs would be delivered at 24B+0D. Some manufacturers and carriers have stretched this limit to seven PRIs. In a configuration involving multiple PRIs, the typical recommendation is that a backup D channel be provisioned in order to maintain signaling and control functionality in the event that the primary D channel fails. While this backup approach diminishes the B-channel count, it is generally considered to be the most prudent approach, for B channels are useless without a D channel. NFAS is required for Switched-1536 data service; because all 24 channels of the T1 PRI line carry user data, the D channel must be on another line.

While designed for transmission over a standard DS-1 trunk, PRI is a significant improvement over T1 or E1, because the channels can be allocated dynamically. In other words, each channel can act as an incoming, outgoing, combination, or DID trunk, as the need arises. The nature of the channel can be determined as required, based on user-definable parameters. Additionally, multiple B channels can be aggregated to serve bandwidth-intensive applications, such as videoconferencing. On the negative side, PRI does not compare favorably with T1 in terms of the raw number of B channels, and PRI can be considerably more expensive, depending on tariff specifics.

7.6.2.2.1 Line Coding PRI requires the same line coding as contemporary T1 and E1. Specifically, the coding technique is Alternate Mark Inversion (AMI) with Bipolar with Eight-Zero Substitution (B8ZS).

7.6.2.2.2 Framing T1 and E-1 framing conventions are exactly the same at the U, T, and S reference points of an ISDN circuit. PRI requires the extended superframe (ESF) format, which supports nonintrusive signaling and control. This approach eliminates any T1 issues of bit robbing and, therefore, supports clear-channel communications. ISDN, of course, accomplishes signaling and control functions over the D channel. PRI uses time slot 24 for the D channel. PRA uses time slot 15, which actually is the 16th time slot, as the E1 time slot numbering is 0–31.

7.6.2.3 H Channels (N × 64)

H channels (*H*igh-speed *channels*) are functionally equivalent to B channels but provide greater aggregate bandwidth in PRI applications. In this $N \times 64$ mode, any number of B channels can be aggregated, or bonded, on a dynamic basis to enable multirate communications through inverse multiplexing. The ISDN approach does have a drawback, however, when compared to traditional inverse muxes, as the connection must be torn down and reinitiated when channels are added or dropped. The feature is known variously as *multirate ISDN, N × 64, channel aggregation*, and *bonding*. H channels find application in fast faxing (Group IV), videoconferencing, high-speed data transfer, high-quality audio transmission, and Frame Relay. Defined H channels include the following:

H_0 channels have an aggregate bit rate of 384 kbps, which is the equivalent of six B channels (6×64 kbps = 384 kbps). This is a common port speed offered by Frame Relay service providers.

H_1 is a full DS-1, with no framing overhead. This channel is sensitive to the specifics of the DS-1 implementation.

- H_{10} operates at 1.472 Mbps, which is the sum of the 23 B channels (23×64 kbps = 1.472 Mbps) in a baseline PRI implementation in which channel 24 is devoted to the D channel. H_{10} applies in North America and Japan and is based on T1 and J1, respectively.
- H_{11} operates at 1.536 Mbps, the sum of all 24 B channels for the North American and Japanese versions, which is based on T1. H_{11} relies on *Non-Facility-Associated Signaling* (NFAS) to provide a D channel an H_{10} facility for signaling and control.
- H_{12} operates at 1.920 Mbps, the sum of all 30 B channels for the European version, which is based on E1.

7.6.2.4 Inverse Muxes

Offered by some manufacturers, inverse muxes enable multiple BRIs to bond, or link, for greater aggregate transmission over multiple BRIs. For example, four BRIs can be linked to support a 512-kbps data transmission. Such an approach competes effectively with Fractional T1 (FT1) where it is available.

7.6.2.5 D-Channel Contention Devices

D-channel contention devices, also known as *ISDN routers* and offered by some manufacturers, permit as many as eight devices to share a BRI circuit, contending for access to the B channel. The individual devices identify themselves to the network through contention for the D channel. The ISDN router also serves traditional switch functions within the context of the small-user domain.

7.6.3 Link Access Procedure—D Channel

The ISDN D channel supports signaling and control messages between user TE and the carrier networks, both circuit switched and packet switched, and between TE devices in a user-to-user private network. The D channel also can be used for user-to-user communications where the carrier supports it for applications such as credit card verification and transaction processing. Messages conveyed over the D channel employ *Link Access Procedure—D channel* (LAP-D), a bit-level protocol that runs at Layer 2, the Data Link Layer. LAP-D evolved from the LAP-B protocol used in X.25 networks, which makes a great deal of sense, as the ISDN SS7 signaling and control network run the X.25 packet format. LAP-B, as previously noted, is a derivative of the *High-level Data Link Control* (HDLC) protocol which, in turn, is based on the IBM *Synchronous Data Link Control* (SDLC) frame.

LAP-D is a balanced protocol that operates in *Asynchronous Balanced Mode* (ABM), referring to the fact that the devices have a *balanced*, rather than a *master/slave*, relationship. Therefore, a device at either end of the link can initiate a dialogue at any time. As illustrated in Figure 7.25, the LAP-B frame format includes the following fields:

- **Flag:** The flag field is a one-octet field that delimits (i.e., establishes the limits or boundaries of) the beginning and end of the frame. The flag is always the specific bit pattern 01111110 (7E in hexadecimal), which is known to all transmitters and receivers.
- **Address:** The address field is a two-octet field known as the *Data Link Connection Identifier* (DLCI), which is divided into two addresses. The first octet is the *Service Access Point Identifier* (SAPI), which identifies the destination service access point, each of which can support multiple terminal devices. The second octet is the *Terminal Endpoint Identifier* (TEI), which is the address of the destination terminal device.
- **Control:** The address field is a one- or two-octet field that identifies the frame type. An *Information* (I) frame carries upper layer (e.g., application) information and some control data. A *Supervisory* (S) frame carries control information such as I-frame acknowledgment, request for retransmission, and flow control. An *Unnumbered* (U) frame carries control data such as disconnection request, acknowledgment frame, and frame reject.

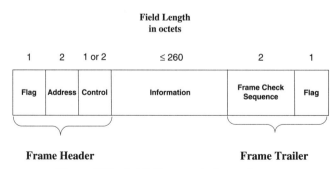

Figure 7.25 LAP-D example frame format.

- **Information:** The information field is a variable-size field with a maximum of 260 octets comprising upper layer information. The size of the field is system dependent. Only information frames include an information field.
- **Frame Check Sequence:** The FCS field is a two-octet field that provides excellent error detection.

7.6.4 ISDN Characteristics and Benefits

ISDN is unusual, if not unique, in that it is undoubtedly the most carefully planned, well-coordinated, and best-documented network technology in history. Despite this fact, or perhaps because of it, ISDN's popularity has lagged due to the previously mentioned issues of cost, availability, and applications. The key characteristics of ISDN include its end-to-end digital nature, which is unusual for a public circuit-switched network. Through a small family of interfaces, a wide range of services can be accessed through the LEC, IXC, or CAP. Rate adaption and channel aggregation permit bandwidth-intensive applications to be supported on a dynamic basis.

The reliance of ISDN on SS7, as is discussed in Chapter 5, offers a number of advantages that include faster call setup and nonintrusive signaling and control. Additionally, SS7 (either with or without ISDN) makes possible a number of interesting CLASS services, including caller ID, name ID, call trace, selective call forwarding, and selective call blocking.

ISDN also is interoperable with X.25, Frame Relay, and ATM. In fact, ISDN standards were developed specifically with these services in mind. X.25 is the packet-level protocol used in SS7. Frame Relay closely aligns with ISDN link-level protocols. Broadband ISDN (B-ISDN) is dependent on ATM network technology, which has been overtaken by the Internet and IP-based networking, in general.

7.6.5 ISDN Characteristics and Drawbacks

ISDN does have some drawbacks. These include limited availability, standards variations, and cost. As discussed earlier in this chapter, availability has always been limited in most nations because ISDN and SS7 software is costly for the carriers to deploy; while ISDN capability can be extended to non-ISDN CO exchanges, that incremental cost is not trivial.

Not surprisingly, cost–benefit considerations dictate the success or failure of technologies, regardless of how compelling they appear at first glance. The cost of an ISDN BRI circuit often is more than twice that of an analog line, and installation costs can be considerably higher. These costs tend to discourage ISDN to some extent, particularly in voice-intensive environments where an ISDN BRI configuration may limit voice to a single channel.

Additionally, many carrier tariffs impose a usage surcharge in the form of a flat rate per minute for circuit-switched connections and a packet surcharge for packet data. The usage charge applies to local as well as long-distance calls. While these additional usage charges tend to discourage ISDN usage, the faster speed of data transfer may serve to reduce call connect time significantly, at least in comparison to dialup modem connections over analog lines.

Hardware costs are additional in support of ISDN. In a BRI environment, such additional equipment might include Terminal Adapters (TAs), inverse muxes, and BRI contention devices. In a PRI application, ISDN software for PBXs and routers represents an additional cost, and older systems generally are not upgradable at any cost.

Notably, ISDN hardware depends on local power. In other words, the phones do not work when the power goes out. Large end-user organizations are accustomed to providing power backup in the form of an *Uninterruptible Power Supply* (UPS) for voice [e.g., PBX, ACD, and KTS) and data (e.g., workstations, servers, hubs, access points, switches, and routers) systems. Residential and small-business users often do not employ UPS systems. Therefore, ISDN BRI is not recommended as a full replacement for analog telephones. Rather, it should be considered a supplemental service to support a mixture of voice and data applications. Also, note that ISDN adds no real value to most voice communications, other than enhanced voice quality and extending functionality from an ISDN-based PBX system. Call centers, however, prefer PRI trunks, as they deliver Calling Line IDentification (CLID) information at no additional cost and are useful in transferring calls within the PSTN.

Also note that ISDN is required on both ends of the connection in order to provide any benefit. An ISDN BRI, for example, is of no value for access to the Internet or Web unless ISDN is in place at the ISP site. Similarly, ISDN BRI is of no value to the telecommuter for corporate intranet access unless the host location also supports ISDN in the form of either BRI or PRI.

It also is worth noting that ISDN can be wasteful of network resources in an Internet or intranet data application. Remember that ISDN is a circuit-switched service. Also you might remember that circuit switching provides temporary, continuous, and exclusive connectivity. Now, consider that interactive data applications such as the Internet, the Web, and intranets are best served by packet-based networks, as discussed earlier in this chapter. Some years ago, several carriers, including Pacific Bell and US West, sought to stimulate IDSN through tariffs that negated usage charges during off-peak hours. Those pricing plans proved highly successful because they encouraged the use of ISDN for purposes of Internet access. However, the increased usage of the PSTN, which is highly inefficient for bursty, low-volume packet data applications, caused substantial network congestion. Since then, manufacturers of carrier-class equipment have developed devices that recognize the telephone number of an ISP and shunt that traffic around the circuit switch to a packet-switched network, thereby eliminating this issue in properly equipped COs.

ISDN certainly has a lot of benefits to offer in comparison to the PSTN, and it has some limitations. Ultimately, the greatest limitation is that of bandwidth. While ISDN offers much greater bandwidth than the PSTN, plus the advantage of digital communications, IDSN BRI bandwidth pales in comparison to DSL, cable modem, and Fiber-To-The-Premises (FTTP) technologies, all of which are capable of running in the Mbps range. At the PRI level, ISDN does not compete well against fiber-optic-based technologies.

7.6.6 ISDN Standards

As mentioned earlier in this chapter, ISDN standards are voluminous. While still under development in some respects, current ITU-T standards for ISDN include the following:

- **I.441/4511:** ISDN Primary Rate Interface (PRI)
- **I.515:** Parameters for ISDN internetworking
- **Q.700:** Signaling System Number 7 (SS7) specifications
- **Q.921:** Layer 2 specification for D channel; Link Access Protocol, D Channel (LAPD)
- **Q.931:** Layer 3 User Network Interface (UNI) specifications
- **V.110:** B-channel procedures (Europe) for Terminal Adapters (TAs)
- **V.120:** B-channel procedures (North America) for TAs

In addition to the ITU-T, other organizations actively develop and promote ISDN standards. For example, ANSI (United States) and ETSI (Europe)each lobby the ITU for the acceptance of their parochial ISDN variations.

7.6.7 ISDN Applications

The applications for ISDN are broad in range. While ISDN was long phrased *a technology in search of an application*, it later opened to applications developers, and aggressively so. There is no *killer app* for ISDN. Rather, there are a number of applications that, in total, enhance its future. A host of applications that benefit from the improved quality of digital networking and are bandwidth intensive are well served by ISDN. Further, ISDN offers an affordable circuit-switched alternative to DDS, Switched 56, and T-carrier, which simply cannot be cost justified in many cases.

Personal office internetworking, remote office internetworking, and *telecommuting* (or *Telework*) all are facilitated by the increased bandwidth and error performance offered by ISDN BRI. In such applications, file transfers and facsimile transmission are accomplished much more quickly and with much greater clarity; the improved quality of the voice communications presents an added benefit.

ISDN also is used for access to packet data networks, including X.25, Frame Relay, ATM, and IP, with users benefiting from the faster call setup and teardown time made available by virtue of SS7. Because either the B channels or D channels can be used for packet data transfer in a BRI implementation, ISDN offers additional flexibility and bandwidth utilization. Additionally, some manufacturers of Terminal Adapters (TAs) offer built-in X.25 PADs and Frame Relay Access Devices (FRADs) for end-to-end error correction [27].

As a replacement or backup for dedicated digital services, ISDN performs well for data and image networking, whether in a host-to-host, LAN-to-LAN, or remote LAN access application. Intensive users of the Internet and Web find ISDN bandwidth to be of great advantage because the speed of call set up and file transfer increases considerably compared to dial-up analog connections. A typical Web page, for example, takes 1.5 min to load over an analog line with a 28.8-kbps modem and a little less with a 33.6-kbps modem but less than 20 s at BRI speed of 128 kbps. As is discussed in Chapter 3, incoming call centers can take advantage of ISDN to increase productivity as well as make use of remote agents working from home.

At least one example merits further discussion, for purposes of illustration. A remote worker might desire to access an application residing on a LAN-connected server. An ISDN call to a local LAN site saves on long-distance charges; that LAN site connects the user to a remote site through another ISDN link. When the remote client workstation is idle, ISDN disconnects the LAN-to-LAN link to save on long-distance charges. Through a process known as *spoofing*, the application remains alive, as it continues to see a logical link over the B channel. The interactive data conversation can quickly be reinitiated due to the fast call setup time of SS7, which is an integral element of ISDN. The remote worker in this scenario might be a telecommuter working from home several days a week.

In terms of vertical markets, ISDN is of particular interest in the health care and education sectors, largely because of its ability to support imaging and video through rate adaption. *TeleMedicine*, for example, enables specialists to diagnose and treat patients in remote areas based on video examination and transmission of X-ray images across error-free and high-speed ISDN links.

The applications for ISDN are virtually unlimited, at least in terms of the network services that you can access. Through a single ISDN local loop, voice, facsimile, data, video, and image information can be accommodated. Additionally, simultaneous access to multiple networks and network services can be accomplished, perhaps including circuit-switched voice, X.25 packet, and Frame Relay. From a user's perspective, ISDN is highly flexible. From a carrier's perspective, ISDN offers the advantage of consolidating access to multiple networks, thereby relieving the strain on local loop, switching, and transport facilities. In other words, the LECs can market ISDN as a single network access solution—sort of a *one-stop shop*.

Notably, ISDN often is used as a backup to Frame Relay. In the event of a Frame Relay network failure, such as the total failure experienced in the AT&T network in 1998, the ISDN circuit responds immediately without dropping the data session. This approach offers diversity of both networks and services, offering greater protection than a backup Frame Relay permanent virtual circuit (PVC), which also would be affected by a total network outage.

7.6.8 Variations on the Theme

Worldwide ISDN has experienced differing levels of success due to various marketing approaches, pricing strategies, and, in some cases, aggressive government support. The Japanese government, for example, has lent strong support to the development and deployment of high-technology networks and network services.

Telecom Australia achieved much success in marketing PRI to large user organizations, in part as an alternative to leased lines. An unusual offering is that of semipermanent circuits within PRI, priced at approximately 50 percent of the cost of a dedicated circuit. In the competitive Australian telecom environment, this approach was successful in countering leased-line networks offered by alternative carriers such as Optus.

Deutsche Bundespost Telekom offers ISDN on a widely available and low-cost basis. In excess of 80 percent of Germany's population has access to ISDN within six weeks. Pricing is very attractive compared to leased lines.

Europe, for years, has offered a service known as *0B+D*. This offering provides access to a solo 16-kbps D channel for low-speed data transmission. Packet data are supported at speeds up to 9.6 kbps, with signaling and control consuming the balance of the capacity; no B channels are involved. This service effectively challenges X.25 packet networking for transaction-oriented applications such as credit card authorization. A number of LECs in the United States now offer ISDN BRI variations

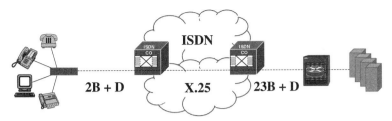

Figure 7.26 Always-on/dynamic ISDN.

such as 1B+D and 0B+D for applications where only a single B channel or a single D channel is required.

7.6.9 Always On/Dynamic ISDN

An interesting variation on the ISDN theme is *Always On/Dynamic ISDN* (AODI), which enables the user to establish a LAN-like *always-on* ISDN BRI connection to an ISP server, corporate intranet server, or corporate video server, for example, through special equipment and using only the D channel. The D channel maintains the always-on logical link between the client and the server systems, enabling the transfer of data (e.g., e-mail, stock quotes, or news bulletins) at rates of up to 9.6 kbps over an X.25 *switched virtual circuit* and without the call set up time required for a circuit-switched connection. As illustrated in Figure 7.26, a portion of the D channel on the client side operates as a link to a X.25 packet network, with the Multilink Point-to-Point Protocol (MPPP) used for access to the network from the TA. The D-channel AODI service employs the X.25 packet protocol that is used for signaling and control purposes in the SS7 network and establishes an SVC over the packet network. On the client side, the TCP/IP suite is encapsulated within the X.25 logical channel carried by the D channel, in support of connectionless data transfer. *Packet handlers* route the data packets around the circuit-switched network and over the X.25 packet-switched network, thereby maximizing efficiency and avoiding the unnecessary use of the circuit-switched PSTN for such an inappropriate application. On the server side, the SVC can be terminated on a BRI D channel, a designated PRI B channel, or a high-speed serial link. Should the need arise, the equipment automatically activates one or both B channels for transfer of large sets of data or to establish a videoconference. Once the need for the B channel(s) has ceased, the equipment automatically terminates those channel connections, the cost of which typically is usage sensitive. The cost of D-channel usage in an AODI application generally is based on a flat rate per month, with surcharges for kilopackets or megapackets applying above a defined usage threshold [28, 29].

REFERENCES

1. Standage, Tom. *The Victorian Internet*. Berkley Books, 1998.
2. Casson, Herbert N., *The History of the Telephone*. A. C. McClurg & Co., 1910.
3. *Engineering and Operations in the Bell System*. Bell Telephone Laboratories, 1977.
4. Brooks, John. *Telephone: The First Hundred Years*. Harper & Row, 1976.

5. *A History of Engineering & Science in the Bell System, The Early Years (1875–1925)*. Bell Telephone Laboratories, 1975.

6. Bates, Regis J. *Introduction to T1/T3 Networking*. Artech House, 1992.

7. http://www.lucent.com/minds/telstar/index.html.

8. Minoli, Daniel. *Enterprise Networking: Fractional T1 to SONET, Frame Relay to BISDN*. Artech House, 1993.

9. Trulove, James E. *A Guide to Fractional T1*. Artech House, 1992.

10. Gelber, Stan. *Introduction to Data Communications: A Practical Approach*. Professional Press Books, 1991.

11. Flanagan, William A. *The Guide to T1 Networking*. Telecom Library, 1990.

12. Doll, Dixon R. *Data Communications: Facilities, Networks, and Systems Design*. Wiley, 1978.

13. Elbert, Bruce R. *Private Telecommunication Networks*. Artech House, 1989.

14. Freeman, Roger L. *Fundamentals of Telecommunications*, 2nd ed. Wiley, 2005.

15. Kleinrock, Leonard. "Principles and Lessons in Packet Communications." In Partridge, Craig. (Ed.), *Innovations in Internetworking*. Artech House, 1988.

16. Heart, F. E., Kahn, R. E., Ornstein, S. M., Crowther, W. R., and Walden, D. C. "The Interface Message Processor for the ARPA Computer Network." In Partridge, Craig (Ed.), *Innovations in Internetworking*. Artech House, 1988.

17. Spohn, Darren L. *Data Network Design*. McGraw-Hill, 1993.

18. Horak, Ray. "ISDN: To Be Delivered as Promised?" *Datapro Communications Analyst*. Datapro Information Services, January 1995.

19. Stallings, William. *ISDN and Broadband ISDN with Frame Relay And ATM*, 3rd ed. Prentice-Hall, 1995.

20. Rybczynski, Antony. "X.25 Interface and End-to-End Virtual Circuit Service Characteristics." In Partridge, Craig (Ed.), *Innovations in Internetworking*. Artech House, 1988.

21. Held, Gilbert. *Understanding Data Communications*. SAMS Publishing, 1994.

22. Levitt, Jason. "Hold The Phone!" *Information Week*, May 15, 1995.

23. Skvarla, Carol A. "ISDN Services in the U.S.: Overview." *IT Continuous Services*. Datapro Information Services, April 14, 1998.

24. Stargess, James. "ISDN D Channel Packet Service: Coming Soon to a Store Near You." *Network World*, March 11, 1996.

25. *ISDN: A User's Guide to Services, Applications & Resources in California*. Pacific Bell, 1994.

26. Flanagan, William A. *ISDN: A Practical Guide to Getting Up and Running*, 2nd ed. CMP Books, 2000.

27. Tredinnick, Ian. "X.25: A New Lease on Life with ISDN." *Telecommunications*, June 1995.

28. LeFevre, Jim. "Intel Taps NetManage for Always On/Dynamic ISDN." *ENT*, May 6, 1998.

29. Minoli, Daniel and Mattern, Peggy. "Integrated Services Digital Network (ISDN) Standards and Implementations." *IT Continuous Services*. Datapro Information Services, March 10, 1999.

CHAPTER 8

LOCAL AREA NETWORKS: CONNECTIVITY AND INTERNETWORKING

Just as computer networks have grown across continents and oceans to interconnect major computing facilities around the world, they are now growing down corridors and between buildings to interconnect minicomputers in offices and laboratories.
Robert M. Metcalfe and David R. Boggs, Xerox Palo Alto Research Center,
Ethernet: Distributed Packet Switching for Local Computer Networks,
Association for Computing Machinery, 1976

Once upon a time, computer networks consisted of mainframes in *glass houses*, so named because you could only watch through the glass window as a highly trained computer operator ran the machine. Input was in the form of punch cards read by card readers, and output was in the form of printed results via local printers. A few local terminals existed for input purposes, mostly control and programming. All processing occurred on a *batch*, rather than an interactive, basis. In other words, the punch cards associated with a job were fed in a batch into a card reader, which transmitted the program and data input in a batch over a short circuit to the processor, which processed the program in a batch, and output the results in a batch to a printer, which printed in a batch. The first true mainframe was the IBM 360, introduced in 1964 [1]. As the first computer capable of both scientific and business computing, it went full circle—hence the designation *360* (degrees) [2]. During my pursuit of several degrees (academic) at the University of Texas at Austin, my computer programs, which seldom ran successfully, were processed in batch mode on a CDC 6600 mainframe computer. That *heavy-metal* machine cost roughly $6 million, occupied a huge room that was at least three times the size of a typical two-bedroom house, was water cooled, required tons of air conditioning to keep the ambient temperature low enough to age meat in the computer room, had enough large cabinets to house the population of a small nation, and possessed far less processing power and storage capacity than a contemporary laptop computer. (I may have exaggerated this a bit, but not much.)

Over time, it became possible for multiple users at dumb terminals to input information to the mainframe. The terminals were connected to ports on the mainframe through *terminal controllers*, or *cluster controllers*. Controllers essentially act as traffic *concentrators* to enable multiple inputs from clusters of slow terminals to share one of a limited number of very expensive ports on the host computer. As time ticked away, *Remote Job Entry* (RJE) was developed to enable groups of users seated at remote clusters of dumb terminals to enter data from a remote location, connecting through a cluster controller to the mainframe over a Wide Area Network (WAN), perhaps over a Dataphone Digital Service (DDS) circuit.

Parallel to the development of data networking, the computers themselves began to change. Computers became more powerful as processor speeds increased with the development of ever-faster microprocessors on silicon chips. Memory became more available as chip technology and hard-drive technology both improved. Additionally, computers became ever smaller and less expensive, to the point that the current typical desktop Personal Computer (PC) is equivalent to an early mainframe that would have filled a moderate-size office building. The PC was legitimized by the introduction of the IBM PC in 1983. By 1993, an estimated 75 percent of professionals in the United States had a workstation on the desktop. In this day and age of the mobile professional, many of those PCs are laptops, and tablet and handheld PCs are increasing in power by the day.

It was logical that all of this computing power and storage capability on all of these desktops would lead to a need to network those devices within the workplace. And it seems as though data traffic is more or less in line with the *Pareto principle*, with an estimated 80 percent of data transfer confined to the workplace, and only 20 percent traveling across the WAN. Whether or not that figure is accurate, PC users clearly have a requirement to share access to hosts, applications, databases, and printers. They also require a means to share access to WANs. Local Area Networks (LANs) provide a solution to those requirements.

Robert M. Metcalfe and his associates at the Xerox Palo Alto Research Center (Xerox PARC) first developed both the concept of a LAN and the enabling technology. That first network originally was known as the Altos Aloha Network because it connected Altos computers through a network based on the University of Hawaii's AlohaNet packet radio system technology. In a memo written May 22, 1973, it became known as *Ethernet*, from *luminiferous ether* [3], the omnipresent passive medium once theorized to pervade all space and to support the propagation of electromagnetic energy, even through a vacuum. The existence of the ether was disproved around 1900 by Albert Einstein, Albert A. Michaelson, Edward W. Morley, and others, but Ethernet thrived. This highly experimental technology supported a transmission rate of 2.94 Mbps over thick coaxial cable. Xerox commercialized the technology, renaming it *The Xerox Wire*. Gordon Bell, Vice President of Engineering at Digital Equipment Corporation (DEC, subsequently acquired by Compaq, which later merged with Hewlett-Packard), hired Metcalfe as a consultant in 1979 for the expressed purpose of developing a LAN technology that would not conflict with the Xerox patent. Metcalfe then facilitated a joint venture of DEC, Intel, and Xerox. Known as *DIX* [4], the venture standardized the technology in 1979 at 10 Mbps, reverting to the name *Ethernet*; it quickly became a de facto standard. LANs were recognized officially in February 1980, when the IEEE established Project 802 at the request of its members. (*Note:* Project *802* took its name from the fact that it was established in the year 19*80* and the month *2* (February). Once again, with the help of my consulting editor in this case, I have proved myself to be the master of the arcane.) In December 1982, the first standard was published and circulated. While IEEE 802.3, to which we commonly refer as *Ethernet*, actually is a variation on the Ethernet standard, I adopt the conversational reference throughout this book and do not dwell on that technical distinction. For our purposes, 802.3 and Ethernet essentially are one. (*Note:* Because the 802.3 and true Ethernet frame formats are dissimilar, they cannot interoperate.)

Ethernet clearly remains the most popular LAN standard. In part, that popularity is due to the fact that 802.3 was the first standard. In part, it also is due to the inherent simplicity of Ethernet, as compared to other standards such as Token Ring. Also, DEC's chip design team sourced chip manufacturing to Intel, Advanced Micro Devices, and Mostek, thereby creating a highly competitive environment that quickly led to low chip prices. Token Ring is considerably more complex and costly. Developed by IBM, Token Ring chips were sourced exclusively to Texas Instruments. The impact of this decision was that of higher prices due to the lack of competition [4]. According to Metcalfe's estimates, in 1994 there were 50 million Ethernet-connected computers, 5 million of which were on 10-Mbps networks [5]. Further, 500,000 Ethernet networks were TCP/IP registered and 50,000 were connected to the Internet. No doubt those numbers have increased considerably in the past dozen years, as have the speeds at which Ethernet runs.

This discussion of the basic concepts of LANs and LAN internetworking serves as the launching pad for discussion of the network technologies of the future. In this chapter, I address the definition, origin, and evolution of LANs and their application. Dimensions of LANs to be explored include media alternatives, physical and logical topology, baseband versus broadband, Medium Access Control (MAC), and standards and standards bodies. I define and illustrate bridges, routers, hubs, switches, and gateways as well as LAN operating systems. This chapter concludes with a discussion of LAN internetworking, remote LAN access, and recent developments in the realm of high-speed LANs and wireless LANs and Personal Area Networks (PANs).

8.1 LANs DEFINED

A *local area network* is a form of local (limited-distance) shared packet network for computer communications. LANs interconnect computers and peripherals over a common medium so users might share access to host computers, databases, files, applications, and peripherals. LANs conform to the *client/server* architecture, which essentially is a distributed computing architecture that takes advantage of the fact that both the client workstations and the servers are intelligent, programmable devices and exploits the capabilities of each. In such a network, *client* applications on microcomputers run against one or more centralized *servers*, which are high-performance multiport computers with substantial processing power and large amounts of memory. Some servers are positioned as devices that control the operational, administrative, and executive functions for the network, including authenticating legitimate users, granting access privileges to a database or perhaps a shared printer, and recording usage data. Some servers are positioned as database engines, that is, application or data repositories, capable of processing client requests for information and managing the resident data. *Note:* LANs also support peer-to-peer communications between clients and between servers.

Generally, LAN specifications are the province of the Institute of Electrical and Electronics Engineers (IEEE), although the American National Standards Institute (ANSI) and other standards bodies are involved, and the regulators are very much involved in spectrum allocation in the Wireless LAN (WLAN) domain. LANs operate at Layer 1, the Physical Layer, and Layer 2, the Data Link Layer, of the Open Systems Interconnection (OSI) Reference Model. Raw bandwidth ranges up to 10 Gbps, although actual throughput often is much less. LANs are limited to a maximum distance of only a few miles or kilometers, although they often operate within a much more confined area measured in feet or meters. LANs support the transmission of data in frame format, with the frames varying in size within specified minimums and maximums.

LANs are used almost exclusively for data communications over relatively short distances such as within an office, office building, or campus environment. LANs enable multiple workstations to share access to multiple host computers, other workstations, applications and databases, printers and other peripherals, and WAN connections. LANs traditionally are used in computer data applications, although they increasingly support video and voice communications as well.

8.2   LAN DIMENSIONS

LANs can be characterized along a number of common dimensions, for ease of understanding. Those dimensions include transmission medium, physical and logical topology, baseband versus broadband, and medium access control method.

8.2.1   Transmission Media

The shared media for LANs can include most of the transmission media discussed in Chapter 2. Although coaxial cable was the original medium, fiber-optic cable has superseded coax in the LAN backbone. Unshielded Twisted Pair (UTP) replaced coax to the desktop beginning in the early 1990s. Radio Frequency (RF) wireless technologies more recently have become extremely popular, particularly in providing the final link to portable and mobile computers. While Wireless LANs (WLANs) generally are limited to special radio technologies, InfraRed (IR) technology is used in certain applications, and microwave and IR systems connect LANs and LAN segments in a campus environment. Satellite rarely is used in any way because propagation delay renders it unsatisfactory for interactive communications. Satellite links also defy the notion of a *local* area network, although they sometimes are used to link LANs and LAN segments in remote areas.

8.2.1.1   Coaxial Cable   Coaxial cable was the transmission medium first employed in LANs. Although coax is expensive to acquire and to configure and reconfigure, its performance characteristics are excellent. Additionally, Data Processing/ Management Information Systems (DP/MIS) managers traditionally were comfortable with coax, which routinely was specified in the mainframe and midrange computer world. In fact, the technology did not exist until fairly recently to make effective use of other options such as twisted pair, fiber optics, and radio systems.

In retrospect, perhaps the use of coaxial cable lessened the resistance of DP/MIS managers to the concept of LANs. Those who lived in the mainframe world (most did) regarded PCs with disdain and sneered at twisted pair, which they referred to as *telephone wire*.

The advantages of coaxial cable include high bandwidth and exceptional error performance over relatively long distances as the thick inner core conductor results in fairly modest signal attenuation. Further, the outer shield rejects ElectroMagnetic Interference (EMI) and Radio Frequency Interference (RFI) as well as providing excellent security. Coax is also highly durable, but the costs of acquisition, deployment, and reconfiguration are high. While the disadvantages of coaxial cable have been mitigated to a large extent through the development of new coax designs, those designs also affect system performance. By way of example, consider three variations on the coax theme: ThickNet, ThinNet, and Twinax.

- **ThickNet:** *Thick* Ether*net*, also known as *10Base5*, was approved by the IEEE in 1983. 10Base5 uses traditional thick coax, often referred to as *goldenrod*, referring to its high cost, high value, and the yellow cable sheath used by some manufacturers. Other manufacturers used orange cable sheaths for thick coax, giving rise to the term *orange hose*. *10Base5* translates to *10* Mbps, *Base*band (one transmission at a time over a single, shared channel), and *500* m maximum segment length. While individual devices can be separated by much greater distances across the network, issues of signal attenuation limit each segment, or link, in the network to approximately 500 m.
- **ThinNet:** *Thin* Ether*net*, also known as *10Base2*, was approved by the IEEE in 1986. 10Base2 uses coax of thinner gauge. The thinner cable is less costly to acquire and deploy, although its performance is less in terms of transmission distance. *10Base2* translates to *10* Mbps, *Base*band, and 200 m maximum segment length (actually 185 m, rounded up).
- **Twinax:** *Twinaxial* cable, resembles ThinNet coax, but with *twin* coaxial conductors, rather than one. Twinax is used in older IBM midrange systems such as Systems 34, 36, and 38 as well as the younger IBM AS/400 and RS/6000. More recently, the IEEE has developed the *10GBase-CX4* standard in support of *10-Gigabit Ethernet* (10GbE). Based on the *Infiniband* high-speed cable assemblies, the specification calls for twinax assemblies operating over distances up to 50 ft. The standard calls for four transmitters and four receivers operating differentially in simplex mode over a bundle of eight twinax cables, with each simplex transmission occurring at 2.5 Gbps at a frequency of 3.125 GHz per channel with *8B/10B* line coding. The cost of this patch cord technology is expected to be approximately $\frac{1}{10}$ th that of comparable 10GBase-optical solutions.

8.2.1.2 *Twisted Pair*   Since the early 1990s, *unshielded twisted pair* has become very popular as a LAN medium. Although its performance characteristics are less appealing than coax, its low cost and high availability certainly are very attractive. As discussed in detail in Chapter 2, UTP of various categories performs very nicely at signaling speeds from 10 Mbps up to 1 Gbps over relatively short distances.

The advantages of UTP include its low costs of acquisition, deployment, and reconfiguration. The disadvantages of UTP include its relatively low bandwidth and poor error performance over long distances. Because the carrier frequency must be high to support a data rate of 10/100 Mbps or more and as high-frequency signals attenuate relatively quickly, error performance suffers considerably over a distance. Therefore, distances are severely restricted. Additionally, the radiated electromagnetic field is considerable at the high frequencies required to support high speeds, which poses security concerns. Security at this level, however, generally is not considered to be a significant issue, as the cabling system is restricted to the premises. More importantly, the radiated electromagnetic field can create noise that affects signals traveling on adjacent pairs in the same cable and in nearby cables. At 10 Mbps, however, Category 5 (Cat 5) cable commonly is used in a *structured wiring plan* to support both voice and data, with two pairs typically pulled to each duplex jack—one pair for voice and one for data.

The disadvantages of UTP have been mitigated to some extent, and the LAN applications have increased through the development and use of Cat 3, 4, and 5 UTP. Since Cat 5 is by far the most capable of these standard options, it currently is the inside wire default for both voice and data. Category 6 is now enjoying application in high-speed LANs and the Cat 7 specification is under development. The following discussions of 1Base5, 10Base-T, 100Base-T, 1000Base-T, and 10GBase-T serve to illustrate the evolution of twisted-pair applications in the LAN domain:

- *1Base5* (IEEE, mid-1980s) translates to *1* Mbps, *Base*band, and *500* m maximum segment length and was the predecessor to 10Base-T. AT&T spearheaded the 1Base5 initiative in support of its StarLAN product. 1Base5 runs over Cat 3, 4, or 5 UTP. 1Base5 is considered obsolete.

- *10Base-T* translates to *10* Mbps, *Base*band over *T*wisted pair (IEEE, 1990) and refers to Ethernet running over Cat 3, 4, or 5 UTP. The maximum segment length between the 10Base-T hub and the attached device (e.g., workstation or printer) is specified at 100 m or less, although good Cat 5 cable will perform well over somewhat longer distances. The 10Base-T hub is a wire hub that serves as a multiport repeater as well as a central point of interconnection.

- *100Base-T* (IEEE, 1995) is similar to 10Base-T Ethernet hub technology, running at 100 Mbps and requiring Cat 5 UTP or better. Distances originally were limited to 100 m over Cat 5 cable but now extend to 350 m over Cat 5e.

- *1000Base-T* (IEEE 802.3ab, 1999) is similar in concept to the predecessor 10/100Base-T. The original specifications called for Cat 5 cable to support Gigabit Ethernet (GbE) over four pairs and distances up to 100 m. Category 6 cabling specifications include UTP, *Shielded Twisted Pair* (STP), and *Screened Twisted Pair* (ScTP) rated at 250 MHz over distances up to 220 m.

- *10GBase-T* (IEEE 802.3an, June 2006) is a specification for 10GbE over Cat 6 cable for distances up to at least 55 m, although distances generally can be extended to 100 m. The expectation is that Cat 7 cable will extend those distances even further. Cat 7 is STP with a combination foil and braided screen construction. Cat 7 supports signaling rates up to 600 MHz, although the usable spectrum can be up to 750 MHz.

Category 3 (Cat 3) UTP also is used for 4-Mbps Token Ring LANs. Category 4 (Cat 4) UTP, developed for 16-Mbps Token Ring LANs, has a bandwidth of 20 MHz. In addition to its application in Cat 6 and 7 cables, STP sometimes is used in high-noise environments in which UTP data transmission is especially susceptible to EMI or RFI. Examples include manufacturing environments where there are large numbers of powerful machines, power plants, and old buildings (e.g., hospitals, or government or military facilities) where it might be impossible when installing a LAN to avoid placing wires close to electric motors or older fluorescent light fixtures.

8.2.1.3 *Fiber-Optic Cable*

Because of its outstanding performance characteristics, optical fiber also is used extensively in contemporary LAN applications. Its cost and fragility, however, generally relegate it to use as a backbone technology in *Fiber Distributed Data Interface* (FDDI) networks, for example.

The advantages of fiber certainly include the combination of high bandwidth and excellent error performance. Additionally, fiber performs well over long distances and offers excellent security. The disadvantages of fiber transmission systems include their high cost of acquisition, as compared to UTP systems. While the fiber, itself, is not significantly more expensive than Cat 5e UTP, the light sources and detectors are considerably more expensive than the metallic interfaces used with UTP. As fiber is very fragile, it must be protected carefully, and redundancy is always a good idea.

As LAN speeds have increased by orders of magnitude during the past several years, however, optical fiber has enjoyed great popularity for the interconnection of hubs, switches, and routers, many of which currently feature direct optical fiber interfaces. As noted in Table 8.1, the interconnection of 100-Mbps Ethernet hubs and switches is supported over much longer distances with fiber than with Cat 5 UTP. Therefore, Cat 5 UTP generally is relegated to terminal connections, although it can be used to interconnect hubs and switches over short distances in the backbone, and fiber generally is used only for backbone applications. While the specifics of the optical fiber system vary widely, Light-Emitting Diodes (LEDs) and Vertical-Cavity Surface Emitting Lasers (VCSELs) traditionally have been used in conjunction with various types of MultiMode Fiber (MMF). As is discussed in Chapter 2, this combination works well at speeds up to 1 Gbps or so over very short distances (300 m or so) and is relatively inexpensive. At higher speeds and over longer distances, the combination of laser diodes and Single-Mode Fiber (SMF) is the clear choice.

TABLE 8.1   Fast Ethernet (100 Mbps) Distance Restrictions [6]

Standard	Cable Type	Duplex	Maximum Nominal Distance
100Base-T4	≥Cat 3 UTP	Full or Half	100 m
100Base-TX	≥Cat 5 UTP	Full or Half	100 m
100Base-FX	MMF	Full	2 km
100Base-FX	SMF	Full	40 km

TABLE 8.2 Gigabit Ethernet (IEEE 802.3z) and 10GbE (IEEE 802.3ae) Media Specifications [6–7]

Standard	Fiber Type	Core Diameter (μm)	Wavelength (nm)	Modal Bandwidth[a] (MHz/km)	Distance Limitations (Maximum)
1000Base-SX	MMF	62.5	850	160	220 m
1000Base-SX	MMF	62.5	850	200	275 m
1000Base-SX	MMF	50.0	850	400	500 m
1000Base-SX	MMF	50.0	850	500	550 m
1000Base-LX	MMF	62.5	1300	500	550 m
1000Base-LX	MMF	50.0	1300	400	550 m
1000Base-LX	MMF	50.0	1300	500	550 m
1000Base-LX	SMF	9.0	1310	Not applicable	5 km
10GBase-SR, SW	MMF	62.5	850	160	300 m
10GBase-LR, LW	SMF	8.3, 9.0, 10.0	1310	Not applicable	10 km
10GBase-ER, EW	SMF	8.3, 9.0, 10.0	1550	Not applicable	40 km
10GBase-LX4	MMF	50.0, 62.5	1310	500	300 m
10GBase-LX4	SMF	10.0	1310	Not applicable	10 km

[a] *Modal bandwidth*, as expressed in MHz/km, is the measure of the capacity of a MMF in Gbps applications. A MMF with a higher modal bandwidth will support data transmission at a given rate over a longer distance. Modal bandwidth is determined by the dispersion characteristics of the fiber, including both modal dispersion and chromatic dispersion. The MMF core diameter (50 μm vs. 62.5 μm) is one of the fiber attributes that influence modal bandwidth.

The advent of Gigabit Ethernet (GbE) standards in 1999 definitely focused attention on optical fiber, although good-quality Cat 5, Cat 5e, and certainly Cat 6 have proven to be acceptable media, and Cat 7 promises to perform well at 1 and 10 Gbps. In support of the interconnection of GbE switches and in order to deal with issues of *modal dispersion* and *pulse dispersion*, which I detail in Chapter 2, several optical fiber standards were developed. 1000BaseSX uses *short-wave* lasers that operate at wavelengths of approximately 850 nm (nanometers). These short-wave lasers are relatively inexpensive and couple efficiently to low-cost MMF for transmission over relatively short distances. 1000Base-LX uses *long-wave* lasers that operate at wavelengths of approximately 1300 nm. 1000Base-LX offers improved performance over longer distances, although the cost of the technology is considerably greater than that of 1000Base-SX [6–8]. Table 8.2 provides a brief comparison of these two standards.

The IEEE 802.3ae standard for 10GbE addresses acceptable fiber media options, as listed in Table 8.2. The MMF standards are 10GBase-SW and 10GBase-SR, both of which use *shortwave* lasers that operate at wavelengths of approximately 850 nm, as is the case with the 1000Base-SX standard for GbE. Again, these short-wave lasers are relatively inexpensive and couple efficiently to low-cost MMF for transmission over relatively short distances. The *S* in *10GBase-SR* refers to the *S*hort range of the system, the *L* in *10GBase-LR* refers to *L*ong range, and the *E* in *10GBase-ER* refers to *E*xtended range. These standards are all intended for *dark fiber* applications, meaning that the fiber is inactive (i.e., dark or not lit) fiber, intended for a specific point-to-point 10GbE application, and not intended for interconnection to the WAN. The letter *X* indicates that the 8B/10B signal encoding technique is used. The letter *R* indicates that 64B/66B signal encoding is used. The number *4* in 10GBase-LX4 indicates the use of *4* wavelengths through *Coarse Wavelength Division Multiplexing* (CWDM). The *W* in *10GBase-SW* refers to the *WAN Interface Sublayer* (WIS) that enables compatibility between 10GbE equipment and SONET long haul equipment in a LAN-to-WAN interface scenario.

8.2.1.4 *Wireless*   Wireless LANs (WLANs) offer the obvious advantage of avoiding much of the time and cost associated with deploying wires and cables. This is especially important in a dynamic environment where portability is desirable, such as an office where cubicles are frequently reconfigured. WLANs have found acceptance in providing LAN capabilities in temporary quarters, where costly cabling soon would have to be abandoned, and in older buildings, where wires are difficult or impossible to run. WLAN technologies include both RF and IR.

The most common approach is that of RF, which involves fitting each device with a low-power transmit/receive radio antenna, which traditionally is in the form of a PC card. Newer laptop, tablet, and hand-held computers boast built-in antennas. Frequency assignments for commercial applications generally are in the 2.4- and 5-GHz bands. The physical configuration involves a hub antenna located at a central point (see Figure 8.1), such as the center or the corner of the ceiling, where Line-Of-Sight (LOS) or near-LOS connectivity can be established with the various terminal antennas. While LOS is not strictly required at these frequencies, it is always desirable and is particularly important at higher frequencies, which suffer greater attenuation from physical obstructions. The hub antenna then connects to the servers, peripherals, and other hosts via cabled connections, which also connect together multiple hub antennas for transmission between rooms, floors, buildings, and so on. In order to serve multiple workstations, *spread-spectrum* radio technology often is employed to maximize the effective use of limited bandwidth. *Frequency-Hopping Spread Spectrum* (FHSS) involves scattering packets of a data stream

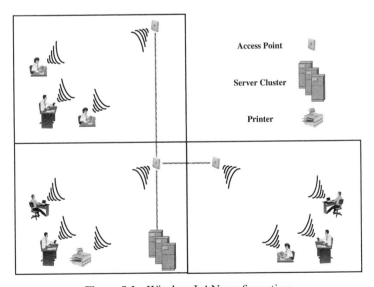

Figure 8.1   Wireless LAN configuration.

across a range of frequencies, rather than using a single transmission frequency. A side benefit of spread spectrum is that of increased security, since the signal is virtually impossible to intercept [9]. While the raw aggregate bandwidth of a wireless radio LAN generally is described as falling into a range (e.g., 2–11 Mbps) sensitive to link quality at any given time, the effective throughput generally is considerably less due to a variety of overhead issues. Some wireless LANs also use *Direct-Sequence Spread Spectrum* (DSSS) transmission, which spreads the original signal across a much wider band of frequencies, thereby yielding a greater likelihood that the signal will get through to the access hub [10]. Regardless of the frequency range employed, metal studs and thermal windows in newer buildings, lead paint in older buildings (e.g., my 1909 farmhouse), and all sorts of other sources of interference can combine to reduce the effectiveness of RF-based wireless LANs.

Most WLANs use unlicensed frequencies in the 2.4- and 5-GHz ranges. This approach avoids the expensive and time-consuming spectrum licensing process. However, the use of unlicensed frequencies creates significant potential for interference from other systems in proximity that use the same frequencies. A wide variety of other devices (e.g., garage door openers, bar code scanners, industrial microwave ovens, and cordless telephones) run in the same frequency ranges, which are in the *Industrial/Scientific/Medical* (ISM) bands. The risk works in both directions, of course, as there is the potential that you could click your mouse and cause garage doors to open and close all over the neighborhood. (That's a joke, just to keep things in perspective, although I suppose it's possible.) The systems that use licensed frequencies avoid the potential for interference but require that the manufacturer carefully police the deployment of such systems on a site-specific basis and under the terms of an omni-license.

A good WLAN example is the IEEE 802.11b Ethernet standard, aka Wi-Fi, which begins at theoretical maximum transmission rates of 11 Mbps, with optional fallback to 2 Mbps in noisy environments. The large number of products supporting that specification generally deliver actual throughput of considerably less, as they are sensitive to factors such as background noise, distance, physical obstructions, and competing systems running in the same band. The more recent 802.11a specification runs at much higher speeds in the 5-GHz range. The 802.11 standards use the same collision-based access protocol as is optional with more traditional wired Ethernet. I discuss WLANs in more detail toward the end of this chapter.

Infrared light also can be used as a WLAN transmission system. As is described in Chapter 2, a *Free Space Optics* (FSO) IR system generally requires a LOS between the light source and receiver. Within a room, however, it is possible to bounce the light signal off of a wall, ceiling, or other surface between the transmitter and receiver using a technique known as *diffused propagation* [5]. This method works well as long as the signal retains sufficient strength but is purely a trial-and-error technique, so results can vary widely. Laptop computers, tablet computer, and *Personal Digital Assistants* (PDAs) and other hand-held computers sometimes are equipped with infrared (IR) ports, as are printers. Microsoft's announcement (April 1999) of IR support for Windows 2000 provided something of a boost to the use of IR in the LAN domain. This turned out to be little more that moral support, however, as Wi-Fi and Bluetooth teamed up and quickly overwhelmed IR. A quick informal poll I conducted of network administrators across the country in June 2006 revealed that very few use IR. A few use it for occasional print jobs when visiting a remote site, but most who use IR do so only occasionally to sync up hand-held computers or cell phones. I also visited a local electronics retail store, where I found that no (i.e., zero) devices had IR ports. A quick search of the Dell and Compaq websites yielded no (i.e., zero) mention of IR.

Wireless LAN technology has matured very quickly and is continuing to evolve. Although acquisition costs are not necessarily low when compared to wired LANs, configuration and reconfiguration costs are virtually nonexistent. Wireless offers the considerable advantages of portability and even mobility. Security remains a concern, but recent developments in encryption technology have largely addressed that issue, at least in the short term. Actually, this sounds so good that I think I'll take my laptop out on the back deck and finish this chapter. (It's a beautiful spring day here in Mt. Vernon.) I can connect to the Web via an 802.11g connection to my Asymmetric Digital Subscriber Line (ADSL) router. Even though I connect over a distance of 30 m or so, around two corners, and through a wall, I can connect to the WLAN access point at no less than 2 Mbps, which is faster than the downstream signaling speed of 1.544 Mbps that I can achieve over my ADSL circuit.

8.2.2 Topology: Physical and Logical Configurations

The physical topology, or layout, of a LAN is in the form of a bus, ring, or star. As trains, ovals, planets, and constellations are not defined, you should avoid vendors promoting such topologies.

8.2.2.1 Bus Topologies   As shown in Figure 8.2, *bus* topologies are multipoint electrical circuits. The original bus topology employed coaxial cable, although contemporary bus systems also can make use of UTP or STP. Data transmission is bidirectional, with the attached devices transmitting and receiving in both directions. While generally operating at a theoretical raw data rate of 10/100 Mbps, actual throughput typically is much less. Bus networks employ a decentralized method of *Medium Access Control* (MAC) known as *Carrier Sense Multiple Access* (CSMA), which

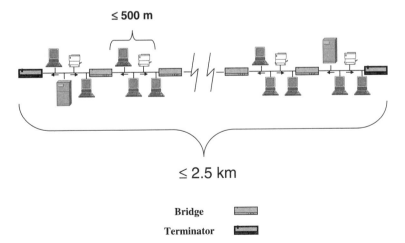

Figure 8.2   Ethernet bus topology.

enables the attached devices to make independent decisions relative to media access and initiation of transmission. Unfortunately, this approach results in data collisions, from which the transmitting device must recover through retransmission, which has a decidedly negative impact on throughput. As bus networks are not controlled from a centralized point, a given device cannot determine when, if ever, it will gain access to the shared bus. Therefore, such networks can be characterized as *nondeterministic* in nature. Bus networks are specified in the IEEE 802.3 standard and generally have a maximum specified length of 1.5 miles (2.5 km).

The original, classic Ethernet was based on a bus topology comprising coaxial cable segments that could be a maximum of 500 m in length. Each segment supports as many as 1024 ($2^{10} = 1024$) network addresses, each of which is associated with an attached device. The maximum segment length was due to issues of attenuation at the relatively high carrier frequency. Ethernet segments could connect through *bridges*, which function as signal repeaters. The total route length of the entire Ethernet was limited to 2.5 km, which is a function of both signal propagation time and MAC mechanisms. I discuss CSMA/CD and CSMA/CA, the Ethernet MAC standards, later in this chapter.

A *tree* topology is a variation on the bus theme, with multiple *branches* off the *trunk* of the central bus. Bus networks also suffer from the vulnerability of the bus—should the bus be compromised, the entire network is compromised. Similarly, tree networks are dependent on the integrity of the *root* bus [2].

8.2.2.2 Ring Topologies   *Ring* networks (refer to Figure 8.3) are laid out in a physical ring, or closed-loop, configuration. Information travels around the ring in only one direction, with each attached station or node serving as a repeater [2]. Rings generally are coax or fiber in nature, operating at raw transmission rates of 4, 16, 20, or 100+ Mbps. Rings are deterministic in nature, employing *token passing* as the method of medium access control to ensure all nodes can access the network within a predetermined time interval. Priority access is recognized. A master control station controls access to the transmission medium, with backup control stations assuming responsibility in the event of a master failure. Throughput is very close to raw bandwidth, as data collisions do not occur in such a carefully controlled

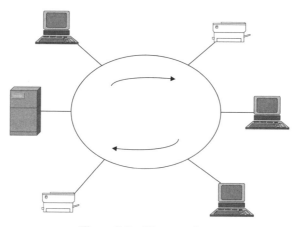

Figure 8.3   Ring topology.

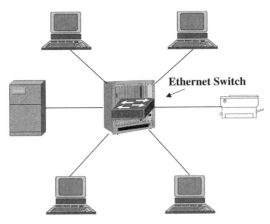

Figure 8.4   100Base-T star topology.

environment and, therefore, retransmissions are unusual. On the negative side, the failure of a single node can compromise the entire network, although many ring networks add a fail-safe mechanism to prevent a total ring failure in the event that a *lobe* is cut. In the case of *Token Ring*, a relay in the wiring hub would actuate if the lobe were cut, thus restoring the balance of the ring to proper operation. Electrical ring networks are specified in the IEEE 802.5 standard; FDDI is an ANSI specification. Token-Passing Ring, IBM Token Ring, and FDDI all are based on ring topologies.

8.2.2.3 Star Topologies   *Star* topologies (Figure 8.4) consist of a central node, or point of interconnection, to which all other devices are attached directly, generally via UTP or STP. The central node is in the form of a hub, switch, or router. Transmission rates vary, with 10Base-T operating at 10 Mbps and 100Base-T at 100 Mbps. Perhaps the greatest advantage of a star is that individual devices can connect to the node via UTP or STP. Another key advantage of a star is the ability to isolate a disruptive or failed station, thereby eliminating any negative effect it may have on overall LAN performance. Additionally, multiple attached devices simultaneously can share the full bandwidth of the LAN, at least in a switched environment. The primary disadvantage is that a hub failure is catastrophic; as all connectivity is provided through the central hub, its failure affects the entire LAN. Examples of star configurations include 10Base-T and 100Base-T. AT&T's StarLAN and DataKit and 100VG-AnyLAN are now considered obsolete. Asynchronous Transfer Mode (ATM), also based on a star topology, enjoyed the LAN switching spotlight briefly but was overwhelmed by switched Ethernet at 100 Mbps and 1 Gbps.

8.2.2.4 Physical versus Logical Topology   A network may be laid out physically in one fashion but operate logically in an entirely different manner. For example, 100Base-T (see Figure 8.4) physically appears as a star configuration because the devices are arrayed around a central node in the form of a hub. But 100Base-T operates as an *Ethernet bus*. The bus still exists, but under the skin of the central *hub* to which all stations in the workgroup connect via UTP. 10/100Base-T collapses the classic Ethernet coax bus into a *collapsed backbone* and places it under the protection of the hub chassis. In this fashion, the network gains the logical advantages of the Ethernet protocol as well as the physical advantages of a UTP-based star. Similarly, a ring network might operate logically as a ring but be supported physically by a collapsed backbone bus.

8.2.3 Baseband versus Broadband

Two LAN transmission options exist: baseband and broadband. Recall from Chapter 2 that a *baseband* signal is a signal in its original form, without being altered in any way, whether by modulation or conversion. A baseband transmission system, therefore, is a single-channel system that supports a single transmission at any given time. Also recall from Chapter 1 that, in the WAN domain, *broadband* is an imprecise term referring to a circuit or channel providing a relatively large amount of bandwidth. In a LAN context, broadband refers to a multichannel system that supports multiple transmissions through Frequency Division Multiplexing (FDM). While broadband LANs were quite common in the 1980s, they are very unusual in contemporary applications.

8.2.3.1 Broadband LANs *Broadband* LANs are multichannel, analog (i.e., RF-based) LANs (see Figure 8.5) typically based on coaxial cable as the transmission medium, although fiber-optic cable also is used on occasion [2]. Aggregate bandwidth may be as much as 500–750 MHz, supporting perhaps 20–30 channels, each with a width of 6 MHz, plus guardbands. The various channels are multiplexed onto the carrier through Frequency Division Multiplexing (FDM). Radio frequency modems accomplish the digital-to-analog conversion process, providing the digital device with access to an analog channel. The modems, which must be tuned and managed carefully, may be either fixed frequency or frequency agile. *Fixed-*

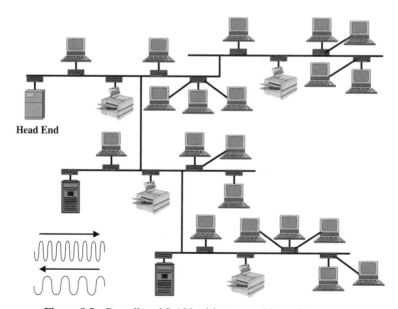

Figure 8.5 Broadband LAN, with tree-and-branch topology.

frequency modems must be tuned to a specific frequency channel, while *frequency-agile* modems can search for an available channel. Although frequency-agile modems are more expensive to acquire, they utilize available bandwidth much more effectively as they mitigate issues of congestion by automatically and dynamically balancing the load among the channels to which they have assigned access privileges.

There are single-cable and dual-cable broadband LANs. A dual-cable LAN involves one cable in support of upstream transmissions from the stations to the head end and a second cable in support of downstream transmissions from the head end to the stations. In a single-cable system, specific frequency channels are designated for upstream transmissions and others for downstream transmissions. The stations connect to the cable through multiport *Media Access Units* (MAUs) that house the RF modems and that connect to the coax cable, as illustrated in Figure 8.5. All transmissions go through the head end. Broadband LANs commonly use 75-ohm CATV–type coax and use CATV-style connectors, taps, filters, and amplifiers in a *tree-and-branch* architecture, which essentially is a variation of the bus with multiple branches off of a main *root bus*, much as there are branches off of the main trunk of a tree.

The IEEE standard for broadband LANs is *10Broad36*, translated as *10* Mbps, *Broad*band (multichannel), with *36*00 m maximum total span. The total span can be divided into multiple segments, each with a maximum distance of 1800 m. The aggregate bandwidth is 550 MHz and the FDM channels are 14 MHz wide with 4-MHz guardbands. The modulation technique is Differential Phase Shift Keying (DPSK), a unibit modulation technique discussed in Chapter 6 and illustrated in Figure 8.6.

The characteristics of broadband LANs, generally speaking, are not endearing, so they are seldom used any longer. But their unique properties do have application. In the mid-1990s, for example, Sea World installed broadband LANs in its theme parks to support analog audio (music and voice paging), closed-circuit analog video (entertainment TV and security), as well as data. As the LAN is analog, it easily supports audio and TV. Further, the application is static, rather than dynamic, because the paging zones and closed-circuit TV channels require fixed amounts of bandwidth, and the associated frequency channel assignments need to be changed infrequently, if ever. Further, the locations of the terminal equipment (music and

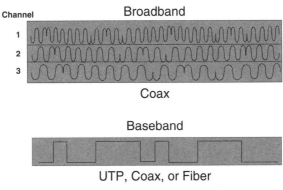

Figure 8.6   Broadband versus baseband.

paging sources and speakers, VCRs, video cameras, and TV monitors) are fixed or seldom change. The data transmissions typically are low speed in nature and in support of transaction processing applications such as cash and credit card transactions initiated from the ticket counters, restaurants, and gift shops. Since the transaction processing applications are low speed, the associated channels are narrowband.

8.2.3.2 Baseband LANs *Baseband* LANs are digital and single channel in nature, supporting one transmission at a time (see Figure 8.6). The full range of media options are available, including both wired (i.e., coax, UTP, STP, and fiber-optic cable) and wireless (i.e., RF and IR). Distance limitations depend on the medium employed and the specifics of the LAN protocol. Baseband LAN physical topologies include ring, bus, and star.

Baseband LANs are by far the most popular and therefore the most highly standardized. Ethernet, token-passing bus, Token Ring, and FDDI are all baseband in nature. While LANs were developed exclusively for computer-to-computer data communications applications, it fairly recently has become quite possible, if not commonplace, to support voice, video, and even videoconferencing over LANs. The support of such isochronous traffic offers clear advantages in support of workgroup communications and lower equipment and cabling costs through a shared infrastructure. As we discussed in Chapter 3, the new generation of PABX systems are client/server IPBX systems running Voice over Internet Protocol (VoIP) over switched Ethernet LANs.

8.2.4 Medium Access Control

Medium access control (MAC) describes the process that is employed to control the basis on which devices can access the shared medium. Some level of control is required to ensure, or at least improve, the ability of all devices to access the network within a reasonable period of time, thereby resulting in acceptable *access times* and *response times*. It also is important that some method exist to either detect or avoid data collisions, which are caused by multiple transmissions placed on the shared medium simultaneously, and to recover from them.

Medium access control takes place at Layer 1, the Physical Layer, and Layer 2, the Data Link Layer, of the OSI Reference Model. Medium access control is accomplished at the *Network Interface Unit* (NIU), or *Network Interface Card* (NIC). A NIU or NIC is at the board level, with the boards typically fitting into an expansion slot of an attached device (e.g., workstation, printer, or server). Alternatively, multiple cards may be contained within a multiport device that supports multiple workstations on a pooled basis. Each NIU/NIC has a unique logical address for purposes of identification, with the address hard coded on a silicon chip at the time of manufacture. Medium access control can be accomplished on either a centralized or decentralized basis and can be characterized as either deterministic or nondeterministic in nature.

8.2.4.1 Centralized Control Centralized control involves a centralized controller that polls devices to determine when access and transmission by each station can occur. Stations can transmit when they are polled or when a station transmission request is acknowledged and granted. This process of polling requires the passing of control packets, which entail overhead and therefore reduce the amount of throughput relative to the raw bandwidth available. Additionally, the failure of the central controller disrupts the entire network; in such an event, the controller is taken offline and a backup controller assumes responsibility. Centrally controlled networks generally employ deterministic access control, whereby for each device the network administrator can establish either the specific point in time at which it is provided access or the maximum interval of time that transpires between access opportunities. The primary advantage of centralized control is that access to the shared network is managed on an orderly (controlled) basis. Access can be provided in consideration of several levels of priority, with the most critical transmissions gaining privileged levels of access. Alternatively, all devices can be of equal priority and therefore can share equally in access privileges. Token Ring and FDDI networks are examples of centralized control.

8.2.4.2 *Decentralized Control* Decentralized control is somewhat anarchistic, as each station assumes responsibility for controlling its access to the shared network. Additionally, each station must assume responsibility for detecting and resolving any data collisions that might occur in the quite possible event that its access to and transmission over the shared medium overlaps with that of other devices. Decentralized control networks generally use a nondeterministic, or contentious, MAC mechanism. By way of example, Ethernet control is decentralized.

8.2.4.3 *Deterministic Access* *Deterministic access* is a MAC convention that enables both the centralized master station, which commonly is in the form of a server, and each slaved station to determine the maximum length of time that passes before access is provided to the network. In other words, each station can be guaranteed the right to communicate within a certain time frame. Additionally, the system administrator can assign access priorities. Deterministic access is also known as *noncontentious* because the devices do not contend for access; rather, access is controlled on a centralized basis.

Deterministic access employs *token passing*. The *token*, which consists of a specific bit pattern, indicates the status of the network—*available* or *unavailable*. The token is generated by a centralized master control station and transmitted across the network. The station in possession of the token controls the access to the network. That station may either transmit or require other stations to respond. Transmission is in the form of a data packet of a predetermined maximum size, determined by the number of nodes on the ring and the traffic to be supported; oversized transmissions are segmented, or fragmented. After transmitting, the station passes the token to a successor station in a predetermined sequence. While the process is complex and overhead intensive, its high level of control over the network avoids data collisions.

Deterministic access is especially effective in high-traffic environments where a lack of control causes chaos in the form of frequent data collisions. It also finds application in environments such as process control, where it is critical that each station have guaranteed access to the network at precise points in time. Oil refineries, for example, employ deterministic MAC through proprietary token-passing LANs. The refining process requires that the specific nature of the raw material be considered in terms of a number of characteristics, including sulfur content, paraffin content, and viscosity. With those factors in mind, the refining process is tailored to act on the crude oil in such a way as to ensure that the end product (e.g., 92 octane unleaded gasoline) remains consistent from one batch to another. In a hypothetical scenario, the master control station addresses tokens to individual devices in the form of various sensors. The individual sensors monitor pressure and temperature conditions as well as the rate of flow of various chemicals through valves throughout the entire process. In order to control the process effectively, the master controller must address each sensor at precise points in time to gather critical data about the processes. At certain times during the process, some devices may require more frequent access. Under specific circumstances, such as an alarm condition, some devices may require high-priority access. In consideration of the data provided by the sensors, the master controller may command other devices to increase or decrease temperature, open or close valves, and so on. Such a network also is highly redundant so that a network or device failure does not compromise the integrity of the process.

Manufacturing Automation Protocol (MAP) is a good example of a deterministic protocol. Developed by General Motors (GM) in the early 1980s for the interconnection of computers and programmable machine tools in factory or assembly line operations, MAP is based on Token Bus (IEEE 802.4) running at 1, 5, 10, and 20Mbps. GM developed MAP as a multivendor solution using off-the-shelf parts, including coax cable, taps, connectors, amplifiers, splitters, and terminators. MAP sometimes is referred to as *Manufacturing Automation Protocol/Technical and Office Protocol* (MAP/TOP). (*Note*: The 802.4 committee disbanded in 2004 as the Token Bus standard has been in hibernation for some years. Token Bus is yet another casualty of Ethernet.)

CATV providers in a convergence scenario are applying deterministic protocols for Internet access, which is provided to large numbers of residences, businesses, and schools over a two-way coaxial cable system terminating in cable modems at each customer premises. As much as 500+ Mbps is reserved for such applications, with as much as 10Mbps available to an individual user. In order to manage contention over such a network, which extends over fairly significant distances, a token-passing MAC technique is employed, with the master station positioned in a *Cable Modem Termination System* (CMTS) physically located at the CATV provider's head end.

General characteristics of token-based networks include a high level of access control, which is centralized. Access delay is measured and ensured, with priority access supported. Throughput is very close to raw bandwidth, as data collisions and therefore retransmissions are avoided. Throughput also improves under load, although absolute overhead is higher than with nondeterministic access techniques. Deterministic access standards include Token-Passing Ring, IBM Token Ring (Figure 8.7), and Token Bus.

Token-based LAN technologies are somewhat overhead intensive, due to the token-passing and management processes. But they can more than compensate for that fact by avoiding data collisions and the retransmissions required to recover from them. Token Ring, for example, comes in 4- and 16-Mbps flavors; in each case, bandwidth utilization is virtually 100 percent under full load [11].

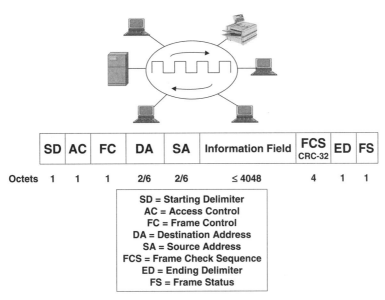

SD	AC	FC	DA	SA	Information Field	FCS CRC-32	ED	FS
Octets 1	1	1	2/6	2/6	≤ 4048	4	1	1

SD = Starting Delimiter
AC = Access Control
FC = Frame Control
DA = Destination Address
SA = Source Address
FCS = Frame Check Sequence
ED = Ending Delimiter
FS = Frame Status

Figure 8.7 Token passing, with 4-Mbps Token Ring frame format.

8.2.4.4 Nondeterministic Access *Nondeterministic access*, or *contentious* MAC, places access control responsibilities on the individual stations. Also known as *Carrier Sense Multiple Access* (CSMA), this approach is most effective in low-traffic environments. There are two variations on the theme: CSMA/CD and CSMA/CA.

CSMA is a decentralized, contentious MAC method used in Ethernet and other bus-oriented LANs. The *carrier* frequency is *sensed* by each of *multiple* stations, or nodes, to determine network availability before *accessing* the medium to transmit data. Further, each station must monitor the network to determine if a collision has occurred because collisions render the transmission invalid and require a retransmission.

CSMA works much like an old telephone *party line*, where there are multiple subscribers with individual logical addresses in the form of telephone numbers, all of which are connected to a single physical circuit. When placing an outgoing call, the subscriber must pick up the telephone to monitor the line for a short while to sense the level of activity. If there is no activity, a call can be placed. If there already is a call in progress, another call attempt cannot be made without causing interference (and hard feelings). Rather, the subscriber must hang up the telephone and subsequently monitor the circuit on some basis in order to determine its availability. Incoming calls are addressed to each party on the party line by varying the number of rings, indicating the unique logical address of each party sharing the circuit, and only the target party can answer the call without violating security and thereby creating ill will. In other words, MAC protocols govern the manner in which the circuit is managed to the satisfaction of all parties. As a result, collisions are less likely. Variations on the theme include Nonpersistent CSMA, 1-Persistent CSMA, and P-Persistent CSMA:

- *Nonpersistent CSMA* allows a machine to transmit data whenever it senses an idle channel. If the channel is busy, the machine backs off the network, calculates a random time interval, and again monitors the channel when that interval expires. This approach mathematically distributes the temporal monitoring of the network, thereby reducing the likelihood that multiple stations will sense its availability at approximately the same time and transmit simultaneously. This is a fairly patient approach to network access.

- *1-Persistent CSMA* also allows a machine to transmit data whenever it senses an idle channel. If the channel is in use, the machine will continuously sense it until the channel becomes free. The protocol gets its name because the machine is *persistent*, that is, tenacious or obstinate, in its monitoring of the channel and transmits with a probability of *1.0*, that is, 100 percent certainty of access success, whenever the channel is idle. If the network includes a large number of stations persistently monitoring the network, a great many of them might sense the availability of the network and begin to transmit simultaneously, virtually guaranteeing a collision. In such a scenario, 1-Persistent CSMA can be characterized as eager, if not downright greedy.

- *P-Persistent CSMA* allows a machine to transmit a frame during an idle time with probability P or lower, based on the length of the idle time as measured by a *time slot*. A time slot is the maximum packet transmission time for a station at one extreme end of the network to send a packet to a station at the opposite extreme end of the network and is based on the physical length of the cable, the physical size of the frames, and the speed of signal propagation through the wire. If a machine senses an idle condition on the channel, it transmits with probability P for one time slot. The machine then delays for worst-case propagation delay for one packet with probability 1-P. If the channel is busy, the machine listens persistently until the channel becomes idle and starts over. For example, P.01 means that there is a probability of 1 percent that the transmission will be unsuccessful. If P is set very low (e.g., .01), throughput is nearly 100 percent, but transmission delays will be very long, as the machine will wait a very long time between idle periods. This highly cautious approach introduces useless delays at low loads but certainly improves the rate of successful transmissions at high loads.

CSMA is implemented in two standard means: CSMA/CD and CSMA/CA. In either case, latency and throughput degrade under heavy loads of traffic; for example, a classic Ethernet network running at the theoretical speed of 10 Mbps typically delivers throughput of no more than 4–6 Mbps, and often much less. Note that CSMA is half-duplex (HDX) in nature, and only one transmission can take place at any given time. While it is less costly than Token Ring networking, it also delivers less efficient bandwidth utilization.

8.2.4.4.1 Carrier Sense Multiple Access with Collision Detection CSMA/CD is the most common MAC method used in bus networks (Figure 8.8). In an Ethernet environment, for example, the transmitting station sends a data packet in both directions of the bus. The 802.3 Ethernet frame (i.e., packet) takes the following form:

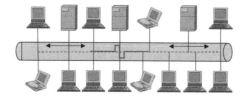

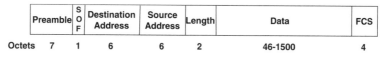

Preamble	S O F	Destination Address	Source Address	Length	Data	FCS
7	1	6	6	2	46-1500	4

Octets

Figure 8.8 CSMA/CD, with 802.3 frame format.

- **Preamble:** The preamble comprises seven octets in an alternating pattern of 1s and 0s that advise the receiving stations that a frame of data is arriving.
- **Start Of Frame (SOF):** The SOF delimiter is a single octet that ends with two consecutive 1 bits that serve to synchronize the receiving stations on the rate of transmission.
- **Destination and Source Addresses:** The destination address and source address fields are the addresses of the target station and the originating station, respectively. Each address comprises each six octets, the first three of which are specified by the IEEE on a vendor-dependent basis and the last three of which are assigned by the vendor. The address is hard coded on the NIC at the time of manufacture. Theoretically, each NIC and therefore each station have a unique address.
- **Length:** The length field of two octets indicates the number of octets of data in the data field.
- **Data:** 802.3 Ethernet frames have a lower limit of 64 octets and an upper limit of 1518 octets. In consideration of the fact that 18 octets are consumed with layer 1 and layer 2 processing, the data field, or payload, must comprise 46–1500 octets. In the event that the payload is less than 46 octets, padding bytes are inserted.
- **Frame Check Sequence (FCS):** The FCS consists of a 32-bit *Cyclic Redundancy Check* (CRC) that is appended to the frame trailer for purposes of error control [12].

Each *transceiver* (*transmitter/receiver*) of each station along the way reads the address in the frame header. If the address matches, the transceiver provides the frame to the target device. If the address does not match, the transceiver forwards the frame to the next transceiver. If any node detects a data collision, that station sends a brief *jamming signal* over a subcarrier (i.e., lower) frequency of the network to advise all stations of the collision. Then all devices back off the network, with each calculating a random time interval before attempting a retransmission. One of the implications of this approach is that transmission is half-duplex (HDX), for full-duplex (FDX) transmission would yield collisions. CSMA/CD is designed to work with frames of specific minimum and maximum sizes. In the Ethernet environment, for example, note that the frame (packet) size varies in length from 64 to 1518 octets, with the application software driver forming frames of the proper specific size. A VoIP frame, for example, would be of the minimum size of 64 octets, and a file transfer most likely would involve frames of the maximum size of 1518 octets.

Several factors help to determine the maximum packet size. First, and as the Ethernet is shared, it is necessary to fragment a file transfer into smaller subsets of data so one transmission does not speak to the entire level of bandwidth available across the network for a long period of time. Second, the maximum frame size is a trade-off between raw efficiency and throughput. In other words, a stream of unfragmented, unframed data is most efficient because it requires little overhead. But a single bit error in a data stream associated with a bulk file transfer might require that the entire file be retransmitted in order to correct for the error. This process might have to repeat an infinite number of times, with the file never making it from transmitter to receiver without error. By fragmenting the file into subsets of data in the form of frames, the level of overhead increases, as each frame requires 18 octets of overhead. An errored bit, however, is confined to a single frame, which easily can be retransmitted without significant likelihood of error and without serious degradation of overall throughput across the LAN.

The minimum packet (frame) size is a direct function of the design of the classic Ethernet and the CSMA/CD control mechanism. In the most extreme case, an Ethernet comprises a number of segments, each of which is up to 500 m in length, supports as many as 1024 addresses, involves a great number of transceivers, and connects to one or more other segments through a bridge. The total route length of the classic Ethernet is up to 2.5 km. Given issues of propagation delays across the segments and through the bridges, it takes a certain amount of time for a frame to traverse the network from the originating device to the target device, which is 2.5 km away in the most extreme possible case. If that frame of data encounters a data collision at the distant end just before it reaches the target device, it takes an identical amount of time for a collision notification to be received by the transmitting device across the subcarrier channel. Then a retransmission is used to adjust to that fact. If such notification is not received in time, the originating device assumes that the data were received in good form when that is not at all the case. Therefore, the minimum frame size in classic Ethernet is 64 octets, which includes 46 octets of payload and 18 octets of overhead.

8.2.4.4.2 Carrier Sense Multiple Access with Collision Avoidance CSMA/CA includes a priority scheme to guarantee the transmission privileges of high-priority stations. CSMA/CA requires a delay in network activity after each completed transmission. That delay is proportionate to the priority level of each device, with high-priority nodes programmed for short delays and low-priority nodes programmed for relatively long delays. As collisions still may occur, they are managed either through *collision detection* or through retransmission after receipt of a *Negative AcKnowledgment* (NAK). CSMA/CA is more expensive to implement because it requires that additional programmed logic be embedded in each device or NIC. CSMA/CA does, however, offer the advantage of improved access control, which serves to reduce collisions and thereby improve the overall performance of the network. Note that CSMA/CA remains HDX in nature.

Wireless LANs, as standardized in IEEE 802.11, employ CSMA/CA. The 802.11 standard uses a *positive ACKnowledgment* (ACK) mechanism which requires that the transmitting station first check the medium to determine its availability. The transmitter sends a short *Request-To-Send* (RTS) packet that contains the source and destination network addresses as well as the duration of the subject transmission. If the shared medium is available, the destination station responds with a *Clear-To-Send* (CTS) packet. All devices on the network recognize and honor this acknowledged claim to the shared network resources. If the source station does not receive an ACK packet from the destination station, it retransmits RTS packets until access is granted.

8.3 LAN EQUIPMENT

In addition to the attached transmit and receive devices, aka *nodes* or *stations*, LANs may make use of other devices to control physical access to the shared medium, extend the maximum reach of the LAN, switch traffic, and so on. Such hardware is in the form of NICs/NIUs, transceivers, MAUs, bridges, hubs, routers, and gateways. As is true of much of the technology addressed in this book, the lines increasingly blur between these devices. Therefore, I focus on the classic definitions, expanding on those concepts and introducing discussion of multifunctional devices as appropriate.

8.3.1 Network Interface Cards

Also known as *Network Interface Units* (NIUs), *Network Interface Cards* (NICs) are chipsets on printed circuit boards that provide physical access from the node to the LAN medium. The NIC is responsible for fragmenting the data transmission and formatting the data packets with the necessary header and trailer. A standard IEEE NIC contains a unique, hard-coded logical address, which it includes in the header of each data packet it transmits. The NIC typically has some amount of buffer memory, which enables it to absorb some number of bits transmitted by the associated device, form the packets, and hold them until such time as the network is available. In the context of the OSI Reference Model, NICs function at the lower two layers, the Physical and Data Link layers. The NIC also may contain a microprocessor that can relieve the attached device of some routine computational functions.

The NIC (refer to Figure 8.9) can take a number of forms, including a circuit board that fits into the expansion slot of a desktop PC, a PCMCIA card, or a stand-alone device. A NIC commonly is embedded in a desktop, laptop, tablet, or hand-held computer at the time of assembly. *Transceivers* (*trans*mitter/re*ceivers*) are used in LANs to receive a carrier signal and then transmit it on its way. They are embedded in NICs/NIUs and MAUs. MAUs (*Medium Access Units*, or *Multistation Access Units*) are stand-alone devices that contain NICs in support of one or more nodes. MAUs are very unusual in contemporary networking.

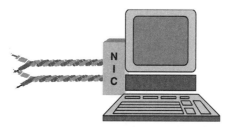

Figure 8.9 Ethernet-attached workstation with NIC.

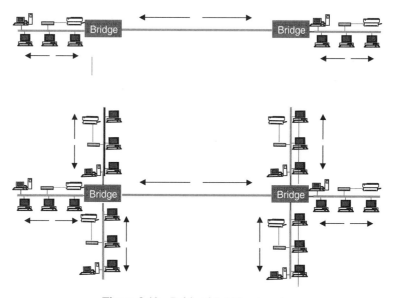

Figure 8.10 Bridged LAN network.

8.3.2 Bridges

Bridges are relatively simple devices that connect LANs of the same architecture (e.g., Ethernet to Ethernet). Bridges operate at the lower two layers of the OSI Reference Model, providing Physical Layer and Data Link Layer connectivity. A bridge, at the most basic level, acts simply to extend the physical reach of a LAN, passing traffic from one LAN segment to another based on the destination address of the frame. In other words, bridges act as LAN repeaters where specified distance limitations are exceeded. Bridges have buffers so they can store and forward frames in the event that the destination link is congested with traffic. Two-port bridges, as illustrated in Figure 8.10, are the most common configuration.

A key advantage of bridges is their inherent simplicity. As protocol-dependent devices, they do not perform complex processes on the data frames traveling through them; neither do they attempt to evaluate the network as a whole to make end-to-end routing decisions. Rather, bridges simply read the destination address of the incoming frame and forward it along its way to the next link. Bridges can be cascaded, or connected in series, link by link. As bridges are so simple, they are

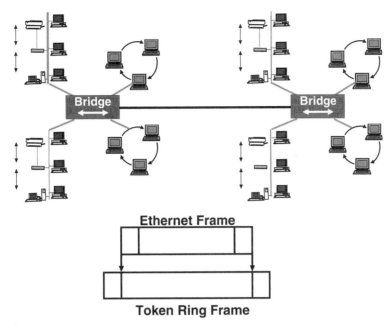

Figure 8.11   Encapsulating bridge, supporting Ethernet to Token Ring.

inexpensive and fast. Such bridges can support multiple LANs and LAN segments connected by multiple media. Essentially, multiple ports are provided with interfaces to an appropriate combination of coax, UTP, STP, RF, infrared, and fiber-optic transmission systems. Figure 8.10 also illustrates a pair of four-port bridges interconnected. In contemporary networking, however, a switch generally would be used in such a situation.

It also is possible for a bridge to support multiple LANs of disparate origin. For example, Ethernet-to-Ethernet and Token Ring-to-Token Ring connectivity can be provided [13]. It also is possible to interconnect disparate LANs, such as Ethernet to Token Ring, through the use of an *encapsulating bridge*. Such a bridge encapsulates the native LAN frame, surrounding it with control information appropriate to the LAN to which the target device is attached (Figure 8.11).

More sophisticated bridges add more functionality, although they are more expensive and slower. Such bridges also can *route* traffic at a simple level between LANs based on the destination addresses. The system administrator can enter the routing table into program logic or it can be *learned* by the bridge as it views the originating addresses of traffic passing through it over a period of time [14]. When initialized, *self-learning bridges* typically broadcast a query to all attached devices. When the devices respond to the query, the bridges associate the originating addresses of the response data frames with the port over which that incoming data were presented. In this fashion, they build address tables on a port-by-port and, by implication, segment-by-segment, basis. Subsequently, the bridges view the destination addresses of transmitted frames, consult the address table, and forward the frames only over the link connected to the proper port. If the frame is intended for a station on the same segment, the bridge simply ignores it, rather than passing it on. Since the frames are not forwarded across other links, such *filtering bridges* do a great deal—at very low cost—to relieve overall congestion on a segment-by-segment basis. This simple level of LAN *segmentation* improves overall access and throughput. Over time, the bridges add new addresses to their routing tables and delete old addresses that have not been viewed in some definable period of time that is much like an expiration date. From time to time, the bridges may repeat the broadcast query process in order to rebuild and resynchronize their address tables.

Medium Access Control (MAC) *bridges* are more sophisticated, still. MAC bridges have the ability to connect disparate LANs (e.g., Ethernet to Token Ring). This is accomplished through the process of *encapsulation*, or *translation*, as depicted in Figure 8.10. When operating in this mode, the bridge alters the frame format by encapsulating, or enveloping, the original frame with control data specific to the protocol of the destination LAN supporting the target device. Such an approach might be used to connect an Ethernet LAN to a Token Ring LAN or where two Ethernet LANs are connected via an intermediate FDDI backbone [15].

Specific bridge protocols include Spanning Tree, Source-Routing Protocol, and Source-Routing Transparent:

- *Spanning Tree Protocol* (STP) bridges, also known as *learning* bridges, are defined in IEEE 802.1 standards. Spanning tree bridges are self-learning, filtering bridges for use in connecting LANs on a point-to-point basis. The bridge is programmed or teaches itself the addresses of all devices on the network; subsequently, the network *tree* of the bridge provides only one *span* (link) for each LAN-to-LAN connection. Some spanning tree bridges also have the capability to provide security by denying access to certain resources based on user and terminal ID. Bridges that support the spanning tree algorithm have the ability to automatically reconfigure themselves for alternate paths if a network segment fails, thereby improving overall reliability [16].
- *IBM Source-Routing Protocol* (SRP) packets are programmed with specific routes (i.e., lists of bridges), based on considerations such as the physical location of the nodes and the capacity of the links involved. The maximum number of bridges *hopped* is 13. SRP bridges are most commonly used in Token Ring networks.
- *Source-Routing Transparent* (SRT) is defined in the IEEE 802.1 standard. It is effectively a combination of STP and SRP. The SRT router can connect LANs by either method, as programmed [2].

8.3.3   Hubs

Hubs reflect the trend toward star and away from bus configurations. Hubs can be either active or passive. *Passive* hubs act simply as cable-connecting devices, while *active* hubs also serve as signal repeaters [2]. The first generation of hubs (1984) acted as LAN concentrators and repeaters, with a single internal collapsed backbone bus for connecting like LANs. The second generation accommodated multiple LAN architectures (e.g., Ethernet and Token-Passing Ring) over separate ports, with rudimentary network management and configuration capabilities included as well [15].

A *collapsed backbone* is a fairly simple concept, and one worth exploring in some detail. LANs traditionally work on the basis of a common electrical bus (i.e., shared cable medium) to which each device or group of devices is directly connected. While this approach works effectively, it requires that the cable be deployed through the entire workplace. The traditional coax medium is expensive to acquire and deploy, as is fiber-optic cable. Additionally, the cable is susceptible to physical damage unless conduits, armoring, or some other means protects it. All of these involve additional cost. Alternatively, the high-speed backbone bus can be collapsed and placed within a hub, with UTP providing the connections between the hub ports and the jacks into which the user plugs various devices such as workstations and printers. Power users can connect to high-speed ports and casual users to low-speed ports. This approach protects the backbone bus and reduces the cost of cabling. Should an individual device (e.g., workstation, NIC, or MAU) create difficulty or should a UTP cable suffer damage, it is a relatively simple matter to disable the associated port and thereby isolate the problem. While the UTP cable is inherently less capable than coax or fiber, equivalent bandwidth can be provided as long as the distances between the devices and the hub are within tolerable limits. This is an accurate description of a 10/100Base-T hub configuration, as illustrated in Figure 8.12. Multiple hubs can be interconnected with various media, depending on bandwidth requirements and distances involved.

This collapsed backbone is analogous to the method by which electrical wiring is run in a home or office. One approach is that of running a thick electrical bus cable all through the walls, floors, and ceilings, splicing in outlets as required. The better approach is that of collapsing the bus and placing it in a centralized circuit breaker box, where the connection to the wide area electrical grid is made and common electrical ground is established. Connections to outlets are made through circuit breakers that snap into the common electrical bus, with some circuit breakers being for power users (e.g., washing machines and dryers) and others for devices drawing less power (e.g., lamps, hair dryers, and food processors). The appliances plug into outlets that are connected to the common electrical bus by relatively thin gauge electrical wiring that is fairly inexpensive and easily installed. In the event that a device causes a problem (e.g., shorts out or draws too much power), it is isolated automatically when the circuit breaker trips.

In the context of the OSI Reference Model, hubs operate at Layer 1, the Physical Layer, with a hint of Layer 2, the Data Link Layer. Since a hub is protocol specific, like a bridge, it works quickly. Actually, a hub is very much like a bridge, except that it provides terminal connectivity on a twisted-pair basis. A hub inherently does nothing internally to control congestion, except for filtering interhub traffic. A 10Base-T hub, for example, runs the Ethernet CSMA/CD protocol over the

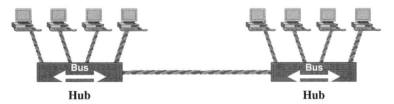

Figure 8.12 The 10/100Base-T hubs, interconnected.

collapsed backbone bus, and the attached devices do the same through UTP NICs. Hubs do have the effect of controlling congestion, however, because they are positioned as workgroup solutions that serve to confine traffic to the users connected to the hub or group of interconnected hubs, much as filtering bridges serve to confine traffic to a coax segment in a classic Ethernet implementation. In other words, a *collision domain* can be defined as single hub or group of interconnected stackable hubs.

Intelligent hubs, the third generation, provide multiple buses for multiple LANs of either the same or disparate architectures, in much the same manner as encapsulating bridges, which are illustrated in Figure 8.11. They can support multiple media (e.g., UTP and fiber), multiple speeds (e.g., 10/100 Mbps), and multiple LAN protocols (e.g., Ethernet and Token Ring). (*Note:* Support for multiple LAN protocols is unusual in contemporary networking, as Ethernet has overwhelmed its competition.) As addressable devices, intelligent hubs can be managed centrally via *Simple Network Management Protocol* (SNMP) or another appropriate network management protocol. Intelligent hubs also provide bridging and basic routing capabilities [15].

Regardless of the generation of the hubs, they serve, at minimum, as central points of interconnection for LAN-attached devices. Additionally, they serve as concentrators of LAN traffic and as repeaters, with multiple hubs interconnected through high-speed media [17]. A number of hub manufacturers offer stackable hubs, which offer the advantage of scalability; in other words, the hubs can be physically stacked and interconnected to increase port and traffic capacity [18]. Hubs, like bridges, do a good deal to reduce congestion through LAN segmentation and at very low cost.

8.3.4 Switches

LAN switches are intelligent hubs with basic packet store-and-forward capabilities that can support multiple simultaneous transmissions. Switches have the ability to read the target addresses of the packets and forward them only and directly to the appropriate port associated with the target device. That device may be directly attached to the switch, may be attached to a lesser workgroup switch, or may be connected to a hub that connects to the switch, as illustrated in Figure 8.13.

The LAN switch architecture can take several forms:

- **Shared Bus:** A *shared bus* switch has a single high-speed bus that is shared by all incoming and outgoing ports on a Time Division Multiplexing (TDM) basis. This is a relatively low cost approach commonly used in smaller workgroup-level switches where issues of congestion typically are relatively modest.
- **Matrix:** A *matrix switch* contains multiple interconnected high-speed internal buses; a multibus switching matrix can provide full bandwidth to multiple, simultaneous transmissions on a port-to-port, point-to-point basis. For example, one workstation can access another over a connection of 100 Mbps, while another has connection to a database server at a full 100 Mbps and still another is passing a file to a print server at 100 Mbps, assuming that the buses are available and can run at that rate [19]. If there are congestion issues in a matrix

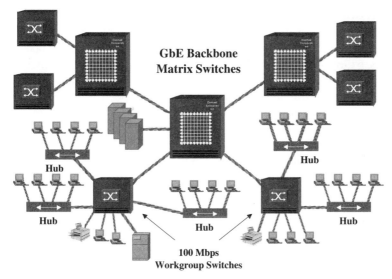

GbE Backbone
Matrix Switches

Hub

Hub

Hub

Hub

Hub

Hub

100 Mbps
Workgroup Switches

Figure 8.13   Interconnected LAN switches.

switch, it may have the ability to subdivide its capacity, with the buses becoming shared buses through a TDM process.

Along another dimension, switches may use several methods of operation:

- **Cut-Through:** A *cut-through* switch quickly reads the address of the data packet and quickly flows the frame through the switching matrix bit by bit.
- **Store and Forward:** A *store-and-forward* switch temporarily buffers, or stores, the frame as it is presented to the incoming switch port, examining the entire frame for errors through a CRC check before forwarding it through the switching matrix to the output port. While cut-through switching is faster and less expensive, it carries with it the risk of the propagation of errored data and the resulting potential for negative impact on overall throughput, as errored frames ultimately must be retransmitted. Therefore, store-and-forward switching generally is preferred over cut-through switching.
- **Fragment Free:** A third, and less common, method is *fragment-free* switching, which is similar to cut-through except for the fact that the switch stores the first 64 octets of the frame before forwarding it. As most errors occur at the beginning of a frame, this approach eliminates the possibility that *runt* frames, that is, truncated frames, will be transmitted [20]. *Note:* Recall that the minimum size of an Ethernet frame is 64 octets. The small frame size is used for VoIP over switched Ethernet.

In order to accomplish these minor miracles, the switch may store the native data packet in buffer memory at the incoming port, examine it for errors, and fragment it into smaller subsets of data. The switch then flows the packet fragments over a path set up across one of the shared buses, directing it only to the designated output port of the switch. At the outgoing port, the switch gathers the data fragments in buffer memory and reconstitutes the packet. For example and depending on the manufacturer, a Token Ring switch may fragment frames of up to 18,000 octets into units of 28, 64, or 4096 octets. The trade-off in size of the data fragment is that of performance, with smaller fragments enabling more users to share the bus at any given time and larger fragments improving switching speed because the switch must analyze and act on fewer packet headers.

Switches operate at the Physical and Data Link Layers of the OSI Reference Model—Layers 1 and 2, respectively. Switches read the destination addresses of the packets, filtering and forwarding as appropriate, based on MAC addresses (Layer 2). Switch logic is relatively simple and is in the form of firmware at the chip level. Therefore, switches are fast and relatively inexpensive. Some switches make routing decisions based on IP addresses (Layer 3). Layer 3 switching (admittedly an arguable term) involves a combination of switching and routing. This involves more complex routing decisions that are made in the context of the network as a whole yet not at the level of complexity that characterizes a router.

As illustrated in Figure 8.13, LAN switches may be positioned either at the workgroup level or in the backbone, or core.

- **Workgroup Switch:** A *workgroup switch* commonly has multiple 10/100-Mbps ports in support of workstations, perhaps connected through hubs, printers, and servers of various kinds.
- **Backbone Switch:** A *backbone*, or *core, switch* serves to interconnect workgroup switches and often provides access to large servers or server clusters. A core switch typically runs at 1 Gbps or perhaps 10 Gbps.

A switch does a great deal to reduce congestion and in a number of ways. First, a switch can support multiple simultaneous transmissions, whether through a matrix, a shared bus, or a combination. Second, switches serve to segment a network through filtering, as they forward traffic only to the port associated with the link to which the target device is connected. Thereby, that traffic does not contribute to congestion on other links or segments. Third, a switch can be equipped to buffer incoming packets until internal bus resources are available to process them. A switch also can be equipped to buffer outgoing packets until such time as the link to the next switch becomes available. Fourth, a switch can exercise a *flow control* mechanism, whereby it can advise a device to stop transmitting when its buffers are in danger of overflowing and then advise the device to resume transmission when the pressure on resources has been relieved. Fifth, store-and-forward and fragment-free switches variously eliminate or reduce the number of errored frames. Finally, a switch supports full-duplex transmission, thereby reducing or eliminating data collisions associated with CSMA in an Ethernet environment, assuming that the station is directly connected to the switch rather than through a hub. This approach is the current *best practice*.

The cost of switches has dropped dramatically in recent years to the point that they often compete effectively against hubs. But switch costs are sensitive to factors such as the type and speed of the transmission media interfaces, the number and speed of the ports, the number and size of the buffers, the number and speed of the internal buses, the complexity of the internal switching matrix, and the complexity of the switching or routing logic.

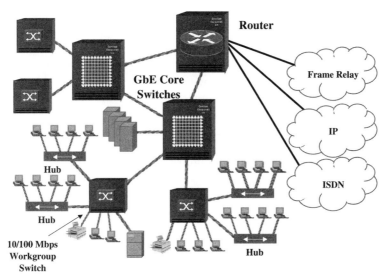

Figure 8.14   High-speed LAN with an edge router for WAN access.

8.3.5   Routers

Routers are highly intelligent devices that can support connectivity between both like and disparate LANs and can provide access to various WANs, such as Frame Relay, IP, and ISDN, as illustrated in Figure 8.14. Router interfaces to ATM also are possible, although other approaches such as *LAN Emulation* (LANE) and *Multi-Protocol over ATM* (MPOA) also may be employed, as is discussed in Chapter 10. Routers are protocol sensitive at the upper layers; they typically support multiple lower layer protocols and large and varying packet sizes such as might be involved in supporting both Ethernet and Token Ring. Routers typically operate at the bottom three layers of the OSI model using the Physical Layer, Link Layer, and Network Layer to provide connectivity, addressing, and switching [17]. Routers also have the capability to operate at all seven layers of the OSI Reference Model, if so equipped.

In addition to supporting filtering and encapsulation, routers route traffic based on a high level of intelligence that enables them to consider the network as a whole. This is in stark contrast to bridges, hubs, and switches, which view the network simply on a link-by-link basis. (*Note:* Layer 3 switches cross the line into the routing domain, although at a relatively simple level.) Routing considerations might include destination address, payload type, packet priority level, least-cost route, minimum route delay, minimum route distance, and route congestion level. Routers also are self-learning, as they can communicate their existence to other devices and can learn of the existence of new routers, nodes, and LAN segments. Routers constantly monitor the condition of the network as a whole, thereby dynamically adapting to changes in the condition of the network from edge to edge. Routers are multiport devices with high-speed ports running at rates up to 155 Mbps or more and with high-speed internal buses that can be on the order of 1 Gbps in the aggregate. Additionally, routers typically provide some level of redundancy so they are less susceptible to catastrophic failure [15].

Routers are unique in their ability to route data based on programmable network policy. Policy-based routers can provide various levels of service based on factors such as the identification of the user, the terminal, and the type of payload. From one edge of the enterprise network to the other, an edge router can select the most appropriate path through the various switches or routers positioned in the core. An important part of this process often is that of dividing the enterprise network into multiple *subnets*. Users associated with a subnet may be afforded access to only a limited subset of network resources in the form of sites, links, hosts, files, databases, and applications. In addition to being limited in terms of access to such a resource, users of another subnet may be prevented from receiving data from it. In effect, even the very existence of those resources is *masked* from view. Creation of such isolated subnets may serve for reasons of security or simply as a means of avoiding unnecessary congestion.

Routers commonly are capable of alternate routing and inverse multiplexing. As noted by Charles Darling [14], what dedicated WAN links lack in cost efficiency they make up for in lack of reliability (e.g., backhoe fade). Routers are available that sense a network failure and reestablish the connection via an alternate means, which may include a totally different network and service. While a backup ISDN Basic Rate Interface (BRI) link may be painfully slow compared with a T1 connection, for example, Darling suggests that *slow* is better than *stop*, and I suspect that most would agree. Additionally, some routers offer inverse multiplexing capabilities over ISDN and T-carrier facilities.

Router protocols include interrouter protocols, serial line protocols, and protocol stack routing and bridging:

- *Interrouter protocols* are router-to-router protocols that can operate over heterogeneous networks. These protocols pass routing information and *keep-alive* packets during periods of idleness. *Routing Information Protocol* (RIP) is an example of in interrouter protocol. The *Internet Engineering Task Force* (IETF) specified RIPv1 in RFC 1058, RIPv2 in RFC 2453, and RFCng for IPv6 in RFC 2080. The IETF specified *Open Shortest Path First* (OSPF) in RFC 1131 (1989), and there have been multiple revisions since.

- *Serial line protocols* provide for communications over serial or dial-up links connecting unlike routers. Examples include *High-level Data Link Control* [HDLC, ISO 3309], *Serial Line Internet Protocol* (SLIP, RFC 1055), and *Point-to-Point Protocol* (PPP, RFCs 1548, 1661, and 1662). Chapter 12, which deals with the Internet, provides more discussion of these specific protocols.

- *Protocol stack routing* and *bridging protocols* advise the router as to which packets should be routed and which should be bridged. Bridging protocols include *Spanning Tree Protocol* (STP), as defined in IEEE 802.1d.

8.3.6 Gateways

Gateways can perform all of the functions of bridges and routers as well as accomplish protocol conversion at all seven layers of the OSI Reference Model. Generally consisting of software residing in a host computer equivalent in processing power to a midrange or mainframe, gateway technology is expensive but highly functional.

Protocol conversion, rather than encapsulation, can serve to fully convert from Ethernet to Token Ring to FDDI or any other standard or proprietary protocol. Additionally, protocol conversion can address higher layers of the OSI model, perhaps through Layer 7, the Application Layer. As the process of protocol conversion is complex, gateways tend to operate rather slowly as compared to bridges and routers. As a result, they impose additional latency on packet traffic and may create bottlenecks of congestion during periods of peak usage. In a large and complex enterprise network, routers tend to be positioned at the edges of the network where they can be used to full advantage. Therefore, they make complex and time-consuming decisions and invoke complex and time-consuming processes only where required. Switches tend to be positioned within the core of the network because they can operate with greater speed, perhaps based on the instructions of the routers.

8.4 LAN OPERATING SYSTEMS

A *LAN Operating System* (OS), or *Network Operating System* (NOS), is software that provides the network with multiuser, multitasking capabilities across the network. The OS facilitates communications and resource sharing, thereby providing the basic framework for the operation of the LAN. The OS consists of modules that are distributed throughout the LAN environment; some NOS modules reside in the servers and other modules reside in the clients.

I want to digress for just a moment. The *client/server* model originated with the development of the U.S. Department of Defense ARPANET in the 1960s. As the cost and size of computer systems decreased and as the capabilities of those systems and the networks increased, the embodiment of client/server changed. A contemporary *client* is an application that generally resides on a microcomputer. Example applications include word processing, spreadsheet, and database software. The client runs against a *server*, which is a multiport computer containing large amounts of memory and enabling multiple clients to share its resources while performing certain functions independently. Servers are database engines capable of processing client requests for information and managing the resident data. For example, client/server continues to be used extensively in the Internet. When accessing America Online, Prodigy, CompuServe, or another service provider, you make use of a *Graphical User Interface* (GUI) and *browser* software that resides on the PC. When initiating an Internet session, that software runs against software installed in the service provider's communications server. Through this approach, the two devices communicate effectively without requiring that the software be downloaded from the server as a part of every Internet session, which, because the graphic files are huge, would cost a great deal of time, and bandwidth would be wasted. Once connected to the communications server, you subsequently can access a large number of database servers on a point-and-click basis, courtesy of the GUI, and with only the target data being transmitted across the Internet.

In addition to supporting multitasking and multiuser access, LAN operating systems provide for recognition of users based on passwords, user IDs, and terminal IDs. On the basis of such information, an OS can manage security by monitoring access privileges. Additionally, LAN OSs provide multiprotocol routing as well as directory services and message services. DOS-based LAN OSs include Microsoft's Windows NT Server, Windows 2000 Server, and Windows Server 2003. Other LAN OSs include Hewlett-Packard HP-UX, Linux, Novell Netware and Open Enterprise Server, and Sun Microsystems Solaris.

8.5 VIRTUAL LANs

Virtual LANs (VLANs) are software-defined LANs that group users by logical addresses into a virtual, rather than physical, LAN through a switch or router (refer to Figure 8.15). The LAN switch can support many VLANs, which operate as subnets [21]. Users within a VLAN traditionally are grouped by physical ports on switches and routers, TCP port address, MAC address, or IP address. Each node is attached to the switch port via a dedicated circuit. A variation on the theme is a *policy-based* VLAN, which can base VLAN membership on such factors as protocol, location, user name, and workstation address [22]. Users also can be assigned to more than one VLAN, should their responsibilities cross workgroup domains.

The LAN switches can be networked, thereby extending the reach of the VLAN. The networking is generally provided through FDDI, 100Base-T, or GbE over optical fiber links. VLANs also can be extended across the WAN through access routers and various services such as dedicated leased lines, Frame Relay, and ATM.

The advantages of switched VLANs include the fact that bridge and router networks can be flattened and simplified, including the elimination of source-routing and bridge hop restrictions. Intelligent segmentation and microsegmentation can serve to reduce congestion, thereby yielding increased accessibility, increased

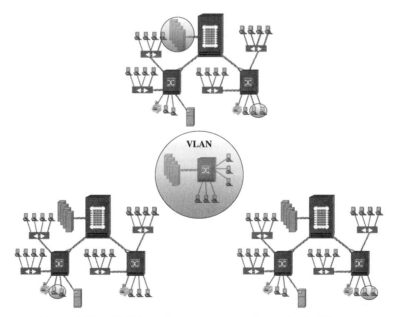

Figure 8.15 VLAN configuration across three physical LANs.

throughput, and shortened response times. Workstations can be provided with full bandwidth at each port, assuming that they connect to nonblocking switches. Particularly in the case of Layer 3 (e.g., IP-based) VLANs, physical *move, add, and change* (MAC) activity is reduced, as many of these activities can be resolved through software changes [23, 24]. As the estimated cost of a move, add, or change varies from $300 to $1000, this approach can offer significant savings in a highly dynamic environment. Additionally, security is much improved through the association in software of users and terminals with subnetworks and hosts. A measure of security also is provided through software firewalls within the confines of each domain [25, 26]. On the downside, VLANs are not easily implemented or managed. It takes a good deal of effort to develop the switch database and identify the various logical subnets.

8.6 REMOTE LAN ACCESS

Remote LAN access is the ability to access a LAN from a remote location. The need for remote LAN access is increasing worldwide, especially in the United States and Western Europe. *Telecommuting* and *telework* now are growing at a fast pace. Various market research firms suggested that 4 million to 9 million employees worked at least 8 h per week from home in 1992. In 1998, an estimated 20 million people worked from home, and an additional 9 million worked at home after normal business hours [27]. Telecommuting estimates now range as high as 40 million in the United States and 16 million in the European Union.

The concept of remote LAN access is one of providing access from remote locations to one or more host computers, which typically are LAN attached. In support of telecommuters, contractors, remote offices, and the Small Office/Home Office (SOHO), remote LAN access often is essential to the operation of the enterprise. Additionally, remote access often is provided to customers, suppliers, trading partners, and so on. The yield is that remote users are provided access to resources with the same level of privilege as though they were on-site [17]. Key components to be examined are the network, the equipment, and the applications supported. Other issues include security management and network management.

The network is clearly a developing enabler. The network can assume a variety of forms, depending on issues such as availability, cost, and bandwidth (as illustrated in Figure 8.16). Network options literally run the full range of conventional data networks, including modem-based communications over the analog PSTN, ISDN, and IP-based *Virtual Private Networks* (VPNs) over ADSL and cable modems. Public Wi-Fi hotspots are now widely available and are often used by traveling employees for access to corporate networks through the Internet. Finally, and in order to support truly mobile remote LAN access, cellular radio data networks often are employed [28, 29].

Equipment required varies according to the nature of the network employed. Improvements in equipment, as well as improvements in network technologies, have enabled cost-effective, remote LAN access. That equipment can include high-performance workstations and high-speed modems for access to the PSTN. ISDN requires Terminal Adapters (TAs), and Frame Relay requires Frame Relay Access Devices (FRADs). Access via a cellular network or a public Wi-Fi hotspot requires

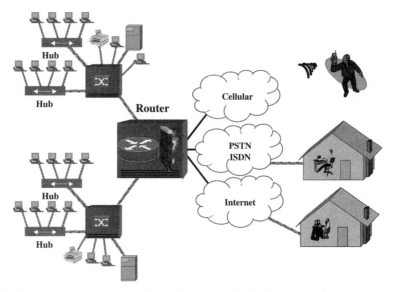

Figure 8.16   Remote LAN access from wireline and wireless networks through an access router with firewall.

the use of appropriate wireless modems. Clearly, the LAN side of the connection requires the installation of modem pools, routers, and various other devices in order to support remote user access. In order to protect internal resources from unauthorized access, security mechanisms must be employed, including intrusion detection devices, firewalls, and VPN (i.e., authentication and encryption) software.

Applications most often supported include e-mail, file transfer, and database access. Additional applications include scheduling, printing, access to online services, client support, and Internet access. The applications and advantages of remote LAN access are clear even to those of us in Mt. Vernon, Washington. Margaret Horak, my lovely bride, is the sole proprietor of The Evergreen Group, an independent consultancy. Among her clients she counts several major banks, Competitive Local Exchange Carriers (CLECs), and one of largest software companies in the world. For those clients she provides a number of services, including developing and maintaining websites and developing professional development curriculum, some of which is Web based. She performs virtually all of her work from right here in Mt. Vernon through remote access to her clients' LANs located on their premises around the country. In one case, her remote access is on the basis of V.90 modem dial-up through the PSTN using an 800 number. That one is really slow because she is downloading multi-megabyte files at 53.3 kbps or less and uploading them at 33.6 kbps or less. In another case, she works a lot faster through the client's VPN, which she accesses over the Internet through our shared ADSL circuit at speeds up to 1.536 Mbps on the download and up to 384 kbps on the upload. Margaret regularly puts virtual project teams together to develop course material. Such a project team typically includes one or more Subject Matter Experts (SMEs) and Instructional Designers (IDs), all of whom are teleworkers. The last several projects included SMEs in Argentina, Australia, Colombia, Norway, Spain, New Zealand, and the United States. Although Internet availability varies widely across these countries, some level of high-speed access generally is available to knowledge workers. Still, it was necessary to send multi-gigabyte program files by courier, as their sheer size made Internet distribution impossible.

8.7 LAN STANDARDS AND STANDARDS BODIES

Standards developed by the *Institute of Electrical and Electronics Engineers* (IEEE) largely have governed the world of LANs since the formation of Project 802 in February 1980. Project 802, working within the scope of the OSI Reference Model, was chartered to deal with the two lower layers. Notably Layer 2, the Data Link Layer, was divided into two subgroups: *Medium Access Control* (*MAC*) and *Logical Link Control* (LLC) [2]. The initial development was based on Ethernet, the embellished version of which was finalized in December 1982 as 802.3, which is commonly referred to as Ethernet. The IEEE membership, the U.S. National Bureau of Standards, and the *European Computer Manufacturers Association* (ECMA) accepted that first release of three standards. The *International Organization for Standardization* (ISO) issued correlating international standards—known as 8802 LAN standards. Since then, the IEEE has continued to develop a broad range of LAN and MAN standards. IEEE standards include the following:

802.1: Architecture and internetworking (high-level interface). Defines architecture layers and rules for interconnection of disparate LAN protocols. Includes data formatting, network management, and internetworking.

802.2: Defines equivalent LLC services, including protocol for data transfer. Largely addresses bridges.

802.3: Defines CSMA/CD Access Method and Physical Layer Specifications. Commonly referred to as the *Ethernet* standard.

802.4: Token Bus Access Method and Physical Layer Specifications.

802.5: Token-Passing Ring Access Method and Physical Layer Specifications. Includes Token Ring.

802.6: Metropolitan Area Network (MAN) Access Method and Physical Layer Specifications. Distributed Queue Dual Bus (DQDB) is defined. Switched Multimegabit Data Service (SMDS), discussed in Chapter 10, was derived from 802.6. This standard has been withdrawn.

802.7: Broadband Technical Advisory Group. Standards for definition of a broadband cable plant design. Established guidelines for LAN construction within a physical facility such as a building. This standard has been withdrawn.

802.8: Fiber Optic Technical Advisory Group. Established to assess impact of fiber optics and to recommend standards. Note that this standard is distinct from that of ANSI's FDDI.

802.9: Integrated Services LAN (ISLAN). Designed for the integration of voice and data networks, both within the LAN domain and interfacing to publicly and privately administered networks running protocols such as FDDI and ISDN.

TABLE 8.3 **Dimensions of Popular LAN Standards**

Dimension\Standard	Ethernet	IBM Token Ring	Fiber Distributed Data Interface (FDDI)
Standard	IEEE 802.3	IEEE 802.5	ANSI X3T9-5
Logical topology	Bus	Ring	Ring
Physical topology	Bus, Star	Ring, Star	Dual ring, Dual bus
Media	Coax, UTP, STP	Coax, UTP, STP	Fiber
Transmission mode	Baseband	Baseband	Baseband
Bandwidth	10/100 Mbps, 1/10 Gbps	4, 16, 20 Mbps	100 Mbps
Medium Access Control	Nondeterministic: CSMA/CD, CSMA/CA	Deterministic: Token passing	Deterministic: Token passing
Payload size	46–1500 bytes	Up to 4048/17,800 bytes[a]	Up to 36,000 bytes
Traffic type	Data	Data	Data, video, voice

[a]These are conventional implementation maximums. According to the IEEE 802.5 specification, however, "although there is no maximum length specified for the information field, the time required to transmit a frame may be no greater than the token holding period that has been established for the station."

802.10: Standards for Interoperable LAN/MAN Security (SILS). This standard was withdrawn in 2004, and the working group is currently inactive. Security for wireless networks is being addressed in 802.11i. VLAN security is addressed in 802.11q.

802.11: Wireless Local Area Networks (WLANs). This is a family of standards describing the over-the-air interfaces for a number of RF-based WLANs.

802.12: 100+ Mbps LANs using demand priority access. The focus was on 100VG-AnyLAN, which standard is considered obsolete.

In addition to the IEEE, other standards bodies are involved in the establishment and promotion of certain LAN and computer networking standards. ANSI (American National Standards Institute), for example, developed the following standards:

X3T9-3: HIgh Performance Parallel Interface (HIPPI)
X3T9-5: Fiber-Distributed Data Interface (FDDI)

Table 8.3 compares Ethernet, Token Ring, and FDDI across a number of critical dimensions.

8.8 LIFE IN THE FAST LAN: THE NEED FOR SPEED

Beginning in the early 1990s, traditional LANs have been pushed to their limits as end-user organizations connect more workstations and users become more active, resulting in more LAN traffic. Increased use of graphics and other more bandwidth-intensive applications developed, adding further to the strain. Collaborative computing increases the demands on existing LAN technologies, especially as voice and videoconferencing are employed to enhance the collaborative experience. Users also have become increasingly impatient, demanding faster response times. In general, LAN users mirror the times in which we live—more is better, bigger is better, and faster is better still. Bandwidth of 10 Mbps and even 16 Mbps just does not do the trick any longer! In response to this requirement, fast LANs began to develop offering bandwidth of 100 Mbps and now an incredible 10 Gbps. Along the evolutionary path, there were some dead ends, as there always are. *Asynchronous Transfer Mode* (ATM) proved too expensive and complex in the LAN domain. *High-Speed Token Ring* (HSTR) failed to gain any traction, as Ethernet overwhelmed it in terms of speed and undercut it in terms of cost. Despite its ability to support virtually any LAN standard, *100VG-AnyLAN* had no market, as Ethernet became the only LAN standard with any following. *Isochronous Ethernet* (IsoEthernet) added an aggregate 6 Mbps in ISDN B channels for voice and video but quickly became irrelevant when switched Ethernet appeared at 100 Mbps. Fast LAN options currently include 100Base-T (fast Ethernet), FDDI, GbE and 10GbE.

8.8.1   100Base-T, or Fast Ethernet

A variation of 10Base-T and standardized as IEEE 802.3u (June 14, 1995), *100Base-T* is a high-speed LAN standard using the CSMA/CD MAC mechanism and operating at 100 Mbps through an Ethernet switching hub. Contemporary 100Base-T hubs and switches can support port speeds of both 10 and 100 Mbps. Acceptable media for 100-Mbps Ethernet include both twisted pair and fiber, as listed in Table 8.1. The predominant version of the standard is 100Base-TX, which calls for two pairs of Cat 5 (or Cat 5e) wiring. 100Base-TX calls for the NICs in the servers, switches, hubs, and workstations to adjust to the capabilities of the medium, much as an analog modem might do. While the obvious choice is 100 Mbps, the devices will fall back to 10 Mbps if the medium will not support the higher speed. The 100Base-T4 standard, now considered obsolete, supports 100 Mbps over four pairs of Cat 3 UTP, with three pairs used for transmission and the fourth pair for signaling and control (CSMA/CD) in half-duplex (HDX) mode [30]. The 100-Mbps media also include fiber for distances up to 32 miles, or 50 km, without repeaters. Remember, however, that Ethernet is collision prone; therefore, 100 Mbps of theoretical bandwidth may yield throughput of only 50 Mbps or so for a hub technology. Switches supporting 100Base-T, however, yield much improved performance through the support of multiple simultaneous transmissions in full-duplex (FDX) mode, yielding a theoretical total bandwidth of 200 Mbps (100 Mbps × 2 = 200 Mbps in FDX), at least from the switch to the attached device [31].

The line-coding technique is sensitive to the medium, of course, but 100Base-TX uses an encoding technique known as *4B/5B*, which refers to the fact that every *nibble* of *4 Bits* of data is encoded into *5 Bits* of signal. (*Note:* The term *nibble* is sometimes used to refer to a four-bit value. The term is a word play on *byte*, which generally is eight bits. 4B/5B also is used in FDDI and 100Base-FX.) Specified by the ANSI X3T9.5 committee, 4B/5B is sometimes referred to as block encoding, as a block of data bits is mapped into a block of signaling bits. This approach increases the number of bit patterns from 16 ($2^4 = 16$) to 32 ($2^5 = 32$), which means that every block of five signaling bits can include at least two 1 bits, even if the original data block of four bits contained either zero or one 1 bit. There are several advantages to this approach. First, the five-bit signal block includes enough clocking pulses and signal transitions to synchronize the network. Second, as only 16 of the possible 32 combinations of five signaling bits are used for data (some are used for signaling

TABLE 8.4   4B/5B Encoding

4-Bit Data Blocks	5-Bit Line Codes
0000	11110
0001	01001
0010	10100
0011	10101
0100	01010
0101	01011
0110	01110
0111	01111
1000	10010
1001	10011
1010	10110
1011	10111
1100	11010
1101	11011
1110	11100
1111	11101

and control), some level of error detection is realized. If one of the valid bit patterns were to be changed to an invalid bit pattern as a result of an error in transit, that fact could be recognized by the receiving terminal and would trigger an error message. Table 8.4 lists the four-bit data blocks and the five-bit line codes into which they map. There are other five-bit line codes, and combinations of two five-bit line codes, used for command (i.e., control) purposes, such as *Start of Stream Delimiter* (SSD) and *End of Stream Delimiter* (ESD). There is a negative side to this approach, as the clock speed and, therefore, the signaling rate of the network must be 125 percent (5/4 = 1.25) of the actual rate. So, the signaling rate must be 125 MHz to support a unipolar code (such as that used with classic 10 Mbps Ethernet) with a signaling rate of 125 Mbps, which in turn supports a data rate of 100 Mbps. However, 100Base-TX uses an intermediate step known as *MultiLevel Transition* (MLT) before placing the signal on the line. This ternary approach cycles through three signal levels in the pattern +1, 0, −1, 0, which is represented as +V, 0 V, −V, 0 V. If the next data bit is a 1, the MLT-3 output transitions to the next signal state in the pattern, which would be +V. (See Figure 8.17.) If the next bit after that is also a 1, the signal state of the output transitions to the next signal state in the pattern, which would be 0 V. If, however, that next bit is a 0, the output does not transition but remains the same. MLT-3 also adds a scrambling step before placing the signal on the line, all of which support a signaling rate of 125 Mbps while placing the main spectral energy at a frequency of only 31.25 MHz. That low frequency results in improved signal quality and a reduced potential for interference.

8.8.2   Fiber-Distributed Data Interface

FDDI is the standard (ANSI X3T9-5; IEEE 802.2) for a fiber-optic, Token-Passing Ring LAN. Bandwidth is pegged at 100 Mbps, although several manufacturers offer 200-Mbps, FDX interfaces. (*Note:* This is the LAN measurement of bandwidth,

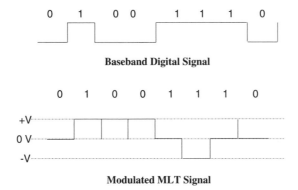

Figure 8.17   Multilevel transmission coding.

which adds the bandwidth in each direction.) The excellent performance characteristics of fiber optics, in general, apply well to the LAN world. Error performance is in the range of 10^{-14} (i.e., one errored bit in every 100 trillion bits transmitted) and devices can be separated by as much as 1.2 miles (2 km) over multimode fiber (MMF) and 37.2 miles (62 km) over single-mode fiber (SMF) [32, 33]. The maximum frame size is 9000 *symbols* (1 symbol = 4 bytes), which easily accommodates the native frame sizes of all standard LAN networks [34]. The line-encoding technique is 4B/5B, as discussed in connection with 100Base-TX.

FDDI largely is deployed as a campus and sometimes a Metropolitan Area Network (MAN) backbone technology for the interconnection of major computing resources such as hubs, switches, routers, and servers. While FDDI can be extended to the device levels, the cost of optoelectric termination is high. The advantages of FDDI, however, can be extended to the workstation through a concentrator that accomplishes the optoelectric conversion process for multiple attached devices. The connection from the concentrator to the workstations is accomplished via UTP over distances of 100 m or less, based on a standard known variously as *Copper Distributed Data Interface* (CDDI) and *Twisted-Pair Distributed Data Interface* (TPDDI).

The fragility of the fiber is a deterrent to the application of FDDI as well. The FDDI specifications provide for a dual counterrotating ring, which provides a measure of redundancy. Should the primary ring fail, a *Dual Attached Station* (DAS) or *Dual Attached Concentrator* (DAC) can communicate with any other device by transmitting in the opposite direction through the secondary ring, which typically is collocated in the same cable sheath as the primary ring (see Figure 8.18). If there is more than one physical failure in the cable plant, however, the ring segments and the network all fail. There are dual-homing solutions to this dilemma, although they involve considerable additional expense, with the designated stations connected via fiber to multiple servers to provide redundancy [35].

Despite its reliability and throughput characteristics, sales of FDDI hubs and switches peaked at $220 million in 1997 [36] and have dropped off to the point that few, if any, analysts currently follow the technology. Not only is FDDI's high cost a detriment, but Ethernet optical fiber backbones currently can be deployed at much higher speeds for the interconnection of hubs, switches, routers, and the like. Further, FDDI is underpowered by current standards, with switched Ethernet running at

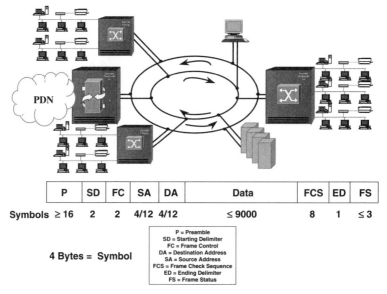

P	SD	FC	SA	DA	Data	FCS	ED	FS
Symbols ≥ 16	2	2	4/12	4/12	≤ 9000	8	1	≤ 3

4 Bytes = Symbol

P = Preamble
SD = Starting Delimiter
FC = Frame Control
DA = Destination Address
SA = Source Address
FCS = Frame Check Sequence
ED = Ending Delimiter
FS = Frame Status

Figure 8.18 FDDI dual counterrotating ring, with frame format.

speeds as high as 10 Gbps. FDDI legacy networks still exist and are sometimes extended but will eventually be replaced by faster, less expensive technologies such as GbE or 10GbE.

8.8.3 Gigabit Ethernet

The standard for *Gigabit Ethernet* (GbE or GigE) was finalized and formally approved on June 29, 1998, as IEEE 802.3z. Although fully compatible with both 10- and 100-Mbps Ethernet, most equipment has to be upgraded (i.e., replaced) to support the higher transmission level. GbE addresses the bandwidth problem in 10/100 Mbps Ethernet networks, which began to feel the stress of bandwidth-intensive, multimedia-based Internet and intranet applications as well as scientific modeling and data warehousing and data backup.

GbE is available in shared and switched versions (see Figure 8.19), both of which support multiple ports that can run at 1 Gbps in full-duplex (FDX). Shared GbE essentially is a much higher speed version of 10Base-T and 100Base-T. Shared GbE is a high-speed hub that uses CSMA/CD for Medium Access Control over the shared bus. Therefore, you can characterize shared GbE as a brute-force attack on congestion. Switched GbE addresses the congestion problem through buffering incoming Ethernet frames and passing them to the output port when the shared bus becomes available. The shared bus can run at a speed of several Gbps. The more substantial switched GbE products offer nonblocking switching through a crossbar switching matrix, which may run at an aggregate of tens of Gbps [37]. The cost of a GbE switch is greater than that of a GbE hub and is sensitive to such factors as port density, buffer placement and capacity, switch matrix complexity, and throughput.

While GbE resembles traditional Ethernet, differences include frame size. As the clock speed of GbE is one or two orders of magnitude greater than its predecessors

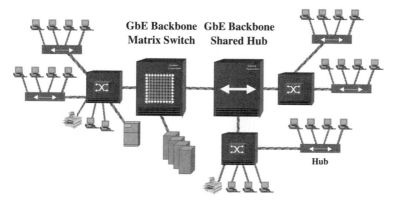

Figure 8.19   Interconnected GbE, shared and switched.

(10/100 Mbps), issues of round-trip propagation delay affect error detection. To avoid potentially incredible collision rates, the minimum frame size has increased from 64 to 512 octets, which generally is equivalent in duration to transmitting a 64-byte frame at 100 Mbps. This larger minimum frame size provides the same time for the transmitting device to receive a collision notification. Although nonstandard, some manufacturers have increased the maximum frame size from 1518 bytes to a *jumbo frame* size of 9000 bytes, which improves the frame throughput of a GbE switch. Since each frame requires switch processing of header information, the fewer frames presented to the switch, the more data the switch can process, switch, and deliver in a given period of time. Where multiple GbE switches are networked, jumbo frames may be passed between them. Where the GbE hub or switch interfaces with lesser, standards-based Ethernet devices, the jumbo frame must be fragmented to effect compatibility [37–40].

Physical transmission media currently focus on fiber optics, as discussed earlier in this chapter and detailed in Table 8.2. MultiMode Fiber (MMF) supports gigabit transmission at distances up to 550 m, and Single Mode Fiber (SMF) up to 5 km. In either case, there is a minimum distance of 2 m because of issues of signal reflection (echo). While UTP, STP, and other electrically based media will support GbE, distances are highly limited. Cat 5 and Cat 5e UTP, for example, can support FDX transmission over distances up to 25 m, with each of four pairs carrying a 125-MHz signal.

GbE generally employs the *8B/10B* line coding technique, which maps 8 data bits into a 10-bit *symbol*, or character. The 8-bit data octet is divided into two groups. The 3 most significant bits, or leftmost bits, are encoded into a 4-bit group (3B/4B). The 5 least significant bits, or rightmost bits, are encoded into a 6-bit group (5B/6B). The two groups are then concatenated, or joined together, and placed on the line. As 8 bits yields 256 possible bit combinations $2^8 = 256$) and 10 bits yields 1024 ($2^{10} = 1024$) bit combinations, each 8-bit data octet can be phrased two different ways, with one being the bitwise inverse of the other. For example, a data octet of 11001010 might be expressed the first time as 1000100111 and the second time as 0111011000. That encoding scheme yields Direct Current (DC) electrical balance on the line, as the number of 1s and 0s will be equal in the long term. This approach also ensures clock recovery, as there are sufficient 1s in sufficient density to serve as clocking pulses. 8B/10B also provides an additional embedded error control mechanism similar to that discussed in 4B/5B, which is used in 100Base-TX. On the downside, 8B/10B adds 25 percent overhead (10/8 = 1.25) to the serial data stream. *Note:* The 10B format also provides for a number of control characters. 8B/10B also is used in *ESCON* and *Fibre Channel*, both of which are standards used in *Storage Area Networks* (SANs), and a number of other protocols, both standard and non-standard. (*Note:* I discuss SANs later in this chapter.)

Clearly, the primary application for GbE in the near future largely will be in the backbone, for interconnecting lesser Ethernet hubs, Ethernet switches, and high-performance servers, rather than connecting individual nodes. However, GbE already is used occasionally in certain bandwidth-intensive desktop applications such as *Computer-Aided Design* (CAD), and GbE hubs and switches commonly offer port speeds of 10/100/1000 Mbps. Gigabit Ethernet supports HDX and FDX interfaces, with FDX offering the advantage of virtual elimination of issues of data collisions. The HDX and FDX declarations are made on a port-by-port basis.

Notably, GbE is not limited to the LAN domain. A number of service providers now offer GbE as a *metropolitan area network* service offering intended for multi-site enterprises confined to a metropolitan area. This service involves centrally positioning one or more GbE switches in the MAN and providing the enterprise with access via a variety of technologies. Although fiber optics clearly is the most attractive access technology from a performance standpoint, alternatives include unchannelized T-carrier, *Very-high-data-rate Digital Subscriber Line* (VDSL), and various wireless options. The real advantage to GbE in the MAN is that all traffic is carried in native Ethernet format, with no requirement for introducing SONET, Frame Relay, ATM, or other Layer 1 or Layer 2 protocols that can increase both complexity and cost.

8.8.4 10-Gigabit Ethernet

As if 1 Gbps is not enough, 10GbE specifications were finalized in 2002 by the IEEE as 802.3ae. As 10GbE uses the same frame format and MAC, it is backward compatible with earlier and slower Ethernets. In addition to the fact that 10GbE runs at 10 times the speed of its fastest predecessor, there are a few other notable differences. 10GbE runs only in FDX mode, which makes collision control unnecessary. The primary line encoding technique used in both GbE and 10GbE is 8B/10B, which carries a 25 percent overhead penalty, forcing the system to run at 125 Gbps. Some 10GbE systems use the more recently developed 64B/66B technique, which is similar but much more efficient, thereby allowing the system to run at close to 10 Gbps. In either case the signaling speeds discourage the use of copper. 802.3ae currently specifies five fiber-optic options, as listed in Table 8.2. The LAN interface specifies MultiMode Fiber (MMF) over distances up to 300 m. The WAN interface calls for Single-Mode Fiber (SMF) over distances up to 40 km and is compatible with SONET long-haul equipment. This WAN interface makes it possible to deploy 10GbE in a *Metropolitan Area Network* (MAN) application, where SONET compatibility will allow enterprises to pass native Ethernet frames as the Layer 2 protocol and with no need for Frame Relay, ATM, or IP. As a result, no reformatting or protocol conversion is required. There also are two options that employ Coarse Wavelength Division Multiplexing (CWDM), running four parallel wavelengths over MMF at distances up to 300 m and over SMF at distances up to 10 km.

Despite the obvious difficulties of running 10GbE over copper, there are several initiatives in this regard. 802.3an is a developing specification for running 10GBase-T over Cat 5e, Cat 6, and Cat 7 at distances up to 100 m using a version of *PAM-16*, a *Pulse Amplitude Modulation* (PAM) technique employing 16 levels of amplitude. There also is an option for twin-axial cable in the form of 10GBase-CX4, which the IEEE specified as 802.3ak in 2004. This standard runs over bundles of twin-axial cables, with the signal split over four cables in each direction running in simplex mode. The signaling speed over each cable is 3.125 Gbps in support of a data rate of 2.5 Gbps, with the difference due to the use of 8B/10B line coding.

While 10GbE certainly has application in support of very bandwidth intensive LAN applications of very large enterprises, it appears at this point to have application primarily as a MAN technology. That said, we always seem to find a way to consume more and more bandwidth closer and closer to the desktop [41–44]. I discuss GbE and 10GbE in the MAN context in Chapter 10.

8.9 WIRELESS LANs

Wireless LAN technology has enjoyed incredible success during the last few years. Offering the obvious advantage of much reduced wiring costs, WLANs can be deployed to great benefit in a dynamic environment where there is frequent reconfiguration of the workplace. They also offer clear advantages in providing LAN connectivity in temporary quarters, where cabling soon would have to be abandoned. Wireless LANs largely are RF based and employ spread-spectrum technology, which was developed during World War II for use in radio-controlled torpedoes. This approach offers significantly increased security and throughput, as I discuss in detail in Chapter 11.

In a typical RF-based WLAN environment (refer back to Figure 8.1), each workstation is fitted with a radio *transceiver* (*trans*mitter/re*ceiver*) with an omnidirectional antenna. The client transceivers, or network adapters, commonly are in the form of PCMCIA cards, although major computer manufacturers have offered laptops with built-in transceivers since 2002. Hub antennas, or *Access Points* (APs), are located at central points, ideally where there is good or at least near Line Of Sight (LOS) between the hub and the workstations. If an AP is located in the center of the ceiling, for example, it will be omnidirectional. If an AP is located in the corner of a room, it will be directional in nature The AP traditionally connects to the servers and peripherals via Cat 5e cabled infrastructure, with multiple hub antennas being placed in strategic locations in rooms throughout each building. As good LOS considerably improves link quality, transmission through walls, floors, ceilings, and other dense physical obstructions should be avoided where possible. In order to serve multiple workstations, spread-spectrum radio technology is employed to maximize the effective use of limited bandwidth. A side benefit of spread spectrum is that of increased security.

Although there exist a number of nonstandard wireless LANs, this discussion focuses on standards-based versions from the IEEE 802.11 Working Group, which began its efforts in 1989. Those standards specify Layers 1 (Physical) and 2 (Data Link), as is the case with all LANs. With respect to transmission media, WLANs operate on infrared (IR) as well as radio frequency (RF) media. The RF specifications include both DSSS (*Direct Sequence Spread Spectrum*) and FHSS (*Frequency-Hopping Spread Spectrum*) and the Ethernet CSMA/CA (Carrier Sense Multiple Access/Collision Avoidance) protocol. This Ethernet Medium Access Control (MAC) protocol involves the establishment of what amounts to a virtual circuit through the positive acknowledgment of the availability of the receiving station through the network. The transmitting station sends a *Request-To-Send* (RTS) packet over the airwaves. If the target device is available, it responds with a *Clear-To-Send* (CTS) packet, which prompts the originating device to begin transmission. Other WLAN-attached devices honor this virtual circuit agreement, thereby avoiding issues of congestion, collision, and packet data loss. This MAC-level protocol works well with the CSMA/CD protocol more typically used in conventional wired Ethernets, thereby supporting physical interconnectivity between the wired and WLANs on a logically indistinguishable basis.

Before proceeding, let us pause to briefly examine some media specifics, building on the discussion of transmission media in Chapter 2. Most WLANs operate in the 2.4- and 5-GHz unlicensed ISM (*Industrial, Scientific, and Medical*) bands. This approach avoids the expensive and lengthy licensing process but carries with it the potential for interference from other such systems in proximity. While the FCC in the United States and other regulatory authorities in other countries initially set aside ISM frequency ranges for unlicensed in-building communications, WLANs running in these ranges are susceptible to interference from other systems such as cordless telephones, microwave ovens, garage door openers, and bar code scanning systems. Spread-spectrum technology generally is used at these frequencies to mitigate issues of interference. As power levels are low, distances generally are limited to 500–800 f or so. (It's comforting to know, therefore, that you are unlikely to cause garage doors to pop up and down all over the neighborhood when you're checking your e-mail over your WLAN. Better yet, garage door openers are unlikely to cause your systems to crash.) WLAN frequency ranges include the following:

- The range *902–928 MHz* is in the original ISM band. At these relatively low frequencies, signals are fairly immune to attenuation caused by dense physical matter such as windows, walls, floors, and ceilings. Those familiar with early cordless phones, analog cellular phones, and pagers, all of which run in this range or in ranges in proximity, are familiar with the advantages of operating in the 900-MHz range.
- The frequency ranges *2.4–2.5 GHz* and *5.8–5.9 GHz* also largely are in the ISM band, which they share with some cordless and cellular phones and other devices. WLANs running in the lower band of 2.4–2.5 GHz certainly are more susceptible to attenuation than those at 902–928 MHz, but much less so than those in the higher 5.8–5.9 GHz.
- The spectrum *5.15–5.35 GHz* and *5.75–5.85 GHz* was made available in the United States by the FCC in January 1997. These bands, which are part of the *Unlicensed National Information Infrastructure* (U-NII) spectrum, are relatively free of interference and offer the potential for transmission at much higher speeds than those available at the lower frequency ranges [45].

• The bands at *18–19 GHz* are sometimes employed in a WLAN environment. As the same frequencies are used in commercial microwave systems, there is considerable potential for interference unless spread-spectrum coding is employed. LOS is critical in this band.

• *Infrared* (IR) light systems currently require no licensing. The potential for interference between systems is very limited as LOS generally is required. IR is seldom employed in contemporary WLANs.

Note that IR generally requires LOS. RF systems generally do not require LOS but certainly benefit from it, particularly at the higher frequencies. Where RF signals must pass through walls, floors, ceilings, and windows, care must be taken with respect to the construction materials used. Metallic foil-backed insulation, for example, can have disastrous effects on the RF signal, as can certain glass windows with UltraViolet (UV) protection afforded by embedded metallic film. In advance of installing a WLAN, it is highly recommended that a site survey be conducted. A number of manufacturers offer highly sophisticated site survey systems that will ensure a satisfactory level of WLAN performance.

8.9.1 IEEE 802.11

The IEEE 802 standards committee formed Working Group 11 to develop a set of specifications for over-the-air (i.e., wireless) LAN/MAN standards. The resulting family of standards, variously referred to in the vernacular as *Wi-Fi* and *Wireless Ethernet*, include IR and RF solutions, although there appear to be no practical applications for IR. The RF standards variously fall into the 2.4- and 5-GHz ISM bands and offer raw bandwidth up to 54 Mbps, at least theoretically. The original 802.11 standard, released in 1997, supported data rates theoretically up to 2 Mbps in the 2.4-GHz band. This early standard included a great number of options, which made interoperability of products of disparate origin difficult, or at least uncertain. As a result, 802.11 never gained any real traction in the market. Soon afterward, however, much improved extensions to 802.11 were finalized, and WLANs quickly gained in popularity.

8.9.1.1 IEEE 802.11a (Wi-Fi5) Dubbed *Wi-Fi5 (Wireless Fidelity 5* GHz) by the *Wireless Ethernet Compatibility Alliance* (WECA), now the *Wi-Fi Alliance*, 802.11a supports speeds up to 54 Mbps in a 300-MHz allocation in the 5-GHz range, which the FCC allocated in support of U-NII (*Unlicensed-National Information Infrastructure*). Specifically, 200 MHz is allocated in the band 5.15–5.35 MHz for in-building applications and 100 MHz in the band 5.725–5.825 MHz for outdoor use.

Rather than using spread-spectrum technology, 802.11a uses *Coded Orthogonal Frequency Division Multiplexing* (COFDM) as the signal modulation technique. COFDM sends a stream of data symbols in a massively parallel fashion, with multiple subcarriers (i.e., small slices of spectrum within the designated carrier frequency band). Each carrier channel is 20 MHz wide and is subdivided into 52 subcarrier channels, each of which is approximately 300 kHz wide. Of those subcarrier channels, 48 are used for data transmission and the remaining 4 for error control purposes. The specified modulation techniques, all of which are explained in Chapter 6, and theoretical data rates include the following:

- BPSK (*Binary Phase Shift Keying*) at 125 kbps per channel for a total of 6 Mbps (125 kbps × 48 channels = 6 Mbps) and 187.5 kbps for a total of 9 Mbps
- QPSK (*Quadrature Phase Shift Keying*) at 250 kbps per channel for a total of 12 Mbps and 375 kbps per channel for a total of 18 Mbps
- 16QAM (*16-level Quadrature Amplitude Modulation*) at 500 kbps per channel for a total of 24 Mbps and 750 kbps per channel for a total of 36 Mbps
- 64QAM (*64-level QAM*) at 1 Mbps per channel for a total of 48 Mbps and 1.125 Mbps per channel for a total of 54 Mbps.

The data rates actually are raw theoretical signaling rates. The actual data rates are more on the order of 40–60 percent of the theoretical due to issues including overhead and medium access control. The *symbol rate* (i.e., the rate of transmission of a *symbol*, or set of bits) is slowed down enough that each symbol transmission is longer than the *delay spread*, that is, the variation in timing between receipt of the signals associated with a given symbol, with the delay spread being caused by *multipath propagation*, which is the phenomenon by which the RF signals carrying a given data symbol arrive at the receiver at slightly different times as a result of their having taken slightly different paths as they bounced off of various physical obstructions.

While the 5-GHz spectrum is relatively clear in the United States, it is not nearly so readily available elsewhere as military and governments use portions of this band overseas. In Japan, only the 5.15–5.25-MHz spectrum is available. In Europe, the 5.725–5.825-MHz spectrum is already allocated for other uses, including HiperLAN, which competes with 802.11a. In any event, the 5-GHz spectrum is highly susceptible to attenuation [46, 47].

8.9.1.2   IEEE 802.11b (Wi-Fi)   802.11b was the first of the 802.11 family to be released. Although the standards development process of 802.11a began first, technical and regulatory difficulties arose and development slowed. WLANs conforming to the IEEE *802.11b* specification now are by far the most common, due not only to the early release of the standard but also to the signal propagation characteristics of the 2.4-GHz band. As a side note, 802.11b has been dubbed *Wi-Fi (Wireless Fidelity)* by the Wireless Ethernet Compatibility Alliance. The term Wi-Fi also has been attributed to the IEEE 802.11 Working Group, with *Wi* referring to the fact that a *wire* traditionally served as the physical medium for LANs and the homonym *Fi* referring to PHY, the PHYsical layer. So, *Wireless PHY* became Wi-Fi. Now, let's get back to work.

802.11b includes three transmission options, one IR and two RF. 802.11b uses *Direct Sequence Spread Spectrum* (DSSS) modulation, which involves the transmission of a bit stream that is modulated with the *Barker code* chipping sequence. Each bit is encoded into a redundant 11-bit Barker code (e.g., 10110111000), with each resulting data *object* forming a *chip*. The chip is put on a carrier frequency in the 2.4-GHz range (2.4–2.483 GHz) and the waveform is modulated using one of several techniques. Systems running at 1 Mbps make use of *Binary Phase Shift Keying* (BPSK), while those running at 2 Mbps make use of *Quaternary Phase Shift Keying* (QPSK). Systems running at 11 Mbps make use of *Complementary Code Keying* (CCK), which involves 64 unique code sequences and supports 6 bits per code word. The CCK code word is then modulated onto the RF carrier using QPSK, which allows another 2 bits to be encoded for each 6-bit symbol. Therefore, each 6-bit symbol contains 8 bits. While all of this may seem highly inefficient, it has considerable advantages in recovering weak signals in wireless transmission.

The FCC limits power output to 1 watt *Equivalent Isotropically Radiated Power* (EIRP). At this low power level, the physical distance between the transmitting devices becomes an issue, with error performance suffering as the distance increases. Therefore, the devices adapt to longer distances by using a less complex encoding technique and a resulting lower signaling speed, which translates into a lower data rate. A system running at 11 Mbps using CCK and QPSK, for example, might throttle back to 5.5 Mbps by halving the signaling rate as the distances increase beyond 30 m or so and error performance drops. As the situation gets worse, it might throttle back to 2 Mbps using only QPSK and 1 Mbps using BPSK. At this lowest rate, link quality generally is acceptable at distances of up to 100 m or so. 802.11b divides the available spectrum into 14 channels, each of which has a width of 25 MHz. In the United States, the FCC allows the use of 11 channels. Four channels are available in France, 13 in the rest of Europe, and only 1 in Japan. There also is overlap between adjacent channels as each has a width of 25 MHz and all share a band that is only 83 MHz (2.4–2.483 GHz) wide, which fact further affects performance. In the United States, for example, only 3 of the 11 available channels are nonoverlapping. Therefore, any given system must maintain maximum channel separation and physical separation from other systems in proximity.

8.9.1.2.1 Operational Mode Wi-Fi can operate in two modes. *Ad hoc mode* allows devices such as laptops to discover each other and communicate directly, without the involvement of an AP. This approach is certainly convenient for ad hoc communications, such as a spur-of-the-moment file transfer when out of range of an AP. *Infrastructure mode* requires that an Access Point (AP) be involved to support a connection. This approach is by far the most common, as it is truly a LAN mode of operation. Access points can be either fat or thin:

- **Fat Access Point:** A *fat* AP is intended to act independently and contains sufficient program logic and processing power to allow it to enforce policies relating to access and usage. Traditionally, multiple APs are hardwired to a switch that serves to interconnect the APs and to provide access to other internal and external resources.
- **Thin Access Point:** A *thin* AP is intended to act under the supervision of a centralized controller that configures, manages, and secures the environment. The centralized controller provides a single point of administration for all APs. This tends to be the preferred approach in contemporary Wi-Fi networks [48–50].

8.9.1.2.2 Mesh Networking A wireless mesh network is quite simply a wireless network in which the nodes interconnect without wires. Along one dimension, there are full- and partial-mesh networks. A *full-mesh* network is one in which all nodes interconnect directly with all other nodes. A *partial-mesh* network is one in which some, but not all, nodes interconnect directly. Along another dimension, there are still two more variations on the theme. A *pure mesh* is a client mesh, meaning that

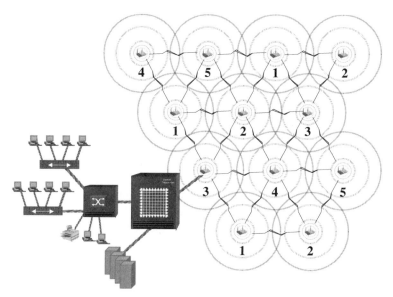

Figure 8.20   WLAN mesh network with overlapping coverage zones and channel separations.

any and all devices can interconnect with any and all other devices without wires. That approach, which is not highly scalable, is more along the lines of a *piconet*, which technologies such as Bluetooth address. An *infrastructure mesh* is a node mesh, meaning that there is no requirement for cabling from the (majority of) APs or wireless routers to a wired port on a switch or for cabling between APs. Rather, the majority of APs interconnect on a peer-to-peer basis through wireless RF links, with only those at the logical edge of the mesh connecting back to the wired LAN domain, as illustrated in Figure 8.20. A mesh configuration requires a considerable level of intelligence, which can be centralized if thin APs are preferred. Alternatively, highly intelligent fat APs can communicate on a peer-to-peer basis to auto-configure the most efficient multihop path for each transmission. The latter approach generally is preferred, as it is more conservative in terms of RF bandwidth, which is always at a premium.

A wireless mesh network autodiscovers topology changes as devices are added and moved. The network also establishes and alters traffic-forwarding paths in order to minimize the number of hops and otherwise optimize bandwidth utilization. Through the use of *cognitive radio* technology, the smart mesh network may also be able to sense sources of potential interference and adjust channel allocations between nodes and their associated coverage areas. Mesh networking also offers considerable redundancy and resiliency, as there are many alternate paths between any two devices. Note that the concept of a full-mesh network implies full coverage. In other words, there must be complete coverage of the area, with carefully placed and closely spaced APs providing overlapping coverage zones, as illustrated in Figure 8.20, in order for the APs to communicate on a peer-to-peer basis. That means that RF channels must be managed carefully, with frequency assignments made in such a way that the same frequency channel is not reused in an adjacent zone, or cell. Power levels must be carefully managed, as well, in order to ensure that the signal from one zone does not inadvertently compete with that of another zone. The IEEE 802.11 Task Group S met in September 2004 to begin developing a standard for interoperable mesh networking. The expectation is that a specification will be released in 2007 [51].

8.9.1.2.3 Power over Ethernet   Speaking of power, all devices comprising a LAN, whether wired or wireless, require electrical power. That includes all clients (even those that can operate on battery power for short periods), all servers, and all APs. If the building was designed with plenty of electrical outlets in just the right places, power is no problem. Since LAN-attached clients now include security cameras, tablet and hand-held computers, and even telephones (more on that later), LAN connectivity and electrical power issues can reach not only into every room, every hallway, and every corner of every building but also into parking garages and outdoor areas. A network designed to provide full coverage at that level can involve considerable costs for electrical wiring, whether it is full-mesh wireless or not. That fact increasingly prompts reexamination of full-mesh Wi-Fi, particularly in consideration of the fact that it can be much less expensive to run Cat 5e LAN cables than to run electrical wiring.

IEEE 802.3af (June 2003) addresses *Power over Ethernet* (PoE), which specifies the method for providing both data and electrical power to low-power devices over Cat 5, Cat 5e, and Cat 6 cable at distances up to 100 m. More correctly known as the *Data Terminal Equipment Power via Media Dependent Interface* amendment to 802.3, PoE defines how power is delivered to devices also using 10Base-T, 100Base-T, and 1000Base-T technologies. (*Note:* Running power and data over the same cable plant is something that has been done in voice telephony since 1876 but is new to the LAN domain.) PoE provides electrical circuits over two separate wire pairs, of course. PoE not only provides an alternative to expensive electrical cabling in hard-to-reach places but also alleviates concerns about power outages if an Uninterruptible Power Supply (UPS) is available. Reliable power is always an issue in the LAN domain, which now extends to security cameras, alarm systems, bar code scanners, smart building controls, and even VoIP telephones, as we discussed in Chapter 3. The 802.3af standard specifies that the *Power Sourcing Equipment* (PSE), or power injectors, provide output of 48 V DC power over the cable plant to terminal units that provide 12 V DC output to PoE-compliant devices known as *Powered Devices* (PDs). The 802.3af standard also specifies four different power draw levels of up to 3.84, 6.49, 12.95, and 15.4 W for attached devices. The PSE automatically senses the power requirements of the PDs. The PoE-compliant IP phones typically consume 3–5 W, wireless access points 6–10 W, and security cameras 9–12 W.

PoE operates in several ways. The purest approach involves power supplied directly from an Ethernet switch to a client device. Alternatively, a midspan device can inject power without interfering with the data signal. These midspan devices are particularly cost effective where PoE is conceptually attractive but existing equipment does not support it. Note that the midspan device does not extend the reach of the LAN, which remains at 100 m, from switch to client. Note also that the midspan device is passive to the data signal, that is, it neither interferes with nor regenerates it. PoE systems are designed to automatically sense whether or not the attached client device is 802.3af comliant. If the device does not present an authenticated PoE signature, the system will not attempt to power it. Device compatibility issues can be resolved by an intermediate PoE-compliant *picker* or *tap* that acts as a splitter, picking off the 48 V DC and making it available to the device at 5, 6, or 12 V DC, for example.

PoE clearly solves a power problem where Cat 5e or Cat 6 is in place and properly installed and over which PoE-compliant endspan or midspan power injectors can connect to compliant client devices. It does, however, also pose a bit of a cabling problem, as it consumes two pairs of what is usually a four-pair cable. Usually Cat 5e cable is run in a four-pair configuration, with two pairs for data and one or two pairs often used for voice. Where PoE is deployed, there is the potential for something of a conflict between voice, data, and power as they all compete for limited pair count. There also can be issues with respect to the total electrical load a PoE system might place on electrical circuits and wiring closet cooling systems in a scenario involving large numbers of PDs.

That said, in November 2004, the IEEE formed the 802.3at Working Group to develop the next level, known as PoE Plus, with the goal of increasing power to 30 W or more. The standard will address Cat 5 or better cable and is expected to support 10Base-T, 100Base-T, 1000Base-T, and 10GBase-T. The 802.3at specification is expected in the 2007–2008 time frame [52–61].

8.9.1.2.4 Security Two security mechanisms are specified in 802.11b. The most basic is the *Extended Service Set IDentifier* (ESSID, or SSID), which is in the form of an identifier code established by the system administrator for each device set up to gain access through each access point. At the next level is *Wired Equivalent Privacy* (WEP), a stream cipher that uses a 40- or 128-bit encryption key to protect data in transit. After WEP was compromised by hackers in 2001, it largely was replaced by *Wi-Fi Protected Access* (WPA), based on the more secure *Advanced Encryption Standard* (AES), which employs a 128-bit block cipher. WPA is included in the 802.11i standard (2004) for Wi-Fi security.

8.9.1.2.5 Public Hotspots 802.11b has become incredibly popular, not only in enterprise, SOHO, and even personal applications but also in public settings. Wi-Fi *hotspots* increasingly are installed in public venues such as airports and coffee shops and even on airplanes, usually on a pay-per-minute basis. A number of municipalities have installed free Wi-Fi access in downtown areas to encourage people to frequent those areas and patronize local merchants, and a great many more municipalities plan to offer such services in the future. A number of Internet Service Providers (ISPs) quite naturally are battling to keep Wi-Fi from attaining the status of a public utility, but their success has been mixed, to say the least.

8.9.1.3 IEEE 802.11g The most recent member of the family is 802.11g (June 2003), which runs at a signaling speed of up to 54 Mbps but is backward compatible with 802.11b. Like 802.11b, 802.11g has access to 11 channels in the United States, only 3 of which can be used in a confined area at any given time without overlap. Like 802.11a, 802.11g uses Orthogonal Frequency Division Multiplexing (OFDM) at data rates of 6, 9, 12, 18, 24, 36, 48, and 54 Mbps, with the attainable speed being highly sensitive to distance and LOS. The modulation technique reverts to the 802.11b Complementary Code Keying (CCK) approach at 5.5 and 11 Mbps. At

2 Mbps, it reverts to DSSS and QPSK and at 1 Mbps to DSSS and BPSK, again defaulting to the 802.11a specification. Dual-band components are widely available, so 802.11a-equipped computers communicate easily with 802.11g APs, and vice versa. Note, however, that when an AP supports both 802.11b and 802.11g simultaneously, the performance of both suffers, especially that of 802.11b systems. This is due to the fact that the AP goes into *protected mode* due to the challenge of supporting two protocols. In any case, roughly one-third of the theoretical bandwidth is consumed by overhead and throughput is likely to be in the range of 50 percent of the signaling rate.

8.9.1.4 IEEE 802.11n   The IEEE formed the 802.11n task group in January 2004 to begin work on the next-generation standards, which are expected to be finalized in late 2006 or early 2007. The focus is on increasing the signaling rate and the data rate to at least 100 Mbps. Achieving that goal can involve several approaches. Additional spectrum just is not available and, if it were, it would be in higher frequency bands, which are less effective due to issues of signal attenuation. More complex modulation techniques would allow more bits to be impressed on each baud, but the nature of radio transmission places practical limits on this approach. So, the focus is on an approach known as *Multiple Input, Multiple Output* (MIMO), which involves multiple transmit antennas and multiple receive antennas operating on the same frequency. 802.11n will operate in the 2.4-GHz band and will be backward compatible with 802.11b/g. As illustrated in Figure 8.21, the transmitter splits the signal across multiple transmit antennas separated by some amount of space, but operating on the same frequency at the same time, as do the receive antennas. MIMO is a LAN variation on the theme of *spatial diversity*, which has been used in microwave and other radio systems for many years as a means of improving communications under circumstances in which physical obstructions cause signals to take multiple paths as they bounce off of one obstruction and then another and perhaps another as they make their way from transmitter to receiver in a phenom-

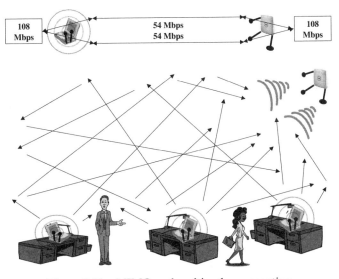

Figure 8.21   MIMO and multipath propagation.

enon known as *multipath propagation*. As we discussed with respect to 802.11a, multipath propagation causes *delay spread*, which is the difference in timing of signal elements caused by the fact that some portions of the signal take longer paths than others from transmitter to receiver and, therefore, arrive at slightly different times. This is all related to distance and the velocity of propagation and is exactly the same set of phenomena as *modal dispersion* and the resulting *pulse dispersion* in multi-mode optical fibers. The MIMO solution to this blurring of the signal is to employ spatial diversity in sets of both transmit (output) and receive (input) antennas. The transmit signal is split across two transmit antennas, thereby doubling the effective transmission rate from 54 to 108 Mbps, for example. As the transmit antennas are separated by some amount of space, the signals will take different paths from transmitter to receiver. The likelihood is that some signal elements will be stronger than others and will arrive ahead of others, since they will take less troublesome paths and suffer less attenuation along the way. Sophisticated signal processing software will take advantage of multipath propagation to combine and correlate many signal elements arriving at different times into one linear combination of a stronger, synchronized, intelligible signal derived from each of the two receive antennas. The signal processor in the receiver will combine the results of the two antennas and reconstitute the original data stream. Although multipath signal propagation generally is considered to be a signal impediment, MIMO actually depends on it to work properly [62–65].

8.9.1.5   *Voice over Wi-Fi*

It was only a matter of time until someone came up with the idea of using a Wi-Fi network for voice communications. After all, Private Branch eXchange (PBX) technology has moved to VoIP over switched Ethernet, wireless voice is commonplace through cellular and cordless technology, Wi-Fi technology offers plenty of bandwidth, and Wi-Fi mesh networks are possible through fat clients and high-speed layer 2 switches. It is not quite that simple, of course. Voice over Wi-Fi (VoWi-Fi) presents a number of technical challenges, including handoffs between APs as the user moves from cell to cell, Quality of Service (QoS), and security.

8.9.1.5.1   Quality of Service   The IEEE addressed QoS issues in 802.11e (2005) through a new coordination function that provides a station with high-priority traffic such as voice with more frequent network access than a station with low-priority traffic such as e-mail. Further, the station with the high-priority traffic is granted a longer transmit opportunity, that is, time window, in which to transmit as many frames as possible. In all, 802.11e defines four access priority classes, which the Wi-Fi Alliance terms to be *Wi-Fi MultiMedia Extensions* (WMMs or WMEs). Those classes are as follows:

- *Voice priority*, the highest level, is defined in support of low-latency voice.
- *Video priority*, the second highest level, prioritizes video relative to other data traffic. One 802.11a/b channel can support three-to-four Standard-Definition TV (SDTV) or one High-Definition TV (HDTV) data streams.
- *Best effort priority* is intended to support traffic from legacy devices and from applications or devices that lack QoS capabilities. Web browsing is an example of best effort traffic.

• *Background priority* is defined in support of low-priority traffic without strict latency and throughput requirements. Examples cited include file downloads and print jobs.

QoS also requires smooth handoffs in order to avoid lapses in conversation and dropped calls as the user moves between cells. This is more than a matter of making a connection with one AP before dropping the connection with another. It is a matter of security as well. When a client seeks access to a Wi-Fi network, it does so through an AP that, either independently if a fat AP or under the direction of a switch if a thin AP, is responsible for authenticating the identity of the client and either granting access or denying it. This process takes a few seconds, which is not noticeable in a data application but is a huge QoS issue for voice calls. As Wi-Fi cells tend to be quite small at a maximum diameter of 200 m or so, these handoffs can be frequent. To mitigate this issue, Wi-Fi switches assume control of large numbers of thin APs, coordinating and controlling their activities much as a Mobile Traffic Switching Office (MTSO) controls a number of dumb cell sites. As long as the user remains within the domain of a single switch, the handoffs can be handled fairly smoothly. If the user moves between switch domains in a large Wi-Fi environment, however, QoS issues develop as the switches struggle to coordinate handoffs. There is an even more complex handoff issue if a user moves from a private VoWi-Fi domain to a public one. The IEEE is addressing these problems through the 802.11r initiative, which it expects to finalize in early 2007. Also, the IEEE 802.21 Task Group is in the early stages of developing specifications to support interoperability and handoff issues between heterogeneous networks, including 802 and non-802 network types. The expectation is that this effort will result in specifications supporting smooth interconnectivity and interoperability between Wi-Fi and cordless telephony and cellular, WiMAX, and other wireless networks. Some dual-mode and even trimode handsets already exist. For example, several ILECs in Africa that also are cellular service providers offer dual-mode phones that support VoWi-Fi and cellular. Since a single service provider owns both the PSTN and cellular networks and both installs and maintains the Wi-Fi LAN, interconnectivity can be achieved more easily than if multiple network operators were involved.

8.9.1.5.2 *Configuration Considerations*
The physical and logical layout of the Wi-Fi network has a real impact on its ability to support voice. A Wi-Fi network in support of laptop users is relatively simple to configure. As computing on such a platform tends to require a flat and stable surface, one looks for places with tables or at least chairs or benches for users with fairly flat laps. Configuring a network for users of tablet and hand-held computers is more challenging, as they can compute in hallways and other unusual places and while on the move. Not only do the coverage areas expand, but cells must overlap, handoffs must be made, and frequency assignments must be carefully administered so that the same channels are not used in adjacent cells. As 802.11b/g offers only 11 channels in the United States, spectrum management can be quite a challenge, particularly where user density is high and, therefore, APs must be tightly spaced and cell sizes must be small. While one obvious solution is that of reducing the power levels of the APs, VoWi-Fi network configuration issues can be tough to solve.

8.9.1.5.3 VoWi-Fi Futures   VoWi-Fi appears to have a bright future. Its key advantage is that of mobility, as is the case with any wireless voice solution. VoWi-Fi involves no airtime charges, unlike cellular, although the handsets are expensive and a Wi-Fi network configured for voice can be costly. The cost of a VoWi-Fi LAN is estimated at roughly double that of a data-only WLAN, as more access points are required, priority mechanisms must be in place, switches and/or APs must be more intelligent, security must be tightened, and so on. The real future of VoWi-Fi is in dual-mode or trimode handsets that can operate as VoWi-Fi handsets over a wireless IPBX at the office and as cellular phones elsewhere. The handsets must be intelligent enough to seek out the lowest cost alternative and to seamlessly switch between them when the need arises [66–70].

8.9.2   HiperLAN

HiperLAN (*Hi*gh *per*formance *radio LAN*) is a high-speed LAN standard running in the 5-GHz range. Approved by the *European Telecommunications Standards Institute* (ETSI) in February 2000, HiperLAN grew out of efforts to develop a wireless version of ATM and a European alternative to 802.11. *HiperLAN1* operates at rates up to 20 Mbps and *HiperLAN2* at rates up to 54 Mbps. As noted in the discussion of 802.11a, in Europe the 5.725–5.825-MHz spectrum is already allocated for HiperLAN. Therefore, ETSI requires that two additional protocols be used in conjunction with 802.11a in order to protect incumbent applications and systems running over previously allocated shared spectrum. *Dynamic Frequency Selection* (DFS) allows the 802.11a system to dynamically shift frequency channels, and *Transmission Power Control* (TPC) reduces the power level. In combination, these protocols serve to eliminate interference issues with incumbent signals. HiperLAN uses *Orthogonal Frequency Division Multiplexing* (OFDM) as the signal modulation technique [46].

HiperLAN has a very small following, as it has been overwhelmed by the popularity of the 802.11 family. HiperLAN is unlikely to survive the next few years as the pressure increases with the release of 802.11n.

8.9.3   Bluetooth

Bluetooth is a specification to standardize wireless transmission between a wide variety of devices such as PCs, cordless telephones, headsets, printers, and PDAs. The initial effort (April 1998) was in the form of a consortium of Intel, Microsoft, IBM, Toshiba, Nokia, Ericsson, and Puma Technology and was code named Bluetooth after Harald Blaatand, the tenth-century Danish king who brought warring tribes together and unified Denmark. Bluetooth is intended to create a single digital wireless protocol to address end-user problems arising from the proliferation of various mobile devices that need to keep data synchronized (i.e., consistent from one device to another). Bluetooth is now formalized in IEEE 802.15.1 (2002) as a *Wireless Personal Area Network* (WPAN) specification. A *personal area network* is defined by the *Personal Operating Space* (POS), that is, the space in the near vicinity of a device or individual [71]. Such a confined area clearly is less than that of a LAN, MAN, or WAN. While Bluetooth has limited use in short-range LAN applications, it generally finds application in providing very short range connectivity between personal productivity tools such as between a computer and a mouse, keyboard, and monitor or between a cell phone and a headphone.

In 2006, Exmovere unveiled a Web-based Bluetooth-enabled biosensor wristwatch designed to provide elderly care assistance. The wearable *Exmocare* sensor system monitors the elderly person's vital signs, including pulse, heart rate, and motion, and allegedly can assess up to 10 different emotional states, including *relaxed, worried,* and *agitated.* The device collects information every 30 min and automatically alerts caregivers of any abnormal status. The wristwatch connects via Bluetooth to an enabled PC, cell phone, or GPS-car kit, which monitors the wearer's location and vehicle speed.

Bluetooth makes use of *Frequency Hopping Spread Spectrum* (FHSS), with devices employing a pseudorandom hop sequence that makes data collisions highly unlikely. Bluetooth operates in the 2.45-GHz range of the ISM band, hopping through a set of 79 (United States and Europe) or 23 (Spain, France, and Japan) channels spaced 1 MHz apart at a rate of about 1600 hops per second, with each hop lasting 62.5 µs. At each hop, the baseband signal is modulated using *Gaussian Frequency Shift Keying* (GFSK), a technique that smoothes out the signals and limits their spectral width in order to reduce the potential for crosstalk in tightly packed radio channels. Gaussian Phase Shift Keying (GPSK) involves a positive frequency shift to represent a 1 bit and a negative frequency shift to represent a 0 bit. GPSK is not highly efficient, yielding a maximum raw link speed of 1 Mbps; however, its simplicity allows for low-cost device implementations. Depending on the transmitted power, Bluetooth devices can be organized into three classes. Class 1 equipment (maximum 100 mW) has a nominal link range of as much as 100 m, and Class 3 devices (maximum 1 mW) are limited to about 10 m [71].

Bluetooth technology supports both *Synchronous Connection–Oriented* (SCO) links for packet voice and *Asynchronous Connectionless Links* (ACLs) for packet data. Bluetooth supports an asynchronous data channel in asymmetric mode of up to 721 kbps in either direction and 57.6 kbps in the reverse direction. Alternatively, the data channel can be supported in symmetric mode of up to 432.6 kbps. As yet another alternative, Bluetooth supports up to three simultaneous synchronous packet voice channels, or a channel that simultaneously supports both asynchronous data and synchronous voice.

Bluetooth supports FDX communications using *Time Division Duplex* (TDD) as the access technique. Voice coding is accomplished using the *Continuously Variable Slope Delta* (CVSD) modulation technique. CVSD yields voice compression of 4:1 (16 kbps) or 8:1 (8 kbps).

8.9.3.1 *Piconets and Scatternets*

Bluetooth operates on a point-to-point and a point-to-multipoint basis. As illustrated in Figure 8.22, current standards allow as many as eight devices to be linked together in a *piconet*, or very small network, with as many as seven devices *slaved* to a single *master*. This relationship is ad hoc and short term in nature, with the role of master assumed by the device initiating the dialogue. The standards also provide for the formation of a *scatternet* comprising overlapping piconets. In such a configuration, a device can serve as master in one piconet and a slave in another or as a slave in both. A device participating in multiple piconets is known as a *bridge* device and can forward packets from one piconet to another. In a true LAN application, Bluetooth is overshadowed by 802.11b, even

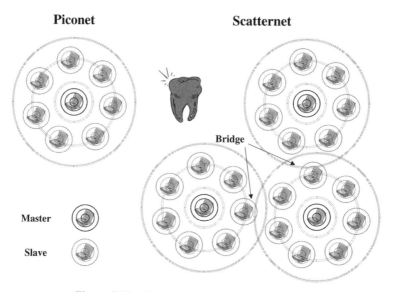

Figure 8.22 Bluetooth piconet and scatternet.

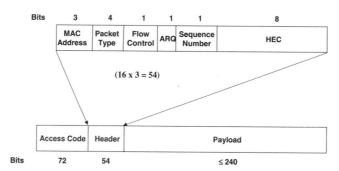

Figure 8.23 Bluetooth packet format.

though Bluetooth is easier to pronounce [72–74]. (*Note:* Score one for the technologists in the ongoing battle with the marketers.)

8.9.3.2 Packet Format As mentioned above, each slot in a Bluetooth connection lasts 62.5 μs and corresponds to a single hop from once device to another. The master always transmits in even-numbered slots and the slave in odd-numbered slots, with one packet transmitted per slot. As illustrated in Figure 8.23, the Bluetooth packet takes the following format:

- **Access Code:** The 72-bit access code is defined by the master and is unique for each piconet. The access code serves to identify incoming packets associated with a given piconet. Devices on the piconet will accept packets with the proper access code and will reject all others. This field also allows synchronization purposes.
- **Packet Header:** A 54-bit packet header contains the following fields, the total of which equal 16 bits. As the header is repeated three times to ensure that

there are no errors in header transmission, the header consumes a total of 54 bits. Bluetooth specifications refer to this redundant error control technique as *1/3 FEC* (Forward Error Correction).

- Medium Access Control (MAC) address: 3 bits, with 000 identifying a broadcast packet. *Note*: As $2^3 = 8$, a piconet is limited to one master and seven slaves.
- Packet type: Four bits indicating the 16 packet types. There are 4 types of control packets and 12 types of data packets. There are 3 types of voice packets, each running at a rate of 64 kbps, including overhead and, in some cases, an FEC mechanism.
- Flow control: One bit.
- *Automatic Repeat reQuest* (ARQ): One bit. This bit is used to request a repeat packet transmission in the event that there is a detected error in the payload. ARQ is not used in SCO links, as error control would increase latency and retransmissions would create jitter.
- Sequence number: One bit.
- Header Error Correction (HEC): Eight bits.
- **Payload:** The variable-length packet is limited to 366 bits, although the theoretical limit is 625 bits (62.5 μs × 1 Mbps = 625 bits). The limit of 366 provides the transmitters and receivers with enough time to hop to the next frequency and stabilize. As the access code and packet header consume 126 bits, the payload cannot exceed 240 bits, or 30 octets. There is a provision for *multislot packets*, which can support a larger payload. The payload includes an error control mechanism in the form of a CRC [71].

8.9.4 ZigBee

ZigBee is a specification from the ZigBee Alliance for a set of high-level communications protocols based on the IEEE 802.15.4 standard for Personal Area Networks (PANs). ZigBee is intended to be simpler, more flexible, and less expensive than either Bluetooth, which also is based on 802.15, or Wi-Fi. ZigBee is designed for connecting devices in ad hoc mesh networks over very short distances with very low power consumption. There are three network topologies possible—star, peer to peer, and mesh. The high levels of redundancy and network resiliency offered by mesh networking make it the preferred approach in consideration of the applications, which include building automation and industrial, medical, and residential monitoring and control. Should a device fail or be removed, the autodiscovery feature of such a mesh network will recognize and register that fact and exercise an alternate path.

ZigBee runs in the ISM band using *Direct Sequence Spread Spectrum* (DSSS) transmission and collision avoidance. Most devices run in the 2.4-GHz range, which is available worldwide. Some run at 915 MHz (Americas) and 868 MHz (Europe), as those bands offer better signal propagation through walls, floors, windows, and so on. Depending on the frequency band selected, raw data rates are 20 kbps (1 channel at 868 MHz), 40 kbps (10 channels at 915 MHz), and 250 kbps (16 channels at 2.4 GHz). Distances range from 10 m to 100+ m, depending on frequency, power output, and environmental characteristics. Security features include access control and encryption based on AES. ZigBee is highly scalable up to 65,536 devices, at least theoretically.

ZigBee is hierarchical in nature, including *ZigBee End Devices* (ZEDs), which are terminal devices, such as sensors, that can perform only a single monitoring or control function. ZEDs communicate with *ZigBee Routers* (ZRs), which can serve as ZEDs in addition to functioning as routers to pass messages to other ZRs or to the *ZigBee Coordinator* (ZC). The ZC initializes the network, coordinates its operation, and is responsible for security.

A ZED comprises a low-cost microprocessor, RAM and ROM memory, a battery, a radio and controller, and the IEEE and ZigBee protocol stacks. The device is small enough to be embedded in a light switch, smoke or carbon dioxide detector, thermostat, security sensor, utility meter, or medical sensor. As battery life is so critical, a ZED goes into sleep mode when not actively performing its functions. Therefore, a battery could last for years if used in a low-duty-cycle application such as meter reading or alarming [75, 76].

The term *ZigBee* refers to the technique, known as the ZigBee principle, that a domestic honeybee uses to communicate the location of a new food source to other members of the colony. The bee dances in a zigzag pattern that communicates information such as distance and direction, at least according to the ZigBee Alliance [76]. Others suggest that it is just a made-up name chosen at random, since independent research, including my own, fails to confirm that ZigBee has anything to do with honeybees. (That's a shame, because it's a really cool story.)

8.9.5 Software-Defined Radio

The FCC has defined *Software-Defined Radio* (SDR) as a generation of radio equipment that can be reprogrammed quickly to transmit and receive on any frequency within a wide range of frequencies and using virtually any transmission format and any set of standards. Theoretically, a device such as a WLAN NIC, cellular telephone, or PDA with an SDR chipset could seek out various available frequency bands and native protocols supported by the networks, lock in on the signals, and negotiate access to the desired network, downloading any necessary supplemental software required to effect network compatibility. In the process, SDR-equipped devices would resolve any conflicts between networks sharing a given band (e.g., 802.11b and Bluetooth overlap in the 2.4-GHz ISM band). The FCC began hearings on SDR in March 2000, with the intent that the development of SDR could promote more efficient use of spectrum, expand access to broadband wireless communications, and increase competition among service providers.

8.10 MINDING YOUR Ps AND Qs

The IEEE recently has taken steps to enable Layer 2 switches to support Quality of Service (QoS), which actually is more in the form of Grade of Service (GoS). In September 1998, the IEEE adopted the 802.1p specification. That specification, in conjunction with the previously adopted 802.1q specification for VLAN tagging, paved the way for standards-based multivendor GoS, much as do *DiffServ* (*Differentiated Services*) and various other protocols in the WAN. The 802.1 specification enables switches and other devices (e.g., bridges and hubs) to prioritize traffic into one of eight classes. Class 7, the highest priority, is reserved for network control data such as *Open Shortest Path First* (OSPF) and *Routing Information Protocol* (RIP) table updates. Classes 5 and 6 can be used for voice, video, and other delay-sensitive traffic. Classes 1–4 address streaming data applications through loss-tolerant traffic such as *File Transfer Protocol* (FTP). Class 0, the default class, is a *best effort* class. Because Ethernet does not provide a mechanism for priority identification in the frame header, 802.1q is employed. That specification defines a 32-bit tag for such purposes. Desktop systems, servers, routers, or layer 3 switches can set the 802.1q tag [77, 78]. Note that 802.1q increases the maximum Ethernet frame size from 1518 to 1522 bytes, which can overload legacy NICs and switches [79].

Ps and Qs are the mechanisms by which layer 2 switches support voice and video as well as LAN traffic in the converged LAN domain. Within the customer premises, these specifications encourage the development of the convergence scenario via such relatively inexpensive technologies as switched Ethernet.

8.11 IEEE 1394 AND FIREWIRE

The IEEE 1394 specification, known as *FireWire* in Apple Computer terminology, is for a data transport bus between a host computer and peripherals (e.g., high-density storage devices and high-resolution still and video cameras) and is designed to eliminate the bottleneck at the serial port of the LAN-attached PC. While 100-Mbps LANs address the bottleneck of the shared medium, they can overwhelm the attached workstation (sort of like drinking out of a fire hose). As videoconferencing and multimedia applications increase in popularity, it is necessary to increase the speed of the *Small Computer Systems Interface* (SCSI) to support them and take full advantage of the speed provided by the LAN. 1394 addresses that requirement through standards for 100, 200, and 400 Mbps. A single 1394 port can support up to 63 peripherals over a six-conductor cable up to 4.5 m in length and as many as 16 cables can be daisy chained to extend the total length to as much as 72 m.

8.12 NONSTANDARD LANs

In addition to standard LANs, there exist a number of nonstandard options. Some are proprietary standards that serve certain applications and vertical markets, such as the oil refining industry described earlier. Others are more widely accepted speci-fications, published by other than a recognized standards body, with *ARCnet* (*A*ttached *R*esource *C*omputer *net*work) being a prime example. Developed in 1977 by Datapoint Corporation, ARCnet is a highly reliable, low-cost token bus system based on a physical star and logical ring and a character-oriented protocol. ARCnet supports data rates up to 2.5 Mbps and as many as 255 attached devices. More recent versions deliver 20 and 100 Mbps, although they have never been in great demand and are not widely available. ARCnet resembles but does not adhere to the IEEE 802.4 specification. RXnet was Novell's implementation of ARCnet.

Manufacturing Automation Protocol (MAP) is another good example of a non-standard LAN protocol. As discussed earlier in this chapter, MAP was developed by General Motors (GM) in the early 1980s for the interconnection of computers and programmable machine tools in factory or assembly line operations.

8.13 BROADBAND OVER POWER LINE

Broadband over Power Line (BPL) is a set of specifications for Power Line Carrier (PLC), a technology that has been used for certain telco local loop applications since 1928. In-house PLC technologies have been used in key telephone and intercom systems since at least the early 1980s, although not particularly successfully. Standards for In-house BPL, a premises networking technology, are a relatively recent development, with HomePlug standards being the most prevalent. Loosely based on Ethernet LAN standards and using a variation of CSMA/CA, In-house BPL allows any device to connect to the LAN directly through the low-voltage electric lines (110 V at 50–60 Hz or 220 V at 50 Hz). HomePlug 1.0 supports up to 16 nodes sharing bandwidth up to a theoretical maximum of 14 Mbps. Some proprietary systems support raw signaling rates up to 85 Mbps, which comes very close to 100Base-T performance.

HomePlug-compatible devices include PCs, routers, bridges, switches, and any other devices that use RF-45 (Ethernet) or USB physical interfaces. The devices plug into a HomePlug adapter that is about the size of a typical low-voltage transformer or power adapter and that plugs into any electrical outlet on the premises. Thereby, every electrical outlet becomes a port into an Ethernet LAN.

HomePlug uses of a version of *Orthogonal Frequency Division Multiplexing* (OFDM) specially tailored for powerline environments. OFDM splits the signal into a stream of data symbols for massively parallel simultaneous transmission over a number of narrowband, low-data-rate subcarrier frequencies. [*Note:* OFDM is the transmission technique used in 802.11a (aka Wi-Fi5) and 802.11g, 802.16, and WiMAX wireless systems. Also known as Discrete MultiTone (DMT), the technique is used in ADSL service as well.] HomePlug 1.0 specifies 84 equally spaced subcarriers within each of which several differential modulation techniques are employed. Security is through 56-bit DES.

Attenuation in HomePlug networks is influenced not only by the propagation of the signal through the copper conductors (commonly 12 or 14 gauge) but also by splices and various components such as fuse boxes, surge suppressors, and circuit breakers. HomePlug currently offers a range of as much as 300 m without repeaters, which compares favorably with the 100 m supported by 10/100BaseT.

ElectroMagnetic Interference (EMI) and Radio Frequency Interference (RFI) present considerable challenges in a HomePlug environment. Sources of EMI include brush motors, switching power supplies, fluorescent lights, and halogen lamps, all of which produce impulse noise that can negatively impact signal integrity over the shared electrical bus. HomePlug deals with these challenges through a combination of Forward Error Correction (FEC) and Automatic Repeat Request (ARQ). As RFI from amateur radio can impact certain frequencies, HomePlug employs spectral density notches around the ham radio frequency bands, thereby reducing the number of OFDM carriers that can be used in the United States. As noise on the powerline can be highly local to the receiver and as the quality of the channel between any two links connecting transmitter and receiver over the common electrical bus can vary considerably, HomePlug 1.0 uses a channel adaptation technique to turn off heavily impaired subcarriers. Tests conducted in 500 or so homes showed that 80 percent of outlet pairs can connect at 5 Mbps or better and 98 percent at 1 Mbps or better, depending on the condition of the inside wire.

The next step in the specification is HomePlug AV, which is intended to support entertainment applications such as HDTV and home theater. HomePlug AV will run in the range of 2–28 MHz, offering a raw signaling speed of up to 200 Mbps through the use of OFDM. Throughput will be more in the range of 100 Mbps, given TCP/IP and other overhead considerations. HomePlug AV will offer inherent Grade of Service (GoS) considerations through the use of IEEE 802.1Q Virtual Local Area Network (VLAN) tags for marking high-priority traffic such as VoIP (Voice over Internet Protocol) and the streaming audio and video components of HDTV. With regard to security, HomePlug AV makes use of the highly secure AES. *Note:* The IEEE is working on the competing P1901 specification for BPL communications, which will address both Access BPL and In-House BPL [80–85].

8.14 STORAGE AREA NETWORKS

As electronic commerce (*e-commerce*) has exploded along with the growth of the Web, so has the requirement for storage. Other applications also are increasingly storage intensive, and it has become clear over the last few years that it is absolutely necessary to maintain highly secure database backups. A number of network storage technologies, which can be quite complex, have developed to satisfy this requirement. The simplest approach is that of *Network-Attached Storage* (NAS), which simply is one or more storage devices (e.g., disk arrays) associated with a server that exists as a node on a LAN. The storage server assumes the responsibility for all data storage and for making the data available to all users on the network who have appropriate access privileges.

A *Storage Area Network* (SAN) generally is in the form of a subnetwork that is part of a larger LAN. SANs generally use the SCSI protocol for communications between computers and storage servers, although they do not use its Layer 1 (Physical Layer) interface. A SAN is much more complex than a simple NAS, as it involves a high-speed, special-purpose dedicated subnetwork designed to transport data-intensive applications such as inventory management, credit and billing management, receivables management, customer relationship management, and supply chain management. SANs provide application users with much faster access to databases, as they avoid congesting the general-purpose LAN with storage traffic, which often involves very active users engaged in very large file transfers. SANs also provide for centralized management of critical data, including accessibility, security, and backup. SAN protocols include 100Base-T, GbE and 10GbE, ATM, IBM's *Enterprise Systems Connectivity* (ESCON) and *Fibre Connections* (FICON), several versions of *Fibre Channel* (FC), *Serial Systems Architecture* (SSA), *Small Computer Systems Interface* (SCSI), and *Internet Small Computer Systems Interface* (iSCSI). The storage technologies include *Just a Bunch Of Disks* (JBOD), *Redundant Array of Inexpensive Disks* (RAID), a cluster of servers on a network, or a more complex and expensive host storage server such as a mainframe computer. SAN applications include disk mirroring, data backup and restoration, data archival and retrieval, data transfer between storage devices, and data sharing between servers [86–91]. SAN technology has been the focus of a great deal of interest during the past few years, and there is every reason to expect that to continue into the foreseeable future. Knowledge workers seem to have insatiable appetites for information, which increasingly is in multimedia format, and they also want access to that information immediately. SANs are a prime solution. There recently has been a great deal of interest centered on the battle between FC and iSCSI.

8.14.1   Fibre Channel

The *InterNational Committee for Information Technology Standards* (INCITS) began work on *Fibre Channel* (FC) in 1988 as a replacement for the *HIgh Performance Parallel Interface* (HIPPI) technology, a highly distance-limited technology that involved 50-pair cable and huge connectors. ANSI approved the resulting specification in 1994. FC is connected at Layer 1, the Physical Layer, by *fibre*, a term the Fibre Channel industry coined to refer to a network comprising a close-knit fabric of access including both optical fiber and copper for large data transfers with low overhead, low-latency switching, and minimal interruptions to the flow of data [92]. The physical media and transceivers are the same as those used in LANs and telecommunications networks. The physical medium of choice is optical fiber, which can be MultiMode Fiber (MMF) of either 62.5 km (300 m, maximum distance) or 50 km (500 m, maximum distance) or Single-Mode Fiber (SMF) (50+ km, maximum distance). The recently developed *Fibre Channel over IP* (FC/IP) technology extends FC to operate through secure tunnels over public IP networks at WAN distances. FC operates at four link speeds (and FDX throughput rates, as measured in bytes per second) as follows: 1 Gbps (200 MBps), 2 Gbps (400 MBps), 4 Gbps (800 MBps), and 10 Gbps (2400 MBps). The line coding technique is 8B/10B, which now also is used in GbE and 10GbE. FC equipment comprises hubs, switching hubs, switches, and routers. Gateway routers are responsible for protocol conversion to support interconnection to telecom networks such as ATM and SONET as well as SCSI SANs and Ethernet LANs.

The FC protocol stack includes a routing protocol similar to Open Shortest Path First (OSPF) and provides the Transport Layer (Layer 4) for upper layer protocols that contain the applications and user interface, with examples being SCSI, ESCON, and FICON. FC is designed to carry IP traffic as well [93].

8.14.2   Internet Small Computer Systems Interface

The *Internet Small Computer Systems Interface* (iSCSI, pronounced *i-scuzzy*) allows SCSI commands and block storage data to travel over Ethernets. An iSCSI *Host Bus Adapter* (HBA) looks to the computer like any storage device (e.g., internal disk or NAS) and looks to the network like a NIC. As it exits the computer headed toward the storage device, HBA converts the data to a SCSI format enclosed in an IP packet and transmitted over an Ethernet network. An advantage of iSCSI is that it is transparent, as the server software sees what looks to be a SCSI controller and the network sees only IP traffic. The protocols employed at Layers 2, 3, and 4 are

Ethernet, IP, and TCP, all of which are well understood and extensively used. Further, iSCSI is intended to run at speeds up to 10 Gbps over 10GbE, at distances that can be very significant through IP, and at lower cost than FC. Critics point to iSCSI's higher overhead rate and processor intensity as issues of significance and points in the favor of Fibre Channel [94–97].

REFERENCES

1. Shelly, Gary B. and Cashman, Thomas J. *Introduction to Computers and Data Processing.* Anaheim Publishing Co., 1980.

2. Nunemacher, Greg. *LAN Primer*, 3rd ed. M&T Books, 1995.

3. Metcalfe, Robert M. and Boggs, David R. *Ethernet: Distributed Packet Switching for Local Computer Networks.* Association for Computing Machinery, 1976.

4. Allocca, Lisa. "Networking for the Masses." *Internet Week*, May 4, 1998.

5. Metcalfe, Bob. "From the Ether." *InfoWorld*, October 31, 1994.

6. Conover, Joel. "Building a Better Infrastructure." *Network Computing*, October 1, 1998.

7. Henderson, Tom. "Gigabit Ethernet Blueprint." *Network Magazine*, September 1998.

8. "Technology Overview: Gigabit Ethernet." Cisco Systems.

9. Williams, Veronica A. *Wireless Computing Primer*. M&T Books, 1996.

10. Larsen, Amy K. "Wireless LANs: Worth a Second Look." *Data Communications*, November 1995.

11. Tolly, Kevin. "Token Ring vs. Ethernet: The Real Cost Story." *Data Communications*, May 21, 1995.

12. Miller, Mark A. *Internetworking: A Guide to Network Communications LAN to LAN; LAN to WAN*, 2nd ed. M&T Books, 1995.

13. Null, Christopher. "Unplugging the LAN." *Network World*, February 8, 1999.

14. Darling, Charles B. "Routers Can Save Your WAN Dollars." *Datamation*, July 1, 1995.

15. Spohn, Darren L. *Data Network Design*. McGraw-Hill, 1993.

16. *Ethernet Tutorial & Product Guide*. LANTRONIX, 1995.

17. Levy, Joseph R. and Hartwig, Glenn. *Networking Fundamentals: From Installation to Application*. MIS Press, 1995.

18. Petrosky, Mary. "Hub Shopping Spree." *Network World*, May 29, 1995.

19. Lopez, Steve. "Why Switch?" *Internetwork*, March 1995.

20. Tyson, Jeff. "How LAN Switches Work." http://computer.howstuffworks.com/lan-switch.htm.

21. Jacobs, Paula. "Virtual LANs: Too Good to Be True?" *Network World*, March 27, 1995.

22. Conover, Joel. "Minding Your Virtual Ps and Qs." *Network Computing*, October 15, 1997.

23. Held, Gilbert. "Virtual LANs Become Reality." *LAN Magazine*, April 1997.

24. Mandeville, Robert and Newman, David. "VLANs: Real Virtues." *Data Communications*, May 1997.

25. Morency, John P. "Do VLANs Really Deliver?" *Business Communications Review*, July 1995.

26. King, Steven S. "VLANs Raise Delicate Design Issues." *Network World*, April 17, 1995.

27. Minoli, Daniel. "Telecommuting Demand Characteristics." Datapro Information Services, November 4, 1998.

28. Allard, Hank. "Remote LAN Access: Strategies & Solutions." *Network World*, Technical Seminars, 1995.

29. Gangler, Barbara. "Remote Possibilities." *Internetwork*, June 1995.

30. Finneran, Michael. "Life in the Fast LAN." *Business Communications Review*, May 1995.

31. Rosen, Ron. "User's Guide to 100Base-T and 100VG-AnyLAN." In Slone, John P. (Ed.), *Handbook of Local Area Networks*. Auerbach Publications, 1995.

32. Peri, Ron. "Life in the Fast Lane." *Communications Week*, January 23, 1995.

33. Sullivan, Kristina B. "Promise of FDDI Holds True." *PC Week*, January 15, 1996.

34. *Implementing FDDI in Enterprise LANs*. SysKonnect, 1995.

35. Heneghan, Hank. "New Configurations Make FDDI More Survivable." *Network World*, November 20, 1995.

36. Korzeniowski, Paul. "Sun Setting on FDDI." *Business Communications Review*, April 1999.

37. Axner, David. "Going Gigabit." *Network World*, July 20, 1998.

38. Fontana, John. "Gig Ethernet Juices Up NT." *InternetWeek*, September 7, 1998.

39. Fontana, John "Jumbo Frames Support Reaches Critical Mass." *InternetWeek*, October 26, 1998.

40. Lounsbury, Al. "Gigabit Ethernet: The Difference Is in the Details." *Data Communications*, May 1997.

41. Ruby, Douglas. "Can Ethernet Go End to End?" *Network World*, September 17, 2001.

42. Allen, Doug. "10 Gigabit Ethernet Reaches for The Metro." *Network Magazine*, October 2001.

43. Rixmer, Rob. "10G Ethernet: We'll All Use It Someday." *Interactive Week*, December 31, 2001.

44. Korzeniowski, Paul. "10 Gigabits and Beyond." *Interactive Week*, December 31, 2001.

45. Karve, Anita. "The Wide World of Wireless." *Network Magazine*, December 1997.

46. Conover, Joel. "802.11a: Making Space for Speed." *Network Computing*, January 8, 2001.

47. Anderson, Fred. "802.11a Speeds Wireless LANs." *Network World*, January 29, 2001.

48. Cox, John. "Switch Broadens Wireless Capabilities." *Network World*, February 17, 2003.

49. Gohring, Nancy. "A Switch in Time." *Network World*, May 19, 2003.

50. Davis, Jeff. "Centralized Wireless LAN: Thin vs. Fat Technology." *Cabling Business Magazine*, November 2004.

51. Wexler, Joanie. "Mesh Moves into the Wireless Office." *Computerworld*, November 29, 2004.

52. Sturdevant, Cameron. "Payoff for Power over Ethernet." *Eweek*, February 24, 2003.

53. Lehr, Amir. "802.3af Powers Up LAN, Lowers Cost." *Network World*, March 10, 2003.

54. Mitchell, Robert L. "Ethernet's Power Play." *ComputerWorld*, May 26, 2003.

55. Hochmuth, Phil. "Power over Ethernet Generates Buzz." *Network World*, November 24, 2003.

56. Mendelson, Galit. "Expanding Existing Capabilities with Power over Ethernet." *Cabling Business Magazine*, June 2004.

57. Schmidt, John. "Power-over-Ethernet Considerations." *Cabling Business Magazine*, August 2005.

58. Gorman, Cheryl. "Power over Ethernet Injectors." *Cabling Business Magazine*, October 2005.

59. Paulov, Steve. "Power over Ethernet Opportunities Defined." *Cabling Business Magazine*, December 2005.

60. Barrass, Hugh and Schindler, Fred. "802.3at Pumps up Power over Ethernet." *Network World*, January 23, 2006.

61. Fluke Networks. "Fundamentals of Power over Ethernet (PoE)." White Paper, www.fluke.com.

62. Jones, V. K., Raleigh, Greg, and van Nee, Richard. "MIMO Answers High-Rate WLAN Call." *EE Times*, December 31, 2003.

63. Cox, John. "Wireless Vendors Try Defining MIMO." *Network World*, August 16, 2004.

64. Mathias, Craig. "My Oh MIMO." *Network World*, August 30, 2004.

65. Mathias, Craig. "Where's MIMO?" *Network World*, October 31, 2005.

66. Lawson, Stephen. "WLAN Quality-of-Service Specification Approved." *Techworld*, October 5, 2005, http://www.techworld.com/mobility/news/index.cfm?NewsID=4520.

67. Watson, Richard. "Wireless VoIP—The PDA Challenge." *Internet Telephony*, September, 2004.

68. Venters, Tracy. "Planning for Voice on Wi-Fi Networks." *Internet Telephony*, September 2004.

69. Horak, Ray. "VoWiFi: What Are They Thinking?" *Commweb*, August 16, 2005, www.commweb.com.

70. Dern, Damiel P. "Voice over Wi-Fi: A Work in Progress." *VON Magazine*, November 2005.

71. Nicopolitidis, Petros, Obaidat, Mohammad Salameh, Papadimitriou, Georgios I., and Pomportsis, Andreas S. *Wireless Networks*. Wiley, 2003.

72. Krasnoff, Barbara. "Bluetooth: Piconets & PANs." *Convergence*, June 2001.

73. Deckmyn, Dominique. "Bluetooth." *Computerworld*, June 12, 2000.

74. Dornan, Andy. "Can Bluetooth Sink Its Teeth into Networking?" *Network Magazine*, November 2000.

75. Kay, Russell. "ZigBee." *ComputerWorld*, May 15, 2006.

76. www.zigbee.org.

77. Bruno, Charles and Tolly, Kevin. "Minding Your QoS p's and q's." *Network World*, April 19, 1999.

78. Sherman, Doug. "Tackling the p's and q's of LAN traffic." *Network World*, September 7, 1998.

79. Green, David. "802.1q VLANs for Better Bandwidth." *Network World*, March 5, 2001.

80. Gardner, Steve. "HomePlug Standard Brings Networking to the Home." *CommsDesign*, December 27, 2004, http://www.commsdesign.com/main/2000/12/0012feat5.htm.

81. HomePlug Powerline Alliance. "HomePlug 1.0 Technology White Paper." White Paper, http://www.homeplug.org/en/docs/HP_1.0_TechnicalWhitePaper_FINAL.pdf.

82. U.S. Department of Commerce, National Telecommunications and Information Administration. "Potential Interference from Broadband over Power Line (BPL) Systems to Federal Government Radiocommunications at 1.7–80 MHz. Phase 1 Study, Volume I." U.S. Government Printing Office, Washington, DC.

83. "Broadband Access Networks Systems—Last Mile." http://cictr.ee.psu.edu/research/bans/index.html.

84. Valdes, Robert. "How Broadband over Powerlines Works." http://computer.howstuffworks.com/bpl.htm.

85. Horak, Ray. "In-House Broadband over Power Line." www.commweb.com.

86. Merian, Lucas. "2Gbit Fibre Channel Boosts Speed in SANs." *Computerworld*, June 11, 2001.

87. Erlanger, Leon. "Storage as Universal as Dial Tone." *Internet World*, August 1, 2001.

88. The Enterprise Storage Group. "Storage Networking: Back to Basics." Special Advertising Supplement to *Computerworld*, April 9, 2001.

89. Lais, Sami. "A Storage Sketchbook." *Computerworld*, October 15, 2001.

90. Ohlson, Kathleen. "Skittish Users Say Hype About SANs is 'Baloney.'" *Computerworld*, August 7, 2000.

91. Clark, Charles T. "Saving Your SAN Dollars." *Computerworld*, November 20, 2000.

92. http://www.fibrechannel.org/OVERVIEW/topology.html.

93. Clark, Elizabeth. "Storage Networking: Fibre Channel, IP, and Beyond." *Network Magazine*, June 2002.

94. Rash, Wayne. "New Standard Set." *InfoWorld*, July 29, 2002.

95. Baltazar, Henry. "iSCSI SAN Plan. *Eweek*, August 12, 2002.

96. Mackig, Bryce. "iSCSO Delivers Storage over Ethernet." *Network World*, September 9, 2002.

97. Baltazar, Henry. "iSCSO Takes on Fibre Channel." *Eweek*, February 13, 2006.

CHAPTER 9

BROADBAND NETWORK INFRASTRUCTURE

Et loquor et scribo, magis est quod fulmine iungo. I speak and I write... but more, it's with light(ning) that I connect.
Poet Giovanni Pascoli, describing his views of the telegraph, 1911, translated from Latin

We appear to have developed a seemingly insatiable appetite for bandwidth in the Wide Area Network (WAN) and Metropolitan Area Network (MAN) as well as the Local Area Network (LAN). A number of broadband network technologies have been developed to address this need for speed. Some of those technologies are specifically oriented toward data applications, and others support the *triple play*, a marketing term derived from American baseball to describe a full range of voice, data, and video services. (*Note:* The term *quadruple play* is used by some service providers who add the wireless component of cellular communications. At this point, the baseball analogy breaks down, however.) Some of these are access technologies that are specific to the local loop while others are transport technologies positioned in the core, or backbone.

Before going further, we should revisit the definition of the term *broadband*. In a LAN context, as we discussed in Chapter 8, broadband refers to a multichannel system that supports multiple transmissions through Frequency Division Multiplexing (FDM). In a WAN context, the definition is all about bandwidth and is oriented toward end-user services.

In 1992, the Regional Bell Operating Companies (RBOCs) defined broadband as a fully symmetrical service running at the T3 nominal rate of 45 Mbps. At the time, they were working to convince the Federal Communications Commission (FCC) and state Public Utility Commissions (PUCs) to support their desires to

Telecommunications and Data Communications Handbook, By Ray Horak
Copyright © 2007 Ray Horak

replace the copper-based local loops with optical fiber [1]. In 1999, the FCC defined *high-speed* services as supporting a data rate of at least 200 kbps in at least one direction and *advanced telecommunications capability* as at least 200 kbps in both directions. The ITU-T defines broadband in Recommendation I.113 as a *transmission rate faster than the primary rate* [referring to Integrated Services Digital Network (ISDN)], which translates into 1.544 Mbps in North America and 2.048 Mbps in most of the rest of the world. Others offer still different definitions designed to support their own agendas. Manufacturers and carriers position Asymmetric Digital Subscriber Line (ADSL) as a broadband local loop service, even though ADSL rates in the United States often run well below 1.544 Mbps. Clearly, broadband is an imprecise, evolving term referring to a circuit or channel providing a relatively large amount of bandwidth. (You may quibble with my characterization of an imprecise term as being clearly defined.) I generally subscribe to the ITU-T's definition and, therefore, use the term to describe capacity equal to or greater than the primary rate, which generically is the DS-1 rate. So, in the context of this book, *pure* broadband is defined as a transmission rate greater than the nominal T1 rate of 1.544 Mbps, or the E-1 rate of 2.048 Mbps, depending on the context. Sometimes, however, I use the term in a more vernacular sense to describe a service that offers a relatively large amount of bandwidth, with ADSL being one example. All of these definitions, of course, are oriented toward end-user services that require more bandwidth than a dial-up connection can provide.

In the network core, or backbone, the levels of bandwidth are much greater and DS-1 has little significance. In this context, broadband is more correctly defined as DS-3 or better and specifically T3 or better. The reason I choose this level is that the T3 rate of 45 Mbps is the foundation for the Synchronous Optical NETwork (SONET), the North American fiber-optic transmission standard, and Synchronous Digital Hierarchy (SDH), the international version of SONET.

Ultimately, of course, the focus has to be on the delivery of end-user services, without which any technology in the core would be totally irrelevant. Manufacturers and carriers are incredibly intent on providing attractive broadband products and services to end users. Governments, in general, and regulatory agencies, in specific, increasingly are centered on ensuring that as many end users as possible have access to broadband services at affordable prices. The FCC, for example, has as a broadband objective, which it has stated as follows [2]:

> *All Americans should have affordable access to robust and reliable broadband products and services. Regulatory policies must promote technological neutrality, competition, investment, and innovation to ensure that broadband service providers have sufficient incentive to develop and offer such products and services.*

Despite this stated level of commitment, the United States does not lead the world in broadband deployment. The ITU-T ranks the United States as 16th in the world in broadband penetration per capita, although 1st in terms of total subscribers. The *Organisation for Economic Co-Operation and Development* (OECD), an organization of 30 member countries sharing a commitment to democratic government and the market economy (i.e., capitalism), conducts a great deal of research across a wide range of subjects. The OECD estimates that broadband subscribers in OECD nations increased by 18 million during the first six months of 2005 to reach a total of 137 million and a penetration rate of 11.8 subscribers per 100 inhabitants. South Korea led with 25.5 subscribers per 100 inhabitants. The Netherlands, Denmark, Iceland, and Switzerland rounded out the top five countries. The United States came in at number 12 at 13.5 percent. The OECD study indicates that DSL is the leading broadband platform in 28 of the 30 OECD countries and that the United States and Canada are the only countries in which broadband cable modem subscribers outnumber DSL subscribers. In the United States, the OECD estimated that 5.5 percent of the population subscribed to DSL service, 8.0 percent to cable modem service, and 1.1 percent to other broadband services, including wireless. [*Note:* CATV and, therefore, broadband cable modem access is largely a North American phenomenon.] The breakdown of broadband technologies in OECD countries, as of June 2005, is as follows:

- DSL: 61.2 percent
- Cable modem: 32.0 percent
- Other technologies: 6.8 percent [3]

The FCC estimated that, as of the end of 2004, there were 37.9 million high-speed access subscribers in the United States. Of that total, there were 13.8 million DSLL installed, compared to 21.4 million cable modem connections. An additional 2.7 million high-speed connections were provided over satellite, terrestrial wireless, fiber, powerline carrier, or other wireline technologies.

Broadband access and transport technologies, according to the FCC, encompass all evolving high-speed digital technologies that provide consumers integrated access to voice, high-speed data, video-on-demand, and interactive delivery services [2]. This chapter deals with a number of access, or local loop, technologies, including Digital Subscriber Line (DSL), Community Antenna TeleVision (CATV), Passive Optical Network (PON), 802.16 and Worldwide interoperability for Microwave Access (WiMAX), and Access Broadband over Power Line (BPL). This chapter also details SONET and SDH, the standards for optical fiber transmission in backbone carrier networks, Dense Wavelength Division Multiplexing (DWDM), Coarse Wavelength Division Multiplexing (CWDM), and Resilient Packet Ring (RPR).

9.1 ACCESS TECHNOLOGIES

Access technologies are absolutely critical. The least capable link defines the greatest level of performance of a circuit or network, and the least capable link is the local loop. There has been a great deal of hyperbole in the United States for a very long time (at least 1992) about Fiber-To-The-Premises (FTTP) but very little investment until very recently. Large user organizations with offices in commercial office parks or high-rise buildings in major markets often have direct access to optical fiber. A privileged few residential and small-business customers have fiber access. A great deal of distribution fiber has been deployed in preparation for FTTP, but most of that fiber has yet to be terminated and activated. So, most of us must contend with the limitations of the copper loop. In the United States and Canada, most broadband users currently subscribe to cable modem service over CATV networks based on coaxial cable. Internationally, most broadband users subscribe to DSL services provided over the Incumbent Local Exchange Carrier (ILEC) twisted-pair local loop. Several terrestrial wireless technologies have been developed to extend broadband capabilities to the premises, although they have yet to achieve any appreciable level of success, and others are under development. Broadband satellite service is an option but usually is limited to remote rural areas where no alternatives exist.

Some of the technologies are incremental in nature. In other words, they are enhancements of existing copper local loop [i.e., Unshielded Twisted-Pair (UTP) or coax] technologies. These incremental approaches substantially improve the performance of the underlying transmission media within the limitations imposed by the regulators and the basic laws of physics. Wireless local loops either overlay the traditional copper network, bypassing it for a variety of reasons that include cost and performance, or are an alternative where wireline loops are either impossible or problematic to deploy. Optical fiber loops generally involve either a complete replacement of trunk facilities in the loop or an overlay. Regardless of whether the access technology is characterized as an enhancement, an overlay, or a replacement, the absolute cost of its deployment is significant.

The cost of UTP local loops in North America ranges from approximately $1000 to $7500, with the upper end of the range applying where terrain is especially difficult (e.g., rocky soil or high water tables) or where population density is low. The cost of an optical fiber loop is much higher, but they are deployed in support of high-volume business in high-density areas. As telco local loop revenues generally are low and costs are high, there really is no stunning profit potential in the telco local loop business. Rather, the profits are in the capabilities of the networks and services to which they provide access and the company that controls the local loop access to those networks and services. Assuming that the environment is a competitive one, a Competitive Local Exchange Carrier (CLEC) or service provider has two basic choices—buy or build:

- **Buy:** In order to encourage competition, the regulator may require that the ILEC lease local loops and other network elements to the CLEC, most likely at some rate that is cost based. The ILEC is unlikely to be in agreement with this requirement and almost certainly will not be especially enthusiastic or helpful in its implementation. However, leasing embedded telco local loops certainly is the fastest way to gain access to large numbers of customers.
- **Build:** Although it is expensive and time consuming to build local loop facilities, the long-term advantages are significant. A local loop built from the ground up, so to speak, likely will be based on the most advanced and appropriate technologies and, therefore, will be optimized to deliver the highest performance at the lowest cost. The CLEC that builds its own network also owns and controls it, which is very significant.

When considering local loop media alternatives, twisted pair is almost out of the question except for an ILEC extending a last-mile legacy network. Coaxial cable is out of the question except for a CATV provider extending a last-mile legacy network. Fiber optics really is the only choice for a wireline local loop, for reasons that include bandwidth, error performance, distance, and resiliency. There are wireless alternatives, although they generally have failed in practice or remain either in development or field trial at the moment.

9.1.1 Generic Digital Subscriber Line

Generic Digital Subscriber Line (xDSL) encompasses a group of digital network access technologies largely intended for the consumer class markets, including residential, Small Office/Home Office (SOHO), and small-business applications. However, several very significant DSL services are oriented toward business class markets. DSL is provisioned over conventional UTP local loop facilities, which comprise the vast majority of network access facilities owned by the ILECs. Beginning in 1880 and continuing through much of the twentieth century, UTP was the only option, or the only viable option, for most local loop applications. UTP inherently is the most limited of all transmission media in terms of the dimensions of bandwidth, error performance, and distance, but a two-wire UTP circuit is quite sufficient in support of analog voice and analog voice-grade data through modems. If properly conditioned, it also is quite sufficient in support of ISDN BRI, a digital service running at a total signaling rate of 144 Mbps. UTP even supports Fractional T1, T1, E-1, and ISDN PRI at signaling speeds up to 2.048 Mbps. The main reason that we continue to deploy UTP is that it is relatively inexpensive and simple to acquire and install as an extension of a legacy UTP-based network.

However, the embedded twisted-pair local loop has become strained under the pressure of increased demand for increased speed of information transfer and improved error performance in support of data communications. In large part, this demand is due to the development of the Web and the resulting growth in popularity of the commercialized Internet. As a result of these factors and competition from CATV providers and others, the ILECs have aggressively sought to develop technologies to extend the life of the embedded UTP local loop through improving its general performance characteristics.

xDSL, or DSL, is a group of Layer 1 (Physical Layer) technologies including ADSL, HDSL, IDSL, SDSL, and VDSL. All of these technologies are digital in nature and make use of sophisticated compression algorithms and multiplexing techniques to derive as much performance as possible from the inherently limited UTP-based local loop. Each technology is loop specific, requiring special equipment on each end of the point-to-point local loop circuit. Most DSL technologies support both voice and data; voice is afforded preferential treatment, and data are supported at relatively high speed. A number of options support video as well, and one supports data only. Most of the technologies involve centralized *splitters*, also called modems or filters, on the customer premises side of the loop. Voice signals, which run at frequencies up to 4 kHz, *cut through* the locally powered splitter in the event of a power failure; therefore, and very much unlike ISDN, the phones still work if the lights go off. A key advantage of all DSL technologies is that they support *always-on* data access because the circuit is always available from the PC through the on-premises splitter to the centralized splitter and *DSL Access Multiplexer* (DSLAM) in the Central Office (CO) or other centralized location and to the Internet. Therefore, there are no dial-up delays such as those you experience when establishing a circuit-switched connection over the Public Switched Telephone

TABLE 9.1   DSL Technology Comparisons

DSL Type	ITU Standard	Maximum Rate Downstream[a]	Maximum Rate Upstream[b]	Maximum Reach; Extendable (Yes/No)	Applications
ADSL	G.992.1 (1999)	7 Mbps	800 kbps	18,000 ft (Y)	Consumer-class Internet
G.lite	G.992.2 (1999)	1.544 Mbps	512 Mbps	18,000 ft (Y)	Consumer-class Internet
ADSL2	G.992.3, G.992.4 (2002)	12 Mbps	1 Mbps	18,600 ft (N)	Consumer-class Internet
ADSL2+	G.992.5 (2003)	24.5 Mbps	1 Mbps	18,000 ft[c] (N)	Consumer-class, SME Internet
ADSL2-RE	G.992.3 (2003)	8 Mbps	1 Mbps	20,700[d] (N)	Consumer-class Internet
IDSL	N/A	144 kbps	144 kbps	18,000 ft (Y)	Consumer-class Internet
Symmetric HDSL	G.991.2 (2003)	4.6 Mbps	4.6 Mbps	6,000 ft (Y)	Business-class Internet
VDSL	G.993.1 (2004)	55 Mbps	15 Mbps	1,000 ft[e] (N)	Voice, data, video
VDSL2: 12 MHz	G.993.2 (2005)	55 Mbps	30 Mbps	1,000 ft[c] (N)	Voice, data, video
VDSL2: 30 MHz	G.993.2 (2005)	100. Mbps	100 Mbps	500 ft[f] (N)	Voice, data, video

[a] Downstream is from the network edge to the customer premises.
[b] Upstream is from the customer premises to the network edge.
[c] Maximum achievable rates downstream and upstream are achievable at distances up to 5000 ft. Considerably reduced rates are achievable at distances up to 18,000 ft.
[d] Maximum achievable rates downstream and upstream are achievable at distances up to 18,000 ft. Rates up to 192 kbps downstream and 96 kbps upstream are achievable at distances up to 20,700 ft.
[e] Maximum achievable rates downstream and upstream are achievable at distances up to approximately 1000 ft, after which rates drop precipitously. Considerably reduced rates are achievable at distances up to 4000–5000 ft.
[f] Maximum achievable rates downstream and upstream are achievable at distances up to approximately 500 ft, after which rates drop precipitously. Considerably reduced rates are achievable at distances up to 4000–5000 ft.

Network (PSTN) through a modem. Table 9.1 provides a comparison of some of the key DSL options.

All xDSL technologies are fairly demanding of the local loop. Some are absolutely unforgiving, and some are adaptive to local loop anomalies. Before examining the specific technologies, I want to examine the problems with electrically based local loops, each of which has definite impact on xDSL performance:

• *Loop length* has a significant effect on signal strength and, therefore, on error performance. This is particularly true at high frequencies, which attenuate much more quickly (i.e., over a shorter distance) than do low frequencies. Further, attenuation especially is an issue given the narrow gauge (usually 24 or 26 gauge) of voice-grade, twisted-pair outside cable plants. As all of the xDSL technologies are designed to support relatively high speed data, they run at relatively high frequencies, at least in comparison to the 4-kHz range over which analog voice is supported. Many of the xDSL technologies are designed to work over local loops as long as 18,000 ft, which describes the vast majority of ILEC loops. Other xDSL technologies require loop lengths in the range of 12,000 ft or less due to the higher frequencies utilized. The reach of some of the technologies can be extended through the use of *loop extenders* or *line doublers*, which essentially are DSL repeaters. Other DSL types cannot be extended and performance is highly sensitive to loop length, with shorter loops offering better performance. Approximately 85 percent of all U.S. residential loops are within 18,000 ft (5.5 km), and approximately 75 percent are within 12,000 ft (3.7 km) [4]. The same holds roughly true for the United Kingdom [5].

- *Splices* and *mixed gauges* can cause echo, or signal reflection, which is a source of noise, or signal distortion. It is not at all uncommon to find mixed 24- and 26-gauge cable in a given local loop. It is not as though the ILEC craftspeople just closed their eyes and reached into the warehouse for a reel of cable, but the effect is the same as if they had. Telcordia Technologies estimates that the average residential local loop involves 22 splices, each of which contributes to the problem [6]. Some of those splices may not be mechanically sound, and some of the splice casings may not be well insulated. The problem is likely to be worse with longer loops, as they generally involve more splices. Echo cancellers and error detection and correction algorithms can compensate for much of this noise, but it is always an issue.

- *Bridged taps* are sections of a cable pair not on the direct electrical path between the CO and the customer premises. They generally are in place because a previous customer between the CO and the subject building was connected to the CO through the same pair. When that customer disconnected service, the splice was not removed. If a telephone is connected, the bridged tap essentially splits the signal, which roughly halves the signal power to the intended customer. If there is no telephone connected to the bridged tap, the open cable acts as a signal reflector, which causes signal distortion. DSL employs high-frequency signals, which attenuate quickly and require a point-to-point local loop for maximum performance. Bridged taps must be removed for some xDSL services to work optimally and for others to work at all.

- *Load coils* are passive devices that filter frequencies above 4 kHz in order to tune the copper circuit and optimize it for mid-voice-band performance. As DSL circuits employ frequencies well above 4 kHz, load coils must be removed.

- *Electronics* such as amplifiers, repeaters, and various types of simple multiplexers and line splitters are installed on some loops. They must be removed. DSL requires a *dry copper pair* (i.e., a pair with no electronics), also known as a *burglar alarm circuit*.

- *Interference* is an inherent problem with unshielded copper conductors. UTP is susceptible to ambient noise from a variety of sources, such as static electricity, radio and TV stations, and electric motors in proximity. Also, in a multipair distribution cable, a high-frequency DSL service on one cable pair can adversely affect Plain Old Telephone Service (POTS), T1 service, or other DSL services on adjacent pairs and even pairs in proximity. As I discussed in Chapter 2, an electrical signal radiates from the center of the copper core, causing the signal to attenuate and creating the potential for *crosstalk* between adjacent pairs in a multipair cable as the signals couple. Attenuation increases over distance, and weakened signals in adjacent pairs are increasingly vulnerable to crosstalk. There are several types of crosstalk that are of particular concern with respect to DSL services:

- *Near-End CrossTalk* (NEXT) occurs at the near end of the circuit, that is, at the end closest to the point of signal origin. Since at that point the outgoing downstream signal is at maximum strength and the incoming upstream signal is at minimum strength, the signals can couple quite easily if the *Attenuation-to-Crosstalk Ratio* (ACR) is not maintained at acceptable levels. NEXT is a particularly significant issue services at the network side of the connection for ADSL services, as the pair count of the cables is quite high at the CO. However, ACR can be managed much more effectively at the edge of the telco network than at the customer premises.

- *Far-End CrossTalk* (FEXT) occurs at the far end of the loop, that is, far away from the point of signal origin. It is at that end that the attenuated downstream signal from the network can experience crosstalk from the strong upstream signal emanating from the customer equipment. FEXT is not a great issue for ADSL services at the customer premises, as the cables are successively smaller, containing fewer and fewer pairs, from the CO to the premises.

In consideration of the phenomenon of crosstalk and the differences between NEXT and FEXT, most DSL services are asymmetric, with the higher frequencies on the downstream side (i.e., from the edge of the telco network to the customer premises) in support of greater bandwidth in that direction. Not only is such asymmetry technically easier to provision, but it is also entirely appropriate for DSL applications that require much more downstream than upstream bandwidth, with Web surfing being a prime example. Note that asymmetric DSL circuits cannot be turned around to provide greater bandwidth upstream and lesser downstream. A Web-based business, for example, would typically not be well served by ADSL as its upstream bandwidth requirements would be much greater than its downstream requirements.

- *Inside wire* systems are of uncertain quality in residential, SOHO, and small-business environments. In many cases, the inside wire has been in place for many years and has been spliced over and over again as buildings were remodeled and as new jacks were added for analog telephones, fax machines, and computer modems. Professionals completed some of this work, and amateurs may have done the rest. My home office, for example, is in a 1909 farmhouse. It's a great place to write a book, but the inside wiring system is of uncertain nature. My lovely wife, Margaret, telecommutes in connection with her consulting practice, so we share the space. We have five lines coming into our SOHO, and I can't even count the number of jacks for our analog telephone sets, fax machine, computer modems (she has to gain access to some client networks on a dial-up basis), and DSL service. Margaret installed some of the jacks, so I have to be careful what I put in print, but suffice it to say that our wiring system is somewhat unusual. In any event, running high-speed data to our workstations could be an interesting, and uncertain, experience unless we recabled the offices. [*Note:* We recabled the connection from the Network Interface Unit (NIU) to our ADSL modem with Cat 5, used a Cat 5 cable to our 802.11 b Wireless LAN (WLAN) bridge/router, used a Cat 5 cable to connect to my desktop, and went wireless to our three laptops.]

- *Voice instrument signaling states* can impact data communications in a DSL environment. While analog voice circuits are standardized and voice terminals are certified, the wide variety of devices vary enough in their electrical characteristics to cause unpredictable problems when integrating voice and data over a common inside wiring plan. For that matter, the electrical characteristics of multiple phones of the same manufacturer and model number can vary considerably, even within the same manufacturing lot. DSL data communications sessions can be dropped when analog telephones ring (voltages can exceed 100 V), when they go off-hook, and when they go back on-hook. The ultimate solution is an adaptive splitter or modem. The interim solution, at least for some versions of DSL, involves equipping all telephone sets, fax machines, and other connected devices with inexpensive high-pass *microfilters* that plug in between the set and the jack.

- *Digital Loop Carrier* (DLC) systems are voice-grade multiplexers embedded in a local loop distribution plant to increase its efficiency. Subscriber Line Carrier (SLC) 96, for example, involves a four-wire copper circuit that runs from the CO to a remote node and provides four T1 circuits in support of 96 voice-grade time division multiplexed (TDM) channels of 64 kbps. At the remote node are concentrated and multiplexed as many as 96 individual voice-grade local loops that terminate at customers' premises, as illustrated in Figure 9.1. SLC-96 and other much more advanced DLC systems are highly cost effective in support of voice-grade communications services to areas that are far removed from the CO. DLC systems, however, filter high-frequency signals such as DSL, shaping and grooming them to fit into 64-kbps channels. Traditional DLCs, therefore, render DSL inoperable, with the exception of IDSL. The only

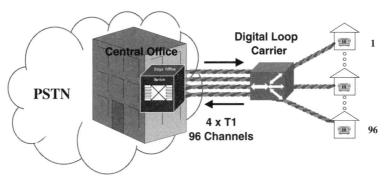

Figure 9.1   SLC-96, illustrating digital loop carrier.

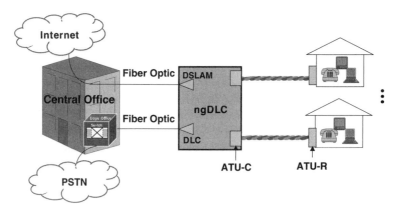

Figure 9.2 Next-generation DLC with built-in DSLAM.

viable solution is to split the voice and data channels through a *next-generation DLC* (ngDLC), as illustrated in Figure 9.2. An ngDLC deals with the voice channels in the traditional manner, digitizing them into *Pulse Code Modulation* (PCM) format and multiplexing them into 64-kbps (TDM) channels. The ngDLC runs the video and data channels through an embedded *DSL Access Multiplexer* (DSLAM), which is a packet multiplexing device designed specifically for that purpose. The ngDLC is a hybrid solution, as it employs optical fiber to the CO but makes use of the existing copper pairs to the premises. In densely packed urban areas, relatively few DLCs are required, while sparsely populated rural areas make extensive use of DLCs.

Also, before exploring the specifics of the DSL technology options, it is important to pause just a moment to examine some of the network elements that generally are involved. Those elements include splitters or modems, local loops, and DSLAMs:

- *Splitters* or *modems* are interface units that must be installed in matching pairs, with one at the customer premises and one at the edge of the network. The splitter at the network edge generally is embedded in a DSLAM. The premises-based unit may be a centralized splitter in the form of an enhanced *network interface unit*, which is the point of termination and demarcation between the carrier-provided local loop and the inside wire and cable system. A centralized NIU generally is attached to the outside of the building or is positioned just inside the basement wall, garage wall, or other point of entry. Alternatively, the splitter may be in the form of a *set-top box*, much like a CATV converter. The splitter functions as a hybrid frequency division multiplexer and time division multiplexer. Additionally, the splitter compresses the data transmissions. The splitter also may function as a codec, digitizing analog voice and fax transmissions as required.
- *Local loops* are the point-to-point dedicated circuits between the customer premises and the edge of the network. As they are dedicated circuits, there are no issues of contention or congestion over the local loops themselves. At the

customer premises, the local loop terminates in an NIU, with the splitter or modem functionally positioned between the NIU and the terminal equipment. A matching splitter/modem is positioned at the edge of the network. If the service provider is an ILEC, that point of interface may be in the form of a DSL card associated with a CO. In the upstream direction, the ILEC splitter splits off the low-frequency voice signals for presentation to the PSTN and the high-frequency data signals for presentation to the Internet or other Public Data Network (PDN) through the DSLAM. A service provider generally dedicates a complete local loop to DSL service. But some DSL services, such as ADSL, can support both data communications for Internet access purposes and traditional analog POTS service over the same local loop through frequency splitting. Under pressure from the end user and competitive service provider communities, the FCC in December 1999 rendered its *line-sharing* decision. That decision requires that the ILECs open the high-frequency portion of qualifying local loops to competing DSL data services at any possible point. That point typically is at the CO, where the CLEC installs its DSLAM. But that point also can be at an ngDLC, in which case a *subloop* (i.e., portion of a loop) is shared. In theory, at least, line sharing is a reasonable solution to the struggles between the ILECs and the CLECs. In practice, however, it rarely seemed to work, and the FCC has since relaxed its position.

- *DSL access multiplexers* provide the interface between the DSL local loop and the service provider's *Point Of Presence* (POP). Most DSLAMs are based on Asynchronous Transfer Mode (ATM), although some earlier models are based on Frame Relay (FR), with the specific technology depending on what the carrier has in place. In any case, the DSLAM and the DSL modem on the user's premises must run matching protocols. In an ADSL scenario, for example, the DSLAM receives upstream DSL traffic and splits, or demultiplexes, the voice and data traffic. The voice traffic then is encoded into PCM format and time division multiplexed over a channelized T1, T3, or, perhaps, SONET link to the PSTN. The data traffic is multiplexed or concentrated in ATM cells over an unchannelized T1, T3, or SONET circuit directly to the Internet backbone or perhaps to an independent Internet Service Provider (ISP). The DSLAM generally represents the first potential point of contention and congestion that affects upstream end-user traffic, as the local loop is a dedicated circuit.

- *Single-Point-Of-Termination* (SPOT) frames typically are used where a CLEC or ISP is leasing dry copper pairs from the ILEC for purposes of provisioning xDSL data services. The SPOT frame, and other hardware, is collocated in the ILEC CO, generally in separately secured leased space. The ILEC cross-connects the individual leased circuits at the *Main Distribution Frame* (MDF) and terminates them in the SPOT frame, where the CLEC or ISP connects them to the DSLAM.

The xDSL technologies discussed first are those intended largely for the residential, SOHO, and small-business markets: ADSL, IDSL, G.lite, and VDSL. The technologies discussed last are those intended primarily for the larger business market: HDSL, HDSL2, SDSL, and SHDSL.

9.1.1.1 Asymmetric Digital Subscriber Line   *Asymmetric Digital Subscriber Line* (ADSL), also known as *full-rate* ADSL, is an advanced, high-bandwidth local loop technology designed to extend the life of existing UTP loops for the transmission of broadband signals. ADSL was developed by Bellcore (now Telcordia Technologies) at the request of the RBOCs in the United States and was later standardized in 1999 by ANSI as T1.413 and the ITU-T as G.922.1. ADSL provides for very high capacity transmission over relatively short local loops in the *Carrier Serving Area* (CSA). The DSL Forum, a Special Interest Group (SIG) dedicated to the promotion of xDSL technologies, promotes this technology [7].

The term *asymmetric* refers to the fact that ADSL bandwidth is provided on an asymmetric (i.e., not symmetric, or unequal) basis, with more bandwidth in the downstream direction than in the upstream direction. This asymmetry is in recognition of the FEXT and NEXT crosstalk issues discussed above and generally suits the applications.

In addition to supporting POTS voice over a separate analog channel running at 4 kHz and below, ADSL supports high-speed data. The upstream data ride in what technically is a bidirectional channel provided in increments of 64 kbps up to 640 kbps in a frequency band from 26 to 140 kHz. Downstream transmission is in increments of 1.536 Mbps up to 6.144 Mbps, based on T1 specifications, in a frequency band that runs from approximately 140–552 kHz. While ADSL is primarily a Layer 1 (Physical Layer) specification, it includes Layer 2 (Data Link Layer) elements. ADSL specifies three modes of operation:

- Synchronous Transfer Mode (STM), which is a bit synchronous mode. In this mode, the ADSL loop is a bit pipe.
- Asynchronous Transfer Mode (ATM).
- Packet Mode is IP-centric, employing both Ethernet and a variation of the High-level Data Link Control (HDLC) framing convention at Layer 2.

ADSL makes use of *Discrete MultiTone* (DMT) modulation, which gives rise to yet another moniker, *G.dmt*. Also known as *Orthogonal Frequency Division Multiplexing* (OFDM), DMT splits the signal into a stream of data symbols for massively parallel simultaneous transmission over 256 narrowband, low-data-rate subcarrier frequencies, each of which is 4 kHz wide. Within each subcarrier, *Quadrature Amplitude Modulation* (QAM) is employed to increase the efficiency of transmission. The DMT modem is dynamically rate adaptive, automatically adjusting the QAM constellation from 1 to 15 bits per baud, depending on the subchannel capacity and the line conditions. Under optimal conditions, therefore, a 4-kHz channel can support a data rate up to 60 kbps (4 kHz × 15-bit QAM = 60 kbps). The ADSL modems can detect any impaired subcarriers and shift data transmission away from them and to the unimpaired subcarriers. *Note:* Variations of the DMT/OFDM technique are used in 802.11 a (aka Wi-Fi5) and 802.11 g, 802.16, and WiMAX wireless systems, as well as Broadband over Power Line (BPL). The OFDM signal is the sum of a number of orthogonal subcarriers. (*Note: Orthogonal* is defined mathematically as *at right angles to* or *perpendicular to*, which in RF terminology means *independent* and *mutually exclusive* to the extent that a receiver can recognize and reject an unwanted signal or signal element.)

During the early stages of development, *Carrierless Amplitude Phase Modulation* (CAP) was the de facto standard modulation technique for ADSL modems. A non-standard variation of QAM, CAP was replaced by DMT even before the ITU released ADSL standards recommendations. Although it is unlikely that any manufacturers still make use of CAP, there likely are some CAP-based DSL modems remaining in use.

ADSL is a broadband local loop technology for broadband access to a broadband WAN. The nature of the WAN is not at issue, although interfaces to the circuit-switched PSTN, the IP-based Internet, and other public networks are supported. The TV signals generally travel via broadcast satellite to a *head end*, which could be in the form of a tandem office, where they are distributed to the lower level COs by microwave or fiber optics. The programming may be stored on tape or disk in a video server at the tandem or CO in support of delayed broadcasting or *Video On Demand* (VOD). The voice input is handled at each end of the connection through a frequency splitter that ensures it is unaffected by the presence or absence of digital data. At the network side of the loop, the video and data inputs are time division multiplexed into an ADSL signal that originates in an *ADSL Transmission Unit— Centralized* (ATU-C). The signal then travels the local loop cable plant, terminating in a matching *ADSL Transmission Unit—Remote* (ATU-R) located on the customer premises, as illustrated in Figure 9.3. The ATU-R acts as a modem, splitting the total bandwidth into three channels: the downstream channel, the bidirectional channel, and a voice channel. The ATU-R can be in the form of a stand-alone unit or a printed circuit board in a PC or other host computer. Alternatively, the ATU-R can be contained within a set-top box in a video or TV application. The Bit Error Rate (BER) of ADSL is 10^{-9}, which compares favorably even with traditional DS-1 at 10^{-7}.

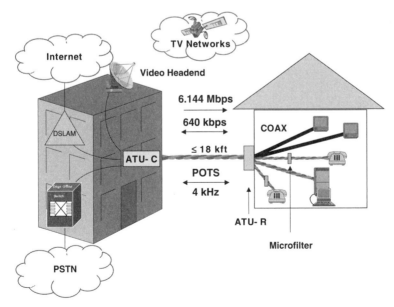

Figure 9.3 ADSL configuration in example residential application, illustrating bandwidth, ATUs, and DSLAM.

ADSL is intended for applications requiring a high-speed downstream channel and a relatively low-speed upstream channel. ADSL is entirely appropriate for residential Internet access. In a *small office/home office* (SOHO) or medium-size business environment, ADSL provides plenty of bandwidth for most applications, including Internet access and file and image transfer. Telecommuters and teleworkers find ADSL highly effective for remote LAN access.

The primary interest in ADSL has resided with the incumbent telcos, as it runs over an embedded twisted-pair cable plant. Where a competitive environment exists, CLECs may be allowed to lease the telco loop and terminate it in their collocated equipment. ILECs and CLECs, alike, find ADSL effective in countering competition from CATV companies for the residential voice and data market. As the CATV providers increasingly compete for that business, they are upgrading existing coaxial cable plant to add bidirectional, switched voice, and data capabilities. In the face of that pressure, the telcos have to consider a method of increasing the bandwidth in the local loop to compete effectively. Their options include laying new cable plant (fiber optic or coax), making use of wireless technology, or increasing the capacity of the existing twisted-pair plant. ADSL offers a relatively inexpensive solution, accommodating entertainment TV, videoconferencing, and higher speed data transfer over an existing cable plant of varying gauges. Many telcos in the United States and other countries with highly developed and competitive telecommunications markets view ADSL as a short-term access solution, filling the bandwidth gap until more substantial fiber-optic cable plant is deployed, usually in the form of Passive Optical Network (PON).

The video component is where ADSL has been disappointing. Video is so bandwidth intensive and, therefore, so demanding of the local loop that few UTP loops qualify at full rate, especially over the longest loop lengths specified. Where ADSL signal impairments such as bridged taps, load coils, and amplifiers or repeaters exist, the loop must be conditioned to remove them. Where the insulation of the cable pair or splice casing is cracked or otherwise compromised, the cable plant must be rehabilitated or replaced. In any case, at loop lengths of 12,000–18,000 ft, there are likely to be 10–20 splices or more, and performance drops as the signal transverses each of them. So the telcos generally avoid video over ADSL and constrain the bandwidth to a maximum of 1.536 Mbps (T1) or 2.048 Mbps (E-1) downstream and 384 kbps upstream. Table 9.2 provides comparative data for 24- and 26-gauge UTP. At these levels, consumer-class and Small-to-Medium Enterprise (SME) requirements for e-mail, Internet access, and remote LAN access are satisfied. Once the cable plant is deemed acceptable, a splitter must be installed at the customer pre-

TABLE 9.2   ADSL Data Rates, Wire Gauges, and Distance Limitations

Data Rate	American Wire Gauge (AWG)	Distance (ft)	Metric Gauge (mm)	Distance (km)
1.544/2.048 Mbps	24	18,000	0.5	5.5
1.544/2.048 Mbps	26	15,000	0.4	4.6
6.1 Mbps	24	12,000	0.5	3.7
6.1 Mbps	26	9,000	0.4	2.7

Source: DSL Forum.

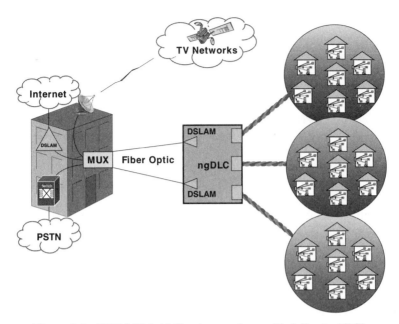

Figure 9.4 FTTN: Hybrid fiber/copper loop with full-rate ADSL.

mises, and the circuit must be tested in order to meet the very tight tolerances. The local loop demands and the installation of a splitter require an ILEC *truck roll*, which makes the provisioning process fairly labor intensive and, therefore, expensive [8].

Some ILECs (e.g., SBC in the United States) are addressing this issue of video performance over copper loops through a hybrid fiber/copper approach known variously as *Fiber-To-The-Neighborhood* (FTTN) and *Fiber-To-The-Curb* (FTTC). FTTN involves one or more high-speed fiber-optic links from the network edge at the CO to an ngDLC that supports both fiber and copper interfaces, as illustrated in Figure 9.4. From the ngDLC to the customer premises, the embedded UTP supports full-rate ADSL. Since this approach reduces the copper portion of the local loop from 12,000–18,000 ft to a few thousand or perhaps a few hundred feet, its performance is improved significantly and it is much more likely to support the full range of voice, data, and video signals. Other ILECs (e.g., Verizon in the United States) consider DSL to be a short-term solution to a long-term problem that they are addressing through PON technology. PON takes fiber optics directly to the customer premises, completely replacing the embedded twisted-pair local loop. Full-rate ADSL development continued well past the introduction of the initial version. Since that time, ADSL2 and ADSL2+ have been released.

9.1.1.1.1 ADSL2 The ITU completed G.992.3 and G.992.4, collectively known as *ADSL2*, in July 2002. Known as *G.dmt.bis* while under development, ADSL2 offers increased data rates of as much as 12 Mbps downstream and 1 Mbps upstream, depending on loop length and quality. ADSL2 achieves the higher downstream data rates by increasing the frequency band from 552 kHz (ADSL) to 1.1 MHz and improving modulation efficiency through the introduction of Trellis-Coded Modulation (TCM) QAM constellations. These modulation techniques yield higher through-put on long loops where the Signal-to-Noise Ratio (SNR) is low. ADSL2 also uses receiver-determined tone reordering of the Discrete MultiTone (DMT) channels to spread out the noise from Amplitude Modulation (AM) radio interference and, thereby, to realize improved coding gain. ADSL2 systems feature reduced framing overhead, enhanced power management, faster startup, seamless rate adaption, and improved diagnostics. ADSL2 also features an all-digital mode, in which the analog voice channel can be used for digital data transmission, thereby increasing aggregate upstream data transmission rates by as much as 256 kbps. ADSL2 adds a packet-mode capability that enables packet-based services such as Ethernet. On long loops, ADSL2 can increase the data rate by as much as 50 kbps and extend the reach by about 600 ft (200 m).

ADSL2 supports bonding in ATM, based on the ATM Forum specification for *Inverse Multiplexing over ATM* (IMA). This allows two ADSL pairs to be bonded together to yield roughly double the single-pair rate across the full range of loop lengths. Bonding has increasing application, as competition from CATV providers and cellular services have led to the disconnection of large numbers of telco-provided copper landlines. The deployment of *pair gain* technologies such as ADSL and the increasing use of fiber optics have combined to further reduce the pair counts required to support primary and secondary lines and trunks and have added to the excess of available copper loops.

ADSL2 further supports channelization, which provides the ability to split the bandwidth into different channels with different link characteristics in support of different applications. For example, a single ADSL2 loop might support a business class videoconferencing application, which is intolerant of both latency and bit errors, while simultaneously supporting e-mail transfers, which are quite tolerant of both latency and bit errors. Channelized Voice over DSL (CVoDSL) allows multiple derived lines of TDM voice traffic to be transported simultaneously over DSL in 64-kbps channels [9].

9.1.1.1.2 ADSL2+   The ITU-T reached agreement on G.992.5 in January 2003, adding *ADSL2+* to the ADSL suite. This specification doubles the downstream data rate to as much as 24.5 Mbps over shorter loops up to approximately 5000 ft (1500 m) in length. The upstream rate remains at a maximum of 1 Mbps. In order to achieve this enhanced data rate, ADSL2+ increases the downstream frequency range from 552 kHz (ADSL) and 1.1 MHz (ADSL2) to 2.2 MHz (ADSL2+) and increases the number of subcarriers to 512. The analog POTS channel remains at 4 kHz (ADSL and ADSL2), and the upstream data channel remains capped at 140 kHz (ADSL2). Table 9.3 provides a quick comparison of the frequency ranges of the various ADSL technologies.

TABLE 9.3   ADSL Frequency Ranges

Channel/ADSL Version	ADSL	ADSL2	ADSL2+
Downstream data	138–552 kHz	138 kHz–1.1 MHz	138 kHz–2.2 MHz
Upstream data	25–138 kHz	25–138 kHz	25–138 kHz
Voice	0–4 kHz	0–4 kHz	0–4 kHz

ADSL2+ also can be used to reduce crosstalk by using only the frequencies between 1.1 and 2.2 MHz and masking those below 1.1 MHz. This can be useful when ADSL signals from a CO and a remote terminal are present in the same 25-pair binder group in a cable [9].

Recent developments in ADSL+ CPE support multiple 10/100 Mbps Ethernet interfaces and optional 802.11 b/g interfaces. Combining these enhancements with the increased bandwidth of ADSL2+ and its channelization capability allows the user organization to derive multiple VLANs from a single ADSL2+ facility, with each VLAN enjoying its own Quality-of-Service (QoS) level. Thereby, separate VLANs can be created for low-latency voice and video as well as latency-tolerant data communications applications in a triple-play scenario. ADSL2 and ADSL2+ certainly relieve the pressure on triple-play bandwidth. MPEG-2 compression reduces the Standard TeleVision (STV) requirement to 3.5 Mbps and that of High-Definition TV (HDTV) to 8.0 Mbps, all of which is downstream, of course. PCM-based voice requires bidirectional bandwidth of only 64 kbps, which leaves plenty of room for high-speed bidirectional data [10].

9.1.1.2 G.lite Also known as *ADSL Lite, Universal ADSL*, and *Splitterless ADSL, G.lite* is an interoperable extension of the ADSL specification ANSI T1.413 and has been standardized by the ITU-T as G.992.2. Three deployment options exist, all of which support simultaneous voice and data communications over a single UTP local loop of one physical pair:

- *Splittered ADSL* resembles the original standardized version of ADSL, in that a single, professionally installed splitter is installed at the point of termination of the local loop at the customer premises. This approach ensures the highest level of data performance, while ensuring that other signals in a multipair cable are not affected by high-frequency ADSL data signals. But this approach also involves an expensive *truck roll*.
- *Distributed Splitter ADSL* involves a number of ADSL splitters, some front-ending PCs and others front-ending analog telephone sets. This approach ensures that the voice and data channels do not interfere with each other and avoids the costs and delays associated with the professional installation of a single splitter, as described in the Splittered ADSL option. This version does not require professional installation.
- *Splitterless ADSL* (see Figure 9.5), which is the most attractive, supports simultaneous voice and data without the requirement for either a centralized splitter or multiple distributed splitters. Rather, high-frequency data communications are supported over an ADSL modem. This version does not require professional installation.

G.lite operates on an asymmetric basis over loops up to 18,000 ft at speeds of up to 1.544 Mbps (T1) downstream and up to 512 kbps upstream, sensitive to loop specifics such as gauge. Upstream and downstream speeds both are selectable in increments of 32 kbps. Shared internal wiring is always an issue but generally is resolved satisfactorily with *microfilters* installed between the jacks and the analog voice sets and fax machines. The filters prevent the analog devices from draining

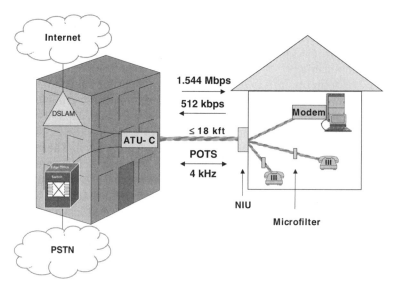

Figure 9.5   Splitterless G.lite.

the high-frequency signal power, thereby improving the signal-to-noise ratio for the modem. Maximum performance can be achieved by upgrading the internal wire to Cat 5, at least for the PC connection, and by home-run cabling the jacks to the NIU [11–13]. (*Note:* This is basically the approach we took at The Context Corporation, and it worked beautifully. With the exception of one desktop computer, however, all computers were interconnected to the G.lite circuit through an 802.11b WLAN.)

G.lite supports both voice and data over the same circuit, just as does full-rate ADSL, but G.lite uses several mechanisms to optimize their coexistence. *Power back-off* dynamically reduces the modem power level at the customer premises in order to eliminate the potential for interference (NEXT) with the analog voice channel. Power back-off occurs automatically and reduces the upstream data transmission rate for the duration of the phone call. *Fast retrain* enables rapid recovery of the upstream when the phone call is terminated. Fast retrain supports the retention in memory of profiles for as many as 16 telephone sets, thereby taking into consideration the specific electrical attributes associated with each. Although early G.lite modems ran Frame Relay as the Layer 2 protocol, contemporary modems mostly run ATM [14]. As G.lite is limited in terms of bandwidth, it never achieved great popularity but did introduce the concept of Splitterless DSL.

9.1.1.3 *ISDN Digital Subscriber Line*   ISDN DSL uses ISDN BRI (Basic Rate Interface) technology to deliver symmetrical transmission speeds of 128 or 144 kbps on digital copper loops as long as 18,000 ft. Like ISDN BRI (Chapter 7), IDSL terminates at the user premises on a standard ISDN Terminal Adapter (TA). At this point, the two concepts diverge. ISDN is a circuit-switched service that connects to a CO at the edge of the carrier domain. IDSL is an always-on data access service that terminates, more or less directly, in a DSLAM. ISDN supports voice, data, video, and any other form of traffic through one or more Bearer (B) channels, each with

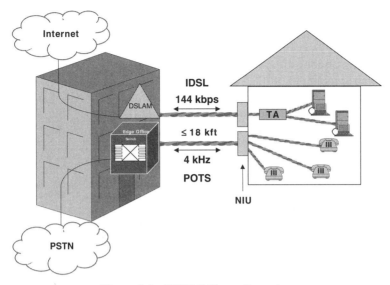

Figure 9.6 ISDN DSL configuration.

a width of 64 kbps. ISDN also can support X.25 packet data over a Delta (D) channel at speeds of up to 9.6 kbps. IDSL supports data access through a contiguous slice of bandwidth with a width 128 or 144 kbps. ISDN can support multiple transmissions, each over a separate channel. IDSL supports a single data transmission at a time over a single, wider channel.

IDSL (refer to Figure 9.6) is a dedicated access service for data communications applications, only. At the Local Exchange Carrier (LEC) CO, the loop terminates in collocated electronics in the form of either an IDSL access switch or an IDSL modem bank connected to a router or DSLAM. In the event that the LEC is not serving as the ISP, the connection is made to the third-party ISP POP via a high-bandwidth dedicated circuit, such as unchannelized T1, T3, or SONET/SDH channel.

IDSL offers several advantages over competing technologies. IDSL is based on ISDN network technologies, which are standardized, well understood, and widely implemented, at least outside the United States. IDSL operates at frequencies that do not interfere with other signals riding over adjacent pairs in a multipair copper cable. IDSL makes use of standard, off-the-shelf Data Communications Equipment (DCE) at the customer premises in the form of a TA with a few modifications. IDSL offers fully symmetric bandwidth, although it admittedly is less than that offered by competing DSL technologies, particularly in the downstream direction. IDSL does not require a *truck roll* for CPE purposes, although the NIU must be installed and the circuit may require conditioning, as is the case with traditional ISDN. Finally, IDSL is the only DSL technology that operates on local loops provisioned with conventional DLCs. On the negative side again, IDSL is application specific, supporting data only and, therefore, requiring an additional local loop. IDSL never gained much traction in the United States, where ISDN never achieved any appreciable level of success and where other broadband access technologies are widely available. Worldwide, IDSL has a market in areas where ISDN is already in place and other broadband access technologies are not available.

9.1.1.4 Very-High-Data-Rate Digital Subscriber Line *Very-high-data-rate Digital Subscriber Line* (VDSL) is a high-speed ADSL technology that was formalized in 2004 as G.993.1. This initial specification provides for downstream data rates up to 55 Mbps and upstream rates up to 15 Mbps over distances up to 1000 ft (300 m) or so, depending on factors such as local loop quality. In order to achieve these data rates, the frequency range increases to approximately 8.8 MHz divided among 2048 subcarriers. As attenuation is a considerable issue at this high frequency, performance drops precipitously beyond 1000 ft. VDSL evolution was rapid. The ITU reached agreement in 2005 on recommendation G.993.2, the specification for two versions of VDSL2.

- **Long Reach:** The *long-reach* version runs at 12 MHz, divided among 2872 subcarriers, to deliver asymmetric data rates of as much as 55 Mbps downstream and 30 Mbps upstream over loops up to 1000 ft, after which data rates drop precipitously. Considerably reduced rates are achievable at distances up to 4000–5000 ft.
- **Short Reach:** The *short-reach* version runs variously at 17.6 MHz and up to 30 MHz, divided among as many as 4096 and 3478 subcarriers, respectively, to deliver symmetric (downstream and upstream) data rates of as much as 100 Mbps over loops up to 500 feet, after which data rates drop precipitously. Again, considerably reduced rates are achievable at distances up to 4000–5000 ft. This version will run in asymmetric mode as well.

VDSL2 employs the same DMT modulation scheme as ADSL but increases the number of subcarriers to 4096 in a typical implementation and up to 3478 in others, as compared to the 512 subcarriers specified in ADSL2+. Trellis-Coded Modulation (TCM) yields higher throughput on long loops where the Signal-to-Noise Ratio (SNR) is low, although data rates drop considerably beyond 500–1000 ft.

A versatile worldwide specification, VDSL2 defines eight profiles for different services, including ATM and Ethernet. Quality-of-Service (QoS) features are integrated into the specification, as is channel bonding for extended reach or data rate. Multiple *bandplans* (i.e., frequency definitions) are specified for different regions. For example, there is a common bandplan for North America, Japan, and the United Kingdom (see Table 9.4) and quite another for other European countries. Each bandplan divides the available spectrum into interleaved, paired upstream and

TABLE 9.4   VDSL Frequency Band Allocations for North America, Japan, and the United Kingdom

Channel Width	Channel Designation
12–28 MHz	Upstream/downstream optional
8.5–12.0 MHz	Upstream
5.2–8.5 MHz	Downstream
3.75–5.2 MHz	Upstream
138 kHz–3.75 MHz	Downstream
25–138 kHz	Upstream
0 Hz–4 kHz	Analog voice

downstream channels. This approach provides some flexibility in terms of the spectrum allocation and overall speed rating under different application scenarios and line conditions. In other words, VDSL2 services can run at several different matched speeds, upstream and downstream.

There are several application scenarios, all of which are based on FTTN hybrid local loops, with fiber optics from the CO to the neighborhood ngDLC and VDSL to the premises. In North America, Europe, and China, there are plans to deploy VDSL2 at downstream rates up to 30 Mbps and upstream rates of 3 Mbps in support of consumer-class triple-play services, including at least three *Digital TV* (DTV) or HDTV channels, Internet access at rates up to 5 Mbps, and VoIP. Clearly VDSL has application in servicing a *MultiDwelling Unit* (MDU), also known as a *MultiTenant Unit* (MTU), where the bandwidth can be shared among multiple tenants. In this *Fiber-To-The-Premises* (FTTP) scenario, the ngDLC is positioned on the premises, and the VDSL service runs over the inside wire and cable system in the apartment building or office building, for example [15].

As VDSL matures and implementations become more prevalent, the pressure increases to increase the bandwidth levels and speed ratings, to alter the bandplan toward symmetrical service, and to increase the distances. In combination, these demands will outstrip the capabilities of the specification. As a result, there has been some movement toward the application of *Multiple-Input, Multiple-Output* (MIMO) technology similar to that used in 802.11 *Wireless Local Area Networks* (WLANs). In a VDSL application, MIMO treats each binder group of 25 pairs as a MIMO channel and the transceiver treats multiple pairs together, thereby separating signal from crosstalk [16]. According to the DSL Forum, some manufacturers have included MIMO in VDSL2 systems in advance of standardization of MIMO technology.

9.1.1.5 High-Bit-Rate Digital Subscriber Line

Bellcore (now Telecordia Technologies) also developed *High Bit-Rate Digital Subscriber Line* (HDSL) at the request of the RBOCs as a more cost-effective means of providing T1 local loop circuits over existing UTP. Standardized by the ANSI T1E-1.4 committee, HDSL eliminates repeaters in the T1 local loop for distances up to 12,000 ft, which can be extended another 12,000 ft through the use of a *line doubler*, which essentially is an HDSL repeater.

Recall from Chapter 7 that conventional T1 uses two pairs, each of which operates in a simplex mode (one upstream and one downstream) at the full T1 transmission rate of 1.544 Mbps (see Figure 9.7). The line coding technique for T1 is *Alternate Mark Inversion* (AMI), which yields 1.544 Mbps at a nominal carrier frequency of 784 kHz, which is exactly half the T1 bit rate, plus some overhead for error control. At such a high frequency, issues of attenuation are significant. As a result, it is necessary that repeaters be spaced at approximately 6000 ft to adjust for distortion and signal loss. Also, the radiated electromagnetic field is significant at T1 frequencies. Therefore, interference between the T1 pairs and other pairs in the cable system is an issue. Also recall that E-1 uses AMI as well, in support of the E-1 transmission rate of 2.048 Mbps at a frequency of 1.168 MHz, which also is half the bit rate, plus some additional overhead.

In the North American implementation of HDSL, the upstream and downstream signals are split across both pairs, with each pair operating in full-duplex (FDX) mode at 784 kbps, which is half the T1 rate plus additional overhead. In the Euro-

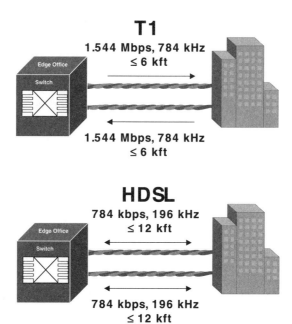

Figure 9.7 Comparison of T1 and HDSL.

pean implementation, each of two pairs operates at 1.168 Mbps, which is roughly half the E-1 rate plus additional overhead. (Some implementations called for three pairs, each operating at 768 kbps.) The yield is that the transmission rate per pair is roughly halved, and so is the frequency level. Therefore, the signal loss is much less and the strength of the radiated electromagnetic field is much less. The obvious yield of this approach is that of longer transmission distances without repeaters and with less distortion.

As a further improvement, HDSL uses the same 2B1Q (*2 Binary, 1 Quaternary*) coding scheme used in ISDN BRI. *Note:* 2B1Q also is known as 4-PAM (*Pulse Amplitude Modulation*). Recall from Chapter 7 that 2B1Q impresses two bits on each symbol (i.e., baud), with each symbol represented by one of four voltage levels. The symbol rate, therefore, is one-fourth the line rate, meaning that an HDSL T1 implementation at a line rate of 784 kbps across each of two pairs requires a carrier frequency of only 196 kHz, at least at the peak power level. At this relatively low frequency, issues of attenuation and crosstalk are further reduced.

On each end of the HDSL circuit, termination gear is installed. On the transmit side, the HDSL modem accepts the T1/E-1 signal in AMI format from the Digital Signal Unit (DSU), splits it, applies 2B1Q line coding, and places the signals on the two pairs. On the receive end of the circuit, the HDSL modem reverses the process and hands the T1/E-1 signal to the DSU. For HDSL to function properly, as one might expect, the UTP cable plant must be in good condition, although HDSL is reasonably tolerant of mixed gauges, bridged taps, and certain other circuit anomalies. Note that HDSL's limitation of circuit length confines it to use in an access environment—HDSL is not a technology to be used in long-haul, private-line, T1 applications.

As HDSL eliminates or reduces the number of repeaters, its incremental cost is mitigated to some extent. Additionally, the cable plant supporting HDSL does not require special conditioning or engineering, further reducing costs, as well as enabling service to be provisioned much more quickly. Finally, HDSL offers error performance of approximately 10^{-10}, as compared to the 10^{-7} level offered by repeatered T1 over twisted pair [17]. The bottom line is that T1 network access can be provisioned much more quickly and much less expensively with HDSL and with improved error performance as compared to traditional T1 or E-1. The carriers, therefore, realize revenues more quickly and at such reduced costs that many pass on those cost savings to end users in the form of lower installation charges and reduced monthly rates for the service. Since its introduction in 1992, the vast majority of T1 and E-1 circuits installed have been HDSL loops.

HDSL2 is an HDSL variant that supports T1 speeds of 1.544 Mbps over a single twisted pair (two conductors) and over 24 AWG loops up to 13.2 kft, 768 kbps at up to 17.7 kft, and 384 kbps at up to 22.5 kft. The international E-1 version supports signaling speeds up to 2.3 Mbps. An HDSL2 regenerator can double the distance for each speed rating. A variation known as HDSL4 can run over two copper loops (four physical conductors) in order to extend the maximum transmission span to as much as 16,500 ft. HDSL2's level of performance is achieved through an advanced coding technique known as *Trellis-Coded Pulse Amplitude Modulation* (TC-PAM). This technique places three bits on a baud, which is an improvement over the two bits per baud realized through the 2B1Q technique used in HDSL. While the bandwidth required to support the HDSL signal is sensitive to the implementation, TCM features inherent *Forward Error Correction* (FEC) to overcome line impairments and issues of Signal-to-Noise Ratio (SNR), thereby offering excellent performance and the same reach as HDSL but requiring only a single pair. With respect to applications, HDSL is essentially a Layer 1 bit pipe that replaces leased full-rate and fractional T1 and E-1 local loops in support of any bit stream, including TDM voice, ATM, and Frame Relay, for example. Although the standard version does not support telephone service over the same line, some equipment manufacturers offer a passive POTS splitter [18].

9.1.1.6 Symmetric Digital Subscriber Line
Symmetric DSL (SDSL), also known as *Single-line DSL*, is an umbrella term for nonstandard variations on the HDSL theme running over a single pair. SDSL uses the same 2B1Q line coding as HDSL and runs at rates from 128 kbps to 2.32 Mbps. Example speeds are 384 kbps for loop lengths of up to 18,000 ft and 768 kbps for loop lengths of up to 12,000 ft (see Figure 9.8). Providers generally limit SDSL to transmission rates of 768 kbps, which is half the payload of a full T1. At that data rate, the signaling rate is 784 kbps, including 16 kbps of error control overhead. At a data rate of 384 kbps, which is one-fourth a full T1, the signaling rate is 400 kbps.

9.1.1.7 Symmetric High-Bit-Rate Digital Subscriber Line
Symmetric HDSL (SHDSL), also known as *G.shdsl*, is a business class DSL technology ratified by the ITU-T in February 2001 as G.991.2. SHDSL supports symmetric rate-adaptive transmission ranging from 192 kbps at 20,000 ft (6 km) to 2.312 Mbps at 10,000 ft (3 km) in increments of 8 kbps over a single pair, sensitive to loop specifics. An optional two-pair (four-wire) mode supports data rates ranging from 384 kbps to 4.624 Mbps

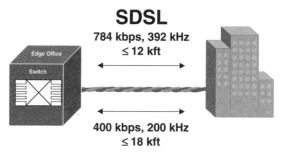

Figure 9.8 Symmetric DSL.

in increments of 16 kbps. At the two-pair mode and at a rate of 2.312 Mbps, the SHDSL reach is targeted at 16,000 ft (5 km). In either mode, repeaters can double the distance. SHDSL uses the same *Trellis-Coded Pulse Amplitude Modulation* (TC-PAM) advanced coding technology as HDSL2 and HDSL4. SHDSL operates in a rate-adaptive mode as well as a fixed-rate mode at 784 kbps and 1.544 Mbps.

SHDSL can transport T1, E-1, ISDN, ATM, and IP signals, with the specific framing requirements negotiated during the handshaking process defined in G.994.1. Although SHDSL is intended for data-only applications, newer techniques will support voice as well [19–21]. Targeted applications include Web hosting, videoconferencing, Virtual Private Network (VPN) services, and remote LAN access. *Note:* SHDSL is specified for North America in ANSI T1E-1.4/2001-174, for Europe in ETSI TS 101524, and worldwide in ITU-T G.991.2.

9.1.1.8 Voice over DSL *Voice over DSL* (VoDSL) refers to a nonstandard technique for supporting voice over various business-class DSL technologies. Specifically, VoDSL runs over SDSL and SHDSL, with ATM as the layer 2 protocol. At the customer premises in a VoDSL implementation, both voice and data connect through a centralized *Integrated Access Device* (IAD). The IAD serves, as necessary, to convert the analog voice into G.711 PCM samples through a codec embedded in a Digital Signal Processor (DSP) contained within the IAD. The DSP then generally compresses the voice streams by using one of a number of available compression algorithms, with G.726 *Adaptive Differential Pulse Code Modulation* (ADPCM) being widely used for this purpose. The IAD then forms the uncompressed voice samples into ATM cells using *ATM Adaptation Layer 1* (AAL1), or the compressed voice samples using *AAL2*, with the voice cells prioritized over the *AAL3/4* data cells in either case. VoDSL also can be supported on an uncompressed G.711 PCM basis through AAL1. (*Note:* ATM and AALs are discussed in detail in Chapter 10.)

At the service provider's CO or other *Point Of Presence* (POP), the ATM-based DSLAM demultiplexes the cells, routing the voice and data streams as appropriate, extracting them from the ATM cell format as appropriate. While the specifics vary by vendor, VoDSL commonly supports 4, 8, 12, or 16 voice conversations over a circuit, depending on the line rate and the compression algorithm employed. Some support as many as 40 voice conversations over full-rate (i.e., 1.544-Mbps) DSL, but there always is an upper limit. VoDSL reserves the remaining bandwidth in support of data communications, thereby ensuring some level of bandwidth is available for both voice and data. Should there be fewer than the maximum number of voice conversations active at any given moment, data can lay claim to that available bandwidth as long as it is available. This dynamic bandwidth allocation serves to optimize the use of the DSL circuit at all times [22–27].

9.1.2 Community Antenna Television

Cable television, formally known as *Community Antenna TeleVision* (CATV), was born in the mountains in the state of Pennsylvania in the United States. In the late 1940s, there were only a few TV stations, all located in major cities. As TV transmit and receive antennas operated exclusively over the air, signal quality was sensitive to Line-of-Sight (LOS), ElectroMagnetic Interference (EMI) and Radio Frequency Interference (RFI), weather conditions, and other factors. Reception was poor, even in the cities, and awful elsewhere. In remote rural areas and particularly in mountainous areas where LOS was not possible, reception was next to impossible, even on the best of days. John and Margaret Walson, owners of the Service Electric Company, a retail appliance store in Mahanoy City, were having a difficult time selling TV sets as reception was so poor in the valley where the town was situated, 90 miles from Philadelphia. In order to demonstrate TV sets to their best advantage, Mr. Walson placed an antenna on top of a tall utility pole on a nearby mountaintop and ran antenna wire to the store. In June 1948, he built some amplifiers to bring the signal to customers who had bought his TV sets, thereby creating the first CATV network. Walson also was the first to use microwave to import TV signals from distant stations and the first to use coaxial cable to improve reception. In 1972, Walson also was the first to distribute Home Box Office (HBO), which marked the beginning of pay TV and the start of the explosive growth of the CATV industry [28, 29].

Until very recently, all CATV systems were analog in nature, and many remain so. The community antenna actually consists of multiple satellite and microwave antennas located at a *head end*, which is the point of signal origin for the CATV network. At the head end, multiple analog TV broadcast signals are interwoven through a frequency division multiplexer and carried over a coaxial cable system to the community or neighborhood to be served. In the United States, each 6-MHz video channel carries a signal in the *National Television Standards Committee* (NTSC) format. [*Note:* In European and many other countries, the *Phase Alternate Line* (PAL) format requires an 8-MHz channel.] Within each 6-MHz channel, approximately 250 kHz is *transition bands*, or *guard bands*, 4.2 MHz is required for the video, another 300 kHz is required for the audio signal, and the balance is due to the *Vestigial SideBand Amplitude Modulation* (VSB-AM) technique utilized as the signal is placed on the Radio Frequency (RF) carrier. Contemporary coax-based CATV systems commonly support aggregate raw bandwidth of 500–850 MHz, thereby supporting a great number of TV channels, all of which reside on the system simultaneously. CATV networks conform to a *tree-and-branch* architecture comprising a *trunk* system, a *distribution* system, and hundreds or perhaps thousands of subscriber *drops*, as illustrated in Figure 9.9. At each subscriber premises, the coax cable drops terminate in *converter boxes*, or *set-top boxes*, which serve as frequency division demultiplexers. As each individual channel is selected, the set-top box selects the appropriate frequency range. Traditionally, these networks have been

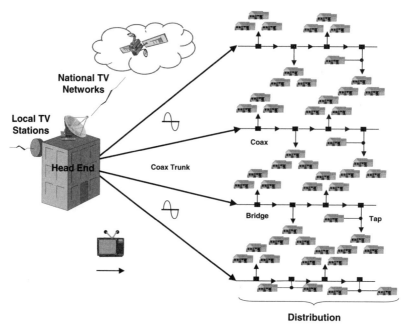

Figure 9.9   Traditional analog CATV network, with tree-and-branch architecture.

analog, coax-based, one-way networks for the downstream delivery of analog enter-tainment TV. Few changes occurred in CATV networks until the 1990s.

Beginning in the mid-1990s, a few large CATV providers began to upgrade their aging coaxial cable systems. As many of those networks were installed in the late 1960s and early 1970s, they were in awful shape. Further, those networks were strained for capacity as the CATV providers sought to increase revenues and profits through the introduction of premium movie channels and *Pay-Per-View* (PPV). The upgrades, in some cases, went beyond simple coax upgrade and replacement to include optical fiber in the trunk facilities from the head end to the neighborhood, where they terminated in an optoelectric conversion box that interfaced with the existing coax for ultimate termination at the premises.

As the telecommunications domain in the United States was deregulated with the Telecommunications Act of 1996, the CATV providers began to consider operat-ing as CLECs. The technology existed to upgrade the coaxial cable system to support two-way communications through frequency splitting and advanced signal modula-tion techniques, much as is done over twisted pair in xDSL technologies. The coax amplifiers and set-top boxes could be upgraded as well. Further, the system could be converted to digital by replacing the amplifiers with repeaters, and TDM channels could run inside the FDM channels, much like xDSL. The coax cable certainly offers much more in the way of bandwidth and distance than does twisted pair, and with an upgrade, error performance could be improved to levels that would make twisted pair pale by comparison. With optical fiber in the long-haul portion of the network (i.e., from the head end to the neighborhood), it became clear that CATV systems could compete very effectively with the ILEC local loop. Further, the largely unreg-ulated CATV providers could not be forced to wholesale their local loops to competitors, unlike the requirements placed on the ILECs by federal and state regulators at the time.

There were several problems with this scenario. Most of the CATV operators were overburdened with debt and, therefore, could not afford to make the necessary capital investments in network upgrades. The requisite investment in the necessary circuit and packet switches added greatly to the problem. Further, the CATV providers lacked the necessary skills, network management systems, and billing systems to support transaction-based switched voice and data services. Also, no standards existed for the set-top boxes, or much of anything else actually. Also, the CATV networks generally were not interconnected—not even those owned by the large *Multiple-System Operators* (MSOs).

Major telecommunications companies began to acquire a number of CATV providers in the late 1990s, addressing many of the cash and management problems associated with this scenario. Most notably, AT&T in early 1999 acquired TCI (Tele-Communications, Inc.) and MediaOne, two of the largest CATV providers in the United States, and invested billions of dollars in system upgrades. The lack of standards for set-top boxes was resolved in March 1997 by the *Data Over Cable Service Interface Specification* (DOCSIS), developed by the limited partnership known as the *Multimedia Cable Network Systems Partners Ltd.* (MCNS) and now administered by *CableLabs*, which has relabeled the initiative *CableLabs Certified CableModem*. Through matching DOCSIS *Cable Modems* (CMs) at the head end and the customer premises, two-way cable paths are provided over a *Hybrid Fiber/Coax* (HFC) system. The head-end portion of the network is in the form of a *Cable Modem Termination System* (CMTS), which supports a packet data connection to an IEEE 802.3 10/100-Mbps Ethernet port on a router. The system runs IP at Layer 3 (Network Layer) in Ethernet frames at Layer 2 (Data Link Layer). Associated with the CMTS are various servers for security, address translation, data caching, video caching, and so on. A CMTS can support as many as 2000 cable modem users on a single 6-MHz channel (8 MHz in Europe), with issues of congestion for shared bandwidth becoming more severe as the number of active users increases. The modem on the customer premises is in the form of a TV/data set-top box, which supports traditional coax connections to multiple TV sets and a 10/100Base-T Ethernet connection to a PC or to a hub serving multiple PCs. Figure 9.10 illustrates a typical CATV network configuration.

DOCSIS is both always on and generally asymmetric in nature, as are most xDSL technologies. DOCSIS 2.0 specifications address downstream and upstream channels:

- **Downstream:** The head end broadcasts downstream transmissions to all premises on the shared point-to-multipoint network. DOCSIS modems carve a downstream channel from the coax cable in the form of one or more 6-MHz channels in the range between 50 and 750–850 MHz, with the exact range being system dependent. DOCSIS 2.0, the current version, specifies *64-point Quadrature Amplitude Modulation* (64-QAM) for the downstream channels, yielding six bits per symbol and yielding a potential of 36 Mbps per 6-MHz channel. (See Table 9.5.) Overhead for framing and Forward Error Correction (FEC) reduces that level to approximately 27–31 Mbps of shared bandwidth per 6-MHz channel. Alternatively, *128-QAM* yields seven bits per symbol,

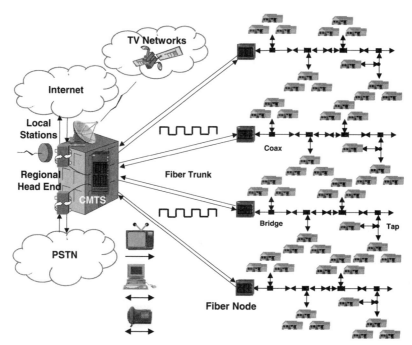

Figure 9.10 Contemporary HFC digital CATV network supporting TV, Internet access, and voice.

TABLE 9.5 DOCSIS Physical Layer

FDMA+[a]	Channel Width (MHz)	Modulation	Bits/ Symbol	Symbol Rate (Msps)	Signaling Speed (Mbps)
		Downstream			
TDMA	6	64-QAM	6	6	36
	6	128-QAM	7	6	42
	6	256-QAM	8	6	48
		Upstream			
TDMA	1.6	QPSK	2	1.28	2.56
	1.6	8-QAM	3	1.28	3.84
	3.2	16-QAM	4	2.56	10.24
	3.2	32-QAM	5	2.56	12.8
	6.4	64-QAM	6	5.12	30.72
S-CDMA	6.4	128-QAM TCM	7[b]	5.12	30.72

[a] Frequency Division Multiple Access.
[b] Six bits payload, 1 bit error control.

respectively, and the signaling speed to 42 Mbps, although it is more sensitive to noise. The standards also provide for the use of *256-QAM*, which increases the raw signaling speed to 48 Mbps. These variations on QAM are compatible with *Moving Picture Experts Group-2* (MPEG-2), the compression technique specified for digital video transmission in CATV networks. Downstream data are encapsulated into MPEG-2 packets of 188 bytes.

- **Upstream:** In early versions of DOCSIS, transmissions upstream from the user premises to the head end are supported either through the contentious CSMA/CD access protocol used in Ethernet LANs or through *Time Division Multiple Access* (TDMA), depending on the specifics of the CATV provider's implementation [30, 31]. According to DOCSIS 1.0 standards, traffic in the upstream direction is supported over 6-MHz channels in the range between 5 and 42 MHz for U.S. systems and 8-MHz channels in the range between 5 and 65 MHz for European systems, each of which can operate at about 12 Mbps through use of the *Quadrature Phase Shift Keying* (QPSK) modulation technique. Again, framing and FEC overhead reduce the level of usable bandwidth, in this case to about 10 Mbps [32–35]. In a field implementation, many systems use QPSK across a 1.6-MHz channel, yielding 2.56 Mbps (2 bits per symbol × 1.28 megasymbols per second). DOCSIS 1.1 added *16-QAM* to the mix, roughly doubling the data rate and doubling the channel width to 3.2 MHz to double it yet again (4 bits per symbol × 2.56 megasymbols per second), for a four fold increase to over 10 Mbps. QAM increases sensitivity to noise, but DOCSIS deals with that through sophisticated noise cancellation techniques. DOSCIS 2.0 triples the upstream rate of DOCSIS 1.1 over a channel of 6.4 MHz using either TDMA in combination with *64-QAM* or *Synchronous Code Division Multiple Access* (S-CDMA) in combination with *128-QAM Trellis-Coded Modulation* (TCM). S-CDMA transmits 128 orthogonal codes simultaneously (i.e., in the same time slot) [36].

There also are *telco-return* modems that use the circuit-switched PSTN for the upstream path, although this approach generally proves less than totally satisfactory. Not only does the telco-return approach require the use of a telco-provided local loop, which adds costs to the equation, but it slows the upstream path to modem speeds. Also, the widely mismatched speeds on the upstream and downstream paths cause the downstream server to slow down, therefore running at less than optimum efficiency. DOCSIS 3.0 standards currently under development are expected to increase upstream data rates to 120 Mbps and downstream rates to 160 Mbps.

CATV networks also support voice in several possible ways. The most common approach at this time is the very same TDM and PCM approach used in the circuit-switched PSTN, although some CATV networks use ADPCM modulation with silence suppression. At a rate of roughly one bit per hertz, a 6-MHz channel supports a considerable number of PCM-based voice channels. (Refer to Chapter 7 for detailed discussions of TDM, PCM, ADPCM, and silence suppression.) The current focus is on VoIP, with protocol conversion from VoIP encoding to PCM encoding occurring at a gateway located at the CATV provider's head end. The ultimate goal is that of end-to-end VoIP, with the CATV provider connecting voice directly from the CMTS to a VoIP backbone. The first approach of TDM-based voice involves separate voice channels, both upstream and downstream. VoIP rides over the shared packet data channels, with the CMTS and cable modems providing priority access in order to support the QoS demanded for toll-quality voice communications. Cable telephony requires local power, unlike traditional telco telephony. In the event that the lights go out, local battery backup provides telephony support for a few hours

[37, 38], which is especially important in support of calls to emergency services (e.g., 911). Various estimates indicate that cable operators in the United States and Canada had signed up approximately 3.2 million VoIP subscribers as of the end of 2005 and that the rate of growth is about 250,000 per month. Including the TDM voice customers, cable operators boasted total voice subscribers numbering about six million [39].

DOCSIS 1.1 provided a number of enhancements, in addition to QoS support. *Fragmentation* allows large Ethernet frames to be fragmented, which naturally improves voice latency. Rather than having to wait in a buffer until a large Ethernet data frame can be fully transmitted, the VoIP packet (once framed) can transit the network after a smaller Ethernet fragment has completed its journey. *Payload header suppression* can serve to reduce overhead, which is especially important for VoIP packets. *Concatenation* is a mechanism that links together multiple packets, which then can be processed as a single entity in a supperrate service context. Concatenation speeds processing, which reduces overall network latency. Dynamic channel allocation allows the CMTS to accomplish dynamically load balancing, shifting traffic flows between channels in order to reduce overall network congestion levels. Security is added in the form of authentication, and support for *Simple Network Management Protocol version 3* (SNMPv3) adds network management capabilities.

The CATV networks remain troubled by several inherent limitations. First, CATV networks are consumer oriented, primarily serving residential neighborhoods and not extending to most businesses. (After all, you are supposed to work at work, not watch TV.) Second, CATV networks are multipoint bus networks, much like a classic Ethernet 10Base5 LAN (Chapter 8). Since your voice and data transmissions pass every other house connected to the coax cable, security is a major concern in the absence of an effective encryption mechanism. Third, much like 10Base5 Ethernet and very much unlike DSL, the CATV local loop network is shared. Therefore, the more active users are on the network, the worse the performance. While the CATV providers speak of total bandwidth of as much as 500 Mbps and user access at rates of as much as 10 Mbps, those are *best case* figures quoted on the assumption that the load on the shared medium is light enough to support transmission rates of 10 Mbps for all active users. DOCSIS cable modems work on an asymmetric basis, as do most xDSL options. Specifically, the maximum aggregate usable bandwidth of a downstream channel currently is 31 Mbps, and the maximum aggregate usable upstream bandwidth ranges from 200 kbps to 10.0 Mbps, depending on the specifics of the field implementation of a network conforming to DOCSIS 1.1 specifications.

Especially with the strength of AT&T (i.e., AT&T prior to the merger with SBC) behind it, it seemed as though CATV network providers had an excellent chance of securing a large segment of the market for high-speed Internet access. The CATV providers did, indeed, enjoy considerable success, although financial problems caused AT&T Broadband to merge with (read *be acquired by*) Comcast in 2001, only two years after it was formed. About the same time, AT&T's primary ISP, At Home Corp., ceased operations, and its Excite@Home customers were left stranded for several days until AT&T and other CATV providers managed to switch them to their in-house ISP networks. Since that time, Comcast and other CATV providers in the United States have enjoyed tremendous success. According to the Telecommunications Industry Association (TIA) *2005 Telecommunications Market Review and Forecast*, there currently are about 22.0 million cable modem subscribers compared with about 15.1 million DSL subscribers in the United States, and cable modems access is expected to continue to surpass DSL access through 2008.

9.1.3 Wireless Local Loop

Wireline local loops are problematic. Cabled media are expensive and time consuming to deploy. Right-of-way must be secured, trenches must be plowed and conduits placed, poles must be planted and crossarms must be hung, conductors must be spliced, amplifiers and repeaters must be connected, and so on. The vast majority of loops are copper twisted pair, which tends to offer relatively little in terms of bandwidth and error performance and is highly distance limited, especially at the higher frequencies required to maximize data rates. The ILECs own most of the conventional local loops, CATV providers own the coaxial cable networks, and neither of them is the least bit inclined to share them, except as required by the regulators. As I discussed earlier in this chapter, the decision boils down to one of *build or buy* (i.e., lease), and *buy* is not always an option.

In building local loops, the advantages of *Wireless Local Loop* (WLL), also known as *fixed wireless,* are fairly obvious at this point in our discussion of communications systems and networks:

- WLL systems often can be deployed much more quickly and much less expensively that wireline systems.
- The aggregate bandwidth of a WLL system can be apportioned to end users on a channelized basis and often very flexibly so.
- The system can be configured and reconfigured remotely, often without a *truck roll.*

The disadvantages of WLL also are fairly obvious:

- Interference always is an issue, and one that in many ways is out of the control of both the service provider and the user organization. While licensed frequency bands are protected from direct interference from others who might covet the same frequencies, unlicensed bands are available for all to share. Regardless of whether the band is licensed or not, Radio Frequency Interference (RFI) and Electromagnetic Interference (EMI) are always issues. Such interference can be caused by electric motors, radio systems transmitting out of their assigned bands or in excess of their prescribed power levels, and forces of nature such as lightning, static electricity, and solar flares. Additionally, the quality of the airwaves is always an issue, as the quality of the signal can be negatively affected by precipitation, fog, humidity, dust, smoke, pollution, and temperature.
- Distances of terrestrial systems are limited due to factors such as quality of the airwaves and the frequencies used. The systems operating in the higher frequency ranges (e.g., 28–31 GHz) suffer from attenuation to a much greater extent than those in the lower frequency ranges (e.g., 2.5–2.7 GHz).
- Line-Of-Sight (LOS) is always preferable and is absolutely required at the higher frequencies.

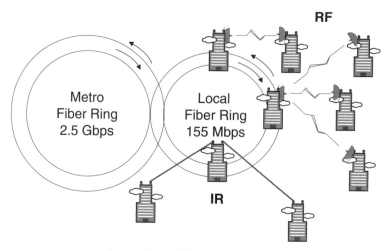

Figure 9.11 WLL configuration.

- The licensing process can be lengthy and expensive, especially given the great demand for licensed spectrum.
- Security is always an issue with RF-based systems.

The WLL options available are numerous and include nonstandard and standard approaches and both licensed and unlicensed frequency bands. This discussion of WLL options includes the most significant and what I consider to be some of the most interesting. Regardless of the specific technology, RF-based WLL configurations include centralized antennas, or base stations, that either are located at the edge of the carrier network or are connected to the network edge via optical fiber, DSL, T1/E-1 facilities, or perhaps microwave. The base stations connect to matching antennas at the customer premises, as illustrated in Figure 9.11. Free Space Optics (FSO) solutions involve infrared (IR) transmitter/receivers.

9.1.3.1 Local Multipoint Distribution Service

Bernard B. Broussard developed *Local Multipoint Distribution Service* (LMDS). Together with Shant and Vahak Hovnanian, Broussard formed CellularVision, a New York wireless cable TV firm that provided 49 TV channels and later added high-speed Internet access. The technical rights to LMDS technology later were spun off into a separate company. In the United States, the FCC first auctioned licenses for LMDS radio in early 1998. The 104 successful bidders yielded revenues of approximately $578 million, which certainly underscored the interest in WLL. Notably, and in consideration of the emphasis on competition in the local loop, the RBOCs and CATV providers were not permitted to participate and further were prevented from holding any LMDS licenses for a period of three years in order to encourage competition. Licenses were awarded in two blocks for each of 492 markets known as *Basic Trading Areas* (BTAs). The *A block* has a width of 1.15 GHz in the frequency ranges of 27.5–28.35 GHz, 29.1–29.25 GHz, and 31.0–31.15 GHz. The *B block* has a width of 150 MHz in the spectrum between 31.15 and 31.3 GHz. Outside of North America, LMDS operates in the 20- and 45-GHz bands. Given the high frequencies used, LMDS requires Line-Of-Sight (LOS) and generally is limited in distance to a cell diameter of 10–15 miles, although cell sizes generally are much smaller. LMDS can carve a 360° cell into four quadrants of alternating antenna polarity (H and V, i.e., Horizontal and Vertical), thereby improving traffic capacity. LMDS offers excellent error performance, with rain fade compensation through the use of adaptive power controls. LMDS supports both point-to-point and point-to-multipoint service configurations. LMDS is flexible enough to support local loops ranging from 1.544 Mbps (T1) to 155 Mbps and even 622 Mbps and in either symmetric or asymmetric configurations. Individual subscribers in office complexes or Multi-Dwelling Units (MDUs) can gain access to bandwidth in increments of 64 kbps (DS-0). Multiplexing access methods include Frequency Division Multiple Access (FDMA), Time Division Multiple Access (TDMA), and Code Division Multiple Access (CDMA). TDMA modulation options include Phase Modulation (BPSK, DQPSK, QPSK, and 8PSK) and Amplitude Modulation (QAM, 16-QAM, and 64-QAM).

As noted above, LMDS generated a lot of excitement in 1998 and again in the 1999–2000 time frame, and hundreds of millions of dollars were spent securing LMDS licenses at spectrum auctions in the United States. Most of the successful bidders (e.g., Teligent, WinStar, and Advanced Ratio Telecom) have since declared bankruptcy. XO Communications (previously Nextlink) invested a total of approximately $900 million in those licenses and now holds licenses in 73 major cities in the United States. XO installed a small number of systems over the intervening years but never was able to achieve any real success with LMDS due to technological problems and high equipment costs. In early 2006, however, XO announced a new LMDS initiative targeted at cellular *backhaul* as well as enterprises and government agencies looking for access alternatives, redundancy, and disaster recovery solutions [40–42]. [*Note:* In this context, *backhaul* refers to the transporting of traffic between distributed sites, such as cellular base stations, and a backbone or centralized network or site, such as a *Mobile Traffic Switching Office* (MTSO), or other point of concentration or switching.] Most of the current interest in WLL is on IEEE 802.16, aka WiMAX, which I discuss later in this chapter.

9.1.3.2 *Multichannel Multipoint Distribution Service*

Operating in five bands in the range 2.15–2.68 GHz in the United States and Canada and in the 3.5-GHz range elsewhere, *Multichannel Multipoint Distribution Service* (MMDS), also known as *wireless cable*, initially was developed for one-way TV transmission. As the name suggests, MMDS is a point-to-multipoint technology that operates from an antenna typically placed on a hilltop or other location in order to maximize Line-Of-Sight (LOS) connectivity. The antenna may be omnidirectional or may be sectorized in order to improve spectrum efficiency, and distances of up to 31 miles (50 km) can be achieved at allowable power levels due to the excellent signal propagation characteristics in this relatively low range of the microwave spectrum.

The first generation of MMDS stalled in the 2001–2002 time frame for several reasons, including technological and cost issues. The first generation of MMDS equipment was geared for one-way TV transmission but was tweaked for two-way applications such as Internet access and required LOS. Competition from DSL, cable modems, and LMDS certainly did not help. Further, MMDS definitely is bandwidth limited at only 200 MHz, which does not compare favorably with most alternative access technologies. Sprint and Worldcom were the most notable providers of first-generation services based on MMDS, having invested a total of approximately $1 billion in MMDS licenses in the late 1990s [43–47]. The deployment strategies were quite aggressive during that time and a number of systems were installed in major metropolitan markets in the United States, but those networks subsequently were capped. Nextel later acquired the Worldcom licenses during bankruptcy proceedings, so when Sprint acquired Nextel, it became by far the largest holder of MMDS licenses. There currently is a resurgence of interest in MMDS, which now works well in Non-Line-Of-Sight (NLOS) applications. Bell-South, Verizon, and Sprint all have conducted tests of second-generation systems. Clearwire offers MMDS service for Internet access in a number of cities in competition with 802.11 b/g commercial hotspot services. A number of small CATV providers employ MMDS in its original *wireless cable* application mode to extend CATV networks in areas where cabled transmission systems are impractical. Elsewhere, Walker Wireless in New Zealand has deployed MMDS in support of high-speed fixed and portable broadband access, including full mobility. In this implementation, MMDS competes directly with 802.11 b/g and 3G cellular systems [48]. Most of the current interest is WLL is on IEEE 802.16, aka WiMAX, which I discuss later in this chapter.

9.1.3.3 IEEE 802.16 and Worldwide Interoperability for Microwave Access

Today, the world is still enjoying the results of new scientific discoveries that are constantly being made, and the one that seemed most marvelous when they were first announced, become commonplace after a few years have passed.... It was a triumph when his [Marconi's] *experiments resulted in communications at will without wires over distance of 250 miles.*

W. J. Jackman et al, *The World's Workshop*, The C. F. Beezley Company, 1911

Worldwide Interoperability for Microwave Access (WiMAX) is a *Broadband Wireless Access* (BWA) solution that has a maximum range of approximately 30 miles, which would hardly impress Marconi's fans but offers bandwidth that they would not have appreciated at the time. WiMAX is based on the standards recommendations from the IEEE 802.16 Working Group and the European Telecommunications Standards Institute (ETSI) HiperMAN group. WiMAX is promoted by the WiMAX Forum, a special-interest group with members from the manufacturing, carrier, service provider, and consulting communities. Although WiMAX and 802.16 are not exactly the same, technically speaking, most people, including myself, use the terms interchangeably. WiMAX is just easier to say and it looks better in print. The same goes for Ethernet and 802.3

The IEEE set up the 802.16 group to standardize LMDS and MMDS, both of which were highly touted technologies that would allow competitive carriers and service providers to provision wireless local loops quickly and inexpensively, bypassing the ILEC copper loops in the process. However, it soon became clear that the first-generation LMDS and MMDS systems were too costly and unreliable. They also suffered from various performance issues, especially given the LOS requirements. About the same time, the economy stumbled. As a result, LMDS and MMDS were commercial failures. While there is a resurgence of interest in both LMDS and MMDS, much of the current focus is on 802.16, aka WiMAX. Officially known as the *WirelessMAN Air Interface for Broadband Wireless Access*, the 802.16 standard evolved over a number of years:

- 802.16 (2001) standardized LMDS. This first specification focused on fixed wireless solutions in both point-to-point and point-to-multipoint configurations. 802.16 specified frequencies in the range 10–66 GHz and required LOS. This first standard garnered little support.
- 802.16a (2003) was based on MMDS and the European HiperMAN system. This extension operates in the range 2–11 GHz, which includes both licensed and license-exempt bands. In the lower end of the range, LOS is not a requirement, although it is always preferable. 802.16a specifications include both point-to-point and point-to-multipoint configurations.
- 802.16d, aka 802.16-2004 (2004), is a compilation and modification of previous versions and amendments 802.16a, b, and c. 802.16d specifies frequencies in the range 2–11 GHz and includes point-to-point, point-to-multipoint, and mesh topologies. This specification recommends, but does not require, LOS and includes support for indoor CPE.
- 802.16e (October 2005), formally known as Mobile WirelessMAN, adds handoff capability, thereby supporting portability and pedestrian speed mobility for users of laptop, tablet, and hand-held computers. Operating in the range 2–6 GHz, it is designed for point-to-multipoint applications and does not require LOS.

9.1.3.3.1 Technical Specifics WiMAX standards address both LOS and NLOS scenarios. Where LOS can be achieved, WiMAX cell radius is as much as 50 km (31 miles). Under NLOS conditions, the maximum cell radius is approximately 9 km (5.6 miles). The fixed wireless standards provide for aggregate raw bandwidth up to about 70 Mbps per Base Station (BS), although the data payload is much less due to overhead. The level of actual throughput depends on LOS, distance, air quality, EMI and RFI, and other signal impairments. Mobile network deployments (802.16e) are expected to provide up to 15 Mbps of aggregate raw bandwidth within a cell radius of up to 3 km. Whether the specification is fixed or mobile, WiMAX data rates and distance are tightly and inversely related. In other words, the maximum signaling rate of 70 Mbps and the maximum throughput of approximately 40 Mbps can be realized over the shortest distance between the BS and the user antenna under LOS conditions. Over the maximum distance of 50 km under LOS conditions, or the maximum distance of 9 km under NLOS conditions, throughput drops considerably. WiMAX specifications address NLOS performance issues through a number of mechanisms, including the following:

- *Orthogonal frequency division multiplexing* subdivides the spectrum into a number of independent, narrowband subcarriers, across which it sends the signal in parallel fashion. The receiving antenna monitors all subcarriers, reassembling the multiple parallel signals into one high-speed serial output.
- *Subchannelization* concentrates signal power into fewer OFDM subcarriers, thereby extending the reach of the system, mitigating the effects of physical obstructions in an NLOS environment and reducing CPE power consumption.

Subchannelization is an option only for the uplink, that is, the link from the remote terminal back to the Base Station (BS) at the head end of the network.

- *Antenna design* clearly impacts signal strength. The fixed wireless specifications call for directional antennas in order to reduce multipath fading and, thereby, improve signal strength and cohesiveness. The directional antenna at the customer premises may be in the form of an adaptive, passive array *pizza box*, so called because it is about the size and shape of a pizza box. Such an antenna possesses beamforming properties that permit it to adjust its logical focus to maximize the strength of the incoming signal from the BS at the network head end. These adjustments in focus are accomplished passively, as no physical (i.e., mechanical) reorientation is required.

- *Multiple Input, Multiple Output* (MIMO) antennas employ space/time coding to compensate for multipath fading over long loops. Copies of the same RF signal are transmitted by multiple antennas separated by physical space and received by multiple antennas also spatially separated. These intelligent antenna systems can compensate for multipath fading and realize *diversity gain*, that is, increase in signal strength. This same technique is specified in 802.11n and VDSL and has been used in long-haul microwave systems for many years.

- *Rate-adaptive modulation* dynamically adjusts the signal modulation technique of each carrier, individually, to compensate for variations in signal quality at that carrier frequency. When the signal is strong, an intelligent WiMAX system can use the most sophisticated modulation scheme, which yields the highest transmission rate. As the signal fades due to factors such as temperature and interference, error performance degrades and link stability is jeopardized. In such a situation, the system can shift to successively more robust modulation schemes, but slower and more overhead intensive, to compensate for those conditions. While transmission rates drop accordingly, at least the link is maintained. The modulation schemes employed, from the most efficient to the least, are specified as 256-QAM, 64-QAM, 16-QAM, QPSK, and BPSK.

- *Reed–Solomon Forward Error Correction* (FEC) is employed to deal with issues of signal quality. While FEC inherently involves some degree of bit-level redundancy, it provides the receiver with enough data to reconstruct a large percentage of frames errored in transit. Automatic Repeat reQuest (ARQ) is employed to request retransmission of any remaining errored frames.

- *Power control* is the responsibility of the base station, which analyzes signal strength and advises the remote terminals. On the basis of that information, the remote terminals can dynamically adjust their transmission levels to conserve power and to minimize the likelihood of cochannel interference with other CPE in proximity.

802.16 specifications include several multiplexing options. Time Division Duplex (TDD) supports Half-DupleX (HDX) communications and Frequency Division Duplex (FDD) supports both HDX and FDX.

Although 802.16 standards address the range 2–11 GHz, spectrum allocations are not completely harmonized throughout the world. WiMAX emphasis in the United

TABLE 9.6   IEEE 802.16, aka WiMAX, Coverage Scenarios

Dimension	Specifics	Standard	Full Featured
Cell radius	LOS	10–16 km	30–50 km
	NLOS	1–2 km	4–9 km
	Indoor self-install CPE	0.3–0.5 km	1–2 km
Maximum throughput per 60° sector	Downlink	11.3–8 Mbps	11.3–8 Mbps
	Uplink	11.3–8 Mbps	11.3–8 Mbps
Maximum throughput per CPE at cell edge	Downlink	11.3–2.8 Mbps	11.3–2.8 Mbps
	Uplink	11.3–2.8 Mbps	0.7–0.175[a] Mbps
Maximum number of subscribers		Less	More

[a]This assumes that a single subchannel is used to extend the range as far as possible.

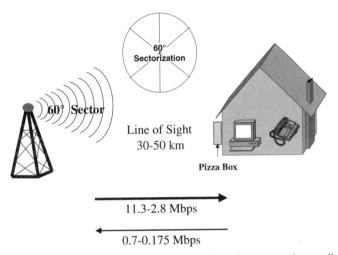

Figure 9.12   Full-featured LOS WiMAX configuration at maximum distance.

States is on the MMDS bands, which mostly are in the range 2.5–2.7 GHz. In Europe and other regions of the world, the emphasis is on the 3.5- and 10.5-GHz bands [49]. The WiMAX Forum describes coverage range in terms of two scenarios. A standard BS includes only mandatory capabilities, including output power. A full-featured BS includes higher RF output power, transmit/receive diversity, subchannelization capability, and ARQ logic. Table 9.6 assumes that the system runs in the 3.5-GHz band, that each channel is 3.5 MHz wide, and that the system is sectorized at 60° to yield six sectors.

Note that the transmission rate in Table 9.6 is stated in terms of *throughput*, in consideration of overhead and assuming reasonable signal performance. In a 3.5-MHz channel and at a maximum of 5 bits/Hz, the raw signaling rate is 17.5 Mbps. In consideration of overhead factors such as framing and error control, the maximum actual bidirectional data throughput is anticipated to be approximately 11 Mbps for a *standard* system operating at short range and 8 Mbps for a *full-featured* system operating at long range, as illustrated in Figure 9.12. 802.16 includes provisions for improved efficiency through header suppression, concatenation, and fragmentation.

Also note that the transmission rate is symmetrical, that is, the same for uplink (upstream) as for downlink (downstream) transmission. This is unlike the typical high-speed modem, cable modem, DSL, and PON implementations, which are asymmetrical. The sole exception to this symmetry is in the case of full-featured CPE at the cell edge, where uplink transmission rates are constrained by power limitations. This forces the invocation of slower subchannelization options in order to avoid crosstalk and adaptive modulation options that reduce the symbol rate to extend the range as far as possible.

The 802.16 security protocol is built on enhancements to the *Privacy-Key Management* (PKM) developed for cable modem communications. The protocol uses X.509 digital certificates with *Rivest–Shamir–Adleman* (RSA) encryption for authentication and key exchange. Traffic encryption options are *Data Encryption Standard* (DES) and *Advanced Encryption Standard* (AES).

9.1.3.3.2 Quality of Service The 802.16 specifications include convergence sublayers designed for mapping services to and from 802.16 connections. The ATM convergence sublayer is for ATM services and the packet convergence sublayer is for packet services such as IPv4, IPv6, Ethernet, and Virtual LAN (VLAN). As 802.16 is connection oriented, all services, including those inherently connectionless (e.g., SMTP and UDP) in nature, are mapped to a connection. This approach provides a mechanism for requesting bandwidth, negotiating service parameters, and establishing Quality of Service (QoS) levels. Downstream transmission is on the basis of a TDM signal, with individual stations allocated time slots serially. As the link is under the control of the BS, downstream QoS is straightforward. Upstream access is by TDMA and QoS is the function of a set of scheduling schemes that the BS has at its disposal in order to optimize performance. An important feature of the Medium Access Control (MAC) layer is an option that allows the BS to grant bandwidth to an intelligent subscriber station, rather than to the individual connection it supports. This allows the station to manage its bandwidth allocation among the users and applications it supports, which provides for more efficient bandwidth allocation in multiuser and multitenant applications [49]. 802.16 defines four polling schedules, as follows:

- *Unsolicited Grant Service* (UGS) is designed for services that periodically generate fixed units of data. TDM services T1 and E-1 are examples.
- *Real-time polling service* is designed for services that are dynamic in nature but require periodic dedicated request opportunities to meet real-time demands. Examples include real-time compressed voice services such as VoIP and IP-based streaming audio and video.
- *Non-real-time polling service* is identical to real-time polling service, except that connections may use random transmit opportunities. Internet access with a minimum guaranteed connection rate is an example of such an application.
- *Best effort service* provides neither throughput nor latency guarantees.

9.1.3.3.3 Strengths and Weaknesses WiMAX offers some real strengths compared to alternative broadband solutions, whether wired or wireless. As a standards-based solution, WiMAX enjoys broad consensus-level support from the manufacturer, carrier, and service provider communities. The existence of specifications for both fixed and mobile systems adds an element of flexibility that is unusual. Line of sight is not an absolute requirement, although it is always desirable. Dynamic link adaptation through adaptive modulation, time–space coding, adaptive antennas, subchannelization, and power control ensures that each link performs optimally, even over long loops in the absence of LOS. Configuration flexibility is an advantage, as point-to-point, point-to-multipoint, and mesh topologies all are defined. QoS, which is extremely important for an integrated voice/data network, is a distinctive feature of WiMAX in the context of wireless systems specifications. Finally, the aggregate bandwidth is considerable, at 70 Mbps per base station.

WiMAX has its share of weaknesses as well. EMI and RFI are always issues with RF-based wireless systems, which places WiMAX at somewhat of a disadvantage compared to DSL and PON. DSL and PON also offer dedicated bandwidth, while WiMAX bandwidth is shared, much like that of cable modems. Also, PON, cable modems, and VDSL all offer higher levels of bandwidth. Competition from the IEEE 802.20 Working Group will also be an issue. That group is chartered to develop similar standards for wireless access systems operating in licensed bands below 3.5 GHz. The technical goal is that of optimizing IP-based data transport, targeting peak data rates per user at over 1 Mbps and supporting vehicular traffic at speeds up to 250 km/h (155 mph). There also will be competition from the IEEE 802.22 Working Group, which is developing a standard intended for wireless data over UHF and VHF spectra currently used for broadcast TV. Also known as *Wi-Fi TV*, 802.22 targets the *Radio Area Network* (RAN), which it defines as having a range of up to 30 miles. Competition extends to Europe and Asia as well. ETSI chartered the *Broadband Radio Access Networks* (BRAN) project. *HiperAccess* is for frequencies above 11 GHz and *HiperMAN* for below 11 GHz. ETSI and IEEE 802.16 cooperate to some extent. South Korea's Electronics and Telecommunications Research Institute (ETRI) developed *Wireless Broadband* (WiBro) for 100 MHz of spectrum allocated in the 2.3-GHz band by the Korean government. WiBro offers aggregate throughput of 30–50 Mbps and has a reach of 1–5 km. Spectrum issues around the world exist, even though WiMAX includes both licensed and license-exempt bands. WiMAX-targeted spectra include license-exempt 2.4- and 5.8-GHz bands, which invite interference. Sprint controls much of the licensed 2.5-GHz band in the United States and that same band already is used for fixed wireless and cable transmission in Mexico [50–54].

9.1.3.3.4 *Applications and Futures*
While there certainly are a number of issues swirling around WiMAX, there is a tremendous amount of interest in it. The numerous applications include full-rate and fractional T1 and E1 services, especially for remote rural or developing areas where broadband telco or cable access is unavailable. ILECs and CLECs, alike, have shown interest in WiMAX for provisioning circuits in urban areas where telco or cable upgrades are not easily, quickly, or cost effectively implemented. Backhaul applications for Wi-Fi hotspots and cellular networks are a natural application, as is disaster recovery. Several manufacturers have announced multimode chips that will allow portable computers to connect via both Wi-Fi and WiMAX, and 802.16e supports portability and mobility for users of laptop and hand-held computers. The 802.16e air interface also likely will be included in trimode voice handsets that will interoperate with Wi-Fi and cellular networks.

Although there are no commercial WiMAX systems in operation at this time (June 2006), there are several trials in progress. AT&T is conducting trials with several customers in New Jersey. Miami University (Oxford, Ohio) is engaged in a trial with NCE Unified Solutions to cover students on campus and living off campus within a radius of 3–5 miles. This application certainly is an interesting broadband private network solution in a campus environment and what is essentially a wireless local loop extension of a campus network [54].

9.1.3.4 *Licensed Microwave*

In the frequency ranges of 24 and 38 GHz, microwave has been licensed by the FCC to a number of service providers for point-to-point WLL applications. The licenses are for aggregate channel capacity of 100 MHz. Some carriers hold multiple licenses, which yield aggregate bandwidth of 400–500 MHz or more. In these microwave frequency ranges, LOS is required and distances are limited to approximately 5 miles. Typically, the cells are much smaller, in consideration of the advantages of frequency reuse. Error performance for these digital microwave systems is in the range of 10^{-13}, which compares very favorably with UTP-based T1 at 10^{-7}, and under optimal conditions, it compares with optical fiber. The paired transmit/receive antennas typically are about 1 ft in diameter. Larger antennas and variable power levels are employed to compensate for *rain fade* in areas where rainfall is heavy and prolonged. As is the case with LMDS, licensed microwave generated a lot of excitement in the 1999–2000 time frame, and hundreds of millions of dollars were spent securing spectrum at FCC auctions in the United States. Most of the successful bidders (e.g., Advanced Radio Telecom, Teligent, and WinStar) since have declared bankruptcy and the licenses have been acquired by others.

9.1.3.5 *Personal Communications Services*

Personal Communications Services (PCS), the U.S. term for the *Personal Communications Network* (PCN) concept originally developed in the United Kingdom, is a service concept that is technology dependent and operates on a set of frequencies set aside specifically for that purpose. PCS ultimately intends to provide a full range of enhanced services through a single device and utilizing one telephone number, which will work *anywhere and anytime, for life. Note:* The term *PCS* also is used in the United States for a digital cellular telephony alternative using analog AMPS bands in the range 800–900 MHz. Actually, at this point, PCS means just about anything. In this section, I explore PCS in the context of the FCC licenses, with a focus on WLL.

While the spectrum allocation varies by country or region, the designated U.S. frequencies include narrowband, broadband, and unlicensed PCS. Actually, much of the PCS spectrum was already spoken for by high-band microwave systems, and clearing that spectrum is the responsibility of the PCS licensees. For the first time in U.S. history, that spectrum was auctioned (and reauctioned in some cases, after default on down payments—a long story) by geographic area. The 1995 auctions of two frequency blocks brought in $7.7 billion, and the 1996 auctions accounted for another $10.22 billion.

- *Narrowband PCS* has been allocated spectrum in the ranges of 900–901 MHz, 930–931 MHz, and 940–941 MHz. That AMPS spectrum is used to extend the capabilities of pagers and cell phones to include acknowledgment paging, two-way messaging, and digital voice.

- *Broadband PCS* is allocated 120 MHz in the ranges 1.85–1.91 GHz and 1.93–1.99 GHz. This spectrum is intended for the delivery of next-generation, high-tier wireless communications, including WLL, voice and data services, and cellular-like services for pedestrian traffic in high-density areas [55].
- *Unlicensed PCS* spectrum serves low-tier applications such as wireless LANs, wireless PBXs, PDAs, and PCS voice and data services within a building or campus environment. Unlicensed PCS spectrum has been set aside in two ranges, with 20 MHz in the range 1.91–1.93 GHz and 10 MHz in the range 2.39–2.40 GHz.

There are no mandatory service offerings associated with PCS licenses. Rather, each license holder defines its own services and applications.

9.1.3.6 *Free Space Optics*   In a local loop context, infrared (IR) light transmission systems also are known as *Free Space Optics* (FSO) and sometimes characterized as *wireless fiber*. FSO systems are airwave systems that use the infrared light spectrum in the terahertz range to send a focused light beam to a small receiver that often resembles a *Direct Broadcast Satellite* (DBS) dish. The transmitter/receivers can be mounted on rooftops or even indoors behind windows.

Contemporary IR systems offer substantial bandwidth at relatively low cost. FSO systems currently operate at rates of 1.544 Mbps (T1), 45 Mbps (T3), 155 Mbps (OC-3), and 622 Mbps (OC-12). Some vendors have advertised systems operating at 1 Gbps in support of GbE (Gigabit Ethernet) and 10 Gbps in support of 10GbE, and systems running at up to 160 Gbps have been demonstrated in the labs. Like microwave systems, FSO systems require LOS. In fact, some systems are so sensitive to LOS as to require autotracking mechanisms where they must adjust to the movements of high-rise office buildings due to wind sway, tremors, and other forces of nature. FSO systems also suffer from environmental interference, particularly fog, which acts like a prism to scatter the light beam. Under optimum conditions, distances are limited to about 2–5 km, although most tests have shown optimum performance at distances 500 m to 1 km. In areas where there is a lot of fog, links more typically are limited to about 200 m, and some manufacturers offer redundancy via RF systems running in unlicensed bands. Under optimum conditions, error performance is in the range of 10^{-8}, which compares favorably with UTP and microwave.

Despite the limiting factors, FSO is enjoying increasing popularity due to its low relative cost when compared to microwave and fiber optics. FSO systems also can be deployed very quickly, as there currently are no FCC licensing requirements and few other regulatory restrictions on its use. The applications are all short haul in nature. FSO, for example, is an attractive alternative to leased lines or private cabled systems for building-to-building connectivity in a campus environment or other short haul point-to-point applications, especially in bridging LANs. Carriers also use FSO as a replacement for various licensed RF-based WLL technologies. In those WLL applications, FSO typically is deployed in a mesh configuration for purposes of redundancy, in consideration of the potential for link failures due to fog and other environmental or LOS issues. More detail on FSO is presented in Chapter 2.

9.1.3.7 *Wi-Fi Hotspots*   Strictly speaking, 802.11b, aka Wi-Fi, does not fall into the realm of broadband access infrastructure, but it certainly merits discussion alongside WiMAX. In fact, WiMAX sometimes is (incorrectly) characterized as the big brother to Wi-Fi. Some suggest that the two compete directly, but I suggest that they are complementary.

Wi-Fi certainly enjoys great popularity, which has increased considerably with the introduction of 802.11g, and 802.11n promises to build on that popularity with increased transmission speeds and ranges due to the incorporation of MIMO technology. In addition to private LAN applications, Wi-Fi works well in public settings. Wi-Fi *hotspots* are now quite commonplace in public venues such as airports and coffee shops and even on airplanes, usually on a pay-per-minute basis. A number of municipalities (e.g., Philadelphia, Pennsylvania) in the United States have installed free Wi-Fi hotspots in downtown areas to encourage people to frequent those areas and patronize local merchants, and great many more municipalities plan to offer such services in the future. Google, one of the leading ISPs, announced in August 2006 that it invested approximately US$1 million to build a Wi-Fi network of 380 radio antennas designed to provide free Internet access to the entire population of its home town, Mountain View, California. The town covers some 11.5 square miles and has a population of approximately 72,000. Commercial ISPs quite naturally are opposed to Internet access as a public utility and are not particularly happy about Google's initiative, but they so far have had only very limited success in battling this trend. Interestingly, groups of residential end users have installed Wi-Fi hotspots in some neighborhoods for their joint use. This approach is reminiscent of some of the old co-op CATV networks.

In any case, a public Wi-Fi hotspot essentially is a WLAN positioned as a highly localized wireless access point. The Wi-Fi network connects to the Internet via a truly broadband local loop technology such as ADSL, WLL, or PON.

9.1.3.8 *Direct Broadcast Satellite*
While it may be a bit of a stretch to characterize satellite TV networks as *wireless local loop* technologies, such networks certainly are wireless and they certainly do offer yet another option for Internet access. Although Teledesic and a number of similar *Low-Earth Orbiting* (LEO) satellite networks (Chapter 11) planned to offer broadband Internet access, currently available offerings are based on *Geosynchronous Earth-Orbiting* (GEO) satellites and use the Ku or Ka band. As discussed at some length in Chapter 2, GEOs are effective for broadcast applications because their footprints, or areas of coverage, are substantial and stable. In terms of the downstream path from the Internet, GEOs offer considerable bandwidth, although it is shared much as is the bandwidth provided over a CATV network. This issue can be mitigated to a considerable extent through the use of highly focused *spot beams* that segment the aggregate footprint of the satellite into smaller areas of coverage. This allows coverage to be segmented and frequencies to be reused, much like the cells of a cellular telephony network or a Wi-Fi network. The upstream channel also is an issue, as two-way satellite dishes are considerably more expensive than one-way dishes. Further, a two-way connection via a GEO in equatorial orbit at an altitude of approximately 22,300 miles imposes round-trip signal propagation delays of at least 0.64 s. While these subsecond delays are tolerable for most applications, they render this Internet access technique unacceptable for users engaged in multiuser *twitch games* involving rapid action. Neither is the service appropriate for those engaged in real-time online financial trading or other activities requiring subsecond response times. Note that

some GEO-based Internet access services use a *telco-return* access technique for the upstream channel. This technique involves dial-up access through an ISP over the PSTN for the upstream channel, with the downstream channel supported over a one-way satellite link. Satellite-based Internet access services available in the United States currently include HughesNet and Starband:

- *HughesNet* (previously known as DirecWay and originally DirecPC) is a service of Hughes Network Systems, which also launched DirecTV, the first high-powered DBS service. While DirecPC and DirecTV historically required separate satellite dishes, the technology quickly evolved to support both through a service dubbed DirecDuo. Subsequently, Hughes sold its interest in DirectTV and now, once again, a separate dish is required. (There is a lesson in there somewhere.) HughesNet offers downstream speeds of up to 1 Mbps, depending on system load in consideration of the total number of users and the total level of activity on the system at any given time. The upstream channel also is via a satellite link and has a maximum speed of 200 kbps. Bandwidth is apportioned among users on the basis of a *fair access policy* that imposes bandwidth restrictions on users who exhibit patterns of system usage that exceed certain thresholds for an extended period of time. The thresholds are sensitive to the nature of the subscription agreement, for example, home, professional, small office, and business Internet. The service supports a Virtual Private Network (VPN), but the associated security mechanisms slow the link speed to that of a dial-up connection [56].
- *StarBand* is an Internet access service of Spacenet offered throughout the United States, Canada, Puerto Rico, the U.S. Virgin Islands, and several Caribbean and Central American countries, providing service running over one of two GEO systems—GE-4 or Telstar 7. Television programming is available through the DISH network via the EchoStar satellite, with the same Starband antenna (24 × 36 in. in the continental United States and 1.2 m in diameter in Alaska, Hawaii, Puerto Rico, and the U.S. Virgin Islands) capable of supporting both applications. Download speeds are up to 1 Mbps, with the target minimum speed being at least 150 kbps and upload speeds are as high as 256 kbps. The service supports a VPN, but the associated security mechanisms slow the link speed to that of a dial-up connection [57].

Internet access speeds via satellite generally do not compare favorably with the speeds offered by DSL, WLL, PON, or other terrestrial services. Satellite-based service also introduces issues of fair use and latency and requires the professional installation of a transmit/receive dish. Satellite access, therefore, generally is restricted to areas where alternative broadband access methods are not available, with examples including rural and remote areas, and developing countries. There are easily a dozen providers in Africa, for example, and about half that number serving Australia.

9.1.4 Passive Optical Network

Passive Optical Network (PON), comprises a family of Physical Layer (Layer 1) access technologies based on the specifications developed by the *Full-Service Access*

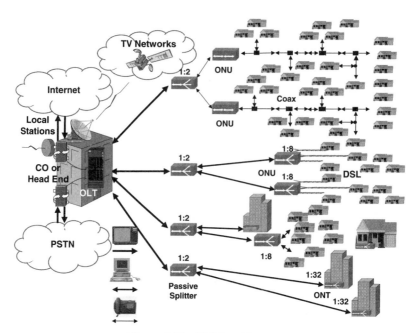

Figure 9.13   PON architecture.

Network (FSAN) initiative for an *ATM-based Passive Optical Network* (APON) scheme developed by an international consortium of vendors and ratified by the ITU-T within the G.983.1 standard (October 1998). A PON is a fiber-optic local loop network without active electronics, such as repeaters, which can be both costly and troublesome. Rather, a PON uses inexpensive passive optical splitters and couplers to deliver signals from the network edge to multiple customer premises. The PON splitters are placed at each fiber junction, or connection, throughout the network, providing a tremendous fan-out of fiber to a large number of end points. By eliminating the dependence on expensive active network elements and the ongoing powering and maintenance costs associated with them, carriers can realize significant cost savings. PON technology generally is used in the local loop to connect customer premises to an all-fiber network.

9.1.4.1   PON Elements and Configurations   Elements of a PON (Figure 9.13) comprise an Optical Line Terminal (OLT), an Optical Network Terminal (ONT), an Optical Network Unit (ONU), a passive splitter, and optical fiber:

- **Optical Line Terminal (OLT):** The OLT is located in the carrier's CO, or head end, where it serves to terminate the optical local loop at the edge of the network. The OLT can either generate optical signals on its own or pass SONET signals from a collocated SONET cross-connect or other device, broadcasting them downstream through one or more ports. The OLT also receives the upstream signals from the ONTs and ONUs.
- **Optical Network Terminal (ONT):** The ONT terminates the circuit at the far end. An ONT is used to terminate the circuit inside the premises in a *Fiber-To-*

The-Premises (FTTP) scenario, also known as *Fiber-To-The-Home* (FTTH) and *Fiber-To-The-Business* (FTTB), where it serves to interface the optical fiber to the copper-based inside wire. This approach maximizes the performance advantages of fiber-optic transmission.

- **Optical Network Unit (ONU):** An ONU is used in a *Fiber-To-The-Curb* (FTTC) scenario, in which the fiber stops at the curb, with the balance of the local loop provisioned over embedded coax in CATV networks or UTP in conventional telco networks. An ONU also is used in a *Fiber-To-The-Neighborhood* (FTTN) scenario, in which it is positioned at a centralized location in the neighborhood, with the balance of the local loop being provisioned over embedded coax or UTP. While this FTTN scenario maximizes the use of embedded cable plant and therefore minimizes the costs associated with cable plant replacement, it compromises performance to some extent.

- **Splitter:** The passive optical splitter sits in the local loop between the OLT and the ONUs or ONTs. The splitter divides the downstream signal from the OLT at the network edge into multiple, identical signals that are broadcast to the subtending ONUs. Optical splitters typically are implemented using cascading 1:2 power splits, each of which more or less evenly splits the incoming signal into two outgoing signals, thereby introducing *insertion loss* (i.e., the loss of signal strength between the inserted signal and the extracted signal) of 3.0 dB, which reduces the power of each by 50 percent. The splitter illustrated in Figure 9.14 comprises three 1:2 splits for a total split ratio of 1:8, with the outgoing signal over each distribution fiber having a power level equivalent to approximately 12.5 percent (one-eighth) that of the incoming signal. As a passive network, PON does not amplify the signal. As each split approximately halves the signal power, trunk lengths and the number of signal splits are limited. A given splitter might have a split ratio of 1:2 (one split), 1:4 (two splits), 1:8 (three splits), 1:16 (four splits), 1:32 (five splits), 1:64 (six splits), or 1:128 (seven splits). The *logical reach* (i.e., the maximum fiber loop length, without regard to the *loss budget*, i.e., the allowable amount of signal attenuation) can be up to 60 km (37 miles), depending on the PON type. In consideration of the *physical reach* (i.e., the maximum fiber loop length in consideration of the loss budget), the loops generally are limited to approximately half that distance.

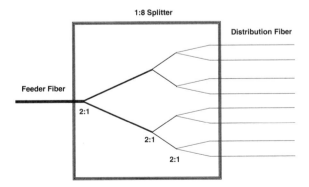

Figure 9.14 Optical splitter.

With respect to upstream transmissions, the splitter serves as a passive signal concentrator.

- **Optical Fiber:** The ITU-T specifies two types of optical fiber to be employed in a PON. G.652 describes a *standard* type of Single-Mode Fiber (SMF). G.652c/d describes low/zero water peak fiber. The fiber link can be organized in either simplex or duplex configuration, the definitions of which are quite different from those used in the context of transmission mode discussion in Chapter 6:
  - *Simplex* refers to a single-fiber configuration supporting transmission in both directions. In this configuration, the downstream transmissions are supported in the 1480–1580-nm window, with voice and data over a wavelength of 1490 nm and video at 1550 nm. The upstream transmissions are supported in the 1260–1360-nm window, with voice and data over a wavelength of 1310 nm.
  - *Duplex* refers to a two-fiber configuration. One fiber supports downstream transmissions and another supports upstream transmissions, both generally in the 1260–1360-nm window.

9.1.4.2 PON Standards There are several variations on the PON theme, each of which specifies Data Link Layer (Layer 2) protocols, transmission speeds, loop lengths, split ratios, multiplexing techniques, maximum loss levels, and distance limitations. Those standards are listed below and compared in Table 9.7:

- *ATM-based Passive Optical Network* (APON) is the term applied to the original specifications set by FSAN and ratified by the ITU-T as G.983.1 (1998). As the underlying bearer protocol, APON specifies Asynchronous Transfer Mode (ATM), which was favored by the ILECs for DSL and for their internal backbone networks. In contemporary terminology, APON generally is known as *Broadband Passive Optical Network* (BPON), which is described in G.983.3 (2001). APON runs in asymmetric mode at 622 Mbps downstream and 155 Mbps upstream or in symmetric mode at 155 Mbps. BPON supports as many as 32 splits over a distance of as much as 20 km (12 miles). The BPON supports voice, data, and video in ATM format.
- *Ethernet-based Passive Optical Network* (EPON) is the term for IEEE 802.3ah (2004). EPON specifies 802.3 (aka Ethernet) at the data link layer. EPON runs

TABLE 9.7   Passive Optical Network (PON) Standards Comparison

Dimension/Protocol	BPON	EPON	GPON
Standard	ITU G.983	IEEE 802.3ah	ITU-T G.984
Data Link Layer	ATM	Ethernet	ATM, Ethernet, TDM
Data rate: Downstream	155 Mbps, 622 Mbps	1.244 Gbps	1.244 Gbps, 2.488 Gbps
Data rate: Upstream	155 Mbps	1.244 Gbps	155 Mbps, 622 Mbps, 1.244 Gbps, 2.488 Gbps
Number of splits	32	16	32, 64, 128
Maximum reach	20 km	10 km, 20 km	30 km, 60 km
Security/Encryption	Churning	None	AES
Video wavelength	Standard	None	Standard

at 1.244 Gbps in symmetric configuration. As many as 16 splits are supported, and the maximum logical reach is approximately 20 km (12 miles). As all services are framed in Ethernet format, EPON inherently supports IP-based applications. EPON is not specifically designed for voice or video but certainly will support VoIP and IPTV. EPON does not specify a video wavelength.

- *Gigabit Passive Optical Network* (GPON), formally known as ITU-T G.984 (2004), describes both asymmetric and symmetric configurations and currently supports transmission rates as high as 2.488 Gbps. GPON supports as many as 32 or 64 splits, with future expectations of 128 splits, and a maximum logical reach of approximately 60 km (37 miles). At full speed of 2.488 Gbps with the maximum of 64 splits, each user has access to sustained bandwidth of more than 35 Mbps, which is far beyond that offered by other access technologies. GPON supports voice, data, and video in ATM format. GPON also supports voice in native PCM/TDM format and data in Ethernet format.

The ITU-T standards for *Wavelength Division Multiplexing* (WDM) certainly apply to BPON and GPON, both of which specify one wavelength for downstream voice and data and another for video. BPON supports 16 wavelengths with 200-GHz spacing and 32 wavelengths with 100-GHz spacing between channels. Upstream transmissions do not make use of WDM, as there is not upstream video requirement. EPON does not make use of WDM at this time.

Security is somewhat of an issue in a PON network, as the OLT is shared among a large number of users. So, PON makes several provisions for enhanced security. First, there are provisions for any number of higher layer encryption mechanisms to act on the downstream transmission. GPON, for example, currently specifies *Advanced Encryption Standard* (AES) for downstream transmissions. Encryption occurs at the OLT, and decryption occurs at the ONT or ONU. Encryption of the upstream transmissions is considered unnecessary as the shared OLT is at the head end, which is under the control of the service provider, which is a trusted party, and other users do not have access to the upstream channel. BPON provides for enhanced security through a technique known as *churning*. Downstream transmissions are encrypted through the use of a byte-oriented *churn key* exchanged between the OLT and the ONU. The ONU or ONT generates the key and sends it to the OLT, which uses it to encrypt downstream transmissions. The key is changed at least once a second—hence the term churning. Again, no security is deemed necessary on the upstream transmissions.

Upstream signals are supported by a TDMA mechanism, with the transmitters in the ONUs operating in burst mode. In this scheme, each subscriber takes turns transmitting upstream, with the length of the transmission window determined by the head end based on the upstream bandwidth subscription. BPON supports only an asymmetric transmission mode, EPON supports only symmetric, and GPON supports both.

9.1.4.3 Deployment Options PON unquestionably is the future of wireline local loops, at least in areas where its considerable costs can be offset by revenue potential. BPON and GPON are both oriented toward the *triple play* of voice, high-speed data (i.e., Internet access and VPN), and entertainment TV applications in the residential market. EPON clearly is oriented toward high-speed data applications in the business market. The cost/revenue equation depends on a number of factors, one of which certainly has to do with demographics, that is, the density of potential subscribers with the requisite applications requirements and the financial means to subscribe to the services that will satisfy them. In other words, if there are enough potential customers with enough bandwidth-intensive needs and enough money to afford to satisfy them, PON is an attractive technology for deployment. Competition must be considered, of course, in countries where it is permitted and where either CATV operators or wireless service providers are active. The service provider, generally in the form of an ILEC, also must consider whether copper facilities are already in place. In this context, the deployment opportunity commonly is characterized in one of three ways—greenfield, brownfield, or overlay:

- *Greenfield* deployments are those in new subdivisions, newly developed areas, or other areas where there is no existing telecommunications wireline infrastructure. The nature of this opportunity is such that, while trenches must be dug, conduits placed, poles planted, and so on, the deployment is engineered specifically for fiber optics. A greenfield deployment most likely would be in the form of FTTP, which takes fiber directly to the premises.
- *Brownfield* deployments take place where there is existing wireline infrastructure in the form of twisted pair, which is removed to create space for PON. A brownfield deployment typically takes the form of FTTN or FTTC, in which case the distribution cables and Digital Loop Carrier (DLC) systems are removed and replaced with optical fiber and PON ONUs. The link from the ONU to the premises is in the form of ADSL or perhaps VDSL over the embedded twisted pair, which yields a hybrid fiber/copper configuration.
- *Overlay* deployments are those in which the new infrastructure parallels that of the existing infrastructure. This approach allows the service provider to construct the new system and provide service to PON subscribers as required, while continuing to serve others subscribing to more basic services from the old cable plant. This phased approach generally is preferred for FTTP deployments where there is existing infrastructure [58, 59].

9.1.4.4 *PON in Practice* The vast majority of the interest in PON is on the part of the ILECs, particularly in the United States, where the CATV providers have captured the greatest share of the broadband access market with cable modem technology. As PON will once again provide the ILECs with a competitive advantage, they are deploying it aggressively in major markets, with the current emphasis on BPON, although GPON will likely supplant it once standards-based equipment becomes widely available. Render, Vanderslice & Associates estimated in a report dated September 2005 that there were FTTH deployments in 652 communities in 43 states, with a total of 2.7 million potential customers passed and 323,000 subscribers [60]. While I find the number of PON subscribers a bit difficult to believe in the context of that time frame, I certainly do not doubt that PON deployment is moving forward quickly.

Verizon is by far the most aggressive of the U.S. ILECs with respect to PON. Verizon's FiOS (*Fiber-Optic Service*) offering is an FTTP architecture based on BPON and using ATM as the underlying bearer protocol for data. Video is on

TABLE 9.8   **Verizon FiOS Internet Access Service Offerings [60]**

Downstream Rate (Mbps)	Upstream Rate (Mbps)	Monthly Rate[a]
5	2	$34.95
15	2	$44.95
30	5	$179.95

[a]Based on a one-year contract.

a separate wavelength, with all channels delivered simultaneously in analog format. FiOS delivers an aggregate of 622 Mbps downstream, with 32 splits, thereby yielding raw bandwidth of over 19 Mbps to each premise. The service offerings are detailed in Table 9.8.

The Verizon service includes up to nine e-mail accounts, Yahoo! or MSN Premium service, and 10 MB of personal Web space. The inside wire to the wired or wireless router must be upgraded to Cat 5 UTP. If a wireless router is used, 802.11b is acceptable for the low-speed service offering but 802.11g is required for the two higher speed offerings. The customer is responsible for providing an electrical outlet close to the location where the ONT is to be mounted in order to provide power to the battery backup unit. The core offering of the English language FiOS Premier TV is available at $39.95 per month, while the Spanish language La Conexion is available for $27.95, assuming that the subscriber also has Verizon voice service. Prices are slightly higher for standalone TV service. Various other packages, add-ons, on-demand content, and Pay-Per-View (PPV) services are additional. As many as 350 TV channels are available in total. Pricing is sensitive to bundling, with the lowest prices offered to those subscribing to the full *triple play* of voice, Internet access, and TV service. While Verizon FiOS is not available everywhere in Verizon territory, the company spent over US$1 billion in 2005 on FiOS-related capital expenditures. At an estimated cost of $1250–$1350 per premises for FTTP, $1 billion can go pretty quickly [61].

AT&T (previously SBC) largely is taking a dual approach in its US$4 billion capital initiative known as *Project Lightspeed*. In most new subdivisions and other greenfield scenarios, the company will employ FTTP. In brownfield scenarios, the company favors a hybrid fiber/copper approach, which it calls *Fiber-to-the-Node* (FTTN), with each node delivering 20–25 Mbps to as many as 300–500 homes. AT&T will use copper to provide ADSL and VDSL services from the node to the customer premises. As the UTP portion of the loop is shortened considerably in an FTTN scenario, many of the associated performance issues are much reduced. BellSouth has taken a relatively passive Fiber-to-the-Curb (FTTC) approach which is quite complementary to the AT&T approach, and that fact quite likely had some impact on the merger discussions between the two companies [62, 63]. FTTN costs are estimated to be in the range of $250 for overlays, which compares favorably with the cost of FTTP.

Elsewhere, EPON is being aggressively deployed in Japan, where it is known as *GigE PON (Gigabit Ethernet PON)*, and is the overwhelming choice throughout Asia. PON spending and penetration levels in Japan currently far exceed those in the United States, while Europe currently lags well behind.

9.1.5   Access Broadband over Power Line

Electricity is the world-power, the most powerful and terrible of nature's hidden forces. Yet, when man has learned how to harness its fiery energies, electricity becomes the most docile and useful of his servants . . . it propels our trains, lights our houses and streets, warms us, cooks for us, and performs a thousand and one other tasks at the turn of a button or at the thrust of a switch.

Joseph H. Adams, *Harper's Electricity Book for Boys*,
Harper & Brothers Publishers, 1907

Broadband over Power Line (BPL) has given fresh spark to an old technology known as *PowerLine Carrier* (PLC), which Bell Telephone Laboratories invented in 1928 as a means of supporting voice and data communications over electric power distribution cabling. Although it proved too unreliable for commercial use at the time, electric utility companies have continued to use PLC for telemetry applications and controlling equipment at remote substations. Telephone companies have used PLC for many years in rural areas to provide *Plain Old Telephone Service* (POTS) to extremely remote customers who had electric service but for whom it would be too costly to provide telephone service over dedicated twisted-pair local loops. BPL is a much advanced broadband version of PLC that represents an alternative to more visible technologies being deployed by the LECs and CATV providers.

As I noted in Chapter 2, PLC uses existing power distribution cabling and inside wire running 120 or 240 V, depending on the electric grid standards in place. In Europe, for example, the standards for the 240-V grid allow for communications at frequencies from 30 to 150 kHz. In the United States, the standards for the 120-V grid allow the use of frequencies above 150 kHz as well. Power utilities use the frequencies below 490 kHz for internal telemetry and equipment control purposes.

The U.S. Federal Communications Commission (FCC) adopted a *Notice of Inquiry* (NOI) on April 23, 2003, expressing enthusiasm in BPL as an alternative broadband access technology. That NOI mentioned two forms of BPL: *access BPL* and *in-house BPL*, with the latter being akin to an Ethernet LAN running over the inside wire and cable system within the premises. I discuss in-house BPL at considerable length in Chapter 8.

9.1.5.1   BPL Elements and Configurations
Access BPL is a form of PLC that uses certain elements of the existing electrical power distribution grid as a broadband local loop. The typical power grid comprises generators, High-Voltage (HV) lines, substations, Medium-Voltage (MV) lines, transformers, and Low-Voltage (LV) lines. The HV transmission lines are unsuitable for BPL as there is too much hum and buzz (i.e., noise) at 155,000–765,000 V, and the power jumps all over the frequency spectrum in a completely unpredictable manner. MV and LV lines are quite usable, however.

Access BPL uses special modems and couplers to transmit over MV lines in the electric power utilities' distribution networks. While the full MV range is defined as 1000–40,000 V, MV distribution lines generally operate at a much more manageable 7200 V, approximately. As illustrated in Figure 9.15, utility substations contain transformers that step the HV power down to the MV level. In a typical access BPL scenario, it is at the utility substation that a fiber-optic network connection termi-

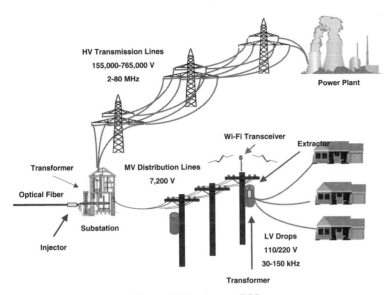

Figure 9.15   Access BPL.

nates in a modem that accomplishes the optoelectric conversion process. The modem connects to an *inductive coupler* that wraps around the MV power line without touching it. The coupler serves as an *injector*, injecting the communications signals onto the distribution lines in a frequency range between 2 and 80 MHz, with the RF carrier sharing the same line with the electrical signals. This is Frequency Division Multiplexing (FDM) of telecommunications and electrical power. From the utility substation to the customer premises, repeaters are spaced every 300 m or so. At the far end, an *extractor* removes the communications signals from the power lines. This typically takes place just ahead of the *transformer*, which typically serves a number of households and steps the voltage down from the MV level of 7200 V to the LV level of 110/220 V used within the premises. The extractor typically bypasses the transformer and couples the BPL signal directly between the MV distribution line and the LV drop to the premises. Alternatively, the extractor connects to a Wi-Fi transceiver, and the final link to the customer premises is via WLAN technology. The purest approach is to use an extractor that amplifies the BPL signal enough that it can travel through the transformer to the LV drop serving the premises, but this technique is problematic. In any case, the extractor not only serves as a demultiplexer for downstream signals from the network to the premises but also as a multiplexer for upstream signals.

The *National Telecommunication and Information Administration* (NTIA) of the U.S. Department of Commerce describes three BPL system types:

- *System 1* injectors and extractors share a common frequency band on the MV power lines for both upstream and downstream communications through the use of a version of *Orthogonal Frequency Division Multiplexing* (OFDM) specially tailored for powerline environments. [*Note:* OFDM also is used in 802.11a (Wi-Fi5) and 802.11g, ADSL, and 802.16 (WiMAX).] The head-end injector couples the fiber to the MV line. Contention for channel access is managed through *Carrier Sense Multiple Access with Collision Avoidance* (CSMA/CA), the same *Medium Access Control* (MAC) technique used in Wi-Fi LANs. Thereby, multiple premises can share the same single-phase MV line. This system may be sufficiently tolerant of interference to permit multiple systems to be installed on adjacent lines in a two- or three-phase MV distribution line.

- *System 2* differs from System 1 only in that it bypasses the LV drop altogether through the use of Wi-Fi technology.
- *System 3* employs *Direct-Sequence Spread Spectrum* (DSSS) in addition to *Carrier Sense Multiple Access with Collision Detection* (CSMA/CD) for data transmission over the MV distribution lines. DSSS is a packet radio technique in which the signal is spread across a wide carrier frequency in a redundant fashion. Thereby, multiple transmissions can share the same wideband, with the transmissions separated from each other through the use of a unique code sequence prepended to each data packet. Multiple systems may be installed on adjacent lines in a two- or three-phase MV distribution line. CSMA/CD is the same MAC used in 802.3 (Ethernet) wired LANs [64].

Access BPL configurations can be either asymmetric or symmetric in nature, with the aggregate amount of bandwidth sensitive to the specific nature of the power grid, for example, one-, two-, or three-phase MV distribution lines. As the BPL local loop is shared, the aggregate bandwidth is shared, and the bandwidth available to each subscriber is highly dependent on system loading and the level of activity at any given moment, much as is the case with CATV cable modem networks and WiMAX networks. Note that as both access BPL and in-house BPL operate over power lines, they cease to work when the lights go out, and there is no point in battery backup. It remains important to have an Uninterruptible Power Supply (UPS) with battery backup to allow graceful shutdown of computer equipment, of course.

Keeping in mind that access BPL is an emerging technology, yet to be standardized, early tests and trials in the United States have been quite positive, and there are a small number of commercial service offerings. In some cases, access speeds are in the range of 200–300 kbps, fully symmetric. Current Communications, which operates in the area of Cincinnati, Ohio, offers fully symmetric access speeds of 1 and 3 Mbps.

9.1.5.2 *Standards, Rules, and Regulations*   The IEEE currently is developing P1675, "Standard for Broadband over Power Line Hardware," which is intended to provide electric utilities with a comprehensive standard for the injectors and extractors used on the MV distribution lines. The standard was targeted for completion in 2006 but that year came and went.

The FCC has shown considerable interest in access BPL, in particular, and most of that interest is in encouraging the deployment of broadband in sparsely populated rural areas where DSL, PON, and cable modem options are unlikely to exist and where the business case cannot be made for WiMAX or other Wireless Local Loop (WLL) technologies. The FCC also is concerned with potential interference issues. As HV and MV power transmission lines largely are unshielded and aerial, they emit considerable electromagnetic fields that potentially can interfere with short-wave and other radio signals. Therefore, the FCC has established certain excluded frequency bands to avoid interference with amateur radio (i.e., ham radio) and aircraft radio. FCC rules also establish exclusion zones in proximity to sensitive operations such as Coast Guard, Navy, and radio astronomy stations. The In-house BPL modems that plug into the wall must comply with FCC Part 15 rules, of course, just as do cordless phones, garage door openers, and Wi-Fi components.

With the exception of the interstate power grid, power utilities tend to be regulated at the state and local levels, at least in the United States. Many power utilities are not only regulated by but are owned by municipalities, which creates a considerable conflict of interest. In recognition of this conflict and the inherent unfairness of a municipality's competing with a private sector telecommunications company, at least 14 states have passed laws preventing municipally owned utilities from offering telecom services, and others are considering similar measures [65–68].

9.2 SONET/SDH

Synchronous Optical Network (SONET) is a set of North American standards for broadband communications over SingleMode Fiber (SMF) optic transmission systems, thereby enabling manufacturers to build equipment that supports full interconnectivity and interoperability. *Synchronous Digital Hierarchy* (SDH) is the internationalized version of SONET, as specified by the CCITT (now ITU-T). As SONET and SDH differ primarily with respect to low-level line rates and some terminology, I lump them together. SONET/SDH uses a transfer mode that defines switching and multiplexing aspects of a transmission protocol, supporting both asynchronous and synchronous traffic in any form on bit-transparent TDM channels. Intended primarily for the carrier networks, SONET/SDH also can be deployed to the user premises, although such implementations are reserved for sites where there are significant bandwidth requirements. The *Network-to-Network Interface* (NNI), also known as *Network Node Interface*, specification allows the blending of national and regional networks into a cohesive global network. The *User Network Interface* (UNI) provides a standard basis for connection from the user premises to SONET/SDH.

SONET/SDH describes the characteristics of a fiber-optic Physical Layer (Layer 1) infrastructure, rather than a set of services. A number of broadband services, however, depend on the bandwidth, error performance, flexibility, and scalability that can be provided best over a SONET infrastructure. Examples of such services certainly include Frame Relay and ATM, which I discuss in Chapter 10. Additionally, T-carrier, DDS, ISDN, X.25, and DSL network traffic benefits from the performance characteristics of the SONET infrastructure. For that matter, even voice traffic gains advantage in terms of improved performance and lower cost of transport.

SONET grew out of the *SYNTRAN* (SYNchronous TRANsmission) standard developed at Bellcore. Designed for operation at rates up to 45 Mbps (T3), SYNTRAN called for multiplexing all signals on the basis of a single master clocking source. Thereby, stuff bits could be eliminated, reducing overhead. Further, DS-0s and DS-1s could be added to (i.e., multiplexed directly into) and dropped from (i.e., demultiplexed directly from), a DS-3 frame, thereby eliminating the intermediate DS-2 level. This synchronous add/drop multiplexing concept formed the basis for SONET, which was proposed as a solution for higher data rate applications. At the time, each manufacturer's products were designed according to proprietary specifications, so systems of disparate origin were incompatible—they simply did not interconnect, much less interoperate. Therefore, each fiber-optic link (e.g., from CO to CO) absolutely was required to have equipment of the same origin (i.e., manufacturer) at both ends. This limitation effectively forced the carriers to select a single equipment vendor, thereby limiting the ability of other manufacturers to compete and stifling technical creativity. Additionally, the interconnecting carriers limited fiber-optic systems either to a single vendor or to optoelectric interfaces that limited the capacity to far less than that actually supported by the individual systems. (The highest common denominator always rules.) Therefore, the economic and technical benefits of a multivendor market were limited and largely unrealized.

Initial SONET standardization efforts began in 1984, when MCI (now part of Verizon) proposed the development of connectivity standards, or *midspan fiber meets*, to the *Interexchange Carrier Compatibility Forum* (ICCF). The ICCF then requested that the *Exchange Carriers Standards Association* (ECSA) develop those standards [69]. In 1985, Bellcore proposed SONET to the ANSI T1X1 committee, and the process continued until final approval for a much-revised SONET was gained in June 1988. The resulting ANSI specification for SONET Phase 1 was released in T1.105-1988 and T1.106-1988, specifying a basic transmission level of 51.84 Mbps, which carries a signal originating as the electrically based T3 of 45 Mbps, plus additional overhead for optical processing and network management.

The CCITT (now ITU-T) began the initial efforts to internationalize SONET as SDH in 1986. In 1988, the CCITT accepted the SONET standards, with modifications that were mostly at the lower multiplexing levels. These differences largely are due to the requirement to accommodate the complexities of internetworking the disparate national and regional networks. The ITU-T Recommendations referenced are G.707, G.708, and G.709.

Standards work continues on SONET/SDH, with the involvement of standards bodies and associations including ANSI, EIA, ECSA, IEEE, ITU-T, and Telcordia Technologies:

- **American National Standards Institute (ANSI):** Founded in 1918, ANSI coordinates and harmonizes private sector standards development in the United States. ANSI also serves as the U.S. representative to the International Organization for Standardization (ISO), the originator of the Open Systems Interconnection (OSI) Reference Model.

- **Electronic Industries Alliance (EIA; previously Electronic Industries Association):** Founded in 1924 as the Radio Manufacturers Association, the EIA is a trade organization representing the interests of U.S. electronics manufacturers. The EIA assists in the development of physical layer interfaces, including optoelectric interfaces and test procedures for SONET. The EIA is best known for its physical layer specification EIA-232, nee RS-232.

- **Exchange Carriers Standards Association (ECSA):** Formed in 1984, ECSA represents the interests of the U.S. IntereXchange Carriers (IXCs). The ECSA T1 committee addresses issues of functionality and characteristics of interconnection and interoperability. The T1X1 committee addresses issues of digital hierarchy and synchronization.

- **Institute of Electrical and Electronics Engineers (IEEE):** The IEEE is a worldwide professional association dealing with SONET only peripherally. The IEEE has significant responsibility for the development of LAN and MAN standards.

- **International Organization for Standardization (ISO):** An organization comprising the national standards organizations (e.g., ANSI) of the various nations and regions, ISO heavily influences international standards set by the ITU-T. In the context of telecommunications, ISO is perhaps best known for its involvement in the Open Systems Interconnection (OSI) Reference Model.

- **International Telecommunication Union—Radiocommunications Sector (ITU-R):** The ITU-R is the sector that deals with issues of radio standards. The ITU-R parallels the ITU-T, which is responsible for the wireline domain.

- **International Telecommunication Union-Telecommunications Standardization Sector (ITU-T):** Previously the CCITT, the ITU-T is an agency of the United Nations. The ITU-T develops international standards in order to promote a world order of interconnectivity and interoperability in the wireline domain. ITU-T sometimes also is abbreviated as ITU-TS or ITU-TSS. The ITU-T predecessor organizations date to 1865.

- **Telcordia Technologies:** Previously *Bellcore (Bell Communications Research)*, Telcordia was formed in 1984 under the terms of the Modified Final Judgement (MFJ), which forced AT&T to divest the Bell Operating Companies (BOCs). Bellcore was the research and development arm of the RBOCs, its client/owners. Bellcore originally focused on standards development, test procedures, and Operations Support System (OSS) development, rather than the physical sciences. Bellcore was privatized and acquired by SAIC in 1998, as the interests of the RBOCs were no longer common in a deregulated, competitive environment. The name was changed to Telcordia Technologies in April 1999, with the stated focus of emerging technologies. Telcordia is now a private, stand-alone organization involved in the development of Operations Support Systems (OSSs) and network management software as well as consulting, testing services, and research services.

9.2.1 SONET Standards Development

SONET standards were developed in three phases. Phase I (1988) defines transmission rates and characteristics, signal formats, and optical interfaces. Phase I also defines *Optical Carrier* (OC) levels and *Data Communications Channels* (DCCs) used for network management purposes in support of midspan meet at the payload level. While Phase I does not support network management from end to end, neither does it preclude that potential. Phase II refines the physical portion of the standards and defines protocols used on data communications channels DS-1 to DS-3. Phase II also defines interoperability parameters for midspan meet, network management, OSI *Common Management Information Service Elements* (CMISEs), and *Add/Drop Multiplexer* (ADM) capabilities. Phase II further defines *Operations, Administration, Management, and Provisioning* (OAM&P) procedures and connectivity to B-ISDN. Phase III provides all network management requirements for midspan meet. Phase III also defines all network management standard message sets (e.g., alarm state, circuit-pack failure, and intermittent failure) and addressing schemes for interconnection. Finally, Phase III provides for ring and nested protection switching standards [69, 70].

9.2.2 SONET/SDH Transmission Hierarchy

SONET defines the *Synchronous Transport Signal level N* (STS-*N*) as the electrical signal, which remains necessary until all-optical switching is developed. When converted to an optical signal for transport over a standard fiber-optic medium, the term *Optical Carrier N* (OC-*N*) is applied. As noted in Table 9.9, the basic building block of the digital hierarchy is OC-1 at 51.84 Mbps and currently tops out at the OC-768 rate of 40 Gbps. Notably, the various STS levels are considerate of the existing digital signal hierarchy, thereby achieving backward compatibility with legacy systems. In other words and by way of example, a T3 frame maps comfortably into an STS-1 signal that becomes an OC-1 frame. Similarly, multiple T1 frames can be aggregated to map into and form an STS-1 signal, which then becomes an OC-1 frame.

At OC-1, for example, a T3 bit stream of 44.736 Mbps is padded and presented as an STS-1 signal of 51.84 Mbps, with an actual maximum payload rate of 49.54 Mbps [72]. When converted from an electrical signal to a fiber-optic photonic signal, the bit stream is known as OC-1. The OC-1 comprises 810-byte frames transmitted at a rate of 8000 frames per second, or every 125 μs, based on the requirement that a PCM-encoded voice byte must be transmitted at that rate. SONET levels fully define the range from OC-1 (51.84 Mbps) to OC-768 (39.813 Gbps), which is composed of 768 OC-1 frames. At OC-768, the entire string of 768 OC-1 frames must

TABLE 9.9   SONET/SDH Signal Hierarchy

Optical Carrier Level[a]	SONET STS Level	SDH STM Level	Signaling Rate	Equivalent DS-3 (45-Mbps) Channels	Equivalent DS-0 (64-kbps) Channels
OC-1	STS-1	—	51.84 Mbps	1	672
OC-2	STS-2	—	103.68 Mbps	2	1,344
OC-3	STS-3	STM-1	155.52 Mbps	3	2,016
OC-4	STS-4	STM-3	207.36 Mbps	4	2,688
OC-9	STS-9	STM-3	466.56 Mbps	9	6,048
OC-12	STS-12	STM-4	622.08 Mbps	12	8,064
OC-18	STS-18	STM-6	933.12 Mbps	18	12,096
OC-24	STS-24	STM-8	1.24416 Gbps	24	16,128
OC-36	STS-36	STM-12	1.86624 Gbps	36	24,192
OC-48	STS-48	STM-16	2.48832 Gbps	48	32,256
OC-96	STS-96	STM-32	4.976 Gbps	96	64,512
OC-192	STS-192	STM-64	9.953 Gbps	192	129,024
OC-768	STS-768	STM-256	39.813 Gbps	768	516,096
OC-1536[b]	STS-1536	STM-512	79.626 Gbps	1,536	1,032,192
OC-3072[b]	STS-3072	STM-1024	159.252 Gbps	3,072	2,064,384

[a] OC-2, OC-9, OC-18, OC-36, and OC-96 are considered to be orphaned rates. OC-3 was defined by the CCITT as the basic transport rate for Broadband ISDN (B-ISDN) [71].
[b] This level is not fully defined.

be transmitted every 125 µs based on that same PCM requirement. It merits reinforcing the point that SONET is based on T-carrier, which is optimized for real-time, uncompressed voice. Fractional speeds are achievable at virtually any level, as subrate transmissions below OC-1 are multiplexed to form an OC-1 channel.

SONET terms of significance include the following:

- *Optical Carrier* (OC) is the definition of the SONET optical signal. The fully defined OC levels begin at OC-1 (51.84 Mbps) and culminate in OC-768 (39.813 Gbps). All SONET/SDH levels are expressed in multiples of OC-1.
- *Synchronous Transport Signal* (STS) is the electrical equivalent of the SONET optical signal; it is known as *Synchronous Transport Module* (STM) in SDH. The signal begins in electrical format and converts to optical format for transmission over the SONET optical fiber facilities. Each STS-1 frame is transmitted in 125 µs, yielding raw bandwidth of 51.84 Mbps. The STS frame includes five elements:
  - *Synchronous Payload Envelope* (SPE) carries the user payload data. It is analogous to the payload envelope of an X.25 packet. The SPE consists of 783 octets (87 columns and 9 rows of data octets).
  - *Transport Overhead* (TO) consists of Section Overhead and Line Overhead:
    - *Section OverHead* (SOH) of nine octets is dedicated to the transport of status, messages, and alarm indications for the maintenance of SONET links between ADMs.
    - *Line OverHead* (LOH) of 18 bytes controls the reliable transport of payload data between any two network elements.
  - *Path Overhead* (PO), contained within the SPE, comprises nine octets for the relay of OAM&P information in support of end-to-end network management.
  - *Payload* is the actual data content of the SONET frame and rides within the SPE. Total usable payload at the OC-1 level consists of up to 49.54 Mbps, into which a T3 frame fits quite nicely. The balance of the 51.84 Mbps is consumed by Transport Overhead and Path Overhead.

Multiplexing is on the basis of direct TDM. Either full SONET speeds or lesser asynchronous and synchronous data streams can be multiplexed into the STS-*N* payload, which then converts into an OC-*N* payload. In other words, an appropriate combination of DS-0, DS-1, DS-2, and DS-3 signals can be multiplexed directly into an electrical STS-1 payload, which then converts to an optical OC-1 payload. The multiplexing process involves byte interleaving, much as described for traditional TDMs.

9.2.3 SONET/SDH Topology

SONET networks are highly redundant. Although they can be laid out in a linear, star, ring, or hybrid fashion, the optimum physical topology is that of a dual counter-rotating ring in which one fiber transmits in one direction and the other transmits in the other direction. Such a layout makes it highly unlikely that any device on the network can be isolated through a catastrophic failure, such as a *cable-seeking backhoe*. Should such a failure occur, the *Automatic Protection Switching* (APS) feature of SONET permits the *self-healing* network to recover. There are two primary implementations of the SONET physical topology: path-switched and line-switched rings:

- *Path-switched rings* employ two active fibers. All traffic moves in both directions, thereby providing protection from network failure. A path-switched approach also improves error performance because the receiving stations examine both data streams, selecting the better signal. Should the primary ring fail, the secondary ring is already active in the reverse direction.
- *Line-switched rings* involve either two or four fibers. The single-ring implementation comprises two fibers, one of which is active and the other of which is held in reserve. Traffic moves in one direction across the active fiber. In the event of a network failure, the backup ring is activated to enable transmission in the reverse direction as well. A four-fiber implementation, which is preferred and typical in carrier-class networks, supports transmission in one direction over one fiber in each of the two rings. By way of example, fibers 1 in rings 1 and 2 might transmit in a clockwise direction, while fibers 2 in rings 1 and 2 transmit in a counterclockwise direction. The second fiber in each ring acts to support transmission in the reverse direction in the event of a failure in the primary ring. Line-switched rings smaller than 1200 km in route distance offer standard restoral intervals of 50 ms or less [73]. Larger rings, such as long-haul transport rings in IXC networks, involve longer restoral intervals due to the increased time of signal propagation.

As illustrated in Figure 9.16, a typical SONET network involves a number of rings operating at multiple speeds. The backbone, for example, might run at the OC-192 nominal rate of 10 Gbps, with subtended metro rings running at the OC-48 rate of 2.5 Gbps and further subtended local rings running at the OC-3 rate of 155 Mbps.

9.2.4 Paths, Tributaries, and Containers

SONET is capable of carrying information at Gbps speeds with excellent performance characteristics, including error performance and network management. The pipe also can carry any variety of asynchronous and synchronous information (e.g., voice, data, video, and image) and present it in a number of frame sizes. While the STS-1 frame is the basic building block, multiple STS-1 frames can be linked together in a process known as *concatenation*. Concatenated STS-*N*c (*N* = Number; c = concatenated) signals are multiplexed, switched, and transported over the network as a single entity. This approach, which currently is defined for STS-3c and STS-12c, offers clear advantages where larger increments of bandwidth are required, because the overall level of overhead is reduced, thereby increasing the payload size. Example applications for concatenation include imaging and HDTV (High Definition TV). The SONET pipe consists of a path, virtual Tributaries, and tributary units:

- **Path:** A *path* is an end-to-end communications path (i.e., route or circuit) that carries traffic from one end point to another.

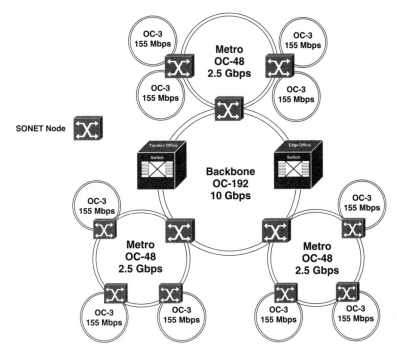

Figure 9.16 Line-switched SONET configuration illustrating an OC-192 backbone ring with multiple subtended OC-48 metro rings with multiple subtended OC-3 local rings.

- **Virtual Tributary and Virtual Container:** A *Virtual Tributary* (VT) carries one form of signal, such as a DS-1, DS-2, or DS-3 signal within a byte-interleaved frame. SONET can map as many as seven *Virtual Tributary Groups* (VTGs) into a single Virtual Path, and as many as four VTs into a VTG, as illustrated in Figure 9.17. (*Note:* The math matches T-carrier, which maps 28 T1s into a T3. Recall that SONET is based on T3.) A VT is a bit-transparent TDM connection that may be channelized (e.g., a 24-channel T1 for voice) or unchannelized (e.g., a clear channel T1 for full-motion video). VTs are sized to accommodate the originating signal and in consideration of the legacy digital hierarchy. VT1.5, for example, operates at 1.544 Mbps (T1), VT2 at 2.048 Mbps (E-1), VT3 at 3.152 Mbps (T1c), and VT6 at 6.312 Mbps (T2). Individual VTs are distinguished by the use of a *pointer*, which identifies the position of the VT within a VTG, which might comprise a group of VT1.5s, within the STS frame. The pointer also provides synchronization in a SONET environment. The SDH equivalent of a VT is a *Virtual Container* (VC).
- **Tributary Unit:** A *Tributary Unit* (TU) is a VT along with a pointer that enables switching and cross-connecting.

9.2.5 SONET Frame Format

The STS-1 frame (see Figure 9.17) is the basic building block for SONET, much like the respective DS-1 frame formats in T-carrier and E-carrier environments. The

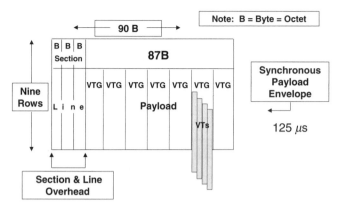

Figure 9.17   SONET frame structure.

STS-1 frame can be considered logically as a matrix of 9 rows of 90 octets, yielding 810 octets in total. The data are transmitted from top to bottom, one row at a time and from left to right. SONET accommodates payloads in increments of 765 octets, logically organized in matrixes of 9 rows by 85 columns. The payload is contained within a *Synchronous Payload Envelope* (SPE) in increments of 774 bytes (9 rows by 86 columns), with the additional column attributable to *Path OverHead* (POH). Where superrate services require more than a single STS-1, they are mapped into a higher level, concatenated STS-*N*c, with the constituent STS-1s kept together [74]. For example, a 135-Mbps B-ISDN H4 frame requires a huge amount of contiguous, unbroken bandwidth. SONET accommodates this requirement by linking three STS-1s into an STS-3c [75].

The SPE, which contains the payload data, actually *floats* within the SONET frame. While SONET is a synchronized transmission system, with all devices relying on a common clocking signal, variations in clocking can occur. Those clocking variations can result from differences in local clocks, thermal expansion and contraction in individual fiber-optic cables, and other phenomena. Rather than buffering individual frames to effect synchronization, floating mode operation enables the network to adjust to frame float with the SPE identified by the pointer. The floating mode reduces cost and delay that would be caused by the use of buffers to synchronize each frame and SPE exactly [75].

The SONET overhead structure mirrors that of the existing digital carrier network for purposes of nonintrusive, end-to-end network management. Overhead layers include transport overhead (TO), which is further divided into section overhead, line overhead, and path overhead.

9.2.5.1 Section Overhead *Section Overhead* (SOH) of 9 bytes provides for management of optical network segments between *Section-Terminating Equipment* (STE), which can be repeaters, Add/Drop Multiplexers, or anything else that attaches to either end of a fiber link. The repeaters can be stand alone or built into switches, such as *Digital Cross-Connect Systems* (DCCSs or DXCs). At the section layer, every repeater in the network performs the SOH functions. These include framing, span performance and error monitoring, and STS ID numbering. These functions resemble those performed by traditional point-to-point protocols such as SDLC and LAP-D. The 9 bytes of SOH include

- 1 byte STS-1 signal ID
- 2 bytes *Bit Interleaved Parity* (BIP) for error monitoring
- 1 byte *Orderwire* (connection request)
- 3 bytes *Data Communication Channel* (DCC)

9.2.5.2   Line Overhead   *Line Overhead* (LOH) of 18 bytes controls the reliable transport of payload data between major network elements. A Digital Cross-Connect System (DXC) performs Line Layer functions, including error control, switching and multiplexing, orderwire, express orderwire (priority connection request), automatic protection switching to back-up circuits, insertion of payload pointers, and synchronization control. The 18 bytes of LOH comprise

- 3 bytes STS-1 pointer
- 1 byte *Bit Interleaved Parity* (BIP) for error monitoring
- 2 bytes *Automatic Protection Switching* (APS)
- 9 bytes *Data Communication Channel* (DCC)
- 1 byte *Orderwire*
- 2 bytes reserved for future use

9.2.5.3   Path Overhead   *Path Overhead* (POH) of 9 bytes comprises all aspects of end-to-end performance monitoring and statistical reporting. Path management is an essential responsibility of the Add/Drop Multiplexers (ADMs). Functions performed at the Path Layer include end-to-end performance monitoring, statistical reporting, STS mapping, and DS-to-OC mapping. The 9 bytes of POH comprise

- 1 byte trace
- 1 byte *Bit Interleaved Parity* (BIP) for error monitoring
- 1 byte Payload ID
- 1 byte maintenance status
- 1 byte user ID
- 1 byte frame alignment
- 3 bytes reserved for future use

9.2.6   SONET/SDH Hardware

The hardware aspects of SONET are difficult to describe in discrete terms because many of the functional elements overlap. Just as manufacturers of traditional voice and data equipment often build multiple functional elements under the skin of a single box, so do SONET/SDH equipment manufacturers. Given that caveat, the following is a description of the discrete functional elements of hardware devices:

- *Terminal multiplexers* are *Path-Terminating Equipment* (PTE) that provide user access to the SONET network (see Figure 9.18), operating in a manner similar

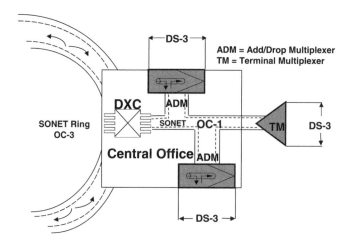

Figure 9.18   DXC and ADM in SONET application.

to a T3/E-3 time division multiplexer. Multiple DS-0s, for example, can be multiplexed to form a VT1.5, several of which would then form a VTG and an STS-1 frame. Terminal multiplexers also accomplish the conversion from electrical STS-*N* signals into Optical Carrier (OC-*N*) signals.

- *Concentrators* perform the equivalent functions as traditional electrical concentrators and hubs. SONET concentrators combine multiple OC-3 and OC-12 interfaces into higher OC-*N* levels of transmission.

- *Add/Drop Multiplexers* do not have exact equivalents in the electrical (Digital Signal, or DS) world, although they perform roughly the same functions as traditional T-carrier TDMs. Generally found in the Central Office Exchange (COE), they provide the capability to insert or drop individual DS-1, DS-2, or DS-3 channels into a SONET transmission pipe. ADMs accomplish the process electrically, with the OC-*N* channel being converted prior to the process and reconverted subsequently. ADMs offer great advantage over traditional DS-*N* MUXs. The T-carrier approach, for example, requires that a DS-3 frame be demultiplexed into its DS-2 and then DS-1 frames, which must be broken down into 24 DS-0 channels in order to extract and route an individual channel. Once that is accomplished, the process must be reversed to reconstitute the DS-3, minus the extracted DS-0, and send it on its way. ADMs perform the additional functions of dynamic bandwidth allocation, providing operation and protection channels, optical hubbing, and ring protection.

- *Digital Cross-Connects* perform approximately the same functions as their electrical equivalents (DACs/DCCSs), providing switching and circuit grooming down to the DS-1 level. They provide a means of cross-connecting SONET/ SDH channels through a software-driven, electronic common control cross-connect panel with a PC user interface. The routing of traffic through a DXC is accomplished through the use of *payload pointers*, which point to the payload in the OC-*N* frame and provide synchronization. Importantly, DXCs also serve to connect the fiber rings, which might take the form of a backbone ring and multiple subtended rings, as illustrated in Figures 9.15 and 9.18. DXCs perform the additional functions of monitoring and testing, network provisioning, maintenance, and network restoral.

- *Regenerators* perform the same function as their traditional electrical equivalents. Often found under the skin of other SONET equipment, they are optoelectric devices that adjust the amplitude, timing, and shape of the signal.

9.2.7 SONET Advantages and Disadvantages

SONET offers a number of advantages, in addition to the inherent advantages of fiber-optic transmission systems, in general. Certainly, the fact that SONET is highly standardized offers the benefits of interconnectivity and interoperability between equipment of different manufacturers. That standardization translates into freedom of vendor choice and yields lower costs through competition. Additionally, SONET/SDH is extendable to the premises on a fully interoperable basis. The increasing availability of SONET local loops provides end-to-end advantages of enhanced bandwidth, error performance, dynamic bandwidth allocation, and network management.

In bandwidth-intensive applications, the high absolute cost of SONET can be offset by virtue of its extraordinarily high capacity. Whether deployed in a carrier network or extended to the user premises, SONET supports the aggregation of all forms of traffic, including voice, data, video, image, facsimile, and multimedia. As a result, a SONET infrastructure can obviate the need for multiple transmission facilities in support of individual services. The simplicity of multiplexing and demultiplexing via ADMs serves to reduce costs, delay, and error. Clearly, SONET/SDH offers the considerable advantage of network resiliency through its inherent redundancy and self-healing capabilities.

Finally, SONET offers tremendous security, as with fiber-optic transmission systems, in general. Fiber is difficult, if not impossible, to physically tap without detection. Perhaps just as important is the fact that it is difficult to identify the one channel for detection out of the thousands of information channels supported in a SONET mode. As Francis Bacon said in 1625, "There is no secrecy comparable to celerity" (*Of Delay*, Essays)—celerity, or speed, is a hallmark of SONET.

SONET also has its disadvantages. As is true of many things, its greatest strengths are also its greatest weaknesses. SONET's redundancy, resiliency, and strong network management all come at a price—hence, SONET's legendarily high cost. Further, SONET's network management strengths translate into overhead intensity, which means that a good deal of raw bandwidth is consumed for management, rather than payload, purposes. SONET's redundancy also means that one-half to three-quarters of the available bandwidth is wasted at any given moment, assuming that the primary ring is operating properly. Finally, and perhaps most importantly, much of SONET's network management overhead and its careful timing mechanisms are considered to be wasted on packet data traffic, which neither expects nor appreciates its elegance.

Generally speaking, incumbent carriers from the heavily regulated voice world of the PSTN tend to place great value in SONET's attributes, which they view as strengths. The insurgent CLECs with a data orientation tend to view those same attributes as weaknesses. Therefore, stripped down, nonstandard versions of SONET have been developed. These versions run at SONET speeds and preserve the framing format. But they do not make full use of the redundancy and network management features provided in standards-based SONET ring architectures.

9.2.8 SONET Applications

SONET primarily is deployed in backbone carrier networks, where its many attributes can be put to full use. Particularly in a convergence scenario, the carriers have the potential to realize considerable cost savings by using a single-fiber infrastructure in support of bandwidth-intensive video and image streams, in addition to voice, facsimile, and voice-grade data traffic. The value of SONET as a backbone technology is ever more evident with the increasing penetration of broadband local loop technologies such as ADSL, cable modems, PON, and WiMAX. In a hybrid network configuration such as FTTN or FTTC, the fiber portion of the loop from the CO to the DLC may be SONET in nature.

It is interesting to track the developments in SONET speeds over the past decade or so using a few select examples. In 1994, Sprint announced the completion of the Silicon Valley Test Track, linking seven companies and several learning institutions to test ATM switching technologies. Plans called for the initial OC-1 ring to be upgraded to OC-3 in early 1995 and to OC-12 later in the year [76]. Shortly thereafter, Sprint deployed the first international SONET ring, linking New York City with three cities in Canada. The capacity of that OC-48 (2.5 Gbps) SONET ring increased to 10 Gbps through the use of Wave Division Multiplexing (WDM) equipment in order to derive four OC-48 channels [77]. In May 1996 AT&T announced plans to beef up its network with WDM equipment to realize 20 Gbps over OC-48 SONET fiber—an eightfold increase [78].

An interesting application of SONET technology is that of the 1996 Summer Olympics in Atlanta, Georgia. BellSouth, the incumbent LEC, deployed 40 OC-48 SONET rings to transport CD-quality voice, broadcast-quality video, and data simultaneously to nine cities in the Greater Atlanta area where the 26 various competitions took place. All information flowed through the SONET pipes, including competition results, accreditation, data, voice, and video. The uncompressed digital video signals traveled to the International Broadcast Center, where the signals were sold to other broadcasters to produce 40 worldwide video satellite feeds for approximately 3000 hours of Olympic coverage [79].

A wide variety of carriers currently deploy fresh SONET networks, and upgrade existing networks, to operate at OC-192 speeds of 10 Gbps. Certain portions of those networks operate at OC-768 speeds of 40 Gbps. These carriers include not only the incumbent voice and data IXCs but also the ILECs and CLECs. Most impressive, perhaps, is the rate at which some of the next-generation carriers deploy SONET. Literally from the ground up, where justified by anticipated traffic, these carriers commonly deploy SONET at speeds up to OC-192 in their long-haul networks. Their overall costs range up to $10 billion and more, in anticipation of huge volumes of voice, data, and video traffic, using various combinations of circuit switching, Frame Relay, ATM, and IP. XO Communications (previously Nextlink), for example, laid optical pipe along its rights-of-way at a rate of as many as 688 fibers per route. Each fiber runs at a speed of as much as 10 Gbps (OC-192) and currently runs WDM (Wavelength Division Multiplexing) through four windows or more, with each wavelength supporting a 10-Gbps data stream. While this currently is a *dim fiber* network (i.e., a network in which some fibers are *lit* only partially, with not all wavelengths activated, and some are left *dark* for future use), the potential is incredible. If all fibers were lit at OC-192 and 32 optical wavelengths were lit, that portion of the network would run at an aggregate rate of approximately 220,160 Gbps, or 220.160 Tbps. At that rate, the cost of billing a customer for the transport of a 1-MB file might well exceed the cost of actually providing the service. The ultimate issue, of course, involves getting the bandwidth where it is needed. That is where access technologies are just as important as transport technologies and where fiber in the local loop becomes very important.

End-user organizations can gain the advantage of SONET local loops with increasing ease through a number of service providers. Particularly where data traffic requirements are significant, SONET offers the advantage of direct and seamless high-speed access to the SONET backbones. For high-speed T3 (45 Mbps) access to Frame Relay networks, SONET local loops have become increasingly advantageous. For ATM at speeds of 155 Mbps and 622 Mbps, SONET access is virtually a requirement, as T1 and even T3 do not address these levels of bandwidth.

SONET also finds application in campus environments, including not only institutions of higher learning but also business campuses. In such an environment, significant volumes of data and image information often are transmitted between buildings, perhaps between mainframes, Gigabit Ethernet switches, or routers. In such applications, optical fiber makes sense (in fact, it may be the only viable solution), and SONET often makes even more sense. Mission-critical environments, such as airports, find the redundancy and resulting resiliency of a SONET infrastructure particularly attractive.

9.2.9 Wavelength Division Multiplexing

Something of a debate continues to swirl around the subject of SONET versus *Wavelength Division Multiplexing* (WDM) and, particularly, *Dense Wavelength Division Multiplexing* (DWDM). The debate centers on issues of cost and complexity; SONET is relatively expensive, overhead intensive, and complex.

As I discussed in Chapter 2, WDM enables multiple wavelengths to be introduced into a single fiber through a process that, simply stated, is Frequency Division Multiplexing (FDM) at the optical level. In order to accomplish this feat, multiple diode lasers transmit optical signals through a *window*, or range of wavelengths, with *wavelength*, or *lambda* (λ), being the inverse of *frequency*. Now, each signal is transmitted at a given nominal wavelength, which would be seen as a visible color of light if our eyes were that finely tuned, over a SingleMode Fiber (SMF). As the system adds a wavelength in a window, the bandwidth of the system is increased as though a new fiber were added to the mix. A system operating at 10 Gbps, for example, enjoys a bandwidth increase of 10 Gbps as each wavelength is added. DWDM is defined in the ITU-T Recommendations G.692 and G.959.1 as supporting eight or more wavelengths in the 1530–1565-nm C-band (Conventional band) and the 1565–1625-nm L-band (Long band). DWDM currently supports channel spacings of 200 GHz (1.6 nm at 1550 nm) and 100 GHz, with spacings of 50 GHz and even 25 GHz expected in the future. As these tight channel spacings require expensive cooled laser sources to prevent wavelength drift, the ITU-T subsequently defined *Coarse Wavelength Division Multiplexing* (CWDM), which specifies 18 wavelengths in the range 1270–1610 nm, with spacing of 20 nm (2500 GHz at 1550 nm). Targeted at networks with a reach of 50 km or less, CWDM offers the advantage of using uncooled laser sources and filters, which are not only less expensive but also consume less power and possess smaller footprints. Although CWDM does not allow channels to be packed as tightly as DWDM, it offers a cost-effective alternative for short-haul metropolitan and local rings supporting applications such as GbE and 10GbE [80, 81].

There also exist a number of nonstandard versions of DWDM. In March 2001, for example, NEC announced a DWDM system capable of supporting 273 lambdas, each modulated at OC-768 rates of 40 Gbps, yielding an aggregate 10.9 Tbps over a single physical fiber [82]. As is the case with FDM in general, the optical signals do not interfere with each, as they are separated by frequency. In terms of bandwidth, therefore, WDM and DWDM offer significant cost advantages over SONET, which requires that the laser diode light sources and the Avalanche PhotoDiode (APD) light detectors be upgraded to run at higher speeds. Currently, 40 Gbps (OC-768) is the maximum speed fully defined for SONET/SDH. At that speed, and certainly at higher speeds, *Polarization Mode Dispersion* (PMD) becomes a limiting factor, so dispersion compensators must be installed. While there are technical limits to the speed at which the laser diodes and APDs can operate and there also are finite speed limitations to optical transmission, at least based on our current understanding of the laws of physics, several higher levels are in the process being defined by the ITU-T. OC-1536 is intended to run at a nominal speed of 80 Gbps and OC-3072 at 160 Gbps. Some suggest that those levels will not be cost effective in comparison to DWDM.

The nature of the repeating devices in the optical network also yields cost advantages for WDM. *Erbium-Doped Fiber Amplifiers* (EDFAs), or *light pumps*, which are discussed at some length in Chapter 2, are used to maximum effect in conjunction with WDM systems because they simultaneously can amplify multiple wavelengths in the operating range of 1550 nm. SONET, on the other hand, specifies optical repeaters, which essentially are back-to-back optoelectric conversion devices tuned to a specific optical wavelength. At the inbound port, the repeater accepts the optical signal, converts it to an electrical signal, boosts and filters the signal, converts it back to an optical signal, and sends it over the outbound port. EDFAs work quite effectively and at lower cost than optical repeaters. EDFAs can be spaced at intervals of 80–120 km, and an optical signal can travel through as many as 10 cascading EDFAs over a maximum distance of as much as 800 km before requiring regeneration by an Optical–Electrical–Optical (OEO) repeater. Optical repeaters, if used as the sole means of signal boosting, often are spaced at intervals of 50–100 km, although current technology allows spacing of as much as 600 km. The exact spacing of the repeaters or amplifiers is sensitive to a number of design factors, as one might expect. A key advantage of an EDFA is its ability to simultaneously amplify multiple wavelengths in both directions. A SONET OEO-based repeater can act only on a single wavelength in a single direction.

Raman amplification requires no fiber doping and usually is accomplished throughout the length of the transmission fiber itself in a configuration known as *distributed amplification*, rather than in a special length of fiber contained within an

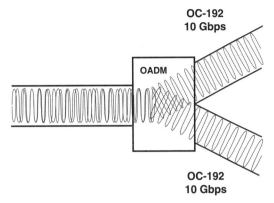

OC-192
10 Gbps

OADM

OC-192
10 Gbps

Figure 9.19   Optical ADM, switching optical signals at the lambda level.

amplification device in a configuration known as *discrete amplification*, or *lumped amplification*. Raman amplification occurs as a high-energy *pump wavelength* is sent in the reverse direction from the output end of the fiber span, where the incoming signal is weakest. The pump wavelength, which generally is in the 1450-nm range, interacts with atoms in the crystalline lattice of the fiber core. The atoms absorb the photons and, when stimulated by the counterflowing signal, quickly release photons with energy equal to the signal photon plus/minus atomic vibration. In other words, a frequency/wavelength shift occurs as the pump wavelength propagates along the fiber in the reverse direction. The energy lost in the pump wavelength is shifted to longer wavelength signals in the forward direction, thereby serving to amplify them. Raman amplifiers offer the advantage of amplifying signals in the broad range extending from 1300 to 1700 nm. Further, they perform better than EDFAs in terms of signal-to-noise ratio (SNR). Raman amplifiers increasingly are used as preamplifiers to enhance the performance of EDFAs in DWDM systems.

Finally, the nature of the multiplexers differs greatly in WDM and SONET. SONET makes use of very sophisticated and expensive ADMs, which can deal with SONET frames at the STS-1 level, VTs within VTGs, or even individual VCs. Some WDM systems, on the other hand, make use of *Optical Add/Drop Multiplexers* (OADMs), which multiplex at the lambda level, as illustrated in Figure 9.19. The OADMs also effectively perform a process of optical switching, or *photonic switching*, at the lambda level, thereby obviating the requirement for complex and expensive SONET DXCs.

The missing element in our discussion of DWDM and its variants is that of network management. SONET offers the advantage of very robust and highly standardized network management, but DWDM does not include any network management mechanisms. The ITU-T addressed this void through its G.709 (2003) specification, "Interface for the *Optical Transport Network* (OTN)," for a *digital wrapper*. This specification includes framing conventions, nonintrusive performance monitoring, error control, rate adaption, multiplexing mechanisms, ring protection, and network restoration mechanisms operating on a wavelength basis. A key element of G.709 is a Reed–Solomon FEC mechanism that improves error performance on noisy links, which enables the deployment of longer optical spans at the expense of some additional overhead. The resulting line rate is approximately 7 percent higher

TABLE 9.10   G.709 Line Rates and Corresponding SONET/SDH Line Rates [83]

G.709		SONET/SDH	
Interface	Line Rate (Gbps)	OC/STM Level	Line Rate (Gbps)
OTU-1	2.666	OC-48/STM-16	2.488
OTU-2	10.709	OC-192/STM-64	9.953
OTU-3	43.018	OC-768/STM-256	39.813

than the corresponding SONET/SDH line rate, which becomes the OTN payload. Table 9.10 lists the G.709 *Optical Transport Unit* (OTU) line rates and the matching SONET/SDH line rates. In addition to the G.709 interface specification, an interface for 10GbE clients exists, utilizing the same overhead structure and FEC mechanism, which results in a line rate of 11.095 Gbps [83].

Some suggest that DWDM, rather than SONET, is the future of optical transmission. Others, including myself, feel that this is not so. Rather, SONET and DWDM will coexist quite nicely into the future. Those promoting DWDM at the expense of SONET assume that a user or carrier can make effective use of a full lambda. Certainly, a large user organization may be able to make effective use of an OC-1 (52 Mbps), OC-3 (155 Mbps), or even OC-12 (622 Mbps) lightstream for access or transport purposes, perhaps in support of integrated voice, data, and video over ATM. But carriers cannot cost effectively dedicate a full lightstream for that purpose when they can run it at OC-192 rates of 10 Mbps in support of a large number of end users and a great variety of native data types. Because SONET is TDM based, it can handle any type of data; in fact and as previously discussed, SONET can support multiple data types over the same facility. Rather, the long-haul carriers typically will use SONET at the access level. For long-haul transport, they will run SONET transparently inside each DWDM wavelength, thereby realizing the advantages of each. The SONET frame format will be supported inside DWDM, EDFAs will be used in conjunction with Raman amplification, OADMs will switch signals at the optical level, and SONET devices will switch signals at the STS-1 level. An additional argument in support of this approach centers on the robust nature of SONET Operations, Administration, and Management (OA&M) features, on which the carriers rely greatly to manage and maintain their networks. Further, WDM is purely a point-to-point technology, with no inherent provisions for recovery in the event of a network failure; the self-healing feature of SONET is of great value in optical networking.

9.2.10 Packet over SONET

Packet over SONET (POS) is a MAN/WAN technology touted as one of the latest threats to ATM. POS offers the advantage of supporting packet data such as IP through either a direct optical interface to a router or through a SONET demarc in the form of a Terminating Multiplexer (TM). POS uses SONET as the layer 1 protocol, encapsulating packet traffic in High-Level Data Link Control (HDLC) frames and using Point-to-Point Protocol (PPP) for layer 2 link control, with the IP packet traffic running at layer 3. The result is that the combined overhead factor

(SONET + PPP + IP) is only approximately 5 percent for a 1500-byte IP datagram. This level of efficiency compares very favorably with ATM, which boasts an overhead factor of about 11 percent for the same IP datagram. This level of performance can be achieved only if packet data are to be transmitted and only if the service is provided over the equivalent of a SONET-based, point-to-point private line, provisioned in the form of a Virtual Tributary (VT). Thereby, POS traffic bypasses any ATM switches that might be in place in the carrier network. If multiple data types (e.g., voice, packet data traffic, SDLC data traffic, video, and fax) require support and if multiple Quality of Service (QoS) guarantees are required, ATM remains the solution [84, 85].

9.3   IEEE 802.17, RESILIENT PACKET RING

The IEEE 802.17 Resilient Packet Ring Working Group published both the base standard and the 802.17a amendment in 2004 and is continuing work on the 802.17b amendment for advanced bridging as well as a maintenance specification and a set of interpretations. *Resilient Packet Ring* (RPR) is a Medium Access Control (MAC) layer protocol that uses Ethernet switching and a dual counterrotating ring topology to optimize the transport of Ethernet/IP packet data traffic over optical fiber rings while maintaining the resiliency of SONET/SDH, but at a much reduced level of overhead. RPR is Layer 1 independent; therefore it can be implemented over existing SONET/SDH physical rings using Layer 1 only or it can run on a stand-alone basis.

RPR calls for dual counterrotating *ringlets* that interconnect nodes where data traffic is intended to drop. The stations connecting through the nodes generally are pure Ethernet, unless TDM circuit emulation or some other option has been added. RPR also uses statistical multiplexing, which allows bandwidth to be oversubscribed, while establishing Committed Information Rate (CIR) and peak-rate thresholds per application. The nodes negotiate bandwidth requirements among themselves based on fairness algorithms and in consideration of a classification scheme that recognizes and provides higher priority access to traffic sensitive to latency and jitter. At the same time, best effort traffic, such as Internet traffic, is ensured equal access and a fair share of the remaining bandwidth. RPR addresses QoS issues through the definition of three Class-of-Service (CoS) levels:

- Class A traffic is intolerant of latency and jitter. RPR addresses class A traffic through a high *Committed Information Rate* (CIR) that ensures the availability of an average suitable amount of bandwidth for high priority traffic. Class A traffic includes real-time voice and video.
- Class B is more tolerant of latency and jitter, yet still has QoS requirements in that regard. RPR addresses class B traffic either through a lower CIR that ensures the availability of an average amount of bandwidth suitable for medium-priority applications that have less stringent QoS requirements or through an *Excess Information Rate* (EIR) option. In the event of network congestion, class B traffic is subject to fairness-based flow control. Class B is intended for business-class data traffic such as transaction processing.

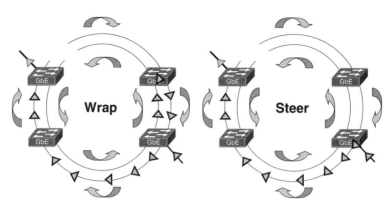

Figure 9.20 RPR protection schemes.

- Class C traffic is best effort traffic with no latency or jitter requirements and therefore is strictly EIR traffic. In the event of network congestion, class C traffic is subject to fairness-based flow control. Class C traffic includes low-priority applications such as consumer-level Internet access.

The RPR protection scheme uses Physical Layer (Layer 1) alarm information and Data Link Layer (Layer 2) communications to detect node and link failures. Once a failure is detected, the RPR switching mechanism can restore the network in 50 ms or less, which is the SONET/SDH benchmark. There are two restoral mechanisms:

- **Wrap:** The wrap option calls for data to travel around the ring until it reaches the node nearest the break, as illustrated in Figure 9.20. That node turns the traffic around and sends it in the reverse direction over the counterrotating ring.
- **Steer:** Using the steer option, the originating station intelligently places the traffic on the ring that retains continuity.

The RPR dual counterrotating ringlets design calls for packet data traffic to travel over both fibers (one in each direction) simultaneously, which is much more efficient than the SONET approach. RPR employs statistical packet multiplexing to optimize the ring for packet data transport. This is unlike SONET/SDH, which employs TDM and, therefore, is optimized for PCM-based voice traffic. RPR further increases efficiency through a spatial reuse mechanism. Older LAN ring protocols such as Token Ring and FDDI require a data packet to traverse the entire ring, even if the destination node is only a single hop away from the originating node. RPR allows data inserted into the ring to be sent directly to the destination node, where it drops off and makes the bandwidth once again available to other stations in the forward direction.

A *Metropolitan Area Network* (MAN) infrastructure technology, RPR is largely oriented toward business-class data applications, with emphasis on Ethernet. Notably, however, RPR traffic classes provide sufficient *Grade-of-Service* (GoS) distinctions to make it viable in appropriate triple-play implementations [86–89].

REFERENCES

1. Kushnick, Bruce and Allibone, Tom. Data Quality Act Complaint from TeleTruth to the Federal Communications Commission, July 26, 2005.

2. http://www.fcc.gov/broadband.

3. Organisation for Economic Co-operation and Development. "OECD Broadband Statistics, June 2005." http://www.oecd.org/document/16/0,2340,en_2649_201185_35526608_1_1_1_1,00.html.

4. "Delivering T1 and E-1 Services over One Copper Pair with HDSL2." Orckit Technologies, June 1999.

5. "UK's Broadband Local Loop Lengths." *ISP Review*, November 2005, http://www.ispreview.co.uk/cgi-bin/news/viewnews.cgi?id=EEFplFyZEkrNkluWdw.

6. "General Introduction to Copper Access Technologies." DSL Forum. www.adsl.com.

7. Paone, Joe. "ADSL: 'The New Kid on the Block'." *Internetwork*, June 1996.

8. Kirby, Rob. "Lesson 145: ADSL and SDSL." *Network Magazine*, August 5, 2000.

9. "ADSL2 and ADSL2plus—The New ADSL Standards." www.dslforum.com, March 25, 2003.

10. Maier, Joe. "Delivering on The Promise." *OSP Magazine*, October 2005.

11. "About the UAWG and UADSL." Universal ADSL Working Group. Originally accessed at www.uawg.org.

12. Cavanaugh, Ken. "Splitting Out the Issues of the ADSL G.lite Initiative." *Telecommunications*, October 1998.

13. Lindstrom, Annie. "G.Lite gets into the starting blocks." *America's Network*. March 1, 1999.

14. Sekar, Richard. "G.lite: Pragmatic, Mass Market, High-Density DSL." *Telecommunications*, April 2000.

15. Lindecke, Sascha. "DSL Tekchnology and Deployment: VDSL2." www.dslforum.org.

16. Lalani, Hamid. "Eeny, Meeny, MIMO, Moe." *OSP Magazine*, January, 2006.

17. "CopperOptics Enhancing the Performance and Application of Copper Cable with HDSL." PairGain Technologies, 1995.

18. Quilici, Jim. "An HDSL2 Primer." *Communication Systems Design*. August 1999.

19. "G.shdsl: New Solution for Businesses." *Telecommunications*. April 2001.

20. Lee, Stephen. "DSL Receives Shot in Arm." *Infoworld*, April 30, 2001.

21. Boyd, Jade. "The Newest DSL: Faster and Further." *InternetWeek*, February 26, 2001.

22. Horak, Ray. "Voice over DSL (VoDSL)." www.commweb.com/techcenters/main/experts/3783/COM20011105S0007.

23. Kaplan, Mark. "Voice over SDSL: Effectively Combining Voice and Data." *Telecommunications*, February 2000.

24. Lee, Seng-Poh. "The Real next-Gen VoDSL." *Telecommunications*, June 2000.

25. Knight, Stefan. "VoDSL: Ready for Prime Time." *Phone +*, November 2000.

26. Perzov, Muni. "Voice over DSL." *A Copper Mountain Networks White Paper*, September 2000.

27. "Voice-over-Digital Subscriber Line (VoDSL) Service—New Revenue from Existing Infrastructure Tutorial." http://www.iec.org/tutorials/vodsl_revenue/index.html.

28. "History of Cable TV." Kansas State University, http://www.telecom.ksu.edu/cable/history.html.

29. Ciciora, Walter S. "An Overview of Cable Television in the United States." http://people. deas.harvard.edu/~jones/csciE-129/nu_lectures/lecturE-13/pdf/CATV.pdf.

30. "Data over Cable Service Interface Specifications FAQ." Cable Television Laboratories. www.cablemodem.com.

31. Nikolich, Paul. "Cable Modems Deliver Fast 'Net access.'" *Network World*, March 1, 1999.

32. Finneran, Michael. "The Cable Modem Picture Comes into Focus." *Business Communications Review*, March 1999.

33. www.cable-modems.org/tutorial.

34. "Coveting Coax Cable." *Network Magazine*, November 2001.

35. Buckley, Sean. "Packet over Cable." *Telecommunications*, March 2001.

36. Fanfelle, Robert. "DOCSIS 2.0: Upping Upstream Performance in Cable Modem Designs." *CommsDesign*, June 19, 2006, http://www.commsdesign.com/design_corner/ OEG20020617S0011.

37. Barthold, Jim. "Cable Telephony: Available in a Variety of Packages." *Cable World*, October 9, 2000.

38. Sweeney, Can. "Cable Telephony Gains Momentum." *America's Network*, August 1, 2000.

39. "Cable Industry Reaches 3 Million VoIP Subscriptions." *OSP Magazine*, March 24, 2006.

40. Sweeney, Daniel. "LMDS: Finally Ready for Prime Time?" *America's Network*, August 1, 1998.

41. Sweeney, Daniel. "LMDS: How Competitive?" *America's Network*, August 15, 1998.

42. Willis, David. "LMDS: Is It a Little Too Much, a Little Too Late?" *Network Computing*, February 8, 1999.

43. Dunlop, Amy. "Wireless Access Enters Real-World Trials." *Internet World*, May 1997.

44. Bernier, Paula. "Carriers Charge Ahead with 38 GHz." *X-Change*, February 15, 1998.

45. Pappalardo, Denise. "Worldcom Adds Wireless MMDS Area." *Network World*, August 20, 2001.

46. Rysavy, Peter. "MMDS Struggles to Find a Foothold." *Network Computing*, October 29, 2001.

47. Gohring, Nancy. "MMDS Shifts Gears." *Interactive Week*, October 1, 2001.

48. Sweeney, Daniel. "A Second Chance for MMDS." *Broadband Wireless Business*, September/October 2003.

49. Nicopolitidis, P., Obaidat, M. S., Papadimitriou, G. I., and Pomportsis, A. S. *Wireless Networks*. Wiley, 2003.

50. Eklund, Carl, Marks, Roger B., Stanwood, Kenneth L., and Wang, Stanley. "IEEE Standard 802.16: A Technical Overview of the WirelessMAN™ Air Interface for Broadband Wireless Access." *IEEE Communications Magazine*, June 2002.

51. Melby, Nathaniel J. "WiMax: Facing the WMAN Challenge." *ACUTA Journal of Communications Technology in Higher Education*, Summer 2005.

52. Barthold, Jim. "Countdown to Plug-and-Play." *Broadband Wireless Business*, May/June 2004.

53. Dornan, Andy. "The WiMAX Anticlimax." *Network Magazine*, December 2004.

54. Horak, Ray. "WiMAX: WLL by the Numbers." Commweb, July 29, 2005, www.commweb. com.

55. Mathias, Craig and Rysavy, Peter. "The ABCs of PCS." *Network World*, November 7, 1994.

56. http://www.hughesnet.com.

57. http://www.starband.com.

58. Horak, Ray. "Passive Optical Network (PON)." *Commweb*, February 19, 2002, www.commweb.com.

59. Horak, Ray. "Passive Optical Network (PON) Redux." *Commweb*, April 13, 2004, www.commweb.com.

60. Kemp, Steve. "Passive Aggressive Providers? NEVER!" *OSP Magazine*, December 2005.

61. www.verizon.com.

62. Hardy, Stephen M. "Not Everyone Loves GPON." *Lightwave*, December 2005.

63. Doiron, Tim. "Got Provisions for Triple Play Provisioning?" *OSP Magazine*, March 2006.

64. "Potential Interference from Broadband over Power Line (BPL) Systems to Federal Government Radiocommunications at 1.7–80 MHz," NTIA Report 04-414, Phase 1 Study, Volume I. U.S. Department of Commerce, National Telecommunications and Information Administration, April 2004.

65. Mears, Jennifer. "Broadband over Power Lines Gaining Steam," *Network World*, August 23, 2004.

66. Gross, Grant. "FCC Action Charges up Broadband over Power Lines," *Network World*, October 18, 2004.

67. Dornan, Andy. "Broadband over Power Lines." *Network Magazine*, April 2005.

68. Horak, Ray. "Access Broadband over Power Line." *Commweb*, June 6, 2005, www.commweb.com.

69. Spohn, Darren L. *Data Network Design*. McGraw-Hill, 1993.

70. Davidson, Robert P. and Muller, Nathan J. *The Guide to SONET: Planning, Installing & Maintaining Broadband Networks*. Telecom Library, 1991.

71. Sexton, Mike and Reid, Andy. *Transmission Networking: SONET and the Synchronous Digital Hierarchy*. Artech House, 1992.

72. Miller, Mark A. *Analyzing Broadband Networks: Frame Relay, SMDS, & ATM*. M&T Books, 1994.

73. Hines, I. J. *ATM: The Key to High-Speed Broadband Networking*. M&T Books, 1996.

74. Minoli, Daniel. *Enterprise Networking: Fractional T1 to SONET, Frame Relay to BISDN*. Artech House, 1993.

75. Kingsley, M. Scott and Amoss, John J. "Synchronous Optical Network (SONET): Overview." *IT Continuous Services*. Datapro Information Services, July 22, 1998.

76. "Sprint's SONET Ring Links Silicon Valley." *New Media*, December 1994.

77. Rendleman, John. "Sprint Readies International Fiber Ring." *Communications Week*, June 10, 1996.

78. Schroeder, Erica. "AT&T's Fiber Network to Get Eightfold Boost." *PC Week*, May 13, 1996.

79. Dziatkiewicz, Mark. "Network of the Olympians." *Convergence*, March 1996.

80. Bovill, Kirk. "Scaling to 10-Gigabit Ethernet and Beyond." *Lightwave*, April 2001.

81. Nebeling, Marcus. "CWDM: Lower Cost for More Capacity in the Short-Haul." *Lightwave*, August 2001.

82. "NEC Tops DWDM Records at 10.9 Tbps per Fiber." *Converge! Network Digest*. Originally accessed at http://www.convergedigest.com/DWDM/archive/010322 NECdwdmrecord.html.

83. Barlow, Guylain. "A G.709 Optical TransportNetwork Tutorial." http://www.innocor.com/pdf_files/g709_tutorial.pdf.

84. Shok, Glen. "Here's Packet over SONET in A Nutshell." *LAN Times*, April 27, 1998.

85. Parente, Victor R. "Packet over Sonet: Ringing Up Speed." *Data Communications*, March 1998.

86. Paripatyadar, Raj. "RPR Supports Efficient Ethernet Data Traffic in Metro Networks." *Lightwave*, May 2005.

87. Vashi, Tejas. "RPR Ups Efficiency of Metro Ethernet." *Network World*, May 31, 2004.

88. "An Introduction to Resilient Packet Ring Technology." White Paper, the Resilient Packet Ring Alliance, October 2001.

89. Hawkins, John. "New Lease on Life." *Telecommunications*, May 2002.

CHAPTER 10

BROADBAND NETWORK SERVICES

The speed of communications is wondrous to behold. It is also true that speed can multiply the distribution of information that we know to be untrue.
Edward R. Murrow (1908–1965), U.S. broadcaster and journalist

The concept of broadband networking has its roots in the early 1970s when the CCITT (now ITU-T) first defined Broadband ISDN in anticipation of the development of highly bandwidth-intensive applications and the demand for them. Since that time, the demand for computer internetworking grew at rates that astounded even the most astute computer and data network pundits. Imaging systems developed and increasingly required networking. The potential for videoconferencing applications became apparent. The development and commercialization of the Internet and the Web truly has been mind-boggling. Finally, entertainment networking now captures the attention of telcos and others as they compete to develop networks that will carry television and movies as well as voice, data, video, image, facsimile, and all other forms of data over a single network infrastructure. These applications—and more—are the drivers of technology.

The focus here is on network service offerings. Frame Relay (FR) and Asynchronous Transfer Mode (ATM) fall under the umbrella term *fast packet services* (see Figure 10.1). Metropolitan Ethernet is the latest addition to this distinguished grouping. *Fast*, of course, refers to the fact that they rely on a broadband transmission infrastructure in the form of fiber optics and on very fast and capable switching systems. *Packet* is a generic term referring to the manner in which data are organized—into packets, frames, blocks, cells, and so on—and variously relayed, transferred, switched, and routed through the network. *Services* is the operative word here. From the Latin *servitium*, it refers to the condition of a slave, whose function

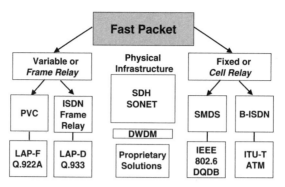

Figure 10.1 Fast packet services networking. (*Source*: Frame Relay Forum.)

is to give good by providing usefulness to others. No matter how elegant the underlying technologies, their ultimate value is in enabling the creation and delivery of services of value. At the bottom of Figure 10.1, note the access standards defined by the ITU-T.

This chapter concludes with a brief exploration of B-ISDN and Advanced Intelligent Networks (AINs). AINs reflect the marriage of computer, database, and network technologies, offering tremendous potential in the creation and delivery of services.

10.1 FRAME RELAY

Gunpowder, since its invention, has been used for signaling, and still is employed in rockets, flares and railroad-track torpedoes. Cannons, since their invention, have been fired as warnings or to announce events. . . . The greatest distance covered by cannon signals was 428 miles from Buffalo to New York City when the Erie Canal was opened, October 26, 1825. Relays of guns got the message through in an hour and twenty minutes.

Telephone Almanac for 1956, Bell Telephone System

Relays of guns may have been effective in 1825 but were obsoleted by the telegraph beginning in 1844. Frame Relay experienced unprecedented growth for at least a decade from the time of its commercial introduction by Wiltel (since acquired by LDDS Worldcom, which was acquired by MCI, which is now part of Verizon) in 1992, before suffering from competition against Virtual Private Networks (VPNs) based on Internet Protocol (IP). Clearly, users in the United States have had the greatest appetite for Frame Relay, although it also enjoyed great popularity in the European Union (EU) and remains widely available at an international level.

10.1.1 Frame Relay Defined

Frame Relay is a network interface, or access, standard that was defined in 1988 by the ITU-T in its I.122 Recommendation, *Framework for Providing Additional Packet Mode Bearer Services*. Access to a Frame Relay network is accomplished using *Link Access Procedure for Frame Mode Services* (LAP-F), as specified in Q.922. LAP-F is an adaptation of the *Link Access Procedure—Data channel* (LAP-D) signaling protocol developed for ISDN, as Frame Relay originally was intended as an ISDN framing convention for a *bearer service*, that is, information-bearing service. As Frame Relay standards address Layers 1 and 2 of the Open Systems Interconnection (OSI) Reference Model, they do not specify internal network operations. As with X.25 packet networks, issues of packet switching and transport are left to manufacturers and service providers [1]. The vast majority of Frame Relay access is at rates of Fractional T1, DS-0 (64 kbps) [2], and below because the vast majority of connected sites are small, remote locations in a larger enterprise network. Access speeds of T1/E-1 are common for connection of centralized data centers, with multiple circuits often aggregated for higher speed access. Most Frame Relay service providers also support T3/E-3, at least in the major metropolitan serving areas. Notably, Frame Relay is backward compatible, as it considers the characteristics of the embedded networks and the standards on which they are based. Electrical and optical interfaces are not rigidly defined for Frame Relay specifically but use established standards such as T1 and V.35 serial interfaces.

Very much analogous to a streamlined and supercharged form of X.25 packet switching, Frame Relay can set up and tear down calls with control packets although no carrier ever activated these features for Switched Virtual Circuits (SVCs) in a public network. Frame Relay also forwards packets of data in the form of frames. Similar to X.25, Frame Relay statistically multiplexes frames of data over a shared network of virtual circuits for maximum network efficiency. The interface is in a *Frame Relay Access Device* (FRAD) that can be implemented on the customer premises and which is analogous to an X.25 *Packet Assembler/Disassembler* (PAD). Like X.25, Frame Relay is intended for bursty data traffic, although it works quite well with fixed-bit-rate applications. While both X.25 and Frame Relay can support voice, video, and audio, the inherently unpredictable levels of packet delay and packet loss over such a highly shared network can yield results that are less than totally satisfactory.

Beyond the basic conceptual levels, the two technologies diverge, as reflected in Table 10.1. Frame Relay is a connection-oriented service working under the assumption of error-free broadband transmission facilities for both network access and transport. While Frame Relay includes an error detection mechanism, it assumes no responsibility for error correction because there are assumed to be no errors in transmission or switching as the frames transverse the network. In the event of a

TABLE 10.1 Comparison of X.25 and Frame Relay

Attribute	X.25	Frame Relay
Infrastructure assumption	Analog	Digital
Maximum payload (octets)	16, 32, 64, 128, 256, 512, 1024	4096
Maximum access speed	DS-1	DS-3
Link Layer protocol	LAP-B	LAP-D/LAP-F
Latency	High	Moderate
Connection(less)	Connection oriented	Connection oriented
Error recovery	Network, link by link	CPE
Primary application	Interactive data	LAN-to-LAN

detected error, the frame is simply discarded and no attempt is made to advise the end-user equipment. Recognition of the loss and recovery of the lost frames is the responsibility of the end equipment. Frame Relay makes no guarantees of frame delivery, there is no sequence numbering, and there are no acknowledgments of any sort provided. Ceding this responsibility to the end user reduces the load on the computational and bandwidth resources of the network, thereby reducing latency significantly and yielding faster processing and relaying of each frame of data and reducing overall network costs.

Frame Relay standards specify a frame payload that is variable in length up to 4096 octets, which supports a payload as large as that of 4-Mbps Token Ring, which generally is considered obsolete. Subsequently, the Frame Relay Forum (FRF) developed an Implementation Agreement (IA) that sets the minimum size at 1600 octets for purposes of interconnectivity and interoperability of FR networks and the switches, routers, and other devices that form the heart of the networks. This frame size easily supports the largest standard 802.3 Ethernet frame of 1518 octets. As Ethernet clearly is the dominant LAN standard and as FR was designed for LAN-to-LAN internetworking, this frame size generally is adequate. Regardless of the minimum frame size, the frames vary in length, just as do LAN frames. This variability results in some level of unpredictability, which contributes to congestion.

10.1.2 Frame Relay Standards

A wide range of manufacturers and carriers, both domestic and international, support Frame Relay. Standards bodies include the American National Standards Institute (ANSI), European Telecommunications Standards Institute (ETSI), and ITU-T. Table 10.2 notes select relevant standards.

The Frame Relay Forum (now merged with the ATM Forum and MPLS Forum to form the MFA Forum), a voluntary group of manufacturers and other interested parties, developed and promoted *Implementation Agreements* (IAs), which address manufacturer interoperability issues. IAs include the following:

- **FRF.1.2:** User-to-Network Interface (UNI) Implementation Agreement (April 2000)
- **FRF.2.1:** Frame Relay Network-to-Network Interface (NNI) Implementation Agreement (July 1995)
- **FRF.3.2:** Multiprotocol Encapsulation Implementation Agreement (MEI) (April 2000)

TABLE 10.2   Frame Relay Standards

Subject Area	ITU-T	ANSI
Architecture and service description	I.233	T1.606
Data Link Layer core aspects	Q.922 Annex A	T1.618
PVC Management	Q.933 Annex A	T1.617 Annex D
Congestion management	I.370	T1.606a
SVC signaling	Q.933	T1.617

Source: DigiNet Corp.

- **FRF.4.1:** SVC User-to-Network Interface (UNI) Implementation Agreement (January 2000)
- **FRF.5:** Frame Relay/ATM PVC Network Interworking Implementation (December 1994)
- **FRF.6.1:** Frame Relay Service Customer Network Management Implementation Agreement (MIB) (September 2002)
- **FRF.7:** Frame Relay PVC Multicast Service and Protocol Description (October 1994)
- **FRF.8.2:** Frame Relay/ATM PVC Service Interworking Implementation Agreement (February 2004)
- **FRF.9:** Data Compression Over Frame Relay Implementation Agreement (January 1996)
- **FRF.10.1:** Frame Relay Network-to-Network SVC Implementation Agreement (September 1996)
- **FRF.11.1:** Voice over Frame Relay Implementation Agreement (May 1997, Annex J added March 1999)
- **FRF.12:** Frame Relay Fragmentation Implementation Agreement (December 1997)
- **FRF.13:** Service Level Definitions Implementation Agreement (August 1998)
- **FRF.14:** Physical Layer Interface Implementation Agreement (December 1998)
- **FRF.15:** End-to-End Multilink Frame Relay Implementation Agreement (August 1999)
- **FRF.16.1:** Multilink Frame Relay UNI/NNI Implementation Agreement (May 2002)
- **FRF.17:** Frame Relay Privacy Implementation Agreement (January 2000)
- **FRF.18:** Network-to-Network FR/ATM SVC Service Interworking Implementation Agreement (April 2000)
- **FRF.19:** Frame Relay Operations, Administration and Maintenance Implementation Agreement (March 2001)
- **FRF.20:** Frame Relay IP Header Compression Implementation Agreement (June 2001)

10.1.3 Frame Relay Access

Frame Relay access is on the basis of a dedicated digital link into a FR node. The speed of the access link can range up to 44.736 Mbps (T3) and can be in the form of Dataphone Digital Service (DDS), Switched 56/64, ISDN BRI or PRI, Fractional T1, T1/E-1 (DS-1), and T3/E-3 (DS-3). Many service providers also offer *Multilink Frame Relay* (MFR) service, which allows multiple T1s to be bundled in support of access speeds between T1 and T3. Essentially a form of *inverse multiplexing*, MFR is an inexpensive and readily available solution compared with T3 or Fractional T3. In addition to offering a bandwidth option between DS-1 and DS-3, MFR offers a significant measure of redundancy as the access speed can throttle down should one or more of the bundled T1s fail. A common MFR example comprises four bundled

T1s running at a nominal speed of 6 Mbps. MFR typically supports as many as eight bundled T1s for an aggregate nominal speed of 12 Mbps. The Frame Relay Forum published the specifications for MFR in its End-to-End MFR (FRF.15) and UNI/NNI MFR (FRF.16), with the latter specifying the various network interfaces.

Customer DCE is the form of a *Frame Relay Access Device* (FRAD). The FRAD can be a stand-alone device that performs the sole function of assembling and disassembling frames, although it generally is embedded under the skin of another device, such as a router or Channel Service Unit (CSU). From the FRAD, access is gained to the link and, subsequently, to the Frame Relay network node at the edge of the carrier network. At the node resides the *Frame Relay Network Device* (FRND). The *User Network Interface* (UNI), as defined by ANSI and ITU-T, defines the nature of this access interface. Many service providers also offer access to Frame Relay on a dial-up modem basis, in which case the FRAD resides at the edge of the carrier network.

10.1.4 Frame Relay Network

The Frame Relay network (Figure 10.2) consists of specified network interfaces in the form of the User Network Interface (UNI) and Network-to-Network Interface (NNI). The specifics of the internal carrier network are based on ISDN, generally making use of Permanent Virtual Circuits (PVCs) between ATM switches. Some carriers also support Switched Virtual Circuits (SVCs), although it is extremely unusual. Frame Relay networks can be public, private, or hybrid.

- *User Network Interface* (UNI) is the demarcation point between the user Data Terminal Equipment (DTE) and the network and is in the form of a Frame Relay Access Device (FRAD) and a Frame Relay Network Device (FRND). The FRAD, which is a Data Communications Equipment (DCE) device, may

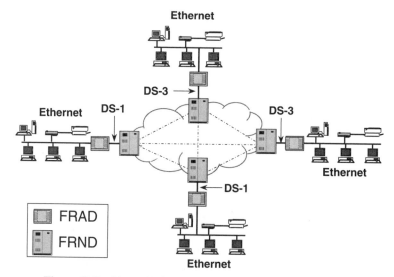

Figure 10.2 Frame Relay network with fully meshed PVCs.

be standalone or may be contained under the skin of another device, such as a router. The FRAD typically is on the customer premises, although it may be included in the FRND for use on a dial-up basis. The FRND is an edge network device, typically in the form of a carrier-class router.

- *Network-to-Network Interface* (NNI) is defined as the interface between FRNDs in different Frame Relay networks and is based on multinetwork PVCs. Most network-to-network connections are provided over digital SONET/SDH optical fiber trunks and through ATM switches. This current method of internetwork connection has implications relative to Implicit Congestion Notification (ICN) and network management. An NNI is required, for example, if a user organization wishes to connect from an intraLATA Frame Relay network provided by a Regional Bell Operating Company (RBOC) to an interLATA network provided by a long-haul carrier. The NNI is provided at additional cost.

- *Integrated Services Digital Network* (ISDN) is the basis of the internal Frame Relay network, as the LAP-F protocol that governs the user links is an adaptation of the ISDN LAP-D protocol. The links generally are not ISDN, however, for reasons that include lack of availability, lack of capacity, and additional cost.

- *Permanent Virtual Circuits* (PVCs) define fixed paths through the network for each source–destination pair. The network administrators *permanently fix* the path definitions in network routing tables, until such time as they *permanently change* them. As bandwidth over the physical path is used only when required, the circuit is virtual in nature. As Frame Relay is a shared packet network, a great many source–destination pairs may share each PVC. Once defined and regardless of the network traffic load, the PVC is always used to serve a given source–destination pair. In order to protect against a catastrophic failure affecting a PVC element (e.g., link or switch), a backup PVC can be provisioned, at additional cost, over a separate physical path. Frame Relay networks generally are based on PVCs, as depicted in Figure 10.2. PVCs typically are fully symmetric, with equal bandwidth in each direction. They also may be asymmetric, with more bandwidth in one direction than in the other.

- *Switched Virtual Circuits* (SVCs) are set up call by call, based on programmed network routing options. These options are triggered by a message sent from the FRAD to the FRND, asking for a connection to a unique address, which is the equivalent of a Frame Relay telephone number. An SVC call is established as the originating FRAD sends a request to the network node, including a destination address and bandwidth parameters for the call. The network node responds with a link designation in the form of a *Data Link Connection Identifier* (DLCI), the originating FRAD responds with an acceptance message, and the data transfer ensues. At the end of the call, the SVCs are torn down. The next SVC provided for the same source–destination pair could differ greatly, depending on network availability. SVCs offer the advantage of improved performance through automatic load balancing, as they are defined on a call-by-call basis. SVCs also offer any-to-any flexibility (like a dial-up call of any type) and resiliency, as the VCs are selected and set up in consideration of the overall performance of the network from end to end. A Frame Relay network based on SVCs offers the option of defining a *Closed User Group* (CUG) for purposes of security. The CUG prevents both the initiation of a call to and the reception of a call from any site not defined in the CUG. *Note:* Few carriers support SVCs due to their complexity.

- *Mesh networking* is easily accomplished with Frame Relay, and on a scalable basis. In other words, the cost of Frame Relay implementation is relatively proportionate to the task at hand. As more sites are added to the network, more FRADs, access links, and FRNDs can be added. As the bandwidth requirement increases at any given site, the capacity of the access link can be increased. The costs grow relatively gracefully as the network expands to provide access to additional sites and as bandwidth increases on a site-by-site basis. While full mesh networking is unusual, it can be accomplished at significantly lower cost than with a private, leased-line network, which requires a separate access link at every site for every other site. Frame Relay networks can be classified as VPNs, as they are virtually equivalent to a private network, at least in many respects.

- *Network processing* is not performed in Frame Relay, at least not to the extent experienced in X.25 networks. A Frame Relay network assumes that the link is error free. As the link is digital (ideally fiber optic) and based on ISDN, there is no compelling requirement to check for errors, although every switch does so. Once the frame is identified as valid and the address is recognized, the frame is relayed through the core of the network to the next node and, ultimately, to the end user. If a frame is errored or somehow mutilated, Frame Relay simply discards it, making no attempt to inform the endpoints. Frame Relay does not support sequence numbers and sends no acknowledgments to endpoints. This is the equivalent of removing OSI Layer 3 functions from the X.25 model. When compared with X.25 and traditional packet switching, this reduction in network processing yields lower cost and reduced packet latency, somewhat counterbalancing the effect of the variable-size frame to reduce issues of overall network congestion.

10.1.5 Frame Relay Equipment

Frame Relay, as a service offering, depends on certain hardware and firmware in order to accomplish the interface. That equipment includes the FRAD, the FRND, and the FR switch:

- *Frame Relay Access Devices* (FRADs) also known as *Frame Relay Assembler/ Disassemblers*, are CPE that are analogous to Packet Assembler/Disassemblers (PADs) in an X.25 packet-switched network. The FRAD essentially organizes the user data into *Protocol Data Units* (PDUs) that can vary in size up to 4096 octets. The FRAD then encapsulates the PDUs into Frame Relay frames, placing the necessary control information around each and appending a Cyclic Redundancy Check (CRC) for error detection. FRADs can be standalone units, serving multiple hardwired devices. FRADs also can be incorporated into X.25 PADs, T-carrier muxes, or even PBXs for dedicated networks. More likely, they are incorporated into high-end routers serving LANs. CO-based FRADs eliminate the end-user investment in such equipment. Although they are limited in functionality as they largely are intended to support the conversion of Synchronous Data Link Control (SDLC) traffic to frame format, CO-based

FRADs offer advantages to user organizations seeking to transition from dedicated networks connecting large data centers [3–5].

• *Frame Relay Network Devices* (FRNDs) are Frame Relay switches at the network side of the connection. The *FRND* (pronounced *friend*) is provided by a friendly service provider, which seems like an amicable arrangement. Frame Relay switches are nodal processors capable of switching frames at very high speed. They contain buffers for flow control purposes. Within limits, the buffers can absorb incoming frames until the switch can act on them as well as absorb outgoing frames until the forward link becomes available. The switches contain very high speed switching matrixes and internal buses. They have sufficient intelligence, in the form of routing tables, to read the control information embedded in the frame header and route the frames correctly over the PVC identified previously in the call set-up process. The nodal processors also have intelligence sufficient to check for errors in the frame. Additionally, the nodal processors may have the intelligence to prioritize certain types of delay-sensitive traffic, such as SDLC and real-time uncompressed voice and video. Since the switches perform no error correction or protocol conversion functions, they act on the frames very quickly, thereby minimizing latency. Frame Relay nodal processors can be managed from a centralized Network Operations Center (NOC), thereby enhancing the scalability of the network [6]. The switch also may contain a voice compression module. While Frame Relay is intended for LAN internetworking, it can accommodate voice with varying degrees of quality.

10.1.6   Frame Relay Protocol: Frame Structure

The Frame Relay protocol involves the relaying of frames of user data across the network. In a typical data application, the preponderance of the frame is payload in the form of user data, with relatively little control information included. As illustrated in Figure 10.3, the five data fields in the ANSI T1.618 frame format comprise

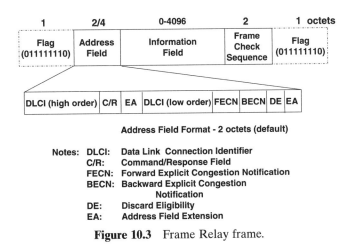

Figure 10.3 Frame Relay frame.

beginning and ending *High-level Data Link Control* (HDLC) flags, an address field, an information field, and a Frame Check Sequence (FCS), which is a CRC. Although the Frame Relay protocol is a subset of the HDLC protocol used in other data communications environments, it lacks control information such as frame sequence number. In the Frame Relay environment, it is the responsibility of the user to identify and correct for frame sequence errors or missing frames that might have been detected by network switches as corrupted or truncated and, therefore, discarded.

- **Flag Field:** One-octet, fixed binary sequence (01111110) is employed to identify and separate frames. Flag fields appear at the beginning and end of the frame. The flags technically are not part of the frame. Rather, they are frame delimiters that precede and follow all frames. The specific bit sequence is recognized by all network devices as a delimiter and nothing more. Essentially, they are much like *idle* bits or *keep-alive* bits. When a network device sees any other bit pattern, it recognizes those bits as being something of significance, rather than a flag field. There may be only a single flag between frames sent in rapid succession or there may be many flags between frames sent at a slow pace.

- **Address Field:** Two octets are used as a default length, although some networks use four octets. Address fields include the necessary control information in the data link connection identifier. The address field also contains a command/response field, address field extension, forward and backward explicit congestion notification fields, and discard eligibility data.

  - **Data Link Connection Identifier (DLCI, pronounced *delsey*):** Ten bits that identify the data link, the virtual circuit (i.e., PVC or SVC), and its service parameters to the network. Those service parameters include frame size, Committed Information Rate (CIR), Committed Burst Size (B_c), Burst Excess Size (B_e), and Committed Rate Measurement Interval (T_c). I discuss the significance of these service parameters later in this chapter.

  - **Command/Response (C/R):** One bit reserved for use of the FRADs, rather than the Frame Relay network. C/R is defined to facilitate the transport of polled protocols such as Systems Network Architecture (SNA), which require a C/R for signaling and control purposes.

  - **Address Field Extension (EA):** Two bits that signal the extension of the addressing structure beyond the two-octet default. The use of EA must be negotiated with the carrier when the service is established.

  - **Forward Explicit Congestion Notification (FECN, pronounced *feckon*):** One-bit field available to the network to advise upstream devices that the frame has experienced congestion. The FRAD clearly recognizes when the frame carrying the FECN survives. It also is advised that subsequent frames may not be so fortunate. Should subsequent frames be discarded or corrupted in transmission, the receiving device is advised that recovery may be required in the form of requests for retransmission. If the upstream device controls the rate of upstream data transfer, it has the opportunity to throttle back.

  - **Backward Explicit Congestion Notification (BECN, pronounced *beckon*):** One-bit field used by the network to advise devices of congestion in the direction opposite of the primary traffic flow. If the target FRAD responds to the originating FRAD in the backward direction, the BECN bit is set in a backward frame. If there is no data flowing in the backward direction, the Frame Relay network creates a frame in that direction, setting the BECN bit. If the originating FRAD is capable of reducing the frame rate, it is well advised to do so, as the network may discard frames once the notification is posted.

- **Discard Eligibility (DE):** One-bit field indicating the eligibility of the frame for discard under conditions of network congestion. Theoretically, the user equipment sets the DE in consideration of the acceptability of the application to packet loss. Should the user equipment not set the DE, the network switches may do so on a random basis, with results that may be less than totally pleasing. SDLC, voice, and video traffic, for example, do not tolerate loss and so typically are engineered to avoid DE status.

- **Information Field:** Contains user information, either in the form of payload data or internetwork control information passed between devices such as routers. Although the information field may be 4096 octets in length, ANSI recommendations dictate a maximum size of 1600 octets. This payload size is more consistent with the frame size (64–1518 octets) of 802.3 Ethernet LANs, which generate the vast majority of LAN traffic. This frame size of 1600 octets also is addressed in Implementation Agreements from the Frame Relay Forum, which were developed to ensure the interconnectivity and interoperability of Frame Relay networks.

- **Frame Check Sequence:** Two-octet Cyclic Redundancy Check (CRC) supporting header error detection in frames with information fields up to 4096 octets in length. Note that no error correction is provided. Rather, error correction for payload data is accomplished by running the TCP/IP suite or some other error-correcting protocol within the information field as part of the payload.

10.1.7 Local Management Interface Protocol

The Local Management Interface (LMI) protocol provides operational support for the UNI. Originally defined by the Frame Relay Forum in 1990, it subsequently was adopted by ANSI and the ITU-T. The LMI is a polling protocol between the FRAD and the network, which periodically verifies the existence and availability of the PVC as well as the integrity of the UNI link.

10.1.8 Congestion Management

Frame Relay, as a highly shared packet network, is extremely efficient. Since all the links and all the switches that comprise the network are shared among large numbers of users and user organizations, the network is subject to variable and somewhat unpredictable levels of congestion. Indeed, it is designed for congestion as a natural occurrence and in support of bursty LAN-to-LAN internetworking applications, which are tolerant of delay and loss and have the time and ability to adjust and recover. This fact serves to reduce the overall cost of the network service, although it compromises the level of performance. This is a classic network optimization scenario in which an appropriate balance is struck between cost and performance. Congestion management is addressed through the following parameters, specified in an addendum to ANSI T1.606 [6–8]:

- **Access Rate:** Maximum data rate of the access channel, as defined by the bandwidth of the access link available for data transmission (e.g., a DS-0 of 64 kbps, a Fractional T1 of 768 kbps, a full T1 of 1.536 Mbps, an E-1 of 2.048 Mbps, or a T3 of 44.736 Mbps). Data can be transmitted or received over the access link at lesser rates, of course.

- **Committed Information Rate (CIR):** The data rate that the network guarantees to handle across the Virtual Circuit (VC), which can be either a PVC or an SVC, under normal conditions over a period of time. While the standards do not specify the time period, the CIR typically is an average data rate over a period of a few seconds. The CIR is based on mutual agreement between the carrier and the customer and should be based on the average maximum amount of traffic required to be supported reliably over the VC during the busiest hour of the day. The VC is defined as a matched source–destination pair, which can be either symmetric or asymmetric in terms of bandwidth. In the event that the CIR is exceeded, the network reserves the option to mark excess frames as Discard Eligible (DE) if the DCE has not already done so. The marking function typically takes place in the entry node at the edge of the network, after which discarding is done by any switch that experiences congestion. In order to obviate any issues of unnecessary congestion in the network core, the first edge device may discard frames rather than just mark them. There is no model for the relationship between the CIR and Access Rate. Some carriers permit *zero CIR*, meaning that the CIR is set at 0 percent. At zero CIR, all Offered Load is handled on a *best effort* basis, with absolutely no commitments or guarantees. Some carriers permit, or even require, the CIR to be set at 100 percent, meaning that all offered load will be guaranteed, with the maximum offered load determined by the Access Rate and Port Speed. Some carriers require that the CIR for a VC be set at no less than 50 percent of the Access Rate and Port Speed, with that relationship capped at T1/E-1 speeds—at E-3 levels of 34 Mbps and certainly at T3 levels of 45 Mbps, the bursts are too wildly variable to be managed effectively. Significantly, multiple VCs can be supported over a single access line.

- **Offered Load:** The data rate offered to the network for delivery measured in bps. The aggregate offered load can be less than the Access Rate supported by the access link and/or the port speed of the FRND but can never exceed whichever is less. As is always the case, the lowest common denominator defines the maximum level of performance.

- **Committed Burst Size (B_c):** Maximum amount of data that the carrier agrees to handle without discard under normal conditions. The B_c and line rate affect the calculation of the CIR for a VC.

- **Excess Burst Size (B_e):** Maximum amount of data that the network will accept in a block from a user without discard if bandwidth is available and over a specified time (T). In recognition of the bursty nature of LAN-to-LAN communications, the transmitting device may burst above the CIR/B_c for a brief period of time; the network attempts to accommodate such bursts within limits of burst size and burst interval. The network reserves the option to mark the excess data above B_c as Discard Eligible (DE) should the user CPE not have done so already. The last parts of larger bursts usually are discarded.

- **Measurement Interval (T):** Time interval measuring burst rates above the CIR and the length of such bursts.

Additional congestion management terminology includes the following:

- **Discard Eligibility (DE):** Indicates the eligibility of the frame for discard, in the event of congestion. Either the FRAD or the FRND may set discard eligibility. Theoretically, at least, the FRAD is programmed to recognize when the CIR is exceeded and to volunteer frames for discard should the network suffer congestion. The FRAD must make DE decisions in consideration of the user-layer application's tolerance for frame loss. As SDLC data, voice, and video, for example, are not tolerant of loss, such data must be confined to the CIR and must not be marked DE. Ethernet LAN data, on the other hand, are tolerant of loss; such data can exceed the CIR for the VC and can be marked DE without compromising the *best effort* expectations of the user-layer application.
- **Explicit Congestion Notification (ECN):** The means by which the network advises devices of network congestion. *Forward Explicit Congestion Notification* (FECN) advises the target device of network congestion so it can adjust its expectations. *Backward Explicit Congestion Notification* (BECN) advises the transmitting device of network congestion so it can reduce its rate of transmission accordingly. Theoretically, it is the responsibility of the various devices on the originating end of the data communication to adjust in some way. While not all FRADs have the ability to adjust, high-end routers can impose a certain amount of flow control through buffering and withholding acknowledgments to sources. The switches and hubs behind the FRAD have no ability to adjust other than filling their buffers, which typically are highly limited in capacity, if they exist at all. The originating terminal devices and peripherals have no ability to adjust, other than filling their very limited buffers, if any, and then simply stopping. If the devices run TCP/IP, then TCP slows down when it detects a lost frame.
- **Implicit Congestion Notification:** Inference by user equipment that congestion has occurred. Such inference is triggered by realization of the user device (e.g., FRAD, mainframe, or server) that one or more frames have been lost. Based on control mechanisms at the upper protocol layers of the end devices, the frames are resent to recover from such loss.

Aside from the congestion mechanisms specified in the frame standards, there are two types of congestion control algorithms. *Open-loop algorithms* permit the acceptance of the frames with no prior knowledge of the likelihood of their successful delivery. *Closed-loop algorithms* prevent frames from entering the network unless there is an extremely high probability of their being accepted, transported, and delivered without discard. The closed-loop algorithm fairly allocates backbone trunk bandwidth among all the PVCs configured on a particular trunk and in proportion to the CIRs [9]. Closed-loop networks clearly are more sophisticated and offer better performance, although at some additional cost.

10.1.9 How Frame Relay Networks Work

Frame Relay networks typically are logically configured as a star, or *hub and spoke*, as illustrated in Figure 10.4. In this example, 96 remote sites connect to a centralized data center. The access links serving the remote sites are defined as DS-0 at 64 kbps, and the access link serving the centralized data center is an unchannelized T1 at a signaling rate of 1.544 Mbps, of which 1.536 Mbps is available for data transmission and 0.008 Mbps is used for signaling and control purposes. Each PVC between the remote sites and the centralized data center is set at a CIR of 32 kbps, which is 50 percent of the access rate for each remote site. The aggregate rate of the CIRs for the 96 PVCs is double the access rate of the link serving the centralized data center; this relationship is a common maximum for Frame Relay service providers.

In this example, 48 of the remote sites can initiate a data session to the centralized data center at a rate of 32 kbps, which is within the CIR of each PVC and is exactly the access rate of the T1 link to the data center. On the average, at least, the carrier supports the CIR without frame loss. Should any of the 48 active remote sites burst above the CIR, the excess frames are marked as DE and are subject to discard if the network suffers congestion. Frames are marked DE by the transmitting FRAD in consideration of the tolerance of the user-layer application for loss. Should the excess frames not be subject to immediate discard, the carrier nodes buffer the excess frames, since the access rate of the T1 is *oversubscribed*. As the level of data destined for the data center diminishes, the excess frames are accommodated, assuming that the buffers in the carrier nodes have not overfilled and the excess frames, therefore, subsequently have been discarded. In a closed-loop scenario, should any other of the 48 remaining sites attempt to initiate a call to the centralized data center, that call is denied because the access link to the data center is oversubscribed. If the public Frame Relay network suffers congestion, overall,

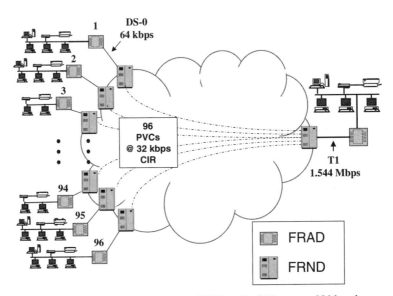

Figure 10.4 Frame Relay network, based on PVCs with CIRs set at 32 kbps between a centralized data center and remote sites in a star configuration.

network routers set FECN and BECN bits in frames flowing in both directions so that both the originating and the destination FRADs are advised of the condition and have the option of adjusting accordingly, if capable of doing so. As the PVCs are susceptible to catastrophic failure, backup PVCs (not shown) also typically are in place. Should both SNA traffic and LAN traffic be supported over the subject network, either the demanding and well-behaved SNA traffic is assigned to the CIR by the FRAD, with the LAN traffic sent in frames marked DE, or a separate PVC is provided in its support.

In addition to public, carrier-based Frame Relay services, a number of end-user organizations run the Frame Relay protocol over their legacy T/E-carrier, leased-line private networks. This approach takes advantage of both the performance characteristics of the legacy network and the protocol independence of Frame Relay, thereby enabling the sharing of the existing infrastructure for a wide variety of data traffic (e.g., SDLC, LAN-to-LAN, and IP-based traffic). As the leased-line network is private and dedicated, rather than public and shared, issues of congestion management are entirely under the control of the user organization. Therefore, desired performance levels are easily established and controlled. In such an environment, issues of congestion that might affect voice and video in a negative way are eliminated. Virtually all the major manufacturers of T/E-carrier nodal MUXs support Frame Relay through special interface cards.

10.1.10   Voice over Frame Relay

The very concept of transmitting voice over any sort of data network, much less a packet data network, would have been viewed as completely bizarre not so long ago. Voice over packet data networks includes *Voice over Frame Relay* (VoFR), *Voice over Internet Protocol* (VoIP), and *Voice over Asynchronous Transfer Mode* (VoATM). Frame Relay certainly was not originally designed for voice traffic, and neither was IP, although ATM was designed from the ground up for all types of traffic. For the moment, let us focus on VoFR, as specified in the Frame Relay Forum's FRF11.1, Voice over Frame Relay Implementation Agreement (May 1997).

While Frame Relay is intended as a data networking service, specifically and originally for LAN-to-LAN internetworking, it also can be used in support of iso-chronous voice and voice-band data (i.e., fax and analog modem) traffic. But you must consider the fact that issues of delay can affect the quality of the voice transmission over time. Even within the constraints of the CIR, the level of bandwidth is measured and ensured only as an average over a period of time, rather than on an instantaneous basis. Because Frame Relay is a packet-switching, rather than a circuit-switching, network service technology, there is no provision for temporary, continuous, and exclusive bandwidth through a Frame Relay network. As a result, voice frames suffer from delay, loss, and error at levels that are variable and unpredictable in nature. The result can be less than aesthetically pleasing. In the event that the received voice stream becomes unintelligible, the listener always can resort to the *Huh?* protocol, as in "Huh? What did you say?"

Assuming, however, that the end-user organization is prepared to accept some level of unpredictability with respect to quality, Frame Relay supports voice and other voice-band data and at very reasonable cost. Generally speaking, you should use the Frame Relay network predominantly for data networking. Assuming, however, that excess capacity exists at the CIR level and that the network is not experiencing congestion at the moment (i.e., no explicit congestion notifications have been delivered and no implicit congestion notifications have occurred), near-toll-quality voice is possible over a Frame Relay network. Why would you even consider VoFR? There really is only one answer: It's FREE—or at least very inexpensive, especially over flat-rate PVCs or SVCs.

Actually, it turns out that there might be another reason, or at least an additional reason. A number of major domestic and international carriers offer, and even promote, VoFR as part of an integrated network solution for the data-heavy enterprise. Those carriers particularly promote VoFR as part of a managed-service offering. Such an offering involves the carrier's managing the service and equipment from end to end, thereby ensuring its ability to strike the proper network design to support business class voice over a packet data network. That design may include separate VoFR PVCs and adequately sized (i.e., oversized) CIRs. Ultimately, however, the only reason to consider VoFR is that it can ride free within the CIR limits of a Frame Relay network intended for LAN-to-LAN communications. Ironically, people hesitant to put voice on Frame Relay eagerly put voice on IP when that VoIP packet may very well be encapsulated in Frame Relay and transmitted over an ATM network.

10.1.10.1 VoFR Theory: Compression Is the Key   In consideration of the fact that the voice information stream is not constant and certainly is not constantly changing at a high rate of speed, VoFR can take advantage of various compression techniques to relax the amount of bandwidth required for the voice stream. Specifically, only the changes in speech patterns from one set of samples to another need to transmit across the network. The natural and predictable pauses in human speech, for example, can be noted, eliminated on the transmit side, and reinserted on the receive side through a process known as silence suppression. Similarly, redundant or repetitive speech sounds can be removed on the transmit side and reinserted on the receive side. Once the silence is suppressed and the redundancy noted, the remainder of the set of voice samples can be further compressed and inserted into relatively small frames. The small frame size allows a number of sound bytes to be presented to and delivered through the network on a frequent basis. While this approach is somewhat overhead intensive, it provides for a relatively normal flow of speech information. There exist a number of low-bit-rate compression algorithms, both standard and proprietary, that are implemented by various router manufacturers. The most popular are those in the CELP (Code-Excited Linear Prediction) family. CELP and other compression algorithms support very reasonable business-quality voice (under conditions of low network congestion or where the voice traffic remains within its CIR) at bit rates as low as 8 kbps. Additionally, the Frame Relay Forum adopted Conjugate Structure-Algebraic Code-Excited Linear Prediction (CS-ACELP) as FRF.11. According to the Frame Relay Forum, only 22 percent of normal speech comprises essential components, with the balance comprising either pauses or repetitive speech patterns [10].

• **Code-Excited Linear Prediction (CELP):** Key to CELP and its derivatives is the creation and maintenance of a codebook, which is a binary description of a

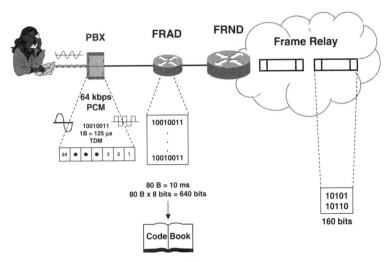

Figure 10.5   CELP compression.

set of voice samples. Specifically, and as illustrated in Figure 10.5, CELP involves the binary description of a set of 80 *Pulse Code Modulation* (PCM) voice samples, representing 10 ms (10 milliseconds, or 1/100th of a second) of a voice stream, gathered in a buffer. Then, the data set is compressed to remove silence and redundancy, the volume level is normalized, and the resulting data set is compared to a set of candidate shapes in the codebook. The data transmitted across the network include the index number of the selected code description and the average loudness level of the set of samples. Every 10 ms, the code is sent across the network in a block of 160 bits, yielding a data rate of 16 kbps, which compares very favorably with PCM voice over circuit-switched TDM networks at 64 kbps. The compression ratio is 4:1. At the receiving end of the transmission, the transmitted code is compared to the codebook, the PCM signal is reconstructed, and, eventually, the analog signal is reconstructed. The reproduction is not perfect but generally is close enough to yield good perceived quality. The devices that perform these processes of compression and decompression are Digital Signal Processors (DSPs), which include the basic codec function.

· **Low-Delay Code-Excited Linear Prediction (LD-CELP):** As defined in ITU-T G.728, LD-CELP also is geared to a rate of 16 kbps, although bit rates as low as 12.8 kbps can be achieved. The lower level of delay suggested by the designation is due to the fact that only five PCM samples, representing 0.625 ms of the voice stream, are accumulated in a block. Considering that each sample is expressed as a 2-bit value and that five samples equal a pattern of 10 adjacent bits, 2^{10} yields 1024 possible combinations, each of which describes a section or shape from the codebook and each of which is part of an overall voice stream. The more frequent transmission of the shorter data blocks yields lower levels of delay through faster processing by the DSPs, and the compression technique yields more efficient use of bandwidth. LD-CELP yields quality that generally is considered to be on a par with *Adaptive Differential Pulse Code Modulation* (ADPCM), which I discussed in Chapter 8.

TABLE 10.3 ITU-T Standard Voice Compression Algorithms

Compression Algorithm, Year	ITU-T Recommendation	Bit Rate (kbps)	Compression Delay (ms)	Mean Opinion Score (MOS)[a]
Pulse Code Modulation (PCM), 1988	G.711	64	0.75	4.4
Adaptive Differential Pulse Code Modulation (ADPCM), 1991	G.721 G.722 G.723 G.726 (FRF.11) G.727	32 64 20, 40 16, 24, 32, 40 16, 24, 32, 40	1.0	4.2
Dual Rate Speech Coder for Multimedia Communications, 1996	G.723, G.723.1 (H.324 umbrella)	6.3, 5.3	30.0	3.5–3.98
Low Delay-Code Excited Linear Prediction (LD-CELP), 1994	G.723.1 G.728	5.3, 6.4 16	3.0–5.0	4.2
Conjugate Structure-Algebraic Code Excited Linear Prediction (CS-ACELP), 1996	G.729 (FRF.11)	8	10.0	4.2

[a]These are representative scores. Actual MOS values are dependent on the skill with which the manufacturer implements the algorithm, the specifics of the DSP, and other factors.
Source: [12–14].

- **Conjugate Structure-Algebraic Code-Excited Linear Prediction (CS-ACELP):** As defined in ITU-T G.729, ACELP improves on the CELP concept through the algebraic expression, rather than the numeric description, of each entry in the codebook. ACELP yields quality that is considered to be as good as ADPCM, but requiring bandwidth of only 8 kbps. CS-ACELP is geared toward multichannel operation. Proprietary versions of ACELP exist as well [11].

Other, nonstandard compression algorithms take the bit rate as low as 4.8 kbps (or lower), but with considerable loss of quality. Current ITU-T standards for voice compression used in various VoFR codecs include those presented in Table 10.3, all of which begin with PCM voice. In addition to the base compression algorithms, various *continuity algorithms* are employed to intelligently fill the void of missing or errored compressed voice frames by stretching the previous voice frames or blending several frames together.

The perceived quality of the voice signal is based on exhaustive tests, the most common of which is defined in the ITU-T P.800 specification, Methods for Subjective Determination of Voice Quality. P.800 involves the subjective evaluation of preselected voice samples of voice encoding and compression algorithms. The evaluation is conducted by a panel of *expert listeners* comprising a mixed group of men and women under controlled conditions. The result of the evaluation is a *Mean Opinion Score* (MOS) in a range from 1 to 5, with 1 being *bad* and 5 being *excellent*. The components of the MOS are as follows:

- *Opinion Scale:* conversation test (bad to excellent), difficulty scale (yes or no)
- *Opinion Scale:* listening test (bad to excellent), listening effort scale ("No meaning understood with any reasonable effort" to "complete relaxation possible, no effort required")
- *Loudness:* preference scale ("Much quieter than preferred" to "Much louder than preferred").

An MOS of 4.0 is considered to be *toll quality*. While P.800 is an effective means of quantifying such a highly subjective perception, it is expensive and time consuming. In the mid-1990s, therefore, the ITU-T began the process of automating the objective measurement and testing of end-to-end voice quality across both circuit-switched and packet-switched networks. The strongest of those techniques include the following:

- Perceptual Analysis/Measurement System (PAMS)
- Perceptual Speech Quality Measurement (PSQM), standardized as P.861
- Perceptual Evaluation of Speech Quality (PESQ), standardized as P.862
- Single Ended Method for Objective Speech Quality Assessment in Narrow-Band Telephony Applications, standardized as P.563

Regardless of the compression technique employed, VoFR suffers from some additional loss of quality due to issues of echo, delay, delay variability (i.e., jitter), and loss. Through network buffering and voice frame prioritization, these factors can at least be mitigated but never entirely be overcome. Additionally, some FRAD manufacturers support the definition of a separate PVC for delay-sensitive traffic such as voice, while others support multiplexing of voice data frames only over the same PVC as that used in support of data. Some manufacturers support both small frame sizes for voice and large frame sizes for data, while others force all data into the smaller frame size more appropriate for voice; the latter approach forces all data into the less efficient (more overhead intensive) frame size. Notably, VoFR FRADs are proprietary in nature, thereby requiring a single-vendor VoFR network [12–20].

In order to mitigate the inherent difficulties of transmitting delay-sensitive voice traffic over Frame Relay, some manufacturers and carriers offer various priority management techniques. Several manufacturers of multifunction FRADs and routers, for example, support the identification of high-priority traffic in the frame header. Some service providers also offer PVCs of varying levels of delay/priority, usually by mapping the Frame Relay connection to an ATM connection with these properties. Priority levels generally are defined as follows [21]:

- **Real-Time Variable Frame Rate:** Top priority; suited to delay-sensitive, mission-critical applications such as voice and SNA; premium pricing

- **Non Real-Time Variable Frame Rate:** No-priority designation; suited to LAN and business-class Internet and intranet IP traffic, which can tolerate some level of delay
- **Available/Unspecified Frame Rate:** Low-priority designation; suited to Internet access, e-mail, file transfer, monitoring, and other low-priority applications that can tolerate relatively significant levels of delay

As it turns out, much of the voice carried over Frame Relay links is, in fact, VoIP traffic. While a more detailed discussion of VoIP occurs in Chapter 12, the subject of VoIP over Frame Relay merits some discussion now. VoIP is very overhead intensive, with a VoIP packet comprising 44 octets of various headers and 20 octets consumed by the payload. As a result, a 64-kbps Frame Relay PVC can support only two G.729 compressed voice conversations. To improve the efficiency of VoFR, the Frame Relay Forum developed FRF.20, Frame Relay IP Header Compression Implementation Agreement. The algorithm described in FRF.20 examines the 44 octets of VoIP header, looking for redundancy and other opportunities to send only a reference to the header, rather than the entire header. This compression technique reduces the header to as few as two to four octets, with the process reversed at the receiving end of the circuit. Thereby, as many as five VoIP calls can be supported over the same 64-kbps PVC [22].

VoFR efficiency can be improved further through subchannel multiplexing. A single data link connection (DLCI) can support multiple VoFR transmissions and multiple data transmissions between two locations in an enterprise over a public or private Frame Relay network, although the voice and data payloads are carried in separate frames. As illustrated in Figure 10.6, multiple VoFR payloads associated with multiple conversations (or even a single conversation) are multiplexed into subframes, or subchannels, each of which adds a header of at least one octet, containing the least significant bits of the *Sub–Channel IDentification* (CID), and extension and length indications. If the *Extension Indication* (EI) bit also is set, the second octet contains the most significant bits of the voice/data channel identification. If the *Length Indication* (LI) bit is set, the third octet contains the payload length, which is sensitive to the compression algorithm employed.

10.1.11 Frame Relay Costs

The costs of Frame Relay vary widely among equipment manufacturers and carriers. It is clear, however, that Frame Relay is highly cost effective for internetworking large numbers of sites at rates up to 44.736 Mbps (T3). Such a network can be configured as a full mesh, a partial mesh, or a star.

Equipment on the premises comprises a FRAD, which can be either stand alone or built under the skin of a bridge or router. A DSU/CSU, which also may be stand

Figure 10.6 VoFR subframes.

alone or built in, provides interface to an electrically based digital access link. In addition to the cost of the FRAD, the carriers' charges typically include some combination of port charge (bandwidth sensitive), access link (bandwidth and distance sensitive), and Committed Information Rate (CIR) per Virtual Circuit (VC), related to the burst size B_c. The B_c is sensitive to time interval T and is calculated as $CIR = B_c/T$.

Frame Relay networks originally were priced on a case-by-case basis. In consideration of the widespread popularity of the service, however, in October 1995 the Federal Communications Commission (FCC) classified Frame Relay as a *basic service* in the United States. This reclassification required that the major carriers file tariffs that included rate structures, which enabled smaller users to comparison shop, which simplified the comparison process and likely reduced their costs. Larger users, on the other hand, lost some bargaining power.

Prices have varied considerably in the United States and worldwide since the FCC decision. In the highly competitive Frame Relay market, mesh or even partial-mesh networking can be accomplished for far less than the cost of a comparable FT1 network. Costs can be mitigated to some extent through the use of compression devices, which permit transmission at lower port speeds and CIRs through the recognition of patterns within data frames intended for multicasting or transmission to multiple addresses.

10.1.12 Frame Relay Attributes and Issues

FR unique characteristics offer some advantages over its predecessor technologies of X.25, Switched 56/64 kbps, and leased lines. Additionally, Frame Relay is widely available both domestically and internationally. Also, the market is highly competitive, which puts it in an advantageous position relative to ATM. However, the recent market shift toward IP has caused not only a slowing in Frame Relay demand but actually a decline in Frame Relay service penetration. The International Data Corporation (IDC) estimated that there were approximately 1.3 million Frame Relay ports in the United States at the end of 2005, but that number was expected to drop to 1.2 million in 2006 and fall to 751,000 in 2008. The IDC reports that not only is Frame Relay decreasing but also carriers are now beginning to eliminate it completely in favor of IP-based VPNs [23].

10.1.12.1 Frame Relay Advantages Advantages of Frame Relay include its excellent support for bandwidth-intensive data and image traffic. Even stream-oriented traffic such as real-time voice and video can be supported with reasonable quality. The absolute speed of Frame Relay, its improved congestion control, and reduced latency certainly are improvements over X.25 networks. Because Frame Relay is protocol insensitive, it can carry virtually any form of data in variable-size frames. Bandwidth on demand, within the limit of the access line, is provided, generally with reasonable costs for high-speed bursts. Additionally, the network is highly redundant, thereby providing improved network resiliency. SVCs are especially resilient, although few carriers support them.

As a VPN service technology, the elimination of dedicated circuits makes Frame Relay highly cost effective when compared to services such as DDS and T/E-carrier, which have cost components calculated based on distance. Frame Relay is reasonably priced due to both its highly shared VPN nature and the fact that the service is widely available and highly competitive, with savings of 30–40 percent over leased lines being quite common. Frame Relay costs also are somewhat scalable, maintaining a reasonably graceful relationship with the needs of the user organization in terms of bandwidth and number of terminating locations, at least in comparison to leased lines. Full-mesh, partial-mesh, and hub-and-spoke networking, therefore, all can be accomplished at reasonable cost on either a symmetric or an asymmetric basis. Frame Relay also offers a considerable measure of security, with each enterprise network essentially forming a Closed User Group (CUG).

10.1.12.2 Frame Relay Issues   Disadvantages of Frame Relay include its latency, which is inherent in any highly shared data-oriented VPN, including X.25 and IP-based networks. Although Frame Relay's latency level is an improvement over X.25, it does not compare well with ATM. The latency issue can be significant when Frame Relay is used in support of SNA traffic, although FRADs can spoof SDLC. The latency issue generally renders it largely unsuitable for highly intensive voice and video, at least when measured in terms of traditional toll quality, although priority mechanisms go a long way toward resolving that issue.

Frame Relay also offers no inherent mechanisms for frame sequencing and error correction for the user payload, unlike its predecessor X.25, which performed these on a hop-by-hop basis at the network nodes. As Frame Relay assumes that these functions are accomplished at higher layers in the end-user domain, Frame Relay is relieved of these responsibilities. Stripping out these functions offers the advantage of reduced processing at the network nodes and, therefore, reduced latency, congestion, and cost. But the elimination of these functions also compromises the reliability of the data stream. In order to compensate for this shortcoming, data applications typically involve running TCP/IP inside the information field as part of the user payload. As a result, the error control is performed on an end-to-end basis in the CPE domain. Since TCP/IP commonly is used in support of LAN communications, this additional step and the additional overhead associated with TCP/IP is not considered particularly burdensome—after all, Frame Relay is intended for LAN-to-LAN internetworking. Where real-time voice and video applications are supported over Frame Relay, User Datagram Protocal (UDP) replaces TCP in the protocol mix because there is no time to recover from loss or error through retransmission, in any event.

Note that Frame Relay originally was marketed as a lower cost replacement for leased lines and generally is provisioned on the basis of PVCs, which are preordained source–destination paths. Like private, leased-line networks, such paths are susceptible to catastrophic failure and require backup. Backup PVCs can satisfy this requirement because they are invoked virtually immediately if the primary PVC fails. SVCs are more reliable, as they automatically seek a reliable path on a call-by-call basis. If the entire network fails, however, neither option is satisfactory. For example, the entire AT&T Frame Relay network failed in April 1998 for a period of up to 23 h, depending on which press releases you believe. The problem apparently had to do with bugs in some software upgrades on a Cascade (now Cisco) switch, again depending on which press releases you believe [24, 25]. A more effective backup strategy in the event of such a catastrophic failure is carrier redundancy, rather than PVC redundancy, or even SVCs. Some carriers also offer redundancy in the form of *redirected PVCs*, which are inactive PVCs that can be activated very quickly to direct traffic around a point of failure in the network or in the access loop. Redirected PVCs also can be used to redirect traffic to a backup data center should the primary data center suffer a failure. Perhaps the most effective backup is in the form of both carrier and service redundancy, such as ISDN, which is a common backup strategy for Frame Relay in environments where data connectivity is both mission critical and time sensitive.

Frame Relay has been criticized for its lack of scalability. Large and complex Frame Relay networks require a considerable number of PVCs, especially in a mesh or partial-mesh configuration. SVCs avoid this requirement but are not universally available. One solution to this scalability issue is that of *IP-Enabled Frame Relay*, also known as *Private IP*, which makes use of virtual routers to route frames at Layer 3 (Network Layer) of the OSI Reference Model. The IP addressing information contained in the data payload and immediately following the Frame Relay header provides the addressing information required for flexible routing on a switched basis. IP-enabled Frame Relay typically makes use of *MultiProtocol Label Switching* (MPLS) to speed frame processing and provide QoS support. I discuss IP and MPLS in considerable detail in Chapter 12.

10.1.13 Frame Relay Applications

Frame Relay was designed to fill the gap between packet-switched networks (X.25), circuit-switched networks (Switched 56/64 kbps), and dedicated data networks (DDS, T/E-carrier). It is intended for intensive data communications involving block-level communications of data and image information. Frame Relay supports voice and low-speed video, although the quality can be spotty due to intrinsic issues of frame latency, jitter, and loss in a poorly designed network.

Frame Relay applications primarily, therefore, are data or image in nature. LAN internetworking is the driving force behind Frame Relay, although controller-to-host, terminal-to-host, and host-to-host applications abound. The more recent availability of dial-up access also makes the service cost effective for bandwidth-intensive telecommuting application. Internet Service Providers (ISPs) and Internet backbone providers make significant use of Frame Relay, both for user access to the ISP and for backbone network application. In Internet applications (indeed, a great many data applications), the TCP/IP protocol suite runs inside Frame Relay in order to ensure the reliability of the data stream.

An excellent example of the application of Frame Relay is that of the airline reservation networks, all of which are transitioning from X.25. In 1995, Apollo Travel Services converted its reservations network to Frame Relay to connect approximately 15,000 travel agency workstations at 1000 sites. Owned by United Airlines, USAir, and Air Canada, Apollo expected response time to improve to 2s from the previous 4s. The network runs at 56 kbps, as opposed to X.25 speeds of 2400–4800 bps. Using AT&T's InterSpan Frame Relay Service, Apollo is able to establish end-to-end TCP/IP connections from each agency to the Apollo reservations center. The service also supports mesh networking between travel agencies, thereby eliminating the requirement for all traffic to go through the Apollo head end [26].

Voice over Frame Relay (VoFR), over both private and public networks, is an option that has created a lot of interest. Although the voice stream is subject to intrinsic Frame Relay delays and although the service is overhead intensive for such an application, it does enable the user organization to take advantage of occasional excess bandwidth to connect voice for free, or at least at very low cost once the cost of the voice FRADs is factored into the equation. Through the use of advanced compression techniques such as standards-based ACELP and LD-CELP, VoFR can support good-quality voice at 16 bps, 8 kbps, and even 4.8 kbps.

10.2 SWITCHED MULTIMEGABIT DATA SERVICE

Switched Multimegabit Data Service (SMDS) is an offshoot of the *Distributed Queue Dual Bus* (DQDB) technology defined by the IEEE 802.6 standard for *Metropolitan Area Networks* (MANs) as a means of extending the reach of the LAN across a metropolitan area. The original work on the DQDB concept was done at the University of Western Australia where it was known as *Queued Packet Synchronous Exchange* (QPSX). Subsequently, the original technology was licensed to QPSX Ltd., formed by the University of Western Australia and Telecom Australia [6]. The commercial success, although highly limited, of SMDS is attributable to further development work by Bellcore (now Telcordia Technologies) at the request of the RBOCs. Bell Atlantic (now part of Verizon) introduced the first commercial offering in 1992 at 5 Mbps based on a test implementation at Temple University [27].

Since then, SMDS enjoyed limited, short-lived success in the United States through deployment by most of the RBOCs and GTE (now part of Verizon). As the RBOCs and their subsidiary BOCs were primarily Local Exchange Carriers (LECs) generally limited to providing service within the confines of the LATA at the time, they exhibited particular interest in network technologies that were appropriate for providing service in confined geographic areas. The geographical limitation of SMDS, therefore, did not pose a practical limitation for the LECs.

SMDS was originally described by Bellcore (now Telcordia Technologies) as a high-speed, connectionless, public, packet-switching service that extends LAN-like performance beyond the subscriber's premises, across a metropolitan or wide area. SMDS is a MAN network service based on cell-switching technology. Generally delivered over a SONET/SDH ring, SMDS has a maximum effective serving radius of approximately 30 miles (50 km). SMDS is a connectionless service that accepts user data in the form of a Protocol Data Unit (PDU) up to 9188 octets in length over access lines of up to 45 Mbps (T3). The user data are segmented into 53-octet cells, 48 octets of which are payload and 5 octets of which are overhead. The segmentation process can occur either at the network node or in the user equipment.

SMDS is designed to support LAN-to-LAN traffic under the IEEE 802.6 standards, although other data applications are supported effectively. SMDS offers excellent performance characteristics, including a guaranteed rate of access, transport, and delivery. SMDS also provides a smooth migration path to ATM. In fact, the 53-octet cell format was chosen specifically for this reason.

While some of the RBOCs deployed SMDS fairly aggressively in the past, their ardor cooled noticeably beginning in 1996. Clearly, the simplicity of Frame Relay contributed to its popularity, which affected SMDS adversely. Further, in their backbones the RBOCs also increasingly deployed ATM, which is not geographically restrained and is much more flexible. SMDS is no longer offered or supported by any carriers in the United States.

SMDS also enjoyed moderate success in Western Europe, where the nations tend to be small in geographic terms and where the population density of large businesses is high in the major metropolitan areas. The European version of SMDS was known as CBDS (*Connectionless Broadband Data Service*). CBDS is available in Australia (Telecom Australia), Germany (DBP), and England (BT).

The history of the SMDS Interest Group (SIG) underscores the overall failure of SMDS. The SIG was a consortium of vendors and consultants that worked to advance SMDS as a standards-based, interoperable solution for high-performance data connectivity. On June 17, 1997, the board of trustees announced that its mission of advancing the cause of SMDS was fulfilled and the group was disbanded, turning all its responsibilities over to unnamed regional organizations. Essentially, the SIG lost the battle, declared victory, and everyone went home a winner. At this point, SMDS is essentially a historical footnote that promoted the cause of high-performance data communications and served as a technological precursor to ATM.

10.3 ASYNCHRONOUS TRANSFER MODE

It seems the only people using the machines were a small number of prostitutes and gamblers who didn't want to deal with tellers face to face.
  Luther George Simjian (1905–1997), inventor of the Bankmatic Automatic Teller
                                          Machine, discussing the slow pace of its early adoption

The automatic teller machine (usually called ATM), also known by at least a dozen other names including *cash machine*, has little to do with Asynchronous Transfer Mode (ATM), except for the fact that the two technologies share a common acronym. Actually, there are some ATMs that interconnect over networks running the ATM protocol, but the real reason that I selected this quote is that I couldn't find any interesting—much less entertaining—quotes about the ATM protocol. This quote also illustrates the dangers of using an Internet search engine when researching a topic for a technical book. In the event that you are interested in the history of the ATM, aka *hole-in-the-wall machine*, Simjian's Bankmatic cash-dispensing machine was installed in a City Bank of New York branch in New York City in 1939. It was removed after only six months, due to lack of acceptance. The modern ATM was invented in 1963 or 1968, depending on which website you believe, by a trio of Docutel engineers, Don Wentzel, Tom Barnes, and George Chastain. The patent for that device was issued in 1973 and the first machine was installed in a New York Chemical Bank [28–30].

Asynchronous Transfer Mode (ATM) was developed in concept by the ITU-T as an outgrowth of ISDN standards that developed in the early 1980s. While the concept of Narrowband ISDN (N-ISDN) was intriguing, it soon became clear that the demand would emerge for a wide range of bandwidth-intensive services that would go beyond the scope of narrowband ($N \times 64$-kbps) transmission facilities and circuit-switched connections. Given that assumption, which has since proven correct, the ITU-T cast an eye toward the future of broadband networking, which it defined as a rate greater than the primary rate, referring to ISDN Primary Rate Interface

(PRI) at 1.544 Mbps or Primary Rate Access (PRA) at 2.048 Mbps. The conceptual result of the ITU-T deliberations is Broadband ISDN (B-ISDN), which is founded on ATM network technology. While B-ISDN has never materialized in full form, ATM has made a considerable mark on the broadband network landscape.

The first ATM network in the United States was a testbed gigabit network known as the *National Research and Education Network* (NREN). Sponsored by the U.S. *Advanced Research Project Agency* (ARPA) and the *National Science Foundation* (NSF), the project began in 1990. In Europe, a consortium of carriers, end users, and universities sponsored the *Research for Advanced Communications in Europe* (RACE) project 1022, which was initialized in 1987 to demonstrate the feasibility of ATM. The result of the RACE initiative was the *R1022 ATM Technology Testbed* (RATT). RACE project 2061, also known as *EXPLOIT*, is a more recent RACE project intended to prove the viability of *Integrated Broadband Communications* (IBC) in the European Community (EC). *Advanced Communications Technologies and Services* (ACTS) was developed as the successor program to RACE and continues that work on ATM networking and some 200 other projects.

Beginning in the early 1990s and for nearly a decade, ATM was highly touted as the ultimate network-switching solution, given its high speed, its ability to serve all information types, and its ability to guarantee each information type an appropriate Quality of Service (QoS). ATM was ultimately expected by many to replace all of the existing circuit- and packet- and frame-switching technologies currently in place, but that now appears highly unlikely. ATM remains widely used as a backbone switching technology, where it was heavily favored by the incumbent LECs and IXCs until just recently, when Internet Protocol (IP) and Multiprotocol Label Switching (MPLS) based networking found support. ATM remains favored in Asymmetric Digital Subscriber Line (ADSL) applications and as the Data Link Layer protocol for Broadband Passive Optical Network (BPON) and Gigabit PON (GPON) broadband local loop.

10.3.1 ATM Defined

Asynchronous transfer mode is a fast-packet, connection-oriented, cell-switching technology for broadband signals. ATM is designed, from concept up, to accommodate any form of information—voice, facsimile, data, video, image, and multimedia —whether compressed or uncompressed, at broadband speeds. Further, all such data can be supported with a very small set of network protocols, regardless of whether the network is local, metropolitan, or wide area in nature. In terms of user access rates, ATM generally operates at minimum access speeds of DS-1 and DS-3. The OC-1 (51.84 Mbps) interface was defined but not commonly implemented, and OC-3 (155 Mbps) is not unusual. The backbone transmission rates are DS-3 and OC-1 at a minimum and generally OC-3 or better.

ATM traffic consists of three basic types. *Constant Bit Rate* (CBR) traffic, such as uncompressed voice based on Pulse Code Modulation (PCM) and Time Division Multiplexing (TDM), requires the presentation of time slots on a regular and unswerving basis. *Variable Bit Rate* (VBR) traffic, such as compressed voice and video and bursty data traffic, requires access to time slots at a rate that can vary dramatically from time to time. *Available Bit Rate* (ABR) traffic, also known as *best effort ATM,* supports bursty LAN traffic and other traffic that can deal with time slot access on an as-available basis. Whether CBR, VBR, or ABR, all ATM traffic enjoys specifically defined QoS parameters.

In any case, the data are presented to and accepted by the network on an *asynchronous* basis, and the ATM switches *transfer* the data from switch to switch in a hop-by-hop *mode*, hence the term asynchronous transfer mode. The optical fiber transmission facilities, of course, are tightly synchronized. The ATM switch and all other network elements are synchronized with the pipe as well.

The user data are sent to the network over a digital facility. At the workstation, router, or ATM switch, data are organized into 48-octet cells. Each cell is prepended with a header of five octets to create a cell of 53 octets. The cells are multiplexed and contend for access to a broadband facility that ideally is SONET in nature. In combination with SONET/SDH, ATM was intended to be the ultimate and complete integrated network solution, supporting a theoretically infinite range of services: B-ISDN.

The small cell size reaps several advantages. First, it can accommodate any form of data—digital voice, facsimile, data, video, and so on. Second, the fixed length of the cell offers the network switches the advantage of predictability, as compared to a variable-length frame. Third, the fixed cell size facilitates the implementation of switching functions in hardware (i.e., silicon), which enables processes to be accomplished at significantly greater speed than does software, especially if processing variable-size frames. These last two considerations yield decreased delay, as data move through the switching systems and across the transmission links in frequent little blasts. Long, and especially variably long, frames occupy the attention of the network for relatively long periods of time, causing delay as other data wait to be processed.

ATM is the first network technology to offer truly *guaranteed bandwidth on demand*, as the bandwidth can vary during the course of the call [7]. True enough, other services offer bandwidth that can vary with each call, but none can offer the ability to adjust the amount of bandwidth required to support a call once the call is established and to guarantee that it will be available when required. A high-quality videoconference, for example, might require 1.544-Mbps (T1) capacity as a rule. Yet, with the sophistication of contemporary compression techniques, that call might require much less bandwidth much of the time. Once that call is set up, a full T1 is dedicated to it, regardless of the actual bandwidth requirement moment by moment. ATM is not so rigid; it can adapt dynamically to the bandwidth actually required.

ATM networks provide for error detection of the header only and not the payload. ATM networks make no provision for error correction. The main concern is to deliver a cell to only the addressee. If the address is corrupted, the cell is discarded, and the endpoint is responsible for determining that fact and recovering from the loss through a request for retransmission. The advantages of this simplified approach to error control are increased speed of switching, reduced latency, and lowered cost, as the ATM switches require less memory and processing power.

10.3.2 ATM Standards

The ITU-T sets ATM standards. The first set of B-ISDN standards recommendations began in 1988, inexorably linking ATM and B-ISDN. In 1992, the ATM Forum formed as a voluntary organization of manufacturers, consultants, and interested parties; that forum developed interoperability specifications based on member consensus. The Internet Engineering TaskForce (IETF) also has gotten involved because ATM has significant implications relative to the Internet, at least at a backbone level. The Frame Relay Forum (FRF) also worked with the ATM Forum in the development and publishing of joint IAs that specify the protocol interworking functions between Frame Relay and ATM networks. (*Note:* The ATM Forum, Frame Relay Forum, and MPLS Forum have since merged into the MFA Forum). ITU-T Standards Recommendations of significance include the following:

- I.113: B-ISDN Vocabulary
- I.121: Broadband Aspects of ISDN
- I.150: B-ISDN ATM Functional Characteristics
- I.211: B-ISDN Service Aspects
- I.311: B-ISDN General Network Aspects
- I.321: B-ISDN Protocol Reference Model
- I.327: B-ISDN Functional Architecture Aspects
- I.361: B-ISDN ATM Layer Specification
- I.362: B-ISDN ATM Adaptation Layer Functional Description
- I.363: B-ISDN ATM Adaptation Layer Specification
- I.413: B-ISDN User-Network Interface
- I.432: B-ISDN User-Network Interface-Physical Layer Specification
- I.555: Frame Relay and ATM Internetworking
- I.610: B-ISDN Operations and Maintenance Principles and Functions

Significant ATM Forum implementation documents include the following:

- ATM User-Network Interface (UNI) Specification for PVCs
- ATM Broadband InterCarrier Interface (BICI) Specification
- ATM Data Exchange Interface (DXI) Specification

Internet Engineering TaskForce (IETF) RFCs (Requests for Comment) include the following:

- RFC 1821: Integration of Real-Time Services in an IP-ATM Network Architecture
- RFC 2225: Classical IP and ARP (Address Resolution Protocol) over ATM
- RFC 2684: Multiprotocol Encapsulation over ATM Adaptation Layer 5

In 1996, the ATM Forum realized that the plethora of ATM standards was creating confusion in the manufacturer community and that issues of backward compatibility were developing as the newer standards leapfrogged the widespread implementations of earlier standards. To alleviate this situation, the Anchorage Accord, a milestone ATM Forum document (April 12, 1998), so named because of the meeting location, outlined which versions of ATM Forum specifications vendors should implement. The approximately 60 baseline specifications designated for successful market entry of ATM products and services include Broadband InterCarrier Interface (BICI), Interim Local Management Interface (ILMI), LAN Emulation (LANE), network management, Private Network Node Interface (PNNI), signaling, Switched Multimegabit Data Service (SMDS), IP over ATM, traffic management, and a number of physical interfaces. The accord also limits the conditions under which specifications are revised in order to reduce the potential for future confusion [31, 32].

10.3.3 ATM Access

Access to an ATM backbone network occurs at rates of DS-1 or greater. Whether access is from an end-user CPE environment, a Frame Relay or X.25 network switch, or an IP-based router, the access rate is DS-1, DS-3, or a higher rate such as OC-3 (155 Mbps) or OC-12 (622 Mbps). Access is provided through a *User Network Interface* (UNI) specified by the MFA Forum:

- *User Network Interface* (UNI) specifically refers to a UNI between a user device and an ATM network. A *private UNI* is employed in a private ATM network (e.g., LAN domain) and for access to a public network. A *public UNI* is used between switches in a public ATM network. As the UNI is cell based, end-user CPE converts the user Protocol Data Units (PDUs) into ATM PDUs and cells.

- *Data Exchange Interface* (DXI) is a private UNI for end-user access to an ATM network from DTE/DCE such as a bridge, router, or ATM DSU. The DXI concept enables the sharing of protocol responsibility between the user and the network provider. At the physical layer, the DXI permits connection via V.35, EIA 449/530, or EIA 612/613 HSSI connection. A variation of the HDLC protocol is used at the data link layer. The user information is encapsulated within a DXI (HDLC) frame and converted to the appropriate, class-specific ATM protocol at the DCE. The DCE assumes responsibility for functions through *Segmentation and Reassembly* (SAR), presenting the data to the ATM network switch in ATM cells. Several DXI modes exist that correspond to AAL 3/4 and AAL 5.

- *The Network-to-Network Interface* (NNI), also known *as B-ISDN InterCarrier Interface* (B-ICI), is a public UNI for interconnection of public networks via PVCs. NNIs exist for interconnection of ATM networks and for interconnection of ATM and Frame Relay networks. Interoperability issues that need to be addressed include protocol conversion, mapping between virtual circuits, alignment of traffic management parameters, and mapping of local network management information.

- *Frame UNI* (FUNI) is a derivative of the DXI standard to extend ATM access to smaller sites at rates from 56 kbps to 1.544 Mbps (T1). Low-speed data enter a router, which forwards the data to the ATM switch as frames similar to Frame Relay frames. Those frames then are converted to cells.

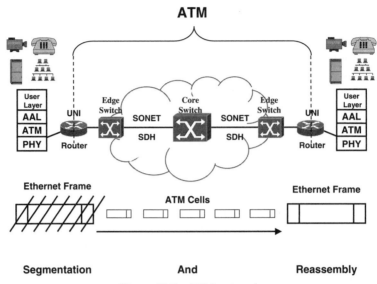

Figure 10.7   ATM network.

10.3.4   ATM Network and Equipment

The ATM network consists of CPE, Broadband Switching Systems (BSSs), and interconnecting transmission facilities (see Figure 10.7) [8, 33]:

10.3.4.1 Customer Premises Equipment   *Customer Premises Equipment* (CPE) comprises Data Terminal Equipment (DTE), Data Communications Equipment (DCE), and voice and video switches, servers, and terminal equipment.

- DTE includes mainframe, midrange, and PC-server host computers connected through the UNI.
- DCE comprises ATM-equipped bridges, switches, routers, and gateways connected through the DXI UNI. A number of vendors offer ATM premises switches. Such switches are used for LAN interconnection, employing a highly redundant cell-switching fabric capable of switching speeds that can reach as high as OC-12 (622 Mbps).
- CPE includes ATM-based PBXs and video servers connected via the UNI or the DXI. These devices were always somewhat unusual and now have been superseded by IP-based devices.

10.3.4.2 Broadband Switching Systems   *Broadband Switching Systems* (BSSs) are carrier exchange switches capable of broadband switching and transport. They employ highly redundant cell-switching fabrics that currently can operate in a range up to 1 Tbps in terms of aggregate bus speed. Maximum port speeds generally are OC-12 (622 Mbps). ATM switches also are highly intelligent, providing buffering, routing, and flow control as well as segmentation and reassembly. The switches are highly redundant and fault tolerant. They fall into two categories, core and edge switches:

- *Edge switches* also are known as *access nodes* or *service nodes*. They are distributed in proximity to ATM users and connect to the core switches via fiber facilities, much like a CO in the voice world. In fact, they often are collocated with a CO switch. Edge switches involve cell-switching fabrics that generally operate at aggregate rates of 5 Gbps or more.
- *Core switches* also are known as *backbone switches*. They generally are housed in a wire center, along with a traditional CO circuit switch or tandem. Core switches involve cell-switching fabrics that generally operate at aggregate rates of up to 1 Tbps.

10.3.4.3 Transmission Facilities Transmission facilities at the Network-to-Network (NNI) level and between network nodes most commonly are SDH/SONET fiber optic in nature, although other media are possible. Local loop facilities (UNI and DXI) can be any medium capable of supporting transmission speeds of DS-1 (T1 at 1.544 and E1 at 2.048 Mbps) or better, although fiber always is preferable. For very short distances in a premises-based ATM environment, Category 5e (Cat 5e) UTP supports ATM speeds up to 155 Mbps. Cat 6 cable supports data rates up to 622 Mbps and beyond.

Inverse Multiplexing over ATM (IMA) commonly is used for high-speed access to a public ATM network when speeds above T1/E-1 are required but T3/E-3 or fractional services are not available. IMA also is used when a T3 or FT3 is not cost effective, with the crossover point generally in the range of 12 Mbps or 8 T1s. With that crossover point in mind, IMA permits as many as 8 T1s to link together to support aggregate bandwidth of up to 12.288 Mbps (1.536 Mbps × 8). The ATM cell stream is spread across the separate T1s in a round-robin fashion and is resynchronized and reconstituted at the edge switch. While the preferred communications link is in the form of SONET optical fiber, it is not always available. Significantly, IMA actually provides some benefit at this bandwidth level in the form of redundancy. As an integrated network service, ATM can support all information types but typically requires an optical fiber link, which is susceptible to catastrophic failure. Multiple T1 links provide some level of redundancy in an IMA scenario, as all are unlikely to fail at the same time, unless one very powerful backhoe manages to rip an entire copper cable out of the ground. And, assuming that the T1 circuits travel different physical routes, even backhoes pose manageable threats [34].

10.3.5 ATM Protocols and Cell Structure

ATM is based on a 53-octet cell structure comprising 48 octets of payload and 5 octets of header, as illustrated in Figure 10.8. Contained within the payload, there also is some amount of overhead in the form of a Convergence Sublayer (CS) header. The choice of 48 octets was a compromise between the U.S. *Exchange Carriers Standards Association* (ECSA) T1S1 committee and the ETSI. The ECSA promoted a cell size of 64 octets, while ETSI favored 32 octets, each reflecting the bandwidth required for the parochial PCM voice-encoding technique. The decision to use 48 octets was a perfect mathematical compromise [35]. It is worth noting that the cell size and composition are very overhead intensive, at about 10 percent, plus Convergence Sublayer and Layer 2 and Layer 3 headers. Standard PCM-encoded voice, at eight bits per sample, can deal effectively with a small cell payload. Data,

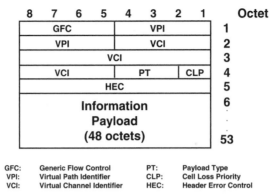

Figure 10.8 ATM cell structure. (*Source:* TA-NWT-00113. Copyright 1993 by Bellcore.)

on the other hand, generally is presented in much larger packets, blocks, frames, and so on. Hence, the data world would have preferred a much larger cell, while many in the voice world actually would have preferred a smaller cell. So, 48 octets was a perfect example of a compromise—none of the parties to the negotiation was ecstatic about it, but all could accept it.

While this level of overhead, commonly known as the *cell tax*, might seem wasteful of bandwidth, the advantages far outweighed the drawbacks—at least initially—a classic trade-off between effectiveness and efficiency. The small cell size offers the advantage of supporting any type of data, including voice, fax, text, image, video, and multimedia—whether compressed or uncompressed. The fixed cell size offers the advantage of predictability, very unlike the variable-length frames of X.25, FR, and Ethernet. This level of predictability yields much improved access control and congestion control. Bandwidth is cheap over optical pipes, especially given the development of DWDM (Dense Wavelength Division Multiplexing) and its variants. ATM switches are geared to process ATM cells in hardware at very high speeds. Finally, ATM switches currently operate at an aggregate internal bus speed in the range of 1 Tbps or more, which makes bandwidth cheap in the switch as well. In any event, the 48-octet payload was set as a standard for the ATM cell.

The cell header provides limited Data Link Layer (Layer 2) functionality, managing the allocation of the resources of the underlying Physical Layer (Layer 1) of the transmission facility, which ideally is SONET/SDH in nature. The ATM cell switches also perform layer 1 functions such as clocking, bit encoding, and physical-medium connection. The header also is used for channel identification, thereby ensuring that all cells travel the same physical path and, therefore, arrive in sequence [36]. The header is structured as follows:

- **Generic Flow Control (GFC):** Four bits that provide local flow control but which field has no significance on an end-to-end basis. The four-bit field supports 16 (2^4) GFC states. Intermediate ATM switches overwrite this field with additional Virtual Path Identifier (VPI) information [8]. In other words and for example, GFC is significant in order to control data flow across a UNI but is unnecessary at an NNI level. Flow control is a congestion control mechanism which requires that the various ATM switches and other equipment in both the network core and at the edges communicate with each other to determine the level of congestion along a defined path. If the network suffers congestion in the path, the ATM switches can buffer finite amounts of data; if the limits of the buffers are exceeded, data will be lost. To avoid that potential loss, the network must communicate with end-user devices to constrain the amount of data entering the network so as not to invite further congestion. A *rate-based* mechanism is an end-to-end flow control scheme that considers resources edge to edge, communicating the level of available resources through a feedback loop. This approach requires that the transmitting end-user device adjust its rate of transmission downward across the UNI in consideration of congestion. A *credit-based* approach either allows or disallows the end-user device to transmit data across the UNI, based on end-to-end consideration of whether sufficient buffer space is available on each link of the network.

- **Virtual Path Identifier (VPI):** Eight bits identifying the *Virtual Path* (VP). The path is determined at the input port and is fixed for each call but is shared among multiple calls. The path is from the switch input port, through the switching matrix, to the output port, and then across a link between any two consecutive ATM entities. The VPI and Virtual Channel Identifier (VCI) jointly can be considered as a label for the allocation of resources across an end-to-end path, realizing that the label may require translation (change in value) from link to link; this translation process would take place at the ATM switching nodes or cross-connect points. As ATM switches, like all switches, work at Layers 1 (Physical Layer) and 2 (Data Link Layer) of the OSI Reference Model, they work on a link-by-link basis. That is to say that the cell address has local significance only.

- **Virtual Channel Identifier (VCI):** Sixteen bits identifying the *Virtual Channel* (VC), which is established each time a call is set up in the ATM network. A VC is a unidirectional channel for transporting cells between two consecutive ATM entities (e.g., switches) across a link.

- **Payload-Type Indicator (PTI):** Three bits distinguishing between cells carrying user information and cells carrying service information.

- **Cell Loss Priority (CLP):** One bit identifying the priority level of the cell to determine the eligibility of that cell for discard in the event of network congestion. Clearly, some applications, such as LAN-to-LAN traffic, are tolerant of loss. Other applications, such as real-time voice, are highly intolerant of loss.

- **Header Error Control (HEC):** Eight bits providing error checking of the header but not the payload. There is no provision for error correction.

The ATM reference model is multidimensional, with three planes and four layers, as illustrated in Figure 10.9. The lower two layers of this reference model loosely compare to the Physical Layer of the OSI Reference Model. As in the OSI model, each layer of the ATM model functions independently to perform its designated functions, all layers are tightly linked, and the functions are highly coordinated. The layers of the ATM reference model are the Physical Layer, ATM Layer, ATM Adaptation Layer, and higher layers and functions. The planes include the Control Plane, User Plane, and Management Plane.

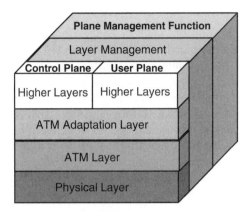

Figure 10.9   ATM protocol reference model. (*Source:* Network VAR.)

10.3.5.1 *Physical Layer* *PHYsical Layer* (PHY) functions are addressed through two sublayers: the Physical Medium (PM) and Transmission Convergence (TC). The ATM Forum's specifications for various User Network Interfaces (UNIs) address the implementation of the Physical Layer. The B-UNI, or Public UNI, is the specification for carrier internetworks. The UNI and DXI are *Private UNIs*, describing the implementation specifics for user access to the ATM network:

- *Physical Medium* (PM) sublayer specifies the physical and electro-optical interfaces with the transmission media on both the transmit and receive sides. Timing functions are provided at this level. SDH/SONET optical fiber is the preferred physical medium.
- *Transmission Convergence* (TC) sublayer handles frame generation, frame adaption, cell delineation, header error control, and cell rate decoupling. The frame generation function takes the frame of data presented by the transmitting device across the PM sublayer for presentation to the ATM Layer and subsequent segmentation into cells. On the receive side, the TC sublayer receives data in cells and decouples it to reconstitute the frame of data, checking all the while for header errors before presenting the data to the PM sublayer, which passes the data to the end-user device.

10.3.5.2 *ATM Layer* *ATM Layer* (ATM) functions include multiplexing of cells, selection of appropriate VPIs and VCIs, generation of headers, and flow control. At this layer, all multiplexing, switching, and routing take place for presentation to the appropriate *Virtual Paths* (VPs) and *Virtual Channels* (VCs) of the SONET fiber-optic transport system, which interfaces through the Physical Layer. Virtual Channels exist within Virtual Tributaries (VTs). For example, VT1.5 (i.e., T1 frame) might carry 24 channels, each of which carries a single voice communication in a time slot or a data communication in multiple time slots. That individual voice or data transmission comprises a VC. As illustrated in Figure 10.10, a VC travels over a VP, which is neither fixed nor dedicated.

VC = Virtual Channel
VP = Virtual Path

Figure 10.10   Relationship of path, VP, and VC.

10.3.5.3 ATM Adaptation Layer   *ATM Adaptation Layer* (AAL) functions are divided into two sublayers: the Convergence Sublayer (CS) and the Segmentation And Reassembly (SAR) sublayer:

- *Convergence Sublayer* (CS) functions are determined by the specifics of the service supported by that particular AAL. Service classes are designated as A, B, C, and D.
- *Segmentation and Reassembly* (SAR) sublayer functions segment the user data into payloads for insertion into cells on the transmit side. On the receive side, the SAR extracts the payload from the cells and reassembles the data into the information stream as originally transmitted. In other words, the process of segmentation takes place at the ingress edge of the ATM domain. Across the entire ATM network, from switch to switch and from edge to edge, data flows in a cell stream. The cells decouple and the data reassemble (reconstitute to original form) at the egress edge of the ATM domain.
- *AAL Types* are supported by the functions of the Convergence Sublayer (CS). There exist defined AAL Types 1, 2, 3/4, and 5, each of which supports a specific class of traffic (see Table 10.4). AAL information is nested within the payload of user information cells.
- *AAL Type 1* supports *Class A* traffic, which is connection-oriented *Constant Bit Rate* (CBR) traffic timed between source and sink. Such traffic is stream oriented and intolerant of delay. Isochronous traffic such as digitized, uncompressed voice is supported via Class 1 AAL, which essentially permits the emulation of a T/E-carrier circuit. All such traffic is carefully timed and must depend on a guaranteed rate of network access, transport, and delivery. Such traffic is marked as high priority in the cell header, as transmission delays could considerably impact presentation quality. Class A traffic is transmitted over a Virtual Path (VP) and in a Virtual Channel (VC) appropriate for such high-priority traffic.
- *AAL Type 2* supports *Class B* traffic, which is connection-oriented, real-time Variable Bit Rate (rt-VBR), isochronous traffic timed between source and sink. Compressed audio and video are Class B. Class B traffic, for example, includes compressed voice using the relatively simple Digital Speech Interpolation (DSI) technique for silence suppression. Compressed video using the Moving Pictures Experts Group (MPEG) compression algorithms also are Class B. Class B traffic is marked as high priority in the cell header and transmitted over an appropriate VP and VC.

TABLE 10.4 ATM Adaptation Layer (AAL) Service Classes/Categories

ITU-T Service Class[a]	Class A	Class B	Class C	Class D	Class X	
AAL type	1	2	3/4; 5 in message mode, only	3/4	5	
ATM Forum service category[b]	CBR	Real-Time VBR (rt-VBR)	Non Real-Time VBR (nrt-VBR)		UBR	ABR
Bit rate	Constant Bit Rate (CBR)	Variable Bit Rate (VBR)			Unspecified Bit Rate (UBR)	Available Bit Rate (ABR)
Timing relationship, source–destination pair	Required		Not required			
Connection mode	Connection-Oriented			Connectionless	Connection oriented or Connectionless	
Traffic contract parameters	Peak Cell Rate (PCR), Cell Delay Variation Tolerance (CDVT)	Peak Cell Rate (PCR), Cell Delay Variation Tolerance (CDVT), Sustainable Cell Rate (SCR), Maximum Burst Size (MBS), Burst Tolerance (BT)			Peak Cell Rate (PCR), Cell Delay Variation Tolerance (CDVT)	Peak Cell Rate (PCR), Cell Delay Variation Tolerance (CDVT), Minimum Cell Rate (MCR)
Quality-of-Service (QoS) parameters	Cell Delay Variation (CDV), Cell Transfer Delay (CTD), Cell Loss Ratio (CLR)		Cell Loss Ratio (CLR)		Not specified	
Example applications	Uncompressed voice, audio, and video; Circuit Emulation Service (CES)	Compressed voice, audio, and video; SNA	X.25, FR, transaction processing	SMDS; LAN traffic; non-real-time buffered video	Signaling and control, network management, e-mail; (FTP); World Wide Web, remote LAN access and telecommuting, LAN internetworking; LAN Emulation (LANE); IP traffic including VoIP (Voice over IP)	

[a] The ITU-T service classes are listed here largely for purposes of historical context. They generally are considered to be obsolete, having been replaced by the MFA Forum service categories.

530

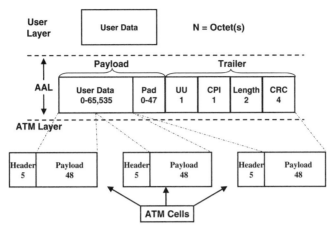

Figure 10.11   AAL Type 5 operation.

• *AAL Type 3/4* supports *Class C* or *Class D* traffic, which is non-real-time
  Variable Bit Rate (nrt-VBR) data traffic with no timing relationship between
  source and sink. Class C traffic, such as X.25 packet data and Frame Relay data,
  is connection-oriented VBR traffic with no timing relationship between source
  and sink. Class D traffic, such as LAN and SMDS data, is connectionless VBR
  traffic that is sensitive to loss but not highly sensitive to delay [8]. AAL Type
  3/4 supports message mode and streaming mode service. *Message mode service*
  is used for framed data in which only one *Interface Data Unit* (IDU) is passed.
  In other words, it is a single-frame message of up to 65,535 octets ($2^{16} - 1$).
  Streaming mode service is used for framed data in which multiple IDUs are
  passed in a stream. The IDUs can be up to 65,535 octets, with a 10-bit CRC
  added at the SAR layer as part of the trailer. As SMDS disappeared, this AAL
  all but disappeared, in favor of AAL 5.

• *AAL Type 5* supports *Class C* traffic in *message mode* only. Such traffic is
  Variable Bit Rate (VBR) traffic with no timing relationship between source
  and sink and consisting of only 1 IDU, as illustrated in Figure 10.11. AAL Type
  5 also is known as *Simple and Efficient AAL* (SEAL), as some of the overhead
  has been stripped out of the Convergence Layer. AAL Type 5 initially was
  intended solely for use in signaling and control (e.g., NNI applications) and
  network management (e.g., Local Management Interface, or LMI). The IDUs
  can vary in length, up to 65,535 octets. A 32-bit CRC is appended to the IDU
  at the Convergence Layer as part of the *trailer*. AAL 5 also supports *Class X*
  traffic, which is either Unspecified Bit Rate (UBR) or Available Bit Rate
  (ABR). Such traffic is VBR and either connection oriented or connectionless
  [37]. AAL 5 is used in support of a wide variety of data traffic, including LAN
  Emulation (LANE) and IP.

10.3.5.4 ATM Service Categories   ATM-defined service categories, which
relate back to AAL types and are reflected in Table 10.4, include CBR, rt-VBR,
nrt-VBR, UBR, and ABR. Guaranteed Frame Rate (GFR) is the most recently
defined service:

- *Constant Bit Rate* (CBR) is a class of service that supports uncompressed voice and video and circuit emulation. CBR traffic is characterized by a continuous rate of data flow, which is intolerant of loss and delay. CBR looks much like a nailed-up circuit. Traffic parameters include *Peak Cell Rate* (PCR) and *Cell Delay Variation Tolerance* (CDVT). QoS parameters include *Cell Delay Variation* (CDV), *Cell Transfer Delay* (CTD), and *Cell Loss Ratio* (CLR). PCM-encoded voice, certain forms of audio, and video encoding intended for TDM channels are examples of CBR traffic.

- *Real-Time Variable Bit Rate* (rt-VBR) traffic is bursty in nature but depends on timing and control information to ensure the integrity of the data stream. Traffic parameters include *Peak Cell Rate* (PCR), *Cell Delay Variation Tolerance* (CDVT), *Sustainable Cell Rate* (SCR), *Maximum Burst Size* (MBS), and *Burst Tolerance* (BT). The QoS parameter is *Cell Loss Ratio* (CLR). Voice, audio, and video encoded at variable bit rates are examples of rt-VBR traffic.

- *Non-Real-Time Variable Bit Rate* (nrt-VBR) traffic is bursty, but its non-real-time nature is not dependent on loss or delay because there is time to recover. Traffic parameters include *Peak Cell Rate* (PCR), *Cell Delay Variation Tolerance* (CDVT), *Sustainable Cell Rate* (SCR), *Maximum Burst Size* (MBS), and *Burst Tolerance* (BT). The QoS parameter is *Cell Loss Ratio* (CLR). Examples include data traffic such as X.25, Frame Relay, transaction processing, LAN-to-LAN, and non-real-time buffered voice and video.

- *Unspecified Bit Rate* (UBR) traffic is a *best effort* type. Traffic parameters include *Peak Cell Rate* (PCR) and *Cell Delay Variation Tolerance* (CDVT). No QoS commitment is made. Traditional computer applications, such as file transfer and e-mail, fall into this category.

- *Available Bit Rate* (ABR) is a best effort category in which the network attempts to pass the maximum number of cells but with no absolute guarantees. Subsequent to the establishment of the connection, the network may change the transfer characteristics through a flow control mechanism that communicates to the originating end-user device. This flow control feedback mechanism is in the form of *Resource Management* (RM) *cells*. In other words, during periods of congestion, the network can buffer cells and advise the sender to throttle back on the rate of transmission. ABR supports VBR traffic with flow control, a minimum transmission rate, and specified performance parameters. Traffic parameters include *Peak Cell Rate* (PCR), *Cell Delay Variation Tolerance* (CDVT), and *Minimum Cell Rate* (MCR). No QoS commitment is made. ABR service is not intended to support real-time applications.

- *Guaranteed Frame Rate* (GFR) is intended to support non-real-time applications that may require a minimum rate guarantee and can benefit from accessing additional bandwidth dynamically as it becomes available. GFR does not require adherence to a flow control protocol. The GFR service guarantee is based on AAL5 Protocol Data Units (PDUs), which also are known as AAL5 frames. During periods of network congestion, GFR attempts to discard entire frames, rather than cells that are segments of frames. Discarding entire frames is highly beneficial in reducing overall network congestion. Traffic descriptors in both the forward and backward directions include *Maximum Frame Size* (MFS), *Burst Cell Tolerance* (BCT), *Minimum Cell Rate* (MCR), and *Peak Cell*

Rate (PCR). Specific applications have yet to be identified. GFR is the most recently defined (August 2001) service category.

10.3.5.5 *ATM QoS Parameters*   The aforementioned references to AAL Service Classes and Service Categories would be incomplete without defining the key QoS parameters. ATM network performance parameters are defined in ITU-T Recommendation I.356. Those parameters are used to measure B-ISDN performance of an end-to-end user-oriented connection in terms of *quality of service* specific to each service class and category. The ATM Forum extended this standard through the definition of QoS parameters and reference configurations for the User Network Interface (UNI). The relevant ATM Forum (now MFA Forum) document is Traffic Management Specification Version 4.0 (af-tm-0056.000, April 1996). QoS objectives are not strictly defined.

Performance parameters defined by the ITU-T address accuracy, dependability, and speed. Accuracy parameters include Cell Delay Variation (CDV), Cell Error Ratio (CER), Cell Loss Ratio (CLR), Cell Misinsertion Rate (CMR), Cell Transfer Delay (CTD), and Severely Errored Cell Block Ratio (SECBR):

- *Cell Delay Variation* (CDV) is the variation in an individual cell's Cell Transfer Delay (CTD) and its expected transfer delay. CTD is a form of *jitter*, which can seriously degrade the quality of voice and video payloads. If cells arrive sooner than expected, the clumping can cause the PCR (Peak Cell Rate) to be exceeded and the excess cells to be discarded. If some cells arrive too late, the result may be gaps in the received information stream. *Cell Delay Variation Tolerance* (CDVT) is a measurement of the maximum allowable CDV tolerance between two end stations. *Peak-to-peak CDV* is negotiated between the end station and the network; peak to peak refers to the best case compared with the worst case, that is, the difference between the earliest and the latest arriving cells on a connection.

- *Cell Error Ratio* (CER) is a dependability parameter expressed as the ratio of the number of errored cells to the total number of transmitted cells sent over a measurement interval. CER is not negotiated.

- *Cell Loss Ratio* (CLR) is a dependability parameter expressed as the ratio of the number of lost cells to the number of transmitted cells. Cell loss can occur for reasons that include misdirection of cells by a switch, a congestion problem causing a discard in consideration of buffer capacity, a station exceeding its PCR resulting in cell discard, or a cell that exceeds the maximum CTD and arrives too late for consideration and processing. CLR is negotiated between the end stations and the network and applies to the lifetime of the connection. CLR applies to all service categories except UBR.

- *Cell Misinsertion Rate* (CMR) is a dependability parameter expressed as the number of cells received over a time interval at a destination endpoint that were not transmitted originally by the source endpoint. CMR is expressed as a rate, rather than as a ratio, because the number of misinserted cells is beyond the control of both the originating and destination endpoints. CMR can result from the corruption of a cell header, which would cause a cell to be misinserted into the cell stream of another source–destination pair of endpoints; in other words, the cell would be misdirected. Also, the fifth octet of the cell header checks the address but may miss a small fraction of changed addresses if another field in the header changes at the same time. CMR is not a negotiated parameter.

• *Cell Transfer Delay* (CTD) is the average time it takes a cell to transverse the network from source to destination across a UNI. CTD is the sum of all delays imposed by coding and decoding, segmentation and reassembly, propagation across transmission media, cell processing at the nodes, queuing of the cell in input and output buffers, and loss and recovery. If a cell arrives too late at the receiving station, it may be considered lost or late and may be disregarded. If the subject cell is a segment of a larger data packet, the entire packet must be discarded and forgotten or retransmitted. *Maximum CTD* (maxCTD) is negotiated between the end stations and the network.

• *Severely Errored Cell Block Ratio* (SECBR) refers to a sequence of some number (*n*) of cells transmitted consecutively (sent in a block) on a given connection, perhaps between OA&M cells. SECBR is a dependability parameter expressed as the number of severely errored cell blocks compared with the total number of cell blocks sent over a period of time, or measurement interval. A severely errored cell block outcome is realized when more than some number of cells in a block is errored, lost, or misinserted. SECBR is not a negotiated parameter.

10.3.5.6 ATM Traffic Contract An ATM traffic contract specifies all characteristics of a connection negotiated between a source endpoint and an ATM network. *Traffic parameters* are descriptions of the traffic characteristics of a source endpoint; they may be quantitative or qualitative in nature. Traffic parameters include (PCR), Sustainable Cell Rate (SCR), Maximum Burst Size (MBS), Maximum Frame Size (MFS), and Minimum Cell Rate (MCR). A traffic descriptor is the entire set of traffic parameters associated with a source endpoint. The traffic descriptors used during connection establishment include the following:

• *Peak Cell Rate* (PCR) is the maximum number of cells per second, in a burst, that is, within a short interval, that the network agrees to accept and transfer for a given UNI. Excess cells may be discarded by the ingress switch or marked as eligible for discard. For CBR service, the PCR is the guaranteed Constant Bit Rate for the virtual circuit. Enforcement of the PCR enables the network to allocate sufficient resources to ensure that the QoS parameters (e.g., Cell Loss Ratio and Cell Transfer Delay) are met. PCR can apply to all service categories.

• *Sustainable Cell Rate* (SCR) is the maximum average rate at which the network agrees to accept cells and support their transfer from end to end for each UNI. In other words, SCR is the average throughput. Enforcement of the SCR enables the network to allocate sufficient resources to ensure that the QoS parameters (e.g., Cell Loss Ratio and Cell Transfer Delay) are met over a period of time. SCR applies to VBR services.

• *Maximum Burst Size* (MBS) is the maximum size of a burst of traffic that can transmit within the PCR, given the *Burst Tolerance* (BT), or *Burst Cell*

Tolerance (BCT), of the network. MBS is expressed as a number of consecutive cells.

- *Maximum Frame Size* (MFS) is the maximum size of a PDU, or frame, supported by the network. MFS relates specifically to the Guaranteed Frame Rate (GFR) service category.

- *Minimum Cell Rate* (MCR) is the minimum number of cells per second that the network agrees to support for a given originating endpoint across a UNI. This ABR service descriptor, expressed in cells per second, is that rate at which the originating endpoint can always transmit during the course of the connection.

10.3.5.7 *Higher Layer Protocols and Functions*   These relate to the specifics of the user Protocol Data Unit (PDU), such as an SDLC frame or Frame Relay frame:

- *Control Plane* functions include all aspects of network signaling and control, such as call control and connection control.

- *User Plane* functions deal with issues of user-to-user information transfer and associated controls (e.g., flow control and error control mechanisms).

- *Management Plane* functions involve the management of the ATM switch or hub. The Management Plane is divided into Plane Management and Layer Management. *Plane Management* acts on the management of the switch as a whole, with no layered approach. Management of and coordination between the various planes is accomplished in Plane Management. *Layer Management* acts on the management of the resources at each specific layer of the model, for example, Operation, Administration, and Maintenance (OA&M) information.

10.3.6 LAN Emulation

LAN Emulation (LANE) is a specification (LANE 1.0, January 1995) from the ATM Forum for an ATM service in support of native Ethernet (802.3) and Token Ring (802.5) LAN communications over an ATM network. Supported by software in the end systems (e.g., an ATM-based host or router, known as a *proxy* in LANE terminology), the ATM network emulates a native LAN environment. LANE acts as a layer 2 bridge in support of connectionless LAN traffic, with the connection-oriented ATM service being transparent to the user application. In the LANE environment, the end system is known as a *LAN Emulation Client* (LEC), which connects to the ATM network over a *LAN Emulation User-to-Network Interface* (LUNI). The network-based *LAN Emulation Server* (LES) registers the LAN MAC (Medium Access Control) addresses and resolves them against (i.e., translates them into) ATM addresses. The LES maps between MAC and ATM addresses through the *Address Resolution Protocol* (ARP). Each LEC (LAN Emulation Client) is assigned to an *Emulated LAN* (ELAN) by a network-based *LAN Emulation Configuration Server* (LECS), which is an optional component of LANE. Each LEC also is associated with a *Broadcast and Unknown Server* (BUS) that handles broadcast and multicast traffic as well as initial unicast frames before address resolution.

The BUS broadcasts queries to all stations on an ELAN in order to identify the MAC and ATM addresses of unknown edge devices; that information then is passed to the LES. Referring back to Table 10.4, LANE traffic generally is Class C Variable Bit Rate (VBR) traffic in Message Mode, and is supported over AAL 5. LANE 2.0 (July 1997) supports up to eight sets of QoS parameters, over different VCs (Virtual Channels), with different priority levels set for LANE traffic as CBR, VBR, ABR, or UBR traffic. LANE can be supported over either PVCs or SVCs. A key advantage of LANE is that it obviates the short-term requirement for wholesale changes to either applications or infrastructure. Rather, it positions the user organization for a relatively smooth future transition to true ATM Wide Area Network (WAN) services.

Note, however, that LANE cannot resolve translational problems between disparate LANs (i.e., Ethernet and Token Ring); rather, a router must accomplish protocol conversion. LANE also supports only Ethernet and Token Ring; FDDI, for example, cannot be supported without the intervention of a router. Also, a router or router function is required to support communications between ELANs, which operate as closed user groups, much like Virtual LANs (VLANs) [31, 38–40].

10.3.7   MultiProtocol over ATM

MultiProtocol Over ATM (MPOA) is a specification (July 1997) from the ATM Forum designed to enhance LANE by enabling interELAN communications without the intervention of a router and the associated packet delay. MPOA provides high-performance, scalable routing functionality over an ATM platform. MPOA expands on LANE, Classical IP over ATM (RFC 1577), and the IETF's Next Hop Resolution Protocol (NHRP) in order to create a standardized notion of a virtual router within an ATM network. Between any two MPOA-capable end devices (e.g., MPOA-enhanced LANE hosts, bridges, or switches), MPOA maps routed and bridged flows of connectionless LAN traffic over cut-through ATM Switched Virtual Channels (SVCs), offloading the packet-by-packet processing steps performed by traditional routers. A route server in the ATM network contains the core intelligence to dynamically track the network topology and performance, thereby providing guaranteed bandwidth, reduced latency, and QoS. Since the routing intelligence is divorced from the packet-forwarding function, MPOA offers a scalable and flexible solution for LAN interconnectivity, corporate intranets, and multimedia applications such as distance learning, desktop videoconferencing, and collaborative work sessions. MPOA supports protocols such as Ethernet, Token Ring, FDDI, and IP. Through the NHRP protocol from the IETF, routing entities within the MPOA network can intercommunicate to determine the most appropriate path for a communication between edge devices on an interELAN basis. MPOA essentially synthesizes bridging and routing of connectionless interLAN traffic over a connection-oriented ATM network [31, 38, 40].

10.3.8   Frame-Based ATM Transport over Ethernet

The Frame-based ATM Transport over Ethernet (FATE) specification from the ATM Forum (February 2000 and July 2002) allows ATM services to be provided over Ethernet media by transporting ATM data within an Ethernet frame. AAL5 applies in FATE applications. At this point, the specification is defined only for the UNI, with FATE for NNI having been targeted for further study. FATE has particular application in the context of an ATM-based ADSL environment interfacing to an Ethernet LAN through a switch or hub on the customer premises [41].

10.3.9 Framed ATM over SONET/SDH Transport

The Framed ATM over SONET/SDH Transport (FAST) specification from the ATM Forum (July 2000) defines the mechanisms and procedures required to support the transport of variable-length datagrams, known as ATM frames, over an ATM infrastructure using SONET/SDH facilities. Interoperability with existing ATM infrastructure is ensured through the use of existing ATM control and management plane functions such as signaling, routing, addressing, and OAM&P. The ATM Forum has specified both *FAST UNI* and *FAST NNI* interfaces. The FAST interfaces are functionally equivalent to their cell-based counterparts with the exception of the maximum size of the *Service Data Unit* (SDU), which is defined as a unit of interface information whose identity is preserved from one end of a layer connection to another.

Two encapsulation formats are defined. *Frame encapsulation* supports the transmission of variable-length frames with a maximum size of 9216 octets and using the AAL5 Adaptation Layer. This approach involves minimal overhead. *Cell encapsulation* involves the encapsulation of ATM cells within a FAST frame, supporting an SDU with a maximum size of 65,535 octets, with the specific size sensitive to the AAL used.

FAST is similar to DXI (Data eXchange Interface) and FUNI (Frame User Network Interface) with several major exceptions. DXI and FUNI are designed for access to an ATM network over relatively low speed plesiochronous transmission facilities, while FAST is designed for access and/or interswitch trunking over very high speed SONET/SDH transmission facilities [42].

10.3.10 ATM Attributes, Issues, and Applications

ATM clearly offered the great advantage of lots of bandwidth, with access rates usually at a minimum of T1/E-1 and backbone interfaces as high as OC-192 (10 Gbps). The advantages of SONET optical fiber, for both access and transport, include bandwidth, error performance, and fault tolerance. ATM is linked closely to SONET/SDH, which is not to say that ATM is not medium independent. While the backbone ATM networks are designed around a SONET/SDH fiber-optic transmission system, UTP and STP can attach workstations to an ATM LAN switch. The local access facilities generally are copper or microwave at the T1 and T3 levels, but fiber is required at higher speeds. ATM over VSATs is available internationally at rates up to 34/45 Mbps (E-3/T3) to leapfrog national wireline networks that do not support ATM.

Further, ATM is the first service to offer truly guaranteed bandwidth on demand. In combination, ATM's strong error performance, access control, and congestion control yield outstanding levels of throughput. ATM also is highly flexible, as it supports any form of data—voice, facsimile, data, image, video, and multimedia—and with guaranteed QoS levels geared to match traffic requirements. Further, the traffic can be asynchronous, synchronous, or isochronous in nature and in any combination. ATM interconnects with X.25, Frame Relay, and IP networks. In fact, ATM commonly is at the core of those networks, which is transparent to the end user, of course.

Much as I discussed relative to Frame Relay, ATM offers mesh networking of virtual circuits with a single access circuit at each site without the need for complex and expensive leased lines. As a result, network configuration and reconfiguration are much simplified. Finally, ATM networks and associated costs are highly scalable, as is the case with any VPN technology. In other words, the cost of the network is very much in proportion to the scale of the network in terms of attached devices, networked locations, and bandwidth requirements.

The incumbent LECs and IXCs, worldwide, aggressively began deploying ATM in the cores of their data networks in the early 1990s. As they experienced growth in voice traffic, some gradually transitioned voice to those backbones, rather than continuing to invest in circuit-switching technology. However, ATM is a forklift upgrade and a very expensive and complex one at that.

Since roughly 2000, Ethernet and IP have greatly increased in popularity at the expense of both ATM and SONET. Ethernet now offers port speeds up to 40 Gbps, a speed that ATM switches likely will never achieve, as development has stopped. The CO remains an ATM stronghold, however, where ATM over OC-3 SONET interfaces is popular as the new highest common denominator for patching together devices from multiple vendors, all of which implement this interface in the same way.

ATM is intended to support any application that the contemporary mind can conceive, and it does an excellent job. The driving force behind ATM development and deployment was primarily that of data applications, where most of the growth was and where it remains, but ATM supports voice, video, and multimedia exceptionally well also and simultaneously with guaranteed QoS.

As a backbone technology, however, the ATM's cost and inherent complexity have not positioned it well against a combination of IP and MPLS. In fact, it was cost and complexity that prevented ATM from taking over the LAN domain, despite its technical elegance. Ultimately, Ethernet's low cost and simplicity prevailed. However, ATM has found its way into the broadband local loop, specifically ADSL and several versions of Passive Optical Network (PON). Notably, the impetus for ATM, ADSL, and PON all originated in the telco domain, where guaranteed QoS is a way of life.

10.4 METROPOLITAN ETHERNET

As you may have discovered, I discussed Ethernet and its variations in quite some detail in Chapter 8 and devoted considerable ink to Gigabit Ethernet (GbE) and 10GbE. Notably, GbE and 10GbE are not limited to the LAN domain. A number of service providers now offer them as *Metropolitan Area Network* (MAN) services intended for multisite enterprises confined to a metropolitan area. This service involves centrally positioning one or more GbE or 10GbE switches in the MAN and providing the enterprise with access via a variety of technologies. Although fiber optics clearly is the most attractive access technology from a performance stand-

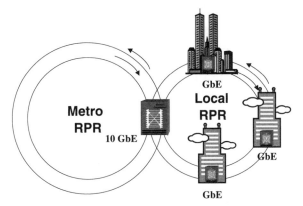

Figure 10.12   GbE and 10GbE in the MAN.

point, alternatives include unchannelized T-carrier, *Very-high-data-rate Digital Sub-scriber Line* (VDSL), various wireless options, and *Ethernet Passive Optical Network* (EPON). The real advantage to GbE in the MAN is that all traffic is carried in native Ethernet format (layer 2), with no requirement for introducing SONET, Frame Relay, ATM, or other Layer 1 or Layer 2 protocols that can increase both complexity and cost while adding overhead.

At 10 Gbps, 10GbE is particularly suitable as a MAN technology, although we always seem to be able to find a way to consume more and more bandwidth closer and closer to the desktop, where Ethernet rules. Estimates are that more than 95 percent of Internet traffic begins life as Ethernet frames [43]. 10GbE will run in full-duplex (FDX) and over fiber only. Fiber options include SONET and *Resilient Packet Ring* (RPR), *Coarse Wavelength Division Multiplexing* (CWDM), and EPON, which is restricted to the local loop level. As I discussed in Chapter 9, RPR runs at SONET rates, either preserves the SONET frame structure or uses Ethernet, and offers SONET-like sub-50-ms cutover times but without the overhead-intensive *Automatic Protection Switching* (APS) functionality of SONET. The LAN interface specifies MultiMode Fiber (MMF) over distances up to 300 m. The WAN interface calls for Single-Mode Fiber (SMF) over distances up to 40 km. In the long term, end-user applications for end-to-end 10GbE include *Storage Area Networks* (SANs), data center disaster recovery, and multimedia transport. In the short term, 10GbE is positioned as a core switch technology for interconnecting GbE edge switches providing native GbE services, as illustrated in Figure 10.12. Internet access via 10GbE is the next logical step, which will not be far behind [44–51].

10.5   BROADBAND ISDN

Potential means that you haven't done anything yet.

Ray Horak, 2006

Broadband ISDN (B-ISDN) was addressed formally in 1988 through the first set of B-ISDN standards from the ITU-T (I.121). Those standards were revised formally in 1990 and continue to experience revision and augmentation. The ATM Forum, formed in 1992 and since merged into the MFA Forum, built on those standards through the development and promotion of specifications for equipment interfaces in the ATM network. As noted previously, B-ISDN builds on the services foundation of Narrowband ISDN (N-ISDN). But B-ISDN is based on cell-switching technology, whereas N-ISDN is a circuit-switched standard. B-ISDN makes use of ATM as the backbone network switching and transport technology, with SDH/SONET as the backbone transmission medium.

Deployment of N-ISDN varies widely from country to country and within each nation on a state, province, and metropolitan area basis. N-ISDN has not been well received in the United States, although it was quite popular in certain countries within Western Europe and was widely deployed in parts of Asia. As the ITU-T anticipated quite early on, N-ISDN proved to be underpowered in the context of the contemporary appetite for bandwidth-intensive applications.

10.5.1   B-ISDN Defined

B-ISDN is defined by the ITU-T as a service requiring transmission channels capable of supporting rates greater than the primary rate. The *primary rate*, of course, is defined in the North American Primary Rate Interface (PRI) as 1.544 Mbps and in the European Primary Rate Access (PRA) as 2.048 Mbps. There are three underlying sets of technologies and standards that are absolutely critical to B-ISDN. First, Signaling System 7 (SS7) is viewed as the signaling and control that supports B-ISDN, just as it supports N-ISDN. Second, Asynchronous Transfer Mode (ATM) is the backbone network switching and multiplexing technology. Third, SDH/SONET is the physical backbone network transmission technology.

10.5.2   B-ISDN Access

B-ISDN user access is broadband in nature, relying on SDH/SONET fiber-optic transmission standards. Currently, there are two access interfaces specified. *User–Network Interface A* (UNI A) operates at OC-3 rates of 155 Mbps, while *User–Network Interface B* (UNI B) operates at OC-12 rates of 622 Mbps. Network-to-Network Interfaces (NNIs) are required for network access to B-ISDN from FR and N-ISDN networks.

10.5.3   B-ISDN Services

The ITU-T defines two types of B-ISDN services, interactive and distribution:

- *Interactive services* involve bidirectional transmission and include three classes of service. *Conversational services* include voice, interactive data, and interactive video. *Messaging services* include compound document mail and video mail. *Retrieval services* include text retrieval, data retrieval, image retrieval, video retrieval, and compound document retrieval.
- *Distribution services* may or may not involve user presentation control. By way of example, interactive TV is a service requiring presentation control. Interactive TV actually enables the viewer to interact with the program, perhaps to

select a product marketed over TV, to influence the ending of a movie, or to change the camera angle to view a football play from a different perspective. Conventional broadcast TV exemplifies a service requiring no presentation control.

10.5.4 B-ISDN Equipment

B-ISDN user equipment is an extension of that described for N-ISDN. *Broadband Terminal Equipment 1* (B-TE1) is defined as B-ISDN-compatible CPE. While the exact nature of the B-TE1 remains undetermined, it is likely that it will take the form of ATM-compatible B-ISDN communications servers. Those servers will combine the functions of a PBX, data switch, and video switch. At the extreme, they will take the form of ATM-based multimedia communications servers.

Broadband Terminal Equipment Type 2 (B-TE2) is defined as terminal equipment that supports a broadband interface other than B-ISDN. Terminal Equipment Type 2 (TE2) continues to be defined as terminal equipment that supports an interface other than ISDN. Both B-TE2 and TE2 equipment will interface with the network through a *Broadband Terminal Adapter* (B-TA). All of this may well be moot now, as IP seems to have relegated B-ISDN to an historical pipe dream.

10.5.5 B-ISDN Attributes and Issues

N-ISDN was intended to be available universally. This did not happen, of course, as many nations and regions did not see the value in upgrading the infrastructure to support it. B-ISDN seems headed for the same fate, although for a different reason. There certainly has been no lack of interest in and commitment to the underlying technologies but, over time, the low cost and inherent simplicity of other technologies seem to have rudely shoved them aside. ATM is on the decline, at least in the backbone, in favor of a combination of IP and MPLS. Ethernet is gaining ground over ATM in PON local loops and in the MAN. SS7 is on the decline in favor of SIP.

There is no question that B-ISDN has a lot of potential, but as a famous author (that would be me) once said, "Potential means you haven't done anything yet." That statement is perhaps a bit harsh when it comes to B-ISDN, as all of the underlying technologies and standards were in place and made considerable impact on telecommunications. It is highly unlikely, however, that B-ISDN will ever be fully specified, much less that it will ever take full form as the ubiquitous integrated network solution, end to end.

10.6 ADVANCED INTELLIGENT NETWORKS (AINs)

Once upon a time, the networks were truly *dumb*. Through the 1960s and even into the 1970s, they remained fairly dumb. In other words, the networks were composed largely of switches and pipes that could do little more than connect calls as directed based on a hierarchical order of limited switching intelligence. That intelligence was in the form of very limited programmed logic housed in databases that interacted at a minimal level. Each switch performed its own job, with little thought of the network as a whole.

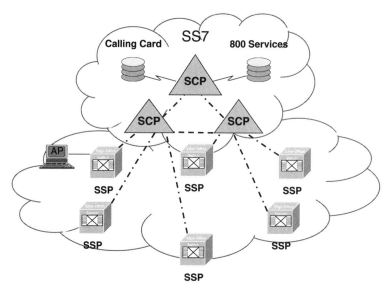

Figure 10.13   IN supporting 800 services and credit card verification.

Intelligent Network Version 1 (IN/1) was conceived at Bell Labs and born in 1976 with the introduction of IN-WATS (800) services and the first Common Channel Signaling (CCS) system. IN/1 provided for the switches to consult centralized, service- and customer-specific databases for routing instructions and authorization code verification, as illustrated in Figure 10.13. IN/1 services include INWATS, calling card verification, and voice Virtual Private Networks (VPNs).

The linchpin of the intelligent network is the *Service Creation Element* (SCE), which is a set of modular programming tools permitting services to be developed independently of the switch, the nature and capability of which can vary by manufacturer and software generic. The SCE divorces the service-specific programmed logic from the switch logic, thereby enabling the independent development of the service. The service, therefore, is available to all switches in the network. The concept of a *sparse network* is one of dumb switches supported by centralized intelligence with connectivity between distributed switches and centralized logic provided over high-speed digital circuits. In the context of a complex network, this concept is highly viable in terms of both performance and economics.

10.6.1   AIN Defined

Bellcore defined AIN in the early 1980s as AIN Software Release 1.0. This release is intended to provide a generic and modular set of tools that enable the creation, deployment, and management of services on a flexible basis. The software tools yield a suite of service offerings that are accessible to all network switches but which operate independently from the switch logic. The services, therefore, can be defined, developed, and deployed quickly and in a multivendor environment.

The depth and complexity of AIN Software Release 1 caused the telcos to move forward with their own various AIN releases, known as Releases 0.X. All Releases 0.X are fully compatible with Bellcore's Release 1.0, which likely will never be implemented in its original and defined form. AIN Release 0.0 addresses basic call modeling functions for *Service Switching Points* (SSPs) and database functions for *Service Control Points* (SCPs). AIN Release 0.1 defines generic call model for interaction between SSPs and SCPs; it includes features in support of N-ISDN services. AIN Release 0.2 adds *Intelligent Peripherals* (IPs) [52].

Characteristics of AINs include service creation toolkits, which enable the creation of centralized logic residing in centralized databases for the development and delivery of features across the network. AINs support all ISDN features, including caller ID, selective call blocking, and distinctive ringing. Finally, AINs are intended to provide support for *Personal Communications Services* (PCS), which permit subscribed features to be supported across networks of all types. PCS, not to be confused with the *PCS* cellular service, currently is conceptual.

10.6.2 Service Creation Environment

The *Service Creation Environment* (SCE) is the key distinction of an AIN. The SCE offers the carrier a toolkit for the development of service offerings, which can be provided on a network basis. The services can be generic or customer specific. Often, you can create the customer-specific services through linking generic services and varying the available options and parameters. Additionally, the SCE can be opened to the user organization, which then can customize the service offering as desired.

10.6.3 AIN Architecture

The architecture of the AIN differs greatly from that of the traditional network and even from that of IN/1. Those differences largely deal with the nature and location of the programmed logic and the databases, which drive the service offerings. The AIN architecture (refer to Figure 10.14) includes SS7, Service Switching Points,

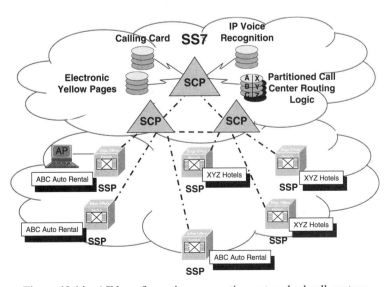

Figure 10.14 AIN configuration supporting networked call centers.

Signal Transfer Points, Service Management Systems, Adjunct Processors, and Intelligent Peripherals.

- *Common Channel Signaling System 7* (SS7) is an out-of-band signaling system for communication between devices in the carrier networks, as is discussed in several previous chapters. SS7 is an absolute requirement for both ISDN and AIN. SS7 is deployed throughout advanced voice networks worldwide.

- *Service Switching Points* (SSPs) are PSTN switches that act on the instructions dictated by AIN centralized databases. SSPs can be end offices or tandem switches, as is defined in the discussion of the PSTN.

- *Signal Transfer Points* (STPs) are packet switches that route signaling and control messages between SSPs and SCPs and between STPs.

- *Signal Control Points* (SCPs) contain all customer information in databases that reside on centralized network servers. SCPs provide routing and other instructions to SSPs, as requested and required.

- *Service Management Systems* (SMSs) are network control interfaces that enable the service provider to vary the parameters of the AIN services. Under certain circumstances, the user organization may be provided access to a partition of the SMS.

- *Adjunct Processors* (APs) are decentralized SCPs that support service offerings limited either to a single SSP or to a regional subset of SSPs. APs might support routing tables or authorization schemes specific to a single switch or regional subset of switches.

- *Intelligent Peripherals* (IPs) provide intelligent peripheral capabilities to enhance the delivery of certain services by offloading processing demands from the SCPs and providing a basic set of services to the SCPs. The role of the intelligent peripheral typically includes collection of digits, collection and playing of voice prompts, collection of voice responses and their conversion to digits, menu services, and database lookups. Voice processing and voice recognition, for example, might be implemented in support of the processing of "collect" calls. Services such as Verizon's 1-800-COLLECT enable the processing of a collect or third-party, billed call without operator intervention. Voice recognition also can be used for *voice printing*, which permits user authentication in highly secure applications. A number of LECs provide automated directory assistance services, accomplishing database lookups based on voice recognition inputs through intelligent peripherals. As the term IP (*Intelligent Peripheral*) has become confused with IP (*Internet Protocol*), the term *Special Resource Function* (SRF) often is used to describe these peripherals and the functions they perform.

10.6.4 AIN Services

AIN services truly are open to the imagination. But the availability of AINs is very uneven within the United States and, certainly, around the world. AIN services are of wide variety, including the following:

- *Find-me service* provides flexible and selective call forwarding. The numbers to which the calls are to be forwarded can be programmed and reprogrammed from any location. Caller priority access can be subject to entry of proper passwords provided by the called party. Such capabilities have appeared in the commercial market, based on Intelligent Peripherals (IPs) known as Personal Assistants. Such systems also provide for scheduling of follow-up calls, provide scheduled reminder messages, and maintain contact lists [53].

- *Follow-me service* provides for call forwarding on a predetermined schedule. A telecommuter, for example, might have the network forward calls to the home office three days a week during normal business hours. Calls would be directed to the traditional office two days a week. Calls clearly outside of normal business hours automatically would be directed to a voice mail system.

- *Computer security service* automatically would serve to provide secure access to networked hosts based on Calling Line ID (CLID) and supplemented by authorization codes. Additionally, the network automatically would keep a log of all access and access attempts, thereby providing an audit trail.

- *Call pickup service* also known as *call notification service*, provides for calls to be answered automatically by a voice processor. The called party can be notified of a deposited voice message by pager, fax, e-mail, or other means. The caller can enter a *privilege code* provided by the called party to distinguish the priority of the calling party. To pick up the call, the called party dials a DISA (Direct Inward System Access) port on the network switch and enters password codes in a manner similar to that used to access contemporary voice mail systems.

- *Store locator service* also known as *single-number dialing*, provides the ability to advertise a single number. The network routes calls to the closest store location in terms of either geography or time zone based on the originating address (i.e., telephone number) of the caller. This service is deployed widely in the United States in support of businesses offering delivery services. It also is used widely for directing calls to networked incoming call centers such as reservation centers (e.g., airlines, auto rental agencies, and hotel chains).

- *Multilocation extension dialing* provides for network routing of calls based on abbreviated numbers. This VPN service resembles a coordinated dialing plan in a networked PBX environment.

- *Call blocking* can work on incoming or outgoing basis. On an incoming basis, the feature allows the subscriber to program some number of telephone numbers from which incoming calls are denied. Incoming call blocking is widely available. On an outgoing basis, this feature typically supports the blocking of calls to international destinations, either in total or to specific country codes. *Content blocking* supports the blocking of calls to specific numbers, such as *900/976* numbers. This capability is deployed fairly commonly in many foreign networks but is not widely available in the U.S. PSTN. Cellular radio providers have made extensive use of such capability on a systemwide basis to avoid cellular fraud involving certain countries in the Middle East, The Bahamas, and South and Central America. Ameritech (now part of SBC) conducted trials of its Call Control Service as early as 1994, enabling residential customers to block calls to specific numbers or to all numbers except those on an allowed list. Widespread deployment of outgoing call blocking is unlikely in the near future.

- *Caller name* is a variation of caller ID, more correctly known as Calling Line ID (CLID), with linkage to directory services. The incoming call is identified on a *Visual Display Unit* (VDU) by originating number and associated directory listing. The VDU can be built into the telephone or can be in the form of an adjunct unit that sits between the telephone and the circuit. In a CT (Computer Telephony) scenario, the VDU can be in the form of a PC or computer workstation.

- *Enhanced call return* enables the subscriber to access what is, in effect, a highly sophisticated, network-based voice mail system. Bell Atlantic (now part of Verizon) trialed an offering that enabled the user to call the system and enter security codes to gain access to the switch and data stores. A voice announcement identified the date, time, and phone number of the last incoming call, at which point the return call could be launched by pressing a number on the telephone keypad [54].

- *Enhanced call routing* is a network-based enhancement to toll-free calling. The callers are presented with options that enable them to specify their needs and then be connected with the offices or individuals best able to satisfy them. MCI (Verizon) and Stentor (Canada) conducted trials on such a system in early 1995, providing seamless interconnection between the U.S. and Canadian networks [55].

- *Call completion service* enables the directory assistance operator to extend the call automatically, perhaps at an additional charge. This capability is offered by cellular providers to avoid accidents caused by "driving and dialing."

- *Number portability* serves to provide portability of telephone numbers such as INWATS and 900 numbers. This function is utilized increasingly as the regulators insist on portability of local numbers between LECs (Local Exchange Carriers). *Local Number Portability* (LNP) is essential to facilitate local competition, as decreed by the Telecommunications Act of 1996.

10.6.5 AIN Futures

The future of the AIN, at least as defined by Bellcore, is a bit cloudy. Many of the large carriers have developed their own versions of IN in support of toll-free services, calling card verification, and various other services. Whether they will move toward a standardized version is not clear, although it appears unlikely. As AIN capabilities continue to roll out on a gradual basis, you likely never will see headlines announcing the arrival of AIN. Rather, it has crept into the networks and will continue to do so. Driving forces over the last decade or so in the United States include FCC (May 4, 1995) and state Public Utility Commission (PUC) decisions in support of caller ID and name ID, which require SS7 and some level of AIN. A number of carriers have offered automated directory assistance; that trend is likely to continue. The Telecommunications Act of 1996 introduced local competition in the United States, and the states then took the initiative of requiring local number portability and equal access at a local exchange level, with AINs being the key to implementing such flexibility. Similar forces and trends have influenced the gradual deployment of AIN services elsewhere in the world. AIN is less of a factor in developing countries, as one might imagine.

More exotic AIN functionality, as illustrated in Figure 10.14, involves a reservation system's taking advantage of processing power embedded in the network; indeed, the following approximate scenario has been followed in several cases. A hotel reservation network might involve incoming call centers in New York, Chicago, and San Francisco. Those call centers are networked so callers might be directed to the closest call center that can handle the call based on criteria such as queue length, average holding time, and priority level of the caller as determined by caller ID or PIN. In cooperation with a third-party vendor, the hotel company might write a generic software program to accomplish the routing of calls. That program would enable inquiries against a relational database in which resides customer profiles so calls can be handled in the most appropriate way. The program and database might reside on a centralized computer platform located in the Chicago call center, in the data center of the third-party vendor, or perhaps even in the carrier's wire center. The centralized network control and routing system would make frequent status checks against the individual Automatic Call Distributors (ACDs) in the call centers to determine traffic load and make performance comparisons against QoS parameters, which are user definable. Based on this process, each call would be routed in the most effective manner. Once this system is deployed and performing effectively, it could be made available on a licensed basis to other hotel chains, rental car agencies, and other reservation networks. Each licensed user then might have similar capabilities, with system administrators remotely accessing physical and logical partitions of the programmed logic and databases to support their own reservation network supported by an appropriate interexchange carrier. Through the development of such a scenario, multiple users take advantage of computer and database technologies embedded in the carrier network and supporting multiple reservation networks. The interaction between the carriers takes place over SS7 links.

These types of services do not necessarily require broadband networks to support them, although broadband is always good. For that matter and in this world oriented toward instant gratification, more is always better and faster is better still. Broadband networks certainly fill that need with more bandwidth that supports faster communications.

REFERENCES

1. Flanagan, William A. *Frames, Packets and Cells in Broadband Networking.* Telecom Library, 1991.
2. "Frame Relay on the Rise." *Network World*, May 17, 1999.
3. Cooney, Michael. "Hypercom Ups Brand Office Ante with New Central FRAD." *Network World*, February 26, 1996.
4. Cooney, Michael. "Trio Targets Central Office FRAD Market." *Network World*, April 1, 1996.
5. Ball, David. "CO FRADs: A New Service Model." *Telecommunications*, April 1998.
6. Minoli, Daniel. *Enterprise Networking: Fractional T1 to SONET, Frame Relay to BISDN.* Artech House, 1993.
7. Spohn, Darren L. *Data Network Design.* McGraw-Hill, 1993.
8. Miller, Mark A. *Analyzing Broadband Networks: Frame Relay, SMDS, & ATM*, 2nd ed. M&T Books, 1997.

9. Hindman, Steve. "SNA over Frame Relay." *Telecommunications*, February 1999.

10. "A Discussion of Voice over Frame Relay." www.frforum.com. October 1996.

11. Flanagan, William A. *Voice over Frame Relay*. Flatiron Publishing, 1997.

12. Minoli, Dan and English, Michelle. "Voice over Frame Relay: Overview." *IT Continuous Services*, March 26, 1999.

13. *Packet Voice Primer*. Cisco Systems. Originally accessed at http://cio.cisco.com/warp/public/cc/sol/mkt/ent/gen/packv_in.html.

14. "Voice over Frame Relay Implementation Agreement," FRF11.1. Frame Relay Forum Technical Committee, June 2001.

15. Matusow, David. "Giving Voice a Boost in Frame Relay Nets." *Network World*, May 4, 1998.

16. Greene, Tim. "Talk Is Cheap with Frame Relay." *Network World*, January 18, 1999.

17. Willis, David. "Listen Up! Cisco Now Does Voice." *Network Computing*, April 1, 1998.

18. Newman, David, Melson, Brent, and Kuman, Siva S. "Imperfect Pitch." *Data Communications*, September 1996.

19. Lindstrom, Annie. "Speaking in Frames." *America's Network*, June 15, 1998.

20. Krautkremer, Todd. "Answering the Call." *Communications News*, October 1998.

21. Adams, Kathleen M. "Infonet International Network Services." *IT Continuous Services*. Datapro Information Services, June 12, 2001.

22. "Frame Relay IP Header Compression Implementation Agreement," FRF.20. Frame Relay Forum Technical Committee, June 2001.

23. Hamblen, Matt. "VPNs Gain with Users; Frame Relay Declines." *Computerworld*, August 29, 2005.

24. Rohde, David and Gittlen, Sandra. "AT&T Frame Relay Service Goes Down for the Count." *Network World*, April 20, 1998.

25. Rohde, David and Gittlen, Sandra. "AT&T Offers Facts about Frame Fiasco." *Network World*, April 20, 1998.

26. Thyfault, Mary E. "Frame Relay Travel Net Set." *Information Week*, December 25, 1995.

27. Anderson, Patricia. "Switched Multi-Megabit Data Service (SMDS)." *Datapro Communications Analyst*. Datapro Information Services Group, August 1995.

28. Bellis, Mary. "Automatic Teller Machines (ATM)." http://inventors.about.com/library/inventors/blatm.htm.

29. "The ATM is 30 Years Old." http://www.cnn.com/TECH/9509/atm/.

30. "Luther Simjian, Inventor of the Week." http://web.mit.edu/invent/iow/simjian.html.

31. Minoli, Daniel. "Asynchronous Transfer Mode (ATM): Technology Overview." *IT Continuous Services*. Datapro Information Services, April 15, 1998.

32. Newton, Harry. *Newton's Telecom Dictionary*, 18th ed. Miller Freeman, February 2002.

33. Minoli, Daniel. "ATM Public Network Switching Technology." *Datapro Communications Analyst*. Datapro Information Services Group, November 1994.

34. Hurwicz, Mike. "ATM for the Rest of Us." *Network Magazine*, November 1997.

35. Minoli, Daniel. *Telecommunications Technology Handbook*. Artech House, 1991.

36. Minoli, Dan. "ATM-Based Voice: An Overview." *IT Continuous Services*. Datapro Information Group, March 25, 1999.

37. Minoli, Daniel, updated by Raymond, Mark. "ATM and Cell Relay Concepts." *Datapro Communications Analyst*. Datapro Information Services Group, September 1995.

38. Amoss, John J. "ATM Standards and Status." *IT Continuous Services*. Datapro Information Group, July 14, 1998.

39. "An Overview of ATM LAN Emulation." Interphase Corporation, www.iphase.com. June 3, 1999.

40. "LAN Emulation and Multi-Protocol Over ATM." Cabletron Systems, www.cabletron.com. April 29, 1998.

41. "Frame-Based ATM Transport over Ethernet (FATE)," AF-FBATM-0139.00. ATM Forum Technical Committee, February 2000.

42. "Frame-Based ATM over SONET/SDH Transport (FAST)," FB-FBATM-0151.000. ATM Forum Technical Committee, July 2000.

43. Ruby, Douglas. "Can Ethernet Go End to End?' *Network World*, September 17, 2001.

44. Sharer, Russ. "Fiber-Optic Ethernet in the Local Loop." *Lightwave*, July 2001.

45. Lecheminant, Greg. "At the Crossroads of 10-Gbit/sec Transmission." *Lightwave*, January 2002.

46. Lancaster, Ken. "Ethernet Makes Its Mark." *Telecommunications*, October 2001.

47. Lee, Tony. "Why Ethernet Is Still the Answer for New Network Requirements." *Lightwave*, January 2001.

48. Bettinelli, Charles. "Fine-Tuning 10-GigE Networks." *Telecommunications*, August 2001.

49. Kennedy, Michael. "Ethernet on the WAN—A Winner?" *Telecommunications*, August 2001.

50. Newman, David. "Ethernet Grows Up—And Out." *Network World*, May 7, 2001.

51. Vickers, Lauri. "Ethernet: The Perfect 10?" *Network Magazine*, June 2001.

52. Briere, Daniel D. and Langner, Mark. "Advanced Intelligent Networks." *Datapro Communications Analyst*. Datapro Information Services Group, November 1994.

53. Briere, Daniel and Heckert, Christine. "Watching the Demise of Calling Cards." *Network World*, March 25, 1996.

54. "Bell Atlantic Enhances Return Call." *Advanced Intelligent Network News*, May 17, 1995.

55. "MCI & Stentor Connect 800 Services as Canadian Telecom Market Heats Up." *Advanced Intelligent Network News*, May 3, 1995.

CHAPTER 11

WIRELESS NETWORKING: EMPHASIS ON MOBILITY

The Bell System has available all of the important inventions with reference to commercial wireless telephony. For the usual commercial telephone services in the United States where wires are or can be provided, wireless from both a service and economic standpoint cannot be used. For the supplemental services which cannot be provided by wires, wireless will be used wherever the commercial demand will warrant such use.
  Telephone Almanac, American Telephone and Telegraph Company, 1922

In 1876, Alexander Graham Bell demonstrated the telephone at the Centennial Exposition of the United States in Philadelphia, Pennsylvania (United States). From that simple demonstration of one-way transmission over a distance of several hundred feet, the copper-based telephone network grew at an astounding rate. In 1880, Bell invented the first wireless communications system using reflected sunlight and photoelectric selenium receivers. Using this technique, however impractical, he was able to transmit intelligible speech a distance of up to 700 ft. Bell named this invention the *photophone,* later renaming it the *radiophone,* which he described as his greatest invention [1, 2]. AT&T continued work on the technology, extending its reach to several miles. The German Nazi military experimented with similar but more advanced systems in tank warfare during World War II [2].

Toward the end of the nineteenth century and not long after Bell's demonstration, a young German scientist named Heinrich Rudolf Hertz discovered the phenomenon of invisible force waves emanating for several meters around an electric spark of sufficient intensity. Classical physicists, at a loss to explain the phenomenon, theorized the existence of an unknown medium, *luminiferous ether,* that conducted that signal. Shortly thereafter, Guglielmo Marconi transmitted these *Hertzian* waves over several kilometers; he named the new technology *radio* because the waves

Telecommunications and Data Communications Handbook, By Ray Horak
Copyright © 2007 Ray Horak

appeared to radiate from the transmitter [2]. In 1886, Marconi was granted a patent for the first practical wireless telegraph, for which he shared in the 1909 Nobel Prize in Physics. In the meantime (1900), Canadian scientist Reginald Fessenden (1866–1932) transmitted his own voice a distance of 1 mile over the first wireless telephone [3]. In what he considered to be his greatest achievement in a long and illustrious career, Fessenden successfully broadcast a short program on Christmas Eve, 1906. Having been alerted to a special event, wireless operators of several United Fruit Company ships in the Atlantic heard the first radio audio broadcast in history when Fessenden transmitted Handel's *Largo* on an Ediphone, played *Oh, Holy Night* on the violin, and read from the Bible before wishing them a Merry Christmas [4].

The first commercial application of radio technology was that of broadcast radio, introduced in the United States in 1920. The first radio station, the Westinghouse station KDKA in Pittsburgh, Pennsylvania, inaugurated service by broadcasting returns of the Harding–Cox presidential election. On July 25, 1922, the first commercial station, WBAY, which was owned by AT&T, began broadcasting from the AT&T Long Lines building in New York City. Its first paying customer, two months later, was the Queensborough Corporation, advertising its Hawthorne Court real estate development in Jackson Heights. In the meantime, AT&T employees supplied the programming, which consisted of vocal selections, piano recitals, poetry recitations, and other content that seems fairly tame by today's standards. Among the performances was a recitation of James Whitcomb Riley's poem *An Old Sweetheart of Mine* by Miss Edna Cunningham; there is no record of the audience reaction [1]. While not an immediate commercial success, radio enjoyed U.S. market penetration of 50 percent of households in 1930, 90 percent in 1940, and essentially 100 percent by 1995.

Television, while first demonstrated in 1926 by John Logie Baier, a Scottish inventor, was refined by yet other Bell Labs employees, Frederick Eugene Ives and Frank Gray. The first public demonstration of color TV took place in the United States in 1927, between Bell Labs in Whippany, New Jersey, and Washington, D.C., where the audience included Secretary of Commerce Herbert Hoover. Vaudeville acts were included. Again, there is no record of the reaction of the audience. Delayed by the Great Depression and then World War II, commercial broadcast TV was introduced in 1946 and enjoyed success similar to that of radio [1]. Market penetration in the United States was estimated at 50 percent in 1955, 90 percent in 1960, and 99 percent in 1995. Wireless radio technology also was deployed early on in maritime communications, or ship-to-shore telephony and telegraphy. The first terrestrial mobile application was a one-way system employed in police radio dispatch trials in 1921 at the Detroit, Michigan, Police Department [2, 4].

The technologies and applications discussed in this chapter include Trunk Mobile Radio (TMR), paging, cordless telephony, Wireless Office Telecommunications Systems (WOTS), cellular, Low-Earth Orbiting (LEO) and Middle-Earth Orbiting (MEO) satellites. I discussed Wireless Local Area Networks (WLANs) and Wireless Local Loop (WLL) technologies in Chapters 8 and 9, respectively.

Wireless is about more than being unplugged or untethered. Wireless, in full form, speaks to a fundamentally different way of business and personal communications. Wireless communications adds the element of portability and even mobility, thereby removing the constraints of constant attachments to a physical space such as an office building or residence. Yet, wireless communications technologies generally are relegated to niche applications, at least for the time being. For many applications, wireless just cannot compete with wired networks. Theodore Vail, twice president of AT&T (1880s and early 1900s), summed it up when he said, "The difficulties of the wireless telegraph are as nothing compared with the difficulties in the way of the wireless telephone" [2]. While wireless telephony may have been quite a trick in those days, it is as nothing compared with mobile wireless data communications.

11.1 WIRELESS DEFINED

Wireless, quite simply, refers to communications without wires. While microwave and satellite communications are without wires, those technologies generally are considered to be high-speed network backbone or access technologies that are point to point, point to multipoint, or broadcast in nature. (See Chapter 2 for discussion of the principles and characteristics of radio transmission.) In the context of this discussion, wireless technologies are local loop or local in nature, with the emphasis on mobility.

11.2 STANDARDS AND REGULATIONS

Standards are very important in telecommunications, and wireless is no exception. Wireless technologies have the dubious distinction of lots of standards, many of which are incompatible and conflicting. Existing formal standards are national, regional, and international in nature. There also are proprietary specifications, which sometimes are characterized as de facto standards. Ultimately, you might like to think that the standards wars will yield clear winners and losers, but that is highly unlikely in the immediate future, as standards often are more about politics and economics than technical purity.

Regulation is extremely important in the wireless world, since there is a high potential of interference between transmissions. To avoid this problem, radio must be managed along several dimensions, including frequency allocation and power level. The ITU-R (International Telecommunications Union—Radiocommunication Sector), originally known as the CCIR (Comité consultatif international pour la radio, or International Radio Consultative Committee) is the branch of the ITU that sets international wireless standards. The Institute of Electrical and Electronics Engineers (IEEE) is in the continuing process of developing standards for Wireless LANs through the 802.11 Working Group. The European Telecommunications Standards Institute (ETSI) sets standards within the European Union (EU). The *Research for Advanced Communications in Europe* (RACE) program, which is directed at the promotion of *Integrated Broadband Communications* (IBC), addresses wireless in its R1043 recommendation document. The Federal Communications Commission (FCC) has handled regulation of the wireless, as well as the wired, world in the United States since 1934. On a periodic basis, the various national regulatory authorities meet to sort out national and international spectrum allocation issues at the *World Radio Conferences* (WRC), previously known as the *World Administrative Radio Conferences* (WARC), which are sponsored by the ITU-R every two years.

Frequency allocation, or s*pectrum management*, involves the designation of certain frequencies in the electromagnetic spectrum in support of certain applications. Examples include AM and FM broadcast radio, UHF and VHF broadcast TV, Trunk Mobile Radio (TMR), cellular radio, and microwave radio. This requirement is essential to avoid interference between various applications using the same, or overlapping, frequency ranges. In limiting each application to a specific range of frequencies, the manufacturers, carriers, and end users of such systems can better be monitored and controlled as well.

The issue of Radio Frequency Interference (RFI), as discussed in Chapter 2, is tied to the specific frequencies employed, the proximity of the transmit/receive antennas, and the power levels involved. As I noted previously, lower frequency signals naturally propagate farther than higher frequency signals, suffering less from attenuation (i.e., loss of signal strength). (*Note:* This admittedly is a generalization. For example, there are notches in the frequency spectrum that work very poorly due to interaction with water molecules.) In consideration of the specific frequency range employed, therefore, the same frequencies cannot be used by multiple antennas within a certain range of proximity without running the risk of mutual signal interference. The power levels of the various transmitters also must be regulated, as stronger signals propagate farther at any given frequency.

11.3 ADVANTAGES AND DISADVANTAGES OF WIRELESS

Deployment of wireless networks certainly can offer advantages of reduced installation and reconfiguration costs. Tremendous costs can be saved by eliminating requirements to secure terrestrial rights of way, dig trenches and plant poles, place conduits and hang cross-arms, splice cables, place repeaters, and so on. Wireless networks work equally well in rocky or soggy terrain where wired networks may be problematic or even impossible and in remote areas of low user density where wired networks may be impractical. Wireless networks also offer great portability, as the antennas quite easily can be disassembled and reassembled at another location. Wired networks, on the other hand, must be abandoned or removed and sold as scrap. Some wireless networks support mobile applications, with examples being cordless telephony, cellular radio, and packet data radio networks.

Wireless also suffers from certain limitations, perhaps the most significant of which is that of spectrum availability—radio spectrum is a finite resource. The laws of physics and Mother Nature state that radio operates between 30 Hz and 30 GHz (Table 2.1), with the spectrum suitable for communications applications at about 30 kHz and above. While that may seem like a lot of spectrum, there also are a lot of applications and users competing for it, and not all frequency bands are suitable for all applications. This limited radio spectrum divides into even more limited ranges of spectrum allocated in support of specific applications (e.g., microwave and cellular radio). Within each slice of allocated spectrum there clearly exists only so much bandwidth; regardless of how cleverly the engineers and mathematicians design compression algorithms to maximize its use, there remains only so much bandwidth available. Airwave transmission systems also suffer from issues of error performance and security.

There are other problems with wireless communications, by the way. Similar to the effect of echo (i.e., reflected energy) in the wired world, wireless communications suffers from *MultiPath Interference* (MPI), also known as *multipath fading*. As radio signals propagate from the transmitter, they naturally diffuse, or spread out, no matter how tightly they are shaped, or focused. If the communication involves a cell phone and a cell site, for example, there really is no shaping. Rather, the signal from the cell site to the terminal is broadcast either on an omnidirectional basis or, more commonly, on a vectored basis. The signal from the terminal back to the cell site always is broadcast on an omnidirectional basis. Assuming that direct line of sight is achievable, some signal elements travel a straight line from transmitter to receiver, while other signal elements bounce off mountains, buildings, cars, trees, your neighbor's dog, and other dense physical objects. This can result in confusion at the receiver. *MPI specter* refers to the *ghosting effect*, which occurs when the path of the echo is relatively long and when some reflected signals, therefore, arrive on a significantly delayed basis in a phenomenon known as *delay spread*. [This ghosting effect is particularly evident in poorly installed coax-based CATV systems, and in traditional antenna-based TV reception in mountainous areas.] To overcome the effects of MPI, the receiver must make comparisons between signals to determine and lock onto the signals of greatest strength, earliest arrival, or both or perhaps to combine the individual weakened signal elements into a single stronger signal.

11.4 CELL CONCEPT: FREQUENCY REUSE

Radio systems are designed for a certain area of coverage, or *footprint*. Even early radio and TV broadcast systems used the concept to define a service area, such as a metropolitan area. Therefore, you could reuse the same frequencies to support service in metropolitan areas some distance away. For example, 98.1 (MHz) on your FM dial might be WXYZ in New York, New York, and KFRC in San Francisco, California. Similarly, Channel 7 on your TV might be WFAA in Dallas, Texas, and KIRO in Seattle, Washington; broadcast TV stations in the United States can reuse frequencies if separated by at least 150 miles [5]. The size of the cells, of course, is sensitive to frequency and power level as well as the height of the antennas, the topography, the time of day or night, solar activity, weather conditions, and other factors.

Now, I want to digress to my youth. Actually, it is not a complete digression, as it illustrates the points I just made. Some of you may remember Robert ("Bob") Weston Smith. Actually, you may remember him as *Wolfman Jack*, a famous rock'n roll Disk Jockey (DJ) of the sixties generation and later a successful star in movies such as *American Graffiti*. (If you don't remember him, just ask your parents.) In any event, as a young man in the 1960s, I would drive hundreds of miles over country roads in Texas to see my girlfriend (which tells you something about the size of the Great State of Texas as well as my difficulty in finding a girlfriend). At 9:00 p.m. on Sunday nights, *The Wolfman* broadcasted from XERF, the AM radio station in Ciudad Acuna, Coahuila, Mexico, and he came through as "clear as a bell" to the little transistor radio that hung from the rearview mirror of my 1962 Volkswagen beetle. It was not much of a radio (I couldn't afford the kind that fit in the dashboard), but it was a heck of a radio show and a heck of a radio station. XERF, you see, blasted *The Wolfman* (and a whole host of radio preachers earlier on a Sunday evening) at 250,000 W—five times the allowable power level for radio stations just across the border in Del Rio, Texas, U.S.A. At that power level, given the right weather conditions, *The Wolfman* could be heard just about as clearly in New York City as I heard him in Corpus Christi. In fact, he overpowered every radio station in North America operating on the same frequency. *The Wolfman* became famous all over North America, at the expense of the poor DJs who tried to make a living on the same frequency band. Having shared that example with you, I can get back to more serious business. By the way, *The Wolfman* passed on to that *big radio station in the sky* on July 1, 1995.

The formal concept of radio cells dates back to 1947, when Bell Telephone engineers developed a radio system concept that included numerous, low-power transmit/receive antennas [1]. Scattered throughout a metropolitan area, this sort of architecture served to increase the effective subscriber capacity of radio systems by breaking the area of coverage into *cells*, or smaller areas of coverage. Thereby, each frequency could be reused in nonadjacent cells. Additionally, the cells can be split, or subdivided, further as the traffic demands of the system increase. In other words, cellular radio networks are highly scalable.

Frequency reuse is sensitive to factors that are covered in the discussion of microwave and satellite systems in Chapter 2 and that can be seen in the *Wolfman Jack* example. Specifically, those factors include frequency, power level, antenna design, and topography. Higher frequency signals always attenuate to a greater extent over distance given the same power level. Antenna design is sensitive to wavelength and other factors. Topography is always an issue, as line of sight is always preferable, especially at the higher frequencies in the spectrum. Lower frequencies signals, such as those used in AM radio, tend to follow the curvature of the earth, while higher frequency signals, such as those used in microwave, tend to follow a straight path—straight into space.

11.4.1 Cell Categories

Cells can be characterized as falling into three broad descriptive categories: *macrocells*, *microcells*, and *picocells*, as illustrated in Figure 11.1. As the cells shrink, the advantages of frequency reuse increase significantly. The increase in traffic-handling capacity can be remarkable, with associated increases in revenue potential. However, the costs of network deployment increase along with the number of base stations and interconnecting links. Also, the complexity of switching mobile traffic between cells increases considerably.

Assume that 12 channels are available for use in a metropolitan area that is 60 miles wide (i.e., has a diameter of 60 miles and a radius of 30 miles) and that the terrain is perfectly flat. Assume a seven-cell reuse pattern, which is typical. Consider the following theoretical scenario:

- *Macrocells* cover relatively large areas, perhaps an entire metropolitan area 60 miles wide. Assume that there is one omnidirectional antenna positioned in the center of the entire metro area and supporting 12 channels. Then only 12 simultaneous conversations can take place within this single macrocell. If the macrocell is divided into seven smaller macrocells in a typical seven-cell reuse

Macrocell

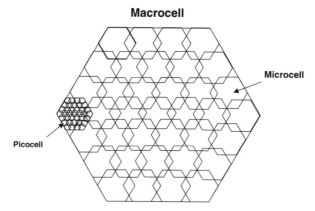

Figure 11.1 Macrocells, microcells, and picocells.

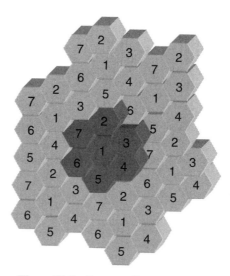

Figure 11.2 Seven-cell reuse pattern.

pattern, as illustrated in Figures 11.1 and 11.2, with each derived macrocell covering a radius of about 11.3 miles, no frequency reuse is possible and no improvement is realized. Only 12 conversations can be supported due to the fact that the cells must overlap, as illustrated in Figure 11.1. (*Note:* Radio cells typically are graphically depicted as cells in a honeycomb pattern. Actually, they are irregular, overlapping areas of coverage that are more or less circular or elliptical in form.) Therefore, conversations on the same frequency channels in adjacent cells interfere with each other due to the phenomenon of cochannel interference.

- *Microcells* cover a smaller area. If each of the derived macrocells were divided into seven equal microcells with a radius of about 4.3 miles, a reuse factor of 7 is realized. In other words, each of the 12 channels could be used seven times and a total of 84 simultaneous conversations could be supported. If the cell size shrinks to 1 mile, the reuse factor is 128, and the same 12 channels could support 1536 simultaneous conversations. (*Note:* The reuse separation ranges from approximately four to six times the radius of the cells.)

· *Picocells* are quite small, covering only a few blocks of an urban area or, perhaps, a tunnel, walkway, or parking garage. In a seven-cell reuse pattern, with each cell covering a radius of approximately one-half mile, the reuse factor climbs to 514. In other words, the same 12 channels could theoretically support up to 6168 simultaneous conversations [2].

11.4.2 Cells, Vectors, and Beams

Within a given cell, the antenna may operate in either an omnidirectional or a vectored mode. The omnidirectional approach involves signal transmission from and signal receipt by the centralized base station antenna in all directions. This approach uses all frequency channels in a 360° beam, and all at the same power level. A vectored antenna can subdivide the coverage area of the cell into multiple vectors of coverage. Cellular antennas, for example, commonly carve a cell into three vectors, each of 120°. This vectoring approach enables the carrier frequencies within each vector to be managed independently from those within the other vectors in terms of dimensions such as channel allocation and signal strength.

Smart antennas can improve significantly on this concept of vectoring. Such antennas can operate within a given cell and its constituent vectors, further subdividing the coverage through the use of as many as 12 beams, each with a beam width of 30°. Thereby, the footprint of the cell can be sculpted to optimize coverage, channel allocation, and interference. The individual beams can rotate from one vector to another and can do so on a dynamic basis in order to adjust to traffic patterns that might vary by time of day, with the process of beam management performed remotely [6].

While I have only a limited amount of space to discuss antenna design, the details of which are the subject for an electrical engineering text, there certainly is at least one other design issue of significance. Code Division Multiple Access (CDMA) systems, which are discussed later in the chapter, deal with issues of multipath interference (MPI) through the use of *rake receivers* in both the Base Stations (BSs) and Mobile Stations (MSs), for example, cell phones. Rake receivers employ *spatial diversity* and *time diversity* in much the same way as *Multiple Input, Multiple Output* (MIMO) technology in the developing 802.11n WLAN standard discussed in Chapter 8. A rake receiver comprises a set of four receivers, or *fingers*, that work in a coordinated way to gather signal elements much like the tines of a garden rake work together to gather leaves. Each finger gathers a faded, or attenuated, signal element at a separate moment in time. The receiver combines and correlates the results of all four fingers to optimize the signal, thereby countering the effects of multipath fading and delay spread.

11.5   MULTIPLEXING AND ACCESS TECHNIQUES

Communications networks take great advantage of the concept of DAMA (*Demand-Assigned Multiple Access*). DAMA enables multiple devices to share access to the same network on a demand basis, that is, *first come, first served*. There exist a number of techniques for providing multiple access (i.e., access to multiple users) in a wireless network. Those techniques generally, but not always, are mutually exclusive.

11.5.1   Frequency Division Multiple Access

At the most basic level, frequency division is the starting point for all wireless communications because all communications within a given cell must be separated by frequency to avoid their mutual interference. *Frequency Division Multiple Access* (FDMA) divides the assigned frequency range into multiple frequency channels to support multiple conversations, as depicted in Figure 11.3. In other words, multiple narrowband frequency channels derive from a wider band of assigned radio spectrum, much as frequency division multiplexers (FDMs) operate in the wired world. Multiple incoming and outgoing calls contend for access to those channels. A given call takes place on one pair of frequencies, with one transmission in the forward direction and another for transmission in the reverse direction. At the same time, another call takes place on another pair of frequencies. The forward and reverse channels in each frequency pair are separated in frequency in order to avoid cochannel interference. The forward channels are separated from each other, as are the reverse channels, in order to minimize the potential for crosstalk. The station equipment must be *frequency agile* in order to search for and seize an available frequency channel, especially as the mobile transmitter/receiver moves from one cell to another in a cellular network.

Analog cellular systems employ FDMA. *Advanced Mobile Phone System* (AMPS), for example, provides for a total allocation of 40 MHz, which is divided into 666 (832 in some areas) frequency pairs. In each of the 734 serving areas defined by the FCC, the available 666 (or 832) channels initially were split equally between *wireline* (i.e., Incumbent Local Exchange Carrier, or ILEC) and *nonwireline* operators, with the carrier in each category initially having been determined on a lottery basis. AMPS separates the 30-kHz forward channel and the 30-kHz reverse channel in each frequency pair by approximately 55 MHz. On average, AMPS cell sites in

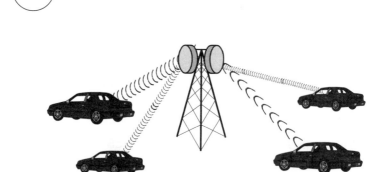

Figure 11.3   Frequency division multiple access.

the United States have a radius of approximately 1 mile. At maximum, based on a 7-cell reuse pattern, AMPS cells support about 56 frequency channels, considering that 21 of the available 416 channels per carrier are reserved for signaling and control purposes [5].

Frequency Division Duplex (FDD) is a means of providing duplex (bidirectional) communications. Forward and backward channels make use of separate frequencies. FDD is used with both analog and digital wireless technologies, including cordless telephony and cellular.

11.5.2 Time Division Multiple Access

Time Division Multiple Access (TDMA) is a digital technique that divides each frequency channel into multiple time slots, each of which supports an individual conversation (see Figure 11.4). This concept is exactly the same as in the wired world, where time division multiplexers perform the same function in a T/E-carrier environment. The total available bandwidth, the bandwidth of the individual channels, and the number of time slots per channel vary according to the particular standard in place, as well as the specific coding technique employed. GSM, for example, involves a carrier channel of 200 kHz, with a channel rate of approximately 200 kbps. The channel is divided into eight time slots of 25 kbps each, easily supporting low-bit-rate digitized voice of 9.6 kbps, plus overhead for framing and signaling. Each set of eight time slots is organized into a logical frame, and the frames are repeated frequently. Each conversation makes use of two time slots, one for the forward channel and one for the reverse channel. Assuming the same level of bandwidth, systems based on TDMA offer roughly three to four times the traffic capacity of those based on FDMA [1]. TDMA was specified first in EIA/TIA Interim Standard 54 (IS-54) and later was included in IS-136, which is an evolved version of IS-54 for use in cellular and PCS systems. TDMA is used in systems based on the D-AMPS and GSM cellular standards, and DECT cordless telephone standards.

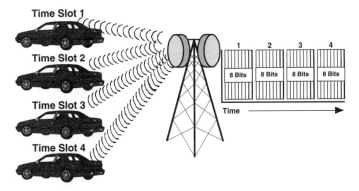

Figure 11.4 Time division multiple access.

A TDMA system, such as GSM, actually is both Frequency Division Multiplexing (FDM) and TDMA. FDM derives multiple carrier channels from a wider band of assigned spectrum. Within each frequency channel, TDMA derives multiple time slots (i.e., digital channels), for which incoming and outgoing calls contend.

E-TDMA (*Enhanced TDMA*), developed by Hughes Network Systems, is an improvement over TDMA, employing *Digital Speech Interpolation* (DSI) compression, also known as *Voice Activity Detection* (VAD), and half-rate *vocoders* (voice coders) operating at 4.8 kbps to enhance bandwidth utilization. E-TDMA provides as much as 16:1 improvement over analog technology.

Time Division Duplex (TDD) is a digital means of providing bidirectional communications. TDD can be employed with both channels using the same frequency, but this *ping-pong* transmission approach can yield poor quality, as it actually is a half-duplex (HDX) transmission mode. (*Note*: In a properly engineered network, this HDX mode can be made transparent.) More commonly, TDD is used in conjunction with FDD, with the forward and backward TDM channels riding over separate frequency channels. Further, the time slots are staggered so the frequency-specific transceivers are not asked to transmit and receive at the same exact points in time. Time slot 1, for example, might be used on the forward channel, or *uplink*, from the user terminal to the centralized cell antenna, while time slot 3 is used on the backward channel, or *downlink*, from the cell site to the user terminal.

In addition to its enhanced traffic capacity when compared to FDMA, TDMA supports data traffic as well as voice. TDMA also offers considerable flexibility, as multiple time slots can be assigned to an application, depending on its bandwidth requirements.

11.5.3   Code Division Multiple Access

Code Division Multiple Access (CDMA) is a relatively new technology that has its roots in *Spread-Spectrum* (SS) radio. Hedy Lamarr, the famous actress and dancer of pre–World War II fame, created the concept of spread spectrum in 1940. As the story goes, Lamarr developed spread spectrum radio in order to remotely synchronize multiple player pianos. Droves of people supposedly paid good money to go to the resulting radio-controlled piano concerts in this much simpler time. The U.S. Patent and Trademark Office issued a patent to Ms. Lamarr and George Antheil, a film-score composer to whom she had turned for help in perfecting the idea, for a *secret communication system* that was, in effect, a spread spectrum radio. In the Pacific Theater during World War II, the Allies used that patented technology extensively to prevent the Japanese from jamming radio-controlled torpedoes. This primitive system used a mechanical switching system much like a piano roll to shift frequencies faster than the Nazis or the Japanese could follow them. Subsequently, spread spectrum has combined with digital technology for spy-proof and noise-resistant battlefield communications. During the 1962 Cuban nuclear missile crisis, for example, Sylvania installed it on U.S. warships sent to blockade Cuba, where the technology provided improved security as well as prevented signal jamming. Ms. Lamarr never asked for, and never received, any royalties from the use of her invention. Ms. Lamarr was quite an innovator, by the way, and across multiple disciplines. She delighted and shocked audiences in the 1930s by dancing in the nude in the movie *Ecstacy* [7].

   As spread spectrum radio spreads the bandwidth of the transmitted signal over a spectrum of radio frequencies that is much wider than that required to support the native narrowband transmission, it commonly is known as a wideband radio technology. Multiple transmissions can occur at the same time in the same frequency domain. Thereby, multiple transmissions can each simultaneously use the entire system wideband, rather than just individual time slots or frequency channels. Spread spectrum uses two techniques: Direct Sequence (DS) and Frequency Hopping (FH).

- *Direct-Sequence Spread Spectrum* (DSSS) is a radio technique in which the narrowband signal is spread across a wider carrier frequency band. Each transmission is assigned a 10-bit pseudorandom binary code sequence, which comprises a series of 1s and 0s in a seemingly random pattern known to both the transmitter and receiver. The original code sequence is mathematically self-correlated to yield a code that stands out from all others, at least on average. The paired transmitters and receivers recognize their assigned and correlated code sequences, which look to all others as *PseudoNoise* (PN). DSSS phase modulates the carrier wave with a continuous string of PN code symbols, or *chips*, each of which has a much shorter duration than a data bit. So, the *chip rate* is much faster than the *bit rate*. Thereby, the noise signal occurs with a much higher frequency than the original data signal and spreads the signal energy over a much wider band (Figure 11.5). *Rake receivers* are used in CDMA systems to improve signal quality.
- *Frequency-Hopping Spread Spectrum* (FHSS), which generally is preferred over DSSS, more closely resembles Ms. Lamarr's original concept. FHSS involves the transmission of short bursts of data over a range of frequency channels within the wideband carrier, with the transmitter and receiver hopping from one frequency to another in a carefully choreographed *hop sequence*. This hop sequence generally is under the control of the centralized base station

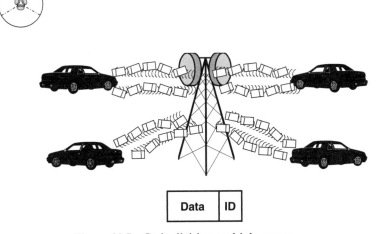

Figure 11.5   Code division multiple access.

antenna. Each transmission dwells on a particular frequency for a very short period of time (e.g., no more than 400 ms for FCC-controlled applications), which may be less than the time interval required to transmit a single data packet, or symbol, or even a single bit. So, again, the chip rate can be faster than the bit rate. A large number of other transmissions also may share the same range of frequencies simultaneously, with each using a different hop sequence. The potential remains, however, for the overlapping of packets. The receiving device can distinguish each packet in a packet stream by reading the various codes prepended to the packet data transmissions and treating competing signals as noise.

CDMA improves bandwidth utilization because a great number of users can share the same wideband radio frequency channel (see Figure 11.5). While initial predictions were that CDMA would improve on AMPS by a factor of up to 20:1, practical results have been more in the range of 15:1 in cellular telephony applications. Particularly through the use of the FHSS approach, CDMA also provides excellent security, as it is virtually impossible to intercept more than a small portion of a transmission. Encryption, of course, can provide additional security [8]. FHSS also offers the advantage of improved overall transmission, since no individual transmission gets stuck with the assignment of a poor-quality channel.

Qualcomm perfected and commercialized CDMA and has gone on to develop, manufacture, market, and license CDMA products. The first commercial CDMA system was placed in service in Hong Kong, where cellular phones have long been considered a necessity and where the networks have suffered terrible congestion. Since then, a great number of manufacturers and providers of cellular, PCS, WLANs, and other systems and networks have licensed CDMA.

11.5.4 FDMA, TDMA, and CDMA Compared: It's Party Time!

The best commonsense means of comparing FDMA, TDMA, and CDMA is to use the *cocktail party* analogy, which has been used by acoustical engineers for many years. This is an international cocktail party with a great number of people all wanting to talk at the same time. The cocktail party is being held in a large ballroom (frequency band).

In an FDMA environment, each conversation takes place in its own space (channel). Therefore, the ballroom must be subdivided into smaller rooms (cells) of suitable size (bandwidth) and then subdivided into smaller cubicles (channels) so there is no interference between the pairs of people engaged in each conversation. At any given time during the conversation, each pair occupies a separate room. The pairs can move from room to room (cell to cell) as they make their way from the entrance to the buffet, although they must interrupt their conversations when they leave one room and must enter another empty room before resuming conversation (*hard handoff*, or *break and make*). The number of simultaneous conversations that can be supported depends on the number of rooms of suitable size that can be derived from the space available in the ballroom. The speakers also must control their volume and frequency levels in consideration of the thickness of the walls (guard bands, or frequency separations) so they do not interfere with others.

In a TDMA environment, multiple pairs of speakers may share an even smaller room, although they have to squeeze (compress) into the physical space. Six people, for example, can hold three separate conversations if they all take turns. The order of the conversations will be carefully controlled, as each pair will get a time slice (time slot) of 20s each minute (frame).

In a CDMA environment, the entire ballroom is open. A larger number of people can fit into the room because the walls have been removed. A larger number of pairs of people can engage in simultaneous conversations, with the conversations overlapping in frequency, volume, and time. Volume and frequency levels still must be controlled, of course. As highly intelligent receivers, the listening parties can distinguish the transmission of their paired speakers by locking on the language (PN code) spoken, even given the high level of background noise (interference) from other conversations. Since the walls have been removed, the pairs can move from one area to another without losing conversational connectivity (*soft handoff*, or *make and break*).

11.6 SPECIALIZED MOBILE RADIO

Two-way mobile radio dates to the very early days of radio. The U.S. Army Signal Corps mounted early spark transmitters in vehicles in 1904 and experimented with air-to-ground communications in 1908. The Detroit, Michigan, Police Department placed the first experimental one-way mobile radio dispatch system into service in 1921, operating in the 2-MHz band with the call letters KOP [9]. The first two-way mobile system was installed by the Bayonne, New Jersey, Police Department in the early 1930s. While this Amplitude Modulation (AM) radio application grew quickly, even as late as 1937 the FCC allocated only 40 channels to mobile radio. In the late 1930s, Frequency Modulation (FM) replaced AM as the method of choice because of its improved quality of reception and lower power requirements. (FM receivers tend to lock in on the stronger competing signal, whereas AM recognizes all competing signals.) In 1949, the FCC recognized two-way mobile radio as a new class of service and began to allocate more spectrum and to regulate its use [2, 10].

In 1946, AT&T was granted the first commercial license for two-way, mobile FM service. That first system in St. Louis, Missouri (United States), employed a centralized antenna with a range of 50 miles. The system not only served to interconnect mobile phones but also provided connection to the Public Switched Telephone Network (PSTN). As the service was fairly inexpensive and extremely convenient, it quickly grew in popularity and soon was oversubscribed. Similar systems were quickly deployed in other major U.S. cities, where they, too, were soon oversubscribed. In fact, it was not uncommon for a provider to load as many as 100 subscribers per channel, which resulted in horrible service quality. In 1976, for example, service in the New York metropolitan area consisted of 20 channels supporting 543 subscribers out of a total population of approximately 20 million. Not surprisingly, there was a waiting list of approximately 3700 [10].

Specialized Mobile Radio (SMR), also known as *Trunk Mobile Radio* (TMR), entered the scene in the 1960s, marketed as *Improved Mobile Telephone Service* (IMTS). This commercially available service made better use of FM bandwidth through narrowband communications involving smaller frequency channels. IMTS

50 Miles

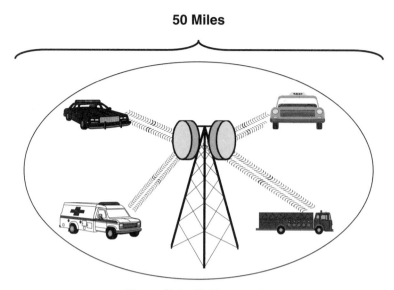

Figure 11.6 SMR network.

also enabled users to manually search multiple frequency channels. Shortly thereafter, intelligent mobile sets were developed that searched channels automatically. The concept of SMR/TMR remains much as it was originally. The provider places a radio tower and omnidirectional transmit/receive antennas on the highest possible point in the area and blasts the signal at the maximum allowable power level. As illustrated in Figure 11.6, this approach provides a coverage area of 50 miles or more, depending on topography. While some SMR systems support full-duplex (FDX) communications, many are only half-duplex (HDX). This HDX communications mode supports transmission in only one direction at a time and, therefore, requires that the parties take turns talking. The talker must depress a key or button on the microphone to talk and must release it to listen. This procedure is commonly known as the *Push-To-Talk* (PTT) protocol and sometimes is referred to as the *press-to-talk* or *Citizens Band* (CB) *radio* protocol. You copy, good buddy?

SMR/TMR largely has been supplanted by cellular service offerings, although it remains widely used in dispatch and fleet applications such as police, fire, and emergency vehicles as well as taxi fleets, utility fleets (e.g., telephone companies, gas and electric utilities, and CATV providers), and courier services. In the United States, 80 MHz has been allocated for SMR.

Enhanced Switched Mobile Radio (ESMR) is a technique developed by Nextel Communications and Geotek Communications for the development of a voice and data, cellular-like network using legacy SMR networks operating in the 800- and 900-MHz range. Nextel acquired and linked a large number of SMR networks throughout the United States, in February 1999 acquired the 191 900-MHz licenses of the bankrupt Geotek, and later added the 1.5-GHz band to the mix. Through the use of TDMA, each frequency channel is divided into multiple time slots to support multiple conversations. ESMR also supports call hand-off so mobile users can maintain connectivity as they travel from cell to cell. The Nextel network offers data throughput of 7.2 kbps, with coverage including most major metropolitan areas in the United States. Nextel terminal equipment supports integrated voice, data, paging, and Internet access. Perhaps the best known ESMR terminal is the Blackberry, a cellular telephone with features including Bluetooth technology, e-mail, PTT *walkie talkie* service, speakerphone, and Global Positioning System (GPS). The Blackberry is manufactured by Research in Motion (RIM) and the PTT service runs over the Nextel ESMR 800-MHz band. (*Note:* The Blackberry is by no means limited to the Nextel band.)

The walkie talkie was invented in 1938 by Al Gross, a high school student in Cleveland, Ohio (United States), at the time. The portable hand-held radio transmitter–receiver caught the attention of the Office of Strategic Services (OSS), predecessor to the Central Intelligence Agency (CIA). The OSS recruited Gross, who then led the effort to develop the walkie talkie for clandestine and military uses. Code named *Joan/Eleanor*, the first walkie talkie system comprised a ground unit, *Joan*, and an airborne unit, *Eleanor*. The system allowed OSS agents behind enemy lines to communicate with aircraft in a manner that virtually defied detection at the time. Gross also is credited with inventing the pager, the CB radio, and the cordless telephone. Gross also lobbied the FCC to create the Personal Radio license spectrum, which later became *Citizens Radio Service Frequency Band*, or CB radio. Gross formed the Citizens Radio Corporation, which manufactured and sold personal two-way radios, mostly to farmers and the U.S. Coast Guard [11].

11.7 PAGING

The paging system was invented by Al Gross as an adaptation of his two-way radio, the walkie talkie. After some early market resistance from doctors who were afraid that the system would upset their patients and disturb their golf games, Gross sold the first system in 1950 to New York's Jewish Hospital [11]. That first system provided a means by which a centralized antenna could broadcast alerts to small, inexpensive pagers, or beepers. A *page* simply transmitted an identification number, which was recognized only by the pager being addressed. If that pager were in range, it beeped, hence the term *beeper*. Response to the page was in the form of a telephone call to the paging company to retrieve a message. The FCC approved pagers for consumer use in 1958. The first consumer pager was the Motorola Pageboy I, which was based on the proprietary protocols including the *GOLAY* standard [10].

During the 1970s, an international team of radio engineers developed a standard set of code and signaling formats. That effort evolved into the POCSAG (*Post Office Code Standardization Advisory Group*) code, the name of which was derived from the fact that the British Post Office (BPO), which was the PTT for the United Kingdom at the time, chaired the effort. The POCSAG standard, which is in the public domain, provides for transmission speed of up to 2400 bps using channels of 25 kHz in the band 150–170 MHz. The CCIR (now ITU-R) standardized that code internationally in 1981, and most nations quickly adopted it. POCSAG can support as many as 2 million individual pager addresses. Tone-only, numeric, and alphanumeric pagers are supported on a one-way basis.

Paging in Europe has been constrained somewhat by the lack of agreement on common standards, although the POCSAG standard generally is recognized. A digital paging system known as ERMES (*European Radio MEssage System*) is supported by ETSI and the EU. In 1990, 26 system operators from 16 countries signed a Memorandum of Understanding (MoU) to create a pan-European system based on this standard. ERMES operates at 6250 bps in the band 169.4–169.8 MHz and uses Frequency Shift Keying (FSK) modulation.

Motorola more recently floated the *FLEX* set of proprietary solutions, which largely have replaced POCSAG in the United States and have become de facto standards throughout most of the world, excepting Western Europe. Those solutions provide two-way messaging, support data transmission, and provide greater bandwidth. FLEX also supports as many as 5 billion addresses, with up to 600,000 supported per channel. The FLEX family of protocols includes the following [12–15]:

- **FLEX:** 1600 bps; 25-kHz channels; one-way
- **ReFLEX:** 1600, 3200, 6400, or 9600 bps; 25- or 50-kHz channels downstream and 12.5-kHz channel upstream; two-way
- **InFLEXion:** Up to 112 kbps; 50-kHz channels in the N-PCS (Narrowband PCS range); two-way; supports compressed voice downstream

11.7.1 Paging Networks

Pagers generally operate over 25-kHz channels in the 900-MHz band. In 1984, the FCC (U.S.) dedicated 1 MHz of 40 channels in this band for nationwide paging purposes. RCCs (*Radio Common Carriers*) and PPOs (*Private Paging Operators*) provide paging services. Regulated by the FCC and the state Public Utility Commissions (PUCs), RCCs make use of FCC-designated frequencies. PPOs are unregulated but must share spectrum with other users in the VHF and UHF bands.

A typical page begins with a message transmitted to a centralized *Network Operations Center* (NOC). Using SkyTel, a Verizon company, as an example, the NOC forwards the page to a satellite with an appropriate footprint, or coverage area. As depicted in Figure 11.7, the satellite forwards the page to a terrestrial network of centralized antennas interconnected through various means, including private microwave, leased lines, and Frame Relay. The terrestrial antenna network forwards the page to the pager. In a two-way paging scenario, the response travels back to the terrestrial base station and then over a Frame Relay network to the NOC, where it is forwarded to the party who originated the page.

Generally speaking, these satellite links are highly beneficial and highly reliable. The May 1998 failure of the Galaxy IV satellite, however, underscored the vulnerability of even the most sophisticated networks. When the satellite lost its orientation, hundreds of terrestrial paging antennas had to be reprogrammed by hand in a process that took over a week. In addition to the 40 million or so pagers that lost service during this outage, 5400 of 7700 Chevron gas stations lost their pay-at-the-pump capabilities, and music-on-hold systems went down, causing callers to hang up when they thought they had been disconnected. Additionally, the results of the multistate Powerball lottery drawing could not be broadcast to the 88 TV stations that regularly carry that programming [16].

The downstream data to the pager originate in several ways. The most common approach involves the paging party dialing a telephone number which often is toll free. That number either may be dedicated to that one pager or may be one of many

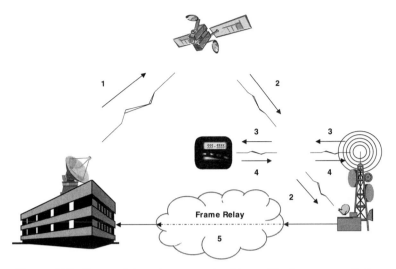

Figure 11.7 Terrestrial paging network with satellite interconnectivity.

associated with the service provider. The telephone number terminates in a voice processor at the central location of the service provider. The voice processor prompts the paging party to use the touchtone keypad of the telephone set to enter a return number. If the telephone number is not dedicated to that one pager, the paging party first is prompted to enter a *Pager Identification Number* (PID), more generically known as a *Personal Identification Number* (PIN), assigned to that one pager. If a textual message is to be sent, the paging service provider provides the option of accessing a human attendant who will answer the call and enter the message (generally limited to 500 characters) for transmission to the alphanumeric pager. Alternatively, many service providers now support Web-based messaging. This approach enables any paging party to access the service provider's website via the Internet and to enter a message of limited length without the intervention of an attendant. Additionally, large organizations with great numbers of employees equipped with pagers may have direct access to the paging system to send alphanumeric messages to their employees. Such access may be on either a dedicated or a dial-up basis and typically involves proprietary software which turns a PC into a paging dispatch terminal. These direct links from the user organization to the service provider also may permit the redirection of e-mail and facsimile transmission from corporate servers once those message formats are converted. As not to overwhelm both the paging network and the terminal equipment, filters must be used to limit the size of the transmitted file.

11.7.2 Paging Equipment

Pagers can be classified as tone only, numeric, alphanumeric, and voice enabled. *Tone-only* pagers cause the device to emit an audible tone and/or to vibrate or blink so it is not disruptive. *Numeric* pagers permit the receipt of numbers only on a display. The vast majority of contemporary pagers are *alphanumeric*, capable of receiving and displaying both alphabetic and numeric characters. Contemporary pagers contain enough memory to support as many as 30,000 characters. But relatively few pagers currently can support the storage of voice messages, which are extremely memory intensive. This voice capability is accomplished through the downloading of compressed voice mail from a centralized voice processor to the pager over a packet network.

Two-Way Paging (TWP) systems have been available abroad for years but were not introduced in the United States until roughly 1995. The simplest and most common version is known as 1.5-way paging. This approach supports *guaranteed message delivery*, as the network does not attempt to download messages until such time as the pager is within range, is turned on, and has enough memory to support the download. The general location of the pager is communicated upstream, so the messages can be downloaded to the antennas supporting that particular geographic area, rather than being broadcast across the entire paging network. Once downloaded successfully, the pager acknowledges to the network the receipt of the page. Such pagers are alphanumeric, supporting the display of a text message and return telephone number, typically for a telephone or pager. Full two-way paging enables the recipient of the page to select and transmit a return message, which most commonly is selected from a small set of predetermined messages, although some pagers support user-definable return messages.

Pagers also have been incorporated into other devices, such as cellular telephones, Personal Digital Assistants (PDAs), watches, and even key chains. Whether standalone or merged into another device, contemporary pager capabilities include some combination of alphanumeric display, two-way communications, message storage, audible and vibration alert, fax receipt (very unusual), and abbreviated e-mail forwarding (very unusual). Many paging service providers also offer information such as weather reports, sports scores, traffic reports, and stock quotes, all of which are provided through a feature known as *Short Message Service* (SMS), which currently is limited to 160 characters [17]. Limited Web content also may be available. Incoming messages generally may be reviewed, erased, and archived. Both incoming and outgoing messages may be date and timestamped.

11.7.3   Paging Applications: Contemporary and Developing

Pagers reached peak popularity in the 1980s and 1990s. In 1994, there were estimated to be well over 61 million pager subscribers worldwide, according to some estimates [18]. In the United States alone, there were over 27 million subscribers to over 2000 paging services [19]. At that point, pagers were overwhelmed by cellular telephones, which offer much more functionality, although at considerably higher cost. As a result, millions of pagers have been disconnected over the past decade or so. Pagers remain used in areas where cellular service is not available and in applications where cellular functionality is undesirable from a cost standpoint.

Paging networks have found new life in a host of applications that were unimagined only a few years ago. CreataLink paging service from Motorola takes paging to the contemporary extreme, in the Americas at least. Through dialing a toll-free number, you can access a menu of eight vehicle-related functions that include turning the lights off and on, vehicle tracking, starting or disabling the engine, locking or unlocking the doors, and opening or closing the trunk. Similarly, into the future, you can unlock the house so that your children have access to the home after school when you are at work. Two-way pagers send information from patients with medical emergencies and receive authorization to automatically increase the dosage of medication from an implanted drug-release system. Two-way pagers also are used in telemetry applications for network-based remote utility meter reading. Tank monitoring supports the remote monitoring of content levels and material corrosion. Lighting control, irrigation control, climate control, and remote monitoring of vending machine inventories are other applications [20]. The Motorola ReFLEX CreataLink transceivers transmit in the range 896–902 MHz at speeds up to 9600 bps and receive in the range 929–942 MHz at speeds up to 6400 bps [21, 22].

Paging costs generally involve a flat monthly fee which includes some number of pages. The cost of pages above the threshold generally is sensitive to their nature (e.g., numeric, alphanumeric, or voice). A few paging service providers offer the option of *paging party pays*, which charges the paging party, rather than the paged party, for the page.

11.8 CORDLESS TELEPHONY AND WIRELESS OFFICE TELECOMMUNICATIONS SYSTEMS

Cordless telephones are quite simply telephone handsets that connect on a wireless RF basis to a base station that connects to a wall jack via an RJ11 or other standard jack. *Wireless Office Telecommunications Systems* generally are in the form of adjuncts that provide cordless telephony communications capabilities behind PBXs, Electronic Key Telephone Systems (EKTS), hybrid's KTS, or Centrex systems. They generally are limited to voice applications, although some also support low-speed data. WOTS involve a wireless master controller, which is hardwired to special ports on the PBX, KTS, Hybrid, or Centrex. As illustrated in Figure 11.8, the master controller is hardwired to subcontrollers and antennas, which are distributed throughout the office complex or campus in a picocell configuration. The terminal equipment is in the form of wireless handsets, which are generally low in cost and limited in

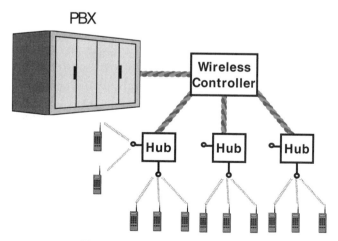

Figure 11.8 WOTS configuration.

range. Some dual-mode handsets exist, which can function as traditional cellular phones when the user is out of range of the WOTS or when otherwise desired.

Generally an extension and application of common cordless telephony, WOTS provides the advantage of mobility for a small group of select employees who must have the freedom to wander around the complex but must have communications capability at all times. However, WOTS PBX adjuncts tend to be highly capacity limited and costs can be easily double those of traditional phones when the additional infrastructure costs are taken into consideration. WOTS falls into the general category of *low-tier* systems, which are intended for pedestrian, in-building, and Wireless Local Loop (WLL) application. There exist multiple cordless telephony standards used to provide WOTS service. The original cordless telephones (c. 1980) in the United States were assigned 1 of 10 channels in the 27-MHz range. As these early analog phones transmitted *in the clear*, that is, with no encryption or signal-scrambling mechanism, and as there were so few channels, it was not unusual to accidentally engage in a three-way conversation with a neighbor who happened to have a phone operating on the same channel. In 1986, the FCC changed the cordless frequency range to the 46- and 49-MHz bands and reduced the allowable power levels, which had the effect of reducing the range slightly but improving signal quality. Contemporary digital versions operate in the 900-MHz and 2.4-GHz bands, which are in the unlicensed Industrial/Scientific/Medical (ISM) band. Contemporary standards include the following, the most significant of which are compared in Table 11.1:

- *CT1* (Cordless Telephony generation 1) was developed in Europe, where it was known as *CEPT-1*. CT1 operates in the 915- and 960-MHz bands over 40 paired channels 25 kHz wide. FDMA and FDD derive two separate channels, one for transmission and one for reception, each of which is 12.5 kHz wide. CT1 is analog and low cost but limited in range to 150 m or so. FM was employed in this voice-only technology. A variation on this standard is *CT0*, which was primarily used in the United Kingdom. CT0 specified eight paired channels, with the base station transmission in the 1.642–1.782-GHz range and portable station transmission in the 47-MHz range. A number of parochial CT0 versions were developed in other countries.

- *CT1+* is a variation on CT1 developed in concert by Belgium, Germany, and Switzerland. CT1+ was intended as the basis for a public wireless service, along the lines of *Telepoint*. CT1+ operated in the 887- and 932-MHz bands over 80 channels 25 kHz wide. FDMA and FDD derive two separate channels, one for transmission and one for reception, each of which is 12.5 kHz wide. Although CT1+ was not successful, it did originate the concept of a *Common Air Interface* (CAI), which enables multiple manufacturers to develop products in support of a public cordless telephony service offering.

- *CT2* was developed in the United Kingdom, where it formed the technology basis for the ill-fated *Telepoint* public cordless service. CT2 is a digital technology using TDMA and TDD and is deployed on a limited basis in Europe, Canada, and the Asia-Pacific. While it originally supported only outgoing calling, contemporary CT2 implementations support two-way calling. As CT2 does not support hand-off, the user must remain within range of the antenna used to set

TABLE 11.1   Cordless Telephony Standards

Standard	CT1	CT1+	CT2	CT2+	DECT	PHS	PCS
Region	Europe	Europe	Europe	Europe	Europe	Japan	U.S.
Frequency band, MHz	915/960	887/932	864/868	944/948	1880–1990	1895–1907	1850–1910, 1930–1990
Carrier spacing	25 kHz	25 kHz	100 kHz	100 kHz	1.728 MHz	300 kHz	300 kHz
Number of FDM carriers	40	80	40	40	10	75	32
Channels/ carrier	1	1	1	1	12	4	8
Access method	FDMA	FDMA	TDMA	TDMA	TDMA	TDMA	TDMA
Duplex method	FDD	FDD	TDD	TDD	TDD	TDD	FDD
Channel bit rate[a]	N/A	N/A	72 kbps	72 kbps	1.152 Mbps	384 kbps	384 kbps
Modulation method	FM	FM	GFSK	GFSK	GFSK	$\pi/4$ DQPSK	$\pi/4$ DQPSK
Speech coding, kbps	N/A	N/A	32	32	32	32	32

Note: Gaussian Frequency Shift Keying (GFSK) is a variation of Frequency Shift Keying (FSK) in which the signal is smoothed with a Gaussian filter before being placed on the carrier. This process reduces the spectral width, which serves to reduce the bandwidth required. The $\pi/4$ *Differential Quaternary Phase Shift Keying* ($\pi/4$ DQPSK) constellation can be viewed as the superposition of two QPSK constellations offset by 45 degrees relative to each other.
[a] Signaling rate.
Source: [23, 24].

up the call. CT2 operates in the range 864–868 MHz, supports 40 channels spaced at 100 kHz, and uses *Gaussian Frequency Shift Keying* (GFSK) as the modulation technique. Dynamic channel allocation requires a frequency-agile handset. CT2 was the first international standard providing a *Common Air Interface* (CAI) for systems operating in the 800- and 900-MHz bands. CT2 supports data communications at rates up to 72 kbps.

- *CT2+* is an improvement on CT2, supporting two-way calling and call hand-off. CT2+ uses 8 MHz of bandwidth in the 900-MHz range. CT2+ is based on dynamic channel allocation, requiring frequency-agile handsets. Encryption is supported for improved security. A common signaling and control channel offers improved call setup times, increased traffic capacity, and longer battery life because the handset must monitor only the signaling channel. CT2+ has been used in applications such as the *Walkabout* public cordless telephony trial in Canberra, Australia.

- *CT3* was developed by Ericsson in 1990 as a proprietary solution designed for high-density office environments. CT3 is based on TDMA and TDD and uses the same frequency bands as CT2+, supporting roaming and seamless hand-off. CT3 supports Telepoint (PCS) and in-building office applications (WOTS).

- *Digital Enhanced (nee European) Cordless Telecommunications* (DECT) is the pan-European standard for digital cordless telephony using TDMA and TDD. Ratified by ETSI in 1992, DECT provides 10 FDM channels in the band 1880–1990 MHz. Channel spacing is at 1.728 MHz, and each channel will support 1.152 Mbps with *Gaussian Frequency Shift Keying* (GFSK) as the modulation technique. Each channel supports 12 users through TDMA, for a total system load of 120 users. Voice encoding is Adaptive Differential Pulse Code Modulation (ADPCM) at 32 kbps. As DECT supports handover, users can roam from cell to cell as long as they remain within range of the system. DECT antennas can be equipped with optional spatial diversity to deal with multipath fading. Security is provided through authentication and encryption mechanisms. In North America, DECT is the basis for the *Personal Wireless Telecommunications* (PWT) standard, which operates in the unlicensed band 1910–1920 MHz. *PWT/E* is an extension into the licensed bands 1850–1910 and 1930–1990 MHz [25].
- *Personal Handyphone System* (PHS) is a digital cordless telephony system developed in Japan and used in China, Taiwan, and throughout Asia. PHS carves out 77 carriers with spacing of 300 kHz in the band 1895–1918.1 MHz. The upper half of the band is used for public systems and the lower half for home office applications. Each carrier supports four TDMA channels employing π/4 *Differential Quaternary Phase Shift Keying* (π/4 DQPSK) to yield a theoretical data rate of 384 kbps. Voice is encoded at 32 kbps. PHS enjoyed considerable market success but lately has declined in the face of competition from low-cost cellular offerings. (*Note*: π is the Greek letter *pi*, which is the symbol for the ratio of the circumference of a circle divided by its diameter. Pi is a mathematical constant with an approximate value of 3.14159. Pi is from the Greek *periphereia*, which translates into *periphery*, i.e., circumference.)
- *Personal Communications Services* (PCS) in the United States is based on the *Wireless Access Communication System* (WACS), which subsequently was modified to an industry standard known as *Personal Access Communications Services* (PACS). PCS employs FDD to carve out 16 paired downstream and upstream carriers with spacing of 300 kHz in the bands 1850–1910 and 1930–1990 MHz. Within each frequency channel, PCS derives eight TDMA channels employing π/4 Quadrature Phase Shift Keying (π/4 QPSK). Voice encoding is ADPCM at 32 kbps and the theoretical data rate is 384 kbps. PCS has never gained any traction in the United States due largely to the prevalence and low cost of cellular telephony [23, 24, 26, 27].

As I discuss in Chapter 3, contemporary PBXs increasingly are Internet Protocol (IP) based, running on a switched Ethernet LAN platform. Mirroring the trend toward WLANs, there currently is a great deal of interest in Voice over Wi-Fi (VoWiFi), which I discuss in Chapter 8. This interest level is heightened in the context of the developing 802.11n recommendation, which includes MIMO antenna technology. Once that standard matures and multimode VoWiFi/cellular handsets become available at reasonable cost, VoWiFi may very well gain the market acceptance that seems to have eluded WOTS cordless telephony.

11.9   CELLULAR RADIO

The basic concept of cellular radio dates back to 1947, when numerous, low-power transmit/receive antennas were scattered throughout a metropolitan area to increase the effective subscriber capacity of SMR/TMR radio systems. This architecture broke the macrocell area of coverage into smaller cells. Thereby, nonadjacent cells could reuse each frequency. Additionally, the carrier could split, or subdivide, the resulting cells further as the traffic demands of the system increased. This cellular concept is highly scaleable. Traditional cellular radio, I should note, involves a circuit-switching mode.

This original concept is the basis of cellular radio. The first prototype system was developed by AT&T and Bell Labs in 1977. In 1979 in Tokyo, Nippon Telephone and Telegraph (NTT) activated the first commercial cellular system, utilizing 600 duplex analog radio channels in the band 925–940 MHz for the uplink channels from the Mobile Station (MS) to the Base Station (BS) and the band 870–885 MHz for the downlink, or reverse, channels [26]. The first system in the United States was activated in Chicago on October 13, 1983. AT&T operated that *AMPS* network for exactly 79 days, at which point the *Modified Final Judgment* (MFJ) took effect and a subsidiary of Ameritech (now part of SBC) assumed ownership [2]. By 1984, the Chicago network already was saturated in some cells and cellular telephones were in the hands of 91,600 people, growing to 19 million by 1994 [2, 3]. A cellular telephony network comprises multiple low-power transmit/receive antennas distributed throughout a geographic area, with each cell site having a relatively small, more or less circular area of coverage (see Figure 11.9). The coverage area of each individual cell overlaps those of neighboring cells, with the cell diameter generally a minimum of about 1 mile and a maximum of about 5 miles, sensitive to factors such as topography and traffic density [8]. As the terminal device moves out of the effective range of one cell, the call switches from one cell antenna to another through a process known as *hand-off* in order to maintain connectivity at acceptable signal strength. The hand-off is controlled through a *Mobile Telephone Switching Office* (MTSO), which is the functional equivalent of the PSTN Central Office (CO). The MTSOs generally are interconnected and connected to the PSTN through either private microwave or leased-line facilities.

The process of hand-off can be accomplished in several ways. The *break-and-make* approach, also known as a *hard hand-off*, breaks the connection with one cell

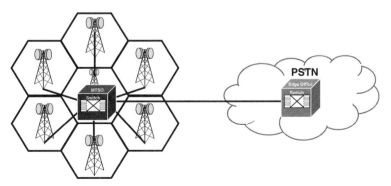

Figure 11.9   Cellular network with MTSO connected to PSTN.

site antenna before connection is reestablished with another preselected cell site. As the duration of the break is very short, it is not noticeable in voice communication. However, it renders data communications difficult at best. *Make-and-break*, also called *soft hand-off*, makes the new connection before breaking the old. This gentler approach offers considerable advantages in the transmission of data, particularly when supporting high-speed vehicular traffic, which may move between many cells in a relatively short period of time. Cellular systems, which support high-speed traffic, generally are known as *high-tier* systems. *Low-tier* systems include those intended for pedestrian traffic or for in-building or WLL applications.

Each cell site supports a limited number of frequency channels in order to take advantage of frequency reuse. The U.S. analog AMPS networks, for example, divide 333 frequencies (416 in certain areas) per carrier among cell sites; the average cell site supports 56 channels. To improve the performance of the network by reducing crosstalk, the original omnidirectional antennas were replaced with vectored versions. Vectoring generally involves three vectors, each with coverage of 120°. Splitting the frequencies into three vectors had the unfortunate effect of reducing by two-thirds the number of channels available to any given user in the cell site. As a result, more cell sites of smaller size were required, which entailed additional cost. For aesthetic reasons, however, many cities and towns placed moratoria on the construction of new antenna sites. In response, antennas were developed that can support the full 333/416 channels, subdividing the area of coverage through the use of 12 beams, each with a 360° sweep. These *smart* antennas communicate continuously with the MTSO so that channel allocation is managed cell by cell and on the basis of the network as a whole and in consideration of shifts in traffic patterns [16].

11.9.1 Cellular Standards

Cellular standards are numerous and largely incompatible. Standards include both early analog and more recent digital solutions. Digital systems offer the advantages of improved error performance, improved bandwidth utilization through compression, and enhanced security through encryption and other mechanisms. Digital systems also support data communications much more effectively. In the United States, carriers are shifting subscribers to digital systems through various marketing enticements. The FCC has authorized U.S. carriers to cease support for analog cellular systems as of March 1, 2008.

11.9.1.1 Generation 1 (1G): Analog Cellular   The first cellular systems were analog in nature. Collectively, these analog solutions are categorized as 1G (1st Generation) systems. They include the following and are compared side by side in Table 11.2.

11.9.1.1.1 Advanced Mobile Phone System (AMPS)   AMPS was the first cellular technology deployed in the United States. Developed by Motorola and AT&T, AMPS is an analog technology operating on 50 MHz in the 800-MHz band and supporting 666 (in some areas 832) channels. In the United States, 25 MHz and 333 (in some areas 416) channels each are provided to the *A-carrier*, or *nonwireline carrier*, and the *B-carrier*, or *wireline carrier* (incumbent telco or telco consortium).

TABLE 11.2   Analog Cellular Standards

Standard	AMPS	ETACS	NTACs	NMT 450	NMT 900
Region	U.S.	UK	UK	Nordic	Nordic
Frequency band, MHz	Tx: 824–849 Rx: 869–894	Tx: 871–904 Rx: 916–949	Tx: 915–925 Rx: 860–870	Tx: 453–458 Rx: 463–468	Tx: 890–915 Rx: 935–960
Carrier spacing, kHz	30	25	12.5	25	12.5
Number of FDM carriers	666/832	1000	400	200	1999
Channels/carrier	1	1	1	1	1
Access method	FDMA	FDMA	FDMA	FDMA	FDMA
Duplex method	FDD	FDD	FDD	FDD	FDD
Channel bit rate[a]	N/A	N/A	N/A	N/A	N/A
Modulation method	FM	FM	FM	FM	FM

Note: Tx = Transmission from mobile station to base station; Rx = Receive by mobile station from base station.
[a] Signaling rate.

Of the total number of channels awarded to each carrier, 21 channels are non-conversational channels dedicated to call setup, call hand-off, and call teardown. The remaining communications channels are split into 30-kHz voice channels, with separation of 45 MHz between the forward and reverse channels. Based on FDMA and FDD transmission, AMPS does not handle data well, with modem transmission generally limited to 6.8 kbps. Although once widely deployed in the United States, Australia, the Philippines, and other countries, AMPS has almost entirely been replaced by digital technology. Australian regulators mandated a cutover from analogue (That's Aussie for analog, mate.) AMPS to digital GSM and CDMA beginning December 31, 1999, in Melbourne, gradually extending throughout the country during 2000. As noted above, the FCC in the United States has authorized carriers to cease support for analog systems as of March 1, 2008.

11.9.1.1.2 Narrowband AMPS (N-AMPS)   Also developed by Motorola, N-AMPS enhances the performance of an analog AMPS system. System capacity is improved by splitting each 30-kHz channel into three 10-kHz channels, thereby tripling AMPS capacity. Very few U.S. carriers deployed N-AMPS.

11.9.1.1.3 Total Access Communications System (TACS)   TACS is a derivative of AMPS developed for use in the United Kingdom in the 900-MHz band. TACS supports either 600 or 1000 channels, each of 25 kHz, compared with the 666/832 channels supported by AMPS. A number of variations were developed, including *Narrowband TACS* (NTACS), *Extended TACS* (ETACS), and *Japanese Total Access Communications System* (JTACS). TACS found acceptance in very few nations, largely has been replaced by GSM, and is considered obsolete in the United Kingdom.

11.9.1.1.4 Nordic Mobile Telephone (NMT)   NMT was developed and placed into service in the early 1980s in Scandinavian countries, including Denmark, Finland,

TABLE 11.3 Digital Cellular Standards

Standard	IS-54/136 (D-AMPS)	IS-95 (cdmaOne)	GSM	PDC
Region	North America	North America	International	Japan
Frequency band, MHz	Tx: 824–849	Tx: 824–849	Tx: 890–915	Tx: 940–956, 1477–1501
	Rx: 869–894	Rx: 869–894	Rx: 935–960	Rx: 810–826, 1429–1453
Carrier spacing, kHz	30	1250	200	25
Number of FDM carriers	832	20	124	1600
Channels/carrier	3	798	8	3
Access method	TDMA	CDMA	TDMA	TDMA
Duplex method	FDD	FDD	FDD	FDD
Channel bit rate[a]	48.6 kbps	1.288 Mbps	270.833 kbps	42 kbps
Modulation method	π/4 DQPSK	BPSK/OQPSK	GMSK	π/4 DQPSK
Speech coding	VSELP: 8 kbps	EVRC: 13 kbps CELP: 8 kbps	RPELPC: 13 kbps VSELP: 8 kbps	VSELP: 9.6 kbps CELP: 5.6 kbps

[a]Signaling rate.
Note: Tx = Transmission from mobile station to base station; Rx = Receive by mobile station from base station. The π/4 *Differential Quaternary Phase Shift Keying* (π/4 DQPSK) constellation can be viewed as the superposition of two QPSK constellations offset by 45° relative to each other. *Offset Quadrature Phase Shift Keying* (OQPSK) is a variant of QPSK that uses a half-symbol timing offset to prevent large amplitude fluctuations in the signal. *Gaussian Minimum Shift Keying* (GMSK) is a variant of Frequency Shift Keying (FSK) in which the signal is smoothed with a Gaussian low-pass filter before being placed on the carrier. This process reduces the spectral width and minimizes cochannel interference. *Regular Pulse Excitation Linear Predictive Coding* (RPELPC) is a speech-encoding technique that uses regular pulses in an *excitation frame* and a long-term predictor, based on long-term correlation of voice samples, to model the speech pitch.

Norway, and Sweden. *NMT 450* operates in the 450-MHz range, which yields excellent signal propagation. Therefore, it is especially appropriate for sparsely populated areas supported by few cell sites. NMT 450 found little acceptance outside of the Scandinavian countries. *NMT 900* operates in the 900-MHz range and is appropriate for more densely populated areas. NMT 900 found acceptance in certain countries in Asia as well as the Nordic countries, although it is not considered a long-term technology. NMT largely has been replaced by GSM [23, 26, 27].

11.9.1.2 *Generation 2 (2G): Digital Cellular* Digital cellular clearly dominates the cellular radio world, having almost completely replaced the analog systems. There are a great number of standards, and a number of standards-based solutions, none of which are compatible. Collectively, these systems are grouped into the 2G (2nd Generation) category. They include the following, the most important of which are compared side by side in Table 11.3.

11.9.1.2.1 *Digital-AMPS (D-AMPS)* Also known as *US TDMA* and *NA-TDMA* (*North American TDMA*), D-AMPS was specified in IS-54 and later evolved into IS-136. D-AMPS is a North American digital cellular standard that operates in the

same 800-MHz band as the earlier analog AMPS. In fact, the two can coexist in the same network. D-AMPS uses the same 30-kHz bands as AMPS and supports up to 416 frequency channels per carrier. Through Time Division Multiplexing (TDM), each frequency channel is subdivided into six time slots, each of which operates at 8 kbps. Each call initially uses two time slots (e.g., 1 and 4, 2 and 5, and 3 and 6) in each direction, for a total of 16 kbps, which supports the data transfer plus overhead for call processing. While the standard recommends speech compression at 8 kbps (actually 7.95 kbps) using *Vector-Sum Excited Linear Predictive Coding* (VSELP), that is an average rate because each call can burst up to 48 kbps. D-AMPS yields a 3:1 advantage over AMPS in terms of bandwidth utilization. IS-136 is known as a *dual-mode* standard because both D-AMPS and AMPS can coexist on the same network, with both using the same 21 control channels for call setup, call hand-off, and call teardown. Thereby, IS-136 offers carriers the advantage of a graceful transition from analog to digital. IS-136 also includes a nonintrusive *Digital Control Channel* (DCCH), which is used for Short Message Service (SMS) and caller ID. SMS supports information transfer for applications such as weather reports, sports scores, traffic reports, and stock quotes, as well as short e-mail-like messages, which may be entered through the service provider's website. Some service providers also allow the cellular user to respond to e-mails via two-way SMS. Data communications is supported at up to 9.6 kbps per channel (paired time slots), and as many as three channels can be aggregated for speeds up to 28.8 kbps. Group 3 facsimile also can be supported. The RF modulation technique is DQPSK (*Differential Quaternary Phase Shift Keying*). Cingular Wireless uses D-AMPS.

11.9.1.2.2 Global System for Mobile Communications (GSM)   Originally known as *Groupe Spéciale Mobile* (French), GSM was adopted by the CEPT in 1987 as the standard for pan-European cellular systems and was first introduced in 1991. GSM operates in the 800-MHz and 900-MHz frequency bands and is ISDN compatible. GSM carves each 200-kHz band into eight TDMA channels of 33.8 kbps, each of which supports a voice call at 13 kbps using *Linear Predictive Coding* (LPC). Data throughput generally is limited to 9.6 kbps, due to FEC and encryption overhead. GSM commonly employs a four-cell reuse plan, rather than the seven-cell plan used in AMPS, and divides each cell into 12 sectors. GSM commonly uses frequency hopping and time-slot hopping, which also is used in CDMA systems. GSM offers additional security in the form of a *Subscriber Identification Module* (SIM), which plugs into a card slot in the handset, much as a PCMCIA card fits into a laptop computer. The SIM contains user profile data, a description of access privileges and features, and identification of the cellular carrier that hosts the home registry. The SIM can be used with any GSM set, thereby providing complete mobility across nations and carriers supporting GSM, assuming that cross-billing relationships are in place. GSM clearly developed to be the international standard of choice. Like D-AMPS, GSM supports SMS text messaging, which generally is two-way. GSM is in place in over 475 networks in more than 190 countries and predominates throughout Europe and much of Asia, supporting full roaming privileges from country to country. With minor modifications, GSM is the basis for *DCS 1800*, also known as PCN (*Personal Communications Network*), in Europe. DCS 1800, in large part, is an *upbanded* version of GSM, operating in the 1800-MHz (1.8-GHz) range. Also with minor modifications, it is the basis for *PCS 1900* in the United States, where it also is known as GSM. PCS 1900 is the ANSI standard (J-STD-007, 1995) for PCS at 1900 MHz (1.9 GHz). Unfortunately, PCS 1900 is not compatible with the original European GSM, due to the difference in frequency bands. T-Mobile (owned by Deutsche Telekom, which explains a lot) has deployed PCS 1900, and Cingular built a GSM network as an overlay to its D-AMPS network.

11.9.1.2.3 Personal Communications System (PCS)   PCS is a U.S. term for cellular systems based on EIA/TIA IS-95a, also known as *cdmaOne* and *CDMA Digital Cellular*. The first commercial systems were installed in South Korea and Hong Kong in 1995 and in the United States in 1996. PCS uses one or more frequency bands of 1.25-MHz converted from the existing AMPS spectrum of each carrier deploying the service. Each 1.25-MHz band is subdivided into 20 carriers, each of which can support as many as 798 simultaneous calls and aggregate bandwidth of up to 1.288 Mbps. Because CDMA is employed, no guard bands are required. IS-95 can support dual-mode communications, operating in the same network as AMPS. As previously noted, CDMA offers the advantages of improved bandwidth utilization as compared to AMPS (as much as 10:1 or even 20:1) and TDMA (as much as 6:1), soft hand-off, variable-rate speech-encoding, and support for both voice and data. The basic user channel rate is 9.6 kbps, although various channel rates can be achieved depending on the carrier implementation. The variable-rate speech-encoding algorithm runs at maximum rates of 8 kbps using *Code-Excited Linear Prediction* (CELP) or 13 kbps using *Enhanced Variable-Rate voCoder* (EVRC) and varies the rate downward to as low as one-eighth rate if the level of speech activity permits. The IS-95-B specification supports symmetric data rates of 4.8 and 14.4 kbps per channel; as many as eight channels can be aggregated to support a data rate up to 115.2 kbps. CDMA offers additional advantages in terms of maximum cell size due to improved antennas sensitivity and battery time due to precise power control mechanisms. IS-95 has been deployed by Airtouch (now part of Verizon), Ameritech Cellular (now part of Cingular), AT&T Wireless (now part of Cingular), Bell Atlantic Mobile (now part of Verizon), GTE MobilNet (now part of Verizon), 360° Communications (now part of ALLTEL), and Sprint Nextel [28–31].

11.9.1.2.4 Personal Digital Cellular (PDC)   Previously known as *Japanese Digital Cellular* (JDC), PDC is a Japanese standard for a digital system operating in the 800-, 900-MHz, and 1400-MHz frequency ranges. PDC derives 1600 RF carriers, each of which supports three TDMA channels at 42 kbps. Full-rate voice encoding is at 9.6 kbps and half-rate at 5.6 kbps. PDC has not found acceptance outside Japan, where it was extremely popular, with the exception of a few Asian countries under Japanese economic influence. PDC is being phased out in Japan, in favor of cdma2000.

11.9.1.3 Cellular Data Communications: The Early Years   Mobile data communications has always been somewhat difficult at best. The first attempt at mobile datacomm, at least as best I can determine, dates to 1907, when the U.S. Cavalry experimented with various solutions to a particular tactical field communications problem. When a cavalry troop was on patrol, it required some means of communicating back to a command-and-control center. At the time, therefore, a single mounted soldier trailed behind the troop, unreeling copper wire from a small wagon as he rode along. When it was necessary to send a message, the troop either stopped until the communications specialist caught up or sent a rider back to meet him. Then the communications specialist reined in his mount, hopped off, planted a metal stake in the ground to complete the circuit, and sent a telegraph message via Morse code to the company command center. This approach clearly slowed the movement of the troop, placing it at a tactical disadvantage. Some brilliant engineers devised a solution which involved shaving a spot on the horse's rump and gluing a copper patch to the horse's bare skin. A telegraph key was mounted on the copper patch, which was connected to the copper wire that connected back to the command center. Because a horse almost always has at least one foot on the ground, the circuit remained complete and the communications specialist could send a message while riding [32]. No mention is made of the attitude of the horse toward this technological breakthrough. Neither is any mention made of the bit error rate (pun intended). In any case, the technique apparently proved less than satisfactory, as it merited only a few references in obscure military documents. Wireless technology certainly offered greater promise.

The 1G analog networks really were not designed with data communications in mind. Over time, a number of manufacturers developed modems that could achieve theoretical data rates of as much as 28.8 kbps under optimum conditions, but conditions were rarely optimum. Factors affecting signal quality include the distance between the mobile station and the base station and Multi-Path Interference (MPI). Given the high error rates typical in cellular networks, the modem protocol either must include Forward Error Correction (FEC), which is overhead intensive, or must provide for repeated transmissions of errored and lost data packets or both. All things considered, throughput generally was more in the range of half the advertised rate at best. As analog systems take a break-and-make approach to call hand-off between cells, dropped connections were routine for the mobile cellular data user.

The 2G digital cellular systems improved on the analog approach, offering all digital data communications advantages enjoyed in the wired world. Those advantages include more usable bandwidth, improved error performance, and therefore enhanced throughput. Issues of inherent error performance and MPI remain, however. The hard hand-offs associated with the break-and-make connection technique persisted in 2G networks, with the exception of GSM, which used the softer approach of make-and-break to maintain connectivity.

In support of data communications over cellular networks, several manufacturers unveiled data-ready cellular phones as early as January 1995. Display screens varied in size from 16 to 160 characters. In addition to voice and SMS, those early data-ready phones supported Internet access, facsimile, and e-mail [33]. Nonetheless, the early cellular networks were not particularly friendly toward data communications.

Cellular Digital Packet Data (CDPD) was formalized in 1993 by a consortium of carriers in the United States to resolve this issue. CDPD operated over existing AMPS networks in the 800-MHz band, taking advantage of the natural idleness in cellular networks in between disconnections and connections and during the break-and-make process to transmit packetized data at rates up to 19.2 kbps using either the Internet Protocol (IP) or the ISO *Connectionless Network Protocol* (CLNP). The frequency-agile CDPD modems searched for available channels over which to send encrypted packets during the periods of channel idleness. As this approach proved expensive, however, the carriers ultimately deployed CDPD over AMPS channels removed from voice service as analog cellular voice subscribers gradually transitioned to digital service. Ultimately, CDPD proved too difficult, too bandwidth limited, and too expensive, especially as data-ready 2.5G and 3G networks made their appearances. The last of the CDPD networks were finally decommissioned around the end of 2005.

11.9.1.4 Generation 2.5 (2.5G) and 3 (3G) and Beyond   Beginning in 1992, the ITU-R began work on *International Mobile Telecommunications-2000* (IMT-2000). Variously dubbed 2.5G (i.e., the generation midway between 2G and 3G) and 3G, some of the many standards are well developed and widely available while others require a bit more work. They all grew out of an attempt to define a single global standard, which effort collapsed in October 1999, when representatives from the various countries agreed to adopt *federal* standards under the IMT-2000 umbrella. Two of the three modes are based on Code Division Multiple Access (CDMA) and one on Time Division Multiple Access (TDMA). cdmaOne, also known as Telecommunications Industry Association Interim Standard-95a (TIA IS-95a), was the first CDMA-based 2G approach to be introduced and is popular with CDMA-based cellular operators in North America and Asia. CDMA2000 3XMC, the high-speed version operating at 2 Mbps, was developed by Qualcomm and has been approved by the ITU-R. Enhanced Data Service for GSM Evolution (EDGE) is the TDMA variant of IMT-2000. General Packet Radio Service (GPRS), an interim step toward EDGE, is the choice of cellular operators with networks based on Global System for Mobile Communications (GSM) and Digital Advanced Mobile Phone Service (D-AMPS). Regardless of the transition approach and the specifics of the technology, all 3G systems will be based on CDMA.

In general, these standards support much increased data rates for both mobile and fixed wireless versions. In general, each has its roots in and is an extension of a 2G technology. So, the incompatibilities roll on, although at faster speeds, in support of voice, *Short Message Service* (SMS) and *Multimedia Messaging Service* (MMS), and Internet access. The 2.5G networks run over the same spectrum as the 2G networks and are backward compatible with them. Hence the transmission rates are lower than those of 3G networks, which run the over the spectrum set aside for their specific purposes. 2.5G and 3G technologies and standards include EDGE, GPRS, IMT-2000, UMTS, and W-CDMA. All of these approaches fit under the umbrella of *International Mobile Telecommunications-2000* (IMT-2000), which replaced *Future Public Land Mobile Telecommunications System* (FPLMTS) as the vision for a single global standard for wireless networks. IMT-2000 is an ITU initiative for a twenty-first century wireless network architecture. As illustrated in Figure 11.10 and compared in Table 11.4, specifications include 128/144 kbps for high-mobility applications, 384 kbps for pedestrian speed (i.e., walking speed) applications, and 2.048 Mbps for both fixed WLL (Wireless Local Loop) and in-building applications such as WLANs (Wireless LANs). IMT-2000 also was intended to operate in the 2-GHz band. Note that *2000* has several meanings: the year 2000, bandwidth of up to 2000 kbps, and frequency range 2000 MHz. As it turned out, the year 2000 was pushed back to at least 2002, bandwidth of 2 Mbps is only for limited in-building applications, and the 2-GHz spectrum has been allocated for other purposes in North America as part of the unlicensed ISM band.

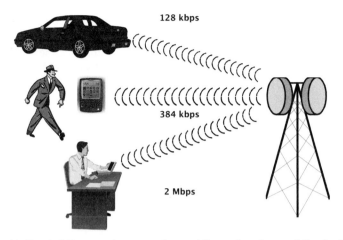

Figure 11.10   A 3G network supporting mobile, pedestrian, and fixed wireless.

TABLE 11.4   IMT-2000

Application	Maximum Bit Rate	Cell Size
Mobile	128/144 kbps[a]	Macro/micro/pico
Pedestrian	384 kbps[b]	Macro/micro/pico
Indoor	2.048 Mbps[c]	Micro/pico

[a] ISDN 2B/2B+D.
[b] 6 × DS-0.
[c] E1 (32 × DS-0).

Terminal equipment across all of these standards includes multifunction cellular telephones, laptop and tablet PCs, and PDAs and other hand-held computers. Generally speaking, a cellular modem is required for a laptop to establish connectivity, although some manufacturers have started selling computers with built-in wireless antennas. In any case, Internet access via a computer requires a separate subscription agreement that is sensitive to airtime.

11.9.1.4.1   High-Speed Circuit-Switched Data   High-Speed Circuit-Switched Data (HSCSD) is a 2G+ upgrade to GSM designed to improve data transmission rates. Recall that each GSM channel of 200 kHz provides capacity of 270.833 kbps through GMSK modulation. Each channel is divided into eight time slots, which yields a theoretical data rate of 33.8 kbps. Actual throughput, however, generally is limited to 9.6 kbps in consideration of overhead for FEC and encryption. HSCSD improves channel throughput to a maximum of 14.4 kbps in GSM host networks operating at 1800 MHz through the use of improved FEC mechanisms. HSCSD also supports the concatenation (linking) of multiple time slots per frame in support of higher speeds. For example, two time slots yield a transmission rate of up to 28.8 kbps, three yield 43.2 kbps, and four yield 57.6 kbps. In GSM networks operating at 900 MHz, the channel throughput remains 9.6 kbps with concatenation yield maximum throughput of 19.2, 28.8, and 38.4 kbps. As GSM provides circuit-switched, rather than packet-switched, connectivity, HSCSD is more suited for connection-oriented applications such as video and multimedia. E-mail and other bursty data communications applications are served more cost effectively by packet data network protocols running over native packet networks. A small number of carriers have deployed HSCSD as an interim step toward EDGE and UMTS networks. HSCSD has not been deployed in PCS 1900 networks, the upbanded GSM version running at 1900 MHz in the United States [26].

11.9.1.4.2 General Packet Radio Service   *General Packet Radio Service* (GPRS) is the 2.5G data service enhancement for GSM host networks. GPRS specifications were developed in 1997 by ETSI, which has since passed that responsibility on to the 3rd Generation Partnership Project (3GPP). GPRS is a packet-switched service that takes advantage of available GSM time slots for data communications and supports both X.25 and TCP/IP packet protocols with QoS. GPRS, an important component in the GSM evolution, enables high-speed mobile datacom usage and is considered most useful for bursty data applications such as mobile Internet browsing, e-mail, and various push technologies. Through linking together as many as eight GSM channels, GPRS has a theoretical transmission rate as high as 171.2 kbps, although it realistically is limited to 115.2 kbps and more typically 30–60 kbps. In practice, however, GSM system operators are unlikely to allow a single user to access eight channels. GPRS maintains the same GMSK modulation scheme used by GSM and provides *always-on* access, although charges apply only for actual data traffic.

Notably, GPRS will support simultaneous voice and data communications over the same wireless link, with voice taking precedence as always. GPRS defines three classes of terminal equipment:

- Class A terminals support simultaneous circuit-switched GSM voice and SMS service as well as GPRS packet-switched data traffic.
- Class B terminals will support nonsimultaneous circuit-switched voice and packet-switched data, automatically switching between the two. A Class B terminal, for example, will suspend an active data session in the event of an incoming voice call or SMS message. Most GPRS terminals are Class B.
- Class C terminals support either circuit-switched voice and SMS service or packet-switched services but must be manually switched from one to the other.

The bit rate is sensitive to the encoding scheme in use, of which there are four. CS-4 supports a bit rate of 20.0 kbps per time slot but can be used only when the Mobile Station (MS) and Base Station (BS) are in proximity. As the distance increases between the MS and BS, the encoding scheme must be more robust to compensate for attenuation and MPI, and the bit rate accordingly must adjust downward. At the edge of the cell, for example, CS-1 supports a bit rate of only 8.0 kbps. The four *Coding Schemes* (CSs) and associated bit rates are as follows:

- CS-4: 20.0 kbps
- CS-3: 12.0 kbps
- CS-2: 14.4 kbps
- CS-1: 8.0 kbps

TABLE 11.5 GPRS Example Service Classes

Multislot Service Class	Downlink Slots	Downlink Speed[a] (kbps)	Uplink Slots	Uplink Speed[a] (kbps)
2	2	40	1	20
4	3	60	1	20
6	3	60	2	40
8	4	80	1	20
10	4	80	2	40
12	4	80	4	80

[a]Maximum speed, assuming CS-4 encoding.

GPRS can run in either the symmetric or asymmetric mode, with the speed in either direction sensitive to the *multislot service class* selected, of which there are 12. The multislot service class determines the number of time slots in each direction, with each time slot supporting a theoretical nominal data rate of 20 kbps (actually 21.4 kbps). The simplest is service class 1, which supports one time slot in each direction. The most capable is service class 12, which supports four time slots in each direction. Generally speaking, the most common service classes are asymmetric in nature, which suits data-oriented Web applications in much the same way as do the asymmetric local loop technologies of ADSL, PON, and WiMAX. The most common GPRS service classes are organized as listed in Table 11.5, which also provides maximum downlink and uplink bit rates based on CS-4 encoding.

The U.S. carriers deploying GPRS include Cingular (800 and 1900 MHz) and T-Mobile (1900 MHz). Upgrades from GPRS to 3G will be in the direction of UMTS, often through EDGE as an intermediate step [26].

11.9.1.4.3 Enhanced Data Rates for GSM Evolution Enhanced Data rates for GSM (nee *Global*) Evolution (EDGE) is a 2.5G standard developed by ETSI in 1999 and touted as the final stage in the evolution of data communications within the existing GSM standards. The only IMT-2000 specification based on TDMA, EDGE supports data transmission rates up to 473.6 kbps over GSM FDD channels 200 kHz wide through an improved modulation technique. *8-Phase Shift Keying* (8-PSK) involves eight levels of phase shift and, therefore, supports three bits per symbol. EDGE supports 124 FDD channels, each of which supports eight time slots/users. EDGE supports two modes of operation:

- *Enhanced GPRS* (EGPRS) is a packet-switched transmission mode that will support data rates as high as 473.6 kbps. EGPRS estimates link quality in order to adapt the *Modulation and Coding Scheme* (MCS), of which there are nine levels, as listed in Table 11.6. If the system estimates that the quality of the link is good, it will select the more efficient 8-PSK modulation technique and, therefore, realize higher signaling rates per time slot and higher data throughput. If the link quality is estimated to be poor, the system will ratchet down to the less capable GMSK. *Incremental Redundancy* (IR) is an enhanced Automatic Repeat Request (ARQ) technique. As transmission begins, IR initially transmits packets with little FEC overhead in an attempt to maximize efficiency, that is, maximize payload by minimizing overhead. If the initial transmission cannot

TABLE 11.6   EGPRS Modulation and Coding Schemes

MCS	Slot Capacity[a] (kbps)	FEC Overhead[b] (%)	Modulation Scheme	Channel Capacity[c] (kbps)
MCS-1	8.8	143	GMSK	70.4
MCS-2	11.2	91	GMSK	89.6
MCS-3	14.8	45	GMSK	118.4
MCS-4	17.6	22	GMSK	140.8
MCS-5	22.4	187	8-PSK	179.2
MCS-6	29.6	117	8-PSK	236.8
MCS-7	44.8	43	8-PSK	358.4
MCS-8	54.4	18	8-PSK	435.2
MCS-9	59.2	8	8-PSK	473.6

[a] Capacity per time slot.
[b] Overhead as a percentage of payload.
[c] Signaling speed assuming eight time slots per channel. Throughput = channel capacity − FEC overhead.
Source: [26].

be successfully decoded by the receiver, IR ratchets up the FEC overhead until it finds a level at which the receiver can successfully decode the transmission.

- *Enhanced Circuit-Switched Data* (ECSD) is an enhancement of the native GSM circuit-switched protocol. ECSD adds 8-PSK as a modulation option, thereby increasing the efficiency of data transmission. A GMSK connection requires four time slots to support a 57.6-kbps data rate, but ECSD requires only two.

EDGE also runs over IS-136 TDMA networks in the United States. In either case, EDGE is an intermediate step between 2G TDMA and 3G WCDMA, although some TDMA-based carriers may stop at EDGE. At the time of this writing (June 2007), the Cingular Wireless network (800 and 1900 MHz) largely supports EDGE and the T-Mobile network (1900 MHz) is fully upgraded [26]. Maximum transmission rates typically are in the range of 75–150 kbps.

11.9.1.4.4   Universal Mobile Telecommunications System   Also known as *Wideband CDMA* (W-CDMA), *Universal Mobile Telecommunications System* (UMTS) is a 3G technology that is seen as a logical upgrade to GSM, although the two are not compatible. UMTS runs over a carrier 5 MHz wide, compared to the 200-kHz carrier used for narrowband CDMA. UMTS specifications provide for both TDD mode and FDD mode, with TDD largely used in Europe and FDD in the United States. The FDD specifications call for the downlink to run in the 2100-MHz range (2110–2200 MHz) and the uplink in the 1900-MHz range (1885–2025 MHz). As is the case with all true 3G systems, UMTS specifications include 128 kbps for high-mobility applications, 384 kbps for pedestrian speed applications, and 2 Mbps (1.920 Mbps) for fixed in-building applications. In reality, UMTS currently caps the transmission rate at a theoretical 384 kbps. UMTS was first deployed in Japan (2000), where it is known as *Freedom of Mobile Multimedia Access* (FOMA). In some locations in the United States, Cingular has deployed UMTS running in the 850- and 1900-MHz bands. T-Mobile has committed to UMTS worldwide.

UMTS networks currently are being upgraded with *High-Speed Downlink Packet Access* (HSDPA), which sometimes is characterized as a 3.5G technology. HSDPA promises to increase theoretical downlink data rates to 14.4 Mbps, although current implementations support speeds more typically in the range of 400–700 kbps, bursting up to 3.6 Mbps for short periods of time using an adaptive modulation technique to throttle bit rates up and down as the link permits. HSDPA has been introduced on a limited basis in Austria, Finland, Japan, South Africa, and the United States (Cingular) [26]. Work has begun in the standards bodies on *High-Speed Uplink Packet Access* (HSUPA). Once increased speeds are in place on both the downlink and uplink, simultaneous voice, data, and even video calls will be quite possible.

11.9.1.4.5 Code Division Multiple Access 2000 Developed by Qualcomm, the company that commercialized CDMA, *Code Division Multiple Access 2000* (CDMA2000) is a 3G system based on earlier CDMA versions (also known as TIA/EIA IS-95a and IS-95b). CDMA2000 (also known as IS-856) has been approved by the ITU-R as part of the IMT-2000 family. The initial version, known as CDMA2000 1 × RTT (*one times Radio Transmission Technology*, with *one times* referring to standard channel width), offers 2.5G capabilities within a single standard 1.25-MHz channel, effectively doubling the voice capacity of the predecessor 2G cdmaOne systems and offering theoretical data speeds up to 153 kbps (throughput in the range of 70–90 kbps) through the use of QPSK modulation. An enhanced 3G version known as 1 × EV-DO (*one carrier EVolution-Data Optimized*) is a *High-Data-Rate* (HDR) version that employs 16-QPSK modulation in support of a peak data rate of 2.4 Mbps on the downlink and 153 kbps on the uplink. 1 × EV-DO supports average aggregate throughput in a fully loaded three-sectored cell of 4.1 Mbps on the downlinks and 660 kbps on the uplinks, with dynamically assigned data rates providing each user with optimum throughput at any given moment. 1 × EV-DO can run in any band (e.g., 450, 800, 1800, and 1900 MHz) and can coexist in any type of network (e.g., CDMA2000, cdmaOne, GSM, TDMA, and AMPS). CDMA2000 runs in the 800-MHz and 1.8–2.0-GHz spectrum. GSM1x is a version designed as a transition specification for GSM operators, involving dual-mode phones. Also known as *IS-2000-A*, 3x is an enhancement that uses three cdmaOne carriers for total bandwidth of 3.75 MHz. This supports data rates up to 2 Mbps by spreading a multicast signal over the three carriers. In North America, Bell Canada, Sprint Nextel, and Verizon Wireless (800 MHz) have committed to CDMA2000 [26, 34–40]. In August 2006, however, Sprint Nextel announced plans to invest up to US$3 billion on an 802.16e, aka mobile WiMAX, network that the company expects to be operational in 2008. Mobile WiMAX is based on Intel technology, rather than Qualcomm's CDMA.

11.9.1.4.6 Time Division Synchronous Code Division Multiple Access *Time Division Synchronous Code Division Multiple Access* (TD-SCDMA) is a 3G mobile telecommunications standard being developed in the People's Republic of China (PRC) by the *Chinese Academy of Telecommunications Technology* (CATT) and adopted by the 3GPP as *UTRA TDD 1.28Mcps Option*. TD-SCDMA is based on CDMA but uses TDMA as well, with the uplink signal synchronized by the base station. The use of TDMA reduces the number of users contending for each time slot, which has the effect of reducing the technical complexity of the system but at the expense of coverage range and mobility. Like European versions of W-CDMA, TD-SCDMA uses TDD rather than FDD. This approach provides additional flexibility in that the uplink and downlink channel capacities can be managed independently. The synchronization of the signal at the base station improves the orthogonality of the coded transmissions, which serves to reduce interference between users and therefore allow increased capacity. TD-SCDMA runs in the 2000-MHz (2-GHz) band, with nominal channel spacing of 1.6 MHz [41]. Asynchronous data rates are highly flexible, ranging from 1.2 kbps to 2 Mbps in both directions. During a field trial in Beijing, it reportedly was possible to make high-quality video calls from a car traveling at 125 km/h at a distance of 21 km from the base station.

While the official reason behind the TD-SCDMA is the avoidance of reliance on Western technology, some suggest that the real reason is the avoidance of patent fees and license fees to Qualcomm and others. Given the fact that there were 190 million users of GSM and CDMA cell phones in the PRC in 2002 and the then current rate of growth was 5 million per month, the PRC certainly had plenty of incentive to avoid those costs. Additionally, it apparently is quite inexpensive to upgrade a cellular network from GSM to TD-SCDMA, while it is very expensive to upgrade a GSM network to UMTS [42]. TD-SCDMA specifications have been finalized, system testing is in progress, and operational systems were expected at the end of 2006.

11.9.1.5 Multimode Cellular Clearly, there are a lot of cellular standards from which to choose, and carriers have chosen them all in various combinations. Many, but not all, of the 2.5G and 3G standards are rooted in GSM, but the upgrades are by no means automatic and the network of any given carrier at any given location typically supports various combinations of 2G, 2.5G, and 3G. Older mobile stations (i.e., cellular telephones) may be able to access only the 2G network, which is ubiquitous until such time as it is decommissioned, as were the 1G analog networks. Newer mobile stations typically are multimode in nature and, therefore, can access the most advanced and most capable generation of technology available, area by area, for a given carrier. Table 11.7 compares carriers in the United States across cellular technologies. Note that the seemingly odd combinations of technologies that do not follow a natural evolutionary path are due to several factors. First, D-AMPS was essentially an evolutionary dead end. Second, many years of acquisitions and

TABLE 11.7 United States Cellular Standards, by Carrier

Generation	1G	2G	2.5G	3G	3.5G
Cingular	AMPS	D-AMPS[a]			
		GSM	GPRS, EDGE	UMTS	HSDPA
		PCS (CDMA)[b]			
Sprint	AMPS	PCS (CDMA)	—	CDMA2000	
T-Mobile	AMPS	GSM	GPRS, EDGE	UMTS	HSDPA
Verizon	AMPS	D-AMPS[a]			
		—	GPRS	UMTS	
		PCS (CDMA)[b]	—	CDMA2000	

[a] IS-54/136.
[b] IS-95.

mergers have created kluged networks that eventually will be sorted out as older generations of technology are phased out and eventually decommissioned altogether. In the meantime, the carriers all offer dual-mode and even trimode terminals. They all now offer international multimode sets that are compatible with the original European version of GSM, so that international travelers can communicate abroad.

11.9.2   Terminal Equipment

Contemporary cellular terminal equipment takes a number of forms today. Cellular telephones are by far the most common and are available in hundreds of styles, most of which are very small form factor models compared to the early *bricks*, so called because they literally were about the size and weight of a clay brick. The small form factor certainly has advantages in terms of size and weight but also presents problems with respect to the user interface, as the visual display and data entry keys are so small. These handset issues are of greatest significance with respect to high-end multifunction sets and with data, rather than voice, applications such as Web surfing and SMS messaging.

As the underlying technologies have evolved and 2.5G and 3G mobile stations have become available, cellular telephones increasingly have been oriented toward data, image, and now video applications. Voice, SMS messaging, and text-based e-mail access have been commonplace on cellular telephones since 2000 or so. Instant Messaging (IM) is much more recent. High-resolution color displays combined with high-speed network access have made Web surfing an integral part of the mobile wireless experience for several years. Digital photography and videography capabilities have now turned the high-end cell phone into an untethered digital camera. In 2004, camera phones made up 36 percent of mobile phone shipments in the United States and more than 50 percent in Europe and dominated the market in Japan [43].

By the way, cell phones do not just ring any longer, they play *ring tones*, and have done so for some years. Ring tones are not just the tones that you hear when your cell phone rings, they now also can be the tones that the caller hears. Just in case the several dozen ring tones that come with a typical cell phone are not to your liking, you can download additional ring tones for a charge. In fact, ring tones are now a billion dollar business in The United States.

PDAs continue to increase in popularity and ever more are wireless enabled. Similarly, laptop and tablet PCs not only can be interfaced with the cellular networks via PCMCIA cards but also now are available with built-in cellular antennas and software, right out of the box, so to speak.

11.9.3   Present and Future of Cellular Radio

In 1994, Global Telecoms Business estimated that there were approximately 35 million cellular subscribers in more than 150 countries and that cumulative growth rates were more than 30 percent per year. The U.S. market grew by almost 50 percent in 1993 and by 325 percent in Pakistan. It went on to say that wireless accounted for nearly 50 percent of Ericsson's sales, compared to a mere 6 percent in 1984. Lehman Brothers estimated cellular penetration of 9.4 percent by the year 2000, equating to 105 million users of a total accessible population of 1.1 billion [44]. Those estimates turned out to be very conservative. In the United States alone, the FCC estimated that there were 33.786 million cellular subscribers on January 1, 1996 [45]. The Cellular Telecommunications Industry Association (CTIA) estimated that there were over 60 million U.S. users as of June 30, 1998 [46]. Now fast-forward to February 2002, when the ITU-T estimated that there were a billion (1,000,000,000) mobile phones worldwide. Fast-forward again to the end of 2005, at which point the CTIA estimated that there were 207.9 million domestic users at year end 2005, sending SMS messages at the rate of 9.8 billion per month and accounting for a total of 1.6 trillion minutes of airtime for the year [47]. Also as of year end 2005, Informa Telecoms & Media estimated that wireless subscriptions reached 2 billion (2,000,000,000) worldwide [47]. That means that the number of mobile phones exceeds the number of wired phones. Specifically, there were more mobile phones than wired phones in several European countries and a number of developing countries where wireline infrastructure is noticeably lacking. Perhaps most intriguing is the fact that in Taiwan there are more mobile phones than inhabitants, according to TeleGeography [48]. International Data Corporation (IDC) estimated that worldwide shipment of mobile phones reached approximately 825 million in 2005, an increase of 165 percent over 2004. Nokia, the manufacturer currently with the largest market share, predicts that there will be 3 billion mobile subscribers by 2008 [49]. Without a doubt, cellular radio growth is unprecedented and will continue to grow at phenomenal rates in the foreseeable future.

While Europe and North America may be reaching saturation levels, at least for cellular voice, the developing world is largely an untapped market. The fastest growth rates are in China and Africa. It is estimated that 80 percent of the world's population lives within range of a mobile network but that only about 25 percent have a mobile phone. A good deal of emphasis these days, therefore, is on low-cost cell phones, including prepaid phones, as this combination is especially attractive to low-income users, particularly if they participate in a cash-only economy. As governments increasingly view communications as a basic human right and as a huge stimulant to entrepreneurial economies, they increasingly are lowering taxes, tariffs, and duties on them.

In any event, we certainly are an increasingly mobile society and we clearly insist on remaining in touch. Cellular radio systems satisfy those requirements. And cellular service increasingly is affordable. While average rates were $0.45 per minute for airtime in the early 1990s (plus toll charges), they have dropped as low as $0.00 per minute in 2002 (with no domestic toll charges), essentially having been flat rated. If you travel as much as I do, you know that this compares very favorably with the cost of a long-distance call from an airport pay phone or hotel room, especially when you add the hotel surcharge. Within the airtime limits of many rate plans, cellular airtime and long distance now essentially are free. Perhaps that explains why, in the United States, approximately 6 percent of households are wireless only, which translates into no wired telephone service [47]. The Yankee Group estimates that since 2003 wireless subscribers in the United States have used their cell phones more than their residential landlines and that, by 2005, personal calling on wireless exceeded that on landlines, even though 35 percent of the population does not have a cell phone [50]. (*Note:* I have two adult children and Margaret, my lovely bride of 10 years, has one. Each of them has a cell phone, but none of them has a residential landline. I don't quite understand that, but it may be that I'm not supposed to. Emergency calls to 911 services are problematic if the cell phone battery runs down, especially if there has been a power failure and there is no way to recharge it. Emergency services personnel cannot locate the caller with nearly the accuracy they can with a landline. There are lots of other advantages to having a landline, of course.) The trend carries over into the business world as well. In January 2005, Ford Motor Company announced that it was replacing the traditional wireline phones of 8000 employees in its product development department with cellular phones from Sprint [51].

The trend toward cellular radio seems unstoppable, at least for voice and mobile data communications applications. It also appears that the trend will only accelerate as 3G, 3.5G, and even 4G cellular become widely available and data rates increase to truly broadband speeds. As manufacturers and carriers collaborate on multimode devices that can seamlessly bridge the gaps between Wi-Fi, WiMAX, and 3G+ cellular networks, we may very well find ourselves in a largely wireless world. By that time, of course, the definition of broadband will have ratcheted up a number of notches, and wireline networks will deliver multimedia content at Tbps speeds.

In the United States, Sprint Nextel announced in August 2006 that it would spend up to US$3 billion to build an 802.16e, aka mobile WiMAX, network slated to begin operation in 2008. The network will be based on technology developed by Intel, which has taken the lead in WiMAX development. Sprint Nextel, with 51.7 million cellular subscribers, expects that the mobile WiMAX network will support connection speeds about five times faster than its current PCS network, which is based on CDMA technology from Qualcomm.

11.9.3.1 *Social Implications*

The ubiquity of multifunctional cell phones and other mobile devices has clearly changed the way that many people conduct their professional and personal lives. Some years ago, you could be reasonably certain that someone walking down the street waving his arms and yelling at thin air was mentally unbalanced, but today that person is more likely just another loud, rude cell phone user with a Bluetooth-enabled headset. Many restaurants, theaters, and other venues have been declared *cell-free* zones in defense of these obnoxious people. (If you are one of those people and I have offended you, I sincerely hope that you will get over it—and mend your ways. ☺)

Cellular technology is just an enabler, of course, but it has been put to some uses that are questionable, to say the least. Cell phones with built-in cameras and video recorders have been used to take *upskirt* or *downblouse* photos and for various other illicit purposes. As a result, some gyms and health clubs have banned cell phones from locker rooms. In the United States, some states have passed laws against such invasions of privacy. In December 2004, the United States enacted the video *Voyeurism Prevention Act*, making it illegal to sneak photos or videotapes of people for lascivious purposes on federal land, such as national parks. A study conducted by Juniper Research and quoted in *The New York Times* indicated that U.S. cellphone adult content was a $1 billion business in 2005 and predicted that it would increase to $2.1 billion in 2006. Camera phones have been used for good as well. The phones have been used to help catch thieves, burglars, rapists, and all sorts of criminals.

As is the case with many things in life, the greatest benefit of cellular telephony is perhaps also its greatest detriment. The ability to communicate anytime and anyplace can be just as much a curse as a blessing, for it robs us of our personal time. *Location-based services* can even track us through *Global Positioning System* (GPS) and *Automatic Location Identification* (ALI) technologies. These services are handy if you are unfortunate enough to need emergency assistance but can be intrusive if you are unfortunate enough to be bombarded with *spim* from retailers along your route. Many of us road warriors who travel extensively for business have found airplanes to be a refuge from this constant contact. Unfortunately for those of us, many airlines plan to offer both mobile phone service and Wi-Fi connectivity in the air.

11.9.3.2 *Religious Repercussions*

As might be expected, religious leaders have taken steps to combat, or at least control, this trend. In 2004, Saudi Arabia's highest religious authority issued an edict banning the use of camera phones, blaming them for *spreading obscenity*. The concern goes quite far in conservative Muslim societies, where religious authorities fear that camera phones are misused to photograph women without their knowledge and without being *properly* covered from view, head to toe. According to the *Associated Press*, a wedding party in July 2004 turned violent after a female guest was caught taking photographs with her phone— scuffles broke out and some guests were hospitalized [52]. (*Note*: In the previous December, the authorities banned dolls and stuffed animals, so a ban on camera phones should hardly be surprising.)

At about the same time, Mirs Communications, the Israeli subsidiary of Motorola, introduced the *kosher phone*, which carries the seal of approval of Israel's ultra-Orthodox rabbinical authorities. The kosher phone is stripped down to its bare functional essentials of making and receiving calls. There is no text messaging, no IM, no Internet access, no camera, and no video. Among the more than 10,000 numbers blocked are those for phone sex and dating services. Arab service providers have sought information from Mirs via backchannels using envoys sent from Jordan, which has a peace treaty with Israel [53].

11.9.3.3 *Safety Issues*

Various safety issues have long swirled around cell phones and their use. Studies conducted over the years have proved and then disproved links between cell phone usage and brain cancer. The most recent of those studies, conducted at the Swedish National Institute for Working Life, indicated that those who heavily used wireless phones had a 240 percent increased risk of a cancerous tumor on the side of the head where they used their phone. *Heavy use* was defined as 2000 or more hours, which roughly translates into one hour of usage a day for 10 years. Kjell Mild, who led the study, said, "The way to get the risk down is to use handsfree". Mild did not offer any suggestions with respect to the hands-free technology, although I suspect that Bluetooth might not do a lot to solve the problem. The U.S. Food and Drug Administration plans to review the study [54, 55].

Studies have long shown links between cell phone usage and vehicle accidents. It seems that driving and dialing, and even driving and talking, are distracting and that even cell phones equipped with hands-free options do not eliminate the problem. As a result, some states in the United States have passed laws banning their use while operating a vehicle. As is always the case, some people deny the link and refuse to abide by the law. (*Note*: They are probably the same people who deny the link between safety and seat belts. You know who you are.)

In June 2006, the *British Medical Journal* published a letter written by doctors at the Northwick Park Hospital claiming a link between cell phone usage during stormy weather and lightning strikes. They suggest that when someone is struck by lightning, the high resistance of the human skin conducts the flash over the body in what is known as a *flashover*. If, however, a metal object such as a cell phone is in contact with the skin, it disrupts the flashover and increases the probability of injury and death. The doctors cited three cases reported in China, South Korea, and Malaysia [56].

11.10 PACKET DATA RADIO NETWORKS

Cellular networks of various descriptions have supported data communications for generations, so to speak, but not very well. The 1G analog and 2G digital networks really were optimized for voice, and understandably so. CDPD offered reasonable data communications speeds, but it really was not a satisfactory solution. Alongside those cellular networks low-speed, packet-switched wireless data networks known as *Mobitex* (*Mobile text*) were deployed in Europe and the United States, and they remain in operation. Developed jointly by Ericsson and Swedish Telecom, Mobitex offers runs in Europe in the band 400–450 MHz and in the United States in the 800- and 900-MHz bands, although it will run in any band.

Mobitex channel spacing of 12.5 kHz supports theoretical data transmission rates of 8 kbps through GMSK modulation. Mobitex packets are up to 512 octets, including overhead, using a *slotted Aloha* protocol. *Slotted Aloha* is an improvement on the original *Aloha* protocol developed at the University of Hawaii in the early 1970s as a contention management mechanism for use in interisland wireless networks. The basic concept behind the Aloha protocol later was incorporated into the Carrier Sense Multiple Access (CSMA) protocols that Bob Metcalf and his associates developed for Ethernet.

Mobitex networks support SMS messaging, corporate e-mail access, and Internet access. Web surfing is supported through proprietary gateways that can filter out graphics or lower their resolution to reduce their bandwidth intensity. Security is provided through various proprietary authentication and encryption mechanisms. Applications include sales agents, maintenance fleets, and truck fleets. Terminal equipment supported includes two-way pagers, PDAs and various other hand helds, and laptops. Mobitex networks largely were retired in Europe in the face of strong competition from GSM but remain in place in 30 countries, including in the United States and Canada. These networks face strong competition from 2.5G and 3G cellular networks that support voice, SMS and MMS, e-mail, and Internet access from traditional cellular phones, laptop and tablet PCs, and PDAs and other hand helds.

11.11 SATELLITE SYSTEMS: LEOs, MEOs, AND GEOs

I discuss satellite systems in quite some detail in Chapter 2, where we explored the basic technology with a focus on GEO (Geosynchronous Earth-Orbiting) systems for point-to-multipoint, or broadcast, applications. As the focus of this chapter is on mobility, I examine satellite systems that support mobile voice and data applications.

Figure 11.11 LEO satellite constellation.

Low-Earth Orbiting and *Middle-Earth Orbiting* satellites (LEO and MEO) satellites operate at low altitudes of several hundred miles or so in a variety of nonequatorial orbital planes. This compares with GEO satellites, which always are placed in equatorial orbital slots at an altitude of approximately 22,300 miles. LEO satellites operate at altitudes of 644–2415 km. Although the term is not tightly defined, *little LEO systems* involve a relatively small number of satellites and operate at frequencies below 1 GHz in support of low-bit-rate data traffic (e.g., telemetry, vehicle messaging, and personal messaging). *Big LEO systems* are bigger networks that operate at higher frequencies in support of voice and higher speed data communications. MEO satellites operate at altitudes of 10,062–20,940 km.

LEO and MEO systems are configured as *constellations* of small, low-power satellites. In combination, the satellites in such a constellation generally provide full coverage of major land masses, and some have been designed to provide full coverage of virtually every square inch of the earth's surface. The various proposals have included as many as 840 satellites and are intended to provide various combinations of voice and data services. These systems also are known as *Mobile Satellite Systems* (MSSs), as opposed to the *Fixed Satellite Systems* (FSSs) in geostatic orbit. Whizzing around the earth like electrons whizzing around the nucleus of an atom and as illustrated in Figure 11.11, LEO and MEO networks are designed so that a satellite is always within reach of a terrestrial terminal.

11.11.1 How LEOs Work: Iridium

The origin of this incredible concept is worth exploring. According to legend, the wife of a Motorola executive was vacationing in the Bahamas during 1987 and was irritated by her inability to place a cellular telephone call (probably from some secluded beach not covered during those early days of cellular service). She complained to her husband and captured his imagination. Motorola named its proposed 77-satellite constellation *Iridium*, after the element Iridium (Ir), which boasts 77

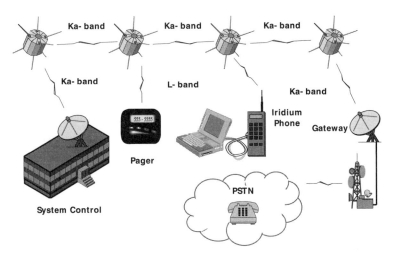

Figure 11.12   Iridium network.

electrons whizzing around its nucleus. Iridium, by the way, gets its name from the Latin and Greek *irid* and *iris*, meaning *rainbow*, after the rainbow colors of the metal when dissolved in hydrochloric acid. (Iris was the Greek goddess of the rainbow and a messenger of the gods.) Subsequently, the proposal was pared down to 66 operational satellites, although the name Iridium stuck. After all, Iridium seems to roll off the tongue better than *Dysprosium* (Dy), the rare earth element with an atomic weight of 66. Dysprosium, by the way is from the Greek *dysprositos*, meaning *hard to get at*, which is hardly a good name for a communications technology. The Iridium constellation is now fully launched and is fully operational. Eleven operational satellites and one spare are placed in each of the six orbital planes at altitudes of 421.5 nautical miles. Connectivity between each satellite and the earth is established via 48 highly focused *spot beams*, each of which has a *footprint* of approximately 30 miles (50 km).

The functioning of the Iridium network, which is the most complex operational constellation at the present, will serve as an excellent example of LEO systems, in general. As illustrated in Figure 11.12, you might initiate a voice call from one Iridium phone to another Iridium phone using a special international telephone number in a block allocated to Iridium. The telephones are slightly supersized versions of cellular phones. Since the frequency band is in the gigahertz range, Line-of-Sight (LoS) is critical, so you either need to be out in the open or you need to stick a little satellite dish on top of your automobile. Your Iridium phone, by the way, is equipped with an antenna the size of a good cigar, which you first must unfold. Your connection directly to the satellite is supported over the L-band at frequencies of 1.616–1.6265 GHz. If you connect from a landline through a local terrestrial gateway, Ka-band frequencies are used, with 29.1–29.5 GHz on the uplink and 19.3–19.7 GHz on the downlink. Once connected to the satellite, the intersatellite links operate in the Ka-band at frequencies of 23.18–23.38 GHz. (See Table 2.5 for a listing of satellite frequency bands.) The satellite constellation finds the target Iridium phone just like a cellular network finds another cell phone. Each active (powered-up) phone maintains a signaling and control link with the satellite network, just as your cell phone keeps in touch with the cellular network when it powers-up. Presto, you're connected, and the quality resembles that of an analog AMPS network, which is pretty good for a satellite call.

Propagation delay is not much of an issue for several reasons. First, the satellites are in low-earth orbits of approximately 485 miles, so the uplink and downlink propagation delays are not significant. Second, the satellites communicate directly over switched intersatellite links, so only one uplink and one downlink are required. As the satellites whiz around the earth like electrons whiz around the nucleus of an atom, however, they have a very short *dwell time*, so you cannot maintain contact with any given satellite for very long, and neither can the other party involved in this telephone call. Therefore, the satellite with which you established the connection must pass that call off to another satellite before it gets out of view, and so must the satellite at the terminating end, and so must every satellite in between. Think of it as a switched cellular network in reverse. In a cellular network, the antennas at the cell sites are stationary. As you whiz through the cellular network in your high-speed vehicle, the cellular network maintains the connection through a hand-off process between cell sites. In an Iridium environment, you are the one who is (relatively) stationary, while the cell sites whiz around you. Iridium also, by the way, supports pagers and airplane communications via the same L-band frequencies. If you call a device that is not on the Iridium network, your call can be connected to the existing PSTN and cellular networks via 12 regional gateways.

Iridium is an incredible network, and it should be, as the total capital investment was in the neighborhood of $4.7 billion. The system went fully operational on November 1, 1998. On May 28, 1999, Iridium LLC received a waiver until June 30 of certain financial covenants from all its lenders to enable it to restructure its capitalization. Those covenants required that the company have at least 27,000 customers by May 31. It seems as though the actual numbers fell far short of that requirement. At a cost of $3000 or so for an Iridium phone, $3795 for a dual-mode phone that also works on existing cellular networks, $500 or so for a pager (one-way, so no guaranteed message delivery), some modest amount of money for a cable and data modem, a considerable monthly charge, and $3–$7 a minute for the connect time (depending on which press releases and articles you believe), Iridium was pretty pricey. Most commercial users just could not justify the cost unless they wanted to call between very remote areas such as the summit of Mt. Everest (true story) to the jungles of New Guinea (it is possible if the canopy of the rain forest is not too thick) [57–67]. Iridium's high cost and relatively poor service did not sit well with prospective users, who failed to appear in numbers even close to those required to make the network financially viable. Iridium filed for Chapter 11 bankruptcy in August 1999 and ceased service in April 2000. At that point, the plan was to let the satellites naturally degrade in their orbits until they simply burned up in the atmosphere in a spectacular Iridium flambé (at least that's the way I like to think of it). Rather than let Iridium flame out, however, a group of investors purchased the network for a paltry $25 million. Shortly thereafter, in what can only be termed an interesting coincidence, the U.S. Department of Defense agreed to a $72 million, two-year contract to provide service to 20,000 government users. Other users can subscribe to Iridium services through one of the 19 or so authorized service partners in various countries around the world. Costs for airtime have dropped to a range of $1.40–$3.00 per minute, and costs for handsets have dropped to $1500 or less.

Analogies such as Icarus and Phoenix aside, suffice it to say that Iridium's close call with a fiery death put a real damper on a number of other proposed LEO projects, and Iridium's long-term financial viability remains in question, to put it mildly. After all, the constellation must be refreshed periodically to avoid future flambé.

By the way, there is a lesson to be learned through Iridium, at least in my humble opinion. The Iridium concept was a great one in 1987 and even through the early 1990s. At that time, the cellular networks were not fully built out and its level of acceptance was not envisioned as ever reaching the current levels, DSL and CATV modems did not exist, and WiMAX and PON were not even gleams in engineer's eyes. Somewhere along the way, however, the folks at Motorola just got too committed to Iridium, even as the competing technologies developed. So, they and others continued to pump billions of dollars into a clever idea that never had a chance. They just breathed their own exhaust (i.e., believed their own marketing hype) until they became so intoxicated that they lost touch with reality. Hindsight is always perfect, but even a little foresight in the mid-1990s would have made it quite clear that Iridium was doomed. Having said that, I also must say that all reliable reports suggest that Motorola made plenty of money off of the idea—at the expense of the other partners. So, maybe Motorola had it right after all, even though other investors lost billions.

11.11.2 How LEOs Don't Work: Teledesic

Let us now turn our attention to the Teledesic LEO network, dubbed the *Internet-in-the-Sky*. *Teledesic* is a combination of *tele*, meaning over a distance, and *geodesic*, meaning the shortest distance between two points that lie on a given surface (i.e., the earth). Teledesic was the brainchild of Bill Gates of Microsoft fame, Craig McCaw of McCaw Cellular (acquired by AT&T Wireless, which then was acquired by Cingular) fame, and Boeing of Boeing aerospace fame. Gates knows a lot about computers and the Internet, McCaw knows a lot about wireless, and Boeing knows a lot about satellites and launch vehicles. Motorola also joined this group at one time. Investors included Bill Gates, Craig McCaw, His Royal Highness Prince Alwaleed Bin Talal Bin Abdul Aziz Alsaud (Saudi Arabia), Abu Dhabi Investment Company, and The Boeing Company. (By the way, they all know a lot about making money.) The idea behind Teledesic was to launch a constellation of 30 (pared down from 288, which was pared down from the original 840) broadband LEOs, plus spares, for Internet access. These two-way satellites were intended to support high-speed Internet access from the subscribers' premises in the Ka-band at rates from 128 kbps to 100 Mbps on the uplink (28.6–29.1 GHz using FDMA) and up to 720 Mbps on the downlink (18.8–19.3 GHz using TDMA). The satellites would whiz around your inexpensive, two-way satellite antenna at low altitudes (about 700 km, or 434 miles), thereby avoiding the aggravating issues of propagation delay that might otherwise have you yelling at your computer. From your computer, you would have a connection to your dish, to the satellite, from satellite to satellite, and down to a portal, which would be your gateway to the Internet and the Web. The costs of your access to this multi-billion-dollar network were expected to be competitive when it would become operational in 2005 or so [57, 58, 62–65]. That was the basic idea, and it was an incredible one—if the technology had fallen into place—and there certainly was no technical reason that it could not. There were a lot of very smart people behind this project, with very deep pockets playing for very big stakes. If you stop to think about Teledesic for just a minute, you will realize that it would have bypassed your modem, your local loop, your Local Exchange Carrier (LEC), your Internet Service Provider (ISP), the wired Internet backbone and all its pieceparts, and your current portal. Teledesic would have bypassed everything conventional. (Thinking back to the definition of Teledesic at the beginning of this discussion, you now understand that the term means that the shortest distance between your computer and the Teledesic portal is the Teledesic network, which seems to have been a codeword for *Microsoft Network, or* MSN.)

Now, I may not be as smart as the Teledesic folks, and I certainly do not have pockets nearly as deep as the least rich of the lot, but I bet my next royalty check from this book that the Teledesic portal will not be optimized for Netscape Navigator. As I said before, it is an incredible idea. Note, however, that Teledesic was conceived in 1986, when there were no optical fiber backbones, no DSL, no cable modems, no PON, and no WiMAX and when Internet access typically involved modems that ran at the blazing speed of 19.2 kbps, best case. In October 2002, Teledesic suspended construction work, and in July 2003, the company gave up its frequency licenses. So, another great idea conceived by engineers went kerplooey.

In the meantime, Craig McCaw and a subset of the principals of the failed Teledesic venture had moved on to ICO, for *Intermediate Circular Orbit*, which they rescued from bankruptcy in 2000. The company reemerged as New ICO, and the system was redesigned. New ICO is a MEO design slated for 10 operational satellites in two orbital planes. After a failed launch or two, New ICO faded into the background for several years. As recently as April 2005, however, New ICO announced that Space Systems/Loral was awarded a contract to build a geostationary mobile satellite system. So, it appears that the system was redesigned yet again and is now a GEO rather a MEO design.

11.11.3   GEOs Work Too: Inmarsat

Inmarsat, for *Int*ernational *Mari*time *Sat*ellite, is the pioneer in the MSS space, having been in service since the early 1990s. Inmarsat also is based on a *bent-pipe* GEO design, with no intersatellite links. Each Inmarsat GEO generates 19 wide spot beams and 228 narrow spot beams. The system reportedly covers 98 percent of the world's population with services including voice, e-mail, Internet access, fax, and videoconferencing at speeds up to 492 kbps [66].

11.12   AND THAT'S NOT ALL

The creativity and inventiveness of the modern human mind should not be underestimated, for there is still more technology to explore, although some of it may seem a bit odd, if not downright wacky. Remember, however, that the telephone seemed like a wacky idea to many people just 130 years ago. Please read on.

Platforms Wireless International plans to develop an unmanned helium-filled aerostat (i.e., static blimp) that will provide inexpensive wireless communications access to rural areas around the world. The 1250-lb antenna payload is mounted on an aerostat roughly twice the size of the Goodyear blimp, tethered to a ground mooring station by a Kevlar cable and floating at an altitude of approximately 3500–4500 m in restricted airspace. The platform technology is well established, as the U.S. Border Patrol uses similar aerostats for surveillance along the Mexican border. The ARC (*Airborne Relay Communications*) system can provide cellular coverage to an area roughly 140 miles in diameter. In the Americas, the system will use TDMA cellular technology, thereby avoiding the issue of expensive terminal equipment. Systems installed elsewhere during subsequent phases will use GSM or other appropriate cellular standards. The initial implementation target is in Brazil, where telephone penetration is very low [67–69].

Along similar lines, Sanswire Networks plans to offer a variety of wireless services via high-altitude airships known as Stratellites. The airships will take up positions in the stratosphere at 65,000 ft (approximately 13 miles, or 20 km) and hold to a single GPS coordinate through the use of solar-powered electrical engines. Each Stratellite will support a payload of several thousand pounds and have clear Line-Of-Sight (LOS) coverage of about 300,000 square miles, an area roughly the size of Texas, according to company information [70].

And there is more. Angel Technologies Corporation plans to develop a high-capacity wireless metropolitan network dubbed HALO (*High-Altitude Long Operation*). The network will use shifts of specially designed aircraft capable of carrying a telecommunications payload of 2300 lb and remaining aloft for 8 h or so at an altitude of approximately 51,000 ft. At that altitude, the coverage area is about 75 miles in diameter. HALO is expected to provide raw aggregate capacity exceeding 16 Gbps, in support of 50,000–100,000 customers requiring symmetric T1 packet data rates [71].

REFERENCES

1. Brooks, John. *Telephone: The First Hundred Years*. Harper & Row, 1975.
2. Calhoun, George. *Digital Cellular Radio*. Artech House, 1988.
3. Williams, Veronica A. *Wireless Computing Primer*. M&T Books, 1996.
4. Reginald Fessenden. http://collections.ic.gc.ca/heirloom_series/volume4/42-45.htm.
5. Baldwin, Thomas F. and McVoy, D. Stevens. *Cable Communications*. Prentice-Hall, 1988.
6. Feuerstein, Marty. "Controlling RF Coverage." *America's Network*, February 15, 1998.
7. Newton, Harry. *Newton's Telecom Dictionary*, 18th Ed. CMP Books, February 2002.
8. Bates, Regis J. *Wireless Networked Communications*. McGraw-Hill, 1994.
9. Lytel, Allan. *Two-Way Radio*. McGraw-Hill, 1959.
10. Paetsch, Michael. *Mobile Communications in the U.S. and Europe: Regulation, Technology, and Markets*. Artech House, 1993.
11. http://web.mit.edu/invent/iow/gross.html.
12. Emmett, Arielle. "Paging FLEXes Its Muscle." *America's Network*, November 1, 1997.
13. "Pager Protocols." *RTS Wireless*. Originally accessed at www.rtsinc.com/pager.html.
14. "FLEX Family of Protocols Q&A." Motorola. Originally accessed at www.mot.com/MIMS/MPSG/CTSD/white_papers/flex_q&a.html.
15. "FLEX Protocol Fact Sheet." Motorola. Originally accessed at www.mot.com/MIMS/MPSG/CTSD/white_papers/FLEX-old/promo/tech_sheet.html.

16. Levy, Doug. "Satellite's Death Puts Millions Out of Touch." *USA Today*, May 21, 1998.

17. Muller, Nathan J. "Wireless PBX Systems." *IT Continuous Services*. Datapro Information Services, September 9, 1998.

18. Bellis, Mary. "History of Pagers." http://inventors.about.com/library/inventors/blpager.htm.

19. Karmer, Matt. "Going Beyond the Beep." *PC Week*, December 11, 1995.

20. Brown, Margaret J. "Major Pagers." *Mobile Computing & Communications*, April 1998.

21. http://www.smartsynch.com/telemetry/creatalink.htm.

22. http://www.smartsynch.com/pdf/creatalinkspecs.pdf.

23. Chang, Kai. *RF and Microwave Wireless Systems*. Wiley, 2000.

24. Freeman, Roger L. *Fundamentals of Telecommunications*. Wiley, 2005.

25. "DECT: The Standard Explained". White Paper, DECT Forum, February 1997.

26. Nicopolitidis, Petros, Obaidat, Mohammad, Papadimitriou, Georgios, and Pomportsis, Andreas. *Wireless Networks*. Wiley, 2003.

27. http://www.ou.edu/engineering/emc/standard.html.

28. "IS-54B Overview." Texas Instruments Incorporated. www.ti.com/sc/docs/wireless/cellsys/is54over.html.

29. "Cellular CDMA Standard: IS-95A." Qualcomm Incorporated. Originally accessed at www.qualcomm.com/cdma/phones/whatiscdma/95a.html.

30. "CDMA Technology & Benefits." Motorola. Originally accessed at www.mot.com/CNSS/CIG/Technology/cdma.html.

31. Sharrock, Stuart. "CDMA/TDMA From Fists to Facts." Originally accessed at www.ericsson.se/Connexion/connexion3-93/techno.html.

32. "Remarks by William E. Kennard, Chairman, Federal Communications Commission, to Personal Communication Industry Association of America." www.fcc.gov/Speeches/Kennard/spwek828.html. September 23, 1998.

33. Moore, Mark. "Data-Ready Cell Phones to Debut." *PC Week*, January 1, 1996.

34. www.qualcomm.com.

35. Dornan, Andy. "Lesson 146: CDMA and 3G Cellular Networks." *Network Magazine*, September 2000.

36. Dornan, Andy. "Broadband from Outer Space." *Network Magazine*, January 2002.

37. Rysavy, Peter. "The Road to a Wireless Future." *Network Computing*, October 30, 2000.

38. Rysavy, Peter. "Clear Signals for General Packet Radio Service." *Network Magazine*, December 2000.

39. McFall, Shaun. "3G Promises New Wireless Applications." *Network World*, June 11, 2001.

40. Stokes, Jeanie. "Building Next-Gen Networks." *Broadband Week*, April 2, 2001.

41. CWTS TSM 11.21 v 3.1.0 (2002-12). China Wireless Telecommunication Standard; 3G digital cellular telecommunications system; Base Station System (BSS) equipment specification; Radio aspects.

42. http://www.palowireless.com/3g/docs/TD-SCDMA-China.pdf.

43. Forsberg, Birgitta. "Camera Phones Wear out Welcome." *Seattle Post-Intelligencer*, May 31, 2005.

44. Dempsey, Michael. "Not-So Saturation Coverage From GSM." *Global Telecoms Business*, October/November 1994.

45. "The Telecommunications Industry at a Glance." Originally accessed at www.itu.int/ti/industryoverview/top20cellular.html.

46. "Frequently Asked Questions & Fast Facts." Originally accessed at www.wow-com.com/consumer/faqs/faq_general.cfm#one.html.

47. *Wireless Quick Facts*, CTIA, April 2006.

48. The Associated Press. "World's Cellular Phones Will Outnumber Fixed Lines within Months, U.N. Predicts." *Seattle Post-Intelligencer*, February 11, 2002.

49. Gohring, Nancy. "Mobile Phone Sales Topped 800 Million in 2005, Says IDC." *ComputerWorld*, January 27, 2006.

50. Kim, Gary. "Wireless Wins the Home." *Fatpipe*, October 2005.

51. Pappalardo, Denise. "Ford to Cut Cord on 8,000 Phones." *Network World*, January 25, 2005.

52. Shihri, Abdullah. "Saudi Arabia Outlaws Camera Cell Phones." *Seattle Post-Intelligencer*, September 2004.

53. Murphy, Brian. "God, Can You Hear Me Now?" *Skagit Valley Herald*, March 31, 2006.

54. Reuters. "FDA to Review Wireless Phone Safety." *ComputerWorld*, April 6, 2006.

55. Reuters. "Study: Long-Term Mobile Phone Use Raises Brain Tumor Risk." *Computer-World*, March 31, 2006.

56. Reuters. "Cell Phone Users Warned of Lightning Strike Risk." *ComputerWorld*, June 23, 2006.

57. Hughes, Tom. "Pie in the Sky?" *Global Technology Business*, January 1999.

58. Ingley, Carol. "Global Vision: Making the Right Connections." *Satellite Communications*, February 1999.

59. Louderback, Jim. "Iridium Phones Give New Ring To Satellite Communications." *PC Week*, February 1, 1999.

60. Louderback, Jim. "Iridium Pager Hits Growing-Pains Stage." *PC Week*, February 1, 1999.

61. Alleven, Monica. "Iridium Gears Up for New Strategy." *Wireless Week*, June 7, 1999.

62. "Fast Facts." Teledesic. Originally accessed at www.teledesic.com/overview/fastfacts.html.

63. Brown, Peter J. "Satellite Companies Launching Dreams at 2 GHz." *Broadband Week*, August 6, 2001.

64. Melnick, Martin and Hadinger, Peter. "Enabling Broadband Satellites." *Satellite Communications*, July 2000.

65. Dornan, Andy. "Waiting for Wireless in the United States." *Network Magazine*, December 2001.

66. www.inmarsat.com.

67. Gohring, Nancy. "Blimps with a Purpose." *Interactive Week*, April 9, 2001.

68. Brewin, Bob. "Giant Aerostats Developed for Rural Cell Phone Service." *Computer World*, March 5, 2001.

69. www.plfm.net.

70. www.sanswire.com.

71. www.angeltechnologies.com.

CHAPTER 12

VIDEO AND MULTIMEDIA NETWORKING

"It's a success, Ned! I've struck it!" cried Tom, in delight.

"Ouch! You struck 'me,' you mean!" replied Ned, rubbing his shoulder, where the young inventor had imparted a resounding blow of joy.

"What of it?" exclaimed Tom. *"My apparatus works! I can send a picture by telephone! It's great, Ned!"*

"But I don't exactly understand how it happened," said Ned, in some bewilderment, as he gazed at the selenium plate.

"Neither do I," admitted Tom.

Victor Appleton, *Tom Swift and His Photo Telephone or The Picture That Saved a Fortune*, Gross & Dunlap, 1914

Tom Swift certainly had a great idea, although he did not understand quite how his *photo telephone* worked. Tom used his invention to save Mr. Damon from Messrs. Peters and Boyman, kidnappers and swindlers who were close to cheating him out of a fortune. While the dastardly pair had poor Mr. Damon kidnapped, they called Tom to arrange for the ransom. Tom took a picture of them with his *photo phone*. That photo ultimately was used as evidence to convict them. While Tom may have saved Mr. Damon a fortune, there is no mention of the cost of the photo phone. However, I must assume that the cost was reasonable, as the young inventor was featured in an entire series of books in which he also invented the motorcycle, the motorboat, the air glider, the airship, the submarine, the wireless telegraph, the electric rifle, and a giant cannon—all funded with his modest allowance. Since Tom invented the photo phone, the concept has been refined by other young inventors and now is used in a variety of applications, including law enforcement. In the right applications, the technology clearly has the potential to save a fortune, but the costs tend to be more than typical teenagers can afford out of their allowances. After all,

Telecommunications and Data Communications Handbook, By Ray Horak
Copyright © 2007 Ray Horak

Tom Swift's invention was just a simple image transmission over an analog telephone circuit. Alexander Graham Bell's *photophone*, on the other hand, was voice only, certainly was not funded out of his allowance, and there is no evidence that it was used to solve any crimes. This chapter deals primarily with interactive video and multimedia networking, neither of which Alexander Graham Bell or Tom Swift could have imagined—or could they?

On the whole, you may wonder how critical interactive video and multimedia networking are to our daily lives. The answer is not clear. While we all understand that a picture can be worth a thousand words and we clearly benefit from one-way video and multimedia, the ability to engage in an interactive video or multimedia communications is not necessarily that important to most of us, either personally or professionally. Such interactive networking currently is expensive in terms of both bandwidth and equipment, although those costs certainly will come down over time. Many of the highly touted benefits have yet to be proven, although a compelling case is developing.

Certainly, video and multimedia have application in education. People tend to have different learning styles. Some are *visual* learners and need to see something in order to process the information. Some are *auditory* and need to hear about it. Still others tend to be more *kinesthetic* and need to touch it. In any event, the addition of the visual information stream certainly assists in the educational process, as it does in the communications process, in general. A fascinating application is in the world of health care. As discussed later in this chapter, telemedicine has proved itself to be of great importance in the delivery of high-quality medical care to remote areas where specialists, and even general practitioners, may not be available.

12.1   VIDEO COMMUNICATIONS: DEFINED AND EVOLVED

The first public demonstration by wire and wireless of Television or "Distant Seeing," as developed by the staff of the Bell System, took place on April 7, 1927. Participating in the demonstration at Washington, D.C., and New York were notable gatherings of leaders in the fields of science, industry and public affairs. Those who talked from the Bell Laboratories in New York were able to see plainly the features of those in Washington with whom they conversed over the long distance circuits of the Bell System. [This] . . . was followed by demonstration of Television by radio in which the audience in New York saw the artists visualized on the screen and heard a varied program from the radio experimental station of the Laboratories at Whippany, N.J.

Things Worth Knowing about the Telephone,
American Telephone & Telegraph Company, 1931

Video (from the Latin *video*, meaning *I see*) communications has its roots in broadcast TeleVision (TV), from the Greek *tele*, meaning *far off*, and the Latin *vision*, meaning *to see*. Broadcast TV is transmitted over the airwaves and remains the primary source of TV in many areas of the United States and in most other countries. The first true TV mechanism was developed in 1884 by Paul Nipkow, a German engineer, using a scanning disk, lenses, mirrors, a selenium cell (as did Tom Swift), and electrical conductors. Doing so, he was able to transmit images in rapid succession to a lamp, which changed in brightness according to the strength of the currents received. Using this mechanical scanning technique, Nipkow demonstrated that portions of a full image viewed in rapid succession (15 images or more per second) created the illusion of viewing the full image [1]. It later was discovered that viewing 15 or more images per second created the illusion of full motion, due to electro-chemical processing delays in the human eye. While this mechanical scanning approach was abandoned in later years, the concept of *persistence of vision* remains valid. Psychologist and film theorist Hugo Munsterberg explained a second process, known as the *phi phenomenon*, in 1916. This process explains the fact that we hal-lucinate, or believe that we see, a continuous action rather than a series of still images—the mind, in effect, fills in the blanks [2]. Modern TV and video systems create the illusion of motion by refreshing screens in rapid succession, a dot at a time and a line at a time. (*Note:* Film projectors refresh the entire screen at once.)

The development of modern TV largely was due to the efforts of Herbert E. Ives, a scientist at Bell Telephone Laboratories. As the son of Frederick Eugene Ives, who in 1878 developed the first practical process for making halftone printing plates, Ives was oriented toward the visual world. In 1923, he and his associates combined the photoelectric cell with the vacuum tube repeater to produce the first commercial system for the rapid transmission of pictures over telephone wires for application by the daily press. Using a scanning beam developed by Frank Gray, another Bell Labs scientist, multiple real-time images of people were transmitted in rapid suc-cession, and TV was born. The first public demonstration, on a black-and-white TV, was conducted in April 1927; color TV first was demonstrated in June 1929 [3]. That experimental color transmission included the transmission of pictures of an Ameri-can flag, a watermelon, and a bunch of roses. The transmitting system had three sets of photoelectric cells, amplifiers, and glow-tubes, with each filtering out one color—Red, Green, or Blue (RGB). At the receiving end, mirrors superimposed the mono-chromatic images to create a single color image [1]. In the 1930s, the first commercial TV stations began operation over the radio waves.

Coaxial cable entered the world of TV in 1936, when the first experimental trans-mission took place between New York and Philadelphia. Jointly conducted by AT&T and the Philadelphia Electric Storage Battery Company (Philco), the experi-ment proved highly successful in terms of transmission performance, as multiple frequencies could be transmitted over the same shielded medium. In 1950, AT&T opened the first coaxial cable for coast-to-coast TV transmission [4]. *Community Antenna TeleVision* (CATV), generally based on coaxial cable, largely has sup-planted broadcast TV in the United States. The popularity of cable is due in large part to the large number and wide variety of available channels; broadcast TV simply cannot compete at this level due to spectrum limitations. More recently, CATV spread to a number of other developed countries, including England and Australia.

Now, only two key evolutionary concepts remain—those of the transmitters and receivers. Vladimir Zworykin, a Russian immigrant to the United States, built on his graduate work in Russia where he had studied the nature of fluorescence under Boris Rosing. On January 1, 1939, Zworykin received patents for his *iconoscope* (transmitting) and *kinescope* (receiving) tubes, which formed the basis for modern cameras as well as *Cathode Ray Tubes* (CRTs)used as display devices in traditional TV sets and computer monitors [1, 5].

Video communications extends well beyond broadcast TV and CATV into videoconferencing, multimedia communications, and, ultimately, interactive TV. In this chapter, I explore the nature of the equipment and networks that support video and multimedia collaborative communications as well as related standards and costs.

12.2 VIDEO BASICS

In order to fully comprehend the nature and implications of advanced video communications, it is necessary to understand the basics of video. The basic concepts are frame rate, scanning, resolution, aspect ratio, luminance, chrominance, and synchronization.

- *Frame rate* refers to the rate at which frames of still images are transmitted. Video is a series of still images that are transmitted in succession to create the perception of fluidity of motion. If transmitted in rapid succession, the perception is one of complete fluidity; 24 *frames per second* (fps) is considered to be motion picture quality, and 30 fps is considered to be broadcast quality. If the frames are transmitted at a slow rate, the result is a poor-quality, herky-jerky video that creates a strobe-light effect reminiscent of 1970s discos. Particularly below 15 fps, quality suffers quite noticeably, as the fluidity of motion is lost even though the image quality may be quite high.
- *Scanning* refers to the process of refreshing the screen; the scanning rate is a function of the power source of the receiver. *Interlaced scanning*, which is used with most analog TV systems, involves two *fields*. Odd lines (field 1) are refreshed in one scan and even lines (field 2) in the next. Each set of odd and even lines refreshed constitutes a frame refreshed. For example, the American NTSC standard provides for 30 fps, involving 60 scans, which relates directly to the 60 Hz of the U.S. power source. The European PAL standard provides for 25 fps, involving 50 scans, which relates directly to the 50 Hz of the European power source. *Progressive scanning* involves displaying all horizontal scan lines in one frame at the same time, which avoids the problem of interline flicker. While the refresh rate varies, most contemporary PC monitors operate at 60–90 fps. The difference between the two techniques is imperceptible to most of us, until we see a broadcast TV news show that includes a video clip of PC monitors.
- *Resolution* refers to the definition, or sharpness, of the image. Resolution is determined by the number and areal density of the *pixels*, or *pels* (picture *elements*), which essentially are dots of picture similar to the dots in half-tone printing. The greater the number and density of the pixels, the better the resolution. If the same number of pixels is spread over a greater area, the result is a grainier picture, as you can readily see by sitting close to a big-screen TV.
- *Aspect ratio* refers to the relationship between the width and the height of the image. The 4:3 (4 wide to 3 high) aspect ratio specified by the American NTSC standard is rooted in the early days of TV, when round picture tubes made effective use of this approach.
- *Luminance* refers to intensity, or brightness, which can vary within an image. An analog video transmission varies the luminance by varying the power level, or amplitude, of the signal, with high power representing black and low power

representing white. (*Note*: Broadcast TV from distant stations often displays as grey due to signal attenuation.)

- *Chrominance* refers to color, with different standards permitting varying levels of color depth. Clearly, the video image is more pleasing and lifelike when the range of color is as broad as possible.

- *Synchronization* between the transmitter and receiver includes vertical and horizontal synch, both of which are critical. Vertical synch is required to keep the picture from scrolling, or flipping. Horizontal synch keeps the picture from twisting.

12.3 ANALOG TV STANDARDS

Television standards are several and incompatible. The initial standards were set in the United States, where broadcast TV originated. In 1945, the Federal Communications Commission (FCC) set the initial VHF (Very High Frequency) transmission standards at 4.5 MHz. The *National Television Standards Committee* (NTSC) was formed in 1948 to standardize the characteristics of the broadcast signal. Ultimately, the Radio Corporation of America (RCA), which was owned by AT&T, lobbied the Electronics Institute of America (EIA) and set the initial black-and-white TV standards. Color TV was commercialized some years later. Among the first live color TV broadcasts were the Cotton Bowl and Rose Bowl football games on January 1, 1954 [1]. As the cost of color TV sets was quite high, color television did not really become popular until the 1960s when sets became affordable for the masses. Table 12.1 compares the major analog standards of NTSC and PAL, which are discussed in more detail in the bulleted list that follows, with digital HDTV.

- *National Television Standards Committee* (NTSC) was established in the United States as the first standard (1953), setting the tone for broadcast TV. NTSC is defined in ITU-R Recommendation 1125. While other standards have since been developed, they all derived from the NTSC baseline. NTSC is characterized as analog in nature, with 525 interlaced scanlines. There are 640 pixels per line, 485 of which are dedicated to the active picture. The frame rate is 30 fps, 60 fields interlaced, and the aspect ratio is 4:3. An early analog standard that is viewed as overly complex and ineffective in the contemporary digital world, NTSC also is said to mean *Never The Same Color* [6].

TABLE 12.1 Television Standards Compared

Standard	NTSC	PAL	HDTV
Analog/digital	Analog	Analog	Digital
Horizontal scanlines	525	625	640, 704, 1280, or 1920
Synchronization	40	49	N/A
Resolution, pixels per line	640	640	480, 720, or 1080
Frame rate,[a] fps	30i	25i	24p, 30p, 60p, and 60i
Aspect ratio	4:3	4:3	4:3 and 16:9

[a] i = interlaced, p = progressive.

- *Phase Alternate Line* (PAL) was established in West Germany, The Netherlands, and the United Kingdom in 1967. PAL addresses problems of uneven color reproduction that plague NTSC due to phase errors associated with electromagnetic signal propagation. PAL inverts the color signal by 180° on alternate lines, hence the term phase alternate line [6]; it currently is used in much of Western Europe, Australia, and Africa. PAL is characterized as analog, with 625 interlaced scanlines. There are 640 pixels per line, with 576 dedicated to the active picture. The frame rate is 25 fps, and the aspect ratio is 4:3.
- *Séquential Couleur Avec Mémoire* (SECAM) (which translates as *sequential color with memory*), a variation of PAL, was developed in France. In addition to its use in France, it also is the standard in regions once under French influence, including areas of the Middle East.

12.4 DIGITAL TV AND HIGH-DEFINITION TV

Digital TV (DTV) transmission offers the same advantages as any other form of digital communications, including enhanced bandwidth efficiency through compression, improved signal quality, and more effective management and control. Additionally, digital video content must be recognized as nothing more than video data. As a result, it offers advantages in terms of processing, storage, and manipulation. Those advantages include editing, alteration (e.g., *morphing*), reproduction, compression, and store-and-forward capability.

TV and most video, for that matter, in their native forms comprise synchronized analog image and analog voice information. Digital video requires that the information be digitized through the use of a video codec. A broadcast-quality video signal is extremely bandwidth and storage intensive. For example, the resolution of a digital video might require 640 horizontal and 480 vertical pixels. As chrominance and luminance require 24 bits per pixel, the video signal, alone, requires 7,372,800 bits per frame. At a frame rate of 30 fps, the bandwidth requirement is 221,184,000 bps. Clearly, it is not within the realm of reason to store and transmit a broadcast-quality video signal requiring 221+ Mbps. It is clear, therefore, that compression is critical if the networks are to support digital TV.

Despite all the advantages of digital technology as applied to TV production, storage, transmission, and reception, all of the TV sets were analog. A conversion from analog to digital broadcast TV is revolutionary and completely unthinkable, unless there exists a compelling reason for the changeout of untold millions of TV sets. According to the manufacturers and the FCC, that compelling reason is in the form of *High-Definition TV* (HDTV).

The definition of HDTV standards caused a debate that raged for a number of years between Japanese analog standards and digital standards proposed by the United States. The U.S. digital standards proposals focused on those offered by the *Grand Alliance*, which was formed by the FCC in May 1993 and comprised AT&T, General Instruments, Zenith, The Massachusetts Institute of Technology (MIT), Thompson Consumer Electronics, Philips Consumer Electronics, and the David Sarnoff Research Center. The efforts of the Grand Alliance led to a set of recommended *Digital TV* (DTV) standards for HDTV and *Standard-Definition TV* (SDTV). Those recommended standards were tested and documented by the

TABLE 12.2   Scanning Formats: ATSC Digital Television Standards

Standard	Vertical Lines	Horizontal Pixels	Aspect Ratio	Frames per Second,[a] fps
HDTV	1080	1920	16:9	24p, 30p, 60i
	720	1280	16:9	24p, 30p, 60p
SDTV	480	704	16:9, 4:3	24p, 30p, 60p, and 60i
	480	640	4:3	24p, 30p, 60p, and 60i

[a]i = interlaced, p = progressive.
Source: [7, 8].

Advanced Television Systems Committee (ATSC) in the summer of 1995 and approved by the FCC in December 1996. Table 12.2 presents the specifics of the ATSC scanning formats.

Considering that an HDTV video signal at 1080×1920 pixels, 24 bits per pixel, and 30 frames per second requires bandwidth of almost 1.5 Gbps, compression becomes very important. The ATSC standard specifies the compression algorithm as MPEG-2, which is discussed later in this chapter and the transport subsystem as ISO/IEC 13818. Packet transport involves a serial data stream of packets of 188 octets, 1 octet of which is a synchronization byte and 187 octets of which are payload. This packet approach is suitable for Asynchronous Transfer Mode (ATM) switching, as each 188-octet MPEG-2 packet fits nicely into the payload of four ATM cells, with only 4 octets of padding required. Forward Error Correction (FEC) is employed in the form of Reed–Solomon coding. Radio Frequency (RF) modulation is accomplished using *8 Vestigial Sideband* (8 VSB), which supports a bit rate of 19.28 Mbps over a 6-MHz terrestrial broadcast channel through the use of eight discrete amplitude levels. Audio compression is based on the *AC-3* specification from *Dolby Digital* and the ATSC. The audio sampling rate is 48 kHz, and the system supports six channels in the Dolby Digital surround format. That format specifies multiple channel outputs, including center, left and right center, left and right surround, and *Low-Frequency Enhancement* (LFE), also known as subwoofer [7]. Despite the fact that most programming is not yet in HDTV format and that most broadcast and cable signals are still analog, Forrester Research estimates that about 15.7 million households had HDTV sets by the end of 2005 and predicts that number to be 50 million by 2009 [9].

The FCC set a timetable for DTV transmission, then changed it several times and softened the requirements after the majority of broadcasters missed the first deadlines. Beginning on November 1, 1998, 42 stations (including noncommercial stations) in the top-10 markets began voluntary DTV transmissions, in addition to analog NTSC transmissions. All stations in those markets were required to transmit DTV by May 1999. After that deadline was missed by a number of broadcasters, the FCC deferred, and then deferred again, its requirement that all commercial TV stations convert fully to digital and fully replicate their analog broadcast signals or lose interference protection. That date ultimately was changed to July 1, 2006, with provisions for waivers. In the meantime, broadcasters constructed and operated lower powered DTV facilities to at least their *community of license*, extending them to their entire service area as the transition progressed. By February 17, 2009, all stations must turn off their analog signals. While the FCC made the DTV RF spectrum available at no cost to the broadcasters, each station is responsible for any costs associated with clearing DTV spectrum of other users. The analog spectrum will be returned for reassignment at the point in time at which 85 percent of the television households in a given market are capable of receiving the digital channels.

Each station also is responsible for the costs associated with upgrades to digital transmission facilities. The costs of converting the several hundred million analog TV receivers are the responsibility of the viewers, although beginning in 2008, the National Telecommunications and Information Administration (NTIA) will provide to eligible households up to two $40 vouchers good toward the purchase of digital-to-analog converters to retrofit existing analog TV sets, *Video Cassette Recorders* (VCRs), and other devices. New TV sets, with screens from 25 to 36 in., were required to have built-in digital tuners as of March 1, 2006. By March 2007, all new TVs, regardless of size, and other devices designed to receive broadcast TV signals must have digital tuners built in. According to U.S. census projections, there were approximately 110 million households in the United States at the end of 2005 [10, 11]. Considering the fact that the average household had 2.6 television sets, the costs of adding 286 million analog TV sets and 100 million or so VCRs to existing landfills will be substantial. Assuming the use of a single landfill, just imagine its size. Just imagine the bewilderment of future archeologists when they excavate it. (This assumes that there will be future generations, which is questionable, since the mercury and lead that leaches out of the TV sets and out of the landfill into the streams, rivers, lakes, and oceans of the world may have killed off all life as we know it.) They will probably think it was some sort of sacrifice to the TV gods, and they will not be far off. In any event and as a byproduct of the conversion from analog to digital broadcast TV, the 750 MHz of analog spectrum will be reclaimed for fixed and mobile commercial wireless services, and that spectrum will be auctioned once the reclamation is complete [12, 13].

You may wonder why I spent so much time discussing broadcast TV technology and standards and their evolution. The reasons are several. First, most of the same technologies and standards apply to interactive videoconferencing and multimedia. Second, although broadcast DTV and CATV do not necessarily have to follow the same rules of the road, it is likely that the CATV providers will carry broadcast TV signals in a digital format since they are upgrading their networks to digital anyway. The FCC has imposed *must-carry* rules on both the CATV and the DBS (Direct Broadcast Satellite) providers, thereby forcing them to carry local broadcast TV channels and to fairly compensate the local stations for that content. Considering that service providers are deploying broadband digital local loops in support of the triple play of voice, data, and TV, that the backbones are digital, that data, image, and movie content are digital, and that the TV broadcast signal will be digital, it is pretty clear that the digital circle will be complete very soon.

12.5   BANDWIDTH AND COMPRESSION

The bandwidth required for video transmission is very significant and is affected by frame rate, resolution, color depth, aspect ratio, and audio. Analog broadcast-quality TV requires about 6 MHz, of which the signal occupies about 4.5 MHz. In the United

States, broadcast channel 2 is assigned frequency band 54–60 MHz, channel 3 is assigned the 60–66 MHz band, and so on, to channel 69, which is assigned the band 800–806 MHz. (*Note:* There is no channel 1, but that story is too long and arcane for this work.) As little can be done to compress an analog signal, analog TV technology is rapidly being replaced with digital technology. Analog airwave broadcast signals in the United States will be turned off in 2009 and CATV networks are being converted quickly. Analog TV is doomed.

Digital video, on the other hand, can be compressed fairly easily. Because uncompressed, broadcast-quality digital video requires between 90 and 270 Mbps, compression is critical. Without compression, a 1-Gbps fiber-optic network could accommodate no more than 11 digitized NTSC channels. To digitize and compress the video information stream, the analog video and data signal first must be digitized through the use of a codec. Clearly, it is possible to reduce the amount of bandwidth required to transmit digital video and the amount of memory required to store it by reducing the frame rate, resolution, or color depth. However, the result is less than pleasing. In order to maintain the quality of the video presentation, therefore, the data must be compressed by using an appropriate and powerful data compression algorithm.

Lossless compression enables faithful reproduction of the video signal, with no data loss, although compression rates are in the range of only 10:1. *Lossy* compression tends to produce *artifacts*, which are unintended and unwanted visual aberrations in the video image that often show up as jagged blockings or tiling effect known as *aliasing*, banding of colors, white spots, and even dropped frames. Although these artifacts result in a degraded picture, this lossy approach achieves compression rates up to 200:1 [14]. Actually, compressed video currently can be transmitted with quite acceptable quality at T1 speeds of 1.544 Mbps or less. MPEG, for example, uses lossy compression in the form of *Discrete Cosine Transform* (DCT). A number of steps are involved in video compression, including filtering, color-space conversion, scaling, transforms, quantization and compaction, and interframe compression.

- *Filtering*, also known as *image decimation*, reduces the total frequency of the analog signal through a process of averaging the values of neighboring pixels or lines. For example, adjoining black-and-white pixels become gray pixels. *Taps* are the number of lines or pixels considered in this process; MPEG, for example, uses a seven-tap filter.

- *Color-space conversion*, also known as *color sampling*, simply involves the reduction of color information in the image. As the human eye is not highly sensitive to slight color variations, the impact is not noticeable. Black and white, however, are prioritized because the human eye is very sensitive at that level.

- *Scaling* addresses the creation of the digital image according to the presentation resolution scale associated with the display device. Rather than digitizing the video signal in large scale, the codec is tuned to the scale of presentation in terms of horizontal and vertical pixels, thereby reducing the amount of data that must be digitized. In consideration of this factor, the aspect ratio must be standardized.

- *Transforms* convert the native two-dimensional video signal into data dimensions. Although they are beyond the scope of this book, the various approaches include *Discrete Cosine Transform* (DCT), *vector quantization, fractal transform*, and *wavelet compression.*

- *Quantization* and *compaction encoding* simply reduce the number of bits required to represent a color pixel. Compaction techniques include *run-length encoding, Huffman coding*, and *arithmetic coding.*

- *Interframe compression* considers and eliminates redundant information in successive video frames. The background of a movie scene, for example, might not change, even though the actors move around the set. While the motion of the actors must be reflected, the background need not be retransmitted over the network. Rather, it can be compressed out of each frame until background changes must be reflected. At the point of decompression at the receive end of the signal, the unchanged background can simply be reinserted over and over again.

12.6 VIDEO STANDARDS

According to an anonymous but very wise man, "The nice thing about standards is that there are so many from which to choose!" Video compression standards are no exception—they are numerous and incompatible. Early standards were developed for specific purposes, such as motion picture production (MPEG) and photographic editing (JPEG). The ITU-T became involved in the development of international standards toward Broadband-ISDN (B-ISDN). In the videoconferencing world, numerous proprietary standards have been developed that require that the network involve equipment from only a single vendor.

As of late, a number of formal standards have developed. Those video compression standards of significance include P × 64, JPEG, and MPEG and its variations. Virtually all manufacturers of significance have embraced these standards, incorporating one or more of them into their systems alongside a proprietary compression technique. Generally involving P × 64, this approach provides at least a minimum level of communication between systems of disparate origin. The world of desktop videoconferencing is much more parochial; proprietary solutions long predominated, yielding incompatibility all too often [15]. As always is the case in networking, the highest common denominator rules. Lacking a common denominator, anarchy rules. The common denominator currently is P × 64, a standards-based solution that virtually all manufacturers support.

12.6.1 P × 64

P × 64 is an ITU-T standard designed to support videoconferencing and various levels of bandwidth, in *p* increments of 64 kbps up to a maximum of 2.048 Mbps (E1). *Note*: In *P × 64*, *p* = 1–30 channels of 64 kbps, with the maximum being the 30 bearer channels supported by E1. P × 64 specifies various frame rates and resolution levels. Most manufacturers of videoconferencing equipment support P × 64 as a common denominator, although they each prefer their own proprietary standards in promotion of their own equipment and unique feature sets. P × 64 more correctly is known as H.320 and is sometimes referred to as H.261, which is the specific ITU-T videocoding standard. The video formats include *Common Intermediate Format*

(CIF), which is optional, and *Quarter-CIF* (QCIF), which is mandatory in compliant codecs. H.261 CIF supports $352 \times 288 = 101,376$ pixels per frame and 30 fps, although lower frame rates also are supported. QCIF supports $176 \times 144 = 25,344$ pixels per frame, exactly one-fourth the resolution of CIF.

Actually, H.261 is an element of the ITU-T H.320 umbrella standard for video telephony over circuit-switched ISDN and addressing narrowband visual telecommunications systems and terminal equipment. Related ITU-T standards include H.221, which defines a frame structure in support of audiovisual teleservices in 64-kbps channels, and H.222, which defines the frame structure for such services in an ISDN environment. While these umbrella standards truly are international in nature, the U.S. and European camps unfortunately (and not surprisingly) are divided over certain implementation aspects of the ITU-T standards. Specifically, those differences deal with the manner in which audio and still-frame graphics are handled. For graphics, for example, the Europeans have adopted JPEG (ITU-T T.81), which is compatible with most PC-based graphics software. That approach is totally incompatible with the H.261 standard implemented in the United States, although it does provide a bridge to true multimedia applications [16]. H.320 is discussed in greater detail later in this chapter.

12.6.2   Joint Photographic Experts Group

The ISO and ITU-T jointly developed *Joint Photographic Experts Group* (JPEG) as a compression standard for editing still images as well as color facsimile, desktop publishing, graphic arts, and medical imaging. A symmetrical compression technique, JPEG is equally expensive, processor intensive, and time consuming in terms of both compression and decompression. *Motion JPEG* is used in the editing of digital video. JPEG is not appropriate for video transmission, as the compression ratio is in the range of only 20:1–30:1. Therefore, JPEG transmission in support of videoconferencing requires bandwidth in the range of 10–240 Mbps, which is far too bandwidth intensive.

12.6.3   Moving Pictures Experts Group

MPEG standards are several and still in final development stages. MPEG standards provide very high compression levels and excellent presentation quality. MPEG is a joint technical committee of the International Organization for Standardization (ISO) and the International Electrotechnical Commission (IEC). MPEG offers the critical advantage of asymmetric compression and decompression. While MPEG compression is time consuming and expensive, the decompression process is rapid and involves relatively inexpensive equipment. MPEG compression is as high as 200:1 for low-motion video of VHS quality, and broadcast quality can be achieved at 6 Mbps. Audio is supported at rates from 32–384 kbps for up to two stereo channels.

- *MPEG-1* was standardized in November 1992 as ISO/IEC IS (International Standard) 11172; it provides VHS quality at 1.544 Mbps and is compatible with single-speed CD-ROM technology. In fact, it was designed as the standard for storage and retrieval of moving pictures and audio on storage media such as

Compact Disc (CD). MPEG-1 integrates synchronous and isochronous audio with video and permits the random access required by interactive multimedia applications such as video games. Intended for limited-bandwidth transmission, it provides acceptable quality and output compatible with standard televisions. Current applications include video games, video kiosks, video-on-demand, and training and education. MPEG-1 supports video compression of about 100:1. MPEG-1 standards are the basis for the MP3 (MPEG-1 audio layer 3) audio data encoding system. Although MP3 can produce CD-quality audio at compression ratios up to 12:1, most MP3 codecs sacrifice considerable quality through compression at ratios as high as 25:1.

- *MPEG-2*, defined in November 1994 as ISO/IEC IS 13818, is the standard for DTV. MPEG-2 was conceived as a broadcast standard for interlaced images at 720×480 pixels at 30 fps and requiring bandwidth at 4–100 Mbps over transmission facilities (e.g., fiber optics, hybrid fiber/coax, and satellite) capable of such support. While MPEG-2 requires much more bandwidth than MPEG-1, it provides much better resolution and image quality and at much greater speed. MPEG-2 has found application in Direct Broadcast Satellite (DBS) services, also known as Direct Satellite Systems (DSS). Such services employ Ku-band satellites and Very Small Aperture Terminal (VSAT) dishes in competition with CATV, running MPEG at rates of about 3 and 7.5 Mbps [16]. In a convergence scenario, MPEG-2 is the standard of choice, supporting compression rates as high as 200:1, depending on content specifics. *MPEG-3*, designed for HDTV application, was folded into MPEG-2 in 1992.

- *MPEG-4* Version 1 was approved in October 1998 as ISO/IEC IS 14496 and as the standard for multimedia applications. Version 2 was approved in 1999. MPEG-4 is a low-bit-rate version intended for application in broadcast TV, videophones, and mobile phones and other small hand-held devices. In a wireless context, MPEG-4 is designed for IMT 2000 applications that transmit at a rate of up to 384 kbps upstream and receive at a rate of up to 2 Mbps downstream, although rates of 64 kbps and lower are supported. Encompassing both client/server and mass storage-based playback scenarios, MPEG-4 deals with the coded representation of audiovisual objects, both natural and synthetic (e.g., computer generated), and their multiplexing and demultiplexing for transmission, playback, and storage.

- *MPEG-7*, approved in September 2001, officially is known as the *Multimedia Content Description Interface*. MPEG-7 is intended to be the content representation standard for multimedia information search, filtering, management, processing, and retrieval. MPEG-7 essentially is metadata standard based on XML (eXtensible Markup Language) for describing multimedia content features in order that one can easily search for multimedia content on the Web. (*Note*: According to the ISO, there apparently is no known reason for the failure to expand the logical numbering sequence and, instead, to jump to MPEG-7 and then to MPEG-21.)

- *MPEG-21*, officially known as *Multimedia Framework*, is an ongoing effort to determine how various multimedia components fit together and to identify new multimedia infrastructure standards that may be required. MPEC-21 also deals with issues of content identification, description, and security. In large part, the focus of MPEG-21 is on the protection of intellectual property through security mechanisms designed to prevent unauthorized access and modification of multimedia content.

12.7   INTERNET PROTOCOL TELEVISION (IPTV)

Several forces have concerned to create a great deal of interest in TV distribution over IP networks. There is an undeniable and irreversible movement towards IP, in general. Telephone companies are aggressively deploying broadband local loop technologies in the local loop to support triple-play services comprising voice, Internet access, and TV. While copper-based ADSL has limited ability to support TV, VDSL certainly can support it over short distances. The telco trend is more towards optical fiber in the loop, either directly from the Central Office (CO) to the premises in a Fiber-To-The-Premises (FTTP) Passive Optical Network (PON) scenario or a hybrid Fiber-To-The-Neighborhood (FTTN) scenario that terminates fiber in a neighborhood node and uses ADSL or VDSL over a short length of embedded copper cable from the node to the premises. Even full-rate ADSL supports enough bandwidth for at least one compressed video channel. VDSL and PON both support multiple channels.

The telco IPTV approach differs considerably from that of traditional CATV networks. CATV networks support analog TV over 6-MHz channels allocated from the total 850-MHz spectrum typically supported over coax or hybrid fiber/coax cable plant. The number of channels clearly is limited to the number of 6-MHz RF channels that can be derived from the available spectrum. As CATV networks are configured as point-to-multipoint networks, each channel is delivered to every connected subscriber premises. The IPTV approach delivers each channel only to the subscriber premises that select to view it, whether one or hundreds do so. Over a PON local loop, the TV channel occupies a separate wavelength from the head end to the premises. In an FTTN scenario, the IPTV signal is converted from optical to electrical format at the neighborhood node and travels over Unshielded Twisted-Pair (UTP) copper cable plant the last few hundred meters or so. AT&T's Project Lightspeed, for example, calls for 20–25 Mbps to each of 300–500 subscriber premises connected to its fiber-optic loops. Verizon's FiOS PON network offers 5, 15, and 30 Mbps, depending on subscriber preferences. Either network offers bandwidth sufficient to support multiple DTV and even HDTV channels, properly compressed. At the moment (9:05 A.M. PDT, May 21, 2007), however, the telco networks run video in uncompressed analog format over a separate wavelength at 1550 nm. The eventual conversion to IPTV speaks to compressed DTV, in packet format, likely still but not necessarily over a separate wavelength.

12.8   THE H.320 FAMILY OF MULTIMEDIA STANDARDS

The ITU-T has developed a number of Standards Recommendations for videotelephony and multimedia communications. These recommendations fall under the umbrella of H.320, which was defined in 1990 for systems operating over Narrowband ISDN (N-ISDN). Table 12.3 details the range of standards under the H.320

TABLE 12.3 Overview of ITU-T Videotelephony and Multimedia Standards

Standard	H.320	H.321	H.322	H.323, V1/V2	H.324
Approval date	1990	1995	1995	1996/1998	1996
Network	N-ISDN, PSTN	B-ISDN, PSTN, ATM LAN	Packet network with guaranteed bandwidth (e.g., IsoEthernet)	Packet Network, with no guaranteed bandwidth (e.g., Ethernet, Token Ring, and Internet)	Analog PSTN
Video compression	H.261, H.263	H.261, H.263	H.261, H.263	H.261, H.263	H.261, H.263
Audio compression	G.711, G.722, G.728	G.711, G.722, G.728	G.711, G.722, G.728	G.711, G.722, G.728, G.723, G.729	G.723
Multiplexing	H.221	H.221	H.221	H.225.0	H.223
Control	H.230, H.242	H.242	H.230, H.242	H.245	H.245
Multipoint	H.231, H.243	H.231, H.243	H.231, H.243	H.323	
Data	T.120	T.120	T.120	T.120	T.120
Communications interface	I.400	AAL I.363, ATM I.361, PHY I.400	I.400 and TCP/IP	TCP/IP	V.34 modem

Source: [17, 18].

umbrella. Each standard was developed for a specific network environment and includes standards for video and audio coding, signaling and control, and multipoint control units.

12.8.1 H.320

Also known as P × 64, *H.320* supports videoconferencing and multimedia communications over N-ISDN B (Bearer) channels at bit rates from 64 kbps to 1.920 Mbps in increments of 64 kbps. Video compression makes use of H.261, which supports image resolution at several levels. As discussed earlier in this chapter, the optional CIF supports resolution of 352 × 288 pixels, and the mandatory QCIF supports resolution of 176 × 144 pixels. Frame rates are 30 fps or lower. Audio coding and compression recommendations include G.711, which is PCM at 64 kbps, thereby requiring a full ISDN B channel. G.722 is high-fidelity ADPCM transmitting up to a 7-kHz range and operating at 48/56/64 kbps. G.728 specifies LD-CELP running in the 3-kHz range and compressing voice at 16 kbps. The balance of the specification (H.221, H.230, and H.242) addresses techniques for call setup and teardown, data framing and multiplexing, and various other operational and administrative functions [17]. Table 12.4 details these various recommendations; the audio compression element largely is a restatement of the data provided in Table 10.3, which supports the discussion of packet Voice over Frame Relay (VoFR).

12.8.2 H.321

H.321 is the ITU-T Standard Recommendation for the adaptation of H.320 visual telephone terminals to B-ISDN environments. B-ISDN depends on ATM switching, which offers the considerable advantage of guaranteed Quality of Service (QoS), as discussed in Chapter 10. H.321 also involves H.310, which is the recommendation for broadband audiovisual communications systems and terminals.

12.8.3 H.322

H.322 is the ITU-T Recommendation for visual telephone systems and terminal equipment for LANs that provide a guaranteed QoS. This specification is limited to IsoEthernet (Chapter 8), which fact renders it little more than a historical footnote.

12.8.4 H.323

H.323 is the ITU-T Recommendation for packet-based multimedia communications systems. Annex D describes real-time facsimile. The recommendation addresses LANs that do not provide a mechanism for guaranteed QoS. H.323 also is used for service over the Internet and other IP-based networks, as illustrated in Figure 12.1. H.323 offers the advantage of supporting various compression techniques for packet-based voice communications, which can yield much more efficient utilization of network resources than does the traditional G.711 approach of PCM over circuit-switched networks. Interoperability of products (e.g., terminals and switches) can be achieved across the LAN and WAN domains. As H.323 is not linked to any

TABLE 12.4   H.320-Related Standards Recommendations

ITU-T Standard Recommendation	Description
G.711	Pulse Code Modulation (PCM) voice coding at 64 kbps
G.722	Adaptive Differential Pulse Code Modulation (ADPCM) voice coding and compression of high-fidelity 7-kHz voice at 64/56/48 kbps
G.723	Dual-rate speech coder at 5.3 and 6.3 kbps for multimedia communications
G.728	Low-Delay Code-Excited Linear Prediction (LD-CELP) coding and compression of 3.3-kHz voice at 16 kbps
G.729	Conjugate-Structure algebraic Code-Excited Linear Prediction (CS-CELP) voice coding and compression at 8 kbps
H.221	Frame structure for channel of 64–1920 kbps in audiovisual teleservices
H.223	Multiplexing protocol for low-bit-rate multimedia communication; annexes address mobile communications over low, moderate, and highly error prone channels
H.225	Call signaling protocols and media stream packetization for packet-based multimedia systems
H.230	Frame synchronous control and indication signals for audiovisual systems
H.242	System for establishing communications between audiovisual terminals using digital channels up to 2 Mbps; addresses call setup and teardown, in-band signaling and control, and channel management
H.245	Call control procedures for multimedia communications
H.261	Video codec for audiovisual services at $p \times 64$ kbps
H.263	Video coding for low-bit-rate communication at rates less than 64 kbps
T.120	Multipoint transport of multimedia data

Source: ITU-T.

specific hardware device or operating system, it can be deployed in a wide variety of devices, including PCs, telephone sets, and cable modems and set-top boxes. H.323 supports multicast communications, thereby avoiding the requirement for specialized Multipoint Control Units (MCUs) in a network where routers assume the responsibility for replicating packets. Version 2, ratified by the ITU in September 1998, provides a means for encryption, includes mechanisms for call transfer and call forward, supports URL-style addresses, and provides the ability for endpoints to set QoS levels through *Resource Reservation Protocol* (RSVP). The four major components specified for H.323 include terminals, gateways, gatekeepers, and Multipoint Control Units (MCUs).

- *Terminals* are the client endpoint devices on the LAN. While all terminals must support voice, data and video are optional. H.245 must be supported for negotiation of channel usage and capability. Q.931 is required for signaling and control. The *Registration/Admission/Status* (RAS) protocol communicates with

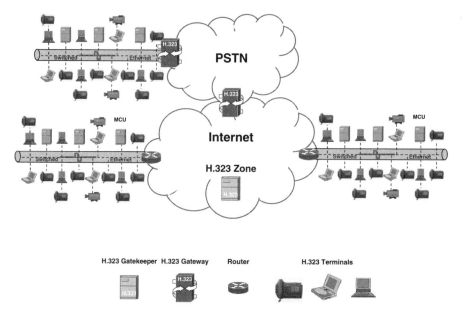

H.323 Gatekeeper   H.323 Gateway   Router   H.323 Terminals

Figure 12.1   H.323 networking over IP-based network.

the gatekeeper. Sequencing of audio and video packets is supported through *Real-time Transport Protocol/Real-time Transport Control Protocol* (RTP/RTCP). Endpoints can set QoS levels through *Resource reSerVation Protocol* (RSVP), although RSVP is unusual. Terminals optionally may include video codecs, T.120 data conferencing capabilities, and MCU functionality.

- *Gateways* are optional elements in the H.323 environment used for various levels of protocol conversion. A gateway comprises a *Media Gateway* (MG) and a *Media Gateway Controller* (MGC), which commonly are physically distinct devices. The MG serves as a protocol converter between devices and networks that have native H.323 capability and those that do not. The gateway also may translate between audio, video, and data formats. Finally, the gateway may perform signaling conversions between the H.225 packet protocol and external protocols such as Signaling System 7 (SS7) and Q.931. Alternatively, signaling conversions may be performed by gatekeepers, call processors, or session border controllers.

- *Gatekeepers* act as the central points in H.323 *zones* (i.e., zones of control). Endpoints may communicate directly, in either a unicast or a multicast environment, if no gatekeeper is present. If a gatekeeper is present, all endpoints in its zone must register with it. The gatekeeper performs the function of admission control, determining if devices are authorized to connect and if there is sufficient bandwidth to support the call. Gatekeepers serve to translate LAN addresses into IP or Internetwork Packet eXchange (IPX) addresses, as defined in the RAS specification. Gatekeepers also can act to route H.323 calls through gateways, if necessary, and monitor the network bit rate capacity, with the ability to deny access to a session if programmable bandwidth thresholds have been reached or exceeded. Gatekeepers also can perform certain administrative functions, such as accounting, billing, directory, and collecting network usage data. Gatekeepers may be distinct network elements, or gatekeeper functionality can be incorporated into (MCUs).

- *Multipoint Control Units* (MCUs) support conferencing among three or more participating terminals. The MCU comprises a *Multipoint Controller* (MC) and optional *Multipoint Processors* (MPs). The MC is responsible for call control negotiation to achieve common levels of communication. The MP may process either a single media stream or multiple media streams, depending on the nature of the conference.

12.8.5 H.324

H.324 is the ITU-T Recommendation for low-bit-rate multimedia communication over the analog PSTN through V.34 modems. As such, modems are limited to maximum transmission rates of 28.8 kbps. Voice must be highly compressed in order to make room for video and other visual information streams [17–20].

12.8.6 T.120

The ITU-T Recommendation for the multipoint transport of multimedia data is T.120, which data can include whiteboarding or binary files. This series of recommendations supports a broad range of underlying network technologies and can work either alone or under the H.320 umbrella. T.120 is entirely platform independent and can run in a variety of network environments, involving either reliable or unreliable data transport. Unicast and multicast modes both are supported [17, 21].

12.9 SESSION INITIATION PROTOCOL

The Internet Engineering Task Force (IETF) defined Session Initiation Protocol (SIP) in its RFC 2543 (March 1999) as an application layer (Layer 7 of the OSI Reference Model) signaling protocol for establishing, modifying, and terminating multimedia sessions or calls over an IP network. Finally approved in June 2002 in RFC 3261, which obsoleted RFC 2543, SIP is a modular component of IP telephony, although it can function over any network. SIP offers considerable advantages over H.323, although it can be used in conjunction with H.323 where appropriate. H.323 commonly is criticized as being too slow in establishing sessions, that is, setting up packet calls. An H.323-compliant client initiating a call must query a gatekeeper for the address of a new destination device. Once the gatekeeper has provided that address, the originating client establishes the session by using the H.225 signaling protocol, and the two clients negotiate features and call control procedures by using H.245. All of this takes time. In fact, it can take much more time than is required to set up a call through the conventional circuit-switched PSTN. The exact amount of time is sensitive to network specifics, and you must consider the fact that the Internet is a network of many networks, each of which is of uncertain nature in many respects. H.323 also is criticized for being overly complex and highly centralized.

SIP addresses all of these problems, and it was built specifically for an IP environment in which intelligence is highly decentralized in a large number of client agent servers. SIP identifies clients through a hierarchical URL similar to an e-mail address, e.g., SIP:ray@contextcorporation.com. (Don't try it, as I'm not SIP-compliant at the moment.) There are two ways that the calling client can initiate the call. If the SIP address of the destination SIP client is known, the calling client simply sends the destination client an *invite* message, in care of (1) a local proxy server. The proxy server sends the invite message to the distant proxy server, inviting the destination endpoint to join the session and providing it with enough information to do so. If the SIP address of the destination SIP client is unknown, the calling proxy server sends (2) the invite message to a *redirect* server, which consults (3) the *location* server for address information. The redirect server passes (4) that information to the calling proxy server, which then issues (5) an invite message to the distant proxy server, including the information required to join the call. If the call is to a call center, such information might include a request to employ H.261 video, G.728 audio, and Japanese as the preferred language. The proxy server on the receive end might consult (6) an optional SIP location server on the receive end to determine the exact location of the called client and connect (7) the call (i.e., ring the multimedia PBX). This approach, as illustrated in Figure 12.2, is a lot simpler and faster than the back-and-forth process involved in H.323, although layers of complexity are being added as the standards process works to enhance SIP to match H.323 and PSTN functionality.

Once the called client receives the invitation to join the session, it can either accept the call or forward it to a messaging system or a user, perhaps a Japanese-speaking call center agent. Assuming that the call is a multimedia call comprising

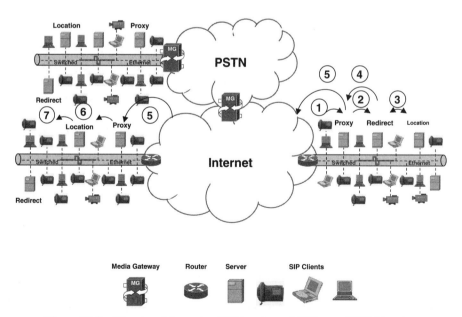

Figure 12.2 IP networking using SIP trunks and SIP over PSTN trunks.

both video and voice, the called client (or messaging system) can elect either to accept the composite call or to accept only one of the data streams, perhaps rejecting the video call but accepting the voice call. SIP also supports *call forking*, or splitting, so that several client extensions can be rung at once. H.323 does not have the same call-forwarding capabilities, although there is a follow-me feature that is similar. H.323 does not support constituent call separation or call forking. The big drawback to SIP is its lack of availability. H.323 is the incumbent protocol, and you know how hard it is to unseat an incumbent. As long ago as early 2001, however, several major carriers announced the commercial availability of limited SIP-based Voice over IP (VoIP) services and their plans to extend that capability in the future. Also particularly noteworthy was Microsoft's decision to incorporate SIP in Windows XP, Windows CE 4.0, Windows.NET server, and other devices embedded with XP [22–29].

A number of carriers now offer SIP trunks that connect the end-user premises directly to the SIP-enabled IP network, as illustrated in Figure 12.2. This approach certainly is the purest, as the enterprise-level SIP clients and servers interconnect directly through the SIP-enabled IP network. The alternative is for the enterprise user to gain access to the IP network through the TDM-based PSTN, with multiple Media Gateways (MGs) making the IP/TDM/IP conversions.

12.10   H.248: MEDIA GATEWAY CONTROL

The Media Gateway Control (Megaco) protocol is a joint standardization effort of the ITU-T (H.248) and the IETF (RFC 3525). Megaco, as it is known at the IETF, evolved from Simple Gateway Control Protocol (SGCP) and Media Gateway Control Protocol (MGCP). Megaco defines the call control protocols employed in a physically decomposed gateway with subcomponents distributed across multiple devices that may be in multiple physically distinct locations. Those subcomponents take the form of a Media Gateway (MG) and a Media Gateway Controller (MGC), also known as a *softswitch* or a *call agent*. A single MGC can control a large number of MGs, each of which is optimized for a particular gateway application function to convert the media format between a packet network and another form of network. Examples include interfaces that accomplish media conversion between IP packet format and an analog PSTN, or DS-0 format in a T/E-carrier-based PSTN interface, or an ATM network, or perhaps a device such as a PSTN-based voice processor.

The call control and signaling logic are centralized in the MGC and can include features such as dial tone, collect dialed digits, call hold, call transfer, call forward, and call conference. Feature changes are made only to the MGC, which simplifies the process of administration. The MGC signals the MGs, which then execute the feature commands and process the call, performing gateway functions as required to interface the incompatible networks or network elements. There is a master/slave relationship between the centralized MGC and decentralized MGs, much like that of a traditional telco network, except for the fact that the MGs that execute the features and perform the switching are distributed across the network. This is in sharp contrast to SIP, which is even more highly decentralized [28–30].

12.11   VIDEOCONFERENCING SYSTEMS

During 1930 two-way Television was demonstrated over a circuit connecting the Bell Telephone Laboratories with the headquarters of the American Telephone and Telegraph Company. Persons in booths two miles apart were enabled to see moving images of each other while they conversed. Television, however, is still in the development stage, and is not available on a commercial basis.

Things Worth Knowing about The Telephone,
American Telephone and Telegraph Company, 1931

Videoconferencing systems consist of cameras, monitors, video boards, microphones, speakers, and software. Videoconferencing can be accomplished in the workplace over a LAN. While it generally is more effective, easier, and less costly to walk across the hall and hold a face-to-face meeting, contemporary LANs often extend across multiple floors and even multiple buildings in a campus setting. Videoconferencing over the WAN clearly offers great benefits in terms of reduced cost and increased availability for meetings. Systems for videoconferencing can be quite substantial or can be PC based. Videoconferencing has increased significantly over the past few years as the cost of equipment and bandwidth have decreased—systems are made up of room systems, rollabout systems, and PC-based systems.

- *Room systems* are complex and quite expensive systems intended for videoconferencing among groups of people. While a specially designed room and equipment easily could cost $250,000 in the 1980s, a room system can be configured for less than $25,000 today. AT&T, MCI, and Sprint have provided room-based conferencing services for years from select locations, and they also provide network-based videoconferencing services for large corporations. Such service offerings include access and transport services as well as network-based MCUs. While many U.S. Incumbent Local Exchange Carriers (ILECs) in the late 1980s or early 1990s announced plans for *video dial tone*, those announcements clearly were way ahead of both the technologies and the market, and all of those plans were abandoned. The proliferation of fiber optics in the network core and advent of PON in the local loop now make such a service technically feasible, at least in terms of transmission capacity. Kinko's Copy Centers (now FedEx Kinko's) began deploying videoconferencing services in 1994 and currently offers such services at 150 locations in the United States and Canada. Kinko's has improved quality to 30 fps at 384 kbps through an alliance with Sprint for high-speed access and transport.
- *Rollabout systems* essentially are portable and much less expensive versions of room systems and also are intended largely for group-to-group conferencing. Such systems account for the preponderance of the stand-alone equipment market.
- *PC-based systems*, also known as desktop or appliance-based systems, are intended for person-to-person conferencing. PC-based systems are enjoying increased popularity, with the systems generally working over LANs or ISDN circuits. Analog transmission over the WAN also is accommodated by some systems, through V.34 and V.34+ modems at speeds up to 33.6 kbps. As you might expect, the cost of the system and the network are directly related to the

frame rate supported, the image resolution, and the sophistication of the compression technique employed. Inexpensive PC-based systems can be had for as
little as several hundred dollars today. The cost is that of camera and codec, the
latter of which fits into the expansion slot of a PC or Macintosh computer, with
the monitor serving as the video presentation device. Inexpensive PC-based
systems are used for Internet videoconferencing, although quality is poor due
to latency and data loss associated with packet switching over the public
Internet.

12.12   VIDEOCONFERENCING EQUIPMENT

Videophones originated with the AT&T Picturephone, which was demonstrated at
the New York World's Fair in 1964. Never intended for practical application, the
Picturephone was extremely bandwidth intensive, requiring bandwidth of about
90 MHz and weighing about 26 lb [6]. During the 1980s, AT&T, BT, and others
developed videophones that sold for less than $1000. As the cost was high, as each
party was required to have a videophone of the same manufacture, and as the
picture quality was poor (2 fps), videophones were stunning failures.

Regardless of the nature of the system, some combination of hardware and software is required. Videoconferencing equipment includes transmit (camera) and
receive (display) equipment that operate in concert with and through various intermediate devices to format the signal properly and otherwise treat it for effective
transmission over a network. Those intermediate devices include codecs, inverse
multiplexers, servers, and control units. Figure 12.3 presents a simple videoconferencing arrangement.

- *Codecs* (*co*ders/*dec*oders) accomplish the process of digitizing, or coding, the
  analog signal on the transmit side and decoding it on the receive end. The
  codecs also accomplish the process of data compression and decompression,

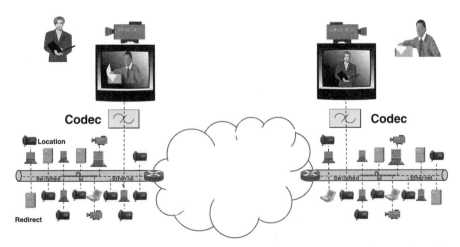

Figure 12.3   Simple videoconferencing network employing cameras, codecs, and monitors.

according to the specifics of the compression algorithm used. Additionally, codecs may include encryption features for security purposes and a mechanism for synchronizing the audio and video elements of the transmission.

- *Inverse Multiplexers* (inverse MUXs) are used in commercial videoconferencing systems where dedicated bandwidth is not available for relatively bandwidth intensive communications. An inverse MUX splits the video signal into two or more component parts that are transmitted over separate circuits or, perhaps, separate channels of multiple multichannel circuits (e.g., T1). The inverse MUX on the receiving end reassembles and resynchronizes the complete video signal for proper presentation.

- *Servers* are extremely high capacity storage devices, containing many gigabytes or even terabytes of memory. Servers store video and audio data for delivery to clients on demand.

- *Multipoint control units* are digital switching and bridging devices that support multipoint videoconferencing, with up to 28 parties (sites) commonly supported. MCUs must be compatible with the compression standards employed with the codecs. H.231, for example, describes ITU-T MCU standards, and T.120 describes generic data conference control functions. MCUs may be found in the carrier network in support of a carrier videoconferencing service or on the end-user premises in support of a videoconferencing network based on leased lines.

12.13 WAN VIDEOCONFERENCING NETWORKS

Video networking can be accomplished over a number of facilities and service offerings, depending on the application and the amount of bandwidth required. Analog circuits will support videoconferencing at low speeds, although the results are less than completely pleasing. The failed videophones offered by AT&T, BT, and others made use of dial-up analog circuits or ISDN circuits. As always is the case, digital circuits offer better performance than do analog circuits.

ISDN circuits are preferable to analog circuits because they provide more bandwidth and better transmission quality. However, the higher cost and lower availability of ISDN have slowed the acceptance of videophones based on ISDN technology in the United States. Note that in Europe and Japan ISDN is not particularly expensive and is much more widely available. Switched 56/64-kbps circuits can be used for videoconferencing—generally aggregated or *bonded* to provide multiple channels. Switched 384-kbps connectivity can be provided on the basis of fractional DS-1 through ISDN PRI channels in a channel group known as HO or over ADSL.

DS-1 facilities support full-motion, high-quality videoconferencing over dedicated networks at rates up to 2.048 Mbps for E-1 and 1.544 Mbps for T1. However, such facilities are costly and not widely dedicated to such applications. Large user organizations with dedicated leased-line T/E-carrier backbone networks make highly effective use of videoconferencing; the video communications contend with voice and data for network access through intelligent MUXs.

Broadband networks are much more capable of supporting the demands of videoconferencing. Frame Relay supports video, although that clearly is not the primary reason for its existence. While Frame Relay performs well under normal circumstances, it is likely to yield herky-jerky video should a poorly designed network suffer severe congestion. ATM was positioned for some years as the network technology of choice in a convergence scenario, at least in the backbone. ATM offers tremendous bandwidth over fiber-optic or hybrid fiber/coax networks and in support of voice, image, facsimile, and data traffic as well as video. ATM offers the unique advantage of supporting all of these traffic types, simultaneously providing each application type with precisely the QoS it expects. Ultimately, ATM fell from grace due to its complexity and high cost. IP has now replaces ATM as the favored protocol du jour, not only for data and voice but also increasingly for video.

12.14   VIDEO OVER IP

I spent a good deal of ink in several previous chapters examining VoIP. Now, I want to shift the emphasis to the other VoIP, Video over IP, which is quite different from IPTV. Video, like voice, is stream oriented and isochronous in nature. That is to say that video, like voice, depends on a continuous flow of interrelated data from the transmitter across the network to the receiver. For that matter, video, in general, usually is a combination of video and voice, and I cannot think of an example of videoconferencing that is not characterized by the same combination. The video and voice information streams are layered together and carefully synchronized so that the voice audio matches the movement of the speaker's lips. Just like stand-alone voice, interactive videoconferencing is very demanding in terms of QoS. It is critical that latency is in the range of 150 ms or less, that jitter be in the range of 50 ms or less, and that packet loss be in the range of 1 percent or less. Issues of latency and jitter are not particularly critical in a one-way streaming video application, as the receiver has time to buffer the incoming data stream and adjust. In a real-time, two-way videoconference, however, these issues are very significant. That's quite a trick over an IP network, but it can be done—and increasingly is being done with quite satisfactory results.

Video is extremely bandwidth intensive. Although video can run at rates as low as 64 kbps, high-quality video requires a high frame rate, good color depth, and good resolution, and those dimensions require a lot of bandwidth. So, the more bandwidth available from end to end, the better the potential quality of the presentation, which of course is also sensitive to the quality of the equipment involved. Given the fact that bandwidth is always limited and considering that the weakest link in the network (i.e., the local loop) always rules, compression is a good idea for both the video and voice elements. H.263, for example, is the choice of many manufacturers for video compression and G.723 (ADPCM) for voice. Because video and voice are stream oriented and, therefore, are intolerant of latency and jitter, it is a good idea to prioritize the data stream throughout the packet network, and that is where *Differentiated Services* (DiffServ) comes in. Considering the fact that there always will be some level of jitter and loss across a highly shared packet network, it is a good idea to both prioritize packets and provide a mechanism for reconstructing and resynchronizing packets at the receiving end, and that is where *Real-time Transport Protocol* (RTP) comes into play. DiffServ and RTP are particularly important in both video over IP and voice over IP, as UDP (User Datagram Protocol), rather than TCP, is used at layer 4. UDP is less bandwidth intensive than TCP but provides no error correction mechanism. Rather, the application assumes all responsibility for error correction. There's no time for retransmission of errored or lost packets in a real-time stream-oriented application, anyway, so it is up to the application to deal with those issues as it sees fit, if at all. Video over IP can work and does work, but all of the right pieces have to be in place for it to work well.

That said, it may surprise you—and it certainly surprised me—to hear that one of the reasons behind the increase in IP videoconferencing is that it actually works better than ISDN. According to Chris DiFiglia, president of the Polycom User Group, the typical 384-kbps ISDN videoconference depends on getting six bonded ISDN B channels to work simultaneously, which can be difficult. DiFiglia states that "IP is a much more fault-tolerant network. With ISDN, if there's a dropped frame or picture the entire call is dropped. An IP world is more forgiving. The worst that would happen is that the video frame would freeze for a second" [31]. IP videoconferencing certainly is less expensive, as ISDN is notoriously costly, and the six B channels carry per-minute charges, which can be particularly significant for international calls. ISDN videoconferencing also requires fairly expensive software and terminal equipment or terminal adapters. IP, on the other hand, generally involves little additional cost.

12.15 MULTIMEDIA CONFERENCING

Software must be in place to support electronic text, image, audio, and video information in a multimedia conference. While the voice and video aspects of the conference are supported in a fairly straightforward manner, the real and distinct advantage of multimedia conferencing is that of enabling multiple parties to collaborate on textual and graphic documents. Special software enables each party to contribute to such documents, in collaboration with the other parties—hence the term *collaborative computing*. During such a collaborative session, the original text document is saved, while each party contributes changes that are identifiable as such, by contributor. Once the parties agree to the collaborative edits and enhancements, the entire text file is refreshed and saved.

Similarly, a design or a concept can be developed graphically and on a collaborative basis through *whiteboarding*, much as the parties would do on a physical whiteboard in a face-to-face meeting. Typically, each party to the conference has access to a special whiteboard pad and stylus which is used to draw. Each party can modify the initial drawing, with each individual's contribution identified by separate color. Again, and once the group has agreed on the final graphic rendition, the graphic is saved and all screens are refreshed.

The clear benefits of such a collaborative process, conducted on a logical basis over a WAN, include reduced travel time, reduced travel expense, and increased speed of collaborative effort. Even in a LAN environment, shoe leather is conserved and productive time is maximized. A number of online Web-based collaborative conferencing tools are now available, offering considerable benefits over audio-only conferencing.

12.15.1   Video and Multimedia Conferencing: Applications and Benefits

There can be little doubt about the value of enhancing a communication with visual information. Pictures add another and very important element to the process of learning and comprehension. Moving pictures enable us to see the person to whom we are talking, thereby creating a more natural and effective person-to-person communication. Taking it one step further, a truly collaborative effort is enhanced greatly when multiple persons can work on a document. Supporting all of this over a network can save a lot of money, time, shoe leather, and gasoline. In a February 2004 report, Yankee Group analysts cited a number of benefits for videoconferencing, particularly in a contemporary multinational, multicultural environment [32]:

- Fifty-five percent of the impact of communications comes from facial expressions and body language, and only 38 percent from vocal inflection. (*Source:* UCLA.)
- Attendees in face-to-face meetings retain 38 percent more information than those in audio-only meetings. (*Source:* Harvard University and Columbia University.)
- Face-to-face meetings increase the power or persuasion by 43 percent over audio-only meetings. (*Source:* 3M Company.)
- Attendees learn 200 percent more in *face-to-face* meetings than in audio-only meetings. (*Source:* University of Wisconsin.)
- Attendees absorb information as much as 40 percent faster in face-to-face meetings than with audio alone. (*Source:* Wharton School of Business.)

Add to these benefits of electronic face-to-face meetings the hard dollar savings in travel expense and employee travel time, and a compelling case can be made for electronic conferencing. The unfortunate events of September 11, 2001, added the element of personal safety to the list of forces driving renewed interest in videoconferencing. Serving to supplement, but never replace, face-to-face in-person collaboration, video and multimedia networking has a legitimate place in the networked world.

The development of videoconferencing has been much slower than many of us projected. Traditional system and network technologies never developed to the point that high-quality videoconferencing was either simple or affordable. The recent development of standards for IP video promises to overcome those issues of complexity and cost, given the high availability and reasonable cost of broadband capacity in both the core and the local loop. Large organizations just cannot deny its value, and the costs of the system technologies are becoming quite reasonable for video-intensive applications. As costs continue to drop, small-business, home business, and even consumer markets for multimedia equipment will expand greatly.

One thriving application is that of *telemedicine*, which supports consultation and even remote diagnosis and treatment. A number of projects have experimented successfully with this concept, largely in support of remote clinics. Through a video-conferencing system and network, a nurse or medical technician in a remote clinic can gain the assistance of a doctor, and even a specialist, located at a major urban hospital. The doctor can guide the technician through the process of diagnosis and treatment, viewing the patient over a videoconferencing network and perhaps viewing x-rays transmitted over the same high-quality, high-speed digital network. For that matter, a multipoint conference can be established so the physician might consult with distant colleagues on a particularly difficult diagnosis and treatment plan. Taking the scenario one step further, the physician can even guide the technician through an emergency surgical procedure. Prescriptions, clearly, can be transmitted electronically to a local pharmacy. While such an application scenario currently is a bit unusual, it is possible, has been accomplished, and is in daily use. Although the typical commercial enterprise might not find videoconferencing or multimedia networking to be a lifesaving application, they will be critical elements of the technology mix for those firms seeking to gain or maintain competitive advantage.

Another interesting application for videoconferencing is in the justice system. Judges around the country, including in my little town of Mt. Vernon, Washington, make use of videoconferencing systems for video arraignments and other court appearances required for suspected criminal defendants. The clear advantage is that a video communication between the judge and lawyers in the courtroom and the defendant in the jailhouse eliminates the costly and sometimes dangerous process of transporting the accused and convicted. Tom Swift would have been proud, indeed, of the evolution and application of his *photo phone*.

Yet another interesting potential application is that of voice- and video-enabled websites. Clearly, the customer's contact experience can be improved tremendously if he or she can click a videoconference button and connect to an agent to establish a videoconference. That assumes, of course, that both parties are properly prepared for the camera.

Despite increased bandwidth, improved compression, and all the various protocols and mechanisms that have improved, and will continue to improve, video quality over the years, the videoconferencing experiencing still leaves us flat—two dimensional, that is. Wouldn't if be nice if video were in a 3-D (three-dimensional) format? Well, there has been a lot of development effort expended to do just that, and there is at least one commercially available system. Teleportec has developed a system that creates a 3-D effect through the use of multiple cameras on the transmit side and on the receive side by projecting the image on a specially treated sheet of glass that acts as a beam splitter, refracting light at different rates. The system builds on the H.320 and H.323 standards and currently runs over ISDN or dedicated circuits. A full T1 is preferred for maximum quality, but 768 and 384 kbps provide acceptable quality [33].

REFERENCES

1. Settel, Irving and Laas, William. *A Pictorial History of Television*. Grosset & Dunlap, 1969.
2. Mast, Gerald, revised by Kawin, Bruce F. *A Short History of the Movies*, 5th ed. Macmillan Publishing Co., 1992.

3. Brooks, John. *Telephone: The First Hundred Years*. Harper & Row, 1975.

4. Mayer, Martin. *About Television*. Harper & Row, 1972.

5. Oslin, George P. *The Story of Telecommunications*. Mercer University Press, 1992.

6. Wilcox, James R. *Videoconferencing: The Whole Picture*. Telecom Books, 2000.

7. ATSC Standard. Doc A/53. September 16, 1995.

8. Outler, Elaine, Baker, Ron and Barr, Tracy. *DTV For Dummies*. IDG Books Worldwide, 1998.

9. Wong, May. "HDTV: Get The Picture?" *The Seattle Times*. February 26, 2006.

10. *Projections of the Number of Households and Families in the United States: 1995–2010*. U.S. Department of Commerce, Economics and Statistics Administration, Bureau of the Census.

11. Feldman, Gary. "Migrating toward PON." *Outside Plant Magazine*, June 2005.

12. "FCC Acts to Expedite DTV Transition and Clarify DTV Build-Out Rules." http://www.fcc.gov/Bureaus/Mass_Media/News_Releases/2001/nrmm0114.html. November 8, 2001.

13. "FCC Reallocates and Adopts Service Rules for Television Channels 52–59." http://www.fcc.gov/Bureaus/Miscellaneous/News_Releases/2001/nrmc0128.html. December 12, 2001.

14. Minoli, Daniel. "Videoconferencing." *Datapro Communications Analyst*, May 1994.

15. Bort, Julie. "Standards Compliance Won't Ensure Interoperability." *InfoWorld*, October 16, 1995.

16. Halhead, Basil R. "Videoconferencing Standards." *Datapro Communications Analyst*, September 1994.

17. Kupst, Shirley, Mehravari, Nader, Olson, Mark, and Rush, Scott. "Designing Virtual Co-Location and Collaborative Environments via Today's Desktop Videoconferencing Technology." http://www.arch.usyd.edu.au/kcdc/conferences/VC97/papers/kupst.html.

18. "A Primer on the H.323 Series Standard. Version 2.0." Databeam. Originally accessed at http://gw.databeam.com/ccts/t120primer.html.

19. "Demystifying Multimedia Conferencing over the Internet Using the H.323 Set of Standards." Intel Corporation, 1999. Originally accessed at http://www.andygrove.com/technology/itj/q21998/articles/art_4a.html.

20. "Videoconferencing Standards." FVC.COM. December 1997. Originally accessed at http://www.fvc.com/whitepapers/stndards.html.

21. "A Primer on the T.120 Series Standard. Version 2.0." Databeam. Originally accessed at http://gw.databeam.com/h323/h323primer.html.

22. Handley, M., Schulzrinne, H., Schooler, E. and Rosenberg, J. "SIP: Session Initiation Protocol," RFC 2543. Internet Society, March 1999.

23. Kraskey, Tim and McEachern, Jim. "Next-Generation Network Voice Services." *Network Magazine*, December 1999.

24. Greenfield, David. "Lesson 138: Inside the Session Initiation Protocol." *Network Magazine*, January 2000.

25. Michael, Bill. "SIP Ascendant." *Communications Convergence*, June 2001.

26. Rosenberg, J., et al. "SIP: Session Initiation Protocol," RFC 3261. Internet Society, 2002.

27. Horak, Ray. "Pure SIPlicity," *Commweb*, February 9, 2004, www.commweb.com.

28. Green, James Harry. *The Irwin Handbook of Telecommunications*, Fifth Edition. McGraw-Hill, 2006.

29. Freeman, Roger L. *Fundamentals of Telecommunications*, 2nd ed. Wiley, 2005.

30. Groves, C., et al. "Gateway Control Protocol Version 1," RFC 3525. Internet Society, 2003.

31. Leung, Linda. "IP Videoconferencing Winning Converts." *Network World*, September 29, 2003.

32. Dern, Daniel P. "Now Your See—And Hear It." *VON Magazine*, October 2005.

33. Connolly, P. J. "Videoconferencing Adds Depth." *InfoWorld*, December 3, 2001.

CHAPTER 13

THE INTERNET AND WORLD WIDE WEB

As a net is made up of a series of ties, so everything in this world is connected by a series of ties. If anyone thinks that the mesh of a net is an independent, isolated thing, he is mistaken. It is called a net because it is made up of a series of interconnected meshes, and each mesh has its place and responsibility in relation to other meshes.

Buddha

From its early beginnings as DARPANET, linking a select few military and Research and Development (R&D) facilities, the Internet has grown to comprise thousands of networks and millions of users in virtually every country in the world. The Internet is the *kudzu* of networks, growing like the uncontrollable weed of Southern infamy.

The Internet truly *was* the *information superhighway*, long before the term was coined. While actually more of a private road, it was characterized by a sense of truly unparalleled freedom. That early network stood apart by virtue of its providing free and open access to information—that characteristic remains today as the hallmark of the Internet. The Internet is a network of networks and has become a *Global Village*, much as envisioned by Marshall McLuhan (1911–1980), education, philosopher, and scholar. In some sense, at least, the Internet *is* the information superhighway.

13.1 THE INTERNET DEFINED

The Internet is a global *network of networks* currently linking over 60,000 networks and spanning over 150 countries. According to the Internet Systems Consortium,

Telecommunications and Data Communications Handbook, By Ray Horak
Copyright © 2007 Ray Horak

the estimated number of connected hosts increased by about 24 percent in 2005, to almost 395 million. While that is a healthy rate of growth, it is much slower than the 36 percent in 2004 and 2003 [1]. According to a 2005 study by IDC, those hosts will account for 84 billion e-mail messages per day worldwide in 2006, about 33 billion of which messages will be *spam*, that is, unsolicited e-mail [2]. International Data Corporation (IDC) predicts that the volume of Internet traffic worldwide will continue to double each year, increasing from 180 petabits per day in 2002 to 5175 petabits per day by the end of 2007 [3]. Others suggest that the Internet is showing signs of maturity, at least in terms of international traffic. TeleGeography's latest survey of Internet backbone providers indicates that cross-border traffic grew by only 49 percent in 2005, which is down considerably from 103 percent in 2004. Tele-Geography pegs the fastest growing regions as Asia (64 percent) and Latin America (70 percent), although their growth rates are way down from previous years [4]. Figure 13.1 provides a graphic view of the growth of the Internet in terms of host computers; notably, there were only four hosts in 1969.

The Internet is grounded in the U.S. Department of Defense *Advanced Research Projects Agency NETwork* (ARPANET), which began in 1969 as a means of linking personnel and systems involved in various computer science and military research projects. ARPANET and its successor DARPANET (*Defense Advanced Research Projects Agency NETwork*), were developed to be totally *fail safe*. The distribution of computing power and the redundancy of the data switches and computer links all were intended to provide a meshed computer network that could withstand a nuclear strike.

In the early 1970s, work began at Stanford University on a set of internetworking protocols designed to prove connectivity among the ARPANET computers. In May 1974, Vinton G. Cerf (Stanford University) and Robert E. Kahn (Advanced Research

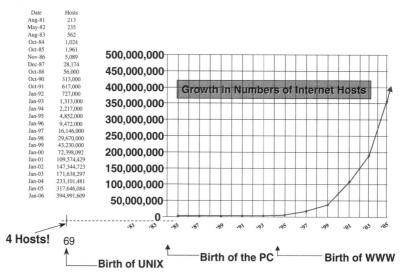

Figure 13.1 Growth in Internet hosts. (*Source:* Internet Software Consortium, http://www. isc.org/.)

Projects Agency, U.S. Department of Defense) published "A Protocol for Packet Network Intercommunication" in the IEEE *Transactions on Communications* [5]. That concept became known as Transmission Control Protocol/Internet Protocol (TCP/IP). They completed development work on the protocols in 1980, and in 1983 the Office of the Secretary of Defense mandated that ARPANET users accept the new set of computer protocols, which became the standard for ARPANET. To encourage colleges and universities to transition to TCP/IP, DARPA eased the implementation process for *Berkeley Software Distribution*, commonly known as *BSD UNIX* or *Berkeley UNIX*, a version of the UNIX operating system that they mostly ran at the time. Toward that end, DARPA funded Bolt Baranek and Newman (BBN, now BBN Technologies) to implement the TCP/IP suite with Berkeley Unix; this approach filled an internetworking protocol void and formed the foundation for the Internet [6].

In late 1983, ARPANET split into two unclassified networks: DARPANET and MILNET (*MILitary NETwork*). Although ARPANET officially was retired in June 1990, the Internet has survived. TCP/IP has been enhanced recently and currently enjoys wide popularity in education, research and development, and commercial applications.

The National Science Foundation founded NSFNET (*National Science Foundation NETwork*) in 1985 to link its supercomputer centers and to provide access to the Internet. NSFNET was a high-speed backbone network consisting of point-to-point links in a mesh configuration. The network was deployed fully in 1988, initially at 56 kbps. In 1986, NSFNET partially funded a number of regional networks, tying into the backbone. In 1992, the backbone was upgraded to T3, operating at 45 Mbps. More recently, major portions of NSFNET were upgraded to OC-3 (155 Mbps) and beyond. NSFNET officially was retired in 1995 and was replaced by the *MERIT* network. MERIT originally was a statewide network operated by the University of Michigan and a regional component of both NSFNET and the Internet.

13.2 INTERNET PHYSICAL TOPOLOGY

The physical topology of the Internet, ultimately, is not that important as long as connectivity is achieved. That topology also changes from day to day and varies considerably from place to place. However, and for those of you who, like me, absolutely *must* know at some level, the Internet physical topology consists of leased lines that connect various major computing centers through switches and routers. This leased-line network is a partially meshed network for purposes of redundancy and network resiliency. Initially, ARPANET consisted of approximately 50 custom-built BBN *Packet-Switching Nodes* (PSNs) scattered around the United States and Western Europe. NSFNET (1985) consisted of 56-kbps leased-line circuits that connected the NSF supercomputer centers in San Diego, California; Boulder, Colorado; Champaign, Illinois; Pittsburgh, Pennsylvania; Ithaca, New York; and Princeton, New Jersey. During this time, a great deal of Internet traffic over the NSFNET backbone became commercial in nature, prompting a number of private commercial Internet backbone providers to establish a point for the exchange of this traffic. This point was established at Wiltel offices in Santa Clara, California, and became known as the *Commercial Internet eXchange* (CIX).

The original NSFNET was replaced in 1988 by an expanded version through a partnership of MERIT, IBM, and MCI. Sites were added in San Francisco, California; Seattle, Washington; Houston, Texas; Ann Arbor, Michigan; and other cities that housed large computer centers. The third version of the NSFNET backbone was deployed in 1989, increasing backbone speed and adding some circuits while deleting others. In 1996, NSFNET officially was retired and replaced with the MERIT network, which has expanded and evolved considerably. That network consisted of *Network Access Points* (NAPs) in San Francisco (Pacific Bell), Chicago (Ameritech), and New Jersey (Sprint) as well as a *Merit Access Exchange* (MAE) in Washington, D.C., and was built by MFS Datanet, which Worldcom later acquired. Subsequently, MAE variously became an acronym for either *Metropolitan Area Exchange* or *Metropolitan Area Ethernet*. (The exact definition is lost in the mists of time, as best I can determine.) and now is just MAE, a registered trademark of MCI, now a Verizon company. Tier 1 MAEs, regarded as national points of interconnection, are located in San Jose, California (MAE West), Vienna, Virginia (MAE East), and Miami, Florida. Tier 2 MAE sites currently are located in Chicago, Illinois; Dallas, Texas; Los Angeles, California; and New York, New York. Also, Terremark Worldwide placed the NAP of the Americas at four locations in and around Miami, Florida, where it provides services between North America, Latin America, and the Caribbean. The NAPs and MAEs are placed at strategic locations (Figure 13.2) where Internet Service Providers (ISPs) can exchange traffic at high speeds. Each MAE comprises Asynchronous Transfer Mode (ATM), Frame Relay (FR), and Gigabit Ethernet (GbE) switching platforms. Synchonous Optical NETwork/Synchronous Digital Hierarchy (SONET/SDH) interfaces include OC-3 (155 Mbps), OC-12 (622 Mbps), and OC-48 (2.5 Gbps). ISPs seeking to interconnect through the NAPs and MAEs do so through their own routers, which may be collocated, and must maintain their own routing tables. ISPs also can exchange traffic at local or regional private peering points [7–11].

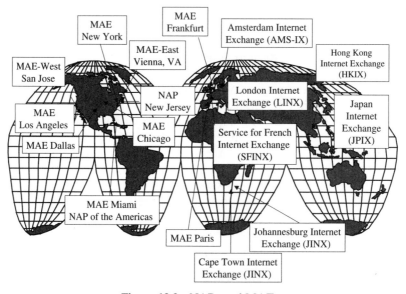

Figure 13.2   NAPs and MAEs.

Aside from the backbone network, NSFNET funded a number of midlevel networks at the state and regional levels. Access networks also have been funded in various networks to support backbone connections for specific organizations or consortia. Campus networks also connect to the backbone through dedicated lines leased from various carriers [6]. In an interesting recent development, universities have become distressed over the degradation in Internet performance as traffic levels have increased because of its commercialization. In fact, a number of them petitioned The National Science Foundation to build a new, separate NSFNET intranet (private Internet) just for them—it seems as though success comes at a price [12]. In October 1996, this concern translated into action in the form of Internet2, which I discuss later in this chapter.

In addition to the basic Internet backbone structure spawned by NSFNET, the major incumbent interexchange carriers—AT&T, MCI, and Sprint—built national backbones. These carriers also have either made a number of acquisitions to enhance those backbones or, more typically, have acquired backbones that they then enhance. A clear example is the acquisition of Metropolitan Fiber Systems (MFS) and its UUNET subsidiary by MCI, which then became MCI Worldcom. MCI also acquired Advanced Networks and Services (ANS), a consortium to which MCI was a party. Ultimately, MCI was acquired by Verizon. Similarly, AT&T acquired Teleport Communications Group (TCG), which previously acquired the CERFnet backbone, and AT&T ultimately was merged into SBC (nee Southwestern Bell), which retained the name AT&T. Competitive interexchange carriers, such as Level 3 Communications and 360 Networks, also offer backbone Internet transport. The incumbent LECs built major regional backbones some years ago and a number of less well-known regional carriers transport Internet traffic as well.

So far, our discussion has focused on the U.S. topology, which is appropriate, as the Internet is a U.S. invention. For all practical purposes, it exclusively was a U.S. network until the 1990s. As the Internet gradually extended outside U.S. borders, all traffic initially was routed through the U.S. infrastructure. That meant, for example, that local traffic from one user in Hong Kong to another in Hong Kong was routed through MAE West in San Jose and back to Hong Kong. In order to confine regional traffic and thereby correct this obvious inefficiency, dozens of NAPs now exist in a number of international locations. Those NAPs include The Hong Kong Internet Exchange (HKIX) in Hong Kong, the Japan Internet Exchange Co. (JPIX) in Tokyo, the Amsterdam Internet Exchange (AMS-IX) in The Netherlands, the Service for French Internet Exchange (SFINX) in Paris, the London Internet Exchange (LINX) in The United Kingdom, and the Johannesburg Internet Exchange (JINX) and Cape Town Internet Exchange (CINX) in South Africa.

To put this physical topology in some sort of reasonable order and context, consider the concept of levels. While these levels are not defined absolutely, they do serve to put the topology of the Internet in perspective. At level 1 are the interconnect points and at level 5 are the end users. In a long-haul, coast-to-coast communication in the United States, you might well work up through all five levels and back down through all five:

- Level 1 includes all of the NAPs and MAEs, both domestic U.S. and international. Again, these NAPs are the major points of interconnection where national, and perhaps regional, carriers exchange traffic. The NAP provider may be independent of other levels (e.g., the international INXs) or may be a level 2 provider (e.g., MCI or Sprint).

- Level 2 is the national backbone level. At this level are the national service providers such as AT&T, MCI, Cable & Wireless, Level 3 Communications, and Global Crossing. Level 2 providers are facilities based, as they own long-haul, high-speed transmission facilities that serve to interconnect their various high-speed switches and routers. (*Note:* Level 3 Communications has built a very substantial IP-based network, primarily in the United States. As IP operates at Layer 3 of the OSI Reference Model, the company originally considered naming itself Layer 3 Communications but felt that name to be less attractive than Level 3, or so I was told by a number of Level 3 employees who claimed to know the underlying logic. I fully understand that this is purely anecdotal, but it is the best I can do.)
- Level 3 comprises the regional carriers. Such carriers typically operate a backbone in a single state or perhaps several states. Level 3 service providers also are facilities based, but at a lesser level than the level 2 service providers.
- Level 4 comprises ISPs, which can be small *mom-and-pop* companies or regional or even national in nature. ISPs may also be Level 3 or even Level 2 providers. They also may function strictly as ISPs and not own or operate any backbone transmission facilities. Examples of pure ISPs include America Online, Earth-Link, CompuServe, Prodigy, and Comcast.
- Level 5 consists of the end users. Level 5 connections to the Internet may be through a Level 4, Level 3, or even Level 2 provider. Access techniques run the full range of network options discussed in previous chapters and discussed again later in this chapter. Such techniques include leased lines, dial-up, Digital Subscriber Line (xDSL), cable modem, and Passive Optical Network (PON) [13].

13.3   INTERNET ACCESS

While the number of Internet users continues to grow at incredible rates, not everyone has access. In this Information Age, there is great concern about creating a new class system of the *information haves* and *have-nots*. In an attempt to address this issue of the *digital divide*, a great debate has arisen about subsidizing some level of *universal service*, much as was assured in the 1930s for the Public Switched Telephone Network (PSTN). With the goal of providing access to every American school by the year 2000, $36 million was raised from private sources in 1995—with matching funds from U.S. government sources and donations. Additionally, various carriers offered special access rates and some manufacturers donated equipment or offered special discounts [14].

Access from the customer premises to the Internet can be accomplished through a variety of means, including a dial-up connection, DSL, CATV modem, or a dedicated circuit. The local loop portion of the access circuit might be provided by the Incumbent Local Exchange Carrier (ILEC), the CATV provider, or a Competitive Local Exchange Carrier (CLEC) and may be either wireline or wireless in nature. The connection might be supported over the circuit-switched PSTN, a packet-switched Frame Relay network, or even a cell-switched ATM network. Access from the user premises to the Internet can be through a level 4 ISP, a Level 3 regional provider, or a Level 2 national backbone provider, with access at Levels 3 and 2 generally on the basis of a dedicated circuit. The performance of the connection between the user premises and the service provider depends on the nature of the underlying technologies and the level of bandwidth provided. Through the service provider and across the Internet, the level of performance depends on the specifics of the networks at Levels 1–4. Congestion, loss, and error may occur anywhere in the network. As is always the case in the networked world, the least capable element of the network defines the maximum level of performance that can be realized end to end.

According to ISP-Planet, top U.S. ISPs at 2005 year end included America Online (AOL) at approximately 18.6 million registered dial-up users (down from 27.7 million at year end 2001), for a market share of 20.1 percent. Comcast reported cable modem customers at 9.0 million, AT&T (includes SBC) at 5.7 million DSL customers, Road Runner at 5.4 million cable broadband users, EarthLink at 5.3 million through a variety of access methods, and BellSouth DSL subscribers at 3.1 million [15].

According to the Federal Communications Commission (FCC), through December 31, 2005, there were approximately 50.2 million subscribers in the United States with *high-speed* Internet access, which it defines as service providing speed of at least 200 kbps in at least one direction. That is an increase of 33 percent over the 37.9 million subscribers at the end of 2004. By comparison, there were 9.6 million at the end of June 2001. Cable modem service accounted for 57.5 percent of the 50.2 million lines, Asymmetric DSL (ADSL) for 40.5 percent, Symmetric DSL (SDSL) or traditional wireline connections for 0.3 percent, fiber optics for 0.5 percent, and other technologies for 1.2 percent. The FCC also reports statistics on *advanced services* lines, which it defines as providing speeds of at least 200 kbps in both directions. According to the FCC, there were 42.8 million advanced services lines as of December 31, 2005, an increase of 48 percent over the previous year [16].

13.3.1 Dial-Up Access

Terminal access via dial-up connection clearly is the most common means of Internet access, certainly for the residential, or consumer, and SOHO (Small Office/ Home Office) markets, worldwide. Whether through an analog modem or an Integrated Services Digital Network (ISDN) modem, dial-up access makes use of the PSTN, which, as is discussed elsewhere in this book, is a circuit-switched, Time Division Multiplexing (TDM)–based network.

The most common dial-up access technique is through an analog modem, as most consumer and SOHO local loops are analog in nature. Chapter 6 explored the process of conventional dial-up modem access in quite some detail, with respect to V.34 (28.8 kbps) and V.34+ (33.6-kbps) symmetric modems and the more recently developed V.90 and V.92 (56-kbps) asymmetric modems. I do not repeat that discussion here, although I must make several points of significance. First, the analog local loop is a source of difficulty because it affects error performance and, therefore, affects throughput in a decidedly negative way. Second, consider the fact that the

PSTN is a circuit-switched network. As such, bandwidth is provided on demand and as available, through the originating Central Office (CO) circuit switch, across the network, and through the terminating CO switch to the ISP. Call set-up takes some amount of time and the quality of the connection can vary considerably from connection to connection, depending on the physical path of the connection and the nature of other activity taking place in local loop cables and other elements of the network. Third, bandwidth is constrained to 2400 baud over the analog local loop and to a maximum of voice-grade 56 kbps (actually 53.3 kbps) in the network cloud and across the channelized T-carrier local loop that terminates at the ISP or other access provider. Finally, the circuit-switched, TDM-based network provides bandwidth on a temporary basis, which is continuous and exclusive in nature. In other words, that capacity is provided for the use of that one connection for the entire duration of the connection, whether it is used or not. This approach is extraordinarily wasteful of limited and, therefore, precious network resources in an interactive, packet-oriented data session such as that associated with Internet access.

ISDN, which I discussed in Chapter 7, offers the advantages of a digital local loop, which certainly improves performance and, therefore, throughput. Further, through a Terminal Adapter (TA), ISDN Basic Rate Interface (BRI) provides fully symmetric bandwidth at levels of either 64 or 128 kbps. ISDN must be supported by the ISP or other access provider, typically in the form of Primary Rate Interface (PRI) at an aggregate rate of 1.544 Mbps in North America and 2.048 Mbps internationally. ISDN certainly is an improvement over analog modem access, but it is also more costly and more limited in availability. As ISDN also is a circuit-switched access technology, it remains a very wasteful approach.

In order to overcome the inherent wastefulness of the circuit-switched PSTN for Internet access, the major CO manufacturers have developed devices that front end the CO. These devices logically are positioned between the local loop access port and the CO's internal switching matrix. Should the terminating telephone number be recognized as one associated with an ISP or other Internet access provider, these devices shunt that traffic away from the CO switch and toward a packet-based data network. This obviates any issues of congestion that might be caused by the decidedly inappropriate use of the circuit-switched PSTN for packet data transfer.

Dial-up access is supported by virtually all ISPs, which focus largely on that method of access. All information service providers (e.g., America Online, CompuServe, Microsoft Network, and Prodigy—national ISPs that also provide content as well as access) support dial-up modem access because they largely are focused on consumer-level services. As I noted earlier, dial-up access through the PSTN to ISPs is by far the most popular means of consumer access to the Internet as it is virtually universally available, worldwide.

13.3.2 xDSL Access

As I discussed in great detail in Chapter 9, xDSL (generic Digital Subscriber Line) is an incredibly powerful network access technology and is widely available throughout metropolitan areas in the United States. The FCC reported that high-speed DSL connections were available to 78 percent of the households to which ILECs cold provide telephone service as of December 31, 2005 [16]. A wide variety of xDSL technologies exist, some of which are standardized, some of which are in the process of standardization, and others of which are proprietary in nature. The most popular version, by far, is ADSL. ISDN DSL (IDSL) is a local loop technology that uses the same mechanisms as ISDN BRI, but without the switch. IDSL is data specific, while ADSL supports voice, data, and video. While the data speeds vary by technology, they are much greater than the speeds offered by even the most sophisticated modem technologies. ADSL access speeds are asymmetric, with much greater bandwidth provided downstream. ADSL speeds are definable, and the cost to the end user is sensitive to speed rating. In addition to the greater level of bandwidth, xDSL access is *always on*. In comparison to a dial-up connection, therefore, xDSL yields the advantages of no delays associated with call setup and no blockage at the originating CO circuit switch. While issues of congestion from the DSLAM (DSL Access Multiplexer) forward remain, congestion is a fact of life in a shared packet network.

13.3.3   Cable Modem Access

Access to the Internet is offered by a number of CATV (Community Antenna TeleVision) providers, as is also discussed in some detail in Chapter 9. Since the mid-1990s, a number of CATV providers have upgraded much of their traditional coax-based networks with optical fiber, thereby increasing overall network performance considerably. Even in the absence of optical fiber upgrades, coax can support not only downstream TV delivery but also two-way Internet access and other data applications. With the upgrade of the electronics to support two-way transmission and the dedication of upstream and downstream data channels, high-speed Internet access can be provided at end-user costs that generally compare quite favorably with xDSL. The development of the DOCSIS (*Data Over Cable Service Interface Specification*) standard for cable modems has encouraged the deployment of this access option, although availability currently remains limited because of the cost of the required network upgrades. Note that older CATV systems use the *telco-return* approach, whereby upstream access to the Internet is provided via dial-up connection to an ISP through the PSTN; downstream access is provided over a shared channel of the coax CATV network. The FCC stated that high-speed cable modem service was available to 93 percent of the households to which CATV operators could provide cable TV service in the United States [16].

As is the case with most xDSL technologies, cable modem access is asymmetric in nature, which generally is quite acceptable for consumer use. Notably, cable modem access usually is limited to consumer or SOHO application, as the CATV networks themselves are oriented toward the consumer, rather than the business, market. Unlike xDSL technologies, cable modem access bandwidth is shared among all users, as the CATV network itself is a shared network. Therefore, the bandwidth available to any individual user at any given time is sensitive to the number of subscribers active on any given CATV network segment at any given time as well as the level of intensity of their usage. The shared nature of the CATV network also poses significant security issues in the absence of effective access control and encryption mechanisms.

Encouraging the deployment and acceptance of cable modem access is not only the development of DOCSIS specifications but also changes in regulation. The Telecommunications Act of 1996 opened the local exchange to competition and clearly stimulated the CATV providers to position themselves as CLECs, initially focusing on the data market, with an emphasis on Internet access. Many CATV providers now offer voice service as well.

13.3.4 Satellite TV Access

Satellite TV networks offer yet another option for Internet access. Currently available offerings are based on Geosynchronous Earth-Orbiting (GEO) satellites, which I discuss at some length in Chapter 2. GEO satellites are effective for broadcast applications because their footprints, or areas of coverage, are substantial and stable. In terms of the downstream path from the Internet, GEO systems offer considerable bandwidth, although it is shared much as is the bandwidth provided over a CATV network. This issue can be mitigated to a considerable extent through the use of highly focused *spot beams* that segment the aggregate footprint of the satellite into smaller areas of coverage. This allows coverage to be segmented and frequencies to be reused, much like the cells of cellular telephony network. The upstream channel also is an issue, as two-way satellite dishes are considerably more expensive than one-way dishes. Further, a two-way connection via a GEO satellite in equatorial orbit at an altitude of approximately 22,300 miles imposes round-trip signal propagation delays of at least 0.64 s. While these subsecond delays are tolerable for most applications, they render this Internet access technique unacceptable for users engaged in applications such as *twitch games* involving rapid action. Note that some GEO-based Internet access services use a telco-return access technique for the upstream channel. This technique involves dial-up access through an ISP over the PSTN for the upstream channel, with the downstream channel supported over a one-way satellite link.

13.3.5 Dedicated Access

Direct connections to the Internet, in this context, are considered to be those that connect to a regional or national backbone provider (i.e., level 1, 2, or 3), bypassing a local ISP (level 4). Dedicated access arrangements are cost effective for large user organizations and for those who use the Internet intensively. Direct access can be on the basis of any number of alternatives, including Dataphone Digital Service (DDS) and T/E-carrier. Direct SONET access also is offered at speeds up to 155 Mbps (OC-3) by select providers. Such direct connections commonly are in the form of leased lines provided by the ILEC, although a facilities-based data CLEC also might provide such access.

13.3.6 Access Anywhere

The Internet is an incredible network of networks that enables you to access from your machine virtually any other machine and any database and any type of information, anywhere in the world—as long as you have access to the Internet. From your residence, your SOHO, and your corporate office, you have that access through a dial-up connection, ADSL, a CATV provider, a fixed wireless link, or perhaps a satellite link. This access involves an ISP with which you have a formal relationship or perhaps a direct connection to a backbone provider. The trick is for the *road warrior* to gain access from some place other than the home or office. The national ISPs solve that problem for U.S. travelers who confine their roaming ways to the domestic United States and Canada, as they have a great many *Points Of Presence* (POPs) that provide local dial-up access from just about anywhere. Smaller independent ISPs do not have the advantage of a national presence and, therefore, cannot provide the same level of access, although they can contract with a service provider that maintains *Remote Access Servers* (RASs) in COs around the country. More difficult is gaining access elsewhere in the world, where local and even national ISPs have no presence and where the plug interfaces can vary as widely as the languages and the currencies. Global modems and plug adapters can solve the basic interface problems. Some ISPs have significant international presence, but most do not. To fill that void, some companies have developed international presence through consortium arrangements with large numbers of local ISPs around the world. But developing countries largely remain unwired for Internet access, as I am constantly reminded during my regular business trips to Africa, where even the best *business-class* hotels often fail to provide dataports in the guest rooms.

Taking *access anywhere* to the contemporary extreme translates into not only portability but also true mobility, and that translates into wireless access in the form of 802.11, Bluetooth, and cellular and packet radio technologies. The most readily available and most effective of these technologies currently is 802.11, as I discussed in Chapter 8.

13.4 INTERNET STANDARDS, ADMINISTRATION, AND REGULATION

You should note that the Internet is most unusual as a network. Virtually any entity can connect to the Internet to offer resources or to access them. Virtually any type of information can transverse the Internet without much in the way of regulatory interference. There is no central authority that regulates the Internet, although there are organizations that set certain fundamental standards and guide its operation. The Internet, by design, is autonomous and even anarchistic; in the end, this is both a strength and a weakness.

There exist a number of organizations that are involved in various Internet administrative and support activities. Those organizations include the following:

- *The Internet Society* (ISOC), a voluntary organization that acts to lend some formal structure to the administration of the Internet, is the organizational home of the IETF, IAB, IESG, and IRTF. The ISOC is active in such areas as censorship/freedom of expression, taxation, governance, and intellectual property. ISOC has granted the IESG formal authority to make decisions on standards.
- *The Internet Architecture Board* (IAB), originally known as the *Internet Activities Board*, is a voluntary board comprising 13 expert individuals who use the resources of their sponsoring companies to further the interests of the Internet. As a technical advisory group of ISOC, the IAB provides oversight for the architecture for the protocols and procedures used by the Internet. The IAB supervises the activities of two task forces: the IETF and the IRTF. In combination, those organizations set policy and direction.

- *The Internet Engineering Task Force* (IETF) identifies, prioritizes, and addresses short-term issues and problems, including protocols, architecture, and operations. Proposed standards are published on the Internet in the form of Requests for Comment (RFCs). Once the final draft of a standard is prepared, it is submitted to the *Internet Engineering Steering Group* (IESG) for approval.

- *The Internet Research Task Force* (IRTF) deals with long-term issues. The work of the IRTF is accomplished in small, focused research groups that work on topics related to Internet protocols, applications, architecture, and technology.

- *The Internet Corporation for Assigned Names and Numbers* (ICANN) is a not-for-profit organization formed in 1999 to assume the responsibilities from the federally funded *Internet Assigned Numbers Authority* (IANA) for assigning parameters for Internet protocols, managing the IP address space, assigning domain names, and managing root server functions. Internet protocol parameters managed by ICANN include the assignment of TCP *ports*, which are logical points of connection in the context of TCP, which is a part of the TCP/IP protocol suite developed for what is now known as the Internet. Port numbers provide a mechanism by which IP host computers can multiplex multiple types of concurrent connections at a single IP address. In combination, the IP address and the port number identify a socket, with a source socket and destination socket defining a TCP connection. Port numbers are 16-bit values that range from 0 to 65,536:

  - *Well-known ports* are numbered 0–1023 for the use of system (root) processes or by programs executed by *privileged users*. Examples of well-known ports include 25 for Simple Mail Transfer Protocol (SMTP), 80 for HyperText Transport Protocol (HTTP), and 107 for Remote Telnet Service. In the TCP/IP-based client/server environment, the server assigns the ports in consideration of the application-level protocol exercised at the client level.

  - *Registered ports*, which are registered by IANA as a convenience to the community, can be used by ordinary user processes or programs on most systems and can be executed by ordinary users. Falling in the range 1024–49151, these same port assignments are used with the User Datagram Protocol (UDP) to the extent possible.

- *Dynamic ports* and/or *private ports* are those from 49,152 through 65,535.

13.5 IP ADDRESSING

The vast majority of networks currently make use of *Internet Protocol version 4* (IPv4), although IPv6 is available and used in some large, recently deployed networks. The specifics of IPv4 and IPv6 are discussed in some detail later in this chapter. Now, I want to focus attention on the specifics of the addressing scheme, with emphasis on IPv4, which was defined in RFC 791.

IPv4 provides an address field size of 32 bits, which yields the potential for 2^{32}, or 4,294,967,296, distinct addresses. The addressing architecture defines five address formats, each of which begins with one, two, three, or four bits that identify the class of the network (Class A, B, C, D, or E). The network ID space identifies the specific

TABLE 13.1   IPv4 Address Classes

Address Class	Start Address	Finish Address	Binary First Digits
A	0.0.0.0	127.555.555.555	0
B	128.0.0.0	191.255.255.255	10
C	192.0.0.0	223.255.255.255	110
D	224.0.0.0	239.255.255.255	1110
E	240.0.0.0	255.255.255.255	1111

network, and the host ID space identifies the specific host computer on the network. Class addresses are defined below and listed in Table 13.1:

- *Class A* addresses are identified by a beginning 0 bit. The next 7 bits identify the specific network, with only a possible 128 (2^7). The remaining 24 bits identify the specific host computer on the network, with as many as 16,777,216 (2^{24}) possible machines. Class A addresses are intended for very large networks supporting a great number of host computers.
- *Class B* addresses are identified by a beginning set of 2 bits in a 10 sequence. The next 14 bits identify the network, with 16,384 (2^{14}) possible networks. The remaining 16 bits identify the specific host computer, with as many as 65,536 (2^{16}) possible machines.
- *Class C* addresses are identified by a beginning set of 3 bits in the binary sequence 110. The next 21 bits identify the network, with 2,097,152 possible networks. The remaining 8 bits identify the specific host computer on the network, with as many as 256 (2^8) possible machines. Most organizations hold Class C addresses.
- *Class D* addresses are identified by a beginning set of 4 bits in the binary sequence 1110. Class D addresses are intended for multicast purposes, with the remaining 28 bits specifying the multicast address.
- *Class E* addresses are identified by a beginning set of 4 bits in the binary sequence 1111. Class E addresses are reserved for experimental use [17].

Dotted decimal notation is the manner in which all IP addresses are written. Each 32-bit address field is divided into four fields, expressed as *xx.xx.xx.xx*, with each field given a decimal number value of 0–255, the range expressed by a single octet in binary form ($2^8 = 256$, or 0–255). Class A addresses begin with 1–127, Class B with 128–191, and Class C with 192–223. As Connect Northwest, my ISP, has an address of 206.40.133.20, it is easily identified as holding a Class C address.

As noted above, ICANN assigns IP addresses to organizations desiring to place computers on the Internet. The IP class, and the resulting number of available host addresses, depends on the size of the organization. The organization assigns the numbers and can reassign them on the basis of either static or dynamic addressing. *Static addressing* involves the permanent association of an IP address with a specific machine. *Dynamic addressing* assigns an available IP address to the machine each time a connection is established. An ISP, for example, may hold one or more Class C address blocks. Given the limited number of IP addresses available, the ISP assigns an IP address to a user machine each time the dial-up user accesses the ISP to seek connection to the Internet. Once the connection is terminated, that IP address becomes available to other users. This dynamic IP address assignment usually is accomplished through a router running the *Dynamic Host Configuration Protocol* (DHCP), as specified in RFC 1541.

There are both public and private IP address spaces. Computers on private Local Area Networks (LANs) running the TCP/IP protocol suite do not require public addresses, at least for internal use within the LAN domain. Private IP addresses are set aside and will never be used publicly. Also known as *Network 10* addresses, in reference to the first field in the first address range, private IP addresses fall into the following ranges:

- 10.0.0.0 to 10.255.255.255
- 172.16.0.0 to 172.31.255.255
- 192.168.0.0 to 192.168.255.255

By convention, routers are not supposed to forward any packets to the public Internet from IP addresses in those private ranges. When a LAN-attached computer requires access to the public Internet, it is necessary that the private IP address be translated into a public IP address. The process takes place in a router that interfaces to both domains and runs the *Network Address Translation* (NAT) protocol, as defined in RFC 3022, which obsoleted RFC 1631. NAT works beautifully in most situations but falls short where the protocol and application demand end-to-end connectivity. In IPsec (IP security), for example, the packet headers are digitally signed and must remain unmodified from source to destination.

As noted earlier, IPv4 provides an address field size of 32 bits, which yields the potential for 2^{32}, or 4,294,967,296, distinct addresses, if it were not for the reserved address spaces. That is not quite enough to address every man, woman, and child on the face of the earth, but it is close, give or take a billion or so. And it seemed adequate for many years. The popularity of TCP/IP, especially given the commercialization of the IP-based Internet, has placed a good deal of strain on the IPv4 numbering scheme, much as the popularity of networked fax machines, cell phones, pagers, computer modems, and even copy machines has strained the E.164 numbering scheme used in the PSTN. Adding to the problem is the fact that the organization of the address space into classes, some of which are reserved for future use, wastes a lot of addresses. Further, very substantial Class A and B address spaces were parceled out to large organizations that really did not need them. As only a small percentage of the addresses were actually used, a huge number of available addresses were wasted. This is not unlike the wasteful way in which telephone numbers were until fairly recently parceled out to U.S. LECs—in CO prefix blocks of 10,000 at a time, regardless of whether they actually needed 1, or 10, or 100, or 10,000.

To alleviate this problem, at least partially, the IETF documented *Classless Inter-Domain Routing* (CIDR) in 1993 in RFCs 1518 and 1519. CIDR builds on the concept of *supernetting*, which allows multiple Class C subnet address blocks to be grouped under a single address. An ISP needing 1000 IP addresses, for example, could get four consecutive Class C blocks, and all addresses in those blocks could be routed to the same host. CIDR simply uses shorthand to specify the *subnet mask*, which also is written in dotted decimal notation. CIDR reduces the number of routes and, therefore, the size and complexity of the routing tables that the Internet switches and routers must support. While CIDR adds flexibility to the IP addressing scheme and, taken together with DHCP and the use of private IP addresses, has gone a long way toward easing the pressure on the IPv4 addressing scheme, it does not solve the basic problem of the lack of IPv4 addresses into the future.

IPv6 resolves this issue through the expansion of the address field to 128 bits, thereby yielding 2^{128} potential addresses. This is a total address potential of 340,282, 366,920,938,463,463,374,607,431,768,211,456. According to Christian Huitema, that is enough to assign 32 addresses per square inch of dry land on the earth's surface, which should just about do the trick [18]. Given the proposals for assigning IP addresses to network cellular phones, Personal Digital Assistants (PDAs), coffee pots, refrigerators, heating and air conditioning systems, automobiles, and virtually every imaginable device, IPv6 clearly adds value. IPv6 also provides additional functionality, although CPE upgrades are required.

13.6 DOMAIN NAME SYSTEM

The Internet is divided into logical domains, which are identified as a 32-bit portion of the total address, under the terms of IPv4. Addresses in the *Domain Name System* (DNS), the administration of which is the responsibility of ICANN, follow a standard convention: user@organization.domain. The vast majority of the 147 million or so registered Top-Level Domains (TLDs) are commercial in nature. TLDs, which are identified as the domain address suffix, are of two types: *generic Top-Level Domains* (gTLDs) and country codes.

13.6.1 Unsponsored Domains

The original gTLDs are *unsponsored*, meaning that they operate under policies established by the global Internet community, directly through the ICANN process. The original gTLDs, their intended use, and the operators responsible for their administration are as follows:

  .arpa—address routing and parameter area, exclusively for Internet infrastructure purposes (IANA, under guidance of the IAB)

  .com—commercial organizations (VeriSign Global Registry Services)

  .edu—accredited degree-granting educational institutions (Educause)

  .gov—U.S. government agencies (U.S. General Services Administration)

  .int—organizations formed under international treaties between governments (IANA.int Domain Registry)

  .mil—U.S. military (U.S. Department of Defense Network Information Center)

  .net—network access providers originally, now unrestricted (VeriSign Global Registry Services)

  .org—noncommercial organizations originally, now unrestricted (Public Interest Registry)

For example, ray@contextcorporation.com is the e-mail address for Ray Horak (user) at The Context Corporation (organization providing the connection), a commercial enterprise. The domain name, for example, contextcorporation, is limited to 63 alphanumeric characters, which is up from the limit of 22 characters that was in place until the fall of 1999.

In November 2000, and after extensive discussion and debate on the subject, ICANN selected seven TLD proposals for inclusion in the gTLD structure. These additions are the first since .com, .net, and .org were included in the mid-1980s. The new unsponsored gTLDs, their intended use, and the operators responsible for their administration include the following:

> *.biz*—**biz**nesses (NeuLevel)
>
> *.info*—**info**rmational sites, unrestricted (Afilias)
>
> *.name*—individuals (Global Name Registry)
>
> *.pro*—certified **pro**fessionals (e.g., doctors, .med.pro; lawyers, .law.pro; and accountants, .cpa.pro) and **pro**fessional companies and associations (RegistryPro)

13.6.2   Sponsored TLDs

A *sponsored* TLD is a specialized TLD that has a sponsor representing the narrower community that is most affected by the TLD. A sponsor is an organization to which ICANN delegates some defined ongoing policy formulation authority regarding the manner in which a particular sponsored TLD is operated for the benefit of a defined group of stakeholders. Sponsored TLDs include the following:

> *.aero*—One of the original sponsored TLDs selected in 2000 and created officially in 2002, .aero is dedicated to **aero**nautical interests and is sponsored by Societe Internationale de Telecommunications Aeronautiques (SITA).
>
> *.cat*—In September 2005, ICANN and Fundació puntCat entered into a Sponsored TLD Registry Agreement under which that organization sponsors the .cat domain, which is intended to support the **Cat**alan linguistic and cultural community. This agreement is most unusual, as Catalan is spoken by less than 16 million people in the world and understood by less than 21 million. Catalan is the language of Catalonia in Spain, the city of Valencia, the Principality of Andorra, and other isolated cities, regions, and islands in Spain, France, Italy, and that general area of Europe.
>
> *.jobs*—In April 2005, .jobs was created through an alliance between Employ Media, the Society for Human Resource Management (SHRM), Verisign, and ICANN. The .jobs TLD is specific for companies that wish to recruit employees and for **job** seekers and companies seeking employees. It is not now permissible for third parties (i.e., corporate recruiters) to use .jobs.
>
> *.coop*—One of the original sponsored TLDs created in 2002, .coop is dedicated to the use of **coop**erative associations and is sponsored by Dot Cooperation.
>
> *.mobi*—In July 2005, ICANN approved .mobi for the use of consumers and providers of **mobi**le products and services. The TLD is sponsored by mTLD Top Level Domain and became available for registrations in May 2006. This TLD is highly controversial, as it partitions the Web along nontraditional lines of mobility and nonmobility.

.museum—One of the original sponsored TLDs created in 2002, .museums is dedicated to the **museum** community and is sponsored by the Museum Domain Management Association.

.travel—In May 2005, ICANN approved .travel as a sponsored domain for those whose primary area of activity is in the **travel** industry. The TLD is sponsored by Tralliance Corporation.

.xxx—In May 2006, ICANN's board of directors voted against a proposed agreement for an .xxx domain, which would have been a TLD for pornography. Some argued that .xxx would serve as a positive, if voluntary, means of segmenting the Internet. ICANN tentatively approved the new TLD before receiving an unprecedented level of correspondence in opposition [19].

13.6.3 Country Codes

Top-level-domains also identify the country, in the form of a neutral two-character *country code*, as established and maintained by the ISO 3166 Maintenance Agency. Country codes, the management of which is delegated to the government of each nation, are appended to the standard address and are necessary only if the target country domain differs from the country domain of origin. The following are examples of the 241 current *country code Top-Level Domains* (ccTLDs) and the countries they designate:

.am—Armenia

.au—Australia

.ca—Canada

.fm—Federated States of Micronesia

.jp—Japan

.nz—New Zealand

.sw—Sweden

.tv—Tuvalu

.us—United States

.za—South Africa

Note that ray@contextcorporation.com does not carry the suffix .us, as this country code generally is assumed. In the event that .us were appended, .us would be defined as the top-level domain and .com would be defined as the *second-level domain*.

The intent is that the country codes are used to designate entities physically located in that country. In the last few years, however, a number of those country codes have been leased to profit-making corporations, yielding substantial revenues to those countries with TLDs corresponding to well-known two-letter expressions. Table 13.2 lists 10 of the most interesting examples.

The story of the .tv website (www.tv) is perhaps the most interesting, so I will share that one with you. In April 2000, the country of Tuvalu licensed .tv to Dot TV Corp., a California-based company since acquired by VeriSign, in a deal that guarantees Tuvalu a payment of at least $4 million per year for at least 10 years. Tuvalu

TABLE 13.2 Commercialized Country Codes

Country Code	Country	Commercial Meaning
.am	Armenia	AM radio
.cc	Cocos Islands	Commerce
.fm	Federated States of Micronesia	FM radio
.la	Laos	Los Angeles (CA)
.md	Moldova	Medical
.nu	Niue	Norwegian for *now*
.tm	Turkmenistan	Trademark
.to	Tonga	Two
.tv	Tuvalu	Television
.ws	Western Samoa	Website

also received a significant minority position in Dot TV. Dot TV markets .tv as a master portal for TV-related content providers, commanding fees in the range of $25–$100,000 for website registrations in the .tv TLD. You might wonder why Tuvalu would be willing to sell its identity. Well, Tuvalu is a small, poor island country located in the western Pacific Ocean. Its closest neighbors are the Fiji Islands and Samoa, both of which are about 650 miles away. Tuvalu comprises nine low-lying coral atolls with a total land mass of about 10 square miles, and the highest point of land is approximately 16 ft above sea level. The main wild animals are rats, lizards, and turtles. The only source of water is rainwater, which is collected in catchment basins. Exports include copra (i.e., dried coconut meat) and postage stamps. In fact, the philatelic bureau issued a series of four stamps to commemorate the .tv deal. The estimated population of 11,992 (July 2001) enjoys little in the way of modern conveniences or infrastructure, as the county is listed by the United Nations as one of the least developed in the world. So, a TLD is of relatively little value to the citizens of Tuvalu but is of great value to others—at least that was the thought in April 2000. In fact, there is great concern among the citizens of Tuvalu that, if global warming continues at the current pace, the entire nation will be under water within 50 years or so [20].

13.6.4 Regional Country Codes

In March 2005, ICANN approved .eu as a regional country code for the European Union. The TLD is administered by EURid, a consortium of the ccTLD registry operators of Belgium, the Czech Republic, Sweden, and Italy. There currently are a number of other proposed regional ccTLDs under active consideration, including the regional TLDs of .asia and .africa.

13.6.5 Internet Registry

There are two aspects to the Internet registry responsibility. The *Internet Assigned Numbers Authority (IANA)* retains responsibility for the administration of IP addresses and root servers under a contract with the U.S. Department of Commerce. IANA delegated the administration of TLDs to ICANN and delegated regional and national registries to allocate IP addresses within their jurisdictions. During the past few years, there has been increasing international pressure for the United States to relinquish these responsibilities to an international body, such as the International Telecommunications Union (ITU). In June 2005, the U.S. government finally made it clear that no such thing would happen.

13.6.5.1 IP Number Assignment IANA is responsible for administering the management of the IP address space. End users get their IP addresses from an ISP, which gets IP address allocations from a *Local Internet Registry* (LIR), which gets allocations from a *National Internet Registry* (NIR) or *Regional Internet Registry* (RIR). There currently are five RIRs, as follows:

- African Network Information Center (AfriNIC)
- Asia Pacific Network Information Center (APNIC)
- American Registry for Internet Numbers (ARIN)
- Latin-American and Caribbean Network Information Center (LACNIC)
- Réseaux IP Européens Network Coordination Center (RIPE NCC)

In addition to general number administration, each RIR is responsible for maintaining one or more of the root servers, maintaining a *Whois* database for IP ownership lookups, deployment of a routing database, coordination of ENUM delegations, and network measurement and statistical reporting.

13.6.5.2 TLD Assignment ICANN assumed the responsibility of administering Internet addresses from IANA. Under IANA administration, the responsibility for TLD assignment was contracted to the *Network Information Center* (NIC) *Internet Registry*, commonly known as *InterNIC*. Now a nonprofit company operating under IANA, for a number of years InterNIC was a commercial enterprise of Network Solutions, now a VeriSign company. For the first few decades of the Internet, domain assignments were free for the asking. InterNIC then began to charge for .com domains, at the rate of $70 for the first two years and $35 for each one-year renewal. In 1999, InterNIC lost its monopoly over domain assignment, as four competing entities were approved in April 1999 for a testbed period to extend through June 25, 1999. Operational responsibility for each TLD now is assigned to an operator chosen through a more or less competitive bidding process, and there now are many hundreds of accredited registrars working under the supervision of the operators. Coordination of domain name assignment is accomplished through the *Shared Registry System* (SRS), so that duplicate names are not assigned. Currently, all domain names are maintained in mirrored databases on 13 *root* (i.e., centralized primary source) *Domain Name Servers* (DNSs) distributed around the world. Thousands of *Domain Name Resolvers* (DNRs) located strategically with ISPs and institutional networks periodically download database updates from the root servers. Through this network of resolvers, translations can be made from domain names and Uniform Resource Locators (URLs) into IP addresses, and vice versa. I discuss URLs in more detail later in this chapter.

13.6.5.3 Language At this point, let us consider the issue of language. While the Internet is an American invention, it clearly is no longer limited to the United

States. While English traditionally was the primary language supported over the Internet and has developed into the worldwide common business language, it clearly is not the only language in the world, and it is natural that non–English speakers would prefer to communicate in their own language. Therefore, there has been increasing pressure over the years to support domain name registration in other languages. Beginning in late 2000, ICANN began registering website addresses in four languages other than English. Currently, a number of accredited registrars support other languages, with Verisign supporting over 60, including Armenian, Bulgarian, Chinese, French, Georgian, German, Greek, Japanese, and Russian. As the ASCII scheme will not support complex alphabets, those domain names are supported by the full UNICODE character set, which then must converted into ASCII for transmission over the ASCII-based Internet. Otherwise, the vast majority of the Internet cannot access many of those websites [21, 22].

13.6.6 Address Translation: Domain Name to IP Address and Vice Versa

The Internet works on the basis of IP addresses, as already discussed in this chapter. You certainly can access another host computer on the basis of an IP address, but IP addresses are hard to remember and hard to type correctly. Domain names are much easier to remember and enter, but a translation must take place to convert domain names to IP addresses; the various routers and switches depend on this in order to route the data to the correct destination device. This process of translation takes place through the *Address Resolution Protocol* (ARP) and through an address resolution database that resides on a server that is accessed by the originating router. As I mentioned earlier in the chapter, all domain names are maintained in mirrored databases on 13 *root* (i.e., centralized primary source) *Domain Name Servers* (DNSs) distributed around the world. Thousands of *Domain Name Resolvers* (DNRs), located strategically with ISPs and institutional networks, periodically download database updates from the root servers. Through this network of resolvers, translations can be made from domain names and *Uniform Resource Locators* (URLs) into IP addresses, and vice versa.

Suppose, for example, that you want to send e-mail to ray@contextcorporation. com thanking me for writing this book and sharing with me the fact that it has changed your life for the better. (This also is an acronym test. The answers are provided at the end of this little tale.) You might access your ISP from your PC at your SOHO through an ADSL line over a UTP local loop, which the ISP has leased for resale from the ILEC. As the DSL line is *always on*, you are assigned an IPv4 address on a static basis. Your client PC runs against the ISP's server, both of which run TCP/IP. Your e-mail message makes use of SMTP, an extension of TCP, and is placed on the UTP local loop in serial mode using PPP. As the message reaches the CO, it is uplinked to the ISP through a DSLAM over an unchannelized T-carrier circuit leased from the ILEC. At the ISP location, an e-mail server receives the e-mail message. It then consults a DNS database on another server, translating ray@ contextcorporation.com into ray@cnw.net, which is the TLD of the e-mail server of Connect Northwest, the ISP that hosts my TLD as a virtual domain on a logical partition of their domain server. That TLD then is translated into the IPv4 address 206.40.133.20, which is the IP address of the Northwest Link server. The DNS at your ISP is updated periodically through downloads from one of the InterNIC root servers. At your ISP, your e-mail message also is converted into a Frame Relay format and forwarded to a NAP over an unchannelized T-carrier circuit, perhaps leased from a CLEC, at which point it enters the Internet backbone. In the core of the Internet backbone, the e-mail message travels in AAL5 format. At the destination edge of the Internet backbone, the process is reversed to Connect Northwest, which translates the IPv4 address into my domain and deposits the incoming e-mail into my mailbox. When I check my e-mail over a dial-up connection and through a 56-kbps modem, which I use as a backup in the event that my ADSL link fails, I also use PPP and a TCP/IP client/server connection-oriented communications mode. As I connect to my ISP, I am dynamically assigned an IPv4 address through DHCP once I enter my password. I then access the e-mail server, and the message is downloaded to my PC at my SOHO. You just put a big ☺. Thank you!

Now for the acronym decoder:

☺	Smile on my face
AAL	ATM Adaptation Layer
ADSL	Asymmetric Digital Subscriber Line
ATM	Asynchronous Transfer Mode
CLEC	Competitive Local Exchange Carrier
CO	Central Office
DHCP	Dynamic Host Configuration Protocol
DNS	Domain Name System
DSL	Digital Subscriber Line
DSLAM	Digital Subscriber Line Access Multiplexer
ILEC	Incumbent Local Exchange Carrier
InterNIC	Network Information Center Internet Registry
IP	Internet Protocol
IPv4	Internet Protocol version 4
ISP	Internet Service Provider
NAP	Network Access Point
PC	Personal Computer
PPP	Point-to-Point Protocol
SMTP	Simple Mail Transfer Protocol
SOHO	Small Office/Home Office
TCP/IP	Transmission Control Protocol/Internet Protocol
TLD	Top Level Domain
UTP	Unshielded Twisted Pair

13.7 INTERNET PROTOCOLS

Internet protocols include IP, TCP, and UDP as well as application-level protocols. TCP/IP is fundamental to the operation of the Internet, while the application-level protocols serve to support specific user applications.

13.7.1 TCP/IP

Transmission Control Protocol (TCP) and *Internet Protocol* (IP) are specific, layered protocols that operate within a protocol stack typically referred to as the TCP/IP protocol suite. TCP/IP is a public domain protocol, as it was developed with public funds for use in a public network, which we now know as the Internet. TCP operates at Layer 4 (Transport) and IP operates at Layer 3 (Network) of the OSI Reference Model. The Internet suite also includes Layer 5 (Session), Layer 6, (Presentation), and Layer 7 (Application) protocols. TCP/IP has been enhanced continuously and used extensively in a variety of computer networks, including X.25 and Ethernet. Additionally, some vendors have layered ISO FTAM, X.400, and X.500 applications protocols (Layer 7) above TCP/IP. The advantages of TCP/IP certainly include its high level of documentation, ease of use, stability, and broad applicability. TCP/IP provides a means of passing datagrams among virtually any networks capable of sending and receiving bits. As such, it is a highly effective *common denominator* protocol.

13.7.1.1 Internet Protocol: Connectionless Datagram Delivery   IP, the basic building block of the Internet, is a Layer 3 (Network) internetworking protocol for the routing of datagrams through gateways connecting networks and subnetworks. Defined in RFC 791, IP is a connectionless protocol, as no true connection is established between the source and destination devices. Rather, the IP packets are presented to the network by the originating device and handled through the network with no advance knowledge of either the existence or the availability of the destination device. IP can be characterized as *datagram oriented* because each IP packet works its way through the network independently, with no thought of an individual packet belonging to a larger stream of packets. IP also can be characterized as a *best effort* protocol, as it offers no guarantees of delivery, no sequencing, and no error detection and correction mechanism.

IP provides for packet segmentation and reassembly and provides specific addressing conventions in the form of dotted decimal notation, as previously described. IP supports routing control as well as status translation and communications. While IP has no concept of the specific content of the packet or of its service requirements, it also supports multiple service types, including low-delay, high-bandwidth, and high-reliability paths. Dial-up IP access protocols include SLIP and PPP.

The total size of the IP datagram, including the IP header, can be up to 65,535 octets in length. At a minimum, all networks must support a packet of at least 576 octets. As illustrated in Figure 13.3, the minimum size of the IP header is 20 octets. Note that the IP datagram is viewed in terms of a 32-bit width, as the original processors that implemented the IPv4 protocol had 32-bit word (i.e., value) lengths [17].

The IPv4 datagram contains the following field:

- **VERS:** Four bits identifying the IP version number. The version number is 4.
- **IHL:** Four bits of Internet Header Length (IHL). The minimum value is five 32-bit words, or 20 octets. The IHL also provides a measurement of where the TCP header, or other higher layer, header begins.

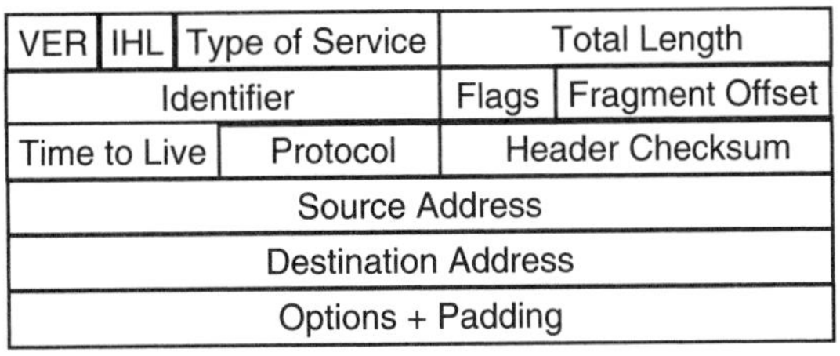

VER	IHL	Type of Service	Total Length	
Identifier			Flags	Fragment Offset
Time to Live		Protocol	Header Checksum	
Source Address				
Destination Address				
Options + Padding				

Figure 13.3   IPv4 datagram format.

- **Service Type:** Eight bits indicating the quality of service requested for the datagram. While TCP/IP networks do not provide guaranteed Quality of Service (QoS) currently, the networks will attempt to honor QoS requests in terms of parameters that include packet precedence (i.e., priority), low delay, high throughput, and high reliability.

- **Total Length:** Sixteen bits describing the total length of the datagram, including the IP header. The maximum size is 65,535 octets ($2^{16} - 1$, with 0 not considered as it has no value), and all network hosts must be able to handle a datagram of at least 576 octets.

- **Identification:** Sixteen bits that are used in fragmentation control. In the event that the receiving network cannot accommodate a datagram of the specified *total length*, that datagram must be fragmented. Each fragment must contain a copy of the *identification* field and certain other fields in the IP header so they can be reassociated and the datagram can be reconstituted.

- **Flags:** Three bits that define the manner in which the fragmentation occurs. The first bit always is set at 0. The second bit defines whether fragmentation is permitted. Fragmentation, for example, may not be permitted in certain applications where only the entire datagram is useful. The third bit is used to identify the last fragment in a series of fragments.

- **Fragment Offset:** Thirteen bits that identify where the fragment fits in the complete set of fragments that comprise the original datagram. This field is used to sequence the fragments correctly, as they may arrive at the destination device out of sequence.

- **Time To Live (TTL):** Eight bits that specify the length of time in seconds that the datagram can live in the Internet system. The maximum length of time is 255 seconds ($2^8 - 1$, with 0 not considered, as it is the official time of death), or 4.25 min. From the instant the IP datagram enters the Internet, each gateway and host that act on the datagram decrement the TTL by at least 1 s, although the time it has possession of the datagram generally is much less. When the TTL reaches 0, the datagram is declared dead and is discarded. The TTL mechanism prevents packets from wandering the Internet for eternity, at which point they would have no value and would only contribute to overall network congestion. Over time, the TTL field has been redefined to indicate, as an option, the number of hops (i.e., routers) through which the packet travels. In effect, the TTL is a hop count, anyway. The default TTL is 64.

- **Protocol:** Eight bits identifying the higher layer protocol that created the message contained in the data field. Examples include TCP and UDP.

- **Header Checksum:** Sixteen bits used for error control in the header. The process is that of *Cyclic Redundancy Check* (CRC).
- **Source IP Address:** Thirty-two bits containing the IP address of the source host.
- **Destination IP Address:** Thirty-two bits containing the IP address of the destination host.
- **IP Options, If Any:** An optional, variable-length field used by gateways to control fragmentation and routing options.
- **Padding:** A variable-length field used only when necessary to ensure that the IP header extends to an exact multiple of 32 bits.
- **Data:** A variable-length field that contains the actual data content [16, 17].

13.7.1.1.1 Serial Line Internet Protocol SLIP is the original and most basic protocol for handling IP packets in a serial bit stream across a voice-grade telephone connection. Installed on both the user's workstation and the provider's server, SLIP forwards packets created by TCP/IP software. SLIP, with origins in the 3COM UNET TCP/IP implementation from the early 1980s, is merely a Layer 2 (Link Layer) packet framing protocol that defines a sequence of characters that frame IP packets on a serial line. SLIP provides no addressing, packet-type identification, error detection/correction, or compression mechanisms. Because the protocol does so little, however, it is usually very easy to implement. SLIP is defined in RFC 1055 as a nonstandard protocol, in the formal sense, although it has become a de facto standard. RFC 1144 defines Compressed SLIP (CSLIP), a method for compressing TCP/IP performance over low-speed (300 bps–19.2 kbps) serial lines by compressing the TCP/IP headers. RFC 1144 does not deal with compressing UDP/IP headers, as they were considered at the time (February 1990) to be too infrequent to be worth the bother.

13.7.1.1.2 Point-to-Point Protocol PPP performs the same basic functions as SLIP. Additionally, it performs fairly sophisticated compression in order to eliminate unused or redundant data in the headers of long sequences of packets in a transmission stream. Further, PPP supports multiple native machine and network protocols as well as subnet routing. PPP installed on a telecommuter's home PC, for example, enables communication with the home office through a router connecting to an Ethernet LAN. PPP also supports IP packet communication through the Internet [23]. RFC 1661 defines PPP. Numerous other RFCs define various PPP implementations.

13.7.1.2 Internet Protocol version 6 (IPv6): Better Yet The IPv6 specification (RFC 1883, replaced by RFC 2460) grew out of the efforts of the IETF IPng (IP next generation) Working Group to define a successor protocol to IPv4. As you discovered in our examination of IPv4, that protocol is highly limited in the context of contemporary packet networking. Limitations of IPv4 include the facts that its addressing scheme is too limited at 32 bits, its address assignment is not flexible enough, application-level protocols are not tightly integrated, QoS is not supported, and security is lacking. IPv6 addresses all of these shortcomings and more through a header of 40 octets (compared to the 20 octets of IPv4) that can be extended as

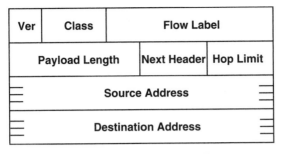

Figure 13.4   IPv6 base header format.

necessary through optional headers. Figure 13.4 illustrates the IPv6 header structure.

The total size of the IPv6 datagram, including the IP header, is changed from that of IPv4. Specifically, datagrams greater than 65,535 octets are allowed, with these *jumbo payloads* identified in the *payload length* field. At a minimum, all network links must support a *Maximum Transmission Unit* (MTU) of at least 1280 octets. The IP header and any extension headers associated with the datagram are in addition to the payload length. This is a departure from IPv4, in which the size of the datagram includes any and all headers. The IPv6 datagram contains the following fields:

- **Version:** Four bits identifying the IP version number. The version number is 6.
- **Class:** Eight bits used by originating nodes and/or forwarding routers to identify and distinguish between different packet classes or priorities as set by upper layer protocols. Originally known as the *priority* field, this field replaces the type-of-service field in IPv4.
- **Flow Label:** Twenty bits used by a source host indicating any special handling requested for a *flow*, or sequence, of datagrams. Each datagram in the flow between an originating host and one or more destination hosts must carry the same flow label. Real-time voice, audio, or video communications are examples of applications involving flows that require non–default handling. The flow label will allow routers to identify and process Voice-over-IP (VoIP) packets more easily.
- **Payload Length:** Sixteen bits describing the total length of the datagram. This field is much like the total-length field in IPv4, although the payload length field does not include the IP header. As mentioned above, IPv6 supports *jumbo payloads* larger than the traditional 65,535 octets. At a minimum, all network links must support a *Maximum Transmission Unit* (MTU) of at least 1280 octets. The recommendation, however, is that network links be configured to support an MTU of 1500 octets or greater in order to support encapsulation of Ethernet payloads without incurring fragmentation. The IPv6 payload length includes any IPv6 header extensions, TCP or UDP headers, and any layer 7 headers that might be associated with the datagram.
- **Next Header:** Eight bits identifying the header immediately following the IPv6 header. Examples (values) include TCP (6), UDP (17), fragment (44), and

authentication (51). IPv6 packets may include one or more extension headers, which I discuss below. This header is similar to the protocol header in IPv4.

- **Hop Limit:** Eight bits that specify the number of hops (i.e., routers) through which the packet can travel. Each router along the path decrements the field value by 1 until the value reaches zero, at which point the packet is discarded. This field is similar to the IPv4 TTL (Time To Live) field, with the exception that the *seconds* parameter has been eliminated and only the *hops* parameter is supported.
- **Source Address:** One hundred twenty-eight bits (hexadecimal) containing the IP address of the source host. (Figure 13.4 depicts this as four 32-bit rows.)
- **Destination Address:** One hundred twenty-eight bits (hexadecimal) containing the IP address of the destination host. (Figure 13.4 depicts this as four 32-bit rows.)

IPv6 supports multiple extension headers. RFC 1883 recommends that they be placed in the following order:

- **Hop-by-Hop Options:** This header carries optional information that must be examined by every node along a packet's path. This header carries information such as the type of extension header immediately following and specific instructions as to what should be done if the processing node does not recognize the option type and whether or not the option data may change en route. This header also identifies the length of the hop-by-hop header and contains padding options.
- **Destinations Options:** This header carries optional information that must be examined by the destination host. This header carries information such as the type of header immediately following, the length of the destination header, and padding options. Optional destination information may be contained in this header or in a separate extension header such as the fragment header or the authentication header.
- **Routing:** This IPv6 source node uses this header to list one or more intermediate modes to be *visited* along the path to the packet's destination.
- **Fragment:** The IPv6 source node uses this header to send a packet larger than the path MTU will accommodate. All fragmentation occurs at the source node, which is a departure from IPv4.
- **Authentication:** As defined in RFC 2402, this header provides a mechanism for ensuring the connectionless integrity and data origin authentication of IP datagrams as well as an anti-replay option. It might also include a nonrepudiation mechanism, which provides the origination node with confirmation of packet receipt. Included in this header may be the authentication algorithm and keys, the encryption algorithm and keys, and other security-related parameters.
- **Encapsulating Security Payload:** As defined in RFC 2406, this header provides confidentiality through encryption and limited traffic flow confidentiality. It also may provide connectionless integrity and data origin authentication of IP datagrams as well as an anti-replay option. *IPsec* (IP security) is a standards-based security suite that operates transparently and may eliminate the need for proprietary firewall mechanisms in some applications. IPsec also provides for

encapsulation of the secured IPv6 packets inside IPv4 datagrams, in consideration of both the increasing need for security and the long-term transition process to IPv6.

The IPv6 address fields merit special discussion at this point. First, the address field of 128 bits yields 340,282,366,920,938,463,463,374,607,431,768,211,456 potential addresses compared to the IPv4 address field of 32 bits, which yields a comparatively paltry potential of only 4,294,967,296 distinct addresses. Despite the best efforts of CIDR, DHCP, and other mechanisms, IPv4 addresses clearly will be exhausted at some point in the foreseeable future, especially given the increased number of IP-addressable devices we have seen over the last few years and expect to see into the future. Other address-related enhancements associated with IPv6 include the following:

- **Address Assignment:** IPv6 offers much improved flexibility of address assignment through two approaches, both of which offer automatic address assignment and discovery. *Stateful* autoconfiguration resembles Dynamic Host Configuration Protocol (DHCP), as the configuration servers dynamically assign unique addresses to devices as they require them, drawing from a pool of such addresses. *Stateless* autoconfiguration employs two IP addresses and is particularly advantageous in mobile applications. One address is assigned permanently to the mobile device, and another address is used to route data to the network to which the mobile device is connected at the time. This stateless approach is much like sending a datagram to a device *in care of* a network and is useful in the context of mobile devices that move among pager, cellular, packet radio, wireless LAN, and other wireless networks.
- **Address Types:** IPv6 supports multiple address types, including unicast, multicast, and anycast (which is a new mode):
  - *Unicast* supports communications between source–destination pairs.
  - *Multicast* involves the communication of data to multiple hosts, each with its own IP address. Rather than the traditional approach of copying a packet stream at the originating router and then transmitting that stream to each device on a sequential, unicast basis, multicast involves a single transmission into the network. Based on its knowledge of the general physical direction in which the individual devices lie, the network fans out the packet stream to its peers, and the process continues until such time as all destination devices are located and the data are presented to them.
  - *Anycast* is a new scheme that supports the communication between a source host and the closest member of a group of destination devices, with the group sharing a single anycast IP address. In this mode, the network routes the packet stream to the *nearest* device in the group sharing the address, based on the routing protocol's measure of distance. That device then assumes responsibility for forwarding the data to the group.

To ease the use of IPv6 addresses, the format is quite different from that of IPv4. In IPv4, as you will recall, the address format calls for four 8-bit binary fields separated by dots and expressed as *xxx.xxx.xxx.xxx*. IPv6 replaces those decimal-separated binary fields with 16-bit hexadecimal fields separated by colons, such as 4ffd:521:195b:2:2e0:91bb:aed8:14f3. IPv6 addresses are more structured than IPv4 addresses. There are various addressing schemes defined and identified by the high-order bits of the block. The most popular scheme splits addresses in half, with 64 bits for the *Regional Internet Registry* (RIR) that parcels out the IP addresses and otherwise manages the address space on behalf of IANA and the other 64 bits for the endpoint device. The high-order bits are composed of 32 bits identifying the RIR (e.g., American Registry for Internet Numbers, or ARIN), 16 bits for the local Internet registry or ISP, and 16 bits for the site to which the address belongs. Similar to a Class B address block, each site supports up to 65,636 devices [24].

In total, IPv6 offers significant advantages over its predecessor IPv4. Through *tunneling* (i.e., the encapsulation of IPv6 packets in IPv4 packets), IPv6 is backward compatible. The real advantages of IPv6 can be realized, however, only if all nodes from source to destination are capable of running IPv6 in native mode. Therein lies the problem. For the most part, IPv6 requires a forklift upgrade. Therefore, IPv6 is finding its way fairly slowly into the domain of internetworking. While the emerging IP-based networks, both private (e.g., Internet2) and public, are building from the ground up, they easily can implement IPv6. They also must run dual protocol stacks to support both versions in native mode. Similarly, new backbone router implementations in large enterprise networks can support both versions easily. Gradually, IPv6 will supplant IPv4, as it works its way from the backbone to the desktop, but the full process may well take decades [17, 18, 24, 25]. IPv6 deployment has been slow to date for reasons including the cost and complexity of conversion from IPv4. Also, the pressure on the IPv4 address space has relaxed due to the success of DHCP and CIDR. The widespread use of private IP addresses within the enterprise and the use of NAT for address translations associated with traffic to the wider Internet also have made IPv6 less of an imperative. On the other hand, NAT can create bottlenecks and, in combination with firewalls, can even deny VoIP traffic. IPv6 can eliminate the need for NAT, at least as far as the addressing scheme is concerned, although network administrators likely will still use NAT for security purposes, but perhaps on a more selective basis.

Many analysts, including myself, see wireless communications as the main driver behind IPv6. The proliferation of IP-enabled wireless cell phones, PDAs, and other hand helds undoubtedly will continue for many years to come and will require IP addresses beyond the limits of IPv4. Additionally, and as noted above, IPv6 offers much improved security and mobility as compared to IPv4. Further, the emergence of 2.5 G and 3 G wireless networks present a ground-up opportunity for IPv6 deployment.

IPv6 opportunities abound in Asia, South America, and Africa, which were fairly late in joining the Internet community and therefore do not have the same heavy investments in IPv4 as do North America and Western Europe. The governments of Japan, Korea, and China all have announced plans for large-scale IPv6 deployment. Of particular emphasis is the need for IP addresses in the People's Republic of China (PRC), where cellular and other wireless technologies are growing at unprecedented rates, encouraged by the strong Chinese economy. When you consider the fact that the population of the PRC is approximately 1.3 billion and that the usage of cell phones and computers is growing, it is not difficult to do a little math and discover that the PRC alone can quickly exhaust the remaining pool of IPv4 addresses. In April 2001, NTT (Japan) was the first carrier to launch a native IPv6 backbone for commercial application [26]. In the United States, the Department of Defense (DoD) mandated in 2003 that all military networks must upgrade to IPv6 by fiscal year 2008. In 2005, the Office of Management and Budget (OMB) mandated that all federal agencies upgrade their backbones to IPv6 by 2008. The high-speed Abilene project, which links approximately 200 universities in the United States, has enabled IPv6 on its network connections and a number of participants use it actively.

In the private sector, Bechtel, the global contracting company, is moving to IPv6 in order to maintain compatibility with the federal government as well as its customers in Asian countries that are deploying the protocol aggressively [27]. Cray Computer announced in April 2005 that it was to be the first to make commercial use of MCI's IPv6 services, citing the requirement for interoperability testing and development of next-generation software and applications for its X1 Supercomputer product line [28]. Despite the obvious advantages of IPv6, it has been very slow in gaining acceptance.

13.7.1.3 *Transmission Control Protocol (TCP)*   Transmission Control Protocol (TCP) is a Layer 4 (Transport) protocol defined in RFC 793. TCP evolved from ARPANET's Network Control Protocol (NCP), which was developed to provide reliable transmission across the essentially unreliable media of analog UTP and packet radio (e.g., AlohaNET). In support of higher layer applications, TCP can be characterized as making use of virtual circuits in support of byte-stream-oriented communications. As a connection-oriented protocol, TCP supports status exchange and synchronization over virtual circuits. TCP provides for file segmentation into packets on the transmit side and for reassembly on the receiving end. TCP also provides for packet sequencing, end-to-end flow control, and error control [29, 30], thereby guaranteeing delivery. Each packet in a stream of packets received by the destination device is either acknowledged as having been received correctly or requested to be retransmitted in the event of corruption. Packets that are not acknowledged or requested for retransmission are considered unacknowledged and are retransmitted by the source host.

The TCP unit of data transfer between two host computers is known as a *segment*. Segments are used to establish connections, transfer actual data, acknowledge packet receipt and request retransmissions, and terminate connections. Figure 13.5 provides a view of the TCP header and its component fields. The standard size of the TCP header is 20 octets, although 4 additional octets may be used to accommodate options.

The TCP header fields are defined as follows:

- **Source Port:** Sixteen bits that define the TCP port number used by the source application program. As discussed earlier in this chapter, TCP *ports* are logical points of connection. *Well-known ports* are numbered 0–1023 for the use of system (root) processes or by programs executed by *privileged users*. Examples of well-known ports include 25 for SMTP (Simple Mail Transfer Protocol), 80 for HTTP (HyperText Transport Protocol), 107 for Remote TELNET Service, and 110 for POP3 (Post Office Protocol version 3).

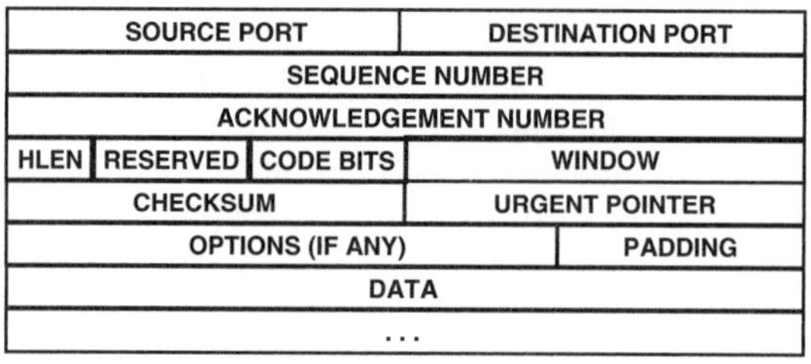

SOURCE PORT		DESTINATION PORT	
SEQUENCE NUMBER			
ACKNOWLEDGEMENT NUMBER			
HLEN	RESERVED	CODE BITS	WINDOW
CHECKSUM		URGENT POINTER	
OPTIONS (IF ANY)		PADDING	
DATA			
. . .			

Figure 13.5   TCP segment format.

- **Destination Port:** Sixteen bits that define the TCP port number used by the destination application program.
- **Sequence Number:** Thirty-two bits that identify the position of the data in the TCP segment relative to the entire originating byte stream. A fundamental notion of TCP is that each octet in a byte stream is numbered. The sequence number refers to the number of the first octet in a given segment. This field is critical if data are to be sequenced properly at the destination host. It also is critical in order that positive acknowledgments can be sent to the originating host and retransmissions of segments can be requested.
- **Acknowledgment Number:** Thirty-two bits that identify the acknowledgment number of the octet that the source expects to receive next. The acknowledgment number explicitly acknowledges that all previous data octets associated with all previous segments were received correctly.
- **HLEN:** Four bits that specify the segment header length, in 32-bit multiples.
- **Reserved:** Six bits reserved for future use.
- **Code Bits:** Six bits that define the purpose and contents of the segment. Examples include acknowledgment, connection reset, and end of byte stream.
- **Window:** Sixteen bits that advertise the size of the sender's sliding receive window (i.e., how much data the host computer is willing to accept, based on buffer size). A window of zero indicates that the receiver is overwhelmed and can not accept any further data until further notice. Large windows indicate that as many as 65,536 unacknowledged bytes can be in transit at a given time. Congestion, which is indicated by expiration of the retransmission timer without an acknowledgment, reduces the window size by half, thereby slowing the transmission rate.
- **Checksum:** Sixteen bits used for error control in the data field as well as the header. The process is that of *cyclic redundancy check*.
- **Urgent Pointer:** Sixteen bits that identify urgent out-of-band data (i.e., data not part of the information stream). Such data are treated on a high-priority basis, in advance of data stream octets that might be awaiting consumption by the destination hosts. Urgent data, for example, might include a keyboard sequence to interrupt or abort a program.
- **Options, If Any:** Twenty-four bits that address a variety of options, such as *Maximum Segment Size* (MSS).
- **Padding:** Eight bits in an optional field used only when necessary to ensure that the TCP header extends to an exact multiple of 32 bits. This field is used only when the *options, if any* field is used.

UDP SOURCE PORT	UDP DESTINATION PORT
UDP MESSAGE LENGTH	UDP CHECKSUM
DATA	
...	

Figure 13.6   UDP format.

- **Data:** A variable-length field that contains the actual data content. As TCP is used in conjunction with IP, the default size of the data field is 536 octets, which is the default size of the IP datagram, less 20 octets each for the standard IP and TCP headers [6].

13.7.1.4 User Datagram Protocol   As defined in RFC 768, *User Datagram Protocol* (UDP) is a Layer 4 (transport) host-to-host protocol that is much simplified in comparison to TCP. Historically used to send datagrams between application programs, UDP offers the same unreliable, connectionless datagram delivery as IP. Like TCP, UDP uses IP for addressing and routing purposes. Unlike TCP, UDP provides no sequencing, error control, or flow control mechanisms. An application program that uses UDP assumes full responsibility for all issues of reliability, including data loss, data integrity, packet latency, data sequencing, and loss of connectivity. UDP is used extensively in VoIP and stream-oriented multimedia applications, where compression techniques are designed to mitigate such issues over a highly shared packet network. UDP also works well where transactions are of such short duration that connection setup overhead comprises a large proportion of overall transaction traffic, with DNS and SNMP exchanges being good examples.

The standard size of the UDP header is eight octets; it comprises the following fields, as illustrated in Figure 13.6:

- **Source Port:** Sixteen bits that define the UDP port number used by the source application program.
- **Destination Port:** Sixteen bits that define the UDP port number used by the destination application program.
- **UDP Message Length:** Sixteen bits that identify the length of the message in the data field.
- **Header Checksum:** Sixteen bits used for error control in the header only. This checksum need not be used; if the value is set to zero, it is disregarded. This lack of regard for header control is possible because it also is accomplished in the IP header. The checksum process is that of CRC.
- **Data:** A variable-length field that contains the actual data content [6].

13.7.1.5 Transmission Framing   Now that I have discussed the header formats and functional characteristics of IP, TCP, and UDP, it is time to view the transmission-framing format used in an Internet context, as illustrated in Figure 13.7. This format generally follows the generic data format in Chapter 6. The IP header comes first, as it is required for routing purposes. Next is the UDP or TCP header. Finally comes the actual application data. The entire transmission frame is considered to be either a UDP datagram or a TCP segment. If you consider this in the context of an Ethernet 802.3 LAN, the datagram or segment becomes the payload of the

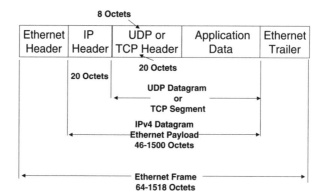

Figure 13.7 IPv4 transmission frame composition for Ethernet 802.3 LAN.

Ethernet frame, which payload cannot exceed 1500 octets. The encapsulating Ethernet header and trailer add another 18 octets, and the framing process is complete.

Note that the entire IPv4 datagram has a maximum length of 65,535 octets and must be fragmented to fit into an Ethernet payload if it exceeds 1500 octets. The IP header consumes 20 octets. If TCP is used, the TCP header is 20 octets. If UDP is used, the UDP header is 8 octets. The balance is available for application layer data subject to any limitations of the local network.

13.7.2 Application-Level Protocols

Application-layer protocols (Layer 7 of the OSI Reference Model) in the IP protocol suite function above TCP/IP in support of specific Internet applications. Examples include the following.

13.7.2.1 *Telecommunications Network* TELecommunications NETwork (TELNET) is perhaps the oldest Internet application protocol. Defined in RFC 854 (May 1983), TELNET provides terminal emulation over a TCP connection, enabling the user to assume control over the applications that reside on a remote system. Virtual network terminal services permit the DTE to emulate other terminal devices, transparently, in a client/server environment.

13.7.2.2 *File Transfer Protocol* File Transfer Protocol (FTP) supports the exchange of files between two hosts across the Internet. Defined in RFC 959, FTP also supports interactive user interface in which humans must interact with a remote host. The specifics of the file type and format [e.g., ASCII, EBCDIC, or binary and compressed or uncompressed] of data can be determined from client to server. FTP also requires clients to satisfy security authorization in the form of login and password. FTP makes use of TELNET for control messages between the hosts and relies on connection-oriented TCP for data transfer.

13.7.2.3 *Simple Mail Transfer Protocol* SMTP operates over UDP, providing the underlying capabilities for networked electronic mail. While SMTP (RFC 821)

does not provide the user interface, it supports text-oriented e-mail between any two devices that support *Message-Handling Service* (MHS). *Multipurpose Internet Mail Extension* (MIME) is a SMTP extension that permits the attachment to textual e-mail of other types of files, including audio, graphics, and video. Thereby, compound mail can be transmitted across the Internet. SMTP is simpler than its predecessor, *Mail Transfer Protocol* (MTP).

13.7.2.4 *Simple Network Management Protocol*

SNMP supports the exchange of network management information between hosts, typically including one or more centralized network management consoles that manage larger numbers of network elements in real time. Defined in RFC 1157, SNMP operates over UDP, thereby avoiding the overhead associated with TCP. There are three versions: SNMPv1, v2, and v3. (*Note:* A host is a computer that runs an application program. Hosts include hand-held computers, such as PDAs, as well as more substantial machines, such as desktops, servers, and mainframes.)

13.7.3 The Quest for IP QoS

All of the protocols discussed so far in the context of the IP protocol suite have been data oriented, but that is by no means the end of the story. The Internet and other IP-based networks increasingly are being used to support real-time voice, audio, and video applications, all of which are extremely demanding in terms of latency, jitter, and packet loss. The Internet and its original underlying protocols were never intended to support Quality of Service (QoS), which is exactly what each of these traffic types requires. As is discussed earlier in the chapter, the IPv4 header contains a Service-Type, or Type-of-Service (ToS), field that is used to request QoS in terms of parameters such as packet precedence (i.e., priority), low delay, high throughput, and high reliability. (Refer back to Figure 13.3.) But that is about as far as it goes, and there certainly are no guarantees. IPv6 goes a bit further (pun intended) with its class and flow label fields. Originating nodes and/or forwarding routers use the IPv6 class field, which replaces the IPv4 ToS field, to identify and distinguish between different packet classes or priorities, as set by upper layer protocols. The flow label field is used by a source host indicating any special handling requested for a *flow*, or sequence, of datagrams, with each datagram in the flow between an originating host and one or more destination hosts carrying the same flow label. Real-time voice, audio, or video communications are examples of applications involving flows that require nondefault handling. The IETF also has developed a number of protocols that are required to support real-time and multimedia applications more effectively. While they do not provide ATM-like QoS guarantees, they can come reasonably close under the right circumstances. The next few sections examine these protocols, which include DiffServ, MPLS, RTP, RTCP, RSVP, and RTSP.

13.7.3.1 *Real-Time Transport Protocol*

In its RFC 1889, the IETF defined the *Real-time Transport Protocol* (RTP) as a mechanism for providing end-to-end network transport functions suitable for applications transmitting real-time data, such as audio, video, or simulation data, over multicast or unicast network services. RTP provides end-to-end delivery services including payload-type identification, sequence numbering, and timestamping. In combination, the sequence numbering and timestamping provide the receiving node with sufficient information to resequence them as necessary. RTP does not address resource reservation. Neither does it guarantee QoS for real-time services, but relies on lower layer protocols to do so. RTP does not either guarantee delivery through the network or prevent out-of-order delivery, and it does not assume that the underlying network is reliable and delivers datagrams in sequence to the receiving machine. RTP does, however, prevent out-of-order delivery to the application. While RTP is designed to be independent of the underlying transport and network layers, applications (e.g., VoIP) generally run RTP on top of UDP, which provides multiplexing and checksum services. The default RTP header comprises 12 octets, although it can be extended to identify contributing payload sources and to provide for individual implementations experimenting with new functions. Although the architecture of the Internet suite does not mesh with that of the OSI Reference Model, RTP would fall into both Layer 5 (Session Layer) and Layer 6 (Presentation Layer) of that model. As defined in RFC 1889, the *RTP Control Protocol* (RTCP) augments RTP. This upper layer companion protocol allows monitoring of the data delivery in a manner scalable to large multicast networks and provides minimal control and identification functionality.

13.7.3.2 *Resource Reservation Protocol*   The IETF defined the *Resource Reservation Protocol* (RSVP) in its RFC 2205 (1997), which it updated in RFC 2750 (2000). RSVP is a Layer 4 (Transport Layer) control protocol that operates on top of IPv4 or IPv6 and depends on existing and future routing protocols. As a control protocol, RSVP does not transport data. Rather, RSVP operates on a hop-by-hop basis in order to signal QoS requirements for unicast and multicast data flows to each node and, thereby, reserve the necessary per-session resources from end to end across an IP network. RSVP can operate in conjunction with other QoS protocols, including DiffServ and MPLS, to effect service discrimination [31–34].

13.7.3.3 *Real-Time Streaming Protocol*   The *Real-Time Streaming Protocol* (RTSP) is defined in RFC 2326 as an *application-level* protocol for control over the delivery of data with real-time properties, such as audio and video and including both live data feeds and stored clips. RTSP is intended to control multiple data sessions, provide a means for choosing delivery channels such as UDP, multicast UDP and TCP, and provide a means for choosing delivery mechanisms based upon RTP. There is no notion of an RTSP connection. Rather, an RTSP server maintains a session labeled by an identifier.

RTSP establishes and controls either a single or several time-synchronized streams of continuous media such as audio and video. It does not typically deliver the continuous streams itself, although interleaving of the continuous media stream with the control stream (e.g., RTCP) is possible. In other words, RTSP acts as a *network remote control* for multimedia servers [33, 35].

13.7.3.4 *Differentiated Services*   *Differentiated Services* (Diff-Serv or DiffServ) is defined by the IETF in RFC 2474 as a framework for enabling the deployment of scalable service discrimination in the Internet. Operating at Layer 3, DiffServ assigns relative priorities to packets on the basis of an eight-bit *code point* in the

Differentiated Services (DS) field in the IP header. This DS field occupies the same position as the IPv4 Type of Service (ToS) octet or the IPv6 traffic class field. At the ingress to each node, the DS field is analyzed and a routing table is consulted in order to determine queuing considerations at the packet's output interface on that node, which considerations reflect the differential level of treatment to be afforded that packet in accordance with a policy that may be based on application, customer, or traffic type or as expressed in a *Service Level Agreement* (SLA). Such policy criteria might include time of day, source and destination address pair, and port number (i.e., application identifier). For example, DiffServ might use RSVP parameters to assign relative priorities to packets, with those priorities being associated with a small number of forwarding classes. There are two primary types of *Per-Hop Behaviors* (PHBs), which effectively represent two service levels, or forwarding classes. *Expedited Forwarding* (EF) provides minimal delay, jitter, and loss. EF traffic exceeding the traffic profile, as defined by the SLA, is discarded. *Assured Forwarding* (AF) comprises four classes, each of which contains three drop precedences and allocates certain amounts of buffer space and bandwidth. AF traffic exceeding the profile may be either dropped or demoted during periods of network congestion. DiffServ operates on a packet-by-packet and hop-by-hop basis [33, 36–39].

13.7.3.5 Multiprotocol Label Switching *MultiProtocol Label Switching* (MPLS) is defined by the IETF in RFC 2702 as a label-swapping framework with Network Layer (Layer 3) routing. Integrating Layer 2 (Data Link Layer) information about network links into Layer 3 routing logic, MPLS is based on several vendor-specific protocols, including Cisco's Tag Switching, Ipsilon's IP Switching, and IBM's ARIS technology. MPLS is designed to work through routers at even higher speed than ATM switches, while realizing much of the flexibility of an IP-based network. MPLS enables routers to make packet-forwarding decisions very quickly on the basis of short labels, comparable to Frame Relay Data Link Connection Identifiers (DLCIs), rather than making complex routing decisions after analyzing lengthy packet headers.

13.7.3.5.1 MPLS in Operation   MPLS works on the basis of *Forwarding Equivalence Classes* (FECs) and flows. A flow consists of packets between common endpoints identified by features such as network addresses and port numbers. An FEC is a class of packets, all of which are treated the same in terms of destination, priority level, and so on. As a flow begins, the first packet exits the user's client workstation, traverses the LAN, goes through an enterprise router, and reaches the ingress *Label Edge Router* (LER) at the edge of the carrier network. The LER identifies the flow based on the IP header, the interface through which the packet arrives, the packet type (e.g., unicast, multicast, or anycast), or perhaps information in the Type of Service (ToS) field.

As illustrated in Figure 13.8, the LER attaches to that packet and to each subsequent packet of the flow a 32-bit MPLS header that includes a 20-bit *label*, or *tag*, as it enters the edge of the MPLS domain. (*Note:* The standards provide for as many as four MPLS headers in a *stack*.) The header can be inserted in several places, depending on the network protocol and the associated packet format. If the Layer 2 protocol is Ethernet or Frame Relay, the header is inserted in a shim between the Data Link Layer header and the IP header. If the network is ATM based, the tag

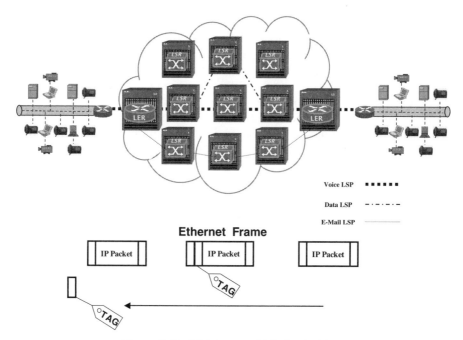

Figure 13.8 Multiprotocol label switching.

populates the *Virtual Path Identifier* (VPI) and *Virtual Channel Identifier* (VCI) fields, which is to say that the TAG is precisely the ATM address. The LER uses a *Label Distribution Protocol* (LDP) to distribute the labels or tags to each intervening *Label Switch Router* (LSR) in the network core, identifying the treatment that should be afforded all packets in the flow on that particular *Label Switched Path* (LSP). If the traffic engineering options are exercised, traffic is balanced between optimum and nonoptimum paths, and congestion is minimized. Otherwise, the traffic takes the same paths that IP packets would take, as MPLS nodes use IP routing protocols [e.g., Open Shortest Path First (OSPF) and Routing Information Protocol (RIP)] to distribute the labels.

From edge to edge through the core of the network, each LSR makes note of the incoming port number and analyzes the label associated with each packet in order to select the appropriate LSP over which the packet is to be forwarded on its way to the next LSR. Thereby, and through a series of links, the end-to-end path is set up and maintained for a given traffic flow. The more complex processes of complete header analysis and routing table lookup are performed only at the ingress edge of the network. In the core of the network, only the abbreviated MPLS label is analyzed in order to make a relatively simple and straightforward packet-forwarding decision. All in all, the routing process is simplified and latency is reduced. At the egress LER, the tag is removed, as it is no longer needed [40–43].

13.7.3.5.2 MPLS Header Structure The structure of the 32-bit MPLS header is as follows and as illustrated in Figure 13.9:

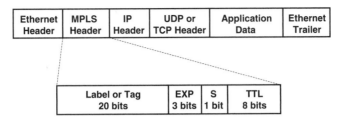

Figure 13.9   MPLS header structure.

- **Label:** The label, or tag, field of 20 bits is structured according to the carrier's requirements and matches the packet to the LSP.
- **Experimental (EXP):** The EXP field of 3 bits is used to indicate the precedence, or packet-queuing priority, for CoS purposes.
- **Stack (S):** The stacking bit is set at 1 to indicate the last (i.e., innermost) MPLS header in a stack of headers. Outer tags carry a 0 bit in this position. MPLS VPNs involve hierarchical routing logic that requires multiple headers. As many as four MPLS headers can be contained within a stack.
- **Time To Live (TTL):** The TTL field is copied from the IP TTL. The TTL is a hop count, with a default value of 64.

13.7.3.5.3 Congestion Management   MPLS addresses congestion issues in a much different way than a typical enterprise Wide Area Network (WAN), in which IP routers use OSPF or some similar routing protocol to interconnect Frame Relay or ATM Permanent Virtual Circuits (PVCs). This approach invariably leads to some level of congestion, which contributes to latency, jitter, and even loss. MPLS supports *constraint-based routing* that considers factors such as bandwidth, hop count, and performance requirements of the traffic flow in selecting the LSP. MPLS also enables traffic engineering, as it allows the carrier to select predetermined LSPs along which the traffic will transit. In a well-designed carrier network, traffic engineering ensures that the performance requirements of the traffic flows are satisfied. MPLS-based traffic engineering also supports backup LSPs that can be invoked quickly in the event of the failure of a physical link or node.

13.7.3.5.4 Class of Service   Through its traffic engineering attribute, MPLS supports the consolidation of multiple traffic types through a single network. However, each traffic type can be associated with a Class of Service (CoS). The carrier can take one of several approaches to honoring the CoS. One approach involves a single LSP between the ingress and egress LSRs. Traffic that flows on that LSP queues in the LSRs in consideration of the precedence bits in the MPLS header, with the high-precedence packets going to the head of the queue and the low-precedence packets remaining in the rear of the queue. Another approach involves multiple LSPs, each of which is engineered to perform according to a set of parameters that considers the performance requirements of a given traffic flow. So real-time voice and video might take one LSP, business-critical data might take another, and best effort e-mail might take yet another. MPLS CoSs are definable by the carrier on a standard and custom basis. AT&T, for example, defines the following standard CoS types:

- **C1:** Real-time CoS intended for applications that are sensitive to availability, latency, and jitter, with high-speed interactive video and voice being examples. This CoS takes precedence over all other options.
- **C2:** Interactive CoS designed for critical applications requiring priority treatment. Examples include one-way video broadcast systems requiring delivery confirmation within a short period of time or database activity.
- **C3:** Enhanced CoS designed for low latency and loss. This CoS is intended for non-mission-critical, yet delay-sensitive, applications such as SNA or streaming video.
- **C4:** Standard CoS designed for bursty traffic with minimal latency requirements. Example applications include e-mail and Internet traffic.

CoS is not at the same level as ATMs' guaranteed Quality of Service (QoS), of course, but it can come reasonably close in a well-engineered network. Consider the following analogy. When you send a personal letter or parcel through the mail via the U.S. Postal Service, you fill out the complete address by longhand or perhaps by typing out a label that you adhere to the letter or parcel. That full address contains the name of the intended recipient, the street address and apartment or suite number, the city, the state, and the ZIP code or other postal code. You also identify whether the letter is to be sent as bulk mail, first-class mail, Priority Mail, or Express Mail. As the letter is first processed at the edge of the system, a computer system scans the ZIP code through the use of *Optical Character Recognition* (OCR) or *Optical Mark Recognition* (OMR), and a postal code is added in the form of a bar code. In the event that the OCR or OMR software cannot translate the handwritten ZIP code, a postal worker adds the bar code manually. As the letter works its way through the core of the system, electronic devices rely solely on the bar-coded ZIP code to forward the letter to the destination edge of the postal processing network. All mail destined for the same local post office is batched and delivered there. At that office, the street address is read in full and the mail is sorted for individual routes and carriers. The carrier verifies the address before depositing the mail in your mailbox. Consider the bar-coded ZIP code as analogous to a tag—it adds additional overhead but speeds up the process of sorting and forwarding in the core of the network of mail-processing centers.

There also are different paths and different levels of precedence for different types of mail. Overnight mail takes an express route by air, taking precedence over priority mail, which also travels by air. First-class mail travels by a surface mode, such as truck and ship, taking precedence over media mail and parcel post, which travel on a space available basis. Such is the case with MPLS.

MPLS works with DiffServ and RSVP and with Frame Relay, ATM, and Ethernet and is used in support of IP-based Virtual Private Networks (VPNs). MPLS works with various routing protocols such as OSPF, RIP, and BGP that can be used to distribute the routing labels. Hence the use of the word *multiprotocol* [33, 44, 45]. MPLS currently operates on only a single domain, as there is no Network-to-Network Interface (NNI) specified for interoperation between carriers. Any such interoperation, therefore, must be by special arrangement.

13.7.3.5.5 Virtual Private Networks MPLS supports several VPN options. The Layer 2 MPLS VPN emulates a point-to-point virtual circuit connection, or *pseudowire*, between two routers or switches. This Layer 2 class of MPLS VPN is commonly referred to as a *Draft-Martini* VPN, or a *PseudoWire Emulation* (PWE) VPN. (*Note:* Luca Martini is a senior architect at Level 3 Communications and a major contributor to the draft standard.) PWE can emulate a variety of Layer 2 protocols, including Frame Relay, ATM, HDLC, and PPP by encapsulating those formats in IP and sending them to the LER. Layer 3 MPLS VPNs offer any-to-any multipoint service and are known as BGP/MPLS VPNs. Membership in the MPLS VPN is defined by a *Virtual Routing/Forwarding* (VRF) table maintained in the LER. As required, the LER adds labels to a label stack in support of the hierarchical routing logic involved in VPN tunneling and routing. Through a combination of tables and labels, the MPLS VPN provides many of the same benefits as a dedicated leased-line network, including security in the form of a *Closed User Group* (CUG). As a Layer 2 protocol with Layer 3 routing support, MPLS does not provide any mechanism for encryption or authentication, however.

13.7.3.5.6 Generalized MPLS   As a final variation on the theme, enter *Generalized MPLS* (GMPLS), once known as *Multiprotocol Lambda Switching* but since extended to other media. GMPLS extends MPLS beyond packet-switched interfaces to include *Time Division Multiplexing* (TDM), *Wavelength Division Multiplexing* (DWDM), and *Add/Drop Multiplexing* (ADM). GMPLS adds the concept of label switching to photonics at the lambda (i.e., wavelength) level in a DWDM system, to time slots in a SONET/SDH system, and to physical fibers in an optical cross-connect or photonic cross-connect. The goal of GMPLS is to speed the provisioning of end-to-end traffic-engineered paths in the TDM and optical domains, much as MPLS has done in the IP domain [40, 46].

13.8 INTERNET APPLICATIONS

Internet applications are growing as fast as the imagination and technology will support. These applications include e-mail, file transfer, Bulletin Board Systems (BBSs), library catalogs, online banking, video and radio broadcasting, and even voice telephony.

13.8.1 E-Mail

Electronic mail (e-mail) clearly is the most popular application. Somewhere in the neighborhood of 84 billion e-mail messages comprising trillions of bytes transverse the Internet every single day. E-mail transport through the Internet makes use of the Simple Mail Transfer Protocol (SMTP). E-mail messages commonly are sent with MIME attachments, which may be in the form of compressed images, video clips, and audio clips.

13.8.2 File Transfer

File transfer is accomplished through the File Transfer Protocol (FTP) and in support of topical research. Through many thousands of FTP servers, Internet attached file servers can be accessed and the resident file resources can be accessed and transferred in ASCII or binary code. While many FTP resources are privileged, others are unrestricted and available for public consumption; you can access such unrestricted resources via the use of the account name *anonymous*. Large files (traditionally equal to or greater than 50 kB) generally are handled in compressed format using compression techniques specific to the computer operating system. Examples include *.Z* or *.tar* (Unix), *.zip* (MS-DOS), and *.hqx* (Macintosh).

13.8.3 Bulletin Board Systems

A *Bulletin Board System* (BBS) is a computer system running software that allows one to connect to what is essentially an electronic bulletin board. BBSs originally were quite local phenomena, as access was on a dial-up basis and only a local call is toll free. BBSs typically were, and often still are, run as a hobby or side interest of the sponsor. Generally, anyone can access the BBS to post messages, reply to messages, post software applications for downloading by others, play games, and otherwise communicate and share with others. BBSs were quite the rage in some Internet circles of interest until the mid-1990s but generally are considered somewhat primitive in the context of the modern World Wide Web (WWW). They remain quite useful, however, for special interest groups such as fans of the Grateful Dead rock band, system administrators struggling with software applications in a specific network environment, or attempting to deal with unusual technical issues.

13.8.4 Library Catalogs

Library catalogs for hundreds of libraries are available through the Internet. Such catalogs include the U.S. Library of Congress, the Research Libraries Information Network, and many major colleges and universities. Through the Interlibrary Loan Program, Internet users can have the document loaned to their local library.

13.8.5 Real-Time Applications

Real-time applications are growing at significant rates. There are a number of highly legitimate applications and many that generally are considered to be less so. Real-time applications include collaborative design and development, interactive role playing, interactive remote education, and chat lines, which operate like simplified real-time BBSs. Other applications include voice and videoconferencing, network games, gambling, and streaming audio and video broadcasting (*cybercasts*).

13.8.6 Financial Services

Online banking has received a good deal of interest. While security issues abound, there is little doubt that this application has gained a broad level of acceptance. In the area of financial services, online stock trading has made a huge impact, having taken significant market share away from conventional brokerage houses. The downside of online stock trading is that several of the Internet-based brokerages have experienced numerous short-term system failures that have resulted in the total loss of their ability to execute trades on behalf of their customers.

13.8.7   Video

Although it is terribly bandwidth intensive and the quality is relatively poor, you can transmit packet video over the Internet. Videoconferencing also is possible with a SLIP/PPP connection and a high-speed modem. A black-and-white video camera can be had for under $100 from a number of manufacturers.

13.8.8   Radio

Radio broadcasting makes use of packet audio systems such as RealAudio from RealNetworks. Although there always are issues of latency and loss in a real-time stream-oriented application, AM quality is quite possible. In other words, the general quality is not necessarily great, but the price is right. Some years ago, a number of students from Asia were enrolled in a graduate-level course I taught at the University of San Francisco. Several students from Japan, in particular, regularly listened to radio broadcasts from Tokyo over the Internet. Again, the quality was not great, but the price was right (read *free*). Additionally, they really had no other means of hearing such a broadcast or getting information from home on a timely basis. My wife, Margaret, also a consultant, several years ago had a long-term engagement with a large wireless company. For the first few weeks, she complained about the fact that the office was as quiet as a tomb. Not only was there none of the usual water cooler conversation, but people seldom ventured out of their cubicles. One evening when we were shopping at the local office supply store, she stopped to buy an expensive headset, which puzzled me. It seemed as though she had figured out the reason for the dead silence in the office. Margaret was about to join the other hundred or so staffers who listened to radio broadcasts on the Internet while working. Bandwidth and quality were not particular problems because the company had several T3 circuits connected directly to the Internet backbone.

13.8.9   Internet Telephony

Internet telephony, also known as *Internet voice*, is a means of transmitting voice over the Internet, bypassing the traditional PSTN and saving money in the process. While I discussed Voice over Internet Protocol (VoIP) at some length in Chapter 5 and dealt with the detailed specifics of IP earlier in this chapter, VoIP merits further discussion here in the context of the public Internet. Internet telephony is accomplished through the use of special software residing on a multimedia PC equipped with a microphone, speaker, and modem or, ideally, a broadband connection to the Internet.

Originally, both parties in the conversation had to install the same software, which worked on a half-duplex basis, enabling only one person to talk at a time using a *push-to-talk, over-and-out* protocol reminiscent of CB radio. Depending on the software in use, it was necessary that both parties schedule the call in advance, typically either via e-mail over the Internet or through a short, preliminary, conventional telephone call. Some software allowed you to determine the status (online or offline) of the other party in much the same manner as you can determine the status of your buddies in an instant messenger application. Current software technology also enables the caller to *ring* the target PC over the Internet or even to connect to a standard telephone set via a gateway between the Internet and the PSTN. In such a scenario, the long-haul portion of the call is over the toll-free Internet. Current technology also supports full-duplex communications.

The quality of a VoIP call over the Internet varies. My lovely bride, Margaret, is the principal of The Evergreen Group, a curriculum company that develops training material for companies all over the world. She recently managed a project involving instructional designers and subject matter experts in Argentina, Australia, Canada, Columbia, New Zealand, Norway, Spain, and the United States. In order to keep long-distance calling costs to a minimum, she used a popular Internet voice service. There were two ways to connect a call:

- **Softphone to Softphone:** She could call from her softphone (i.e., computer equipped with the VoIP software) to the other person's softphone through the service provider's website. In this scenario, the call was entirely over the Internet, and both parties were required to have broadband, always-on access. The quality was sometimes quite acceptable and sometimes awful due to latency, jitter, and loss. The call was free, but usually not free enough.
- **Softphone to PSTN:** She could call from her softphone to the other person's PSTN hardphone or cell phone. In this scenario, the international portion of the call was over the Internet to a local gateway in the distant city, where the connection to the PSTN (and cellular network) was established. The quality generally was quite good and the cost of the call compared quite favorably to the cost of a traditional PSTN call.

As you might expect, some of the traditional PSTN carriers are very upset about this application because it threatens to shift large volumes of traffic away from the conventional PSTN and to the Internet. Some carriers in the United States have even blocked VoIP calls. Many nations, particularly developing countries, have even outlawed the practice, at least for end users, as it so negatively affects the associated revenue stream, which oftentimes is a significant source of hard currency. On the other hand, many of those same nations have contracted, through their monopoly national carriers, with IP-based international carriers that use both private IP networks and the Internet.

Other issues abound, including reduced tax revenues at the local, state, and federal levels. In the United States, the continued viability of the *Universal Service Fund* (USF) is a major source of concern. Traditionally, the Internet has been considered a tax-free zone. Therefore, VoIP calls over the Internet were not taxed and did not contribute to that fund. ISPs long have contended that they support the Universal Service Fund through charges embedded in the circuits they and their subscribers lease from the ILECs. Clearly, however, these fees do not come close to matching the USF contributions that apply to voice, fax, and data traffic over the PSTN. The Internet Tax Freedom Act became law in October 1998, placing a three-year moratorium on taxation of Internet access at the state and local levels. The act also protected e-commerce from taxation (e.g., sales taxes) for out-of-state transactions. In November 2001, the Internet Tax Non-Discrimination Act extended this tax moratorium through November 1, 2003, reaffirming the Internet and the Web as a tax-free zone. In February 1999, the FCC ruled that dial-up Internet calls are interstate in nature, rather than local, which ruling gave the U.S. federal government jurisdiction over them, which set the stage for future taxation of the Internet. Finally, in June 2006, the FCC voted to require taxation of all VoIP calls that connect to the PSTN, with that revenue going to support the USF. Peer-to-peer calls traveling exclusively over the Internet remain untaxed, at least for the time being.

Internet telephony actually offers more than just low cost in return for low quality. Consider a convergence scenario in which you access the Internet and the Web through your ISP to view a website to make a reservation or order a product or service. As you click your way around the website through your browser, you have a question about the specifics of a product. You click the *Talk to An Agent* button and are connected to a real person with whom you can carry on a conversation over the public Internet using VoIP technology. While you are talking, you both view the same information on the website. The agent can take control of the website to present you with the proper information, which saves you a lot of mouse clicks, thereby improving communications and enhancing your Internet experience. The voice quality may be less than toll quality, but the convergence of voice and data has added considerable value overall.

Now, take this experience to the next level. If you access your ISP over an ADSL local loop and connect to a private next-generation IP-based carrier, rather than through the public Internet; then your converged voice and data communications might transport over an improved IP-based backbone optimized for voice as well as data. Your voice communication will be improved. Your data communication probably will be more satisfactory as well. At this level, Internet telephony really is not Internet telephony at all, except perhaps at the very edges. Rather, the majority of your experience is one of voice and data communications over a finely tuned, IP-based network.

There currently are various proposals to impose a tiered, content-sensitive pricing arrangement for Internet usage whereby real-time, stream-oriented applications such as voice and video would be charged at a premium. These proposals are the source of considerable controversy, of course.

13.10 INTERNET2

Universities have become distressed over the degradation in Internet performance as traffic levels have increased because of its highly successful commercialization. In fact, a number of them petitioned the National Science Foundation to build a new, separate NSFNET intranet (private Internet) just for them. In October 1996, this concern translated into action in the form of Internet2, which is a project of the *University Corporation for Advanced Internet Development* (UCAID), a not-for-profit entity created specifically to develop and manage the network. Internet2 is a not-for-profit consortium of the NSF, the U.S. Department of Energy, more than 200 U.S. research universities, and over 60 private companies. Internet2 is intended as a private Internet for the benefit of its member organizations, although it is not a separate physical network, and it does connect to the present Internet, as required. The Internet2 and its members are in the process of developing and testing technologies such as IPv6, multicasting, and QoS mechanisms in support of what they characterize as revolutionary Internet applications such as digital libraries, virtual laboratories, distance-independent learning, and tele-immersion. *Tele-immersion* is intended to allow multiple, geographically distributed users to collaborate in real time in a shared, simulated hybrid environment through a synthesis of media technologies such as 3D environment scanning, projection and display, tracking, audio, video, robotics, and haptics (i.e., touch) technologies. In other words, tele-immersion creates a multimedia virtual meeting space. The university-led Internet2 initiative is parallel and complementary to the federally led *Next-Generation Internet* (NGI) initiative. While only member organizations have access to Internet2, the underlying technologies being developed to support these applications are intended to form the basis for the next generation of the public Internet. In 2001, the Internet began the *Sponsored Education Group Participation* (SEGP) program that allows states to tap into the network under the sponsorship of an Internet2 member.

Initially, Internet2 made use of existing networks such as the *very-high-speed Backbone Network Service* (vBNS), a network provided under a cooperative agreement between the NSF and Worldcom (now MCI, which is part of Verizon) in support of NSF-approved institutions of higher learning. vBNS initially ran over an ATM/SONET backbone at 155 Mbps (OC-3) and later was upgraded to 622 Mbps (OC-12). More recently, the *Abilene Project* was developed as a high-performance network by the *University Consortium for Advanced Internet Development* (UCAID) in cooperation with Qwest Communications, Nortel Networks, Cisco Systems, and Indiana University. Abilene supports the Internet2 infrastructure through high-speed routers connected to several dozen *GigaPOPs* (i.e., Gbps Points of Presence, routers positioned as access points) interconnected over optical fiber transmission facilities running at speeds up to 10 Gbps (SONET OC-192) nationwide [47, 48]. Abilene is now being phased out in favor of a DWDM backbone that initially will support ten 10-Gbps lambdas and eventually will scale to eighty 10-Gbps lambdas [49].

Géant2, a similarly ultrafast European backbone, was lit in Milan in June 2005, with the 186 million Euros funded half by the European Union and half by member nations. *Géant2*, which translates from French as *Giant2*, connects some 5000 institutions across Europe, including many high schools, in the European Union, plus Russia, Switzerland, Turkey, and Israel. Anyone connected with any of those institutions has access to the network [50].

13.11 WORLD WIDE WEB

The *World Wide Web*, also known as *the Web*, is the Internet's first *killer application*. Tim Berners-Lee developed the Web at the l'Conseil Européen pour la Recherche Nucléaire (CERN), which translates from French as *The European Council for Nuclear Research* and is generally known as the European Laboratory for Particle Physics in Geneva, Switzerland. The Web is a multiplatform operating system that supports multimedia communications on the basis of a *Graphical User Interface* (GUI). The GUI provides *hypertext*, which enables the user to click a highlighted text word and search related files across Web servers and through *hot links*; in other words, the Web is *hyperlinked*. In addition to text, the Web supports graphics, audio, and video, with various levels of quality and speed depending on the bandwidth available.

While CERN served to conceive the Web, its home has moved to the *World Wide Web Consortium* (W3C). W3C is a cooperative venture of CERN, The Massachusetts Institute of Technology (MIT), and *l'Institut National de Recherche en Informatique et en Automatique* (INRIA), which translates from French as the *National Institute for Research in Computer Science and Control*. The primary focus of W3C is that of leading the technical evolution of the Web by promoting interoperability and providing an open forum for discussion. Since it was organized in 1994, W3C has published over 30 technical specifications. You can monitor progress through the W3C website (of course).

13.11.1   Websites and Home Pages

Organizations or individuals can develop a *website*, consisting of a computing platform (server) connected to the Web on a full-time basis. Although the typical business approach involves the deployment of a dedicated server, smaller users may achieve the same end through renting website capability from a hosting ISP. The ISP logically partitions a server in support of multiple users and multiple home pages.

Home pages effectively provide a POP on the Web for businesses, typically supporting advertising and informational purposes. Within each website is a home page, or multimedia informational document, which may contain graphics, animated graphics, video clips, and audio clips as well as text. Individuals may develop personal home pages, which may offer updates on personal life along the lines of a cyberspace version of the ever popular, *What I did on my summer vacation.* (Don't laugh. We've done the same thing here at the Horak household.)

13.11.2   Uniform Resource Locator

A *Uniform Resource Locator* (URL) is a type of *Uniform Resource Identifier* (URI) that consists of a *uniform* address that both identifies an abstract or physical *resource* on the Web and indicates how to *locate* it. As specified in RFC 3986, the syntax follows a standard convention that is *scheme://authority/path?query#fragment*. Consider the example http://info.cern.ch/hypertext/WWW/MarkUp/MarkUp.html, where *http = hypertext transport protocol* and *html = hypertext markup language*.

The *method*, or *scheme* name, indicates the network protocol used to assign identifiers. Examples of schemes include the following:

- http: HyperText Transfer Protocol
- https: HyperText Transfer Protocol (HTTP) over Secure Sockets Layer (SSL)
- ftp: File Transfer Protocol
- news: Usenet newsgroups
- Telnet: *Tel*communications *net*work protocol

The *authority* is preceded by a double slash (//) and is terminated by the next slash (/), question mark (?), or number sign (#) or the end of the URI. The authority can contain user information followed by a commercial *at* sign (e.g., ray@) and host information in the form of an IP address (e.g., IPv4 dotted decimal notation) or registered name (e.g., contextcorporation.com). The authority also may contain an optional port number, which is unnecessary if the number is the same as the scheme's default (e.g., *80* would be redundant with *http*).

The *path* component contains data, usually organized in a hierarchical form that serves to identify a resource within the scope of the URI's scheme and naming authority. Within the path, a slash (/) is used as a delimiter between components. The path is terminated by the first question mark (?), number sign (#), or the end of the URI.

The *query* component contains nonhierarchical data that, along with data in the hierarchical path component, serve to identify a resource within the scope of the URI's scheme and naming authority.

The *fragment* component allows the indirect identification of a secondary resource that may be some portion or subset of the primary resource, some views or representations of the primary resource, or some other resource defined or described by those representations [51].

HTTP is the default protocol for transmitting HyperText Markup Language (HTML) content over the Internet. It advises the browser to use the HTTP protocol in accessing Web documents. HTML was the first programming language for creating compound documents for websites and for supporting hot links to other sites. HTML has the clear advantage of device independence; in other words, the specifics of the user terminal (e.g., Macintosh, IBM-compatible PC, or Sun workstation) do not affect the presentation of the file [24]. As I discussed earlier in the context of domain names, English traditionally was the native language of the Internet and Web. In 2000, URL support was announced for over 60 other languages. This support requires the full UNICODE character set, rather than the ASCII character set traditionally used in support of English.

Clearly, URLs are of great value in the world of e-commerce. For example, Wiley, the publisher of this book, has the website www.wiley.com, which is very important to its online business. (Unabashedly self-serving hint 1: You order large quantities of this book from that website.) Microsoft reportedly paid $5 million for the rights to www.internetexplorer.com, which was the property of another company that made legitimate use of it [52]. Some companies have filed lawsuits against *squatters*, who had registered websites mimicking the copyrighted, trademarked, or service-marked names of other companies specifically for the purpose of selling them to the rightful owners for substantial sums of money.

Some companies and individuals have secured rights to attractive, unclaimed URLs with the intention of selling them. Some years ago, Marc Ostrofsky of Houston, Texas, put a number of his employees to work searching for URLs that had e-commerce potential. They searched for just about any work in the English language preceded with *e-* or *e*. As a result of that effort, he identified and registered www.eflowers.com. In 1999, he rejected an offer of $1 million from Flowers Direct, preferring to sell it to them for $25,000 plus $0.50 for every transaction generated over the website—plus free flowers for his wife (now ex-wife), Sarah, for the rest of her life. Given the projections of 500,000 transactions per year, that is not a bad return on an investment of $70. Subsequently, Ostrofsky sold the rights to www.business.com for $7 million, a URL he had acquired a few years earlier for $250,000 for use in connection with a business he later sold for many millions of dollars.

13.11.3 Standards

Standards for the WWW are set by the *W3 Consortium*, which is run by The Massachusetts Institute of Technology (MIT), *Intitut National de Recherche en Informatique Et En* (INRIA) in Europe, and in collaboration with CERN. Tim Berners-Lee, creator of the Web, serves as the group's director.

13.11.4 Applications

Web applications are in a world of their own, including advertising, publishing, micromarketing, catalogs and direct sales, entertainment, and e-commerce. The great hope is that the Web will become a significant tool for commerce, enabling customers to access a home page on a website, gain information about a product or service, and actually make a purchase online.

13.11.5 Advertising: Home Page Sponsorship

Home page sponsorship is offered by several directories and frequently hit portals and home pages. Virtually all browser portals and home pages for publishers of periodicals, for example, sell sponsorships. In return for a rather princely sum, your home page banner will appear on their home page; users can click your banner and hyperlink to your home page.

13.11.5.1 Personal Home Pages and Blogs   Personal home pages exist by the millions, at least. Text describing collections of airsick bags, photos of the children, audio clips of the growls and screeches of family pets, and video clips of summer vacations—all are possible and all are present on the Web. The latest variation on this theme is the *blog*, a contraction of W*eb log*. A blog is a website where an individual maintains a personal journal or even an interactive forum much like a personal newsgroup.

   The vast majority of blogs, or so it seems, are pretty silly electronic diaries posted by adolescents. Those who do not want to take the time and go to the expense of developing their own website and registering their own URL can use a site such as MySpace.com. That site is an online community that allows you to post a personal profile, including photos, keep a journal of your daily activities and thoughts, and so on. You can invite your friends to join and share in your *personal network*, as the site terms it. You can then *view the connections you create between your friend and their friends. Some people have 1000 people in their extended network* [53]! Some (mostly young) people have done fun (read *stupid*) things like post descriptions, including photos, of their underage drinking and sexual exploits and other dumb and irresponsible things, only to find that prospective employers, as well as their extended network of friends and their friends, have read them. Sexual predators prowl the same sites, looking for likely prospects to engage in conversation, which they hope will lead to unimaginable horrors. As you can imagine, the impacts have been considerable. Windows Live Spaces (nee MSN Spaces) currently boasts over 120 million unique users, according to the website at least. Users can choose a number of different page layouts, which they can customize. Windows Live Spaces supports textual blogs, photos, and video clips.

Many blogs, however, are quite serious. Some companies maintain public blogs to foster dialogue among employees with respect to projects, strategies, and other matters of interest. Some blogs take the form of well-researched personal opinion columns on politics or other controversial and weighty subjects. In August 2006, U.S. Senator Joe Lieberman, the Democratic senator from Connecticut and vice-presidential candidate in 2000, lost the Connecticut senatorial primary election. According to mainstream media news sources, activist left-wing bloggers were largely responsible for his narrow defeat.

Popular in the blog community is a software application program known as a news aggregator, which maintains contact with selected sites in the *blogosphere* area in *cyberspace* that make content available in RSS format. *RSS* is a *push* technology, that is, a *metadata* technology, that can identify changes in data and initiate a content push to the end user without the user having to search it out and *pull* it from the site. The term RSS is an umbrella term variously used to describe a number of versions of several data Web feed formats specified in eXtensible Markup Language (XML) and used for syndication of Web content. Those standards include:

- Really Simple Syndication
- Rich Site Summary
- RDF Site Summary. Resource Description Framework (RDF) is a W3C specification that integrates a variety of applications using XML as an interexchange syntax.

More recently, the IETF adopted the *Atom Publishing Protocol* (APP), which builds on the previous RSS work but corrects some RSS deficiencies, according to some. The Atom format is documented in the *Atom Syndication Format* (RFC 4287).

13.11.6 E-Commerce

Electronic commerce, or *e-commerce*, has grown to incredible proportions over the past several years. The U.S. Census Bureau defines e-commerce as "sales of goods and services where an order is placed by the buyer or price and terms are negotiated over an Internet, extranet, Electronic Data Interchange (EDI) network, electronic mail, or some other on-line system. Payment may or may not be made online." According to that definition, e-commerce retail sales in the United States reached US$85.993 billion in 2005, which is approximately 2.2 percent of the total retail sales of US$3.861 trillion [54].

Examples of Web-based businesses include online booksellers, some of which have expanded into sales of music, videotapes, toys, and electronic games. Online auction houses recently have gained a good deal of attention as well; online auctions of airline tickets, cruises, and hotel rooms currently are especially popular. Application software routinely is sold over the Web and then downloaded over it. Using the *Moving Pictures Experts Group, Audio Layer 3* (MP3) compression algorithm, near CD-ROM quality music also can be purchased over the Web and then downloaded over the Internet, although there are significant concerns relative to copyright infringements and avoidance of royalty payments. Some companies even distribute the music for free, in anticipation of increased revenues from concerts and T-shirts. Electronic brokerage houses have made a significant dent in the overall stock brokerage market; although plagued by occasional router and server failures, electronic trades are accomplished at a fraction of the commissions paid to conventional stockbrokers.

As noted earlier in this chapter, the Internet Tax Freedom Act became law in October 1998. One of the provisions of that act protects e-commerce from taxation at the state and local levels, in effect declaring the Internet and the Web a tax-free zone in terms of sales taxes for interstate transactions. In other words, your online purchases from a company in Rhode Island are not taxed, assuming that you live outside the state. Technically speaking, no taxes apply as long as the seller does not maintain a physical presence in the buyer's state. This step is grounded in tradition, as mail-order catalog companies long have enjoyed the same advantage. For that matter, the purchases that you make in New York City are exempt from state and local sales taxes if you have them shipped to me in Mount Vernon, rather than taking them back to your hotel and hauling them back home yourself. And, as most people know, the cost of postage easily can be less than the sales taxes. Now, buyers are required to pay levies on their online and mail-order purchases when they file their state income tax returns, assuming that there is an income tax in the state of residence, but it is unlikely that any but the most right-minded consumer actually would do such a thing. The risk, of course, is that the increasing popularity of e-commerce can undermine the state and local tax bases, which fund a wide variety of social and other services.

13.11.7 Search Mechanisms and Browsers

The huge number of servers and incredible amount of information available on the Internet and Web quickly made it difficult to find the desired resource. In other words, you could not find the information unless you knew pretty much where to find it—sort of like having to know at least the approximate spelling of a word in order to look it up in the dictionary to determine the exact spelling. This problem was addressed in the early 1990s through the development of search engines and, more recently, Web browsers.

Search engines are in the form of client/server software application programs that support the search for Internet informational resources through the development and maintenance of resource directories. In contemporary terminology, search engine commonly is used to describe systems such as Alta Vista, Ask Jeeves, Excite, Google, LookSmart, Lycos, Infoseek, and Yahoo! (Yet Another Hierarchical Officious Oracle).

Web browsers are software application programs used to locate, display, and interact with Web pages. The first primitive text browser appeared in 1991, courtesy of CERN. In 1993, the first graphical browsers appeared: Viola for X Windows, Mac browser from CERN, and Mosaic for X Windows [55]. Early graphical browsers that also support hyperlinks include Archie, Gopher, JUGHEAD, VERONICA, and WAIS. Currently, the most popular browsers include Microsoft's Internet Explorer, Mozilla Firefox, Apple Safari, AOL's Netscape Navigator, and Opera. Select search engines and browsers include the following:

- *Archie*, a corruption of *archive*, is an FTP search mechanism first deployed in 1991. Archie enables you to search for a file (exact name unknown) on a file server (name unknown) somewhere on the Net. Archie servers contain directory listings of all such files updated on a monthly basis through a process of file server polling. Archie provides a user-definable number of file hits, as well as file names, server names, and directory paths to access each listed file. Archie capabilities are limited to specific search *strings*, thereby providing little flexibility. Currently, Archie is often integrated into Gopher or Web clients, to be activated when the user accesses an Archie server.

- *Gopher* was developed at the University of Minnesota, where the Golden Gopher is the school mascot. Developed as a user interface to provide easy access to server resources in educational institutions, Gopher has become a de facto user interface standard. Gopher servers enable the user to access a directory of over 1800 Gopher server sites, click the name of the server, and browse its file resources on the basis of nested menus. Gopher requires that the user know the server on which the subject file is located, somewhere in *gopherspace*.

- *Very Easy Rodent-Oriented Netwide Index to Computerized Archives* (VERONICA) is an Archie variation that supports an index of gopherspace titles on which a search can be performed. The selected resources are then delivered to the user in the form of a Gopher menu.

- *Jonzy's Universal Gopher Hierarchy Excavation And Display* (JUGHEAD) is similar in operation to VERONICA, although it limits the search to a specific organization. JUGHEAD also delivers a custom menu of available resources located on the basis of the *keyword* search. Playing off Archie and Gopher, subsequent developers of search mechanisms tried to stay with the Archie comic book/rodent theme, proving once and for all that even acronyms can be fun. Alas, the more recent and more powerful browsers were not named with a noticeable sense of humor.

- *Mosaic* is a browser developed by a team led by Marc Andreessen at the National Center for Supercomputing (NCSA) at the University of Illinois Urbana-Champaign campus. Mosaic provides a consistent user interface available in versions to support Macintosh, Microsoft Windows, and Unix-X Windows. Mosaic can be used over dedicated or dial-up Internet connections; the dial-up access provider must support either SLIP or PPP. Mosaic enables the easy browsing of Web resources through menus that support hypertext. Through a simple process of mouse clicking, the user can select menu options. Selected files can include audio and graphics, both of which can be viewed without the requirement to download the subject file. Mosaic technology is licensed by the NCSA for commercial application. Spyglass licensed well over 12 million copies of Mosaic to IBM, DEC, and others who intended to resell the company's Enhanced Mosaic, which included enhanced security mechanisms based on *Secure HTTP* (S-HTTP). Ultimately, Spyglass Mosaic was licensed to Microsoft, where it formed the basis for Internet Explorer.

- *Wide Area Information Service* (WAIS) servers enable the user to specify the databases requested for search and to conduct a subject matter search on the basis of *keywords*.

- *Internet Explorer* (IE) is a highly capable browser which has the advantage of being packaged with Microsoft's Windows suite of software. While at one point IE included AltaVista, GoTo.com, Infoseek, Lycos, and MSN (Microsoft Network) browsers, all but MSN were either eliminated or subjugated as IE rose to market dominance.
- *Netscape Communication*, which includes *Netscape Navigator*, was developed by Netscape Communications, since acquired by AOL. The software, built by a team led by Marc Andreessen (creator of the original Mosaic), features simultaneous image loading and continuous document-streaming speed performance. Navigator quickly became the top browser choice, before suffering a meltdown beginning in the late 1990s.
- *Mozilla* was originally a code name for Netscape Navigator and was a contraction of *Mo*saic killer, referring to the hope that it would unseat Mosaic as the top browser, and God*zilla*, referring the fictional monster of Japanese science fiction movies. Mozilla now refers to an open-source application suite based on the Netscape Navigator source code, which was released by Netscape in 1998 under an open-source license.

13.11.8 Access Anywhere Revisited

Earlier in this chapter, I discussed the concept of access anywhere in the context of the Internet. Our focus there was on e-mail messaging. Well, that is by no means the end of the story. Contemporary cell phones, PDAs, and other hand helds increasingly are Web enabled through built-in microbrowsers. As hand helds just do not offer the same capability as a laptop or desktop in terms of processing power and display technology, and as wireless networks just do not offer the same bandwidth or error performance as wired networks, some adjustments have to be made to support an effective Web experience. Those adjustments largely involve tailoring the content so that it fits comfortably on a smaller display that does not support colors other than black and white. Further, the complexity of the website must be adjusted in consideration of the limitations of the user interface of a hand held, which currently does not support point-and-click capability easily, as it does not support a mouse. There are two competing approaches: Internet Mode (i-Mode) and Wireless Access Protocol (WAP):

- *Internet Mode* is a proprietary service developed by NTT DoCoMo, initially for the Japanese market. i-Mode supports text, graphics, audio, and video over the Japanese cellular network. In consideration of the inherently limited bandwidth of the cellular network, i-Mode employs *Compact HTML* (C-HTML), a simplified version of HTML similar to *Wireless Markup Language* (WML) used in WAP networks. Transmission between the hand helds and the i-Mode-enabled cell sites is via packet mode, using packets of 128 octets, at rates up to 9.6 kbps. Since its introduction in February 1999, i-Mode has grown to include over 32 million users and thousands of CWML-coded websites. The popularity of i-Mode is attributable to at least two factors. First, the Japanese are notoriously enamored with technology. Second, it is at least partially a personal-space issue, as the typical Japanese home does not have enough space for a desktop or even a laptop. In other words a Japanese SOHO would have to be very small indeed.

- *Wireless Access Protocol* is a carrier-independent, device-independent, transaction-oriented protocol employed in cellular networks outside of Japan, also in support of text, graphics, and audio. While the best performance is achieved when accessing websites written in WML, which is similar to but different from HTML, this requires that the content be rewritten. The alternative is transcoding from HTML to WML, which is accomplished through gateways. A much simpler but much less attractive technique is *Web clipping*, which strips the graphic content out of Web pages. Security over the wireless link is provided through *Wireless Transport Layer Security* (WTLS), pronounced *witless*.

The most popular applications for wireless Web access generally are thought to include services such as stock quotes, weather forecasts, airline and train schedules, and weather reports. Location-based services, such as finding the closest restaurant or bar, are not only possible but are also highly attractive to wireless users. Wireless banking also holds great promise, with the eventual hope being that hand helds could replace cash, checks, and credit cards altogether [56–60].

13.12 INTRANETS AND EXTRANETS

Intranets are the concept of the Internet turned inward—an unexpected turn, perhaps, but a very significant one. Intranets essentially are mini-Internets deployed within organizations or groups of organizations. They can be in the form of internal Internets, functioning to provide access to information resources within the company, university, or other organization. They can be confined to a campus environment or can extend across the wide area to link together multiple, geographically dispersed locations. They also can function as a closed subnet of the Internet, much as is intended for Internet2 in the college and university market. Although conceived as recently as 1995, intranets have spread quickly to the point that most medium to large user organizations have them in place. Intranets use the same browsers as used for Internet application, thereby avoiding the training and support requirements imposed by another application software package such as groupware.

Intranets can be used for communications to and between employees for just about any purpose imaginable. A number of corporations use intranets to keep their employees advised of company policies, job postings, company events, product literature, press releases, and so on. With the proper password for security purposes, of course, privileged users can access sensitive internal company information, including customer billing records and network usage data. As is the case with the Internet and Web, images, video clips, and sound clips associated with textual information are supported. Hypertext links can be included to *hot link* to other sites and databases and even to the Internet and the Web.

Health care organizations have made fairly extensive use of intranets to link remote clinics, reducing paperwork and abbreviating communication time. Kaiser Permanente has put the intranet concept to use in order to keep employees abreast of changes in health legislation and insurance law as well as to provide access to company telephone directories and human resource manuals.

Extranets are intranets opened to select groups of users outside the company. Access generally is provided to groups of vendors, suppliers, customers, and others who have a requirement to access select databases and processes, perhaps for *Electronic Data Interchange* (EDI) applications. Extranets, for example, can enable customers to place orders electronically and to track them to fulfillment, and vendors can track retail sales of their products, perhaps store by store. Security clearly is a major issue with extranets.

13.13 INTERNET SECURITY: A SPECIAL ISSUE

Not a cow, nor a gift of land, nor yet a gift of food, is so important as the gift of safety, which is declared to be the great gift among all gifts in this world.
Panchatantra, fifth century B.C.

The Internet is inherently insecure. It is, after all, an open network. Its openness certainly is one of its major strengths and, at the same, perhaps its major weakness. While e-commerce is promoted as the future of the Internet, this lack of security certainly has slowed development of commercial applications. It is worth noting, however, that providing your credit card number over the Internet is probably as secure as giving your credit card to a server at a restaurant.

Beyond security concerns over Internet commerce, you must remember that access to the Internet is a two-way door—just as users can get out, others can get in. All too often, those others have no legitimate right to do so. Although few organizations admit to having had their systems breached via the Internet, it is clear that such occurrences are all too common.

13.13.1 Security Risks and Countermeasures

The Internet is rife with risks. Hackers, crackers, saboteurs, and other unsavory characters abound, eagerly attacking the Net and its users at every opportunity. The risks certainly include system intrusion, unauthorized data access, system sabotage, planting of viruses, theft of data, theft of credit card numbers, and theft of passwords. While the Internet and the Web cannot be blamed for the concepts and practice of mischief, fraud, theft, and other socially unacceptable forms of behavior, they certainly provide another high-tech cyberalley on the information superhighway.

The *Computer Emergency Response Team* (CERT) at Carnegie-Mellon University comprises a group of experts which are responsible for overseeing security issues on the Internet. While it is highly doubtful that a single security measure or standard will prevail in the near and distant future, there exist a number of options, including message encryption, authentication, and authorization. Firewalls, incorporating much of the above, recently have gained the spotlight in terms of a defense mechanism.

13.13.1.1 Encryption *Encryption* involves scrambling and compressing the data prior to transmission; the receiving device is provided with the necessary logic in the form of a *key* to decrypt the transmitted information. Encryption logic generally resides in firmware included in stand-alone devices, although it can be built into virtually any device. Such logic now, for example, is incorporated into routers, which can encrypt/decrypt data on a packet-by-packet basis. Encryption comes in two basic flavors:

- *Private-key encryption*, also known as *single-key* or *secret-key* encryption, uses the same key for both encryption (encoding) and decryption (decoding). This approach requires that the key be kept secret through some form of secure key transmission prior to the ensuing data transfer.

- *Public-key encryption* involves the RSA encryption key that can be used by all authorized network users. The key for decryption is kept secret. Public-key encryption is much slower than private key, but the dissemination of the key is accomplished much more quickly. Public-key encryption is available freely on the Internet via a program known as *Pretty Good Privacy* (PGP), developed by Philip Zimmerman. PGP was under a cloud for some time because there was concern that it was so powerful as to violate U.S. technology export laws. By the way, encryption technology technically is classified under U.S. law as a form of *munitions*. The commercial version of PGP is known as ViaCrypt PGP, offering an improved user interface.

13.13.1.2 *Data Encryption Standards*   Data encryption standards include *Data Private Facility* (DPF), DES, RSA, and Clipper. *Data Encryption Standard* (DES), which uses a challenge–response approach and intelligent tokens, was formulated by the U.S. National Bureau of Standards. RSA, named after its developers, Rivest, Shamir, and Adleman, is for public key encryption. *Clipper*, an encryption standard developed by the U.S. government, uses escrowed keys to permit government deciphering through a *back door*. Clipper, which is nonexportable, is used extensively by the U.S. government and those who wish to do business with it. Encryption programs used on the Net include SSL, S-HTTP, and combinations of them.

- *Secure Sockets Layer* (SSL) from Netscape negotiates point-to-point security between client and server, including type of encryption scheme and exchange of encryption keys. SSL sends messages over a socket, which is a secure channel at the connection layer and existing in virtually every TCP/IP application. While SSL can accommodate a number of encryption algorithms, Netscape has licensed RSA Data Security's BSafe to provide end-to-end encryption as well as key creation and certification. Netscape's Netsite Commerce Server technology, including SSL, has been licensed by the likes of DEC (now part of Compaq, which is part of Hewlett-Packard), Novell, the Bank of America, and Delphi. *Socket*, by the way, is an operating system abstraction that permits application programs to access communications protocols automatically. Bolt Beranek and Newman (now BBN Technologies) developed this concept in conjunction with the company's early work on TCP/IP.

- *Secure HyperText Transport Protocol* (S-HTTP) from Enterprise Integration Technologies also negotiates point-to-point security between client and server, although at the application layer. EIT has licensed RSA Data Security's BSafe *Toolkit for Interoperable Privacy-Enhanced Messaging* (TIPEM). S-HTTP is a superset of HTTP and, therefore, is specific to the Web; several manufacturers of Web servers have announced plans to include S-HTTP in their products. S-HTTP has gained the support of the W3C and looks to be moving toward acceptance as a de facto standard.

13.13.1.3 Authentication *Authentication* provides a means by which network managers can confirm the identities of those attempting access to computing resources and the data they house. Authentication consists of password protection and intelligent tokens.

- *Password protection* is imposed to restrict individuals on a site, host, application, screen, and field level. Passwords should be of reasonably long length, alphanumeric in nature, and changed periodically. There is a current trend toward the use of dedicated password servers for password management. *Password Authentication Protocol* (PAP) is a commonly used mechanism for password protection in support of remote users. While PAP is easy to use, passwords typically are sent to the Remote Access Server (RAS) in *plain text* (i.e., *in the clear*, or unencrypted).
- *Intelligent tokens* are hardware devices that generate one-time passwords to be verified by a secure server. They often work on a cumbersome *challenge–response* basis. *Challenge Handshake Authentication Protocol* (CHAP) is an example of this improved approach. CHAP involves the RASs challenging the remote user with a random number. The user responds with a *digest*, which is an encrypted password based on the random-number challenge. The RAS then decrypts the password using that same random-number key to verify the identity of the remote user.

13.13.1.4 Authorization *Authorization* provides a means of controlling which legitimate users have access to which resources. Authorization involves complex software that resides on every secured computer on the network; ideally, it provides *single sign-on* capability. Authorization systems commonly used in support of Internet security include Kerberos, Sesame, and Access Manager:

- *Kerberos* draws its name from *Kerberos* (also known as *Cerebrus*), the three-headed monster that guarded the entryway to the infernal regions in Greek mythology. Perhaps the best known authorization software, it was developed by the Massachusetts Institute of Technology (MIT) and is available free, although more powerful commercial versions exist as well. As Kerberos uses DES, it is not easily exportable. IBM's Kryptoknight is a weaker but exportable Kerberos variant. Although, according to Greek legend, Hercules defeated Kerberos, a hacker of Herculean proportions has yet to emerge victorious over this powerful software.
- *Sesame* (*Secure European System for Applications in a Multivendor Environment*) was developed by the ECMA (European Computer Manufacturers Association). It is flexible, open, and intended for large, heterogeneous network computing environments. It also is highly complex and not effective for smaller applications.
- *Access Manager* is an authorization mechanism approved by the IETF. Access Manager uses an API for application development, employing scripting. *Scripting* involves a process of mimicking the log-on procedures of a program, providing basic levels of security for small networks.

*13.13.1.5 **Firewalls*** *Firewalls* comprise application software that can reside in a communication router, server, or some other device. That device physically and/or logically is a first point of access into a networked system. On an active basis, the device can block access to unauthorized entities, effectively acting as a *security firewall.* Firewalls provide logging, auditing, and *sucker traps* to identify access attempts and to separate legitimate users from intruders. Firewalls can be in the form of a programmable router or a full set of software, hardware, and consulting services.

13.13.2 Virtual Private Networks

The term *Virtual Private Network* (VPN) has many definitions, all of which are valid. In Chapter 5, I explored voice VPNs, also known as Software-Defined Networks (SDNs). In Chapter 7, I explored classic data VPNs. Both of these conventional VPNs are circuit switched in nature. In other chapters, I examined X.25 and packet switching, Frame Relay, and ATM, all of which also are characterized as VPNs. In contemporary usage, the term most commonly refers to the creation of a virtually private network over the public Internet or over a public IP-based network.

At the center of all definitions is the fact that the VPN has some of the characteristics of a private, leased-line network without being one. True private networks are distinguished by the fact that dedicated circuits, or channels, or at least channel capacity interconnect multiple sites in an enterprise network. Therefore, the bandwidth always is available and without any usage charges, and there are no issues of access and congestion control, at least not at the network level. Performance essentially is guaranteed, and security never is an issue. The disadvantages of private networks include long configuration and reconfiguration times, high installation costs and recurring charges, lack of scalability, and susceptibility to catastrophic failure. Private networks are optimized on the side of performance, rather than raw efficiency.

VPNs reverse these factors to some degree. VPNs are fast and easy to configure and reconfigure and are highly scalable, with a solid relationship between cost and functionality. As VPNs make use of a public network that is shared in terms of access, switching, and transport and is highly redundant, issues of catastrophic failure are much reduced. Their shared nature, however, creates ever-present issues of access control and congestion control. Therefore and particularly in the case of IP and Frame Relay packet networks, issues of latency, jitter, loss, and throughput always must be considered. Further and particularly in the case of the IP-based Internet, security is a considerable concern. In the current context of the Internet-based VPN, security issues are mitigated through the use of a combination of authentication, encryption, and tunneling.

*13.13.2.1 **Authentication*** Authentication, as previously discussed, is a means of access control that ensures that users are who they claim to be. Whether through password protection or intelligent tokens, the authentication process is intended to avoid the possibility that unauthorized users might gain access to internal computing or network resources.

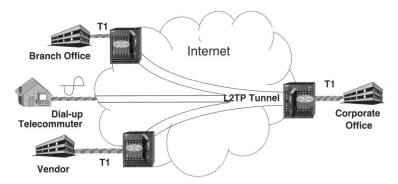

Figure 13.10   VPN through the Internet, illustrating the use of tunneling and firewalls.

13.13.2.2   Encryption   Encryption, also as discussed previously, is the encoding, or scrambling, of the data for security purposes. In order to decode the data, the destination device must have the correct key. In the VPN context, data encryption and decryption can occur either at the user endpoints or at the edge of the service provider's network. In the first case, the user endpoint devices can include workstations or other host computers, routers, or *Remote Access Servers* (RASs). In the second case, the carrier's or service provider's (e.g., ISP) equipment is responsible for encryption and decryption.

13.13.2.3   Tunneling   *Tunneling* is the process of encapsulating an encrypted data packet in an IP packet for secure transmission across the inherently insecure Internet, as illustrated in Figure 13.10. The four leading tunneling protocols are SOCKSv5, PPTP, L2TP, and IPsec:

- *SOCKSv5* is an authentication technique that runs at the Session Layer (Layer 5) of the OSI Reference Model. Through the use of secure sockets negotiated between client and server over a virtual circuit and on a session-by-session basis, SOCKv5 supports the security of UDP datagrams as a data stream, rather than on a packet-by-packet basis. SOCKSv5 also supports protocol-specific communications, such as SMTP, to prevent hackers from extracting e-mail data through the use of an alias. SOCKSv5 is a cross-platform technique, working across multiple operating systems and browsers. SOCKSv5 also interoperates on top of IPv4, IPsec, PPTP, L2TP, and other lower level protocols.
- *Point-to-Point Tunneling Protocol* (PPTP) operates at the Data Link Layer (Layer 2) of the OSI Reference Model. Initially conceived by Ascend and developed by Microsoft and embedded in Windows NT, PPTP is a proprietary technique that encapsulates Point-to-Point Protocol (PPP) frames with IP packets. Packet filters provide access control end to end and server to server.
- *Layer 2 Tunneling Protocol* (L2TP) is an IETF (RFC 2661) standard that evolved from a combination of PPTP and Cisco's *Layer 2 Forwarding* (L2F) protocol. L2TP is used for secure, node-to-node communications by ISPs and other VPN service providers, in support of multiple, simultaneous tunnels in the network core. End users gain access to the service provider on an

unencrypted basis; the service provider assumes that responsibility at the edge of the packet network.

- *IP Security* (IPsec) as I noted earlier in this chapter, is the security mechanism developed for IPv6 and is used in IPv4 as well. In a dual-stack mode, IPv4 frames are encapsulated for transport within encrypted IPv6 frames. IPsec runs at the Network Layer (Layer 3). IPsec is a standards-based solution from the IETF and is defined in RFC 2401.

13.13.2.4 *Applications Scenarios*   VPNs really are all about cost effectiveness, with the emphasis on cost rather than raw performance. Especially in the context of the IP-based VPN, the reduction in costs can be very significant indeed. Most such VPNs are based on the public Internet, although the IP-based networks from carriers such as Level 3 and Qwest also support VPNs. Access techniques include all of the options I have detailed, including dial-up modems, cable modems, ISDN, xDSL, Frame Relay, and ATM. The application scenarios include remote access, intranets, and extranets. Note that VPNs can be incredibly cost effective in support of the multinational enterprise, as truly private networks often are prohibitively expensive, if even available, on an international basis.

- *Remote access VPNs* are highly cost effective in support of telecommuters and mobile and remote workers. Assuming that the worker can reach the Internet or other IP-based network over a dial-up or other form of connection, reasonably secure communications can be accomplished with the home office or any branch office on the VPN. The level of bandwidth provided, of course, depends on the speed of the remote access link, the speed of the link to the home or branch office and the total volume of traffic over that link, and the level of congestion currently experienced over the shared IP backbone network. Significant cost savings can be realized when you compare VPNs to dial-up PSTN costs, whether based on Direct Distance Dialing (DDD) or toll-free access numbers. Each client PC, of course, must be equipped with software that supports the necessary tunneling protocols. Savings also can be realized at the branch, regional, or corporate location, as both T-carrier circuits connected to the PSTN and associated *Remote Authentication Dial-In User Service* (RADIUS) routers can be either eliminated or consolidated, in favor of shared T-carrier circuits connecting to the VPN.
- *Intranet VPNs* serve to link branch, regional, and corporate offices. Access to the VPN generally would be on the basis of dedicated circuits, usually in the form of unchannelized T-carrier, which might run the Frame Relay protocol or, perhaps, ATM. In this scenario, the elimination or consolidation of access circuits can yield significant cost benefits at all connected sites.
- *Extranet VPNs* serve to link vendors, customers, affiliates, and distributors into the main corporate office. An extranet VPN often must be extremely flexible in terms of access techniques and security mechanisms, as the corporate sponsor may have little control over the specifics of the external users' systems and processes.

Cost savings can be extreme if access to and usage of the VPN is based on a flat rate, with no usage charges. Large ISPs offering enhanced VPN services typically offer service-level guarantees, as stated in contracts known as *Service Level Agreements* (SLAs). SLAs address performance parameters such as downtime and overall network throughput, which can be crucial in a mission-critical and time-sensitive application environment. While such enhanced performance contracts commonly are based on an additional usage-based pricing algorithm, the overall costs still compare quite favorably with those of DDD and toll-free, dial-up access.

13.14 MISUSE AND CONTENT

There have been a large number of highly publicized cases of the Internet and the Web being misused for illicit and immoral purposes. The Internet has been used for transmitting stolen credit card numbers and cellular telephone ID numbers, for example. While any communications medium can be used for such purposes, the Net creates another set of difficulties for law enforcement because communications are virtually instantaneous and multiple parties can gain access to the illegal data through a bulletin board. Further, it is difficult, if not impossible, in many cases to track down the offenders.

More significant from a social perspective is the fact that Internet *chat rooms* have been used to lure minors into the clutches of pedophiles and others who wish to take advantage of them. The offenders generally seem to mask themselves as minors, striking up an electronic conversation and suggesting a meeting somewhere. While there have been only a relative few such cases uncovered and publicized, the risks exist and the consequences can be terrible.

Also of great concern is the issue of content and access to it. Numerous websites contain sexually explicit material, including photographs. While adults have a constitutional right to view such material, minors generally do not; the Internet and the Web really have no effective means of controlling access, since the users are anonymous and their ages and other characteristics are unknown. Most of the truly explicit websites (so the author is told) offer very little for free, with full access to the offending material generally being provided only on the basis of subscription paid by credit card.

The FCC and Congress, however, have and continue to consider regulating the content of the Internet. While I think we all would agree that censorship can go too far, I am equally certain that we all would agree that there are numerous and clearly documented cases of beasts who prowl the Internet. (Author's humble opinion: At the risk of lapsing into a discussion of morals, there should be a method for constraining those who would use the Internet and the Web for immoral purposes, as defined by law. At the very least, there should be a solid means of blocking access of minors to such material. There also should be stiff penalties imposed on violators.) The Electronic Frontier Foundation and numerous other organizations promoting privacy are continuously battling the FCC and Congress in this regard, citing freedom of speech; petitions are passed (you guessed it) over the Internet. Although a Communications Decency Act was passed in 1996, it was overturned in federal district court, citing violation of free speech as guaranteed by the First Amendment to the Constitution. While we may never see Internet content censored, there do exist a number of commercially available software filters that enable parents to deny access to Internet sites that might contain unsavory content, as defined by the filter developers. Those filters run against Web servers, which are updated on a regular basis in order to keep pace with the dynamics of the websites and their content.

Outside the United States, very tight content controls have been exercised in some countries. As you might expect, those nations include Singapore and the Peoples Republic of China (PRC). In January 1999, an Internet e-mail broker in the PRC was sentenced to two years in prison for *inciting the overthrow of state power* by supplying 30,000 e-mail addresses to a prodemocracy magazine in the United States. The sentence was considered by many to be light, as three leaders of the China Democracy Party received sentences of 11, 12, and 13 years after being found guilty of the same charge in December 1998. Since that time, there have been a number of similar occurrences in various countries.

13.15 INTERNET ODDITIES, SCREWBALL APPLICATIONS, AND SOME REALLY GOOD IDEAS

As mentioned previously, only technology and the human mind can limit the applications for the Internet and the Web. There have been a number of recent announcements of Internet oddities and screwball applications as well as some really good ideas. I've got my own ideas about which are which, and I'm sure you have yours.

- *Package tracking* is offered on the Web by FedEx, UPS, USPS, and other carriers. Customers can access the websites and get the latest status of packages, based on package tracking numbers. The savings easily are in the millions of dollars, compared to the costs of handling voice calls through human agents or even interactive voice response applications in an incoming call center.
- *Document delivery* over the Internet has been promoted by UPS, through its document exchange service known as UPS OnLine Courier. Senders upload documents to the secure UPS server, at which point UPS sends e-mail notifications to the recipients, advising them of documents awaiting delivery. The recipient uses the URL provided in the notification to download the document from the secure server via a Web browser. An option enables the sender to require that the recipient use a password known only to the sender and recipient. Security is provided via 128-bit encryption on the server and 40-bit Secure Socket Layer (SSL) encryption during transport. Pricing is attractive, especially when compared to U.S. Postal Service (USPS) Registered mail and traditional courier services. Although UPS discontinued OnLine Courier effective September 1, 2002, it is a good idea that surely will find a market at some point in the future. Over the past few years, the USPS has offered a number of e-mail services, some of which have experienced modest levels of success and some of which have been discontinued.
- *Distribution of software, upgrades, and bug fixes* is a really good idea. In many cases, software is provided free of charge as a public domain release, with subsequent, enhanced versions provided commercially.
- *Online publishing* has really taken off in recent years. Most of the technology publications, of course, have developed websites for online access to published articles and late-breaking news. While associated advertising revenues are not

made public, they are very substantial. Even my local *Skagit Valley Herald* has a website, and began charging for access in 2006, much to my dismay, as I enjoy keeping up to date on the local news while traveling.

- *Cyberfairs* are job fairs that enable companies to post job listings and recruit applicants online. Interactive communications are supported.

- *Cybergambling* (Now there's a constructive idea!) is now offered by casinos. The casinos can circumvent U.S. gambling laws by placing their servers offshore. Users are required to have offshore bank accounts, as well, to deposit their winnings. imho (that's netspeak for *in my humble opinion.* note the lack of capitalization. that's also an e-mail convention. it's in the style of e.e. cummings.), we do not need a Virtual Vegas. The U.S. government appears to agree. Actually, the State of Washington outlawed cybergambling in 2006.

- *Cybersex* is something I mention only because its popularity demands that I do so. It is not worthy of further comment, at least not in my humble opinion.

- *Music distribution* over the Internet currently makes use of MP3 (MPEG-1, Audio Layer 3) compression algorithm. As each minute of uncompressed digital music requires a file of roughly 10 MB, it takes a very long time to download an entire CD-ROM album, even over an ADSL connection. MP3 compresses a 600-MB CD-ROM music album down to about 50 MB. Issues of copyright infringement are significant with respect to this application.

- *Online auction* houses offer electronic auctions for just about everything imaginable, from airline tickets, hotel rooms, and cruises to sporting goods, antique china, and fabric remnants for quilting. The risks are several: You do not get to see the goods before you buy, and you have no real assurance of delivery. In other words, you buy solely on faith. Personally, I don't get it. I generally can get much better fares, much better flight schedules, and much better service from my travel agent. As it seems that the airlines are doing everything they can to put the travel agents out of business, I may not have that option forever. Margaret, my lovely wife, on the other hand, loves this sort of thing. Apparently, a lot of other people love this sort of thing as well. Beware, however. The U.S. Internet Crime Complaint Center (IC3) received 231,493 complaints in 2005. Internet auction fraud was by far the most reported offense, comprising 62.7 percent of referred complaints [61].

- *Online grocers* attracted a lot of attention several years ago. WebVan and HomeGrocer, the most prominent members of this group, offered the convenience of shopping from your PC over the Internet. In other words, you could shop from home or office for exorbitantly overpriced groceries that would be delivered right to your doorstep. Believe it or not, I actually know people who used these services, and they were very disappointed when both WebVan and HomeGrocer went bankrupt. Frankly, it did not surprise me a bit, although it does surprise me that at least one online grocer has survived. Peapod delivers to the Chicago, Illinois, Milwaukee, Minnesota, and southeast Wisconsin areas. Through an alliance or strategic relationship of some sort with Stop & Shop, it also supplies portions of Connecticut, Massachusetts, Rhode Island, New Jersey, and New York. Through the Giant grocery chain, it also offers service in the greater Washington, D.C., area. Is it any wonder that the latest government studies show that 62.5 percent of Americans are overweight or obese?

- *Bill presentment*, or *bill presentation*, is the rendering of a bill on a website. A number of large voice and data communications providers offer this service via secure websites, also providing for electronic payment in the form of authorizations for wire transfers or credit card charges in a variation known as *Electronic Bill Presentation and Payment* (EBPP).

- *Click-and-smell* technology certainly is a high-tech step above scratch-and-sniff marketing, and it is a generation beyond Smell-O-Vision, for you movie-going baby boomers who remember that obnoxious and obscure technology from the early 1960s. The various permutations of click-and-smell involve devices that attach to your PC, or perhaps your IPTV set. When you access a website designed to support this application, the device downloads the necessary commands and reproduces a particular scent by mixing several base chemicals in the proper proportions. So, you theoretically could sample a new perfume or cologne before ordering it online or you could smell a pizza or a get a whiff of the odor of the interior of a new car. I really have to scratch my head over this one, but if you are really interested, you can get more information at www.aromajet.com and www.trisenx.com.

- *SETI@home* is a means of harnessing the power of hundreds of thousands of home PCs to assist in the SETI (Search for ExtraTerrestrial Intelligence) project. SETI involves sifting through billions and billions of radio signals captured by large radiotelescopes in the hopes of finding a pattern that could represent a message from an intelligent life form on another planet. SETI@home screen saver can be downloaded across the Internet, along with samples of radio signals for analysis. During periods of idleness, the software searches for patterns. Once the analysis is complete, the results are uploaded to the SETI server, and another sample is downloaded for analysis. This is a cool idea! Carl Sagan would be proud. No question about it! Check it out at http://setiathome. berkeley.edu/.

- *Application Service Provider* (ASP) is a class of Internet-based company that first appeared in 1999. An ASP provides access to software over the Internet, or perhaps over a private network, for a fee that generally is based on the number of users. ISPs, systems integrators, and software vendors all have adopted this term, which describes what essentially is an Internet-based outsourcing approach that allows the user company to avoid the cost of acquiring, installing, supporting, securing, and upgrading expensive software applications and the platforms on which they reside. A *Managed Service Provider* (MSP) is an ASP that delivers and manages network-based services, applications, and equipment for a fee. An MSP may load a company's application data on its servers, customizing the data as necessary, and operating the service at a remote data center. There are clear advantages to this approach for the client companies, as they can focus on their core businesses, leaving the complex technical details of transaction-based e-commerce to the ASP, which theoretically knows that aspect of the business best and can take advantages of the economies of scale. Applications served by ASPs include human resources, accounting and other financial services, and retail sales and services [62–66].

- *Map sites*—There are several very cool websites that variously provide maps and driving directions to general points of interest and specific physical addresses. MapQuest (www.mapquest.com), for example, provides excellent maps, with zoom capabilities and directions that you can download to your computer and print out for future reference. There is an option for Web-enabled cellular telephones, so that you can access the information en route. There also is a GPS-enabled service that lets you identify your location and share it with trusted others as long as all parties are on the Sprint-Nextel cellular network. (I presume the GPS service is so that your friends can find you in the event that the directions are wrong and you can not find them.) Google Earth is a free-of-charge, downloadable virtual globe program that completely maps the entire Earth by pasting together satellite imagery, aerial photography, and *Global Information System* (GIS) technology. Google Earth provides zoom capabilities, map rotation, and even 3D with tilt views, all controllable by mouse. Both of the sites are very cool.

- *Unclaimed property*—There may be more than one of these, but I am familiar with the National Association of Unclaimed Property Administrators (http://www.unclaimed.org), which allows you to search unclaimed property records in each state in the United States to determine whether there is unclaimed property being held in your name. Common forms of unclaimed property include savings or checking accounts, stocks, uncashed dividends or payroll checks, refunds, traveler's checks, trust distributions, unredeemed money orders or gift certificates (in some states), insurance payments or refunds and life insurance policies, annuities, certificates of deposit, customer overpayments, utility security deposits, mineral royalty payments, and contents of safe deposit boxes. I found $25.00 that the State of California had been holding for me since my days in Bakersfield 25 years before.

- *Myth busters*—There are several sites dedicated to investigating and either authenticating or debunking those unbelievable e-mail stories and photos that circle the Internet. My personal favorite is Snopes (www.snopes.com). The Computer Incident Advisory Capability (CIAC) Hoaxbusters site (http://hoaxbusters.ciac.org/), another good one, is maintained by the U.S. Department of Energy, Office of the Chief Information Officer, Office of Cyber Security. By the way and just in case you were wondering, Bill Gates will not pay you $245 for every address to which you forward a certain e-mail message just to test the market penetration of Internet Explorer.

- *Dictionaries and encyclopedias*—The Skeptic's Dictionary (http://skepdic.com/) provides lengthy definitions based on skeptical analysis of terms, concepts, people, and apparently anything else that interests the author, Robert Todd Carroll. Examples include Noah's Ark, Nostradamus, Occam's razor, occult, and Yeti. Wikipedia (http://en.wikipedia.org) is an online reference written collaboratively by just about anyone with Internet access who wishes to write, edit, correct, or improve information. There are thousands of detailed definitions, with many providing links to other information sources. In my experience, Wikipedia is well done, for the most part, although many topical categories are incomplete, and some information is incomplete or even incorrect, but such is the nature of such an open forum. The portal refdesk.com (http://www.refdesk.com) is a good source of facts on a wide variety of subjects.

- *Municipal and regional*—A number of municipalities and regions have established websites, including my little town of Mount Vernon, Washington (http://www.ci.mount-vernon.wa.us/), and Skagit County (http://www.skagitcounty.net). There also are commercial and not-for-profit websites that serve to promote a town, city, or region and that often take positions on issues of local significance. One good example is East Texas Towns Online (http://easttexastowns.com/), which characterizes itself as *a gadfly website with a strong environmental bias*.

- *Photos and videos*—There are a number of sites that allow users to post digital photos for storage and viewing by friends and family over the Internet. There also are several sites that allow users to post video clips. In fact, the most popular website currently is YouTube (http://www.youtube.com/), with over 100 million video downloads per day from a selection that grows by 65,000 videos per day. The site was started in 2005 by three young entrepreneurs in their twenties, and now reportedly is worth as much as US$1 billion.

Truly bored Netheads can while away their time on the following sites:

- *Mr. Potato Head*—An online version of everyone's favorite starchy vegetable toy (http://www.cs.utk.edu/~ffowler/javahtml/potato/Potato.html).
- *Strawberry Pop-Tart Blow-Torches*—A step-by-step guide to the process of using this popular breakfast food as an incendiary device. (www.sci.tamucc.edu/%7Epmichaud/toast). I do not recommend that you try this at home.
- *Lost in Space*—A guide to every episode of this popular TV show. (www.lostinspacetv.com).

13.16 THE DARK SIDE: AN EDITORIAL

It has been said that the Internet is the electronic equivalent of the Gutenburg printing press in terms of its impact on the information age. That quite likely is an understatement. Certainly, the Internet and the Web support an unprecedented level of information access for the electronically privileged. In contrast to the printed word, however, much of the content on the Internet and the Web essentially is self-published. Therefore, there are no guarantees of either its objectivity or its accuracy. It is up to the reader to sort out the bias, the subjective, the self-serving, the inaccurate, and the outright lie in order to get the truth. Similarly, it is up to the reader of the text and the viewer of the image or video to sort out the ugliness of racial hatred and pornography that assault the senses and sensibilities of the vast majority of us in the global society.

Take this book, for example. It is based on 30 years of experience, countless hours of research, and an unyielding commitment to the objective truth. It has been written with all the skill at my disposal. A highly skilled and knowledgeable team, including a consulting editor, a great technical editor, a development editor, and a copy editor, reviewed it. For what it is, it is the best that we can make it. If my drafts contained any obviously and blatantly biased, subjective, self-serving, inaccurate, or untrue statements, at least one member of the team corrected them or suggested that I eliminate them. If my drafts overly emphasized the ugliness of websites that promote racial, ethnic, or religious hatred or that of pornography, my editors would have struck it, and rightfully so. Compare that with your own experience on the Internet and the Web. I make that comparison every day, as I do my research and sort through the tens of thousands of sources of information and misinformation.

Having said all that, I also have to tell you that the first edition of this book, along with a number of other books, was once published on the Internet by a university in a nation that once was one of the Soviet republics. It was an obvious case of copyright infringement that took a good deal of time, effort, and expense to stop. It was outright thievery that I did not appreciate a bit, and neither would you, were you in my position.

You may recall the horror of the TWA Flight 800 incident. You also may recall the statements of Pierre Salinger, who was the press secretary for President John F. Kennedy. Salinger claimed to know on national news that Flight 800 was, in fact, shot down by an errant missile fired from a U.S. Navy warship. Salinger got this bogus information from a website.

You may remember the *Good Times* virus and a host of other viruses. While there are many real viruses that pose real threats, many of them are absolute hoaxes, the rumors of which spread like wildfire over the Internet. Virtual panic resulted.

You may remember that in June 1998 an Associated Press reporter mistakenly posted a prepared obituary for Bob Hope on that wire service's website. While the error was discovered virtually immediately and the obituary was removed within 15 min or so, a U.S. congressman saw it and eulogized Bob Hope from the floor of the House of Representatives. Mr. Hope's response was somewhat along the lines of Mark Twain's, "Reports of my death have been greatly exaggerated." Perhaps Sir Walter Scott said it best: "Oh, what a tangled web we weave when first we practice to deceive."

The bottom line is that it is your responsibility to seek the true and the beautiful. There are no filters on the Internet or the Web that truly will protect you. Protect yourself if you can, and please protect your children, for they often cannot—or will not—protect themselves.

REFERENCES

1. "Internet Domain Survey, Number of Hosts Advertised in the DNS." Internet Systems Consortium, http://www.isc.org/index.pl.

2. "IDC Examines the Future of Email as It Navigates Security Threats, Compliance Requirements, and Market alternatives." IDC Press Release, December 22, 2005, http://www.idc.com/getdoc.jsp?containerId=prUS20033705.

3. "IDC Predicts Internet Traffic Will Continue to Double." IDC Press Release, February 28, 2003. http://lw.pennnet.com/Articles/Article_Display.cfm?Article_ID=169837&CFID=6008592&CFTOKEN=92937694.

4. "Internet Traffic Growth Slows by Half in 2005." TeleGeography Press Release, August 23, 2005, http://www.telegeography.com/press/releases/2005-08-23.php.

5. Cerf, Vinton G. and Kahn, Robert E. "A Protocol for Packet Network Interconnection." *IEEE Transactions on Communications*, Institute of Electrical and Electronics Engineers, 1974.

6. Comer, Douglas E. *Internetworking with TCP/IP*, 2nd ed. Prentice-Hall, 1991.

7. "MCI Worldcom MAE Services." Originally accessed at www.wcom.com/tools-resources/ about...services/mae_services_defined/tiers.shtml.

8. Garfinkel, Simson. "Where Streams Converge." *Hotwired.* www.hotwired.com/packet/ garfinkel/96/37/geek.html.

9. Pappalardo, Denise and Gittlen, Sandra. "Is It Too Late for WorldCom's MAEs?" *Network World*, August 24, 1998.

10. Pappalardo, Denise and Marsan, Carolyn Duffy. "How Ready Are the Nation's Networks?" *Network World*, November 26, 2001.

11. Pappalardo, Denise. "New Florida NAPs to Improve Connectivity to Latin America." *Network World*, July 2, 2001.

12. Metcalfe, Bob. "Coming Internet Collapse Spurring Shortsighted Proliferation of Intranets." *InfoWorld*, May 20, 1996.

13. Rickard, Jack. "Internet Architecture." *Boardwatch Magazine*, 1996. Originally accessed at www.boardwatch.com/isp/fallisp/archi.html.

14. Carl, Jeremy. "Universal Access to Internet: Who Pays?" *Web Week*, May 20, 1996.

15. Goldman, Alex. "Top 22 U.S. ISPs by Subscriber: Q1 2006." http://www.isp-planet.com/ research/rankings/usa.html.

16. "Federal Communications Commission Releases Data on High-Speed Services for Internet Access." Federal Communications Commission, July 26, 2006. http://hraunfoss.fcc. gov/edocs_public/attachmatch/DOC-266593A1.doc.

17. Miller, Mark A. *Implementing IPv6*. M&T Books, 1998.

18. Miller, Mark A. "Finding Your Way Through the New IP." *Network World*, December 16, 1996.

19. Jesdanun, Anick. "Red Light Flashed for '.xxx' Domain." *Seattle Post-Intelligencer*, August 7, 2005.

20. http://encarta.msn.com/find/Concise.asp?z=1&pg=2&ti=761574118.

21. Dornan, Andy. "Can the Internet Move Beyond Dot-Com?" *Network Magazine*, March 2001.

22. Weiss, Todd R. "Domain-Name Registration Gets 64 More Languages." *Computerworld*, February 27, 2001.

23. Miller, Robert and Keeler, Elissa. *Internet Direct: Connecting through SLIP and PPP*. MIS Press, 1995.

24. Sihl, Karl. "IPv6 Addresses Demand for Space." *Network World*, May 16, 2005.

25. Karve, Anita. "Lesson 115: IP Security." *Network Magazine*, February 1998.

26. Smetannikov, Max. "IPv6: So Far, Few Takers." *Interactive Week*, May 21, 2001.

27. Garretson, Cara. "Bechtel Says Move to IPv6 Is All About Business." *Network World*, October 10, 2005.

28. Marsan, Carolyn Duffy. "IPv6 Fears Seen Unfounded." *Network World*, December 15, 2003.

29. Saho, L. Michael. "Transmission Control Protocol/Internet Protocol (TCP/IP)." *Datapro Communications Analyst*. Datapro Information Services Group, March 1994.

30. Muller, Nathan J. "Using the Internet." *Datapro Communications Analyst*. Datapro Information Services Group, December 1994.

31. Clark, Elizabeth. "Lesson 157: The Resource Reservation Protocol." *Network Magazine*. August 2001.

32. Treece, Greg, Steinman, David, and Bittle, Ben A. "QoS Challenges, Opportunities for Next-Generation Optical Internet." *Lightwave*. April 2001.

33. Miller, Mark A. *Voice over IP Technologies*. M&T Books, 2002.

34. Branden, R., Zhang, L., Berson, S., Herzog, S., and Jamin, S. "Resource ReSerVation Protocol—Version 1 Functional Specification," RFC 2205. Internet Society, September 1997.

35. Schulzrinne, H., Rao, A., and Lanphier, R. "Real Time Streaming Protocol (RTSP), RFC 2326. Internet Society, April 1998.

36. Nichols, K., Blake, S., Baker. F., and Black, D. "Definition of the Differentiated Services Field (DS Field) in the IPv4 and IPv6 Headers," RFC 2474. Internet Society, December 1998.

37. Blake, S., Black, D., Carlson, M., Davies, E., Wang, Z., and Weiss, W. "An Architecture for Differentiated Services," RFC 2475. Internet Society, December 1998.

38. Ma, Tao and Bingxin, Shi. "Integrated Services and Differentiated Services Team Up to Deliver True QoS to the Internet." *Network Magazine*. November 2000.

39. Clark, Elizabeth. "Lesson 158: Differentiated Services." *Network Magazine*, September 2001.

40. Greenfield, David. "Lesson 163: MPLS in Brief." *Network Magazine*, February 2002.

41. Metzler, Jim. "Innovation in MPLS-Based Services." *IT Innovation Report*, March 2006, http://www.webtorials.com/abstracts/Kubernan2.htm.

42. Fineberg, Victoria. "QoS Support in MPLS Networks." MPLS/Frame Relay Alliance White Paper, May 2003, http://www.mfaforum.org/tech/MPLSQOSWPMay2003.pdf.

43. Horak, Ray. "MPLS: FR RIP." *CommWeb*, January 21, 2005, www.commweb.com.

44. Awduche, D., Malcolm, J., Agogbua, J., O'Dell, M., and McManus, J. "Requirements for Traffic Engineering over MPLS," RFC 2702. Internet Society, September 1999.

45. Treece, Greg, Steinman, David, and Bittle, Ben A. "QoS Challenges, Opportunities for Next-Generation Optical Internet." *Lightwave*, April 2001.

46. "Generalized Multiprotocol Label Switching (GMPLS)." International Engineering Consortium, http://www.iec.org/online/tutorials/gmpls/.

47. www.internet2.edu.

48. Stapleton, Richard M. "Bigger, Better, Faster: Here Comes Internet2." *InterActive Week*, August 28, 2000.

49. Pappalardo, Denise. "Internet2's Network to Get a Facelift." *Network World*, May 1, 2006.

50. Shannon, Victoria. "The New Net: A Wide Effort." *International Herald Tribune*, August 20–21, 2005.

51. "Uniform Resource Identifier (URI): Generic Syntax." Internet Society, 2005.

52. Driscoll, Paul A. "What's in a Name? For 'Internet Explorer,' $5 million." *Seattle Post-Intelligencer*, July 2, 1998.

53. http://www.myspace.com/Modules/Common/Pages/AboutUs.aspx.

54. "Quarterly Retail E-Commerce Sales, 4th Quarter 2005." *U.S. Census Bureau News*. U.S. Department of Commerce, February 17, 2005, http://www.census.gov/mrts/www/data/html/05Q4.html.

55. Blum, Adam. *Building Business Web Sites*. MIS Press, 1996.

56. Railsback, Kevin. "Transcoding Renews Apps for Mobile Access." *Infoworld*, May 29. 2000.

57. Boswell, Rebecca. "Location-Based Technology Pushes the Edge." *Telecommunications*, June 2000.

58. Dornan, Andy. "Pulling the Internet's Plug." *Network Magazine*, July 2000.

59. Ostergaard, Bernt. "WAP vs. I-Mode." *Mbusiness*, January 2001.

60. Dornan, Andy. "WAP Reaches the Second Generation." *Network Magazine*, September 2001.

61. http://www.ic3.gov/media/annualreports.aspx.

62. Carr, David F. "The Rise of the ASP." *Internet World*, July 15, 1999.

63. Hayward, Mark and Strong, Deborah. "The Future of ASPs." *TelOSSource Magazine*, March 2000.

64. Cone, Edward. "What Is an ASP?" *Inter@ctive Week*, September 27, 2000.

65. Toigo, Jon William. "ASPs Hit the Mainstream." *Network Computing*, May 28, 2001.

66. Collett, Stacy. "MSPs: The New Hosts." *ComputerWorld*, November 14, 2005.

CHAPTER 14

NETWORK CONVERGENCE

We have now reached the stage when virtually anything we want to do in the field of communications is possible. The constraints are no longer technical, but economic, legal, or political.

Arthur C. Clarke, United Nations Telecommunications Day, 1983

While Arthur C. Clarke may have overstated things a bit in 1983, his statement seems to be pretty accurate in 2006. Virtually anything we want to do is possible, short of *teleportation*, and scientists are working on that, or so I'm told. The limitation may well be that we just do not want to do enough. In other words, our imaginations are probably more limited that our abilities to satisfy them through the invention of new technologies. Ultimately, of course, technology really is just an enabler of applications, and people keep dreaming up new and exciting applications, many of which are highly demanding technically. Many of these applications are extremely bandwidth intensive, including high-speed data, video, and even multimedia. Billions of dollars have been spent on the development and deployment of new network infrastructure in an attempt to satisfy our seemingly insatiable desires to communicate instantly and in a variety of formats, from audio to data to video to multimedia.

Arthur C. Clarke certainly was correct in saying that the constraints are largely economic, legal, or political, but he failed to include cultural and religious. It seems that there is something of a technology backlash taking place in some parts of the world. Internet access is constrained in the People's Republic of China, for example, as the general population is denied access to websites that offer content contrary to official government policies. It takes little imagination to create a mental image of the restrictions on Internet access in North Korea, where the totalitarian regime is even more oppressive. Media restrictions are excessive in much of the Middle East, where violations of cultural and religious codes of conduct often are punished severely. So, I maintain that we could do so much more if we would only get out of our own way. Oh, well, back to the technology.

Analog has given way to digital technology. Copper has yielded to glass in the backbone and is finding its way into the local loop. Wireline networks have given way to wireless, at least in support of mobile communications, and wireless technologies do a wonderful job of supplementing the wireline backbone networks. Wireless Local Area Networks (WLANs) have been standardized and are enjoying great popularity. Circuit switching is challenged by packet switching across all applications types. Satellite constellations support communications anywhere on the face of the earth. The Internet and the Web provide access to virtually any type of data in any database residing on any networked computer anywhere in the world—issues of security not withstanding and assuming that economics, laws, politics, culture, and religion do not get in the way. Electronic commerce has changed the way we shop for everything from books to music to clothing to automobiles to groceries. Telephone calls are so inexpensive that we no longer give any thought to picking up the phone and calling across the country and even across continents. Telephone calls, snail mail, and faxes have largely yielded to electronic mail, which is virtually free, assuming Internet access. Plain old telephone sets are yielding to softphones and cell phones, and the size of the cell phones we use seems to be related inversely to the size of the SUVs (*Sports Utility Vehicles*) we drive. By the time you buy your next Chevy Subdivision, your cell phone may well fit in your ear and be activated by your brain waves, so be careful what you think. *Note:* That cell phone likely will cost less than a tank of gas. (Actually, my cell phone was free, although it doesn't quite fit in my ear.)

The U.S. economy has changed in the last 200 years from agrarian to industrial to informational. Ten years ago, we lamented the fact that we no longer *make* anything in this country and that our having lost our industrial edge would be the ruination of the American way of life. In fact, generation X could look forward to being the first generation in our history to enjoy a lower standard of living than the previous generation. While we no longer make every part that goes into every car that we assemble, buy, and drive, we do create much of the information technology that drives the rest of the world that makes the parts that go into the Chevy Subdivisions and other vehicles that are assembled in Mexico or Canada or who knows where and shipped to the United States and bought over the Web that runs on technology that was developed mostly in the United States and is embedded in machines that we were at the forefront of inventing and which include components we invented and bought over the Web; and the same goes for the cell phones that soon will fit in our ears so that we are not distracted while driving our Subdivisions—and on and on and on.

In any event, this network infrastructure is being developed to deliver something that a few years ago was known widely as the *information superhighway*. Initially conceived in the United States as the *National Information Infrastructure* (NII), heavy sponsorship was proposed by the federal government. That government commitment was withdrawn in favor of commercial development of the concept, which still enjoys government endorsement and encouragement. Internationally, the concept also goes under the names *International Information Infrastructure* (III) and *Global Information Infrastructure* (GII) and typically is heavily subsidized by national and regional governments.

Although the term information superhighway has fallen out of favor and is even considered quaint, the concept remains sound. The applications are exciting and even compelling. Many of the enabling technologies have been invented, many standards are in place, and billions of dollars are being invested in infrastructure. Technology is moving forward and propelling us along with it at ever-increasing speed. Progress marches on. Resistance is futile. At this point, it is worth reflecting on the content of a letter reportedly sent in 1829 from Martin Van Buren, Governor of New York, to President Andrew Jackson:

Dear Mr. President:

The canal system of this country is being threatened by the spread of a new form of transportation known as railroads. The federal government must preserve the canals for the following reasons.

One: If canal boats are supplanted by railroads, serious unemployment will result. Two: Boat builders would suffer and tow-line, whip and harness makers would be left destitute. Three: Canal boats are absolutely essential to the defense of the United States.

As you may well know, Mr. President, railroad carriages are pulled at the enormous speed of 15 miles per hour by engines which, in addition to endangering life and limb of passengers, roar and snort their way through the countryside, setting fire to crops, scaring the livestock, and frightening women and children. The Almighty certainly never intended that people should travel at such breakneck speed.

(signed)
Martin Van Buren
Governor of New York

The above letter apparently is a fabrication. According to Snopes.com, the earliest sighting was in an advertisement for Virginia Coal Pipeline Associates that ran in *The Washington Post* in 1983. Oh well, I suppose it was too good to be true. True or not, it serves to illustrate the fact that the Luddites lost their struggle against the Industrial Revolution and will lose the battle again the information revolution as well. It also serves to underscore the fact that you really have to be careful of your information sources.

Actually, the information revolution is more of an information evolution, although certainly a rapidly developing one, that has developed and will continue to develop in various ways and with various levels of functionality. In India, for example, it might mean placing a single solar-powered, satellite-based payphone in every rural village while building state-of-the-technology call centers in Mumbai to handle technical support for software products. Many millions of business and residential users in affluent areas of developed countries now have access to one or sometimes multiple broadband networks in support of voice, data, video, entertainment, and multimedia at speeds measured in Mbps. Schools and libraries, with government support in the United States and many other countries, increasingly have universal broadband access to the wisdom of the scholars since the beginning of recorded history. Even in rural markets in the United States and many other countries, broadband access is available via satellite. As Arthur C. Clarke noted in 1983, "The constraints are no longer technical, but economic, legal, or political" —to which I add cultural and religious.

14.1 CONVERGENCE DEFINED

Convergence is defined as *the moving toward union or one another*. In the context of the information superhighway, revolution, evolution, or whatever you choose to call it, the concept of convergence cuts across a number of dimensions, including a wide range of applications and the underlying technologies. In full form, convergence represents the coming together of every technology and application discussed in the previous 13 chapters.

14.1.1 Applications

Applications, truly, are at the very crux of convergence. The only conceivable business reason for investing billions of dollars in network technologies is to serve revenue-producing, profit-generating applications. (There are, of course, a number of social, cultural, and political reasons for either encouraging or discouraging such investments.) There appears to be no single *killer app* driving the information superhighway—although the Web comes close, it is really a means of serving a vast collection of applications. Certainly, a number of interesting and productive niche applications exist which, in various user-specific combinations, constitute a killer app suite. One of the most compelling applications is that of Web-enabled call centers, as I discussed previously. Multimedia, in full form, currently is viewed as the ultimate in terms of presentation mode, although compelling applications of real substance have yet to be defined.

14.1.2 WAN Technologies

Analog switches and transmission facilities rapidly are being replaced with digital network elements in the Wide Area Network (WAN). This process largely is complete in the carrier backbone networks and the Central Office Exchange (COE) networks, at least in developed countries. The local loop, of course, remains largely copper based and analog in nature, at least in the residential and small-business market segments, although Passive Optical Network (PON) is now being deployed aggressively by several Incumbent Local Exchange Carriers (ILECs) in the United States and Wireless interoperability for Microwave Access (WiMAX) is nearing the commercial deployment phase. Integrated Services Digital Network (ISDN) never enjoyed much success in either the residential or business markets in the United States but made considerable inroads in many other developed countries, especially in Asia and the European Union. T/E-carrier is used widely as an access technology in the medium and large business markets. Synchronous Optical NETwork/ Synchronous Digital Hierarchy (SONET/SDH) optical fiber transmission systems remain unusual in local loop applications but are used widely in the carrier backbone networks. Dense Wavelength Division Multiplexing (DWDM), used both in conjunction with SONET/SDH and on a stand-alone basis, supports multiple optical data streams over a single fiber. At the edge, the incumbent voice carriers' networks are based on digital circuit-switching and time division multiplexing (TDM). A great deal of Asynchronous Transfer Mode (ATM) is now in place in the PSTN core, but the interest in ATM has abated. The Public Switched Telephone Network (PSTN) core is shifting toward Internet Protocol (IP)–based packet switching, and some ILECs are beginning to offer Voice over Internet Protocol (VoIP) over softswitches in response to competitive pressures.

Data networks are digital in the backbone and have been so for years. While residential and small-business users typically access the network on a dial-up basis through modems, larger users typically make use of digital local loops in the form of T-carrier. The generic Digital Subscriber Line (xDSL) local loop technology is making a huge impact, although availability remains something of an issue. Frame Relay (FR) is on the decline, no major new investments are being made in ATM, and the undeniable shift to IP is taking place quickly.

Community Antenna TeleVision (CATV) networks also are experiencing significant upgrades. The old one-way analog coaxial cable distribution systems are being converted to two-way digital and the coaxial distribution cable systems are being replaced with fiber. The CATV networks also are easing out legacy protocols in favor of IP. Additionally, some of the satellite TV networks have been upgraded to support two-way Internet access, and IP is the protocol of choice.

Cellular networks have almost completely shifted from analog to digital. Speeds have increased considerably with the introduction of 2.5G and 3G technologies that support not only voice but also data and even video. Cellular networks are one of the few domains in which the trend is not toward IP, although IP gateways certainly support the necessary internetwork interfaces.

14.1.3 LAN Technologies

Local Area Network (LAN) technology has made incredible strides in the past decade or so. Ethernet overwhelmed Token Ring, Fiber-Distributed Data Interface (FDDI), and other protocols, switches replaced bridges and hubs, and speeds increased from 10 Mbps to 100 Mbps, 1 Gbps, and even 10 Gbps. Priority mechanisms developed to support Grade of Service (GoS), if not guaranteed Quality of Service (QoS). In combination, the speed and GoS mechanisms have made switched Ethernet entirely appropriate for voice as well as data. In fact, switched Ethernet LANs have become the platform for integrated voice/data Private Branch eXchanges (PBXs), which, by the way, run the Internet Protocol.

WLANs became commonplace with the standardization of *wireless Ethernet* as 802.11a/b/g, and WLAN speeds now reach 54 Mbps, at least theoretically. 802.11n is soon to make an appearance, doubling theoretical speeds to 108 Mbps. Wi-Fi WLANs have now replaced wired LANs in many cases, and Voice over Wi-Fi (VoWiFi) is beginning to make serious sense.

14.1.4 Terminal Technologies

Terminal devices are evolving at a rapid pace as well. Multimedia PCs are widely available to support synchronized data, video, image, and audio applications. Cell phones, Personal Digital Assistants (PDAs), and other hand-held devices increasingly are multimedia capable and Web enabled. Dual-mode VoWi-Fi/cellular handsets are a true crossover technology that has the potential to knock down the walls that separate cordless and cellular technology and make true mobility a seamless reality.

14.2   DRIVING FORCES

The evolution of the network moved at a relatively glacial pace for the first hundred years or so. With the exception of Step-by-Step (S × S) electromechanical switches and rotary dial telephones, little in the way of technology was introduced for the first 50 years. The introduction of Crossbar (Xbar) switching in 1937, Direct Distance Dialing (DDD) in the 1950s, and tone dialing in the 1960s were considered absolutely revolutionary. The pace accelerated with DDS, Switched 56 kbps, and T-carrier services in the 1960s, 1970s, and 1980s. Toward the mid-1980s, the pace picked up considerably, and ISDN deployment began to make an impact in the early 1990s. In the last 10 years, the pace picked up even more and became a full-tilt race through April 2000. At that point absolutely awful fiscal policy in the United States, combined with irrational exuberance in the technology sector of the stock markets, general weakness in the worldwide economy, and a number of other factors created a worldwide recession. Things got back on track in 2004 or so, although the business of telecommunications changed a good deal in the meantime. Forces driving the development of the convergence scenario include deregulation, privatization and competition, applications, and technology.

14.2.1   Deregulation and Competition

Deregulation, perhaps, was the primary driving force. Beginning with the Federal Communications Commission (FCC) Carterfone decision in 1968, end users were presented with a wide variety of options for terminal equipment. The Modified Final Judgment (MFJ), which took full effect on January 1, 1984, was the next step, dictating the breakup of the AT&T Bell System and ending what had constituted a virtual monopoly over communications in the United States—from research and development, to network equipment manufacturing, to service delivery. Competition developed as a result of the actions of the FCC and the federal courts. That level of competition has intensified by orders of magnitude since the passage of the Telecommunications Act of 1996 and now includes competition for local service and broadband Internet access as well as equipment and long-distance services.

Customer Premises Equipment (CPE) and long-distance competition developed quickly after the Carterfone decision and the MFJ. CPE competition is widespread, with literally thousands of manufacturers competing for the voice, data, video, and image systems markets. Competition in the long-distance business is intense in the United States. In 2002 there were well over 400 facilities-based carriers vying for the interLATA (Local Access and Transport Area) market, although most of them have since gone out of business or been acquired by the incumbent carriers. The ILECs also acquired IntereXchange Carriers (IXCs) and now compete against each other intensely.

Local loop competition did not exist to any great extent in the United States or other developed nations until the late 1990s. The primary exception, of course, was that of the Alternative Access Vendors (AAVs), which extended their optical fiber facilities directly to the customer premises in major markets in the United States. The Telecommunications Act of 1996 required that the ILECs lease local loops to the Competitive Local Exchange Carriers (CLECs), and provide space in their Central Offices (COs) so the CLECs might collocate their termination facilities in a convenient and cost-effective Point of Presence (POP). The act also required the ILECs to *unbundle* the cost of those loops, thereby charging the ILECs only for the loop and not for the various Operations Support Systems (OSSs) and other network elements that are bundled into the overall cost of a loop for purposes of calculating the rate base for regulatory purposes. CATV providers also are providing voice service and Internet access in an increasingly large number of areas. A few electric power utilities also compete in the local loop through the deployment of optical fiber networks and in support of voice, data, and entertainment TV services. Access Broadband over Power Line (BPL) is out of the trial stages and in commercial application in a few select areas of the country. Through one means or another, local service competition exists, at some level, in most states.

14.2.1.1 Cost The cost of an integrated terminal device certainly can be less than the cost of multiple devices. Assuming that you have a requirement for a telephone set, a computer, a videoconferencing unit, and a television set, the cost of a single, multifunctional device can be less than the total cost of the individual devices. The cost of a single, multifunctional local loop can be less than the cost of multiple, application-specific loops. And the cost of a single, multifunctional network can be less than the cost of multiple, application-specific networks. A basic assumption is that the technology is in place to enable such an integrated scenario—and that enough people are sufficiently interested to buy enough integrated terminals, rent enough loops, and subscribe to enough network services to enable the manufacturers to build and sell enough devices and develop and sell enough application software to bring the unit cost down to an affordable level. Only then can the network providers offer such services at affordable cost.

The public Internet clearly is being positioned as the converged network, at least for the great unwashed masses of us who cannot afford or justify private IP networks. The volume of Internet traffic has grown to incredible proportions and increasingly is burdened with ever more bandwidth-intensive applications such as graphics and image transmission as well as QoS-sensitive applications such as voice and video. The end result is an Internet that is increasingly stressed at all levels. So, the backbone service providers and ISPs alike are expected to add more capacity in order to meet end-user expectations, but it is not clear how they will recover their costs. One proposal is for a tiered, content-sensitive, and discriminatory pricing structure that charges end users more for QoS in support of applications such as VoIP, streaming video, Virtual Private Networks (VPNs), and even the Web, in general. This has sparked considerable controversy, particularly in the United States, where the Internet has always been a public and content-neutral medium. The controversy over *network neutrality* is unlikely to be resolved in the near future.

14.2.1.2 Applications Applications, clearly, are the primary force driving the concept of convergence. Users have developed a real appetite for bandwidth-intensive applications, lacking only the network infrastructure to support them. The capability of a single provider to deliver a full range of voice, data, video, image, and even multimedia services across the full spectrum of meaningful applications is a compelling feature of a converged network. Basic telecommunications applications certainly must be supported, including voice and data communications as well as Internet access. A full convergence scenario also adds TV to the basic service mix, resulting in the blending of voice, data, and entertainment applications. The more exciting applications include videoconferencing, distance-delivered learning, music and video on demand, home shopping, publishing, and integrated messaging.

14.2.1.2.1 Voice Communications Voice communications, including both the provisioning of local loops and the delivery of local service, certainly not only is part and parcel of convergence but also represents its very foundation. While local service is not stunningly profitable, the successful local service provider has a competitive edge with the consumer with respect to long-distance service. Bundled with other voice services such as enhanced custom calling features and broadband Internet access, such a package can be most attractive to the user and highly profitable to the service provider, especially when delivered over a single Asymmetric Digital Subscriber Line (ADSL) or perhaps PON local loop.

14.2.1.2.2 Data Communications Data communications services tend to be highly profitable, even in these highly competitive times. The historical growth of—and growth potential for—bandwidth-intensive data services is well documented. In the competitive market for data communications services, emphasis is on highly profitable dedicated services such as native LAN-to-LAN connectivity as well as on enhanced and unregulated services such as Frame Relay and IP-based VPNs.

14.2.1.2.3 Internet Access Internet access is a natural for a convergence scenario. A great number of large end-user organizations have dedicated Internet access, often provided over unchannelized T/E-carrier service perhaps delivered to the premises through a SONET/SDH local loop. Tens of millions of business and residential users have broadband Internet access over ADSL or cable modem network. Additionally, millions of small businesses and individuals have dial-up access to the Internet via conventional modems. Depending on the specifics of the technology, those applications can include voice, LAN to LAN, entertainment TV, and videoconferencing. For the most part, Internet access is not regulated and can be highly profitable. The Internet and Web, of course, offer access to an incredible array of applications at another level. As I discussed in Chapter 13, those applications include the following:

- Home pages and blogs
- Home shopping
- Auction sites
- Distance-delivered learning
- Newspapers and magazines
- Dictionaries and encyclopedias
- Weather forecasts
- Stock quotes and trading

- Tax reporting and filing
- Airline, hotel, and rental car reservations
- Games
- Gambling

14.2.1.2.4 Television Currently, CATV providers offer television on a highly profitable basis, especially since rate regulations were lifted in recent years. In fact, CATV providers essentially are unregulated at the federal level and only marginally regulated at the local level. In addition to basic offerings, premium channel subscription, Pay-Per-View (PPV), and Video-On-Demand (VOD) are extremely popular and profitable. As I discuss in previous chapters, CATV networks rapidly are being upgraded in support of Internet access and voice communications. Videoconferencing, multimedia, and other applications also can be supported over CATV networks.

While CATV is likely to continue as the primary means of TV delivery in the United States, Direct Broadcast Satellite (DBS) has grown in popularity, offering greater choice of channels at reasonable cost. Employing digital satellites and MPEG-2 compression, transmission quality is excellent, although propagation delay remains an issue with traditional Geosynchronous Earth-Orbiting (GEO) satellites.

Telco PON networks currently in the early stages of deployment in the United States are capable of supporting entertainment TV as well as voice and broadband Internet access. While the telcos have yet to offer TV programming over those networks to any appreciable commercial extent, that certainly is the intent in the longer term under the triple-play concept of converged voice, Internet access, and entertainment TV. Verizon is already promoting entertainment over its FiOS PON offering.

14.2.1.2.5 Multimedia The performance of an integrated suite of applications certainly is an improvement over that of multiple, disconnected applications. This is the essence of multimedia. We do not necessarily need to view text, or image, or video information in connection with every telephone call, but it often is advantageous to do so. When you talk to a realtor in another city about an impending house-hunting trip, it is helpful to see images of houses that meet your criteria, see a map of the city to get a sense of the house's location, view the proposed contract with the realtor, and even see the realtor through a videoconference. You certainly can do most of these things through a combination of a telephone call, a fax transmission, and an e-mail transfer, but combining them all together in a single, interactive, multimedia presentation enhances the overall performance of the communication.

14.2.1.2.6 Videoconferencing Videoconferencing expands to *video dial tone*, or *visual dial tone*, in a full-on broadband convergence scenario. This application appeals greatly to some and borders on the ridiculous to others. Affordable bandwidth and terminal equipment remain major issues. Once the cost becomes reasonable and the CPE configuration issues are resolved, videoconferencing will become commonplace.

14.2.1.2.7 Music on Demand *Music On Demand* (MOD) provides access to a wide variety of music over high-quality local loops. MOD involves access to an audio server, much like a CD-ROM jukebox. For some years, this application was forecast to be highly attractive, but the technology did not evolve sufficiently until 1999 when the MP3 (MPEG-1 audio layer 3) compression algorithm was developed. MP3, as discussed in Chapter 12, supports the downloading of near CD-ROM quality music over the Web. Significant concerns remain relative to copyright infringements and avoidance of royalty payments. Audio programming has been available for a number of years over CATV networks. In the more recent past MOD also has become available over CATV networks in the form of music VOD, which is even better.

14.3 CONVENTIONAL CONVERGENCE: WIRELINE NETWORKS

Several versions of a converged network appear in Figure 14.1, at least at the wireline local loop level. (This figure may seem familiar, as I *leveraged* it from Chapter 9. Leverage is a euphemism for *reuse*.) In one converged local loop scenario, a Hybrid Fiber/Coax (HFC) CATV uses PON technology to the neighborhood node and embedded coax to the premises. One telco scenario illustrates pure PON Fiber-To-The-Premises (FTTP). The other telco scenario illustrates Fiber-To-The-Neighborhood (FTTN), aka Fiber-To-The-Node, and uses embedded Unshielded Twisted Pair (UTP) to the premises. In each case, the local loops support the full *triple play* of voice, Internet access, and entertainment TV. Other scenarios include WiMAX and access BPL, although they are limited to voice and data.

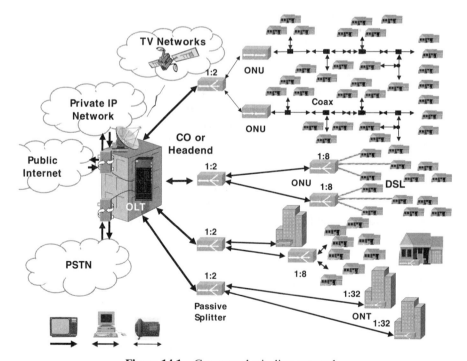

Figure 14.1 Converged wireline network.

At the edge of the carrier network, the telco CO or CATV head end provides switched interfaces to three networks:

- **PSTN:** The conventional circuit-switched PSTN offers unyielding voice quality and will be in place for many years to come. Ubiquitous access to this legacy public voice network is essential for the foreseeable future.
- **Internet:** The public Internet will continue to be the network by which we access the Web and by which we send public e-mail. It also increasingly will serve as a low-cost alternative for voice using VoIP.
- **Private IP Network:** Carriers will offer access to well-designed and carefully managed private IP networks for business-class voice, VPN service, and other business-class or premium services that benefit from increased bandwidth, improved congestion management, and enhanced security.

14.4   THE RACE IS ON: MERGERS AND ACQUISITIONS (M&As)

The race is on and here comes pride up the backstretch. Heartaches are goin' to the inside. My tears are holdin' back. They're tryin' not to fall.... And the winner loses all.
> *The Race Is On*, composed by Don Rollins and first recorded by
> George Jones in 1965

ILECs, CLECs, IXCs, CAPs/AAVs, CATV providers, PCS licensees, satellite service providers, electric utilities, and others all have jockeyed for the pole position to lay the grid and provide the services that will make them and their stockholders wealthy. In recognition of the fact that convergence is the name of the game and that no one company has all of the answers, hardware manufacturers, software developers, and carriers all have been gobbling each other up at a record pace. The race definitely is on! The following is a small but representative sample of the scope and scale of such activity. I find it relevant in the context of convergence for several reasons. First, there has been a lot of consolidation (i.e., convergence) in the corporate world as companies have merged and acquired each other. Within the domain of the original AT&T Bell Systems, there has been a great deal of fragmentation (i.e., divergence) and reassembly (convergence). All of that activity has led to increased levels of competition at times and decreased levels at other times. One way or another, this activity has defined, at least in part, the nature and form of convergence.

14.4.1   Evolution of the Bell System

Break up to make up, that's all we do. First you love me, then you hate me. It's a game for fools.
> *Break Up To Make Up*, The Stylistics, 1973

The evolution the old Bell System is an interesting case in point. It may not be a game for fools, but there has been an awful lot of breaking up and making up. Now, I want to reflect on the breakup of the Bell System, which I discuss at length in the context of the PSTN (refer to Chapter 5). On January 1, 1984, AT&T spun off its

TABLE 14.1   Bell System Operating Company Organizational Structure Before and After the MFJ and to Present

Bell Operating Companies (Primary States of Operation), Predivestiture	Regional Bell Operating Companies (Headquarters), Postdivestiture
Illinois Bell (IL), Indiana Bell (IN), Michigan Bell (MI), Ohio Bell (OH), Wisconsin Telephone (WI)	Ameritech (IL), acquired and absorbed by SBC Communications (October 1999)
Bell of Pennsylvania (PA), Diamond State Telephone (DE), The Chesapeake and Potomac Companies (DC, MD, VA, WV), New Jersey Bell (NJ)	Bell Atlantic (PA), now Verizon Communications
South Central Bell (AL, KY, LA, MI, TN), Southern Bell (FL, GA, NC, SC)	BellSouth (GA), acquired by AT&T (nee Southwestern Bell Corporation) (December 2006)
New England Telephone (MA, ME, NH, RI, VT), New York Telephone (NY)	NYNEX (NY), Acquired by Bell Atlantic (August 1997), now Verizon Communications
Pacific Bell (CA), Nevada Bell (NV)	Pacific Telesis (CA), acquired and absorbed by SBC (April 1997)
Southwestern Bell (AR, KS, MO, OK, TX)	Southwestern Bell Corporation (TX), renamed SBC and later renamed AT&T (November 2005)
Mountain Bell (AZ, CO, ID, MT, NM, UT, WY), Northwestern Bell (IA, MN, ND, NE, SD), Pacific Northwest Bell (OR WA)	US West (CO), acquired and absorbed by Qwest (June 2000)

Note: Official U.S. Postal Service (USPS) abbreviations are used for states.

22 wholly owned *Bell Operating Companies* (BOCs) under the terms of the *Modified Final Judgment* (MFJ), also known as the *Divestiture Decree.* These BOCs were reorganized into seven *Regional Bell Operating Companies* (RBOCs), also known as *Regional Holding Companies* (RHCs), as noted in Table 14.1. Over time, the RBOCs fully absorbed the individual BOCs, creating a single legal entity with a centralized management structure. Cincinnati Bell and Southern New England Telephone (SNET) were not affected by the MFJ, as they were not wholly owned subsidiaries of AT&T.

14.4.2   BOCs Break Out of the Box

While the MFJ had incredible impact on the telecommunications environment in the United States, the resulting landscape was neat and orderly. That is no longer quite the case. Southwestern Bell Telephone Company changed its name to Southwestern Bell Corporation and then to SBC Communications. The last name change was a bit puzzling until the merger (read *acquisition*) with Pacific Telesis was announced—*Southwestern* no longer had any positive value. Pacific Telesis ceased to exist, but Pacific Bell and Nevada Bell retained their identities. Subsequently, SBC also merged with Southern New England Telecommunications (SNET), a Connecticut Local Exchange Carrier (LEC) once partially owned by AT&T. SNET also retained its identity. A key advantage to SBC of the SNET merger is the fact that

SNET was not affected by the MFJ; therefore, SNET could freely develop a small but successful position as an IXC. SBC felt it could potentially build a much more substantial IXC business on the SNET foundation once the terms of the Telecom Act of 1996 were lifted. SBC also merged Ameritech into the fold in October 1999. SBC acquired AT&T in 2005 and adopted the name AT&T.

Bell Atlantic got busy, merging with NYNEX in August 1997. NYNEX ceased to exist. In July 1999, Bell Atlantic acquired GTE, and renamed the company Verizon.

US West reached a definitive agreement in 1999 to merge with Global Crossing, an upstart international submarine fiber-optic carrier. Global Crossing also reached a definitive agreement to merge with Frontier, a LEC and IXC that began life as Rochester Telephone. Qwest, an upstart IP-based IXC, then made a hostile bid for both US West and Frontier and a war broke out. Actually, it was more of a skirmish, as Qwest very quickly (July 1999) was declared the winner with respect to the US West merger. Global Crossing walked away with the Frontier merger agreement remaining intact. Global Crossing and Qwest were attracted to US West because a good deal of long-distance traffic either originates or terminates in US West territory. While Frontier's properties in Rochester, New York, were not particularly attractive, its status as a second-tier IXC was very attractive. The bottom line is that both Global Crossing and Qwest spent billions of dollars *laying pipe* (i.e., building backbone infrastructure), which they needed to fill with minutes of traffic. It is faster and easier and often less expensive to buy the minutes than it is to take them away from the competition. Qwest has a history of this sort of activity, having merged with LCI in June 1998; LCI previously (September 1997) acquired USLD Communications. US West previously spun off its CATV business, US West Media Group, which became MediaOne, which was subsequently merged into AT&T. By the way, Global Crossing sold the ILEC portion of Frontier to Citizens Communications in July 2000.

14.4.3 So What Was AT&T Up To?

Who owns AT&T? The stock of American Telephone & Telegraph Company, parent company of the Bell System, is held by more than 750,000 shareholders. These stock-holders are about equal to the population of the city of Pittsburgh or the state of Rhode Island.

Telephone Almanac, Bell Telephone System, 1949

Speaking of AT&T, the company reorganized into two business units at the time of divestiture in 1984. AT&T Long Lines became AT&T Communications, operating as an IXC. AT&T Technologies was formed of Western Electric, the manufacturing arm of AT&T, and AT&T Bell Telephone Laboratories (Bell Labs), the research and development organization. For the next 13 years, AT&T did very well focusing on its core businesses, although it did acquire NCR in a failed attempt to get into the computer business. With all of the hype that surrounded the merging of voice and data at the time, it apparently seemed to AT&T management that AT&T Technologies and NCR would make a great match and a great launching pad for computer telephony systems. That seemingly great idea just did not work. IBM previously experienced a similarly dismal failure with its acquisition of ROLM, an almost

TABLE 14.2   AT&T, Postdivestiture and Postspinoff

Postdivestiture	Postspinoff
AT&T Communications Services (formerly AT&T Long Lines): *Role*—long-distance service, universal card, AT&T McCaw Cellular (acquired 1994)	AT&T Corp.: *Role*—long distance, universal card, AT&T McCaw Cellular, wireless, Internet services, AT&T Laboratories. Revenues: US$51 billion. Assets: US$56 billion. Employees: 127,000
AT&T Bell Telephone Laboratories: *Role*—research & development AT&T Technologies (formerly Western Electric): *Role*—Manufacturer of CPE/DTE	Lucent Technologies: *Role*—research and manufacturing of carrier equipment such as circuit switches and transmission equipment, and CPE/DTE. Includes Bell Laboratories and AT&T Technologies. Revenues: US$21 billion. Assets: US$20 billion. Employees: 131,000
AT&T Global Information Solutions (formerly NCR Corp.): *Role*—manufacturer of computers, automatic teller machines, electronic cash registers	NCR Corp.: *Role*—data processing systems, ATMs, and electronic cash registers. Revenues: US$8 billion. Assets: US$5 billion. Employees: 38,000

Source: AT&T 1995 Annual Report.

legendary PBX manufacturer, which it subsequently sold to Stromberg-Carlson at a substantial loss.

On January 1, 1997, AT&T effected the largest voluntary breakup in history. The US$75 billion company split into three market-focused companies, also selling AT&T Capital, its captive financing business. Approximately 8500 employees, all in the Global Information Solutions (GSI) computer business, lost their jobs fairly immediately. GSI resulted from the NCR acquisition, which did not live up to expectations. Hundreds of thousands of others lost their jobs over time. The postdivestiture AT&T boasted assets of US$79.2 billion, annual revenues of US$75.1 billion, and a total workforce of 303,000, which was down from over 1,000,000 prior to divestiture. Table 14.2 provides a view of AT&T, both postdivestiture and immediately postspinoff.

After the voluntary breakup, AT&T kept busy, of course. In February 1996, AT&T announced its Internet access service, which was an instant hit. AT&T World-Net service suffered considerably when faced with competition from the likes of AOL, MSN, United Online, and Prodigy and never recovered. The brand still exists and is marketed outside the AT&T (nee SBC) franchise service areas but is not in the top 22 Internet Service Providers (ISPs) in the United States, according to ISP Planet.

In January 1996, AT&T purchased a 2.5 percent stake (US$137 million) in DirecTV, a provider of entertainment TV via DBS satellite. AT&T marketed DirectTV to its customer base, billing for the monthly services through its standard long-distance billing system. Purchase and installation of the dishes also could be financed through AT&T credit cards. Subsequently, AT&T divested its stake in DirecTV.

In July 1998, AT&T merged with TCG (Teleport Communications Group), a large CAP/CLEC that previously (January 1997) acquired Cerfnet, a large Internet backbone provider. AT&T acquired TCI (Tele-Communications Inc.) in 1999 in an all-stock deal worth about US$48 billion at the time. Also in 1999, AT&T acquired MediaOne, which previously had been spun off from US West, in a bidding war against Comcast. The *winning* bid was in the form of AT&T stock worth US$58 billion at the time plus the assumption of US$4.5 billion in debt. Together, these acquisitions formed AT&T Broadband, the largest CATV provider in the United States. Under extreme financial pressure due to the inflated cost of its acquisitions and the high costs of upgrading its CATV systems, AT&T Broadband agreed to merge with Comcast to form AT&T Comcast in a deal that initially valued AT&T Broadband at US$72 billion and later shrunk to US$53 billion, which is quite a discount from the US$110.5 billion AT&T spent to form the company.

In July 1998, AT&T and British Telecom (BT) announced the formation of Global Venture, an international alliance that expected revenues of US$10 billion in its first year of operation. BT contributed to Global Venture its Concert services, which it previously linked with MCI. BT, at one time, owned 20 percent of MCI, which subsequently merged with Worldcom to become MCI Worldcom. AT&T's link with BT required that it dissolve its WorldPartners WorldSource international alliances, which involved a large number of partners all over the world. Global Venture subsequently was renamed Concert. On October 16, 2001, AT&T and BT announced their decision to dismantle Concert but to honor existing contracts and service-level agreements for three years.

In 2006, the tattered remnants of AT&T were acquired by SBC for approximately US$16 billion, which named the combined entity AT&T. In just over 20 years, one of the oldest, largest, and most respected companies in the world was reduced to a ghost, at least partially due to the sheer incompetence of some members of its senior management. On a personal note, I thank my lucky stars that I left the Bell System of my own free will long, long before AT&T collapsed. Heck, I never did fit in, anyway. I reckon I didn't have enough of a *Bell-shaped head*, as we used to say.

14.5 ONE POTATO, TWO POTATO, THREE POTATOE, FOUR . . .

Potato(e) can be spelled more than one way, as then-Vice President Dan Quayle illustrated several years ago. There also is more than one way to build an information superhighway. There is a very serious question as to just how many wires (hardwires and *wireless wires*) should be extended to provide access to the networks in a convergence scenario. After all, the very word *convergence* means *coming together*. You have to wonder how much redundant infrastructure is reasonable and at what cost. Perhaps the electric utilities have the right idea—build one information grid and provide access to a wide variety of service providers. Of course, this idea is not new. Utilities traditionally have provided a single grid for telecommunications, data communications, CATV, electric power, gas, water, and sewer services. The fundamental economic concept of a *natural monopoly* served by a *common carrier* still has merit.

Rochester Telephone (Rochester, New York), well known as an innovator, several years ago hit upon a variation of the same theme. The New York Public Utility Commission (PUC) approved the separation of the company into R-Net, the grid provider, and R-Comm, the service provider. R-Net acted as an information grid wholesaler, or common carrier, providing access to service providers, including R-Comm. On January 1, 1995, the people of Rochester became the first U.S. citizens since 1919 to have a choice of local telephone service providers [1]. Rochester Telephone was acquired by Frontier, which now is part of Citizens Communications.

It now seems as though a *natural duopoly* or *triopoly* will prevail in many areas at the local loop level. Telecommunications and CATV local loop networks already exist and have been upgraded in many areas. While the level of capital investment is not trivial in either case, it is quite likely that two grid providers can survive quite comfortably and profitably. If you add a wireless carrier or two to the equation through Wi-Fi or WiMAX technology, perhaps that number expands to three or four, given the flexibility of wireless network configuration. Broadband over Power Line (BPL) conceivably adds another wireline carrier to the mix, although BPL likely will be reserved for remote rural applications. But it is highly unlikely in the foreseeable future that there will be more than two or three wireline grids and perhaps a wireless bypass alternative except in small geographic areas where there exists a high concentration of substantial business users.

That is all at the Physical Layer (Layer 1), of course, but that is a grid issue. The service providers using the grid is another issue altogether.

- **CATV Loops:** The CATV networks will likely be closed to competition—they always have been and there is no reason to expect that will change in the foreseeable future. It also is highly unlikely that another CATV provider will build a competing wireline network. The telcos, however, will increasingly compete head on with triple-play services over PON.
- **Telco Loops:** The telco local networks were closed for 120 years, until the Telecom Act of 1996 opened them to competition. Now they seem to be closed again, for all practical purposes, with the recent designation of DSL as an *information service*. The RBOCs continue to wholesale DSL loops to competing ISPs, but that will not necessarily continue forever. PON local loops are similarly closed and will remain so. Full-on PON is an overlay network, which creates an interesting situation. Verizon is building overlay PON networks in New York City and a number of other major cities in the northeast United States. Those PON loops, of course, will be much more capable, much less maintenance intensive, and much more profitable that the UTP loops right alongside them. The PON loops likely will have unregulated status at the state as well federal level while the UTP loops will remain regulated, except for DSL, which Verizon will transition to PON. The PON loops likely will be maintained by the nonunion, or quasi-union, contractors who built them, while the regulated UTP voice loops will continue to be maintained by the solidly embedded members of the Communications Workers of America (CWA) and the International Brotherhood of Electrical Workers (IBEW). So, I am willing to bet the royalties from this book that Verizon will sell off the legacy regulated local exchange business, including the copper local loop, and will retain the nonregulated PON loops, over which it will deliver the triple play of IP voice, Internet access, and IPTV.
- **Wireless Local Loop:** WLL competition will become a reality in select areas through the commercial introduction of WiMAX, even though it will be limited

to voice and Internet access. Commercial public Wi-Fi access will increasingly expand into the neighborhood level in select areas but generally will be limited to Internet access. Some service providers will support VoWiFi.

· **Access Broadband over Power Line:** Access BPL largely will remain a niche solution for remote rural situations. Access BPL will remain targeted primarily at broadband Internet access and voice.

While some of us remember certain aspects of the Bell System days of yore with a certain fondness, I think we all agree that competition has been a good thing for the most part. Competition may have gotten out of hand at some point, but the impacts largely were positive. The failures of the (old) AT&T, MCI, and Sprint as CLECs were undoubtedly their own fault, but it is a shame, for their losses are our own as well. Three once powerful competitors effectively disappeared.

14.6   NEXGEN CONVERGENCE: WIRELINE AND WIRELESS NETWORKS

Next-Generation (NexGen) convergence goes beyond the local loop alternatives of CATV, telco UTP and PON, WLL, and Access BPL. NexGen convergence also goes beyond the core network alternatives of the circuit-switched PSTN, the Internet, and private IP networks. NexGen convergence folds 2.5 G and 3 G cellular networks into the mix. Such a converged network is more than a gleam in some engineer's eye or a spark in some marketer's mind; it is taking form through the *IP Multimedia Subsystem* (IMS) architectural concept.

IMS originated in the 3rd-Generation Partnership Project (3GPP), which was seeking a common means by which GSM cellular operators could deliver data services. IMS subsequently transcended the cellular domain and is now being embraced by both wireless and wireline service providers. Industry groups such as the *Multiservice Switching Forum* (MSF), European Telecommunications Standards Institute (ETSI), and Alliance for Telecommunications Industry Solutions (ATIS) have adopted IMS as the foundation for their next-generation infrastructure strategies.

In the pre-IMS world, the PSTN and the Internet are very different, as we have discussed. The circuit-switched PSTN conforms to an architecture that centralizes the intelligence in the network. The Advanced Intelligent Network (AIN) places each service in an intelligent node and shares applications and communicates instructions to less intelligent switching platforms through Signaling System 7 (SS7). PSTN terminal devices are dumb. The packet-switched Internet distributes intelligence among clients and servers in support of peer-to-peer communications. Internet signaling and control are through the TCP/IP stack, with movement toward the Session Initiation Protocol (SIP) for applications such as VoIP. The cellular networks generally are circuit switched in nature and conform to the PSTN model of centralized intelligence and dumb terminals, with SS7 signaling and control.

IMS is built around a packet core and provides an environment in which a user can access a wide range of multimedia services using any device and any type of network connection. IMS supports IP sessions between devices over any type of connection and protocol, whether wireline (e.g., ADSL, PON, CATV, or Ethernet) or wireless [e.g., 802,11a/b/g/n, GSM, EDGE, EV-DO, CDMA, GPRS, or WiMAX] in nature. IMS will support sessions between devices in the PSTN, Internet, and cellular domains, recognizing each device and each network for what they are and what their capabilities are.

An IMS-capable multimedia client connects over an IP-based core network to application and content servers that provide various value-added services such as multimedia conferencing and unified messaging. The client can operate over a PSTN connection, a wired Ethernet port, a Wi-Fi channel, a WiMAX channel, or a cellular channel and can roam between any of them. The IMS network will recognize the changing nature of the network and adjust accordingly, perhaps enabling video when the user connects to an 802.11 g Wi-Fi network that will support the bandwidth requirement and disabling it when the user gets out of range of the Wi-Fi network but maintains connectivity through a handoff to a GSM cellular network that will not support the video bandwidth requirement. IMS manages internetwork handoffs, bandwidth negotiation and QoS, while it keeps peers engaged in the session advised via SIP as to the specifics of the level of multimedia presence [2–6].

So, that all translates into one device with one address that can be used just about anywhere at just about any time under just about any circumstances and that communicates with its peers through a network that can adapt to just about anything. That seems like convergence to me.

REFERENCES

1. "Rochester Telephone's New Frontier." *Convergence*, April 1995.
2. Lawson, Stephen. "Primer: Here's the Lowdown on IMS." *Network World*, September 25, 2005.
3. Buckley, Sean. "IMS: A New Spin on Convergence." *Telecommunications Americas*, March 2005.
4. Duffy, Jim. "RBOCs Tout Promise of Next-Gen Services." *Network World*, October 31, 2005.
5. Grigonis, Zippy. "Multimedia and IMS." *VON Magazine*, November 2005.
6. Wallace, Bob "IMS." *Telecommunications Americas*, January 2006.

CHAPTER 15

REGULATION: ISSUES AND (SOME) ANSWERS

Order marches with weighty and measured strides; disorder is always in hurry.
Napoleon I (Napoleon Bonaparte), *Maxims*, 1804–1815

Regulations, from the Latin *regula*, meaning *rule*, are rules or orders established by governmental bodies and having the force of law. As generally is the case in matters of government, regulations are not developed quickly. Like standards, they develop slowly, based on human experience and setting the stage for the next cycle of experience. Generally written by groups of lawyers, they are characteristically lengthy and complex to the extent that it takes other groups of lawyers to interpret them. Where there are disagreements or disputes, lawyers appeal to the regulators and all too often litigate those matters through the courts, where judges, who usually are lawyers, make the final decisions, which often are appealed to higher courts, where eventually some judge or panel of judges makes the final decision. As noted by Karl Shapiro in *Reports of My Death* (1990), "Lawyers love paper. They eat, sleep and dream paper. They turn paper into gold, and their files are colorful and their language neoclassical and calligraphically bewigged." While current regulations also are available electronically, the size of the files to be downloaded reflects the size of the official paper documents, and you will wind up printing them out anyway. Whether in hard copy or soft copy, attorneys always have the final word, and so it is with this book. Chapter 15 is the final chapter.

This chapter isolates discussion of recent developments in U.S. regulation as well as pending regulatory issues. As I have dealt with specific regulatory events and issues in a number of previous chapters to the extent that they were germane to the particular network at hand, I do not repeat those discussions. Rather, I focus on the Telecommunications Act of 1996 (the act) and how it has shaped the future of com-

Telecommunications and Data Communications Handbook, By Ray Horak
Copyright © 2007 Ray Horak

munications in the United States. I also discuss subsequent legislation and regulations and remaining issues of decency, rates, tariffs, number portability, and universal service. You can obtain information relative to the act and other regulatory and legislative information from the Federal Communications Commission (FCC) and Congress over the Internet. Although regulation and legislation seem to move at a glacial pace, an iceberg can break loose at any time, and with considerable impact. The websites I suggest are as follows:

- Federal Communications Commission: http://www.fcc.gov
- U.S. House of Representatives: http://www.house.gov
- U.S. Senate: http://www.senate.gov

Regulations tend to build on one another, for without the experience of history the future tends to be a bit uncertain. Over time, a complex fabric of regulation has been woven at the international, national, state, and local levels. While the seams are generally well stitched, there are significant overlaps in some cases and the patterns do not always match up quite right. In the United States, the weave of the fabric traditionally has been tight and has covered the body of communications from head to toe. The clear trend is one of a much lighter and looser weave and for less coverage. In other words, the trend has been toward *deregulation*. As deregulation takes hold and the competition heats up, comfort and freedom of movement become increasingly important.

Full deregulation, however, is unlikely. At the risk of stretching the fabric analogy to the tearing point, the telecom world of New Zealand is virtually threadbare. In the late 1980s, the government completely deregulated telecommunications, abolishing the regulator in the process. As a result, any legal disputes among carriers must be addressed through the courts, as must any other commercial civil dispute. Over the years, the carriers have engaged in several such disputes. The issue of interconnection, for example, worked its way through the entire national court system and then was appealed to the Privy Council in England for final resolution. Australia took much the same approach as New Zealand during the late 1990s when it abolished *AUSTEL*, which was the equivalent of the FCC.

For the most part, the United States sets the regulatory model for the rest of the world. This has been the case traditionally and continues to be so, with notable exceptions such as those of New Zealand and Australia. Following the U.S. lead, CPE and long-distance competition has been introduced in much of the world, and regulation generally has been relaxed in favor of permitting market forces to determine which products and services will find success and at what price. Quite a few countries have privatized their government-owned networks, either completely or partially. Numerous others have plans in that regard, and in some cases the proceeds of privatization have been earmarked for specific purposes such as education. Table 15.1 provides a chronology of selected key regulatory and legal events in telecommunications history.

15.1   TELECOMMUNICATIONS ACT OF 1996

On February 8, 1996, the administration signed into effect a law that passed both houses of Congress by overwhelming majorities and that, once again, dramatically

TABLE 15.1 Summary of Selected Key Regulatory and Legal Events in U.S. Telecommunications History

Year	Event
1865	International Telegraph Union (ITU) created to establish rules for interconnectivity of national telegraph networks
1876	Alexander Graham Bell granted patent for telephone
1910	*Mann-Elkins Act* grants Interstate Commerce Commission (ICC) interstate regulatory authority
1913	Department of Justice (DOJ) considers antitrust action against Bell System, based on commitment by President Woodrow Wilson to break up monopolies *Kingsbury Commitment* causes DOJ antitrust action to be dropped in return for AT&T's agreement to interconnect with independent telcos, stop acquiring them, and divest its stock in Western Union
1918	Post Office assumes interstate regulatory authority for telephone and telegraph by executive order of President Wilson
1921	*Graham-Willis Act* establishes telephone companies as natural monopolies
1924	ITU forms CCIR (International Radio Consultative Committee, now ITU-R) to set standards for international connectivity of radio networks
1927	*Radio Act of 1927* establishes Federal Radio Commission to regulate all radio spectrum, except bands owned by federal government
1934	*Communications Act of 1934* establishes the FCC to regulate interstate, international, and maritime communications, with universal service stated as the goal; DOJ begins major antitrust action against Bell System, which is delayed due to issues of national interest during World War II;
1935	First state Public Utility Commissions (PUCs) formed to assume intrastate regulatory authority from municipal and city governments
1945	Supreme Court rules in *Ashbacker Radio Corporation vs. the FCC* that radio spectrum allocation is to be on the basis of comparative hearings
1949	DOJ files antitrust action against AT&T, which action had been delayed by World War II; this action results in 1956 Consent Decree
1955	FCC's *Hush-a-Phone Decision* supports AT&T's contention that even acoustically coupled foreign (non-telco-provided) devices cannot be connected to the network without special arrangement
1956	*Consent Decree* negotiated as settlement between AT&T and DOJ, allowing AT&T to retain ownership of Western Electric if it manufactures only for Bell companies; also prevents Bell System from offering data processing services and other services not related to functions of a common carrier; requires that Bell System patents be licensed to others on basis of reasonable fees
1959	FCC's *Above 890 Decision* grants private microwave access to a dedicated portion of radio spectrum; also permits construction of such networks, regardless of economic impact on the established common carrier
1962	*Communications Act of 1962* places authority with FCC to assign commercial satellite frequencies; act establishes *Communications Satellite Corporation (Comsat)* to act as a carriers' carrier (wholesaler) for international satellite service and in conjunction with Intelsat; *Intelsat (International Telecommunications Satellite Organization)* established as international financial cooperative that owns and operates satellites for international communications
1963	Microwave Communications (MCI) files application to operate as a Specialized Common Carrier (SCC)

TABLE 15.1 (*Continued*)

Year	Event
1968	FCC's *Carterfone Decision* counters Hush-a-Phone decision, allowing interconnection of foreign equipment through standard protective coupling device provided by telco
1971	FCC's *SCC Decision* clears way for MCI and other SCCs to construct and operate networks
1972	FCC's *Domsat Decision* permits domestic satellite market; AT&T excluded from market for three years
1975	FCC establishes *Part 68* registration program for certification of foreign equipment, eliminating requirement for coupling devices
1978	FCC's MTS/WATS *Decision* permits MCI and others to offer switched MTS voice services
1980	Second Computer Inquiry (aka Computer Inquiry II) requires AT&T to offer enhanced communications and (CPE/DTE) through separate subsidiary
1982	*Modified Final Judgment* (MFJ) negotiated between DOJ and AT&T as a modification to 1956 Consent Decree; MFJ forces divestiture of Bell Operating Companies (BOCs) and establishes equal access; also removes restrictions on AT&T against computer and related businesses; AT&T retains Long Lines (long distance), Bell Telephone Laboratories (R&D), and Western Electric (manufacturing); AT&T retains embedded CPE base; AT&T files reorganization plan with Federal District Judge Harold H. Greene
	Congress grants FCC authority to award radio spectrum licenses on basis of lottery, rather than comparative hearings; cellular radio licenses granted on this basis
1983	Judge Greene files order based on AT&T reorganization plan; order specifies seven (RBOCs), with Bellcore for common R&D support; (LATAs) established, with AT&T and other (IXCs) permitted to provide interLATA service; BOCs and other (LECs) granted exclusive rights to provide local and intraLATA long-distance services
1984	MFJ takes full effect January 1
1985–1996	MFJ relaxed through series of waivers (e.g., permitting RBOCs to offer enhanced services such as voice mail); several PUCs permit intraLATA and local competition
1993	*Omnibus Budget Reconciliation Act* provides for creation of new class of wireless services, (PCS); FCC authorized to auction PCS spectrum
1994	First PCS auctions held, raising $8.3 billion
1996	*Telecommunications Act of 1996* passed by Congress and signed into law February 8, allowing full and open competition across all dimensions, including manufacturing, local service, and long distance; conditions established for RBOC entry into long-distance market within home states; FCC begins process of establishing rules for implementation of act; implementation specifics in process of full definition; ultimate impact remains unclear
	Communications Decency Act (CDA) enacted to hold both creators of content and service providers responsible for access of minors to indecent or offensive material over the Internet; the act was ruled unconstitutional, in violation of *free speech* guaranteed by the First Amendment

Sources: [1–6].

redrew the landscape of telecommunications. The *Telecommunications Act of 1996 (The Act)* effectively superseded the 1982 MFJ, removing line-of-business restrictions and promising to permit full and open competition in virtually every aspect of communications, from radio broadcasting to Community Antenna Television (CATV) to local exchange and long distance. Ownership restrictions in large part were lifted, enabling the carriers to invest, relate, merge, and acquire. The plan is that, over time, rates will be deregulated, tariffs may be eliminated, and the regulator may well take a back seat. The Act also effectively deregulated CATV, opened spectrum for broadcast TV stations to introduce High-Definition TV (HDTV), and formally established a Universal Service Fund (USF) to keep rates low in rural areas and to subsidize telecommunications and Internet services to schools and libraries.

15.1.1 Lines of Business

The Act effectively removed the line-of-business restrictions imposed by national regulations since the early part of this century and strengthened by the MFJ. Some of the major reversals included allowing the RBOCs to enter the long-distance business, permitting AT&T and others to compete as LECs, enabling them all to own and operate CATV businesses both within and outside of their home states, and permitting them all to manufacture products. In some cases, restrictions apply, at least until certain tests are satisfied.

15.1.2 Mergers and Acquisitions

The Act largely lifted restrictions on ownership. In other words, LECs and IXCs, CATV, satellite, radio, TV, and wireless companies became free to merge, acquire, and otherwise invest in each other without much in the way of restriction, although both the FCC and the DOJ must be satisfied that such activities are in the public interests. As is discussed in Chapter 14, there has been a flurry of such activity since The Act was signed into law. SBC acquired (whoops, I mean *merged with*) Pacific Bell, Nevada Bell, SNET, and Ameritech. Bell Atlantic merged with NYNEX and added GTE to the fold in order to create Verizon. Qwest acquired US West, Global Crossing acquired Frontier, only to sell the ILEC portion of that business to Citizens Communications shortly thereafter, and the list goes on.

AT&T merged with Teleport Communications Group (TCG), a large CAP/CLEC with a considerable Internet backbone. AT&T acquired TCI and MediaOne at a cost close to US$100 billion to form AT&T Broadband, the largest CATV provider in the United States. AT&T also acquired a number of wireless providers, including McCaw Cellular (now Cingular). In total, the optical fiber facilities of TCG, the coax-based CATV networks of TCI and MediaOne, and its Wireless Local Loop (WLL) spectrum (including its PCS licenses) presented AT&T with considerable local loop options that would enable AT&T to largely avoid having to deal with its disowned children, the RBOCs, for local loop access to its prospective customers. The promise was that AT&T's position as a Competitive Local Exchange Carrier (CLEC) would be enhanced considerably once these acquisitions were developed fully and the necessary network upgrades had been accomplished. Unfortunately, it did not quite work out that way. AT&T's PCS-based WLL solution never made it to commercial release, and that business unit was folded in 2002 at a cost of approximately $1.2 billion. AT&T Broadband, formed of acquisitions during the high-priced high-tech heydeys, subsequently merged into Comcast on a stock swap valued at about US$53 billion. Some years before, AT&T spun off its equipment manufacturing business as Lucent Technologies. In 2006, the remnants of AT&T were acquired by SBC, which named the combined entity AT&T. At least the name survived.

Media ownership restrictions were eased as well. However, a company was restricted, in a single local market, from owning two TV stations, or a newspaper and TV station, or a newspaper and CATV operation, or a CATV operation and broadcast TV station. On the other hand, The Act did increase the number of TV stations that any company could own, up to a maximum of 35 percent of the total national market. In 2004, the FCC replaced the newspaper/broadcast and radio/television cross-ownership rules with a single set of cross-ownership rules. The FCC reviews ownership restrictions every four years.

15.1.3 Rules and Implementation

The FCC, despite budget cuts, has a truly overwhelming task, made more difficult by the fact that many of the provisions of The Act remain under the cloud of constant legal challenges. Included in the issues of significance are the Universal Service Fund (USF), rules under which the LECs are permitted to offer long-distance service in their home states, interconnection between incumbent and competing carriers, and rights of way.

15.1.3.1 Universal Service

> The Bell System is, in effect, a vast switchboard providing universal service, nation-wide and world-wide in reach. Every Bell telephone may be connected with more than 35,000,000 American or foreign telephones, or about 93 per cent of all the telephones in the world.
>
> *Telephone Almanac*, American Telephone & Telegraph Company, 1938

Universal service remains an issue without a complete answer or certainly a complete answer that is agreeable to everyone. The Communications Act of 1934 established the concept of universal service. The *Universal Service Fund* (USF) was formally created some years later, although it was not codified until the Telecommunications Act of 1996. The USF traditionally was intended to subsidize the cost of providing service to *high-cost* areas, defined as areas where the cost of providing service is at least 115 percent of the national average. Thereby, the USF ensured that even the most remote, sparsely populated, and impoverished areas of the United States had access to good-quality basic voice telephone service at reasonable cost. In effect, it was a national cost-averaging scheme designed for the benefit of society in general. The USF extended over time to support the provisioning of *lifeline service* to those end users who cannot afford the cost of basic telephone service. The Telecommunications Act of 1996 codified the USF, extending its benefits to subsidize Internet access to schools and libraries and telecommunications networks to link rural health care providers to urban medical centers in order to provide access to advanced diagnostic and other medical services. The USF currently involves an estimated US$7.3 billion to be paid into the fund by carriers and end users in 2006. Under the direction of the FCC, the *National Exchange Carrier Association* (NECA) governs the USF, which actually is administered by the *Universal Service Administrative Company* (USAC), a NECA subsidiary. USAC accepts the collected funds from the LECs and the IXCs. The LECs collect USF fees from both IXCs and subscribers. Those fees are embedded in the access charges to the carriers and generally are billed to subscribers as a separate line item. The LECs net out their approved USF requirements, retain an inflated percentage as reimbursement for billing and administrative costs, and pass any remaining monies to USAC. The IXCs, under orders from the FCC, have added to customers' bills a surcharge calculated as a percent of interstate and international revenues. USF charge to the carriers was set at 10.5 percent of interstate revenues for the third quarter of 2006. (The USF charge changes each quarter, depending on the needs of the USF and the antici- pated carrier revenues from interstate commerce.)

Cellular providers also contribute to the USF, usually on the basis of an average assumed percentage of interstate and international traffic, which currently is set at 37.1 percent. As of June 2006, Voice over Internet Procal (VoIP) providers also contribute to the fund for VoIP to PSTN calls; peer-to-peer VoIP-to-VoIP calls remain untaxed. Interconnected VoIP providers contribute to the USF on the basis of the 64.9 percent of their revenue that is assumed to be interstate and international in nature. The carriers then pass those surcharge revenues to USAC, which redistributes the funds in support of USF objectives. Specifically, USAC distributes designated funds in support of the following programs and organizations set up under The Act:

- **Schools and Libraries Program:** The USF disburses up to US$2.25 billion (2006) to the *Schools and Libraries Corporation* (SLC) to make tele- communications service, Internet access, and internal connections affordable for eligible schools and libraries. Discounts for support depend on the level of poverty and the urban/rural status of the population served and range from 20 to 90 percent of the costs of eligible services. This program is also known as the *E-rate Program.*
- **Rural and High-Cost Program:** The USAC expects to disburse US$4.2 billion in USF funds in 2006 in support of subsidies to affordable basic telephone service in high-cost areas of the United States and its territories. Those funds are parceled out directly to LECs deemed qualified by the state commissions or by the FCC directly.
- **Rural Health Care Program:** Through the *Rural Health Care Corporation* (RHC), the USF provides subsidies of approximately US$45 million (2006) in support of telecommunications and Internet service to eligible rural health care providers in high-cost areas. Health care providers apply for the discounts, which are provided directly to the service providers.
- **Low-Income Consumers Program:** Lifeline subsidies of approximately US$820 million (2006) are parceled out to LECs to reduce the installation and monthly costs of basic telephone service for low-income consumers.

In addition to serving its normal, ongoing purposes, the Universal Service Fund recently came in very handy in a national emergency when Hurricane Katrina devastated the Gulf Coast of the United States in 2005. In response, the FCC modified certain USF rules as follows:

- **Schools and Libraries Program:** The commission opened a new application window to allow schools and libraries in the affected area and those serving increased numbers of students and patrons to resubmit their requests for E-rate funds for 2005. The commission also assigned the E-rate program's highest level of funding priority (i.e., 90 percent) to schools and libraries in the Hurricane Katrina disaster area for funding years 2005 and 2006.
- **Rural and High-Cost Program:** The FCC clarified that, under the rules for receiving high-cost support, carriers in the disaster area may use US$240 million in high-cost support funds received for high-cost areas in Alabama, Louisiana, and Mississippi to assist in reconstructing facilities damaged by the hurricane.
- **Rural Health Care Program:** The commission adopted rules to allow rural and nonrural public and nonprofit health care providers, including American Red Cross shelters providing health care services to disaster victims, to apply for support for advanced telecommunications and information services used for telemedicine applications in temporary quarters.
- **Low-Income Consumers Program:** The FCC adopted *Lifeline* rules to provide households eligible for individual housing assistance under rules of the *Federal Emergency Management Agency* (FEMA) with temporary wireless telecommunications service. The commission also adopted *Link-Up* rules and authorized US$211 million to help pay the costs of reestablishing connections, both temporary and permanent, to the telecommunications network [7].

Note that especially in a deregulated competitive environment some carrier must be designated as the *provider of last resort*, which concept is part and parcel of universal service. In other words, some carrier must serve even the least desirable residential end user, even if that customer lives in the middle of nowhere, requires only the most basic service, has neither the willingness nor ability to pay, and so on. Traditionally, the designee was what we currently refer to as the ILEC. Into the future, it is unclear how this will be handled.

By the way, each state in the United States also has a Universal Service Fund of some sort. In the case of the state of Washington, with which I am most familiar because I happen to have the great pleasure of living there, the *Telephone Assistance Program* (TAP) derives funding through an excise tax at a rate of 14 percent that applies to all switched access lines. The tax appears on the customer bill as a separate line item. TAP revenues go toward providing telecommunications services for low-income customers. If the resident is homeless, the fund will provide access to a community voice mailbox.

Universal service is no longer confined to the United States. Australia, for example, formally established a *Universal Service Obligation* (USO) in 2001, when the government finalized an agreement with Telstra, the ILEC, for the provision of untimed local calls and upgraded services in the *extended zones* of Australia. The agreement provides A$150 million for the infrastructure upgrade necessary to provide these services and was funded from the proceeds of the second sale of shares in Telstra, which originally was a government agency. The extended zones are located in the most sparsely populated areas of Australia. They cover about 80 percent of Australia's land mass and contain approximately 40,000 *telephone services* [8].

15.1.3.2 LEC Long Distance
Effective with the MFJ, the RBOCs and GTOCs (*GTE Operating Companies*) had the exclusive right to provide local service and intraLATA long-distance service. Effective immediately on passage of The Act, they also were free to offer interLATA long-distance services, but only outside their home states of operation. Several did just that, initially reselling long-distance service to their cellular customers. Otherwise, the ILECs made little effort to develop interLATA long-distance networks. SBC's acquisition of SNET, which was unconstrained by The Act, clearly signaled its intentions. Bell Atlantic's acquisition of GTE to form Verizon was a similarly clear signal, as GTE had a very substantial long-distance network.

The Act prevented the RBOCs from competing in the interLATA market within their home states of operation, however, until such time as they could satisfy a 14-point competitive checklist. Although a detailed discussion of those 14 points is beyond the scope of this book, each point on the list was intended to prove that the ILECs offer competitors easy access to local switching and transmission facilities and local databases and other resources at fair and reasonable rates that are cost based and that compensation is reciprocal where appropriate. At this point in time, those restrictions largely have been lifted. In the 22 years since the MFJ took effect and the 10 years since The Act was passed, the once lucrative long-distance market has eroded considerably, largely due to three factors:

- The US$11 billion fraud perpetrated by Bernie Ebbers, Chief Executive Officer, and other Worldcom executives sparked an irreversible long-distance price war. MCI Worldcom slashed long-distance prices while reporting inflated earnings. At the same time that AT&T, Sprint, and other long-distance competitors were in disbelief and shock over Worldcom's financial performance, they were slashing prices to match Worldcom's.
- About the same time, cellular telephony providers slashed airtime costs and began to eliminate domestic long-distance costs altogether, in favor of flat-rate service plans that bundled monthly service and airtime allowances at very attractive rates. Despite the fact that the quality of cellular communications does not compare favorably with landlines, many people now choose to use cellular telephones exclusively.
- VoIP service has been nibbling away at traditional PSTN long-distance revenues since the late 1990s and now presents serious competition.

15.1.3.3 Interconnection: The Quid Pro Quo
Interconnection is an issue of great significance. The voice world has been fully interconnected for most of its long history, while the data world, with the notable exception of the Internet, has been much less so, and the CATV world never had much use for the concept. In a full-convergence scenario, all competing carriers will offer some combination of voice and Internet access, and most will offer entertainment TV as well. Voice and Internet access, of course, absolutely requires interconnection through local loop facilities.

Interconnection issues cut several ways. Clearly the IXCs, CATV providers, and other competing carriers are concerned about the terms under which they can connect to the existing local loop infrastructure in order to reach the customer quickly. While many of them are building or rebuilding infrastructure to do that independently, that process takes time and is accomplished on a highly selective basis, addressing areas where the market potential justifies the investment, which clearly is not trivial. In the meantime, the thought was that competing carriers must have access at some reasonable cost to the embedded local loop plant owned by the incumbent carrier.

15.1.3.3.1 Telco Local Loops As I previously discussed, the RBOC ILECs had to satisfy a checklist of 14 points in order to provide interLATA long distance within their home states. Most especially, they were required to satisfy the state PUCs, DOJ, and FCC that interconnection was a significant reality or, at least, was significantly available on reasonable terms. Further, the costs of interconnection were required to be at levels that reflected real costs so the competing carriers could buy local loop access at compensatory wholesale rates that enabled them to retail those loops at competitive prices. Those costs were required to be unbundled so switch costs, for example, could not be applied to nonswitched facilities and the costs of support systems could not be included.

The act specified three ways by which CLECs could provide competing local phone service:

- **Build and Interconnect:** CLECs could build their own wireline or wireless local loops and interconnect with the ILEC and IXC networks
- **Bundled Wholesale Purchase:** CLECs could purchase bundled local telephone service from the ILECs at government-controlled wholesale prices, which typically were 15–25 percent below retail prices. *Bundled* service includes the following:
  - Local loops
  - Local and tandem switches (including software features)
  - Interoffice transmission facilities
  - Signaling and call-related database facilities
  - Operations Support Systems (OSSs) and information
  - Operator and directory assistance facilities
- **Unbundled Service:** The CLECs can purchase the very same network elements on an unbundled basis, which essentially is a menu of *Unbundled Network Elements* (UNEs) from which the CLECs can choose to purchase (actually lease) only what they need on a case-by-case basis. When all elements from the menu are chosen, this approach is known as *Unbundled Network Elements—Platform* (UNE-P) and the effect is the same as the bundled wholesale purchase, but the total price typically is at a discount of 40–60 percent from retail prices.

The comparison between bundled service and UNE-P is interesting. Think of buying bundled service as an automobile dealer buying an automobile from the manufacturer at the wholesale price, which is the retail price less 15–25 percent. So, an automobile with a retail value of US$50,000 costs the dealer as little as US$37,500. Think of UNE-P as the same dealer buying all of the parts required to make the same automobile and contracting with the manufacturer to assemble the parts into an automobile, with the price for parts plus assembly being the retail price less 40–60 percent. The UNE-P wholesale cost to the dealer for the same automobile is as low as US$20,000. The ILECs complained that the UNE-P costs were unrealistically low, while the CLECs, of course, complained that they were still too high.

Aside from considerable cost issues, the CLECs complained that the LEC processes were too slow in providing the loops to the CLECs; meanwhile, the ILECs claimed that they were doing all they could to cooperate. The CLECs complained that they had to fax their orders to the ILECs, which then processed them manually, rather than providing direct interfaces into the appropriate OSSs for order placement and tracking and for trouble reporting. The ILECs were concerned about system security, of course, as their OSSs intentionally were closed systems.

The CLECs also complained that the ILECs made unreasonable demands with respect to the collocation of termination facilities, switches, DSLAMs, and other equipment. Most of the ILECs required that collocated POPs be physically distinct and secure, so neither the ILEC nor the CLEC would have free access to the other's facilities. That requirement for physical partitioning sometimes translated only into separate equipment cages, while in other instances the ILECs insisted on separate entrance facilities (i.e., doors). In either case, the CLECs bore the costs, which sometimes reached levels in the area of $400,000 per Central Office (CO) POP. In 1999, after a great deal of pressure from competitive service providers, the FCC expanded its ruling to allow shared cage arrangements and even cageless collocation. Despite the relaxed collocation requirements, space remained unavailable in some LEC COs, the associated costs were relatively high, and the relationships often were far from cordial. As a result, a number of third parties established neutral collocation sites, also known as *telco hotels*, where fully equipped space could be rented by ILECs and CLECs alike [9, 10].

With respect to the local loop itself, there have been two standard approaches with respect to competitive DSL. One approach is for the ILEC to provide the DSL local loop and to terminate it in a DSLAM or other termination facility owned by an independent ISP and collocated in the ILEC CO. That approach, of course, meant that your DSL service is provided by two separate entities, that you receive two separate bills, that you must deal with two separate maintenance and support organizations, and so on. The cleanest approach is for the ISP to lease the local loop from the ILEC, thereby providing a single point of contact for all installation, billing, and maintenance and support issues. The FCC offered a third approach in its December 1999 *line-sharing* decision. Line sharing allows both the ILEC and the independent ISP to share the same local loop, with the ILEC retaining the low-frequency voice-grade portion of the spectrum and the ISP gaining access to the high-frequency portion of the spectrum for high-speed Internet access, and on an unbundled cost basis. There initially was a lot of interest in this approach. In fact, SBC and US West (now part of Qwest) even offered line sharing to ISPs at no charge. Over time, however, technical difficulties and generally poor relationships between the various parties seem to have cooled everyone's ardor for this approach. Note that line sharing is appropriate only for ADSL, as the other versions (e.g., HDSL and SDSL) all require access to the full spectrum supported over the local loop.

The battle lines were drawn. The PUCs, the DOJ, and the FCC attempted to sort it all out, and the courts got involved, but the issue remained unresolved. In

February 2002, the FCC adopted a rule promoting greater deployment of broadband services, concluding that wireline broadband Internet access services are *information services*, much like voice mail and e-mail, with a telecommunications component, rather than *telecommunications services*. Appeals to the FCC and challenges in the courts delayed the implementation of this decision for several years, but it finally became the law of the land in August 2005. The ILECs now are no longer required to share broadband local loops with CLECs. DSL is now unregulated, and the ILECs are no longer required to file DSL tariffs or otherwise seek regulatory approval for DSL prices. Broadband PON and WLL local loops are similarly deregulated, at least with respect to data and entertainment TV. At this point, however, the ILECs still wholesale DSL to CLECs and ISPs, and retail prices have dropped in fact. Most other nations of the world that have opened the network to competition have also adopted *Local Loop Unbundling* (LLU). Some nations heavily regulate wholesale prices of UNEs, although the trend seems to be toward deregulation.

15.1.3.3.2 CATV Local Loops   Note that the CATV providers generally are not required to provide open access, that is, interconnectivity, to competing ISPs. While federal regulations do not require that they do so, AOL and Time Warner were required to support open access over the Time Warner CATV networks as a prerequisite to their merger, which created AOL Time Warner in 2001. You may recall from detailed discussion of CATV networks in several previous chapters that the CATV providers are regulated largely at the local level. At the federal level, they largely are unregulated, except for issues of content, which were restated and redefined by the Telecommunications Act of 1996. The act focused on opening the RBOC's local networks to competition, never considering the CATV monopolies. As a result and for example, CATV networks are not required to be opened to CLECs of any sort. In January 1999, AT&T and TCI filed suit against Portland, Oregon, Multnomah County, and eight local cable regulators. Those local regulators placed an open-access condition on their approval of the AT&T/TCI merger, insisting that competing ISPs be provided open access to the network. This first challenge to the closed nature of CATV networks was not the last, but they all effectively were dismissed by the FCC's March 2002 classification of cable modem service as an *information service*, which effectively deregulated it. The Supreme Court affirmed that decision in June 2005 in the *Brand X* decision, when it ruled against Brand X Internet LLC, a Santa Monica (CA) based ISP that had challenged the FCC's position. CATV networks remain closed to competition.

There recently has been some movement with respect to CATV franchising, however. Such franchises traditionally have the province of municipal and local governments, which has caused the telcos considerable concern as they have sought to expand their services to include entertainment TV as part of the *triple-play* convergence scenario. On August 31, 2006, the California Assembly passed the Digital Infrastructure and Video Choice Act, which will allow new entrants to seek statewide permission, rather than approvals at the local level. If signed into law by the governor, California will join Indiana, Kansas, New Jersey, North Carolina, South Carolina, Texas, and Virginia in enacting statewide CATV franchise laws. The California law requires that providers continue to pay local franchise fees and adhere to local rights-of-way requirements.

15.1.3.4 Right-of-Way Right-of-way traditionally has been the domain of state and local governments, which are responsible for building and maintaining streets, roads, most highways, sidewalks, and so on. The act, however, gives the FCC authority to preempt any state or local government from action that may inhibit the ability of an entity to compete effectively in providing telecommunications services. That provision essentially stripped the state and local governments of their authority to limit the various carriers and service providers in their trenching to deploy new transmission facilities. Streets and roads in towns and cities all across the country have been trenched time and time again over the last few years, causing significant disruptions in traffic flows and compromising the structural integrity of the roadbeds, thereby increasing maintenance costs for the sake of competition in the local loop.

The Act contains provisions relative to pole attachments. The FCC amended those provisions in its August 1, 1996, preliminary implementation rules. Specifically, The Act and its FCC interpretation provide for nondiscriminatory access by CATV providers and telcos to poles, ducts, conduits, and rights-of-way owned by power utilities and incumbent LECs. Despite these provisions, the ILECs, CATV providers, and power utilities continue to battle over implementation specifics such as reciprocal pole attachment fees.

15.2 RATES AND TARIFFS

Rates and tariffs have been part and parcel of telecommunications regulation for many years, as previously discussed in Chapter 5. Beginning with the MFJ, the posture of the FCC gradually relaxed to the point that interstate long-distance tariffs largely are optional. Actually, the FCC prefers that they not be filed at all and is considering eliminating them altogether. In fact, there is discussion about eliminating local service tariffs as well, in favor of permitting market forces to dictate prices in the soon-to-be competitive market. Such a move would enable carriers to bundle services, perhaps including CPE in a package of cellular, PCS, long-distance, and local service, all of which would be unregulated. CATV providers have been up and down the regulatory path—first regulated at the federal level, then deregulated, then reregulated, and then deregulated again in 1999.

15.3 THE INTERNET

The Internet and, especially, the Web remain both hot topics and open issues. As part of the government's initiatives in telecom reform, Congress passed the *Communications Decency Act* (CDA) in February 1996. That act provided for sentences up to two years and levies of fines up to $250,000 for those who made indecent or offensive material available to minors on the Internet. Not only were the creators of such content liable, but also the Information Service Providers (e.g., America OnLine, CompuServe, and Prodigy). A number of coalitions and groups led by the *American Civil Liberties Union* (ACLU) and *Electronic Frontier Foundation* (EFF) challenged the law. A panel of federal judges struck down the CDA in June 1996, citing violation of *free speech* as guaranteed by the First Amendment to the

Constitution. Judge Stewart Dalzell, who wrote the opinion, stated, "The Internet may fairly be regarded as a never-ending worldwide conversation. The government may not, through the CDA, interrupt that." Numerous appeals have been unsuccessful so far.

Regardless of the outcome, the issue of censorship of the Internet and the Web will continue unabated for many years. It is clear that measures of some sort must be taken to limit access of minors to certain material. It may well be that the government will require installation on newer PCs of something akin to the *V-chip (Violence-chip)*, which was required to be included in new TV sets under provisions of The Act. Software filters currently are available that enable parents to impose some level of restriction, although the cost is additional and the filters must be updated continuously.

15.4 NUMBER PORTABILITY

As noted in Chapter 5, the PSTN relies on a standard logical addressing scheme known as a numbering plan. The ITU-T E.164 numbering plan includes country codes, area codes, CO prefixes, and line numbers. Various combinations of leading 1s and 0s advise the network of things such as a request for operator assistance or the crossing of an area code boundary or national border. These logical addresses always were oriented geographically in the voice world and oriented by customer and carrier in the data world; further, those numbers were not portable. The sole exception to this rule was that of toll-free (800/888/877/866) numbers in the United States, which became portable some years ago.

The lack of portability of local telephone numbers became a considerable issue with the introduction of deregulation and competition in the local exchange domain. As discussed in Chapter 14, the Telecommunications Act of 1996 came to the rescue with a provision for Local Number Portability (LNP) between LECs within a given geographic area. In July 1996, the FCC mandated that LNP be in place by January 1, 1998. Generally speaking, LNP takes advantage of the Advanced Intelligent Network (AIN) infrastructure, as discussed in Chapter 10. The FCC mandated that the various wireline and wireless LECs cooperate in the implementation of LNP through the establishment and synchronization of regional number databases that reside in Signal Control Points (SCPs) in the AIN control network. Neustar currently is responsible for running the *Number Portability Administration Centers* (NPACs), which serve as national clearinghouses for local service operators in North America. (*Note:* Neustar also is the primary administrator for the *North American Numbering Plan*, or NANP. Details on the NANP are available at www.nanpa.com.)

Wireless LNP took effect in November 2003, allowing users to change phone companies, wireless-to-wireless, wireless-to-wireline, or wireline-to-wireless, within a local area and keep their phone number.

15.5 LAWS AND SAUSAGES

Laws are like sausages, it is better not to see them being made.

Otto von Bismarck

Well, I have never watched sausages being made, but I did watch a lot of laws being made as I worked my way through graduate school as an aide to a Texas state senator. It was a great experience, but I have to agree with Bismarck. I suppose that's why I'm a technologist, rather than a politician. Actually, I'm not much of a politician anyway, which explains why I'm now an independent consultant, rather than something more similar to the CONTEL vice president I was in the mid-1980s.

In any event, a lot of folks do not like laws or regulations of any sort, whether they have to do with communications or with life in general. For those folks who have a problem with laws and regulations as they apply to the Internet, there is a place. *Sealand* was founded in 1967 as a sovereign principality in international waters about 6 miles off the coast of Britain. Sealand is an island nation consisting of an abandoned antiaircraft gun battery that looks much like an offshore drilling platform. The British built it during WW II (that's World War II for you youngsters who were playing games on your cell phones when you were supposed to be learning something in history class), never even considering the possibility that it would become a sovereign nation, much less that it would serve as the home of HavenCo, which has established a data haven hosting site. HavenCo boasts that it is a secure hosting site not only in the traditional sense but also in the sense that it is secure from legal action. Sealand, you see, has no laws governing data traffic, and the terms of HavenCo's agreement with Sealand specify that none shall ever be enacted. Please note, however, that HavenCo's acceptable use policy prohibits spam, hacking, and child pornography. While HavenCo has no restrictions and Sealand no laws regarding copyrights, patents, libel, restrictions on political speech, nondisclosure agreements, cryptography, or much of anything else, please note that individuals and corporations are still subject to the laws of their home countries. It may be comforting to you to know that your website will still be up and running when you're in the Graybar Hotel (that's jail), but I think I'll just mind my manners, abide by the laws of the good old US of A, and leave my website right where it is, in Mount Vernon, Washington [11–13].

Hello, Shreeve! Hello, Shreeve! And now, Shreeve, good night. The first (October 21, 1915) wireless transatlantic telephone call, between H.R. Shreeve, a Bell Telephone engineer at the Eiffel Tower in Paris, France and B.B. Webb in Arlington, Virginia.
Telephone: The First Hundred Years by John Brooks

REFERENCES

1. Williams, Veronica A. *Wireless Computing Primer*. M&T Books, 1996.
2. Parker, Edwin B. and Hudson, Heather E. *Electronic Byways: State Policies for Rural Development through Telecommunications*. Aspen Institute, 1995.
3. Bates, Bud and Gregory, Donald. *Voice & Data Communications Handbook*. McGraw-Hill, 1996.
4. Tunstall, W. Brooke. *Disconnecting Parties*. McGraw-Hill, 1985.
5. Hudson, Heather E. *Communications Satellites*. Free Press, 1990.
6. Snow, Marcellus S. *Marketplace for Telecommunications*. Longman, 1996.

7. "What Is the Universal Service Obligation?" http://internet.aca.gov.au/ACMAINTER. 131180:STANDARD::pc=PC_2491.

8. "Hurricane Katrina Universal Service Fund Relief Information." http://www.katrina-usf. org/katrina/.

9. Blacharski, Dan. "Dial C for Competition." *America's Network*, May 2001.

10. Fishel, Alan and Coleman, Sana. "What's Next for the FCC and Colocation." *America's Network*, May 2001.

11. www.havenco.com..

12. www.sealandgov.org.

13. Mullen, Theo. "A Haven for Net Lawbreakers?" *Internet Week*, June 19, 2000.

APPENDIX A

ACRONYMS, ABBREVIATIONS, CONTRACTIONS, INITIALISMS, AND SYMBOLS

"The question is," said Alice, *"whether you can make words mean so many different things."*

Lewis Carroll, *Through the Looking Glass and What Alice Found There*, 1872

*+$^*&@%&**$!	Encryption (Get it?)
1G	1st Generation
1 × EV-DO	One times EVolution—Data Optimized
1 × RTT	One times Radio Transmission Technology
2G	2nd Generation
2.5G	2nd Generation + but not quite 3rd Generation
2B1Q	2 Binary, 1 Quaternary
3B/4B	3 Bits/4 Bits
3G	3rd Generation
3GPP	3G Partnership Project
4B/5B	4 Bits/5 Bits
4B3T	4 Binary, 1 Ternary
4G	4th Generation
5B/6B	5 Bits/6 Bits
8B/10B	8 Bits/10 Bits
8-PSK	8-Phase Shift Keying
8 VSB	8 Vestigal SideBand
10GbE	10-Gigabit Ethernet
64B/66B	64 Bits/66 Bits
AAL	ATM Adaption Layer
AAV	Alternative Access Vendor

ABM	Asynchronous Balanced Mode
ABR	Available Bit Rate
AC	Alternating Current
ACD	Automatic Call Distributor
ACELP	Algebraic Code-Excited Linear Prediction
ACK	ACKnowledgment
ACL	Asynchronous Connectionless Link
ACLU	American Civil Liberties Union
ACR	Attenuation-to-Crosstalk Ratio
ACTS	Advanced Communications Technologies and Services
ADCCP	Advanced Data Communications Control Procedures
ADM	Add/Drop Multiplexer
ADML	Asymmetric Digital Microcell Link
ADPCM	Adaptive Differential Pulse Code Modulation
ADSL	Asymmetric Digital Subscriber Line
AES	Advanced Encryption Standard
AF	Assured Forwarding
AfriNIC	African Network Information Center
AGC	Automatic Gain Control
AIN	Advanced Intelligent Network
aka	also known as
ALI	Automatic Location Identification
AM	Amplitude Modulation
AMI	Alternate Mark Inversion
AMIS	Audio Messaging Interchange Specification
AMPS	Advanced Mobile Phone System
AMR	Automated Meter Reading
ANI	Automatic Number Identification
ANSI	American National Standards Institute
AO/DI	Always On/Dynamic ISDN
AP	Adjunct Processor, Application Processor
APD	Avalanche PhotoDiode
API	Applications Programming Interface
APNIC	Asia Pacific Network Information Center
APON	ATM-based Passive Optical Network
APP	Atom Publishing Protocol
APPN	Advanced Peer-to-Peer Networking
APS	Automatic Protection Switching
ARC	Airborne Relay Communications
ARCnet	Attached Resource Computer network
ARIN	American Registry for Internet Numbers
ARP	Address Resolution Protocol
ARPA	Address Routing and Parameter Area, Advanced Research Projects Agency
ARPANET	Advanced Research Projects Agency NETwork
ARQ	Automatic Repeat reQuest
ARS	Automatic Route Selection

ASCII	American Standard Code for Information Interchange
ASIC	Application-Specific Integrated Circuit
ASK	Amplitude Shift Keying
ASP	Application Service Provider
ASR	Automatic Speech Recognition
AT&T	American Telephone & Telegraph
ATM	Asynchronous Transfer Mode, Automated Teller Machine
ATMS	Advanced Traffic Management System
A-to-D	Analog-to-Digital
ATSC	Advanced Television Standards Committee
ATU-C	ADSL Termination Unit—Centralized
ATU-R	ADSL Termination Unit—Remote
AUSTEL	Australian Telecommunications Authority
AWG	American Wire Gauge
b	bit
B	Bit (rarely), Byte (typically)
B_c	Burst size, committed
B_e	Burst size, excess
B-ICI	Broadband ISDN InterCarrier Interface
B8ZS	Bipolar with Eight-Zeros Substitution
BBN	Bolt Beranek and Newman
BBS	Bulletin Board System
BCC	Block Checking Character
BCD	Binary-Coded Decimal
BCM	Bit Compression Multiplexer
BCT	Burst Cell Tolerance
BECN	Backward Explicit Congestion Notification
BEL	BELl (Bell)
Bellcore	Bell Communications Research
BER	Bit Error Rate
bit	*bi*nary digi*t*
BFT	Binary File Transfer
BHCA	Busy-Hour Call Attempt
BHCC	Busy-Hour Call Completion
BHCCS	Busy-Hour Centum Call Second
BHT	Busy-Hour Traffic
BICI	Broadband InterCarrier Interface
BIP	Bit-Interleaved Parity
B-ISDN	Broadband Integrated Services Digital Network
blog	we*b log*
BLT	Bacon, Lettuce, and Tomato (sandwich)
BMP	Basic Multilingual Plane
BOC	Bell Operating Company
BPL	Broadband over Power Line
BPO	British Post Office
BPON	Broadband Passive Optical Network
bps	bits per second

Bps	Bytes per second
BPSK	Binary Phase Shift Keying
BPV	BiPolar Violation
BRA	Basic Rate Access
BRAN	Broadband Radio Access Network
BRI	Basic Rate Interface
BS	Base Station
B&S	Brown and Sharp
BSC	Binary Synchronous Communications
BSD UNIX	Berkeley Software Distribution UNIX
BSS	Broadband Switching System
BT	Burst Tolerance
BTA	Basic Trading Area
B-TA	Broadband Terminal Adapter
B-TE	Broadband Terminal Equipment
B-UNI	Broadband User Network Interface
BUS	Broadcast and Unknown Server
BWA	Broadband Wireless Access
c	constant, velocity of light in a vacuum
C	Conventional
CAC	Carrier Access Charge
CAD	Computer-Aided Design
CAI	Common Air Interface
CALC	Customer Access Line Charge
CAN	CANcel
CAP	Carrierless Amplitude Modulation, Competitive Access Provider
CAPI	Common ISDN Applications Programming Interface
CAS	Centralized Attendant Service
Cat	Category
CATT	Chinese Academy of Telecommunications Technology
CATV	Community Antenna TeleVision
CB	Citizens Band
CBDS	Connectionless Broadband Data Service
CBR	Constant Bit Rate
CCIR	Comité consultatif international des radiocommunications (International Radio Consultative Committee)
CCITT	Comité Consultatif International Télégraphique et Téléphonique (International Telephone and Telegraph Consultative Committee)
CCK	Complementary Code Keying
CCLC	Carrier Common Line Charge
CCS	Centum Call Second, Common Channel Signaling
ccTLD	country code Top-Level Domain
CDA	Communications Decency Act

CDDI	Cable Distributed Data Interface
CDMA	Code Division Multiple Access
CDPD	Cellular Digital Packet Data
CDR	Call Detail Recording
CDSU	Channel Digital Service Unit
CDV	Cell Delay Variation
CDVT	Cell Delay Variation Tolerance
CE	Circuit Emulation
CELP	Code-Excited Linear Prediction
Centrex	*Cent*ral/ex*change*
CEPT	Conférence Européenne des Administrations des Postes et des Télécommunications (European Conference of Postal and Telecommunications Administrators)
CER	Cell Error Ratio
CERN	Conseil Européen pour la Recherche Nucléaire (European Council for Nuclear Research)
CERT	Computer Emergency Response Team
CES	Circuit Emulation Service
CHAP	Challenge Handshake Authentication Protocol
C-HTML	Compact HyperText Markup Language
CICS	Customer Information Control System
CID	Channel IDentification
CIDR	Classless InterDomain Routing
CIF	Common Intermediate Format
CIR	Committed Information Rate
CISC	Complex Instruction Set Computing
CIX	Commercial Internet eXchange
CLASS	Custom Local Access Signaling Services
CLEC	Competitive Local Exchange Carrier
CLID	Calling Line IDentification
CLNP	ConnectionLess Network Protocol
CLP	Cell Loss Priority
CLR	Cell Loss Ratio
cm	centimeter
CMISE	Common Management Information Service Element
CMR	Cell Misinsertion Rate
CMTS	Cable Modem Termination System
CO	Central Office
codec	*co*de/*dec*ode
COE	Central Office Exchange
COFDM	Coded Orthogonal Frequency Division Multiplexing
CoS	Class of Service
CPE	Customer Premises Equipment
CPU	Central Processing Unit
CR	Carriage Return

C/R	Command/Response
CRC	Cyclic Redundancy Check
CRT	Cathode Ray Tube
CS	Coding Scheme, Convergence Sublayer
CSA	Carrier Serving Area
CS-ACELP	Conjugate Structure-Algebraic Code-Excited Linear Prediction
CSMA	Carrier Sense Multiple Access
CSMA/CA	Carrier Sense Multiple Access with Collision Avoidance
CSMA/CD	Carrier Sense Multiple Access with Collision Detection
CSTA	Computer-Supported Telephony Applications
CSU	Channel Service Unit
CT	Computer Telephony, Cordless Telephony
CTD	Cell Transfer Delay
CTI	Computer Telephony Integration
CTS	Clear To Send
CTX	CenTreX
CUG	Closed User Group
CVoDSL	Channelized Voice over DSL
CVSD	Continuously Variable Slope Delta
CWDM	Coarse Wavelength Division Multiplexing
DAA	Data Access Arrangement
DAC	Dual Attached Concentrator
DACS	Digital Access Cross-connect System
DAMA	Demand-Assigned Multiple Access
D-AMPS	Digital Advanced Mobile Phone System
DARPANET	Defense Advanced Research Projects Agency NETwork
DAS	Dual Attached Station
dB	deciBels
DBS	Direct Broadcast Satellite
DC	Direct Current
DCC	Data Communications Channel
DCCH	Digital Control CHannel
DCCS	Digital Cross-Connect System
DCE	Data Communications Equipment
DCT	Discrete Cosine Transform
DCTE	Data Circuit Terminating Equipment
DDD	Direct Distance Dialing
DDN	Defense Data Network
DDS	Dataphone Digital Service
DE	Discard Eligibility
DECT	Digital Enhanced Cordless Telecommunications, nee Digital European Cordless Telecommunications
demarc	demarcation point
DES	Data Encryption Standard

DFB	Distributed FeedBack
DFS	Dynamic Frequency Selection
DG XIII	Directorate General 13
DHCP	Dynamic Host Configuration Protocol
DID	Direct Inward Dial
DiffServ	Differentiated Services
DISA	Direct Inward System Access
DLC	Digital Loop Carrier, Data Link Control
DLCI	Data Link Connection Identifier
DMT	Discrete MultiTone
DNA	Digital Network Architecture
DNIS	Dialed Number Identification Service
DNR	Domain Name Resolver
DNS	Domain Name Server, Domain Name System
DOCSIS	Data Over Cable Service Interface Specification
DOD	Direct Outward Dial
DOJ	Department Of Justice
Domsat	*Dom*estic *sat*ellite
DoS	Denial of Service
DP	Data Processing
DPCM	Differential Pulse Code Modulation
DPF	Data Private Facility
DPLC	Digital Port Line Charge
DPMA	Demand Priority Media Access
DPSK	Differential Phase Shift Keying
DQDB	Distributed Queue Dual Bus
DQPSK	Differential Quaternary Phase Shift Keying
DS	Digital Signal (level), Direct Sequence
DSA	Digital Switched Access
DSF	Dispersion-Shifted Fiber
DSI	Digital Speech Interpolation
DSL	Digital Subscriber Line
DSLAM	Digital Subscriber Line Access Multiplexer
DSMA/CD	Digital Sense Multiple Access/Collision Detect
DSP	Digital Signal Processor
DSS	Direct Satellite System
DSSS	Direct-Sequence Spread Spectrum
DSU	Data Service Unit, Digital Signal Unit
DTE	Data Terminal Equipment
DTMF	Dual-Tone MultiFrequency
D-to-A	Digital-to-Analog
DTV	Digital TeleVision
DWDM	Dense Wavelength Division Multiplexing
Dy	Dysprosium
DXC	Digital Cross-Connect
DXI	Data eXchange Interface
EA	Extended Address
EBCDIC	Extended Binary Coded Decimal Interchange Code

EBPP	Electronic Bill Presentment and Payment
EC	European Community
E-Carrier	European Carrier
ECC	Electronic Common Control
ECM	Error Control Mode
ECMA	European Computer Manufacturers Association
ECN	Explicit Congestion Notification
ECSA	Exchange Carriers Standards Association
ECSD	Enhanced Circuit-Switched Data
ECTF	Enterprise Computer Telephony Forum
EDFA	Erbium-Doped Fiber Amplifier
EDGE	Enhanced Data rates for Global Evolution
EDI	Electronic Data Interchange
EEB	Extended Erlang B
EF	Expedited Forwarding
EFF	Electronic Freedom Foundation
EGPRS	Enhanced General Packet Radio Service
EHF	Extremely High Frequency
EI	Extension Indication
EIA	Electronic Industries Alliance
EIEIO	EIEIO
EIR	Excess Information Rate
EIRP	Equivalent Isotropically Radiated Power
EKTS	Electronic Key Telephone System
ELAN	Emulated Local Area Network
ELF	Extremely Low Frequency
EM	End of Medium
EMI	ElectroMagnetic Interference
EMS	Element Management System
EOC	Embedded Operations Channel
ESD	End of Stream Delimiter
ENQ	ENQuiry
ENUM	Electronic NUMber
EOT	End Of Text, End Of Transmission
EPABX	Electronic Private Automatic Branch eXchange
EPON	Ethernet-based Passive Optical Network
EQEEB	Equivalent Queue Extended Erlang B
ERMES	European Radio MEssage System
ESC	ESCape
ESCON	Enterprise Systems CONnectivity
ESF	Extended SuperFrame
ESMR	Enhanced Specialized Mobile Radio
ESS	Electronic Switching System
ESSID	Extended Service Set IDentifier
ETACS	Extended Total Access Communications System
ETB	End of Transmission Block
E-TDMA	Enhanced Time Division Multiple Access
ETSI	European Telecommunications Standards Institute

ETX	End to Text
EU	European Union
EVRC	Enhanced Variable-Rate voCoder
FAST	Frame-based ATM over SONET/SDH Transport
FATE	Frame-based ATM Transport over Ethernet
FC	Fibre Channel
FCC	Federal Communications Commission
FC/IP	Fibre Channel over Internet Protocol
FCS	Frame Check Sequence
FDD	Frequency Division Duplex
FDDI	Fiber Distributed Data Interface
FDM	Frequency Division Multiplexing
FDMA	Frequency Division Multiple Access
FDX	Full DupleX
FEC	Forward Error Correction, Forwarding Equivalence Class
FECN	Forward Explicit Congestion Notification
FEP	Front End Processor
FEX	Foreign EXchange
FEXT	Far-End CROSSTalk
FH	Frequency Hopping
FHSS	Frequency-Hopping Spread Spectrum
FICON	FIbre CONnections
FIGS	FIGureS
FM	Frequency Modulation
FOD	Fax-On-Demand
FoIP	Fax over Internet Protocol
FOMA	Freedom of Mobile Multimedia Access
FOTS	Fiber Optic Transmission System
FPLMTS	Future Public Land Mobile Telecommunications System
fps	Frames per second
FR	Frame Relay
FRAD	Frame Relay Access Device, Frame Relay Assembler/Disassembler
FRF	Frame Relay Forum
FRND	Frame Relay Network Device
FSAN	Full-Service Access Network
FSK	Frequency Shift Keying
FSO	Free Space Optics
FSS	Fixed Satellite System
FTP	File Transfer Protocol
FTTC	Fiber-To-The-Curb
FTTH	Fiber-To-The-Home
FTTN	Fiber-To-The-Neighborhood, Fiber-To-The-Node
FTTP	Fiber-To-The-Premises
FUD	Fear, Uncertainty, and Doubt
FUNI	Frame User–Network Interface

FWM	Four-Wave Mixing
FX	Foreign eXchange
G	Generation, Giga
GbE	Gigabit Ethernet
Gbps	Gigabits per second
GEO	Geosynchronous Earth Orbiting
GFC	Generic Flow Control
GFI	General Format Identifier
GFR	Guaranteed Frame Rate
GFSK	Gaussian Frequency Shift Keying
GHz	GigaHertz
GIF	Graphics Interchange Format
GII	Global Information Infrastructure
GMPLS	Generalized MultiProtocol Label Switching
GOF	Glass Optical Fiber
GO-MVIP	Global Organization for MultiVendor Integration Protocol
GoS	Grade of Service
GPB	Grand Pooh-Bah
GPON	Gigabit Passive Optical Network
GPRS	General Packet Radio Service
GPS	Global Positioning System
GPSK	Gaussian Phase Shift Keying
GSM	Global System for Mobile communications, nee Groupe Spéciale Mobile
GSTN	Global Switched Telephone Network
GTE	General Telephone and Electric
gTLD	generic Top-Level Domain
GTOC	GTE Operating Company
GUI	Graphical User Interface
HALO	High-Altitude Long Operation
HAN	Home Area Network
HBA	Host Bus Adapter
HCV	High-Capacity Voice
HDB3	High-Density Bipolar 3-zeros
HDLC	High-level Data Link Control
HDSL	High-bit-rate Digital Subscriber Line
HDTV	High-Definition TV
HDX	Half DupleX
HEC	Header Error Control, Header Error Correction
HF	High Frequency
HFC	Hybrid Fiber/Coax
HiperLAN	High-performance radio Local Area Network
HIPPI	HIgh-Performance Parallel Interface
HLEN	Header LENgth
HomeRF	Home Radio Frequency
HSCSD	High-Speed Circuit-Switched Data
HSDPA	High-Speed Downlink Packet Access
HSTR	High-Speed Token Ring

HSUPA	High-Speed Uplink Packet Access
HTML	HyperText Markup Language
HTTP	HyperText Transport Protocol
HV	High Voltage
Hz	Hertz
IA	Implementation Agreement
IAB	Internet Architecture Board
IAD	Integrated Access Device
IANA	Internet Assigned Numbers Authority
IAP	Internet Access Provider
IBC	Integrated Broadband Communications
ICANN	Internet Corporation for Assigned Names and Numbers
ICF	Interexchange Compatibility Forum
ICN	Implicit Congestion Notification
ICO	Intermediate Circular Orbit
IDSL	ISDN Digital Subscriber Line
IDU	Interface Data Unit
IEC	InterExchange Carrier, International Electrotechnical Commission
IEEE	Institute of Electrical and Electronics Engineers
IETF	Internet Engineering Task Force
IFP	Internet Fax Protocol
IGF	International Gateway Facility
IHL	Internet Header Length
III	International Information Infrastructure
ILEC	Incumbent Local Exchange Carrier
ILMI	Interim Local Management Interface
IM	Instant Messaging
IMA	Inverse Multiplexing over ATM
IMAP	Internet Message Access Protocol
imho	in my humble opinion
i-Mode	internet Mode
IMPP	Instant Messaging and Presence Protocol
IMS	IP Multimedia Subsystem
IMT	InterMachine Trunk, International Mobile Telecommunications
IMT-2000	International Mobile Telecommunications-2000
IMTC	International Multimedia Teleconferencing Consortium
IMTS	Improved Mobile Telephone Service
IN	Intelligent Network
INCITS	InterNational Committee for Information Technology Standards
Inmarsat	*In*ternational *Mar*itime *Sat*ellite
INP	Interim Number Portability
INRIA	Institut National de Recherche en Informatique et en Automatique
intercom	intercommunication (system)

InterNIC	Network Information Center Internet Registry
INWATS	INward Wide Area Telecommunications Service
IOR	Index Of Refraction
IP	Intelligent Peripheral, Internet Protocol
IPBX	Internet Protocol (Private) Branch eXchange, Intranet Private Branch eXchange
IP PBX	Internet Protocol Private Branch eXchange
IPsec	IP security
IPTV	Internet Protocol TeleVision
IPX	Internet Packet eXchange
Ir	Iridium
IR	Incremental Redundancy, InfraRed
IRC	International Record Carrier, Internet Relay Chat
IRTF	Internet Research Task Force
IS	Interim Standard, International Standard
ISCSI	Internet Small Computer Systems Interface
ISDN	Integrated Services Digital Network
ISM	Industrial/Scientific/Medical
ISO	International Organization for Standardization (not an acronym)
ISOC	Internet SOCiety
ISP	Internet Service Provider
ISSI	Interswitching System Interface
ISU	Integrated Service Unit
IT	Information Technology
ITU	International Telegraph Union, International Telecommunications Union
ITU-D	International Telecommunications Union— Development Sector
ITU-R	International Telecommunications Union— Radiocommunication Sector
ITU-T	International Telecommunications Union— Telecommunication Standardization Sector
IVD	Integrated Voice/Data
IVR	Interactive Voice Response
IWU	InterWorking Unit
IXC	IntereXchange Carrier
JBOD	Just a Bunch Of Disks
JDC	Japanese Digital Cellular
JPEG	Joint Photographic Experts Group
JTACS	Japanese Total Access Communications System
JTAPI	Java Telephony Application Programming Interface
JUGHEAD	Jonzy's Universal Gopher Hierarchy Excavation And Display
k	kilo
kbps	kilobits per second
kHz	kiloHertz
km	kilometer

KSU	Key Service Unit
KTS	Key Telephone System
KTU	Key Telephone Unit
λ	lambda
L	Long, Loran
L2TP	Layer 2 Tunneling Protocol
LACNIC	Latin-American and Caribbean Network Information Center
LAN	Local Area Network
LANE	Local Area Network Emulation
LAP-B	Link Access Procedure—Balanced
LAP-D	Link Access Procedure—Data channel
LAP-F	Link Access Procedure—Frame mode services
laser	*l*ight *a*mplification by *s*timulated *e*mission of *r*adiation
LATA	Local Access and Transport Area
LCI	Logical Channel Identifier
LCR	Least Cost Routing
LDAP	Lightweight Directory Access Protocol
LD-CELP	Low-Delay Code-Excited Linear Prediction
LDP	Label Distribution Protocol
LEC	LAN Emulation Client, Local Exchange Carrier
LECS	LAN Emulation Client Server
LED	Light-Emitting Diode
LEO	Low-Earth Orbiting
LER	Label Edge Router
LES	LAN Emulation Server
LF	Low Frequency
LFE	Low-Frequency Enhancement
LI	Length Indication
LIR	Local Internet Registry
LLC	Logical Link Control, Limited Liability Corporation
LLPOFYNILTATW	Liar, Liar Pants On Fire, Your Nose Is Longer Than A Telephone Wire
LLU	Local Loop Unbundling
LMDS	Local Multipoint Distribution Service
LMI	Local Management Interface
LNP	Local Number Portability
LNPA	Local Number Portability Administration
LOH	Line OverHead
LOS	Line Of Sight
LPC	Linear Predictive Coding
LRC	Longitudinal Redundancy Check
LRN	Local Routing Number
LSB	Least Significant Bit
LSR	Label Switched Router
LTRS	LeTteRs

LU	Logical Unit
LUNI	LANE User-to-Network Interface
LV	Low Voltage
μ	micro, micron
μm	micrometer
m	meter, milli
M	Mega
MAC	Medium Access Control; Move, Add, and Change
MAE	Merit Access Exchange, Metropolitan Area Exchange
MAN	Metropolitan Area Network
MAP	Manufacturing Automation Protocol
MAPI	Messaging Application Programming Interface
MAT	Maintenance and Administration Terminal
MAU	Medium Access Unit, Multistation Access Unit
Mbps	Megabits per second
MBS	Maximum Burst Size
MC	Multipoint Controller
MCR	Minimum Cell Rate
MCS	Modulation and Coding Scheme
MCTD	Mean Cell Transfer Delay
MCU	Multipoint Control Unit
MDF	Main Distribution Frame
MDP	Motorola Data Protocol
Megaco	*Media ga*tewa*y co*ntrol
MEMS	MicroElectroMechanical System
MEO	Middle-Earth Orbiting
MF	Medium Frequency
MFJ	Modified Final Judgment
MFR	Multilink Frame Relay
MFS	Maximum Frame Size, Metropolitan Fiber System
MG	Media Gateway
MGC	Media Gateway Controller
MGCP	Media Gateway Control Protocol
MHS	Message-Handling Service
MHz	MegaHertz
MILNET	MILitary NETwork
MIME	Multipurpose Internet Mail Extension
MIMO	Multiple Input, Multiple Output
MLT	MultiLevel Transition
mm	millimeter
MMDS	Multichannel Multipoint Distribution Services
MMF	MultiMode Fiber
MMS	Multimedia Message Service
MNP	Microcom Networking Protocol
Mobitex	*Mobi*le *tex*t
MOD	Music On Demand
modem	*mo*dulate/*dem*odulate
MoH	Modem on Hold

MOS	Mean Opinion Score
MoU	Memorandum of Understanding
MP	Multipoint Processor
MP3	Moving Pictures Experts Group, audio layer 3
MPEG	Moving Pictures Experts Group
MPI	MultiPath Interference
MPLS	MultiProtocol Label Switching
MPOA	MultiProtocol Over ATM
MPOE	Minimum Point Of Entry
MPPP	Multilink Point-to-Point Protocol
MS	Mobile Station
MSF	Multiservice Switching Forum
MSN	MicroSoft Network
MSS	Mobile Satellite System
MTP	Mail Transfer Protocol
MTS	Message Telecommunications Service
MTSO	Mobile Telephone Switching Office
MTU	Maximum Transmission Unit
mux	multiplexer
MV	Medium Voltage
MVIP	Multivendor Integration Protocol
NAK	Negative AcKnowledgment
N-AMPS	Narrowband Advanced Mobile Phone System
NANC	North American Numbering Council
NANP	North American Numbering Plan
NAP	Network Access Point
NAS	Network-Attached Storage
NAT	Network Address Translation
NA-TDMA	North American Time Division Multiple Access
NCC	Network Control Center
NDSF	Non-Dispersion-Shifted Fiber
NE	Network Element
NEC	National Electrical Code
NECA	National Exchange Carrier Association
NEXT	Near-End CROSSTalk
NFAS	Non-Facility-Associated Signaling
ngDLC	next-generation Digital Loop Carrier
NGI	Next-Generation Internet
NHRP	Next Hop Resolution Protocol
NIC	Network Information Center (Internet Registry), Network Interface Card
NID	Network Interface Device
NII	National Information Infrastructure
NIR	National Internet Registry
N-ISDN	Narrowband Integrated Services Digital Network
NIST	National Institute of Standards and Technology
NIU	Network Interface Unit
NLOS	Non-Line-Of-Sight
nm	nanometer

NMT	Nordic Mobile Telephone
NNI	Network-to-Network Interface
NOC	Network Operations Center
NOS	Network Operating System
NPA	Numbering Plan Administration
NPAC	Number Portability Administration Center
NREN	National Research and Education Network
nrt-VBR	non-real-time Variable Bit Rate
NSA	National Security Agency
NSFNET	National Science Foundation NETwork
NT	Network Termination
NTACS	Narrowband Total Access Communications System
NTSC	National Television Standards Committee
NTU	Network Termination Unit
NUL	NULl (null)
NZDF	Non-Zero-Dispersion-shifted Fiber
OADM	Optical Add/Drop Multiplexer
OA&M	Operations, Administration, and Maintenance
OAM&P	Operations, Administration, Management, and Provisioning
OC	Optical Carrier
OCC	Other Common Carrier
OCR	Optical Character Recognition
OD	Outside Diameter
OECD	Organisation for Economic Co-operation and Development
OEO	Optical–Electrical–Optical
OFDM	Orthogonal Frequency Division Multiplexing
OLT	Optical Line Terminal
O&M	Operations and Maintenance
ONT	Optical Network Terminal
ONU	Optical Network Unit
OOO	Optical–Optical–Optical
OPS	Off-Premises Station
OPTIS	Overlapped PAM Transmission with Interlocking Spectra
OPX	Off-Premises eXtension
OQPSK	Offset Quadrature Phase Shift Keying
OS	Operating System
OSI	Open Systems Interconnection
OSP	OutSide Plant
OSPF	Open Shortest Path First
OSS	Operations Support System
OTN	Optical Transport Network
OUT	Optical Transport Unit
π/4 DQPSK	Pi/4 Differential Quaternary Phase Shift Keying
π/4 QPSK	Pi/4 Quadrature Phase Shift Keying
p	peta

P	Probability
PA	PreArbitrated
PABX	Private Automatic Branch eXchange
PACS	Personal Access Communications Services
PAD	Packet Assembler/Disassembler
PAL	Phase Alternate Line
PAM	Pulse Amplitude Modulation
PAMS	Perceptual Analysis/Measurement System
PAN	Personal Area Network
PAP	Password Authentication Protocol
PBX	Private Branch eXchange
PC	Personal Computer
PCB	Printed Circuit Board
PCM	Pulse Code Modulation
PCMCIA	Personal Computer Memory Card International Association
PCN	Personal Communications Network
PCR	Peak Cell Rate
PCS	Personal Communications Services
PD	Powered Device
PDA	Personal Digital Assistant
PDC	Personal Digital Cellular
PDH	Plesiochronous Digital Hierarchy
PDN	Public Data Network
PDU	Protocol Data Unit
pel	*pi*cture *el*ement
PESQ	Perceptual Evaluation of Speech Quality
PGP	Pretty Good Privacy
PHB	Per-Hop Behavior
PHS	Personal Handyphone System
PHY	PHYsical layer
PID	Pager IDentification number
PIN	Personal Identification Number, Positive–Intrinsic–Negative
PISN	Private Integrated Services Network
PKM	Privacy-Key Management
PLC	PowerLine Carrier
PLP	Packet Layer Protocol
PM	Phase Modulation, Physical Medium
PMD	Polarization-Mode Dispersion
PMI	Property Management Interface
PN	PseudoNoise
PNNI	Private Network-to-Network Interface
POCSAG	Post Office Code Standardization Advisory Group
PoE	Power over Ethernet
POF	Plastic Optical Fiber
POH	Path OverHead
PON	Passive Optical Network

POP	Point Of Presence, Post Office Protocal
POS	Packet Over SONET, Personal Operating Space
POTS	Plain Old Telephone Service
P-phone	Proprietary phone
PPO	Private Paging Operator
PPP	Point-to-Point Protocol
PPTP	Point-to-Point Tunneling Protocol
PPV	Pay-Per-View
PRA	Primary Rate Access
PRI	Primary Rate Interface
prosigns	procedural signals
PSC	Public Service Commission
PSD	Power Spectral Density
PSE	Power Sourcing Equipment
PSK	Phase Shift Keying
PSN	Packet-Switching Node
PSQM	Perceptual Speech Quality Measurement
PSS1	Private Signaling System no. 1
PSTN	Public Switched Telephone Network
PTE	Path-Terminating Equipment
PTI	Packet-Type Identifier, Payload-Type Indicator
PTT	Post, Telegraph, and Telephone; Push To Talk
PU	Physical Unit
PUC	Public Utility Commission
PVC	Permanent Virtual Circuit, PolyVinyl Chloride
PWE	PseudoWire Emulation
PWT	Personal Wireless Telecommunications
PWT-E	Personal Wireless Telecommunications—Enhanced
QA	Queued Arbitrated
QAM	Quadrature Amplitude Modulation
QCIF	Quarter-Common Intermediate Format
QoS	Quality of Service
QPSK	Quadrature Phase Shift Keying, Quaternary Phase Shift Keying
QPSX	Queued Packet Synchronous eXchange
QSIG	Q SIGnaling
RACE	Research for Advanced Communications in Europe
RADIUS	Remote Authentication Dial-In User Service
RADSL	Rate-Adaptive Digital Subscriber Line
RAID	Redundant Array of Inexpensive Disks
RAM	Random-Access Memory
RAN	Radio Area Network
RAS	Registration/Admission/Status, Remote Access Server
RATT	R1022 ATM Technology Testbed
RBOC	Regional Bell Operating Company
RCC	Radio Common Carrier
RDF	Resource Description Framework

RF	Radio Frequency
RFC	Request For Comment
RFI	Radio Frequency Interference
RFID	Radio Frequency IDentification
RG	Radio Guide
RGB	Red, Green, Blue
RHC	Regional Holding Company, Rural Health Care Corporation
RIP	Routing Information Protocol
RIPE NCC	Réseaux IP Européens Network Coordination Center
RIR	Regional Internet Registry
RISC	Reduced Instruction Set Computing
RJE	Remote Job Entry
RM	Resource Management
RPC	Remote Procedure Call
RPELPC	Regular Pulse Excitation Linear Predictive Coding
RPR	Resilient Packet Ring
RS	Recommended Standard
RSA	Rivest–Shamir–Adleman
RSS	RDF Site Summary, Really Simple Syndication, Rich Site Summary
RSVP	Resource reSerVation Protocol
RTCP	RTP Control Protocol
RTP	Real-time Transport Protocol
RTS	Request To Send
rt-VBR	real-time Variable Bit Rate
Rx	Receive
S	Short
SAA	Systems Application Architecture
SAIC	Science Applications International Corporation
SAN	Storage Area Network
SAPI	Service Access Point Identifier
SAR	Segmentation And Reassembly
SAT	Subscriber Access Terminal
SATAN	Security Administrator Tool for Analyzing Networks
SBE	Small-Business Enterprise
SCADA	Supervisory Control And Data Acquisition
SCAI	Switch-to-Computer Applications Interface
SCC	Specialized Common Carrier
S-CDMA	Synchronous Code Division Multiple Access
SCE	Service Creation Element, Service Creation Environment
SCO	Synchronous Connection Oriented
SCP	Signal Control Point, Service Control Point
SCR	Sustainable Cell Rate
SCSA	Signal Computing System Architecture
SCSI	Small Computer Systems Interface

ScTP	Screened Twisted Pair
SDDN	Software-Defined Data Network
SDH	Synchronous Digital Hierarchy
SDLC	Synchronous Data Link Control
SDN	Software-Defined Network
SDR	Software-Defined Radio
SDSL	Symmetric Digital Subscriber Line
SDTV	Standard-Definition TeleVision
SDU	Service Data Unit
SEAL	Simple and Efficient ATM Adaptation Layer
SEC	Symmetrically Echo Canceled
SECAM	Séquential Couleur Avec Mémoire
SECBR	Severely Errored Cell Block Ratio
SEGP	Sponsored Education Group Participation
Sesame	*S*ecure *e*uropean *s*ystems for *a*pplications in a *m*ultivendor *e*nvironment
SETI	Search for ExtraTerrestrial Intelligence
SFTP	Shielded Foil Twisted Pair
SGCP	Simple Gateway Control Protocol
SHDSL	Symmetric High-bit-rate Digital Subscriber Line
SHF	SuperHigh Frequency
S-HTTP	Secure HyperText Transport Protocol
SIG	SMDS Interest Group, Special Interest Group
SIM	Subscriber Identification Module
SIMPLE	Session Initiation Protocal (SIP) for Instant Messaging and Presense Leveraging Extension
SIP	Session Initiation Protocol, SMDS Interface Protocol
SLC	Schools and Libraries Corporation, Subscriber Line Carrier, Subscriber Line Charge
SLIP	Serial Line Internet Protocol
SMDI	Simplified Message Desk Interface
SMDR	Station Message Detail Recording
SMDS	Switched Multimegabit Data Service
SMDSU	Switched Multimegabit Data Service Digital Service Unit
SME	Small-to-Medium Enterprise, Subject Matter Expert
SMF	Single-Mode Fiber
SMR	Specialized Mobile Radio
SMS	Service Management System, Short Message Service
SMTP	Simple Mail Transfer Protocol
SNA	Systems Network Architecture
SNET	Southern New England Telephone Company
SNI	Subscriber Network Interface
SNMP	Simple Network Management Protocol
SNR	Signal-to-Noise Ratio

SOH	Section OverHead, Start Of Header
SOHO	Small Office/Home Office
SONET	Synchronous Optical NETwork
SPC	Stored Program Control
SPE	Synchronous Payload Envelope
SPIM	SPam over Instant Messaging
SPOT	Single Point Of Termination
SPX	Sequenced Packet eXchange
SRC	Spiral Redundancy Check
SRDL	SubRate Digital Loop
SRF	Special Resource Function
SRP	Source-Routing Protocol
SRT	Source-Routing Transparent
SS	Spread Spectrum, Switching System
$S \times S$	Step-by-Step
SS7	Signaling System 7
SSA	Serial Systems Architecture
SSD	Start of Stream Delimiter
SSID	Service Set IDentifier
SSL	Secure Socket Layer
SSP	Service Switching Point
STDM	Statistical Time Division Multiplexing
STM	Synchronous Transfer Mode
STP	Shielded Twisted Pair, Signal Transfer Point, Spanning Tree Protocol
STS	Synchronous Transport Signal
STX	Start of TeXt
SUB	SUBstitute
SUV	Sport Utility Vehicle
SVC	Switched Virtual Circuit
SYNTRAN	SYNchronous TRANsmission
T	Tera
T	Measurement interval (time)
TA	Terminal Adapter
TACS	Total Access Communications Systems
TAP	Telephone Assistance Program
TAPI	Telephony Application Programming Interface
TASI	Time-Assigned Speech Interpolation
Tbps	Terabits per second
T_c	Time—committed (rate measurement interval)
TC	Transmission Convergence
T-carrier	Terrestrial carrier
TCM	Trellis-Coded Modulation
TCP	Transmission Control Protocol
TC-PAM	Trellis-Coded PAM
TDD	Telephone Device for the Deaf, Time Division Duplex
TDM	Time Division Multiplexing

TDMA	Time Division Multiple Access
TD-SCDMA	Time Division Synchronous Code Division Multiple Access
TE	Terminal Equipment
TEI	Terminal Endpoint Identifier
TELNET	TELecommunications NETwork
THz	TeraHertz
TIA	Telecommunications Industry Association
TIB	Tag Information Base
TIFF	Tagged Image File Format
TIPEM	Toolkit for Interoperable Privacy-Enhanced Messaging
TLD	Top-Level Domain
TM	Terminating Multiplexer
TMR	Trunk Mobile Radio
TO	Telecommunications Organization, Transport Overhead
TOH	Transport OverHead
TOP	Technical and Office Protocol
TP	Transport Protocol, Twisted Pair
TPC	Transmission Power Control
TPDDI	Twisted-Pair Distributed Data Interface
transceiver	*trans*mitter/re*ceiver*
transpond	*trans*mit/res*pond*
TSAPI	Telephony Services Application Programming Interface
TSR	Terminate-and-Stay-Resident
TTL	Time To Live
TTS	Text-To-Speech
TTY	TeleTYpe
TV	TeleVision
TWP	Two-Way Paging
TWX	TeletypeWriter eXchange
Tx	Transmit
UBR	Unspecified Bit Rate
UCAID	University Corporation for Advanced Internet Development
UCD	Uniform Call Distributor
UCS	Universal Character Set
UDP	Uniform Dialing Plan, User Datagram Protocol
UGS	Unsolicited Grant Service
UHF	UltraHigh Frequency
UIFN	Universal International Freephone Number
UMTS	Universal Mobile Telecommunications System
UNE	Unbundled Network Element
UNE-P	Unbundled Network Elements—Platform
UNI	Universal Network Interface, User–Network Interface
Unicode	Universal code

U-NII	Unlicensed National Information Infrastructure
UPS	Uninterruptible Power Supply
URL	Uniform Resource Locator
USAC	Universal Service Administrative Company
USB	Universal Serial Bus
USDC	U.S. Digital Cellular
USF	Universal Service Fund
USO	Universal Service Obligation
US TDMA	U.S. Time Division Multiple Access
UTF	Unicode Transformation Format
UTP	Unshielded Twisted Pair
UV	UltraViolet
UWB	Ultra-WideBand
v	variable, velocity of light in a medium
V	Volt
VAD	Voice Activity Detection
VAN	Value-Added Network
VBNS	Very-high-speed Backbone Network Service
VBR	Variable Bit Rate
VC	Virtual Channel, Virtual Circuit, Virtual Container
V-chip	Violence chip
VCI	Virtual Channel Identifier
VCSEL	Vertical-Cavity Surface-Emitting Laser
VDSL	Very-high-Data-rate Subscriber Line
VDU	Visual Display Unit
VERONICA	Very Easy Rodent-Oriented Net-wide Index to Computerized Archives
VERS	VERSion number
VG	Voice Grade
VGE	Voice-Grade Equivalent
VHF	Very High Frequency
VLAN	Virtual Local Area Network
VLF	Very Low Frequency
VoATM	Voice over Asynchronous Transfer Mode
VOD	Video-On-Demand
VoDSL	Voice over Digital Subscriber Line
VoFR	Voice over Frame Relay
VoIP	Voice over Internet Protocol
VoWi-Fi	Voice over Wireless Fidelity
Vp	Velocity of propagation
VP	Virtual Path
VPI	Virtual Path Identifier
VPIM	Voice Profile for Internet Mail
VPN	Virtual Private Network
VQC	Vector Quantizing Code
VQL	Variable Quantizing Level
VRC	Vertical Redundancy Check
VRF	Virtual Routing/Forwarding
VSAT	Very Small Aperture Terminal

VSB	Vestigial SideBand
VSELP	Vector Sum-Excited Linear Prediction
VSM-AM	Vestigial Side-band Amplitude Modulation
VT	Virtual Tributary
VTAM	Virtual Telecommunications Access Method
VTOA	Voice and Telephony Over ATM
W	Watt
WACS	Wireless Access Communication System
WAIS	Wide Area Information Service
WAN	Wide Area Network
WAP	Wireless Application Protocol
WARC	World Administrative Radio Conferences
WATS	Wide Area Telecommunication Service
WCDMA	Wideband Code Division Multiple Access
WDM	Wavelength Division Multiplexing
WECA	Wireless Ethernet Compatibility Alliance
WEP	Wired Equivalent Privacy
WiBro	Wireless Broadband
Wi-Fi	Wireless Fidelity
WiMAX	Worldwide interoperability for Microwave Access
WIS	WAN Interface Sublayer
WLAN	Wireless Local Area Network
WLL	Wireless Local Loop
WMBTOTCITBWTNTALI	We May Be the Only Telephone Company In Town But We Try Not To Act Like It
WME	Wi-Fi MultiMedia Extension
WML	Wireless Markup Language
WMM	Wi-Fi MultiMedia extension
WOSA	Windows Open Services Architecture
WOTS	Wireless Office Telecommunications System
WPA	Wi-Fi Protected Access
WPAN	Wireless Personal Area Network
WRC	World Radio Conference
WTLS	Wireless Transport Layer Security
WTO	World Trade Organization
WWW	World Wide Web
WWDM	Wide Wavelength Division Multiplexing
x	generic, unknown
XBar	CROSSBar
xDSL	generic Digital Subscriber Line
XML	eXtensible Markup Language
XMPP	eXtensible Messaging and Presence Protocol
XNS	Xerox Networking System
YCDBSOYA	You Can't Do Business Sitting On Your Ass
ZDSF	Zero-Dispersion Shifted Fiber
ZC	Zigbee Coordinator
ZED	Zigbee End Device
ZR	Zigbee Router
ZWPF	Zero Water Peak Fiber

APPENDIX B

STANDARDS ORGANIZATIONS AND SPECIAL INTEREST GROUPS (SIGs)

There is no useful rule without an exception.

Thomas Fuller, Gnomlogia, 1732

Standards are rules, principles, or measures established as models or examples by authority, custom, or general consent. Standards take several forms. *De jure* standards are formal standards that do not have the force of law but often come close as they are set by formal standards bodies that generally are established by governmental or regulatory bodies or at least by industry consensus. Such formal bodies include the ANSI, CableLabs, and the ITU. *Du jour* standards are the standards of the day, that is, those that are popular at the moment. For example, Asynchronous Transfer Mode (ATM) was the standard du jour for a number of years, until it fell from grace and Internet Protocol (IP) became the favorite. (Actually, there really is no such thing as a du jour standard, at least not formally. I made it up. Pretty funny, don't you think?) *De facto* standards and specifications take on the effect of formal standards simply because they become so widely accepted. De facto standards are not established by such formally constituted bodies and may even be established by a dominant vendor in its own self-interest and often for its own internal use in the context of an ad hoc solution. Hayes, IBM, and Microsoft, for example, have developed numerous specifications that have become de facto industry standards. Standards generally are in the form of baseline specifications according to which manufacturers can develop products with the assurance that they will interconnect and interoperate with those of other manufacturers. Standards also commonly allow for options that manufacturers can exercise in various fashions peculiar to their own product development philosophies, strategies, and so on, thereby distinguishing those products from others. While standards have been

---

criticized as common denominator solutions that stifle creativity, they in fact provide a common framework of technical specifications within which manufacturers can exercise a considerable level of creativity. The yield is a competitive market that offers buyers a choice while ensuring interconnectivity and interoperability. In other words and within limits, buyers can implement multivendor solutions as they see fit.

FORMAL STANDARDS ORGANIZATIONS

3GPP
3rd Generation Partnership Project
ETSI
Mobile Competence Centre
650 Route des Lucioles
06921 Sophia-Antipolis Cedex, France
Tel: +33 (0) 4-92-94-42-00
Fax: +33 (0) 4-93-65-47-16
www.3gpp.org

3GPP2
3rd Generation Partnership Project 2
2500 Wilson Boulevard, Suite 300
Arlington, VA 22201
Tel: (703) 907-7700
Fax: (703) 907-7728
www.3gpp2.org

ANSI
American National Standards
  Institute
1819 L Street, NW, Suite 600
Washington, DC 20036
Tel: (202) 293-8020
Fax: (202) 293-9287
www.ansi.org

ATIS
Alliance for Telecommunications
  Industry Solutions
1200 G Street NW, Suite 500
Washington, DC 20005
Tel: (202) 628-6380
Fax: (202) 393-5453
www.atis.org

ATSC
Advanced Television Systems
  Committee
1750 K Street NW, Suite 1200
Washington, DC 20006
Tel: (202) 872-9160
Fax: (202) 872-9161
www.atsc.org

CableLabs
Cable Television Laboratories, Inc.
858 Coal Creek Circle
Louisville, CO 80027-9750
Tel: (303) 661-9100
Fax: (303) 661-9199
www.cablelabs.com

CEN
Comité Européen de Normalisation
European Committee for
  Standardization
36 Rue de Stassart, B
1050 Brussels, Belgium
Tel: 32-2-550-08-11
Fax: 32-2-550-08-19
www.cenorm.be

CENELEC
Comité Européen de Normalisation
  Electrotechnique
European Committee for
  Electrotechnical Standards
35 Rue de Stassartstraat, B
1050 Brussels, Belgium
Tel: 32-2-519-68-71
Fax: 32-2-519-69-19
www.cenelec.org

CSA International
Canadian Standards Association
   International
5060 Spectrum Way
Mississauga, Ontario L4W 5N6, Canada
Tel: (416) 747-4000 or (800) 463-6727
Fax: (416) 747-2473
www.csa.ca

Ecma International
(nee ECMA, European Computer
   Manufacturers Association)
114 Rue du Rhone
CH-1204 Geneva, Switzerland
Tel: 41-22-849-6000
Fax: 41-22-849-6001
www.ecma-international.org

EIA
Electronic Industries Alliance
2500 Wilson Boulevard
Arlington, VA 22201
Tel: (703) 907-7500
Fax: (703) 907-7501
www.eia.org

ETSI
European Telecommunications
   Standards Institute
650 Route des Lucioles
06921 Sophia Antipolis Cedex, France
Tel: +33 (0) 4-92-94-42-00
Fax: +33 (0) 4-93-65-47-16
www.etsi.org

FCC
Federal Communications Commission
445 12th Street, SW
Washington, DC 20554
Tel: (888) 225-5322
Fax: (202) 418-0232 or (866) 418-0232
www.fcc.gov

IEC
International Electrotechnical
   Commission
3 Rue de Varembe
CH-1211 Geneva 20, Switzerland
Tel: 41-22-919-02-11
Fax: 41-22-919-03-00
www.iec.ch

IEEE
Institute of Electrical and Electronics
   Engineers
445 Hoes Lane
Piscataway, NJ 08854-1331
Tel: (732) 981-0060
Fax: (732) 981-1721
www.ieee.org

ISO
International Organization for
   Standardization
1 Rue de Varembe
Case Postale 56
CH-1211 Geneva 20, Switzerland
Tel: 41-22-749-01-11
Fax: 41-22-733-34-30
www.iso.ch

ISOC
Internet Society
1775 Wiehle Avenue, Suite 102
Reston, VA 20190-5108
Tel: (703) 326-9880
Fax: (703) 326-9881
www.isoc.org

ITU
International Telecommunication
   Union
Place des Nations
CH-1211 Geneva 20, Switzerland
Tel: 41-22-730-5111
Fax: 41-22-733-7256
www.itu.ch

NIST
National Institute of Standards and
Technology
(nee National Bureau of Standards,
NBS)
100 Bureau Drive
Gaithersburg, MD 20899
Tel: (301) 975-2000
www.nist.gov

NTIA
National Telecommunications and
Information Administration
U.S. Department of Commerce
1401 Constitution Avenue, NW
Washington, DC 20230
Tel: (202) 482-7002
www.ntia.doc.gov

Telcordia Technologies
(nee Bellcore)
1 Telcordia Drive
Piscataway, NJ 08854-4157
Tel: (732) 699-2000
Fax: (732) 336-2320
www.telcordia.com

TIA
Telecommunications Industry
Association
2500 Wilson Boulevard, Suite 300
Arlington, VA 22201-3837
Tel: (703) 907-7700
Fax: (703) 907-7727
www.tiaonline.org

UL
Underwriters Laboratories, Inc.
333 Pfingsten Road
Northbrook, IL 60062-2096
Tel: (847) 272-8800
Fax: (847) 272-8129
www.ul.com

W3C
World Wide Web Consortium
Massachusetts Institute of Technology
32 Vassar Street, Room 32-G515
Cambridge, MA 02139
Tel: (617) 253-2613
Fax: (617) 258-5999
www.w3.org

CONSORTIA, FORA, AND SPECIAL INTEREST GROUPS (SIGs)

ACM
Association for Computing Machinery
1515 Broadway
New York, NY 10036
Tel: (212) 626-0500 or (800) 342-6626
www.acm.org

ATM Forum
See MFA Forum

BICSI
Building Industry Consulting Service
International
8610 Hidden River Parkway
Tampa, FL 33637-1000
Tel: (813) 979-1991 or (800) 242-7405
Fax: (813) 971-4311
www.bicsi.org

Bluetooth Special Interest Group
500 108th Avenue NE, Suite 250
Bellevue, WA 98004
Tel: (425) 691-3535
www.bluetooth.com

CDG
CDMA Development Group
575 Anton Boulevard, Suite 560
Costa Mesa, CA 92626
Tel: (714) 545-5211 or (888) 800-2362
Fax: (714) 545-4601
www.cdg.org

CEA
Consumer Electronics Association
2500 Wilson Boulevard
Arlington, VA 22201-3834
Tel: (703) 907-7600 or (866) 858-1555
Fax: (703) 907-7675
www.ce.org

CompTIA
Computer Technology Industry
    Alliance
1815 South Meyers Road, Suite 300
Oakbrook, IL 60181-5228
Tel: (630) 678-8300
Fax: (630) 628-1384
www.comptia.org

CTIA
The Wireless Association
(previously Cellular Telecommunications
    & Internet Association)
1400 16th Street, NW, Suite 600
Washington, DC 20036
Tel: (202) 785-0081
Fax: (202) 785-0721
www.ctia.org

DSL Forum
39355 California Street, Suite 307
Fremont, CA 94538
Tel: (510) 608-5905
Fax: (510) 608-5917
www.dslforum.org

EFF
Electronic Frontier Foundation
454 Shotwell Street
San Francisco, CA 94110-1914
Tel: (415) 436-9333
Fax: (415) 436-9993
www.eff.org

Ethernet Alliance
P.O. Box 200757
Austin, TX 78720
Tel: (512) 363-9932
Fax: (512) 532-6894
www.ethernetalliance.org

Fibre Channel Industry Association
www.fibrechannel.org

Frame Relay Forum
See MFA Forum

IEC
International Engineering Consortium
300 West Adams Street, Suite 1210
Chicago, IL 60606-5114
Tel: (312) 559-4100
Fax: (312) 559-4111
www.iec.org

IMC
Internet Mail Consortium
127 Segré Place
Santa Cruz, CA 95060
Tel: (831) 426-9827
Fax: (831) 426-7301
www.imc.org

IMTC
International Multimedia
    Telecommunications Consortium, Inc.
Bishop Ranch 6
2400 Camino Ramon, Suite 375
San Ramon, CA 94583
Tel: (925) 275-6600
Fax: (925) 275-6691
www.imtc.org

IPv6 Forum
www.ipv6forum.com

IrDA
Infrared Data Association
P.O. Box 3883
Walnut Creek, CA 94598
Tel: (925) 943-6546
Fax: (925) 943-5600
www.irda.org

MFA Forum
(merged MPLS, Frame Relay, and
  ATM Fora)
39355 California Street, #307
Fremont, CA 94538
Tel: (510) 608-5910
Fax: (510) 608-5917
www.mfaforum.org

MPLS Forum
See MFA Forum

MSF
Multiservice Switching Forum
39355 California Street, #307
Fremont, CA 94538
Phone: (510) 608-5922
Fax: (510) 608-5917
www.msforum.org

NAB
National Association of Broadcasters
1771 North Street, NW
Washington, DC 20036-2891
Tel: (202) 429-5300
Fax: (202) 429-4199
www.nab.org

NARTE
National Association of Radio and
  Telecommunications Engineers
167 Village Street
Medway, MA 02053
Tel: (508) 533-8333 or (800) 896-2783
Fax: (508) 533-3815
www.narte.org

NARUC
National Association of Regulatory
  and Utility Commissioners
1101 Vermont Avenue, NW
Washington, DC 20005
Tel: (202) 898-2200
Fax: (202) 898-2213
www.naruc.org

NCTA
National Cable & Telecommunications
  Association
1724 Massachusetts Avenue, NW
Washington, DC 20036
Tel: (202) 775-3550
www.ncta.com

NECA
National Exchange Carrier Association
80 South Jefferson Road
Whippany, NJ 07981-1009
Tel: (973) 884-8000 or (800) 228-8597
Fax: (973) 884-8469
www.neca.org

NTIS
National Technical Information Service
Technology Administration
U.S. Department of Commerce
5285 Port Royal Road
Springfield, VA 22161
Tel: (703) 605-6000
Fax: (703) 321-8547
www.ntis.gov

OMA
Open Mobile Alliance
4275 Executive Square, Suite 240
La Jolla, CA 92037
Tel: (858) 623-0740
Fax: (858) 623-0743
www.openmobilealliance.org

OMG
Object Management Group, Inc.
1410 Kendrick Street
Building A, Suite 300
Needham, MA 02494
Tel: (781) 444-0404
Fax: (781) 444-0320
www.omg.org

The Open Group
Previously the Electronic Messaging
  Association (EMA)
44 Montgomery Street, Suite 960
San Francisco, CA 94104-4704
Tel: (415) 374-8280
Fax: (415) 374-8293
www.opengroup.org

PCIA
Personal Communications Industry
  Association
500 Montgomery Street, Suite 700
Alexandria, VA 22314-1561
Tel: (703) 739-0300 or (800) 759-0300
Fax: (703) 836-1608
www.pcia.com

PCMCIA
Personal Computer Memory Card
  International Association
2635 North First Street, Suite 218
San Jose, CA 95134
Tel: (408) 433-2273
Fax: (408) 433-9558
www.pcmcia.org

SAI
Satellite Industry Association
1730 M Street, NW
Suite 600
Washington, DC 20036
Tel: (202) 349-3650
Fax: (202) 349-3622
www.sia.org

SBCA
Satellite Broadcasting and
  Communications Association
1730 M Street, NW, Suite 600
Washington, DC 20036
Tel: (202) 349-3620 or (800) 541-5981
Fax: (202) 349-3621
www.sbca.com

SCTE
Society of Cable Telecommunications
  Engineers
140 Philips Road
Exton, PA 19341-1318
Tel: (610) 363-6888 or (800) 542-5040
Fax: (610) 363-5898
www.scte.org

SIIA
Software & Information Industry
  Association
1090 Vermont Ave NW, Sixth Floor
Washington, DC 20005-4095
Tel: (202) 289-7442
Fax: (202) 289-7097
www.siia.net

SIP Forum
(Session Initiation Protocol Forum)
Tel: (978) 824-0111
www.sipforum.org

SMPTE
Society of Motion Picture & Television
  Engineers
3 Barker Avenue
White Plains, NY 10601
Tel: (914) 761-1100
Fax: (914) 761-3115
www.smpte.org

SNIA
Storage Networking Industry
  Association
500 Sansome Street, Suite 504
San Francisco, CA 94111
Tel: (415) 402-0006
Fax: (415) 402-0009
www.snia.org

TM Forum
TeleManagement Forum
240 Headquarters Plaza
East Tower, 10th Floor
Morristown, NJ 07960-6628
Tel: (973) 944-5100
Fax: (973) 944-5110
www.nmf.org

Unicode Consortium
Attn: Magda Danish
1065 L'Avenida Street
Microsoft Building 5
Mountain View, CA 94043
Tel: (650) 693-3921
Fax: (650) 693-3010
www.unicode.org

USTA
United States Telecom Association
607 14th Street NW, Suite 400
Washington, DC 20005
Tel: (202) 326-7300
Fax: (202) 326-7333
www.ustelecom.org

WCA International
Wireless Communications Association
   International
1333 H Street, NW
Suite 700 West
Washington, DC 20005-4754
Tel: (202) 452-7823
Fax: (202) 452-0041

Wi-Fi Alliance
Previously Wireless Ethernet
   Compatibility Alliance (WECA)
3925 West Braker Lane
Austin, TX 78759
Tel: (512) 305-0790
Fax: (512) 305-0791
www.wi-fi.com

WiMAX Forum
2495 Leghorn Street
Mountain View, CA 94043
Tel: (503) 712-2206
www.wimaxforum.org

WTO
World Trade Organization
Centre William Rappard
154 Rue de Lausanne
CH-1211 Geneva 21, Switzerland
Tel: 41-22-739-51-11
Fax: 41-22-731-42-06
www.wto.org

ZigBee Alliance
2400 Camino Ramon, Suite 375
San Ramon, CA 94583
Tel: (925) 275-6607
Fax: (925) 886-3850
www.zigbee.org

INDEX

The corporations which will excel in the 1980's will be those that manage information as a major resource.

John Diebold, as quoted in Infosystems, October, 1979.

In an attempt to help you manage the information in this book, this index points you to the pages on which topics are discussed. Each term and subject is spelled out completely, with the acronym or abbreviation following in parentheses. If you know only the acronym or abbreviation, please consult the Acronyms and Abbreviations appendix, which serves as a cross-reference to this index. **Boldface** page numbers denote in-depth coverage of a topic. *Italic* page numbers denote the definition of a term. Page numbers followed by "f" denote associated figures. The numbers not in **boldface** or *italics* point you to the pages on which the term or subject is discussed either in lesser detail or is placed in an application context. Pages on which the term or subject is mentioned only peripherally are not listed.

Telecommunications and Data Communications Handbook, By Ray Horak
Copyright © 2007 Ray Horak